Continental Intraplate Earthquakes: Science, Hazard, and Policy Issues

edited by
Seth Stein
Department of Earth and Planetary Sciences
Northwestern University
1850 Campus Drive
Evanston, Illinois 60208
USA

and

Stéphane Mazzotti
Geological Survey of Canada
Pacific Geoscience Centre
9860 West Saanich Road
Sidney, British Columbia V8L 4B2
Canada

THE
GEOLOGICAL
SOCIETY
OF AMERICA®

Special Paper 425

3300 Penrose Place, P.O. Box 9140 ▪ Boulder, Colorado 80301-9140 USA

2007

Published by The Geological Society of America, Inc.
3300 Penrose Place, P.O. Box 9140, Boulder, Colorado 80301-9140, USA
www.geosociety.org

Printed in U.S.A.

GSA Books Science Editors: Marion E. Bickford and Abhijit Basu

Library of Congress Cataloging-in-Publication Data

Continental intraplate earthquakes : science, hazard, and policy issues / edited by Seth Stein,
  Stéphane Mazzotti.
     p. cm. — (Special paper ; 425)
  Includes bibliographical references and index.
  ISBN 978-0-8137-2425-6 (pbk.)
     1. Plate tectonics. 2. Seismology. 3. Earthquake prediction. 4. Geology, Structural. I. Stein,
  Seth. II. Mazzotti, Stéphane, 1972–.
QE511.4.C655 2007
551.22—dc22

                                                                                2007012988

Cover: Comparison for eastern North America of locations of intraplate seismicity (dots) to postglacial vertical motion (colors) interpolated from GPS measurements (Sella, G.F., Stein, S., Dixon, T.H., Craymer, M., James, T.S., Mazzotti, S., and Dokka, R.K., 2007, Observation of glacial isostatic adjustment in "stable" North America with GPS: Geophysical Research Letters, v. 34, L02306, doi: 10.1029/2006GL027081), including those at the site shown. Photograph by S. Mazzotti.

10 9 8 7 6 5 4 3 2 1

Contents

Preface

Earthquakes within continental plates are an embarrassing stepchild of modern earthquake seismology. The discovery of plate tectonics explained why the overwhelming majority of earthquakes and seismic moment release occurs on plate boundaries. As plate motions became understood, they explained the mechanisms of interplate earthquakes and constrained their recurrence interval from the rate at which these motions were released seismically. More recently, the recognition of some plate boundaries as broad deformation zones allowed us to explain earthquakes within them in the context of the plate tectonic model.

In contrast, plate tectonics provide no direct insight into intraplate earthquakes, beyond the trivial prediction that they should not occur within ideal rigid plates. Hence, earthquake occurrence gives insight into the internal deformation of the major plates and provides a scientific challenge to address. Moreover, although large intraplate earthquakes are rare, they can do significant damage when they occur in populated areas. Thus, study of those intraplate earthquakes that occur within continents has societal significance.

Progress on understanding intraplate earthquakes has, until recently, been difficult. Because deformation within plates is slow, at most a few mm/yr compared to the generally much more rapid plate boundary motions, seismicity is much lower and harder to study. This difficulty is compounded by the fact that, unlike at plate boundaries where plate motions give insight into why and how often earthquakes occur, we have little idea of what causes intraplate earthquakes and no direct way to estimate how often they should occur. As a result, progress in understanding these earthquakes is much more difficult than on plate boundaries, and key issues may not be resolved for a very long time.

In recent years, however, new techniques and approaches have started to yield important new insights into these issues. Among these, three stand out to date. Space geodesy can measure the slow intraplate deformation that causes the earthquakes and thus constrain the stresses involved and the rates at which they accumulate. Paleoseismology can extend the sparse instrumental record backward in time, constraining the recurrence history, and thus exploring whether seismicity remains concentrated on specific structures or migrates between them. Numerical deformation modeling makes it possible to test hypotheses for the stresses causing the earthquakes and how they would cause seismicity to vary with time.

As a result, it seemed timely to compile a sampling of research addressing issues of continental intraplate earthquakes. We convened a series of special sessions addressing these issues at the spring 2004 Joint Assembly of the American and Canadian Geophysical Unions in Montreal, the papers from which became the core of this volume. We also invited relevant papers at other meetings and via an Internet discussion group. We asked authors to, as appropriate, look at the topic in a worldwide rather than North American context.

The papers in the book address the broad related topics of the science, hazard, and policy issues of large continental intraplate earthquakes.

A group of these papers addresses aspects of the primary scientific issue—where are these earthquakes and what causes them? We need to know whether there is something special about the areas where our limited record shows that such events occurred. Answering this question is crucial to determining whether they will continue there or migrate elsewhere. Part of the same question is understanding when they started in a given area, and how often (if ever) they will recur.

A second group of papers addresses the challenge of assessing the hazard posed by intraplate earthquakes. Although it may be a very long time before the scientific issues are resolved, the progress being made is helping attempts to estimate the probability, size, and shaking of future earthquakes, and the uncertainty of the results.

Finally, a third group of papers explores the question of how society should mitigate the possible effects of future large continental intraplate earthquakes. Although we will not know for hundreds or thousands of years how well the hazards were estimated, communities around the world face the challenge of deciding how to address this rare, but real, hazard, given the wide range of other societal needs.

Continental intraplate earthquakes will remain a challenge to seismologists, earthquake engineers, policy makers, and the public for years to come. Still, our sense is that significant progress toward understanding and addressing this challenge is now being made.

We thank Bill Dickinson for suggesting this book, and the authors for their efforts in preparing papers suitable for the book's intended general audience and accepting the delays in publication (compounded by the December 2004 Sumatra earthquake) that result from publishing in a book rather than a journal. We also thank the reviewers for their assistance. We have enjoyed putting together this volume and hope it provides readers with a flavor of this intriguing field.

Seth Stein
Stéphane Mazzotti

The Geological Society of America
Special Paper 425
2007

Approaches to continental intraplate earthquake issues

Seth Stein[†]

Department of Earth and Planetary Sciences, Northwestern University, Evanston, Illinois 60208, USA

"We choose to go to the moon in this decade and do the other things, not because they are easy, but because they are hard."—John F. Kennedy, 1962

ABSTRACT

The papers in this volume illustrate a number of approaches that are becoming increasingly common and offer the prospect of making significant advances in the broad related topics of the science, hazard, and policy issues of large continental intraplate earthquakes. Plate tectonics offers little direct insight into the earthquakes beyond the fact that they are consequences of slow deformation within plates and, hence, relatively rare. To alleviate these problems, we use space geodesy to define the slowly deforming interiors of plates away from their boundaries, quantify the associated deformation, and assess its possible causes. For eastern North America, by far the strongest signal is vertical motion due to ice-mass unloading following the last glaciation. Surprisingly, the expected intraplate deformation due to regional stresses from plate driving forces or local stresses are not obvious in the data. Several approaches address difficulties arising from the short history of instrumental seismology compared to the time between major earthquakes, which can bias our views of seismic hazard and earthquake recurrence by focusing attention on presently active features. Comparisons of earthquakes from different areas illustrate cases where earthquakes occur in similar tectonic environments, increasing the data available. Integration of geodetic, seismological, historical, paleoseismic, and other geologic data provides insight into earthquake recurrence and the difficult question of why the earthquakes are where they are. Although most earthquakes can be related to structural features, this explanation alone has little predictive value because continents contain many such features, of which a few are the most active. It appears that continental intraplate earthquakes are episodic, clustered, and migrate. Thus on short time scales seismicity continues on structures that are active at present, perhaps in part because many events are aftershocks of larger past events. However after periods of activity these structures may become inactive for a long time, so the locus of at least some of the seismicity migrates to other structures. Analysis of the thermo-mechanical structure of the seismic zones gives insight into their mechanics: whether there is something special about them that results in long-lived weak zones on which intraplate strain release concentrates, or as seems more likely, that they are not that unusual, so seismicity migrates. Accepting our lack of understanding of the underlying causes of

[†]E-mail: seth@earth.northwestern.edu.

Stein, S., 2007, Approaches to continental intraplate earthquake issues, *in* Stein, S., and Mazzotti, S., ed., Continental Intraplate Earthquakes: Science, Hazard, and Policy Issues: Geological Society of America Special Paper 425, p. 1–16, doi: 10.1130/2007.2425(01). For permission to copy, contact editing@geosociety.org.

**the earthquakes, the limitations of the short instrumental record, and the possibility
of migrating seismicity helps us to recognize the uncertainties in estimates of seismic
hazards. Fortunately, even our limited knowledge can help society develop strategies
to mitigate earthquake hazards while balancing resources applied to this goal with
those applied to other needs.**

Keywords: intraplate earthquakes, continental deformation, seismic hazards.

INTRODUCTION

The papers in this book represent a range of ongoing research addressing the related topics of the science, hazard, and policy issues of large continental intraplate earthquakes. As summarized in the preface, addressing these issues is more difficult than for the far more common earthquakes on plate boundaries, for two reasons. First, we lack a model like plate tectonics that gives insight into the causes, nature, and rate of the earthquakes. Second, because intraplate earthquakes are much rarer owing to the slow deformation rate, we know much less about these earthquakes and their effects.

As a result, probably none of the authors would claim to be an "expert" on intraplate earthquakes. After all, an expert should know why, where, and when such earthquakes occur, what their effects will be, and how society should address them. Because none of these issues is well understood at present, the authors are simply researchers exploring these messy issues.

These issues involve both fundamental science and societal implications. The challenge is to understand the nature and causes of these relatively rare but sometimes very destructive earthquakes and use what we learn to assess the hazard they pose and help society formulate sensible policies to address the resulting risk. In doing so, it is useful to distinguish between hazards and risks. The hazard is the intrinsic natural occurrence of earthquakes and the resulting ground motion and other effects. Although we can define it in various ways for different purposes, and our estimates of it have large uncertainties, the hazard is a natural feature. In contrast, the risk is the danger the hazard poses to life and property, and can be reduced by human actions. Hence, we seek to estimate the hazard and choose policies consistent with societal goals to reduce the resulting risk.

An underlying theme is that many of the scientific and societal issues differ significantly from those posed by the far more common earthquakes at plate boundaries. Figure 1 illustrates this point by comparing a type example of a continental intraplate seismic zone, the New Madrid seismic zone in the central United States, with southern California, part of the boundary zone between the Pacific and North American plates. New Madrid seismic zone earthquakes of a given magnitude are ~30–100 times less frequent because southern California earthquakes result from the ~46 mm/yr motion within the plate boundary zone, whereas New Madrid is within the interior of the North American plate, which is stable to better than 2 mm/yr. However, shaking from New Madrid seismic zone earthquakes

is thought to be comparable to that from California earthquakes one magnitude unit larger because rock in the stable continental interior transmits seismic energy more efficiently. Because earthquakes of a given magnitude are ~10 times more frequent than those one-magnitude-unit larger, the shaking difference reduces the effect of the difference in earthquake rates by about a factor of 10. The precise net effect of these differences depends on the recurrence rate of large earthquakes and the resulting ground motion, neither of which are well known. Even so, the comparison indicates that different approaches to mitigating the seismic hazard are likely to make sense.

The hazard posed by large continental intraplate earthquakes is a small, but still significant, fraction of the threat posed by all earthquakes. Earthquakes, in turn, are just one of many challenges societies face. In the United States, on average, fewer than ten people per year are killed by earthquakes (Fig. 2), and intraplate events make up less than 10% of the total. Hence earthquakes are at the level of in-line skating or football, but far less than bicycles, for risk of loss of life (Stein and Wysession, 2003). Similarly, the approximately $5 billion average annual earthquake losses for the United States, though large, is ~2% of that due to automobile accidents. Nonetheless, large earthquakes occasionally cause many fatalities and major damage. Similarly, on a global basis, earthquakes cause an average of ~10,000 deaths per year, significant but relatively minor compared to other causes. For example, malaria causes about a million deaths per year. The challenge to societies is to thus to develop strategies that balance resources allocated to earthquake hazard mitigation with other needs.

Papers in this volume explore many of the issues in these examples. Although written by different authors addressing various geographic areas, and hence often taking different views, they illustrate approaches that are becoming increasingly common and offer the prospect of making significant advances. The goal of this introduction is to highlight some of these approaches, using North America and New Madrid as examples for comparison with some of the results and ideas presented in this volume.

DEFINING PLATE INTERIORS

Although the discovery of plate tectonics explained why the overwhelming majority of earthquakes and seismic moment release occurs on plate boundaries, it remained unclear for some time how to define plate boundaries and distinguish them from plate interiors. Although early papers defined narrow plate boundaries between idealized rigid plates, for example, treat-

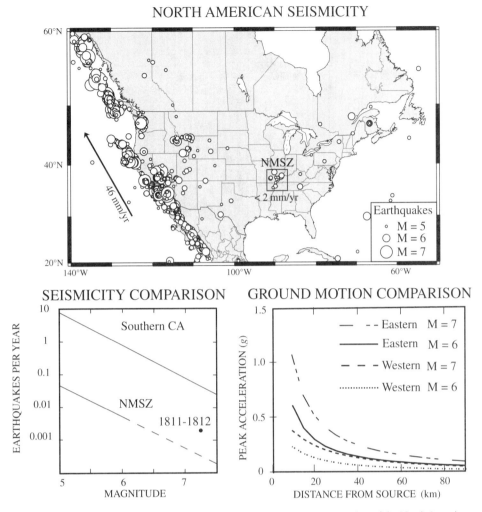

Figure 1. Top: Seismicity (M 5 or greater since 1900) of the continental portion of the North American plate and adjacent areas. Seismicity is concentrated along the Pacific–North America plate boundary zone, reflecting the relative plate motion. The stable eastern portion of the continent, approximately east of 260°, is much less active, with seismicity concentrated in several zones, notably the New Madrid seismic zone (NMSZ). Bottom left: Comparison of the annual rates of earthquakes greater than a given magnitude for Southern California and the New Madrid seismic zone. Solid lines are computed from recorded seismicity, whereas dashed are extrapolated. Dot indicates paleoseismically inferred recurrence for the largest New Madrid seismic zone earthquakes, assuming M 7.2. Bottom right: Comparison of the predicted strong ground motion from M 7 and 6 earthquakes in the eastern and western United States (Stein et al., 2003).

ing the San Andreas fault as the boundary between the Pacific and North American plates, they recognized that not all plates were perfectly rigid. Morgan (1968, p. 1960), for example, noted that noted that "such features as the African rift system, the Cameroon trend, and the Nevada-Utah earthquake belt are most likely the type of distortion denied in the rigidity hypothesis."

As understanding of motions at plate boundaries and within plate interiors grew, ideas about the distribution of earthquakes and deformation away from idealized boundaries became more specific. Hence, we now would regard Morgan's three examples as illustrating three different types of slowly deforming regions. The seismically active East African rift system is now regarded

as a slowly opening plate boundary between the Nubian (East African) and Somalian (West African) plates (Chu and Gordon, 1999). The Nevada and Utah earthquakes are regarded as part of the deformation associated with the broad plate boundary zone between the Pacific and North American plates (Bennett et al., 1999). In contrast, the earthquakes associated with the Cameroon volcanic line (Sykes, 1978) are considered to be within the Nubian plate.

This view came about because plate motions became better understood, both from geological plate motion models (e.g., Chase, 1972, 1978; Minster et al., 1974; Minster and Jordan, 1978; DeMets et al., 1990, 1994) and space-based geodesy

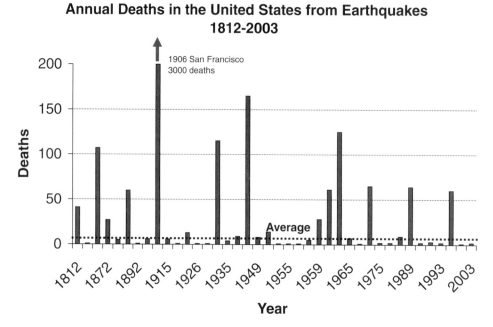

Figure 2. Earthquake deaths in the U.S. Data are from http://earthquake.usgs.gov/regional/states/us_deaths.php.

(e.g., Sella et al., 2002), which made it easier to distinguish plate boundaries from plate interiors. A key to doing so was quantification of deviations from rigid plate behavior, first by using plate motion data at plate boundaries (Stein and Gordon, 1984; DeMets et al., 1990), and later using space geodesy to measure deformation within plates. This process is illustrated by Figure 3, which shows the motions of global positioning system (GPS) sites in North America. Because the motion of a rigid plate is described by a rotation about an Euler pole, sites on the rigid North American plate should move along small circles about the pole, at rates that increase with the sine of the angular distance from the pole. This is the case in eastern North America, whereas motions in the west are quite different, showing that they are part of a broad plate boundary zone.

The deviations of GPS site velocities from those expected for a rigid plate can be used to quantify the deformation of the plate interior, which causes the intraplate earthquakes. Successive studies using increasing amounts of data from the growing number of continuous GPS sites yield increasingly precise velocities. The resulting root-mean-square (rms) misfit of site velocities to those predicted by a single Euler vector if the plate were perfectly rigid is now less than 1 mm/yr (Table 1).

The misfit is strikingly small, given that it reflects the combined effects of intraplate deformation due to tectonics and glacial isostatic adjustment, uncertainties in the positions of geodetic monuments due to the GPS techniques, and local motion of the geodetic monuments. The result seems plausible because similar values emerge from very long baseline radio interferometry studies (Argus and Gordon, 1996). Hence,

sites that move faster with respect to the stable interior of the plate than a specified rate, perhaps 2–3 mm/yr, can be viewed as within the boundary zone, whereas those that move more slowly can be viewed as within the plate interior.

This process can be formalized using the GPS data to distinguish a plate boundary zone from deformation within a plate interior, just as plate motion data are tested to see whether they are statistically better fit by assuming the existence of two distinct plates (Stein and Gordon, 1984; Gordon et al., 1987). In such cases, Euler vectors can be derived and used to describe the motion of the two plates, which occurs primarily at their boundaries. Such analyses have shown that North and South America (Stein and Gordon, 1984), India and Australia (Wiens et al., 1985), and Nubia and Somalia (Chu and Gordon, 1999) should be regarded as distinct plates, often with seismicity along their boundaries, rather than single plates with distinct zones of intraplate seismicity. Conversely, application of such analysis to GPS data on opposite sides of the New Madrid seismic zone shows that treating eastern North America as two distinct blocks is not statistically justified (Dixon et al., 1996; Newman et al., 1999). As a result, the New Madrid seismic zone is regarded as a zone of deformation within the North American plate, which contains several others (Mazzotti, chapter 2; Swafford and Stein, chapter 4). Similarly, the earthquakes in the Rhine Graben of northwest Europe (Camelbeeck et al., chapter 14; Hinzen and Reamer, chapter 15) are regarded as intraplate because no significant motion across it has yet been resolved with GPS (Nocquet et al., 2005).

Hence, adequate GPS data can identify the extent of a plate boundary zone and distinguish between it and the plate

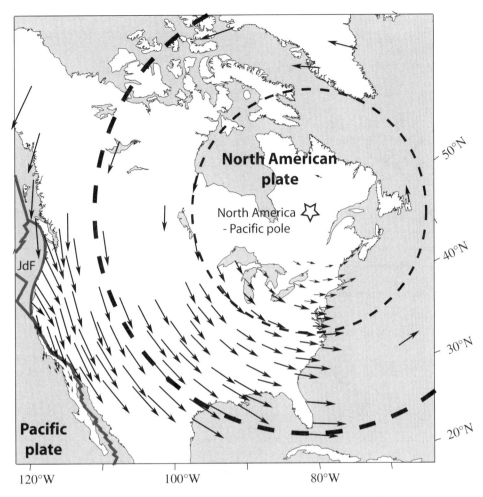

Figure 3. Global positioning system (GPS) site motions (arrows) show the difference between the interior of the North American plate and the Pacific–North America plate boundary zone. Within the plate interior, sites move in small circles about the plate rotation pole (star) at a rate increasing with distance, whereas motions in the boundary zone differ noticeably. These data show that the plate is stable to better than 2 mm/yr, it can be described by a single Euler vector, and it shows no significant motion across the New Madrid seismic zone (Stein and Sella, 2002). JdF—Juan de Fuca plate.

interior. For example, Bada et al. (chapter 16) use GPS data to map deformation in the broad Adriatic deformation region, part of the plate boundary zone between Nubia and Eurasia. However, in areas where adequate GPS data are not yet available, an earthquake can be regarded as either part of the plate boundary zone or within the plate interior. For example, the 2001 Bhuj, India (Mw 7.7), earthquake has been interpreted as a continental intraplate earthquake with analogies to the New Madrid seismic zone in the central United States (Abrams, 2001; Beavers, 2001; Bendick et al., 2001; Ellis et al., 2001). However, it occurs within the broad zone of seismicity and deformation that forms the Indian plate's diffuse western boundary with Eurasia (Fig. 4) (Stein et al., 2002; Li et al., 2002). In western U.S. terms, this location corresponds to Nevada, within the deforming plate boundary zone, where the earthquakes reflect the kinematics and dynamics of the boundary zone (Flesch

TABLE 1. GLOBAL POSITIONING SYSTEM (GPS) SITES AND ROOT-MEAN-SQUARE (RMS) FITS

Study	Number of sites	Rms misfit (mm/yr)
Dixon et al. (1996)	8	1.3
Newman et al. (1999)	16	1.0
Sella et al. (2002)	64	0.86
Calais et al. (2006)	119[†]	0.70
[†]Sites with best-determined velocities.		

et al., 2000). In contrast, the New Madrid seismicity is ~2400 km from the San Andreas fault, the nominal boundary, with no obvious relation to the Pacific–North America boundary zone (Li et al., chapter 11). This view of the Bhuj event as part of a plate boundary zone is consistent with Sarkar et al.'s (chapter 20) suggestion that the basement there shows evidence of long-term deformation.

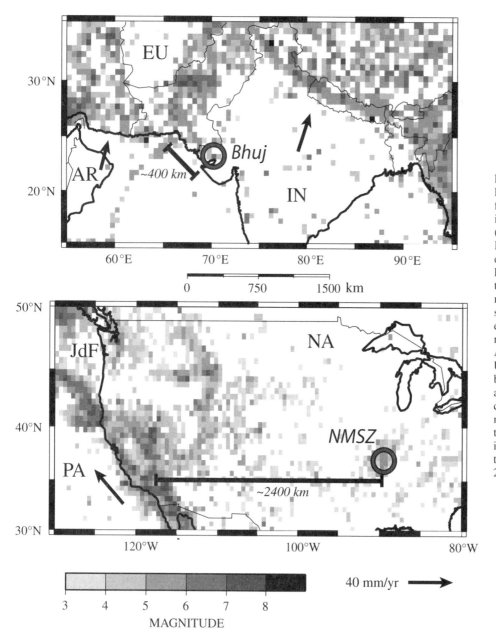

Figure 4. Earthquake magnitude release (1900–1999, depths <100 km) for part of Indian plate and surroundings (top) and the western United States (bottom), plotted at same spatial scale. In each pixel, cumulative seismicity is estimated by summing the moment release inferred from published magnitudes and reinterpreting its sum as the magnitude of a single event; shaded as shown by the horizontal bar. The Bhuj earthquake is ~400 km from the nominal boundary (EU—Europe, IN—India, AR—Arabian plate), a distance that in U.S. terms is about halfway across the boundary zone between the Pacific (PA) and North American (NA) plates, in the central Nevada seismic belt where magnitude 7 earthquakes occur. In contrast, the New Madrid seismic zone (NMSZ) is in the plate interior, ~2400 km from the nominal boundary (Stein et al., 2002). JdF—Juan de Fuca plate.

DESCRIBING AND MODELING INTRAPLATE DEFORMATION

Space geodetic data have dramatically improved our view of intraplate deformation beyond what was previously possible with sparse earthquake, paleoseismic, and other geologic data. The results can be surprising (Sella et al., 2006; Calais et al., 2006). Figure 5 shows a vertical velocity field for eastern North America that is clearly dominated by the effects of glacial isostatic adjustment from ice-mass unloading following the last glaciation. Vertical velocities show upward rebound (~10 mm/yr) near Hudson Bay, the site of maximum ice load at the Last Glacial Maximum, that decreases to slower subsidence

(1–2 mm/yr) south of the Great Lakes. Also shown is a residual horizontal velocity field derived by subtracting the best-fit rigid plate rotation model. These data show coherent deformation associated with the Cascadia subduction zone. The scattered motions in eastern North America are interpreted as showing motions directed outward from Hudson Bay and secondary ice maxima in western Canada. In addition, the motions show a pattern of southeast-directed flow in southwestern Canada that rotates clockwise to southwest-directed flow in the central-western United States. Some of the horizontal scatter is presumably a combination of local site effects (noise for these purposes) and intraplate tectonic signal, but no coherent pattern beyond the glacial isostatic adjustment signal is obvious.

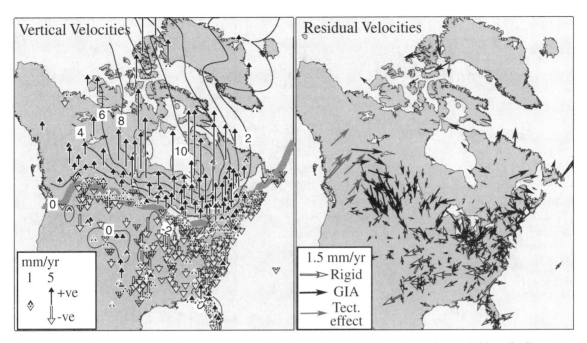

Figure 5. Left: Vertical global positioning system (GPS) site motions. Solid line shows observed "hinge line" separating uplift from subsidence. Sites in the plate boundary zone are not shown. Right: Horizontal motion site residuals after subtracting best-fit rigid plate rotation model (after Sella et al., 2006). GIA—glacial isostatic adjustment.

Such data thus provide powerful new constraints on the intraplate deformation field and the stresses causing it. They are being used to improve models of the effects of glacial isostatic adjustment (e.g., Peltier, 2004; Wu and Mazzotti, chapter 9) via more accurate descriptions of the ice load and laterally variable mantle viscosity. The data will address the long-suspected role of glacial isostatic adjustment as a possible cause or trigger of seismicity in eastern North America and other formerly glaciated areas (e.g., Stein et al., 1979, 1989; Hasegawa and Basham, 1989; Mazzotti and Adams, 2005). Previously, assessing the significance of this effect has proven difficult because the predicted velocities and hence strains vary significantly among glacial isostatic adjustment models, which until recently could not be well constrained. Hence, James and Bent (1994) and Wu and Johnston (2000) found that glacial isostatic adjustment may be significant for seismicity in the St. Lawrence valley but not the more distant New Madrid zone, whereas Grollimund and Zoback (2001) favored glacial isostatic adjustment as the cause of New Madrid seismicity.

Surprisingly, the data show no clear evidence for the plate-wide compression inferred from stress data and interpreted as a consequence of platewide stresses (e.g., Zoback and Zoback, 1989). Moreover, as will discussed shortly, there is no clear evidence of strain accumulation across the New Madrid zone. Hence the data provide strong upper bounds on both platewide and local deformation.

The increasingly high-quality intraplate velocity fields are now providing data that can be combined with earthquake mechanisms and other data to improve our understanding of intraplate deformation. They can be used to test numerical models of deformation, such as those shown by by Liu et al. (chapter 19) and Wu and Mazzotti (chapter 9). The approach has provided new insights in plate boundary zones, where rates are higher (e.g., Flesch et al., 2000; Liu et al., 2000, 2002). It will become increasing useful within plate interiors for understanding the stresses that cause earthquakes and the rheology of the plate interior, assessing what fraction of the deformation occurs seismically, and providing information on the location and recurrence time of future earthquakes.

TAKING A GLOBAL VIEW

A key to the development of plate tectonics was the formulation of a global synthesis by concentrating on similarities between different areas. The same approach is increasingly being taken in studies of continental intraplate earthquakes. Hence, papers in this book discuss earthquakes in regions outside North America, including Antarctica (Reading, chapter 18), Australia (Leonard et al., chapter 17), China (Liu et al., chapter 19), Europe (Bada et al., chapter 16; Camelbeeck et al., chapter 14; Hinzen and Reamer, chapter 15) and India (Sarkar et al, chapter 20).

Such earthquakes are increasingly viewed not only in terms of specific locations, but also in terms of their tectonic environments (e.g., Gangopadhyay and Talwani, 2003; Schulte and Mooney, 2005). For example, a significant fraction of continental intraplate seismicity occurs along passive continental margins, presumably due to reactivation of fossil structures, including those associated with postglacial rebound (Stein et al., 1979,

Stein

1989; Mazzotti et al., 2005). As a result, studies are exploring common features that may contribute to the seismicity (e.g., Mazzotti, chapter 2), such as fault geometry (Gangopadhyay and Talwani, chapter 7) and the effects of postglacial rebound (Wu and Mazzotti, chapter 9; Jacobi et al., chapter 10).

A similar global view is also increasingly being taken in addressing seismic hazards, illustrated by the recent Global Seismic Hazard Map (Giardini et al., 2000). Figure 6 compares earthquake recurrence rates among continental intraplate seismic zones discussed in this volume. On average, a magnitude 6.5 or greater earthquake is expected in Australia about every 20 yr, whereas an earthquake of this size is expected about every 350, 500, and 800 yr in the Pannonian Basin, New Madrid seismic zone, and northwestern Europe, respectively. Hence, some of these areas face similar challenges in assessing the earthquake hazard (Atkinson, chapter 21; Camelbeeck et al., chapter 14; Hinzen and Reamer, chapter 15; Leonard et al., chapter 17; Wang, chapter 24) and developing sensible mitigation strategies (Crandell, chapter 25; Lomnitz and Castanos, chapter 26; Searer et al., chapter 23).

CONFRONTING THE SHORT EARTHQUAKE RECORD

A major difficulty for continental intraplate studies is the short history of instrumental seismology compared to the time between major earthquakes. As a result, inferences drawn from the earthquake history can have serious limitations and leave many questions unanswered. This problem arises even at some plate boundaries. For example, modern seismicity maps show little activity on the segment of the southern San Andreas fault on which the Mw ~7.9 1857 earthquake occurred. The segment of the Sumatra trench on which the great (Mw 9.3) December 2004 earthquake occurred was not particularly active seismically, was not considered particularly dangerous, and was not high risk on seismic gap maps. However, because intraplate deformation is typically much slower (<1 mm/yr) than at most plate boundaries, the recurrence times for large earthquakes in individual parts of the seismic zones are longer, making the recorded seismicity an even worse sample.

This situation gives rise to a number of difficulties. Almost every aspect of hazard estimation faces this challenge, because hazard estimates seek to quantify the shaking expected during periods of time (once in 500 yr in California and most other countries, once in 2500 yr in the central and eastern United States) that are much longer than the seismological record.

One issue is deciding where large earthquakes are likely. Seismic hazard maps for places like the North African coast, North America's eastern continental margin, or the St. Lawrence valley sometimes show bull's-eyes of high predicted hazard where we know from instrumental or historic records that moderate to large earthquakes have occurred. These bull's-eyes result from the assumption that the sites of recent seismicity are more likely to have future large earthquakes than other sites on

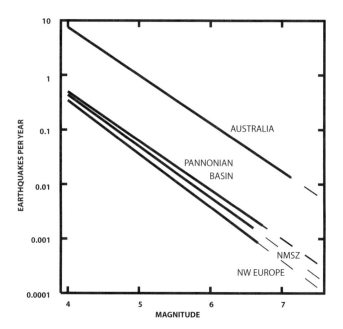

Figure 6. Comparison of the annual rates of earthquakes greater than a given magnitude for several seismic zones discussed in this volume. Solid line shows reported data; dashed line is extrapolated. Sources: New Madrid (Stein and Newman, 2004), Australia (Leonard et al., chapter 17), Pannonian Basin (Bada et al., chapter 16), northwest Europe (Camelbeeck et al., chapter 14). NMSZ—New Madrid seismic zone.

the same structures. However, one could also assume that the risk is comparable in similar environments for which the short record does not show earthquakes, or higher in these locations due to stress transfer from previous earthquakes. Aspects of this issue are also explored in this book (Mazzotti, chapter 2; Kafka, chapter 3; Swafford and Stein, chapter 4; Li et al., chapter 11; Atkinson, chapter 21).

A related issue is inferences of the maximum size and recurrence interval of future earthquakes in a given area from the short earthquake history. This involves estimating the frequency-magnitude (b value) curve for an area (Okal and Sweet, chapter 5). A crucial question is how well the rate and size of the largest earthquakes can be inferred from the small earthquakes (Fig. 6), even when historical and paleoseismic data are added (Camelbeeck et al., chapter 14; Hinzen and Reamer, chapter 15; Bada et al., chapter 16; Leonard et al., chapter 17). Some insight comes from plate boundary segments with long records, which show variability in the size and recurrence time of large earthquakes. Hence, a short earthquake record from an area with long recurrence times is likely to either miss the largest earthquakes entirely or preferentially detect large earthquakes with recurrence times shorter than the average. As a result, frequency-magnitude (b value) studies are likely to either underpredict the size of the largest earthquakes or conclude that they are characteristic, i.e., more common than expected from the rate of smaller earthquakes (Fig. 7). Moreover, whether characteristic earthquakes appear can depend on the portion of a seismic zone samples (Wesnousky, 1994; Stein et al., 2005).

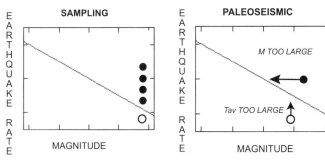

Figure 7. Possible apparent deviations from a log-linear frequency-magnitude relation due to a short earthquake record. Left: Due to sampling bias, the largest earthquakes can seem more common (characteristic, solid circles) than their long-term average recurrence interval, Tav. Alternatively, they can be missed or seem less common (uncharacteristic, open circles) than their long-term average. Right: Apparent characteristic earthquakes occur if paleoseismic data yield overestimates of magnitudes. Apparent uncharacteristic earthquakes occur if paleoseismic data yield overestimates of recurrence intervals (after Stein and Newman, 2004).

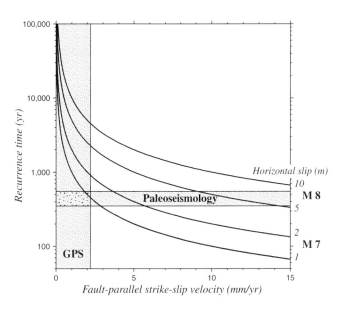

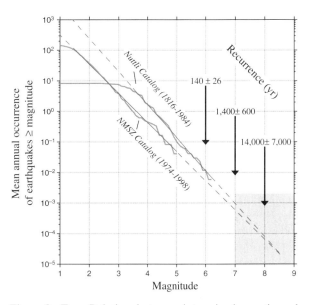

Figure 8. Top: Relation between interseismic motion observed from global positioning system (GPS) and paleoseismic estimates for the recurrence interval of large New Madrid earthquakes. The paleoseismic and geodetic data are jointly consistent with slip in 1811–1812 of ~1 m, corresponding to a magnitude 7 earthquake. Bottom: Earthquake frequency-magnitude data for the New Madrid seismic zone (NMSZ). Both the recent and historic (1816–1984) data have slopes close to one and predict recurrence intervals of ~1000 yr for magnitude 7 earthquakes and 10,000 yr for magnitude 8 earthquakes. Estimates are shown with 2 sigma uncertainties (after Newman et al., 1999).

Although additions of historical and paleoseismic data are valuable, combining these data with seismological data is tricky. Historical studies add events with known dates but with considerable uncertainty in magnitudes. For example, magnitude estimates for the 1906 San Francisco earthquake based on early seismological data have been as high as 8.3, compared to the typical current value of 7.9. The challenge is even greater for pre-instrumental data; recent results suggest low M 7 magnitudes for the largest 1811–12 New Madrid earthquakes (Hough et al., 2000), but published estimates range from low M 7 to over M 8. Paleoseismic studies have uncertainties both in the estimated dates and in recurrence times due to possibly missed events and even larger uncertainties in estimated magnitudes. For example, paleoliquefaction analysis for New Madrid seems to have overestimated the size of paleoevents, producing apparent characteristic earthquakes (Stein and Newman, 2004). Conversely, some paleoearthquakes may not yet have been identified in the nearby Wabash seismic zone, making the implied recurrence interval for large events too long and causing apparent uncharacteristic earthquakes (earthquakes less frequent than expected from the small earthquakes).

INTEGRATING GEODETIC, SEISMOLOGICAL, HISTORICAL, AND PALEOSEISMIC DATA

Geodetic data provide crucial insights into the issues raised by the short earthquake record because they measure the strain accumulating that will be released in future earthquakes. Hence, combinations of the geodetically observed deformation rate with the earthquake history give insight into the size and recurrence time of future large earthquakes.

This approach is illustrated in Figure 8 for New Madrid zone, where GPS data show less than 1–2 mm/yr of motion across the seismic zone (Newman et al., 1999; Gan and Prescott, 2001; Calais et al., 2005, 2006; Stein, 2007; Newman, 2007). Large earthquakes occurred in 1811 and 1812, and earlier such events have been inferred from the distribution of paleoliquefaction features. Wesnousky and Leffler (1992) did not find paleoliquefaction features comparable to those attributed to the 1811–1812 earthquakes and, hence, suggested that such large earthquakes are less common than implied by the instrumental and historic seismicity. In contrast, Tuttle (2001) interpreted paleoliquefaction features as showing that earthquakes comparable to or perhaps somewhat smaller than those in 1811–1812 occurred ca. 1450 ± 150 A.D. (M ≥ 6.7) and 900 ± 100 A.D. (M ≥ 6.9). Taken together, the GPS and paleoseismic data indicate that large earthquakes ~500 yr apart that release 1–2 mm/yr of interseismic motion would have magnitude ~7, consistent with the frequency-magnitude data from smaller earthquakes (Stein and Newman, 2004). Earthquakes with magnitude 8 would require motion across the seismic zone much faster than observed. These constraints are improving as the precision of the GPS site velocities increases (Calais et al., 2005, 2006).

A point worth noting is that such analyses relate the long-term seismicity to the presently observed deformation. The results can thus be biased by transient postseismic deformation. For example, if motions near the fault are dominated by transient strain after the 1811–1812 earthquakes (Rydelek and Pollitz, 1994; Rydelek, 2007), the interseismic strain accumulation rate is even smaller. Alternatively, Kenner and Segall (2000) proposed that a weak zone under the New Madrid seismic zone has recently relaxed, such that, for a few earthquake cycles, strains can be released faster than they accumulate. This hypothesis suffers from the fact that there is no evidence for such a weak zone (McKenna et al., chapter 12) and no obvious reason for why the proposed weakening occurred.

Geodetic data are being integrated similarly with seismological, historical, paleoseismic, and other geologic data in other intraplate seismic zones (Mazzotti, chapter 2; Camelbeeck et al., chapter 14; Leonard et al., chapter 17). Among the best such data at present are those presented by Bada et al. (chapter 16) for the Pannonian Basin, where the GPS shortening rate is well constrained and consistent with the seismicity. As for New Madrid, longer series of higher-quality GPS data will make this approach progressively more powerful.

This approach is also starting to shed light on the question of what fraction of the intraplate deformation is released seismically, because geodetic strain rates can be compared to those inferred from the seismic moment release. It appears that essentially all of the expected motion occurs seismically on the San Andreas fault (Stein and Hanks, 1998) and in continental interiors, as implied in Figure 8 and by the Pannonian Basin results (Bada et al., chapter 16). In contrast, trenches (Pacheco et al., 1993), oceanic transforms (Kreemer et al., 2002), and some (but not all) continental plate boundary zones (e.g., Jackson and McKenzie, 1988; Klosko et al., 2002; Pancha et al., 2006) appear to have significant aseismic motion. At present, it is unclear how well these variations are known, and whether they reflect differences in rheology and deformation, or are artifacts of the short earthquake history, which is crucial because most of the slip occurs in the infrequent largest events.

INVESTIGATING THE MECHANICS AND LONGEVITY OF SEISMIC ZONES

A fundamental question about continental intraplate earthquakes is why they are where they are. Although most earthquakes can be related to some structural feature, the explanation has limited predictive value, because continents contain many such features, of which a few are the most active. Hence, it is important to know whether over time seismicity continues on the structures that are most active at present, or is episodic and migrates between many similar structures. This issue is both of scientific importance and is crucial for assessing seismic hazards.

One approach to the question is to compare seismological, historical, paleoseismic, and other geological data. This approach increasingly finds that continental intraplate earthquakes are episodic, clustered, and migrate. Faults seem to go through cycles of activity punctuated by long periods of inactivity (Crone et al., 2003). Sarkar et al. (chapter 20) examine basement structure near the site of the Bhuj earthquake for evidence of long-term deformation. Studies for Australia (Leonard et al., chapter 17) and northwest Europe (Camelbeeck et al., chapter 14) consider the role of faults that appear to have been active in the past, although the short seismic record sometimes shows no activity on them. The idea that seismicity migrates is consistent with results for North America—these results indicate that the New Madrid zone became active recently (Schweig and Ellis, 1994, Newman et al., 1999; Holbrook et al., 2006), and they also show evidence of Holocene surface faulting that appears to be seismically inactive at present (Crone and Luza, 1990). What mechanism makes faults "turn on," "turn off," or change sense of motion remains unclear. Possible factors include stress changes due to regional tectonics (Bada et al., chapter 16; Liu et al., chapter 19), postglacial rebound (Stein et al., 1979, 1989; Mazzotti et al., 2005; Wu and Mazzotti, chapter 9; Jacobi et al., chapter 10), and denudation (Van Arsdale et al., chapter 13).

Another approach is to explore spatial and temporal correlations in seismicity. Kafka (chapter 3) finds that portions of seismic catalogs predict later seismicity well. An interesting question is: Does the fact that seismically active areas are likely places for continued small earthquakes make future large earthquakes more likely there than in other regions that may be equally or more susceptible to strain concentrations? Part of the challenge in answering this question involves understanding the role of static (Li et al., chapter 11) and dynamic (Hough, chapter 6) stress triggers in controlling future earthquake locations. A related question is whether much of the present seismicity reflects aftershocks of large past earthquakes (Stein and Newman, 2004).

A third approach explores the thermo-mechanical structure of the seismic zones to assess whether there is something special about them that results in long-lived weak zones on which intraplate strain release concentrates. Mazzotti (chapter 2) considers various models for the relations among lithospheric strength, strain distribution, and seismicity. Gangopadhyay and Talwani (chapter 7) propose that fault geometry favors earthquake occurrence. McKenna et al. (chapter 12) use heat-flow data to infer that the New Madrid zone is not significantly hotter and weaker than its surroundings, although such weakness has been postulated. These results argue against the New Madrid seismic zone being a long-lived weak zone on which intraplate strain release concentrates, and they favor a model of migrating seismicity.

RECOGNIZING THE UNCERTAINTY IN SEISMIC HAZARD ESTIMATES

Given the limitations of our present knowledge about continental intraplate earthquakes, it is not surprising that estimates of the hazard they pose have considerable uncertainties (Atkinson,

chapter 21; Wang, chapter 24). These uncertainties result from the fact that we do not understand the underlying causes of the earthquakes and have a limited earthquake history, typically without seismological records of the largest earthquakes of concern. Hence, their magnitudes and recurrence intervals are difficult to reliably infer, and the resulting ground motion must be extrapolated from smaller earthquakes (Bent and Delahaye, chapter 22).

As a result, a wide range of hazard estimates can be made. These are illustrated by comparison of maps for the New Madrid region (Fig. 9) that show the maximum predicted acceleration expected approximately once every 2500 yr for different assumptions. As shown, the areas of significant hazard (0.2 g corresponds approximately to the onset of major damage to some buildings) differ significantly. The differences are even

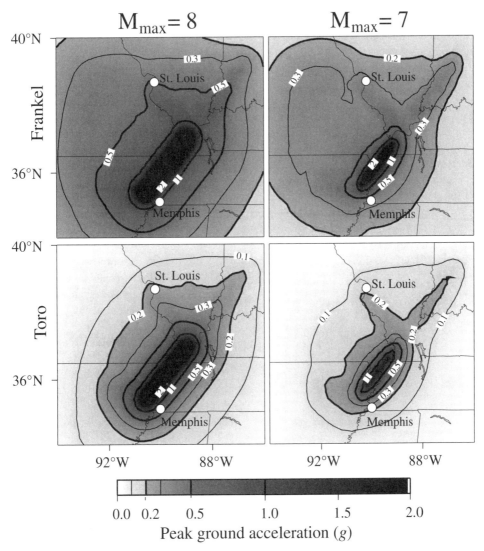

Figure 9. Comparison of the predicted seismic hazard (peak ground acceleration expected at 2% probability in 50 yr) from New Madrid seismic zone earthquakes for alternative parameter choices. Columns show the effect of varying the magnitude of the largest earthquake every 500 yr from 8 to 7, which primarily affects the predicted acceleration near the main faults. Rows show how different ground motion models affect the predicted acceleration over a larger area (after Newman et al., 2001).

greater for longer-period ground motion, which poses the threat to tall buildings. These uncertainties will remain unresolved at least until the next major earthquake.

An important additional contributor to the uncertainty, discussed earlier, is the question of whether to view the hazard as highest where recent seismicity has been concentrated or as essentially uniform within regions of similar structure. This question relates to the issue of whether locations of large future earthquakes are well predicted by the short seismic record or if instead seismicity migrates such that faults that seem aseismic from the earthquake record may be the next to generate a damaging earthquake. Depending on the assumptions made, quite different hazard estimates arise (Atkinson, chapter 21; Swafford and Stein, chapter 4). Put another way, we can assume that earthquakes are most likely in parts of a seismic zone where they have happened recently, more likely where they haven't happened recently, or equally likely throughout the zone. The predicted hazards vary: time-independent models predict the same probability of a large earthquake regardless of the time since the last one, whereas time-dependent models predict lower probabilities for the first two-thirds of the mean recurrence interval, and then higher probabilities as the earthquake is "due" (Fig. 10; Stein and Wysession, 2003; Stein et al., 2003). There is no standard choice: some California maps have been based on time-dependent probabilities, whereas the central U.S. maps (Frankel et al., 1996) are based on time-independence. In each region, these opposite assumptions chosen tend to predict higher probabilities than the alternative, due to the longer recurrence time in the central United States.

A final crucial issue is how to define the hazard. This issue is crucial in discussions of the appropriate codes to specify the earthquake resistance of buildings for intraplate areas. For example, the U.S. Federal Emergency Management Agency (FEMA) has proposed a new building code that would increase the earthquake resistance of new buildings in the New Madrid zone to levels similar to those in southern California. This proposal derives from an argument (Frankel, 2004) that the seismic hazard, defined as the maximum predicted acceleration expected at 2% probability in 50 yr, or approximately once every 2500 yr, is comparable for sites in the New Madrid zone to that for sites in California.

The utility of this criterion, which is much more stringent than the 500 yr one used for other natural disaster planning, is debatable. Searer et al. (chapter 23) show that the long time window makes the assumed hazard in the New Madrid seismic zone and California comparable, whereas use of a 500 yr window (as is used in California or most other countries) yields much higher hazard in California. Similarly, by taking a sufficiently long time, the hazard anywhere can be defined as comparable to California's (Stein, 2004a). This situation arises because the hazard is defined as the maximum shaking at a geographic point over a period of time rather than what would be experienced by a typical structure during its much shorter (50–100 yr) life. The difference is illustrated in Figure 11, which contrasts the fractions of the regions that might be shaken strongly enough to seriously damage some buildings. In 100 yr (upper panels), much of the California region will be shaken seriously, whereas a much smaller fraction of the New Madrid seismic zone would be. After 1000 yr (lower panels), much of the New Madrid seismic zone has been shaken once, whereas most of the California area has been shaken many times. Although the maximum shaking at a given location in the New Madrid seismic zone over thousands of years may be comparable to that in California, a building in California is much more likely to be seriously shaken during its ~50–100 yr life. Thus, over the life of a new building in Memphis, there is a reasonable probability of low to moderate shaking, but a significantly lower probability of severe shaking. Similar issues arise in other areas of intraplate seismicity.

DEVELOPING MITIGATION STRATEGIES

The final theme in this book, explored by Crandell (chapter 25), Lomnitz and Castanos (chapter 26) and Searer et al. (chapter 23) is the use of our knowledge to formulate policies that address the societal risk posed by continental intraplate earthquakes. Several approaches are used, all of which are equally applicable to mitigating the effects of other natural disasters. These include site restrictions that exclude certain structures from hazardous areas, building codes that require levels of earthquake resistance, insurance that compensates for losses and provides funds for reconstruction, and emergency preparedness for response during and after an earthquake.

Society must decide how much to accept in additional present costs in order to reduce both the direct and indirect losses in future earthquakes. This involves tradeoffs between present uses of resources and the use of those same resources for other applications that also do societal good. For example, funds

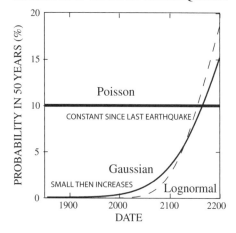

Figure 10. Predicted probabilities of a large New Madrid earthquake in the next 50 years as a function of time since the last one in 1812, for different models assuming a recurrence interval of 500 ± 100 yr. The predicted probability is much higher for the time-independent Poisson model than for the two-time-dependent models (Stein et al., 2003).

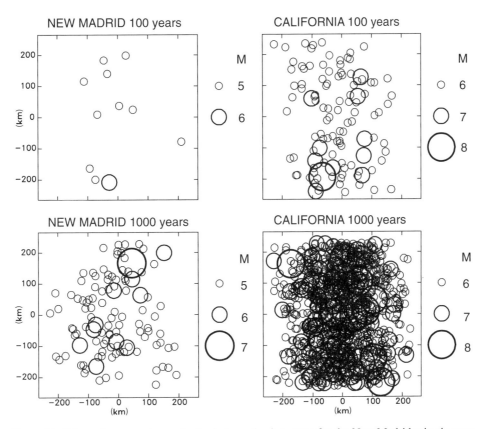

Figure 11. Schematic comparison of seismic hazard using maps for the New Madrid seismic zone and southern California on two time scales. Seismicity is assumed to be random, with California 100 times more active but New Madrid earthquakes causing strong shaking over an area equal to that for a California earthquake one-magnitude-unit larger. Areas of shaking with acceleration >0.2g are shown by circles (Stein and Wysession, 2003).

spent strengthening schools are not available to hire teachers, and stronger hospitals may come at the expense of providing health care. Similarly, imposing costs on the private sector can cause reduced economic activity (firms don't build or build elsewhere) and impose other costs, which in turn affect society as a whole. Choosing a mitigation strategy thus requires estimations of the costs and benefits of various possible strategies. Surprisingly, these have often been proposed and even implemented without this crucial analysis. For example, the 2500 yr hazard definition for the central United States was adopted without economic analysis, making its justification questionable (Searer et al., chapter 23).

Fortunately, there is an increasing trend to explore these issues. FEMA (2001) has developed estimates of annualized earthquake losses for various cities and states in the United States that can be used for comparison with the costs of potential mitigation strategies (Stein, 2003; Crandell chapter 25; Searer et al., chapter 23). Leonard et al. (chapter 17) illustrate how the probabilistic seismic hazard estimates and their uncertainties can be used to study potential earthquake losses. An important challenge is estimating how much various mitigation strategies would reduce losses, which is the benefit that needs

to be compared to their cost. A tricky aspect of this challenge is that it involves seismologists and earthquake engineers working together and appreciating each group's approach and the associated uncertainties.

Decisions on mitigation strategies involve tough choices that are ultimately economic and societal (Stein et al., 2003; Stein, 2004b, Crandell, chapter 25). Although these decisions are hard for earthquake hazard mitigation in any setting, it is especially difficult for the rarer intraplate earthquakes, the recurrence and effects of which are even less well understood. Helping to make these choices, given our imperfect knowledge, will be an increasing challenge for earth scientists in years to come as the population in earthquake-prone areas continues to grow.

ACKNOWLEDGMENTS

Much of the fun involved with puzzling over intraplate earthquakes and tectonics comes from the stimulating interchanges that this puzzling phenomenon generates. I have benefited from discussions over the years with many researchers including Andrew Newman, Tim Dixon, Mian Liu, Sue Hough, John Schneider, Giovanni Sella, Stephane Mazzotti, and Eric Calais.

REFERENCES CITED

Abrams, D., 2001, Will Gujarat's problem be ours: Mid-America Earthquake Center Newsletter, April, v. 4, p. 2–3.

Argus, D.F., and Gordon, R.G., 1996, Tests of the rigid-plate hypothesis and bounds on intraplate deformation using geodetic data from very long baseline interferometry: Journal of Geophysical Research, v. 101, p. 13,555–13,572, doi: 10.1029/95JB03775.

Atkinson, G., 2007, this volume, Challenges in seismic hazard analysis for continental interiors, *in* Stein, S., and Mazzotti, S., eds., Continental Intraplate Earthquakes: Science, Hazard, and Policy Issues: Geological Society of America Special Paper 425, doi: 10.1130/2007.2425(21).

Bada, G., Grenerczy, G., Toth, L., Horvath, F., Stein, S., Windhoffer, G., Fodor, L., Fejes, I., Pinter, N., and Cloetingh, S., 2007, this volume, Motion of Adria and ongoing inversion of the Pannonian Basin: Seismicity, GPS velocities and stress transfer, *in* Stein, S., and Mazzotti, S., eds., Continental Intraplate Earthquakes: Science, Hazard, and Policy Issues: Geological Society of America Special Paper 425, doi: 10.1130/2007.2425(16).

Beavers, J., 2001, Kutch earthquake: What we now know: Mid-America Earthquake Center Newsletter, February, v. 4, p. 2–4.

Bendick, R., Bilham, R., Fielding, E., Gaur, V.K., Hough, S.E., Kulkarni, M.N., Martin, S., Mueller, K., and Mukul, M., 2001, The 26 January 2001 "Republic Day" earthquake, India: Seismological Research Letters, v. 72, p. 328–335.

Bennett, R.A., Davis, J.L., and Wernicke, B.P., 1999, Present-day pattern of Cordilleran deformation in the Western United States: Geology, v. 27, p. 371–374, doi: 10.1130/0091-7613(1999)027<0371:PDPOCD>2.3.CO;2.

Bent, A., and Delahaye, E., 2007, this volume, H/V at short distances for four hardrock sites in eastern Canada and implications for seismic hazard assessment, *in* Stein, S., and Mazzotti, S., eds., Continental Intraplate Earthquakes: Science, Hazard, and Policy Issues: Geological Society of America Special Paper 425, doi: 10.1130/2007.2425(22).

Calais, E., Mattioli, G., DeMets, C., Nocquet, J.-M., Stein, S., Newman, A., and Rydelek, P., 2005, Tectonic strain in plate interiors?: Nature, v. 438, doi: 10.1038/nature04428.

Calais, E., Han, J., and DeMets, C., 2006, Deformation of the North American plate from a decade of continuous GPS measurements: Journal of Geophysical Research, v. 111, doi: 10.1029/2005JB004253.

Camelbeeck, T., Vanneste, K., Alexandre, P., Verbeeck, K., Petermans, T., Rosset, P., Everaerts, M., Warnant, R., and Van Camp, M., 2007, this volume, Relevance of active faulting and seismicity studies to assess long-term earthquake activity in Northwest Europe, *in* Stein, S., and Mazzotti, S., eds., Continental Intraplate Earthquakes: Science, Hazard, and Policy Issues: Geological Society of America Special Paper 425, doi: 10.1130/2007.2425(14).

Chase, C.G., 1972, The *n*-plate problem of plate tectonics: Geophysical Journal of the Royal Astronomical Society, v. 29, p. 117–122.

Chase, C.G., 1978, Plate kinematics: The Americas, East Africa, and the rest of the world: Earth and Planetary Science Letters, v. 37, p. 355–368, doi: 10.1016/0012-821X(78)90051-1.

Chu, D., and Gordon, R.G., 1999, Evidence for motion between Nubia and Somalia along the Southwest Indian ridge: Nature, v. 398, p. 64–67, doi: 10.1038/18014.

Crandell, J., 2007, this volume, Policy development and uncertainty in earthquake risk in the New Madrid seismic zone, *in* Stein, S., and Mazzotti, S., eds., Continental Intraplate Earthquakes: Science, Hazard, and Policy Issues: Geological Society of America Special Paper 425, doi: 10.1130/2007.2425(25).

Crone, A.J., and Luza, K.V., 1990, Style and timing of Holocene surface faulting on the Meers fault, southwestern Oklahoma: Geological Society of America Bulletin, v. 102, p. 1–17, doi: 10.1130/0016-7606(1990)102<0001: SATOHS>2.3.CO;2.

Crone, A.J., De Martini, P.M., Machette, M.N., Okumura, K., and Prescott, J.R., 2003, Paleoseismicity of two historically quiescent faults in Australia: Implications for fault behavior in stable continental regions: Bulletin of the Seismological Society of America, v. 93, p. 1913–1934.

DeMets, C., Gordon, R.G., Argus, D.F., and Stein, S., 1990, Current plate motions: Geophysical Journal International, v. 101, p. 425–478.

DeMets, C., Gordon, R.G., Argus, D.F., and Stein, S., 1994, Effect of recent revisions to the geomagnetic reversal time scale on estimates of current plate motion: Geophysical Research Letters, v. 21, p. 2191–2194, doi: 10.1029/94GL02118.

Dixon, T.H., Mao, A., and Stein, S., 1996, How rigid is the stable interior of the North American plate?: Geophysical Research Letters, v. 23, p. 3035–3038, doi: 10.1029/96GL02820.

Ellis, M.E., Gomberg, J., and Schweig, E., 2001, Indian earthquake may serve as analog for New Madrid earthquakes: Eos (Transactions, American Geophysical Union), v. 82, p. 345–350, doi: 10.1029/01EO00211.

FEMA (Federal Emergency Management Agency), 2001, HAZUS 99 Estimated Annualized Losses for the United States: Federal Emergency Management Agency Publication 366.

Flesch, L.M., Holt, W.E., Haines, A.J., and Shen-Tu, B., 2000, Dynamics of the Pacific–North American plate boundary zone in the western United States: Science, v. 287, p. 834–836, doi: 10.1126/science.287.5454.834.

Frankel, A., 2004, How can seismic hazard in the New Madrid seismic zone be similar to that in California?: Seismological Research Letters, v. 75, p. 575–586.

Frankel, A., Mueller, C., Barnhard, T., Perkins, D., Leyendecker, E., Dickman, N., Hanson, S., and Hopper, M., 1996, National Seismic Hazard Maps Documentation: U.S. Geological Survey Open-File Report 96-532, 110 p.

Gan, W., and Prescott, W., 2001, Crustal deformation rates in central and eastern U.S. inferred from GPS: Geophysical Research Letters, v. 28, p. 3733–3736, doi: 10.1029/2001GL013266.

Gangopadhyay, A., and Talwani, P., 2003, Symptomatic features of intraplate earthquakes: Seismological Research Letters, v. 74, p. 863–883.

Gangopadhyay, A., and Talwani, P., 2007, this volume, Two-dimensional numerical modeling suggests preferred geometry of intersecting seismogenic faults, *in* Stein, S., and Mazzotti, S., eds., Continental Intraplate Earthquakes: Science, Hazard, and Policy Issues: Geological Society of America Special Paper 425, doi: 10.1130/2007.2425(07).

Giardini, D., Grunthal, G., Shedlock, K.M., and Zhang, P., 2000, The GSHAP Global Seismic Hazard Map: Seismological Research Letters, v. 71, p. 679–686.

Gordon, R.G., Stein, S., DeMets, C., and Argus, D.F., 1987, Statistical tests for closure of plate motion circuits: Geophysical Research Letters, v. 14, p. 587–590.

Grollimund, B., and Zoback, M.D., 2001, Did deglaciation trigger intraplate seismicity in the New Madrid seismic zone?: Geology, v. 29, p. 175–178.

Hasegawa, H.S., and Basham, P., 1989, Spatial correlation between seismicity and postglacial rebound in eastern Canada, *in* Gregerson, S., and Basham, P., eds., Earthquakes at North Atlantic Passive Margins: Neotectonics and Postglacial Rebound: Dordrecht, Kluwer, p. 483–500.

Hinzen, K., and Reamer, S., 2007, this volume, Seismicity, seismotectonics, and seismic hazard in the northern Rhine Area, *in* Stein, S., and Mazzotti, S., eds., Continental Intraplate Earthquakes: Science, Hazard, and Policy Issues: Geological Society of America Special Paper 425, doi: 10.1130/2007.2425(15).

Hough, S., 2007, this volume, Remotely triggered earthquakes following moderate mainshocks, *in* Stein, S., and Mazzotti, S., eds., Continental Intraplate Earthquakes: Science, Hazard, and Policy Issues: Geological Society of America Special Paper 425, doi: 10.1130/2007.2425(06).

Hough, S., Armbruster, J.G., Seeber, L., and Hough, J.F., 2000, On the modified Mercalli intensities and magnitudes of the 1811/1812 New Madrid, central United States, earthquakes: Journal of Geophysical Research, v. 105, p. 23,839–23,864, doi: 10.1029/2000JB900110.

Jackson, J., and McKenzie, D., 1988, The relationship between plate motions and seismic moment tensors, and the rates of active deformation in the Mediterranean and Middle East: Geophysical Journal of the Royal Astronomical Society, v. 93, p. 45–73.

Jacobi, R., Lewis, C.F., Armstrong, D., and Blasco, S., 2007, this volume, Popup field in Lake Ontario south of Toronto, Canada: Indicators of late glacial and post-glacial strain, *in* Stein, S., and Mazzotti, S., eds., Continental Intraplate Earthquakes: Science, Hazard, and Policy Issues: Geological Society of America Special Paper 425, doi: 10.1130/2007.2425(10).

James, T.S., and Bent, A.L., 1994, A comparison of eastern North America seismic strain-rates to glacial rebound strain-rates: Geophysical Research Letters, v. 21, p. 2127–2130, doi: 10.1029/94GL01854.

Kafka, A., 2007, this volume, Does seismicity delineate zones where future large earthquakes are likely to occur in intraplate environments?, *in* Stein, S., and Mazzotti, S., eds., Continental Intraplate Earthquakes: Science, Hazard, and Policy Issues: Geological Society of America Special Paper 425, doi: 10.1130/2007.2425(03).

Kenner, S.J., and Segall, P., 2000, A mechanical model for intraplate earthquakes: Application to the New Madrid seismic zone: Science, v. 289, p. 2329–2332, doi: 10.1126/science.289.5488.2329.

Klosko, E.R., Stein, S., Hindle, D., Kley, J., Norabuena, E., Dixon, T., and Liu, M., 2002, Comparison of GPS, seismological, and geologic observations of Andean mountain building, *in* Stein, S., and Freymueller, J., eds., Plate Boundary Zones: Washington, D.C., American Geophysical Union, p. 123–133.

Kreemer, C., Haines, J., and Holt, W.E., 2002, The global moment rate distribution within plate boundary zones, *in* Stein, S., and Freymueller, J., eds., Plate Boundary Zones: Washington, D.C., American Geophysical Union, p. 173–190.

Leonard, M., Robinson, D., Allen, T., Schneider, J., Clark, D., Dhu, T., and Burbidge, D., 2007, this volume, Toward a better model for earthquake hazard in Australia, *in* Stein, S., and Mazzotti, S., eds., Continental Intraplate Earthquakes: Science, Hazard, and Policy Issues: Geological Society of America Special Paper 425, doi: 10.1130/2007.2425(17).

Li, Q., Liu, M., and Yang, Y., 2002, The 01/26/2001 Bhuj earthquake: Intraplate or interplate?, *in* Stein, S., and Freymueller, J., eds., Plate Boundary Zones: Washington, D.C., American Geophysical Union, p. 255–264.

Li, Q., Liu, M., Zhang, Q., and Sandoval, E., 2007, this volume, Stress evolution and seismicity in the central-eastern USA: Insight from geodynamic modeling, *in* Stein, S., and Mazzotti, S., eds., Continental Intraplate Earthquakes: Science, Hazard, and Policy Issues: Geological Society of America Special Paper 425, doi: 10.1130/2007.2425(11).

Liu, M., Zhu, Y., Stein, S., Yang, Y., and Engeln, J., 2000, Crustal shortening in the Andes: Why do GPS rates differ from geological rates: Geophysical Research Letters, v. 27, p. 3005–3008, doi: 10.1029/2000GL008532.

Liu, M., Yang, Y., Stein, S., and Klosko, E., 2002, Crustal shortening and extension in the Andes from a viscoelastic model, *in* Stein, S., and Freymueller, J., eds., Plate Boundary Zones: Washington, D.C., American Geophysical Union, p. 325–339.

Liu, M., Yang, Y., Shen, Z., Wang, S., Wang, M., and Wan, Y., 2007, this volume, Active tectonics and intracontinental earthquakes in China: The kinematics and geodynamics, *in* Stein, S., and Mazzotti, S., eds., Continental Intraplate Earthquakes: Science, Hazard, and Policy Issues: Geological Society of America Special Paper 425, doi: 10.1130/2007.2425(19).

Lomnitz, C., and Castanos, H., 2007, this volume, Disasters and maximum entropy production, *in* Stein, S., and Mazzotti, S., eds., Continental Intraplate Earthquakes: Science, Hazard, and Policy Issues: Geological Society of America Special Paper 425, doi: 10.1130/2007.2425(26).

Mazzotti, S., 2007, this volume, Geodynamic models for North America intraplate earthquakes, *in* Stein, S., and Mazzotti, S., eds., Continental Intraplate Earthquakes: Science, Hazard, and Policy Issues: Geological Society of America Special Paper 425, doi: 10.1130/2007.2425(02).

Mazzotti, S., and Adams, J., 2005, Rates and uncertainties on seismic moment and deformation in eastern Canada: Journal of Geophysical Research, v. 110, doi: 10.1029/2004JB003510.

Mazzotti, S., James, T., Henton, J., and Adams, J., 2005, GPS crustal strain, postglacial rebound, and seismic hazard in eastern North America: The Saint Lawrence valley example: Journal of Geophysical Research, v. 110, doi: 10.1029/2004JB003590.

McKenna, J., Stein, S., and Stein, C.A., 2007, this volume, Is the New Madrid seismic zone hotter and weaker than its surroundings?, *in* Stein, S., and Mazzotti, S., eds., Continental Intraplate Earthquakes: Science, Hazard, and Policy Issues: Geological Society of America Special Paper 425, doi: 10.1130/2007.2425(12).

Minster, J.B., and Jordan, T.H., 1978, Present-day plate motions: Journal of Geophysical Research, v. 83, p. 5331–5354.

Minster, J.B., Jordan, T.H., Molnar, P., and Haines, E., 1974, Numerical modeling of instantaneous plate tectonics: Geophysical Journal of the Royal Astronomical Society, v. 36, p. 541–576.

Morgan, W.J., 1968, Rises, trenches, great faults, and crustal blocks: Journal of Geophysical Research, v. 73, p. 1959–1982.

Newman, A., Stein, S., Weber, J., Engeln, J., Mao, A., and Dixon, T., 1999, Slow deformation and lower seismic hazard at the New Madrid seismic zone: Science, v. 284, p. 619–621, doi: 10.1126/science.284.5414.619.

Newman, A., Stein, S., Schneider, J., and Mendez, A., 2001, Uncertainties in seismic hazard maps for the New Madrid seismic zone: Seismological Research Letters, v. 72, p. 653–667.

Nocquet, J.M., Calais, E., and Parsons, B., 2005, Geodetic constraints on glacial isostatic adjustment in Europe: Geophysical Research Letters, v. 32, doi: 10.1029/2004GL022174.

Okal, E., and Sweet, J., 2007, this volume, Frequency-size distributions for intraplate seismicity, *in* Stein, S., and Mazzotti, S., eds., Continental Intraplate Earthquakes: Science, Hazard, and Policy Issues: Geological Society of America Special Paper 425, doi: 10.1130/2007.2425(05).

Pacheco, J., Sykes, L.R., and Scholz, C.H., 1993, Nature of seismic coupling along simple plate boundaries of the subduction type: Journal of Geophysical Research, v. 98, p. 14,133–14,159.

Pancha, A., Anderson, J., and Kreemer, C., 2006, Comparison of seismic and geodetic scalar moment rates across the Basin and Range Province: Bulletin of the Seismological Society of America, v. 96, p. 11–32, doi: 10.1785/0120040166.

Peltier, W.R., 2004, Global glacial isostasy and the surface of the ice-age Earth: The ICE-5G (VM2) model: Annual Review of Earth and Planetary Sciences, v. 32, p. 111–149, doi: 10.1146/annurev.earth.32.082503.144359.

Reading, A., 2007, this volume, The seismicity of the Antarctic plate, *in* Stein, S., and Mazzotti, S., eds., Continental Intraplate Earthquakes: Science, Hazard, and Policy Issues: Geological Society of America Special Paper 425, doi: 10.1130/2007.2425(18).

Rydelek, P.A., 2007, New Madrid strain and postseismic transients: Eos (Transactions, American Geophysical Union), v. 88, p. 60–61.

Rydelek, P.A., and Pollitz, F.F., 1994, Fossil strain from the 1811–1812 New Madrid earthquakes: Geophysical Research Letters, v. 21, p. 2303–2306, doi: 10.1029/94GL02057.

Sarkar, D., Sain, K., Reddy, P.R., Catchings, R.D., and Mooney, W.D., 2007, this volume, Seismic reflection images of the crust beneath the 2001 M = 7.7 Kutch (Bhuj) epicentral region, western India, *in* Stein, S., and Mazzotti, S., eds., Continental Intraplate Earthquakes: Science, Hazard, and Policy Issues: Geological Society of America Special Paper 425, doi: 10.1130/2007.2425(20).

Schulte, S.M., and Mooney, W.D., 2005, An updated global earthquake catalogue for stable continental regions: Reassessing the correlation with ancient rifts: Geophysical Journal International, v. 161, p. 707–721, doi: 10.1111/j.1365-246X.2005.02554.x.

Schweig, E.S., and Ellis, M.A., 1994, Reconciling short recurrence intervals with minor deformation in the New Madrid seismic zone: Science, v. 264, p. 1308–1311, doi: 10.1126/science.264.5163.1308.

Searer, G., 2007, this volume, Does it make sense from engineering and economic perspectives to design for a 2475-year earthquake?, *in* Stein, S., and Mazzotti, S., eds., Continental Intraplate Earthquakes: Science, Hazard, and Policy Issues: Geological Society of America Special Paper 425, doi: 10.1130/2007.2425(23).

Sella, G.F., Dixon, T.H., and Mao, A., 2002, REVEL: A model for recent plate velocities from space geodesy: Journal of Geophysical Research, v. 107, no. B4, doi: 10.1029/2000JB000033.

Sella, G.F., Stein, S., Dixon, T., Craymer, M., James, T., Mazzotti, S., and Dokka, R., 2006, Observations of glacial isostatic adjustment in stable North America with GPS: Geophysical Research Letters, v. 34, doi: 10.1029/2006GL02708.

Stein, R.S., and Hanks, T.C., 1998, M > 6 earthquakes in Southern California during the twentieth century: No evidence for a seismicity or moment deficit: Bulletin of the Seismological Society of America, v. 88, p. 635–652.

Stein, S., 2004a, Comment on "How can seismic hazard in the New Madrid seismic zone be similar to that in California?" by A. Frankel: Seismological Research Letters, v. 75, p. 362–363.

Stein, S., 2004b, No free lunch: Seismological Research Letters, v. 75, p. 555–556.

Stein, S., 2007, New Madrid GPS: Much ado about nothing?: Eos (Transactions, American Geophysical Union), v. 88, p. 58–60.

Stein, S., and Gordon, R.G., 1984, Statistical tests of additional plate boundaries from plate motion inversions: Earth and Planetary Science Letters, v. 69, p. 401–412, doi: 10.1016/0012-821X(84)90198-5.

Stein, S., and Newman, A., 2004, Characteristic and uncharacteristic earthquakes as possible artifacts: Applications to the New Madrid and Wabash seismic zones: Seismological Research Letters, v. 75, p. 170–184.

Stein, S., and Sella, G.F., 2002, Plate boundary zones: Concept and approaches, *in* Stein, S., and Freymueller, J., eds., Plate Boundary Zones: Washington, D.C., American Geophysical Union, p. 1–26.

Stein, S., and Wysession, M., 2003, Introduction to Seismology, Earthquakes, and Earth Structure: Oxford, Blackwell, 498 p.

Stein, S., Sleep, N.H., Geller, R.J., Wang, S.C., and Kroeger, G.C., 1979, Earthquakes along the passive margin of eastern Canada: Geophysical Research Letters, v. 6, p. 537–540.

Stein, S., Cloetingh, S., Sleep, N., and Wortel, R., 1989, Passive margin earthquakes, stresses, and rheology, *in* Gregerson, S., and Basham, P., eds., Earthquakes at North Atlantic Passive Margins: Neotectonics and Postglacial Rebound: Dordrecht, Kluwer, p. 231–260.

Stein, S., Sella, G.F., and Okal, E.A., 2002, The January 26, 2001, Bhuj earthquake and the diffuse western boundary of the Indian plate, *in* Stein, S., and Freymueller, J., eds., Plate Boundary Zones: Washington, D.C., American Geophysical Union, p. 243–254.

Stein, S., Newman, A., and Tomasello, J., 2003, Should Memphis build for California's earthquakes?: Eos (Transactions, American Geophysical Union), v. 84, p. 177, 184–185.

Stein, S., Friedrich, A., and Newman, A., 2005, Dependence of possible characteristic earthquakes on spatial sampling: illustration for the Wasatch seismic zone, Utah: Seismological Research Letters, v. 76, p. 432–436.

Swafford, L., and Stein, S., 2007, this volume, Limitations of the short earthquake record for seismicity and seismic hazard studies, *in* Stein, S., and Mazzotti, S., eds., Continental Intraplate Earthquakes: Science, Hazard, and Policy Issues: Geological Society of America Special Paper 425, doi: 10.1130/2007.2425(04).

Sykes, L.R., 1978, Intraplate seismicity, reactivation of pre-existing zones of weakness, alkaline magmatism, and other tectonism postdating continental fragmentation: Reviews of Geophysics and Space Physics, v. 16, p. 621–688.

Tuttle, M.P., 2001, The use of liquefaction features in paleoseismology: Lessons learned in the New Madrid seismic zone, central U.S.: Journal of Seismology, v. 5, p. 361–380, doi: 10.1023/A:1011423525258.

Van Arsdale, R., Bresnahan, R., McCallister, N., and Waldron, B., 2007, this volume, The upland complex of the central Mississippi River valley: Its origin, denudation, and possible role in reactivation of the New Madrid seismic zone, *in* Stein, S., and Mazzotti, S., eds., Continental Intraplate Earthquakes: Science, Hazard, and Policy Issues: Geological Society of America Special Paper 425, doi: 10.1130/2007.2425(13).

Wang, Z., 2007, this volume, Seismic hazard and risk assessment in the intraplate environment: The New Madrid seismic zone of the central United States, *in* Stein, S., and Mazzotti, S., eds., Continental Intraplate Earthquakes: Science, Hazard, and Policy Issues: Geological Society of America Special Paper 425, doi: 10.1130/2007.2425(24).

Wesnousky, S.G., 1994, The Gutenberg-Richter or characteristic earthquake distribution, which is it?: Bulletin of the Seismological Society of America, v. 84, p. 1940–1959.

Wesnousky, S.G., and Leffler, L.M., 1992, The repeat time of the 1811 and 1812 New Madrid earthquakes: A geological perspective: Bulletin of the Seismological Society of America, v. 82, p. 1756–1785.

Wiens, D.A., DeMets, C., Gordon, R.G., Stein, S., Argus, D., Engeln, J.F., Lundgren, P., Quible, D., Stein, C., Weinstein, S., and Woods, D.F., 1985, A diffuse plate boundary model for Indian Ocean tectonics: Geophysical Research Letters, v. 12, p. 429–432.

Wu, P., and Johnston, P., 2000, Can deglaciation trigger earthquakes in North America?: Geophysical Research Letters, v. 27, p. 1323–1326, doi: 10.1029/1999GL011070.

Wu, P., and Mazzotti, S., 2007, this volume, Effects of a lithospheric weak zone on postglacial seismotectonics in eastern Canada and northeastern USA, *in* Stein, S., and Mazzotti, S., eds., Continental Intraplate Earthquakes: Science, Hazard, and Policy Issues: Geological Society of America Special Paper 425, doi: 10.1130/2007.2425(09).

Zoback, M.L., and Zoback, M.D., 1989, Tectonic stress field of the continental United States, *in* Pakiser, L.C., and Mooney, W.D., eds., Geophysical Framework of the Continental United States: Geological Society of America Memoir 172, p. 523–539.

MANUSCRIPT ACCEPTED BY THE SOCIETY 29 NOVEMBER 2006

The Geological Society of America
Special Paper 425
2007

Geodynamic models for earthquake studies in intraplate North America

Stéphane Mazzotti

Geological Survey of Canada, Pacific Geoscience Centre, 9860 West Saanich Road, Sidney, BC V8L 4B2, Canada

ABSTRACT

A common view of continental intraplate seismicity is that large earthquakes occur in areas where peculiar local conditions favor lower lithospheric strength and/or higher stress concentration compared to typical intraplate settings. Although there are numerous explanations for these local strength reduction and stress increase effects, their application to seismic hazard assessment is limited to the few specific regions for which these explanations were developed. In this paper, I present four general models that can be used to define seismic hazards based on the associated geodynamic frameworks and their implications for earthquake locations, sizes, and recurrence rates. The four models are defined by the relationships among lithospheric strength contrasts, strain distribution, and earthquake characteristics, and they may apply to different intraplate regions. (1) The random model, defined by the lack of significant lithospheric structure and the spatial and temporal randomness of seismicity, may be applicable to Precambrian cratons and shields. (2) In the plate-boundary model, earthquakes concentrate along lithospheric-scale tectonic structures under low intraplate strain rates, which may apply to eastern North America. (3) The localized weak zone model postulates that large earthquakes are limited to small areas of crustal weakness and high strain concentration (e.g., New Madrid seismic zone in the central United States). (4) The large-scale weak zone model is characterized by high crustal strain concentration in major paleotectonic structures, along which large earthquakes are spatially confined but susceptible to migration with time. This last model may apply to Paleozoic and Mesozoic rift and basin regions, such as the St. Lawrence valley in eastern Canada. Because all four models are built on the relationship between lithospheric strength, strain distribution, and earthquake characteristics, they can be used as a framework for experiments designed to test their validity. I discuss two lines of studies that address the relationship among strength, strain, and earthquakes. The first type deals with strength of the crust and upper mantle using rock rheology, thermal profiles, and average strain rates in intraplate seismic regions. The second type is based on geodetic measurements of intraplate strain rate patterns and amplitudes.

Keywords: earthquakes, intraplate, geodynamics, tectonics.

Mazzotti, S., 2007, Geodynamic models for earthquake studies in intraplate North America, *in* Stein, S., and Mazzotti, S., ed., Continental Intraplate Earthquakes: Science, Hazard, and Policy Issues: Geological Society of America Special Paper 425, p. 17–33, doi: 10.1130/2007.2425(02). For permission to copy, contact editing@geosociety.org. ©2007 The Geological Society of America. All rights reserved.

INTRODUCTION

Although not as frequent, and possibly not as large, as their plate-boundary cousins, large continental intraplate earthquakes can pose a significant risk to population and infrastructure centers. The most common mechanisms invoked to explain the occurrence of intraplate seismicity are based on the concepts of lithospheric weakness and local stress concentration compared to a typical intraplate state of strength and stress. As a mechanism for promoting earthquakes, local reduction of crustal and/or upper mantle strength can be caused by thermal (e.g., Liu and Zoback, 1997), mechanical (e.g., Talwani and Acree, 1984), or chemical anomalies (e.g., Costain et al., 1987). Local increases in differential stresses can be related to density contrasts (e.g., Zoback and Richardson, 1996), local topography (e.g., Assameur and Mareschal, 1995), or kinks and intersections in a fault system (e.g., Talwani, 1988). In all cases, the common underlying assumption is that seismicity is promoted by local physical perturbations in an otherwise less-seismic "typical" intraplate environment.

Because they have been developed only for a few areas with relatively good geological and geophysical data coverage and earthquake record, these explanations for intraplate earthquake occurrence are difficult to include in large-scale seismic hazard models. If the frequent large earthquakes in the New Madrid seismic zone, central United States, are induced by the combination of a slightly warmer geotherm, the presence of a dense rift pillow, and the transient perturbation of Holocene postglacial rebound (e.g., Pollitz et al., 2001), what are the implications for seismic hazard in the rest of the central and eastern United States, eastern Canada, and intraplate regions in general?

In this regard, the current version of the Canadian seismic hazard model (Adams and Halchuk, 2003) has adopted a new geodynamic approach to the problem of describing earthquake spatial distribution. One of the concepts used for defining the hazard model is that large earthquakes may occur anywhere along the paleotectonic zones encompassing the Late Proterozoic–Cambrian Iapetan rift margin and associated aulacogens (Adams et al., 1995; Adams and Halchuk, 2003). In this paper, I extend this approach and present four simple geodynamic models that can be used to describe the location, rate of occurrence, and magnitude of intraplate earthquakes. The four models are compared with current seismicity and deformation data, mostly using the eastern North America example. These models are described in terms of three principal characteristics: correlation of earthquakes with zones of lithospheric weakness, amplitude and spatial pattern of crustal strain rate, and spatial and temporal distribution of earthquakes. These three characteristics are obviously interrelated and form the basis of the geodynamic framework. Each of the four models has specific implications for the location, size, and recurrence of large earthquakes, leading to specific impacts on seismic hazard assessments.

The concept of weak zones in the lower crust and/or upper mantle acting as a potential control on earthquake location, size, and recurrence is central to the four models presented here. The existence, or absence, of such weak zones in the lithosphere can be studied through different perspectives. I discuss two lines of study that can help to constrain the characteristics of potential lithospheric weak zones and thus discriminate among the four models. The first line is based on a standard rheology approach the crustal and upper-mantle strength and how strength variations relate to strain rate concentration. The second, and complementary, approach is based on global positioning system (GPS) measurements of crustal strain with a specific emphasis on how strain rate patterns would vary with different loading and geodynamic model assumptions.

INTRAPLATE EARTHQUAKE CHARACTERISTICS

Before discussing the geodynamic models that may apply to intraplate regions, I review the main characteristics of intraplate seismicity that provide the principal constraints in the development of these models: the correlation of seismicity with paleotectonic structures, the lack of evidence for significant recent deformation structures or active faults, and the low level of intraplate strain rates. This short review is by no mean comprehensive and only serves as a background for discussing the merits of the proposed geodynamic frameworks. The review is mostly based on eastern North America seismicity with some examples from other major intraplate regions.

Eastern North America is one of the most studied examples of continental intraplate seismicity. Typically, background seismicity has been monitored and recorded down to $M = 2$–3 for the last 20–30 yr, and the relatively long history of settlements provides information on large earthquakes back to the seventeenth century in some areas of New England and southeastern Canada. Figure 1 shows a compilation of seismicity east of 100°W, excluding the Canadian Arctic, with the historical $M = 6$–8 earthquakes indicated.

One of the main characteristics of intraplate earthquakes is their high spatial correlation with paleotectonic zones and in particular with paleo–rift systems. As pointed by Johnston et al. (1994), about two-thirds of continental intraplate earthquakes occur within regions characterized by Mesozoic or older structures of crustal extension: i.e., rifted margins, aulacogens, and extensional basins. This correlation is particularly striking in eastern North America, where most of the known $M = 6$–7 earthquakes in the past 350 yr have occurred along the Atlantic and Iapetan rift basins, rifted margin, and aulacogens (Fig. 2; e.g., Adams and Basham, 1991). This distribution is also reflected in the background earthquake statistics. Averaged over the same unit area, the rate of $M = 3$–6 earthquakes in the paleotectonic zones of eastern Canada is ~50–100 larger than that of the Canadian Shield (Fig. 3).

Although there is a clear tendency for intraplate earthquakes to occur within paleotectonic regions, there are very few clear associations between known earthquake locations and active faults. In many cases, earthquake locations are relatively poorly known (especially depth estimates), precluding any definitive association

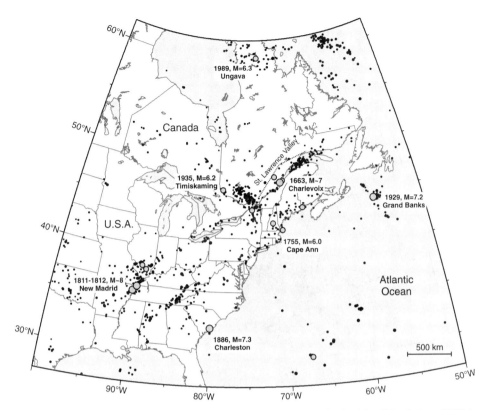

Figure 1. Intraplate seismicity in eastern North America. Background seismicity (M ≥ 3 since 1973) is shown by small black circles, and historical large earthquakes (M ≥ 6) are shown by large gray circles. Data are from the Geological Survey of Canada and the U.S. Geological Survey online databases.

with fault scarps or other geological or geophysical lineaments. In a few cases where the earthquake locations are well constrained, the association with active faults is still dubious. The Charlevoix seismic zone of eastern Canada (Fig. 1) is a good example of this unclear relation. The largest recorded event (1925 M_w = 6.2 earthquake) occurred along a fault plane similar in strike and dip to the major faults associated with the Iapetan rift system (Bent, 1992). However, moderate size M = 4–5 earthquakes show a more complex distribution of fault-plane mechanisms (Lamontagne, 1999), and small events appear to cluster away from the projected locations of the main Iapetan faults (Vlahovic et al., 2003).

The Charlevoix and New Madrid seismic zones, the two most active zones of eastern North America (Fig. 1), lack mapped fault traces that would show significant relative displacements associated with the earthquake activity. In both regions, extrapolation of current earthquake statistics over the last 1 m.y. would suggest total seismic deformation on the order of a few kilometers (e.g., Mazzotti and Adams, 2005). In contrast, detailed seismic-reflection studies have indicated that the total accumulated deformation on any individual structure in these two regions cannot exceed a few tens of meters (e.g., Hamilton and Zoback, 1981; Lamontagne, 1999; Van Arsdale, 2000). This paradox leads to the conclusion that seismic activity must be transient and represent either the recent (Holocene) onset of significant deformation or the spatially migrating character of long-term deformation.

The lack of significant deformation structures and the low level of seismicity, compared to plate-boundary regions, may be an expression of the intrinsically low strain rates that characterize continental intraplate regions. Typical intraplate strain rates are less than ~10^{-10} yr^{-1} (~10^{-17} s^{-1}, equivalent to ~100 m per 1000 km strain over 1 m.y.). This upper bound on average intraplate strain rates is well constrained by plate tectonic reconstructions (e.g., Gordon, 1998), geodetic measurements (e.g., Dixon et al., 1996; Calais et al., 2006), and earthquake statistics (Anderson, 1986; Mazzotti and Adams, 2005). Estimates of long-term intraplate strain rates also suggest that high strain rates and high seismic activity are transient features. Examples of higher strain rate regions are discussed in more detail in the next section. Temporal and spatial variations of intraplate strain rate, and their link with the lack of clear active deformation structure, serve as the basis for the definition of the geodynamic models discussed here.

GEODYNAMIC MODELS

Seismic hazard assessment requires essentially three types of information: (1) the location of future earthquakes; (2) estimates of the magnitude and frequency of these events; and (3) ground motion relations to relate earthquake magnitude to shaking estimates. The latter component is a critical aspect of hazard estimates but mostly relates to wave propagation theory

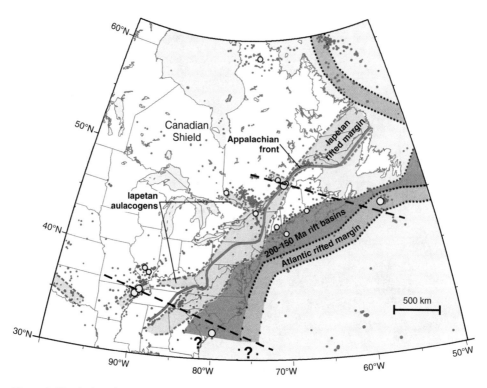

Figure 2. North America seismicity and paleotectonic structures. Background seismicity and large earthquakes are shown by orange and large yellow circles (data as in Fig. 1). Paleotectonic zones (from west to east): Green shade: 650–600 Ma Iapetan rifted margin and aulacogens; blue line: westward limit of 450–300 Ma Appalachian thrust nappes; purple shade and blue shade: 200–150 Ma rift basins and rifted margin of Atlantic Ocean. The thick dashed lines show the locations of the cross sections in Figure 7.

(c.f. discussion in Boore and Atkinson, 1987). The first two components are directly related to the dynamics of intraplate deformation. The four models proposed in this section can be used as a framework to describe the location, magnitude, and recurrence rate of large earthquakes (Fig. 4). Each model is simply defined by three parameters: crustal/lithospheric weakness along geological structures, long-term strain rate, and spatial and temporal distribution of earthquakes. In all four models, I emphasize the importance of the reduction of crustal and/or upper mantle strength as the main constraint on earthquake location. This view assumes that "stress concentrators," such as fault intersections or local topography, are second-order features that might control earthquake locations within a preweakened crust but cannot trigger significant seismic deformation in a typical strong intraplate crust. This assumption is discussed in more details in the Mechanical Control section.

Random Seismicity and Low Strain Model

This framework is characterized by a lack of large-scale paleotectonic weak structures, a low uniform intraplate strain rate, and a random distribution of earthquakes in space and time (Fig. 4A). In other words, large earthquakes can occur anywhere, but their magnitudes and recurrence intervals are limited by the low strain rate. Assuming that a $M = 7$ earthquake is associated with ~1 m of slip over a 50×20 km fault dimension, a typical intraplate strain rate of 10^{-10} yr^{-1} is equivalent to about one $M = 7$ earthquake per 10,000 yr per 1000 km. In three dimensions, this low strain corresponds to about one $M = 7$ earthquake per 1000 yr per million km^2 (i.e., about one $M = 7$ per 100 yr over the entire central and eastern North America area).

In terms of earthquake hazard, the implications of this model are straightforward. The lack of association between large earthquakes and significant weak structures under low intraplate strain rates implies that $M > 6$ events can occur at any location but with a very low return frequency. In this case, the hazard may be best treated as a probabilistic uniform value across the entire region considered.

The random model may be appropriate for stable continental cratons, which are characterized by extremely low strain rates and seismic activity. Based on worldwide compilation of instrumental seismicity, Fenton et al. (2006) showed that the average recurrence rate of large earthquakes in Precambrian cratons and shields is about one $M = 6$ earthquake per 250 yr per million km^2. Apparent variations exist between low-rate regions such as Siberia and Arabia (no known $M \geq 6$ earthquake) and high-rate regions such as the western Australian craton ($M \geq 6$ return rate of about one per 30 yr per million km^2). These recur-

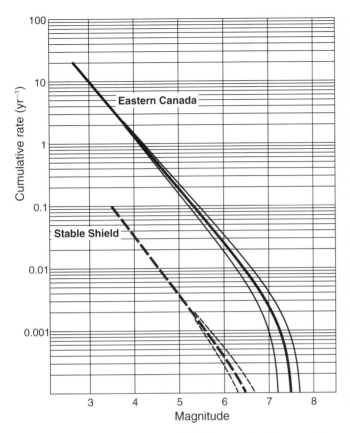

Figure 3. Earthquake recurrence statistics for central and eastern Canada showing cumulative recurrence rate versus earthquake magnitude normalized by area (in million km²). The eastern Canada curve corresponds to all earthquakes east of the western limit of the Iapetan structures (cf. Fig. 2). The stable shield curve corresponds to all earthquakes between the eastern Canada zone and the Rocky Mountains.

rence statistics are compatible, within one order of magnitude, with general estimates of intraplate strain rates.

On the other hand, the random model is likely not valid in paleotectonic regions such as eastern North America or eastern Australia, where the seismicity correlates with major tectonic structures. Instrumental, historical, and paleoseismic records in these regions indicate that earthquakes tend to cluster along large-scale structures with, to a first order, return periods of large events compatible with small earthquake statistics. Examples from eastern Canada (Mazzotti and Adams, 2005), eastern United States (Anderson, 1986), or southeastern Australia (Sandiford et al., 2004) also indicate seismic strain rates locally as high as 10^{-9}–10^{-8} yr^{-1}, about two orders of magnitude larger than craton values.

Plate-Boundary Zone Model

Under this model, intraplate earthquakes are associated with large-scale lithospheric structures and are symptomatic of deformation of the entire lithosphere section, as in a plate-boundary zone (Fig. 4B). The major difference with a typical plate-boundary region is that strain rates, although focused along specific structures, are typical of intraplate environments (~10^{-10} yr^{-1} or less) and are several orders of magnitude smaller than typical plate-boundary rates. Under this framework, tectonic structures that extend over thousands of kilometers and cut through the entire lithospheric thickness constrain the spatial location of earthquakes. The time distribution of large earthquakes is such that, over the long-term, the spatial distribution is more or less uniform, whereas it appears random and discontinuous in short-term instrumental record (e.g., Swafford et al., this volume, chapter 4).

This concept may appear absurd in that, by definition, intraplate regions are far from tectonic plate boundaries. However, this discrepancy may just be regarded as a matter of strain amplitude. Typical plate-boundary regions are characterized by high, mostly localized, strain rates associated with high, mostly localized, seismic activity. Strain patterns and earthquake distributions delineate the major fault systems that mark the plate boundaries. This is obviously not the case in intraplate regions, but, in the eastern North America example, one could regard the easternmost U.S. and Canadian seismic activity as reflecting a large deformation zone between two rigid blocks, namely the North America craton and the western Atlantic Ocean (Fig. 5). To a first order, the edges of this higher seismicity area correspond to the eastern limit of the cold continental craton and lowland (e.g., as defined by tomography; Van der Lee and Frederiksen, 2005) and the western limit of strong oceanic lithosphere.

The difficulty in resolving independent North America craton and western Atlantic blocks may be related to the very slow relative motion and the limited data set. A differential motion between these two blocks of a few tenths of millimeters per year would be compatible with seismic strain rates (e.g., Anderson, 1986; Mazzotti and Adams, 2005) and just at the limit or below the resolution of geodetic techniques (e.g., Mao et al., 1999; Calais et al., 2006). For example, discussions about independent microplates in the northeast Russia–north China–Japan region illustrate the difficulty of resolving slow-moving tectonic plates, even when these plates are surrounded by geodetic data (e.g., Apel et al., 2006).

A direct implication of this model on seismic hazard is that large earthquakes can occur at any point along the main geological structures accommodating the current relative plate motion, e.g., Iapetan and Atlantic rifted margins (Fig. 5). The magnitude-recurrence relation of large events is limited by the low strain rates, e.g., roughly a few M = 7 events per 500 yr over the eastern U.S.–Canada area (Fig. 5).

Localized Weak Zone Model

The third possible model is at the opposite of the first two. Under this view, intraplate earthquakes occur along small, local structures associated with a weak zone in the otherwise strong lithosphere (Fig. 4C). Typical dimensions for the weak zone and

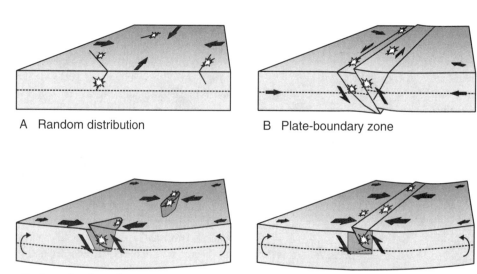

Figure 4. Geodynamic models described by the relationship among earthquake characteristics, strain rates, and tectonic structures. (A) Random: Large earthquakes can occur anywhere, but their magnitudes and recurrence intervals are limited by the low strain rate. (B) Plate boundary: Earthquakes are associated with large-scale lithospheric structures and are limited by the low strain rates. (C) Large-scale weak zones: Earthquakes focus along long paleotectonic structures associated with a weak layer in the lower crust and/or upper mantle. Lateral variations in lithospheric strength concentrate strain along the structures. (D) Localized weak zones: Earthquakes occur along small, local structures associated with a weak zone in the otherwise strong lithosphere. Strain concentration results in higher seismicity in the weak zones.

the associated tectonic structures are a few hundreds of kilometers horizontal and a few tens of kilometers vertical. The model is characterized by a strain rate concentration in the weak zones, which leads to high rates and possibly high-magnitude earthquakes, and by a distribution of earthquakes limited in space and time to the weak structural zones.

This model has been proposed to explain the earthquake concentration in the New Madrid area, central United States. Numerical simulations of stress and strain concentration within a weak crustal or upper-mantle zone show that, locally, strain rates can reach 10^{-8}–10^{-9} yr^{-1} (Grollimund and Zoback, 2001), associated with a recurrence rate of large M = 7–8 earthquakes of ~1/250–1/4000 yr within this small area (Kenner and Segall, 2000). These weak zone models require a very low viscosity (~10^{19} Pa s or less) for the lower crust and/or the upper mantle, which may be achieved by a locally hotter geothermal gradient (Liu and Zoback, 1997) (see section Weak Zones as Stress Concentrators).

A possible example for this model is the lower St. Lawrence valley, eastern Canada (Fig. 1), where GPS measurements and seismicity suggest a lateral variation in crustal strain (Mazzotti et al., 2005). Seismic activity varies significantly in this region (Fig. 6): the Charlevoix region is characterized by five M = 6–7 earthquakes since 1650; the lower St. Lawrence seismic zone, ~150 km downstream, has only been affected by a few M = 4–5 events during the last century; the rest of the valley shows no significant earthquake activity. GPS-derived strain rates are consistent both in style and in amplitude with the seismic deformation

and suggest that strain rates across the two most active seismic zones are ~2–3 times larger than the regional average (Fig. 6). The correlation between seismic and GPS strain may indicate that large earthquakes in this region tend to cluster within small local zones of lithospheric weakness. However, these spatial variations are not significant at the 66% confidence level and do not disprove the hypothesis that current GPS strain rates and long-term seismic strain rates could be uniformly distributed along the entire St. Lawrence valley.

Large-Scale Weak Zone Model

The last geodynamic model represents a large-scale extension of the localized weak-zone model. Under the large-scale weak zone model, intraplate earthquakes focus along long geological structures associated with a weak layer in the lower crust and/or upper mantle. These geological structures, with a typical length scale of ~1000 km, correspond to paleotectonic zones, mostly Paleozoic to Mesozoic rifts or aulacogens. As for the localized weak zone model, lateral variations in the lithospheric strength concentrate stress and strain along the paleotectonic structures, leading to local strain rates up to 10^{-8}–10^{-9} yr^{-1}, a few orders of magnitude larger than background intraplate strain rates, and this allows for earthquakes with higher magnitude and shorter recurrence periods. Under this model, earthquakes are spatially confined to the paleotectonic zones, but their temporal distribution can vary significantly along these zones.

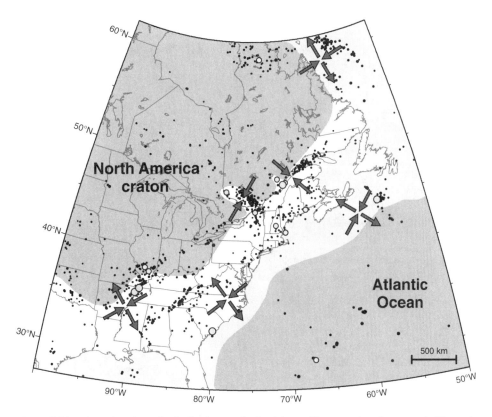

Figure 5. Plate-boundary hypothesis. Background seismicity and large earthquakes are as in Figure 1. The gray shaded areas show the low seismicity, rigid regions of the North America plate (Canadian and northern U.S. craton) and Atlantic Ocean. The dark gray strain crosses show the style of deformation associated with the relatively higher seismic activity between the more rigid blocks.

The strongest supporting evidence for this framework is the global correlation between intraplate earthquakes and paleo–rifted margins or aulacogens (Johnston et al., 1994). As discussed in the Intraplate Earthquake Characteristics section, eastern North America provides a good example of this high correlation (Fig. 2). Figure 7 shows two schematic geological cross sections through southeastern Canada and the southeastern United States that highlight the relationship between large earthquakes and zones of crustal weakness associated with extension tectonics (Adams and Basham, 1991). There are only a few examples of medium or large earthquakes (M ≥ 5) occurring along the thrust structures of the Appalachian orogen. In regions such as the lower St. Lawrence valley or the eastern Tennessee seismic zones, where the Appalachian thrust nappes account for the first ~5 km of the crust, most of the seismicity occurs within the deeper Cambrian to Late Proterozoic structures of the Iapetan rifted margin (Adams and Basham, 1991; Adams et al., 1995).

THE LITHOSPHERIC WEAKNESS HYPOTHESIS

The main factor underlying the four geodynamic models presented in the previous section is the concept of weak zones in the strong intraplate lithosphere. The presence or absence of such zones, their correlation with earthquake activity, and their lateral extent constrain the location of intraplate earthquakes and possibly how frequent and how big they can be. The impact on seismic hazard can be summarized in the following hypothesis: "The spatial and temporal distribution of large intraplate earthquakes is controlled by lithospheric strength contrasts along paleotectonic structures." Disproving this hypothesis would validate the random distribution model. Alternatively, constraints on the correlation between zones of weakness and earthquake distribution would discriminate between the other three models (plate boundary, large, and small weak zones). In this section, I discuss two aspects of the weak zone hypothesis: the mechanical definition and the geodetic signature.

Mechanical Control

From a mechanical perspective, the presence of weak zones in the intraplate lithosphere can be addressed through the relatively standard strength envelope approach (e.g., Kohlstedt et al., 1995; Ranalli, 1995). The strength of the crust and upper mantle is described in terms of brittle behavior and plastic flow rheology and depends principally on the local geotherm, background strain rate, presence of fluids, and composition of the lithospheric layers.

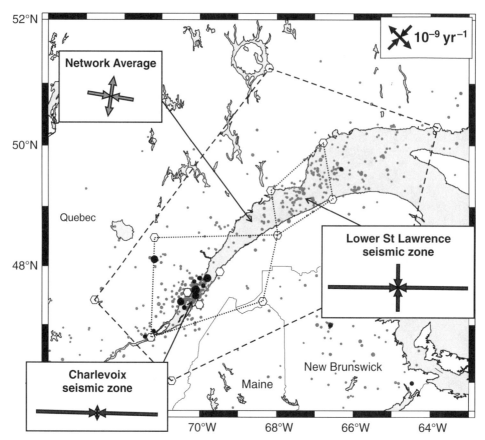

Figure 6. Lower St. Lawrence valley seismicity and global positioning system (GPS) strain rates. Background seismicity and large earthquake are as in Figure 1. The open hexagons show the locations of the 16 GPS campaign sites used to define crustal strain rates. Strain rates (gray shaded crosses) were estimated for three subnetworks shown by the dotted dashed lines. Figure was modified from Mazzotti et al. (2005).

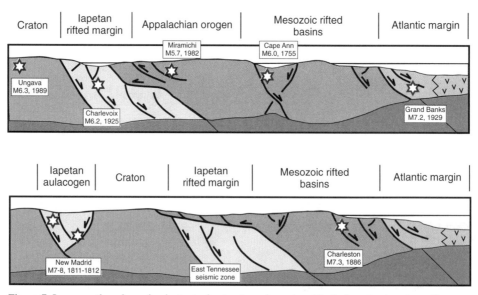

Figure 7. Large earthquake and paleotectonic structures in eastern North America in schematic cross sections of the crustal structure of eastern Canada and central eastern United States (see Fig. 2 for locations). The black arrows indicate fault displacements during their main tectonic phases (e.g., extension during Iapetus Ocean rifting). The red stars show the schematic location of large earthquakes. Figure was modified from Adams et al. (1995).

Examples of applications of the strength envelope approach to intraplate regions include studies of the relation between lithospheric strength and geotherm evolution, elastic thickness, or seismicity (e.g., Burov and Diament, 1995; Deverchere et al., 2001; Cheng et al., 2002). In particular, studies by Liu and Zoback (1997) and Zoback and Townend (2001) indicate that, given the amplitude of plate-boundary forces, a "typical" intraplate lithosphere can only deform at relatively fast strain rates (larger than ~10^{-10} yr^{-1}) in the presence of high fluid pressure and/or a hot geotherm. In this section, I extend the results presented by Zoback and Townend (2001) to a more general set of intraplate conditions and discuss their implications for the importance of lithospheric weak zones.

Cratons Cannot Deform at Seismic Strain Rates

The study of intraplate strength presented in Zoback and Townend (2001) assumed a typical continental surface heat flow of 60 mW/m^2, equivalent to a Moho temperature of ~600 °C, and a strike-slip stress regime. Although these conditions are representative of parts of the continental United States, most of the North America intraplate region is under lower temperatures and/or a compressive stress regime. The average stress regime in

central and eastern North America is characterized by a change from a compressive regime in Canada to a strike-slip regime in the central United States (Zoback and Zoback, 1980). Figure 8 shows a compilation of heat-flow measurements for eastern North America. In the Canadian Shield, a heat-flow average of ~40 mW/m^2 and relatively low crustal heat production indicate a cold geotherm, with temperatures at the Moho of 400–500 °C (Mareschal et al., 1999, 2000). Although the average heat flow is slightly higher in the Appalachian region (~55 mW/m^2), crustal temperatures along the eastern seaboard are similar to those in the shield due to the higher crustal heat generation in the Appalachian orogen (Mareschal et al., 2000).

In the following, I assume a lithosphere section composed of a felsic upper crust, an intermediate middle crust, and a mafic lower crust over a uniform upper mantle (Table 1). The brittle behavior of the lithosphere follows Byerlee's friction law, and the plastic behavior follows a dislocation creep law, except for the upper mantle, where glide-controlled dislocation is assumed for stresses larger than 200 MPa (see Kohlstedt et al., 1995; Evans and Kohlstedt, 1995, for details). Semibrittle and nonlocalized brittle behaviors are considered to be second-order effects and are approxi-

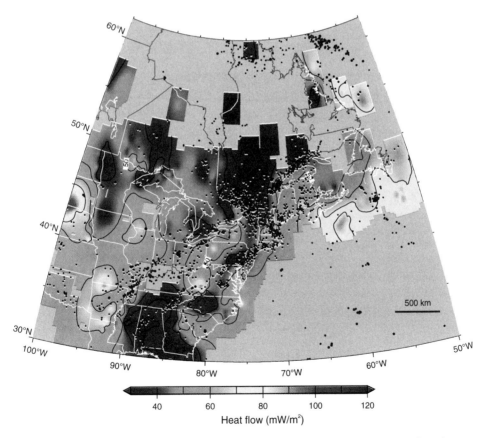

Figure 8. Surface heat flow and seismicity of eastern North America. Background heat flow data are from Blackwell and Richards (2004). Heat flow data for eastern Canada are from Mareschal et al. (1999, 2000). Regions without data points within 3° range are masked by gray shading. The thick contour line shows the 50 mW/m^2 isoline; contour lines are every 10 mW/m^2. Background seismicity is as in Figure 1.

TABLE 1. RHEOLOGY PARAMETERS OF CONTINENTAL LITHOSPHERE MODELS

Layer	Depth range (km)	A $(Pa^{-n}\,s^{-1})$	n	Q $(10^3\,J\,mol^{-1})$	Rock name	Ref.
Upper crust	0–13	1.04×10^{29}	3.4	139	Dry Westerly granite	1
		7.94×10^{16}	1.9	137	Wet Westerly granite	1
Middle crust	14–26	5.18×10^{18}	2.4	219	Dry quartz-diorite	1
		1.26×10^{16}	2.4	212	Wet quartz-diorite	2
Lower crust	27–43	1.20×10^{26}	4.7	485	Dry diabase	3
		7.94×10^{25}	3.4	260	Wet diabase	4
Lithospheric mantle	43–250	4.85×10^{17}	3.5	535	Dry dunite	5
		4.89×10^{15}	3.5	535	Wet dunite	5

Note: References: 1—Hansen and Carter (1982); 2—Carter and Tsenn (1987); 3—Macwell et al. (1998); 4—Shelton and Tullis (1981); 5—Hirth and Kohlstedt (1996).

mated. Strength envelopes are derived for a strong lithosphere case (near-hydrostatic pore fluid pressure and dry rheologies) and for a weak lithosphere case (near-lithostatic pore fluid pressure and wet rheologies) under a cold and a mild geotherm (Table 2).

Under a cold geotherm, a compressive stress regime, and typical intraplate strain rates (10^{-13}–10^{-10} yr^{-1}), the integrated strength of a lithospheric section is 5–10 times larger than typical plate-boundary forces (10–80×10^{12} N m^{-1} versus 2–5×10^{12} N m^{-1}; Figs. 9A and 9B). Such a uniform lithosphere cannot deform at these rates under these driving forces. For typical plate-boundary forces, the cold North America lithosphere can only deform at rates slower than ~10^{-18} yr^{-1}. Under these extremely slow rates, deformation is accommodated by plastic flow in most of the lithosphere (including the upper crust), and the average deformation is equivalent to one M = 7 earthquake per 1 m.y. per 1 million km^2.

Assuming milder continental U.S. conditions, Zoback and Townend (2001) argued that the central and eastern U.S. lithosphere is kept strong (i.e., deforms slowly) by the near-hydrostatic pore fluid pressure in the upper crust. Figure 9 shows that the lithosphere of the Canadian Shield and the Appalachians is strong regardless of the pore fluid pressure or the amount of water in the minerals.

Weak Zones as Stress Concentrators

The deformation rates compatible with plate-boundary forces and the strong rheology of the northeast U.S. and Canadian lithosphere are several orders of magnitude smaller than intraplate strain rates derived from seismicity and GPS (<10^{-18} yr^{-1} versus ~10^{-10} yr^{-1}, see Intraplate Earthquake Characteristics section). In other words, plate-boundary forces distributed over a 100-km-thick lithosphere only account for a few tens of MPa and cannot bring the crust near failure (hundreds of MPa at mid-crustal levels; Fig. 9). Similarly, stress "concentrators" (e.g., local topography,

density contrasts, fault intersections) only produce differential stress increases on the order of tens of MPa (e.g., Assameur and Mareschal, 1995). When distributed over a typical strong intraplate lithosphere, tectonic and local stresses are about an order of magnitude too small to trigger failure. Thus, locally weaker zones are required in regions of significant seismic activity. Weak zones can be due to conditions that reduce the strength of the entire lithospheric section, thus allowing it to deform faster, or to conditions that allow for high stress and strain concentration in a weak layer, thus driving high strain rates in the seismogenic upper crust.

The hypothesis of a weak lithosphere was proposed by Liu and Zoback (1997) for the New Madrid seismic zone. They argued that the high surface heat flow and low Pn wave velocities under this region indicate a locally hot geotherm (700–800 °C at the Moho) and a very low plastic strength of the lower crust and upper mantle. Under these conditions, strike-slip deformation of the entire lithosphere in the New Madrid area can occur at rates similar to the estimated seismic strain rates. However, the existence within the New Madrid seismic zone of a geotherm 200–300 °C higher than that of the surrounding lithosphere is debated both on the grounds of heat-flow interpretations (Swanberg et al., 1982; McKenna et al., this volume, chapter 12) and mechanisms capable of generating such a small-scale, large-amplitude anomaly (e.g., Pollitz et al., 2001).

Hot geotherm conditions may prevail in other regions with high seismic activity, such as the Charlevoix seismic zone (Guillou-Frottier et al., 1995). Figure 9 shows that a temperature increase of ~200 °C of the geotherm can reduce the integrated lithospheric strength by a factor of 2–10. However, because the Charlevoix region is primarily under a compressive state of stress, other factors are required to weaken the entire lithosphere sufficiently to deform it at seismic strain rates with typical tectonic forces. Figure 9D shows an example of a combination of mild geotherm, high pore fluid pressure, and wet

TABLE 2. THERMAL MODEL PARAMETERS

Model	Q $(mW\,m^{-2})$	Upper crust		Middle crust		Lower crust		Lithospheric mantle	
		A $(10^6\,W\,m^{-3})$	K $(W\,m^{-1}\,K^{-1})$	A $(10^6\,W\,m^{-3})$	K $(W\,m^{-1}\,K^{-1})$	A $(10^6\,W\,m^{-3})$	K $(W\,m^{-1}\,K^{-1})$	A $(10^6\,W\,m^{-3})$	K $(W\,m^{-1}\,K^{-1})$
Cold	40	1.5	3.0	0.6	3.0	0.2	2.5	0.02	3.5
Mild	55	1.0	3.0	0.4	3.0	0.2	2.5	0.02	3.5

Note: *Q*—surface heat flow; *A*—heat generation; *K*—heat conductivity.

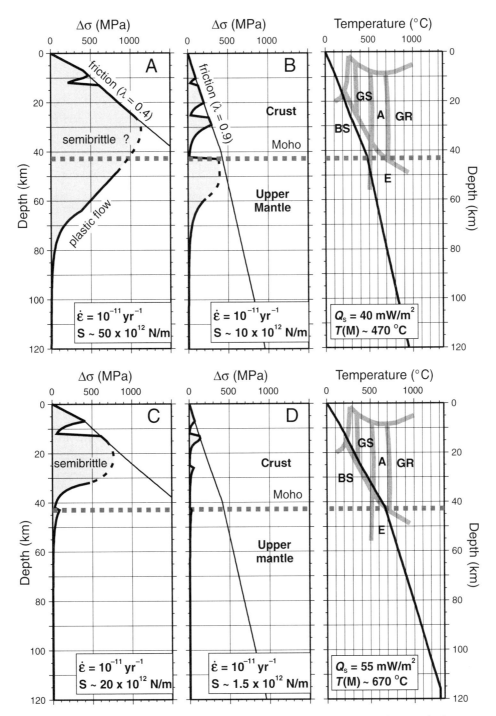

Figure 9. Strength envelope examples for a cold intraplate lithosphere. The geotherm varies between cold (upper panels) and warm (lower panels). (A) and (C): Strong lithosphere, dry rheologies, and near hydrostatic pore fluid pressure ($\lambda = 0.4$). (B) and (D) Weak lithosphere, wet rheologies, and near lithostatic pore fluid pressure ($\lambda = 0.9$). The background strain rate is 10^{-11} yr^{-1}. The integrated lithospheric strength (S) varies between 50×10^{12} N m^{-1} and 1.5×10^{12} N m^{-1} (compared to typical tectonic forces of 1–5×10^{12} N m^{-1}). The dashed portion of the strength envelope approximates the effect of a semibrittle deformation zone. BS—blue schist, GS—green schist, A—amphibolite, GR—garnet, E—eclogite (metamorphic facies).

rheologies that reduces the Charlevoix lithospheric strength to ~1.5 × 10[12] N m[-1] under low seismic strain rates (10[-11] yr[-1]). Similar conditions also allow for the faster strain rates associated with the current seismicity in this region (~10[-8] yr[-1]). Although possible, such a combination of drastic conditions is somewhat extreme for a Precambrian lithosphere and remains largely untested.

In contrast to a reduction of the bulk lithospheric strength, weak material may also act as stress and strain concentrators, allowing for relatively fast seismic deformation of the upper crust in response to the high stress accumulation in the lower crust and/or upper mantle. A weak lower crust has been proposed to explain the increase in seismicity in the New Madrid region after the Holocene melting of the Laurentia ice sheet (e.g., Kenner and Segall, 2000; Grollimund and Zoback, 2001; Pollitz et al., 2001). These models, based on a visco-elastic approach, generally require an effective viscosity for the weak layer of 10[18]–10[20] Pa s, a few orders of magnitude smaller than the effective viscosity of the surrounding, slowly deforming crust and lithospheric mantle.

Low viscosity values are inferred for lower-crustal rocks in active tectonic settings and are commonly associated with factors promoting plastic flow under low stresses, such as a very hot geotherm or the presence of fluids (e.g., McKenzie and Jackson, 2002). In contrast, typical lower-crustal rocks in intraplate environments are mostly mafic rocks, likely dry and depleted, with low temperatures (Rudnick and Fountain, 1995; Cheng et al., 2002). Thus, atypical conditions are required to create a weak, low-viscosity layer in the lower crust (or the upper mantle). The mechanism most commonly used is high temperature. As an alternative to the high-geotherm hypothesis, Figure 10 shows how the presence of weak material can result in low effective viscosity values in the lower crust, even under a cold craton geotherm. In this example, the lower crust is represented by a wet quartz-diorite rheology (similar to the mid-crust layer). Effective viscosity for the lower crust is 10[20]–10[21] Pa s, about two orders of magnitude smaller than the viscosity estimates for the uppermost mantle. These low-viscosity values are associated with a strain rate of 10[-8] yr[-1], representative of the very active seismic zones such as Charlevoix, eastern Canada.

Whether mechanical weakness results in bulk strength reduction of the lithosphere or in stress concentration in the lower crust/uppermost mantle, exceptional local conditions are required to produce a significant effect and allow the lithosphere to deform at strain rates compatible with the observed seismic rates. This strength reduction/stress concentration state may be achieved more easily in regions characterized by strike-slip deformation regimes and mild geotherms, such as the central United States. Under a cold geotherm and compressive regime, such as the Canadian Shield or Appalachian regions, the conditions for local weak zones are more extreme. The association of frequent large earthquakes with the presence of a weak lithospheric layer requires a very high local geotherm anomaly, the presence of a mechanically weak crust, or the presence of a very weak fault zone. Weak faults have been proposed for plate-boundary systems such as the San Andreas fault (e.g., Zoback, 1987) or the Cascadia subduc-

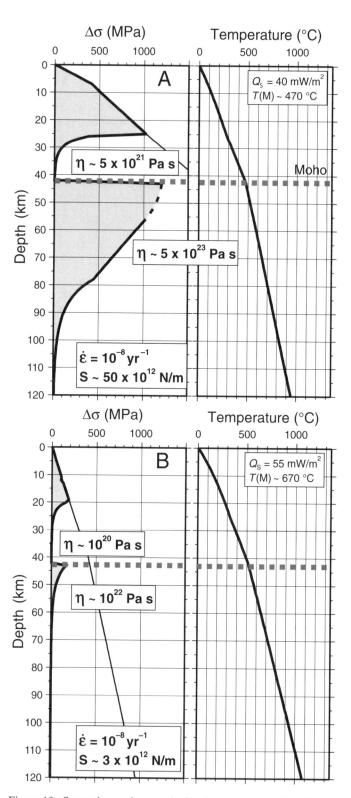

Figure 10. Strength envelope and effective viscosity of the lower crust. (A) Strong cold lithosphere. (B) Weak warm lithosphere. In both cases, the lower crust layer is represented by a wet quartz-diorite rheology, leading to effective viscosities of 10[21]–10[20] Pa s for a background strain rate of 10[-8] yr[-1].

tion fault (Wang et al., 1995). Although the nature of conditions promoting low-stress plate-boundary faults is subject to debate, similar conditions may apply to some intraplate fault zones.

Geodetic Signature

In active tectonic regions, GPS measurements of crustal deformation have revolutionized the study of strain buildup and release during the earthquake cycle of active faults. In intraplate environments, the major challenge with geodetic measurements of crustal strain is that the expected signal is about the same level as the uncertainty. Typical horizontal uncertainties for GPS surveys are ~0.1–1 mm/yr (10^{-9}–10^{-8} yr^{-1}) depending mostly on the type of survey (campaign or continuous) and length of the time series (e.g., Mao et al., 1999). In the last ten years, geodetic studies of intraplate deformation across the stable North America plate have shown that, east of the Rocky Mountains, horizontal deformation of the continental plate occurs at less than ~10^{-9} yr^{-1} (less than ~1 mm/yr over 1000 km) (e.g., Dixon et al., 1996; Calais et al., 2006; Sella et al., 2007). A possible exception is the present-day strain rates associated with postglacial rebound in

Canada. Figure 11 shows an example of horizontal contraction rates predicted by one particular postglacial rebound model (ICE3G-VM1; Peltier, 1998) for eastern North America. Although significant variations are expected with different models, typical postglacial rebound horizontal strain rates range between ~10^{-10} yr^{-1} and ~10^{-8} yr^{-1}, with the higher rates mostly in Canada.

As mentioned earlier, strain rates associated with intraplate seismicity are about the same order of magnitude as, or slightly smaller than, postglacial rebound and geodetic upper bounds of intraplate deformation. In eastern North America, seismic strain rates vary between ~10^{-12} yr^{-1} in the low seismic regions to ~10^{-9}–10^{-8} yr^{-1} in the more seismically active zones, such as New Madrid or Charlevoix (Anderson, 1986; Mazzotti and Adams, 2005). Similar rates, 10^{-12}–10^{-9} yr^{-1}, have been derived from intraplate seismicity in Australia (e.g., Sandiford et al., 2004).

The difficulty in measuring crustal strain rates associated with intraplate seismic activity is reflected in the New Madrid case. Attempts to measure and interpret geodetic strain across the New Madrid seismic zone have been inconclusive due essentially to the relatively large uncertainties in the measurements (e.g., Newman et al., 1999; Calais et al., 2006), and this problem precludes dis-

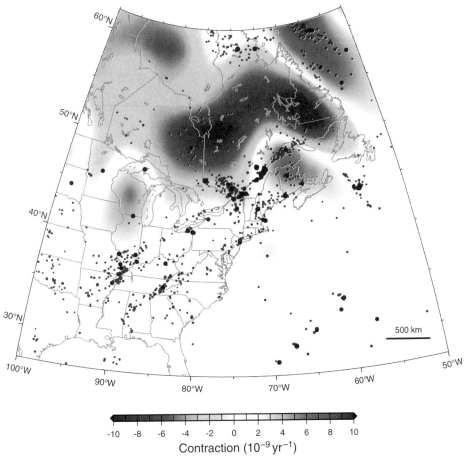

Figure 11. Postglacial rebound strain rates and seismicity of eastern North America. Example of present-day surface contraction rates (first invariant of strain rate tensor, contraction positive) predicted by the ICE3G-VM1 postglacial rebound model (Peltier, 1998). Seismicity is as in Figure 1.

crimination among different models of seismic strain accumulation (Stuart, 2001). To illustrate this point, I use recent estimates of crustal strain in the lower St. Lawrence area, which encompasses the Charlevoix seismic zone (Figs. 1 and 6). The GPS velocity data for this region show significant crustal shortening, and the rate and direction are consistent with the earthquake mechanisms and sta-

tistics (Mazzotti et al., 2005). The velocity data are associated with relatively small uncertainty (0.6 mm/yr on average) and provide one of the best-constrained examples of intraplate strain patterns.

Figure 12 shows cross sections of the eastward velocity field (direction of maximum horizontal strain, Fig. 6) projected along an E-W section centered on the St. Lawrence River. Five

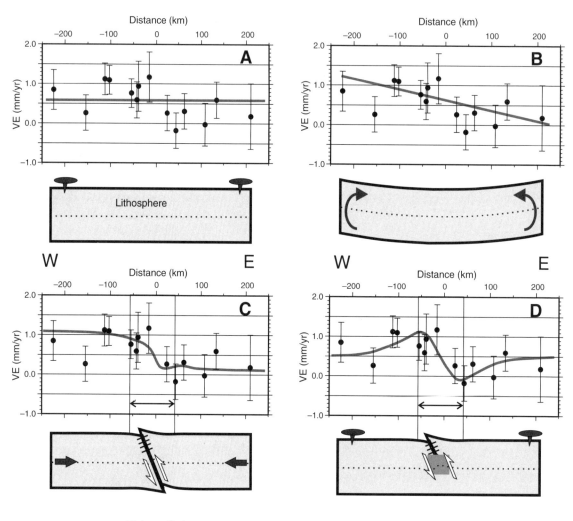

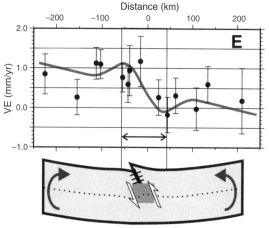

Figure 12. Global positioning system (GPS) velocity and strain loading model for the St. Lawrence valley, Quebec. East-west cross section of eastward velocity component centered along the St. Lawrence River (cf. Fig. 6). The GPS velocities can be explained by different loading models: (A) No strain (less than ~10^{-11} yr^{-1}); (B) purely elastic postglacial rebound strain; (C) plate-boundary–type elastic loading of a locked thrust (barbed thick line) by far-field displacements; (D) loading of locked thrust by strain concentration within a weak layer (gray shaded square) in a low strain background; and (E) loading of locked thrust by strain concentration within a weak layer in a high strain background.

possible models of deformation and associated strain rate patterns are tested against this data set. The models range from no deformation (Fig. 12A), to long wavelength postglacial rebound strain (Fig. 12B), to local strain concentration in relation to a locked thrust fault with a long-term slip rate of 1 mm/yr (Figs. 12C to 12E). None of these models can be ruled out at the 95% confidence level.

Although it is compatible with the data set at the 95% confidence, the no-strain model (Fig. 12A) is not realistic because most postglacial rebound models predict horizontal shortening rates for the St. Lawrence valley area of $\sim 10^{-9}$ yr^{-1} (equivalent to ~ 1 mm/yr of shortening across the whole network). As shown in Figure 12B, the GPS data are consistent with this purely elastic postglacial rebound deformation but cannot resolve the detailed shape and small-scale variations of the strain pattern. In both models, the lack of significant strain accumulation along a locked fault or fault system would suggest a very small seismic hazard.

Within their resolution, the GPS velocities can also be explained by a typical elastic loading model, in which the seismogenic portion of a thrust is locked and loaded by continuous slip on the deep lithospheric-scale portion (Fig. 12C). This model corresponds to the plate-boundary model. The long-term far-field relative motion between two rigid blocks is accommodated along a major fault zone that cuts through the entire lithosphere. Applied to the St. Lawrence valley GPS data, this interpretation would suggest that up to 1 mm/yr of relative motion is presently loaded on a locked thrust (Fig. 12C). However, the resolution of the data can only loosely constrain the locked fault characteristics (location, geometry, downdip extent, etc.). Under this model, the magnitude-recurrence relation of large earthquakes can be estimated using relationship between earthquake statistics and long-term strain developed for plate-boundary regions (e.g., WGCEP, 1995).

The last possible interpretation of this GPS data set is related to the weak zone model (local or large-scale). Strain concentration in a weak, low-viscosity layer results in the loading of an upper-crust fault system and a surface velocity gradient amplified compared to the regional velocity pattern. Figure 12 shows two end-member examples where the amplitude of the local strain anomaly is large compared to an insignificant regional strain field (Fig. 12D) or is about the same level as the regional strain (e.g., postglacial rebound strain, Fig. 12E). As for the plate-boundary model, the GPS data only provide a loose constraint to the characteristics of the loaded fault. The interpretation of this type of system in terms of earthquake hazard is more complex than for the plate-boundary case. The same low strain rate measurements can correspond to a low or high loading rate (and corresponding hazard) depending on the geometry of the weak zone and on the time constant of the system (e.g., Kenner and Segall, 2000; Stuart, 2001).

CONCLUSION

Different geodynamic models have significantly different implications for seismic hazard assessment in an intraplate environment. For example, the large-scale weak-zone model applied to the St. Lawrence valley, eastern Canada, predicts peak ground acceleration values that differ by up to a factor of two from the predictions of a small-scale weak-zone model (Adams and Halchuk, 2003). In both models, the hazard is high along the paleotectonic structures of the St. Lawrence valley and low elsewhere. By opposition, the random model would predict a uniform hazard over the entire region.

The four models presented in this paper are built on the relationships among lithospheric strength, strain, and earthquake characteristics. The concept of a weak layer concentrating stress and strain in region of higher seismicity is at the center of these geodynamic frameworks. Except for the random model, which may apply to Precambrian shield regions, large earthquake locations, recurrence times, and magnitudes are likely controlled by the contrasts in strength and strain between the "typical" strong intraplate lithosphere and local weak zones. Thus, identification and characterization of the conditions for lithospheric strength and strain variations have the potential to produce very significant improvements to the science and hazard assessment of continental intraplate seismicity.

In this paper, I present two lines of study that can address the question of relationships among strength, strain, and earthquakes. A mechanical approach can be used to identify the conditions required for creating a weak zone in the otherwise strong intraplate lithosphere. Local anomalies such as a hot geotherm, high pore fluid pressure, weak bulk rock composition, low-friction fault gouge material are examples of conditions that promote lithospheric weakness and strain concentration. In some regions, such as eastern Canada, a combination of these factors is probably required to allow for the observed level of seismic activity. Detailed studies of the velocity structure, heat flow, and state of stress suggest that specific mechanical conditions may exist under high seismicity regions like the New Madrid seismic zone (e.g., Liu and Zoback, 1997). However, characterizing the "typical" mechanical state of the low seismicity areas is as important as characterizing the peculiar conditions in the active regions. High-resolution seismic experiments combining passive and active sources can provide detailed information on the mechanical state of the crust and upper mantle. With an ~ 70 km spacing, the USArray component of EarthScope may only provide the backbone for such experiments in the central and eastern United States. In combination with seismic surveys, a better understanding of geotherm variations would provide critical constraints for mechanical models. This requires precise estimates of surface heat flow, crustal heat generation, and potential superficial perturbations (e.g., hydrothermal circulation).

The second line of study discussed here is the use of geodetic techniques to measure the patterns and variations of strain rates in high and low seismicity areas. Compared to studies in plate-boundary regions, intraplate GPS measurements have been relatively unsuccessful in constraining seismotectonic models, mostly because the strain level is at the limit of GPS resolution. However, four factors may contribute to making GPS measurements a critical tool for understanding the dynamics of earthquake concentration in intraplate regions: (1) GPS resolution improves with the duration of the measurements. Results obtained with 6–9-yr-long time series indicate a resolution of $1–4 \times 10^{-9}$ yr^{-1}, with a few campaign stations over ~100 km scale (Mazzotti et al., 2005). (2) Improvements in the definition of the reference frame provide a more stable background for local strain measurements (Blewitt et al., 2004). (3) Next generation instruments combining the GPS and European Galileo systems will likely improve the resolution of future measurements. (4) Better understanding of the spatial and temporal strain patterns will help in designing experiments specifically targeted for measuring these patterns. Such experiments could include a combination of dense campaign stations (e.g., sites spaced 10–20 km apart surveyed once per year) and backbone permanent stations (e.g., ~50 km spacing). With special care given to monument and station design, such a network would likely resolve variations of crustal strain related to seismic loading within a 5–10 yr period, thus significantly improving our knowledge of intraplate dynamics.

ACKNOWLEDGMENTS

I thank S. Halchuk for providing the Geological Survey of Canada earthquake catalog and statistics for eastern and central Canada, T. James for providing the postglacial rebound model predictions, and J.-C. Mareschal for his database and clarification of heat flow in eastern Canada. This work greatly benefited from numerous discussions with J. Adams, R. Hyndman, and T. James, although they are not responsible for the views presented here. Reviews by J.-C. Mareschal, S. Van der Lee, and S. Stein helped improve the final manuscript. This is Geological Survey of Canada contribution 2005058.

REFERENCES CITED

Adams, J., and Basham, P., 1991, The seismicity and seismotectonics of eastern Canada, *in* Slemmons, D.B., Engdahl, E.R., Zoback, M.D., and Blackwell, D.D., eds., Neotectonics of North America: Boulder, Colorado, Geological Society of America, Decade of North American Geology Map, v. 1, p. 261–276.

Adams, J., and Halchuk, S., 2003, Fourth generation seismic hazard maps of Canada: Values for over 500 Canadian localities intended for the 2005 National Building Code of Canada: Geological Survey of Canada Open File 4459, 155 p.

Adams, J., Basham, P.W., and Halchuk, S., 1995, Northeastern North America earthquake potential—New challenges for seismic hazard mapping, *in* Current Research, 1995-D: Ottawa, Ontario, Geological Survey of Canada, p. 91–99.

Anderson, J.G., 1986, Seismic strain rates in central and eastern United States: Bulletin of the Seismological Society of America, v. 76, p. 273–290.

Apel, E.V., Bürgmann, R., Steblov, G., Vasilenko, N., King, R., and Prytkov, A., 2006, Independent active microplate tectonics of northeast Asia from GPS velocities and block modeling: Geophysical Research Letters, v. 33, doi: 10.1029/2006GL026077.

Assameur, D.M., and Mareschal, J.C., 1995, Stress induced by topography and crustal density heterogeneities: Implications for the seismicity of southeastern Canada: Tectonophysics, v. 241, p. 179–192, doi: 10.1016/0040-1951(94)00202-K.

Bent, A.L., 1992, A re-examination of the 1925 Charlevoix, Québec, earthquake: Bulletin of the Seismological Society of America, v. 82, p. 2097–2113.

Blackwell, D.D., and Richards, M., 2004, Geothermal Map of North America: Tulsa, Oklahoma, American Association of Petroleum Geology, scale 1:6,500,000.

Blewitt, G., Bennett, R.A., Calais, E., Herring, T.A., Larson, K.M., Miller, M.M., Sella, G., Snay, R.A., and Tamisiea, M.E., 2004, First report of the Stable North America Reference Frame (SNARF) working group: Eos (Transactions, American Geophysical Union), v. 85, Joint Assembly Supplement, abstract G21C-01.

Boore, D.M., and Atkinson, G.M., 1987, Stochastic prediction of ground motion and spectral response parameters at hard-rock sites in eastern North America: Bulletin of the Seismological Society of America, v. 77, p. 440–467.

Burov, E.B., and Diament, M., 1995, The effective elastic thickness (Te) of continental lithosphere: What does it really mean?: Journal of Geophysical Research, v. 100, p. 3905–3927, doi: 10.1029/94JB02770.

Calais, E., Han, J.Y., DeMets, C., and Nocquet, J.M., 2006, Deformation of the North American plate interior from a decade of continuous GPS measurements: Journal of Geophysical Research, v. 111, doi: 10.1029/2005JB004253.

Carter, N.L., and Tsenn, M.C., 1987, Flow properties of continental lithosphere: Tectonophysics, v. 136, p. 27–63, doi: 10.1016/0040-1951(87)90333-7.

Cheng, L.Z., Mareschal, J.C., Jaupart, C., Rolandone, F., Gariepy, C., and Radigon, M., 2002, Simultaneous inversion of gravity and heat flow data: Constraints on thermal regime, rheology and evolution of the Canadian Shield crust: Journal of Geodynamics, v. 34, p. 11–30, doi: 10.1016/S0264-3707(01)00082-5.

Costain, J.K., Bollinger, G.A., and Speer, J.A., 1987, Hydroseismicity: A hypothesis for the role of water in the generation of intraplate seismicity: Seismological Research Letters, v. 58, p. 41–64.

Deverchere, J., Petit, C., Gileva, N., Radziminovitch, N., Melnikova, V., and San'kov, V., 2001, Depth distribution of earthquakes in the Baikal rift system and its implications for the rheology of the lithosphere: Geophysical Journal International, v. 146, p. 714–730, doi: 10.1046/j.0956-540x.2001.1484.484.x.

Dixon, T.H., Mao, A., and Stein, S., 1996, How rigid is the stable interior of the North America plate?: Geophysical Research Letters, v. 23, p. 3035–3038, doi: 10.1029/96GL02820.

Evans, B., and Kohlstedt, D.L., 1995, Rheology of rocks, *in* Rocks Physics and Phase Relations, A Handbook of Physical Constants: American Geophysical Union Reference Shelf, v. 3, p. 148–165.

Fenton, C.H., Adams, J., and Halchuk, S., 2006, Seismic hazards assessment for radioactive waste disposal sites in regions of low seismic activity: Geotechnical and Geological Engineering Journal, v. 24, p. 579–592, doi: 10.1007/s10706-005-1148-4.

Gordon, R.G., 1998, The plate tectonic approximation: Plate non-rigidity, diffuse plate boundaries, and global plate reconstructions: Annual Review of Earth and Planetary Sciences, v. 26, p. 615–642, doi: 10.1146/annurev.earth.26.1.615.

Grollimund, B., and Zoback, M.D., 2001, Did deglaciation trigger intraplate seismicity in the New Madrid seismic zone?: Geology, v. 29, p. 175–178, doi: 10.1130/0091-7613(2001)029<0175:DDTISI>2.0.CO;2.

Guillou-Frottier, L., Mareschal, J.C., Jaupart, C., Gariepy, C., Lapointe, R., and Bienfait, G., 1995, Heat flow variations in the Grenville Province, Canada: Earth and Planetary Science Letters, v. 136, p. 447–460, doi: 10.1016/0012-821X(95)00187-H.

Hamilton, R.M., and Zoback, M.D., 1981, Tectonic features of the New Madrid seismic zone from seismic reflection profiles: U.S. Geological Survey Professional Paper 1236-F, p. 55–82.

Hansen, F.D., and Carter, N.L., 1982, Creep of selected crustal rocks at 1000 MPa: Eos (Transactions, American Geophysical Union), v. 63, p. 437.

Hirth, G., and Kohlstedt, D.L., 1996, Water in the oceanic upper mantle; implications for rheology, melt extraction and the evolution of the lithosphere: Earth and Planetary Science Letters, v. 144, p. 93–108, doi: 10.1016/0012-821X(96)00154-9.

Johnston, A.C., Coppersmith, K.J., Kanter, L.R., and Cornell, C.A., 1994, The earthquakes of stable continental regions: Assessment of large earthquake potential, *in* Schneider, J.F., ed., Electric Power Research Institute Report TR-102261: Palo Alto, California, Electric Power Research Institute, 309 p.

Kenner, S., and Segall, P., 2000, A mechanical model for intraplate earthquakes: Application to the New Madrid seismic zone: Science, v. 289, p. 2329–2332, doi: 10.1126/science.289.5488.2329.

Kohlstedt, D.L., Evans, B., and Mackwell, S.J., 1995, Strength of the lithosphere: Constraints imposed by laboratory experiments: Journal of Geophysical Research, v. 100, p. 17,587–17,602, doi: 10.1029/95JB01460.

Lamontagne, M., 1999, Rheological and geological constraints on the earthquake distribution in the Charlevoix seismic zone, Québec: Geological Survey of Canada Open File D3778.

Liu, L., and Zoback, M.D., 1997, Lithospheric strength and intraplate seismicity in the New Madrid seismic zone: Tectonics, v. 16, p. 585–595, doi: 10.1029/97TC01467.

Mackwell, S.J., Zimmerman, M.E., and Kohlstedt, D.L., 1998, High-temperature deformation of dry diabase with application to tectonics on Venus: Journal of Geophysical Research, v. 103, p. 975–984, doi: 10.1029/97JB02671.

Mao, A.C., Harrisin, G.A., and Dixon, T.H., 1999, Noise in GPS coordinate time series: Journal of Geophysical Research, v. 104, p. 2797–2816, doi: 10.1029/1998JB900033.

Mareschal, J.C., Jaupart, C., Cheng, L.Z., Rolandone, F., Gariepy, C., Bienfait, G., Guillou-Frotier, L., and Lapointe, R., 1999, Heat flow in the Trans-Hudson orogen of the Canadian Shield: Implications for Proterozoic continental growth: Journal of Geophysical Research, v. 104, p. 29,007–29,024, doi: 10.1029/1998JB900209.

Mareschal, J.C., Jaupart, C., Gariepy, C., Cheng, L.Z., Guillou-Frotier, L., Bienfait, G., and Lapointe, R., 2000, Heat flow and deep thermal structure near the southeastern edge of the Canadian Shield: Canadian Journal of Earth Sciences, v. 37, p. 399–414, doi: 10.1139/cjes-37-2-3-399.

Mazzotti, S., and Adams, J., 2005, Rates and uncertainties on seismic moment and deformation in eastern Canada: Journal of Geophysical Research, v. 110, doi: 10.1029/2004JB003510.

Mazzotti, S., James, T.S., Henton, J., and Adams, J., 2005, GPS crustal strain, postglacial rebound, and seismic hazard in eastern North America: The St. Lawrence valley example: Journal of Geophysical Research, v. 110, doi: 10.1029/2004JB003590.

McKenna, J., Stein, S., and Stein, C.A., 2007, this volume, Is the New Madrid seismic zone hotter and weaker than its surroundings?, *in* Stein, S., and Mazzotti, S., eds., Continental Intraplate Earthquakes: Science, Hazard, and Policy Issues: Geological Society of America Special Paper 425, doi: 10.1130/2007.2425(12).

McKenzie, D., and Jackson, J., 2002, Conditions for flow in the continental crust: Tectonics, v. 21, doi: 10.1029/2002TC001394.

Newman, A., Stein, S., Weber, J., Engeln, J., Mao, A., and Dixon, T., 1999, Slow deformation and lower seismic hazard at the New Madrid seismic zone: Science, v. 284, p. 619–621, doi: 10.1126/science.284.5414.619.

Peltier, W.R., 1998, Postglacial variations in the level of the sea: Implications for climate dynamics and solid-earth geophysics: Reviews of Geophysics, v. 36, p. 603–689, doi: 10.1029/98RG02638.

Pollitz, F.F., Kellog, L., and Burgmann, R., 2001, Sinking mafic body in a reactivated lower crust: A mechanism for stress concentration at the New Madrid seismic zone: Bulletin of the Seismological Society of America, v. 91, p. 1882–1897, doi: 10.1785/0120000277.

Ranalli, G., 1995, Rheology of the Earth (2nd edition): London, Chapman and Hall, 413 p.

Rudnick, R.L., and Fountain, D.M., 1995, Nature and composition of the continental crust: A lower crustal perspective: Reviews of Geophysics, v. 33, p. 267–309, doi: 10.1029/95RG01302.

Sandiford, M., Wallace, M., and Coblentz, D., 2004, Origin of in-situ stress field in south-eastern Australia: Basin Research, v. 16, p. 325–338, doi: 10.1111/j.1365-2117.2004.00235.x.

Sella, G.F., Stein, S., Dixon, T.H., Craymer, M., James, T.S., Mazzotti, S., and Dokka, R.K., 2007, Observation of glacial isostatic adjustment in "stable" North America with GPS: Geophysical Research Letters, v. 34, L02306, doi: 10.1029/2006GL027081.

Shelton, G.L., and Tullis, J., 1981, Experimental flow laws for crustal rocks: Eos (Transactions, American Geophysical Union), v. 62, p. 396.

Stuart, W.D., 2001, GPS constraints on M7–8 earthquake recurrence times for the New Madrid seismic zone: Seismological Research Letters, v. 72, p. 745–753.

Swafford, L., and Stein, S., Limitations of the short earthquake record for seismicity and seismic hazard studies, *in* Stein, S., and Mazzotti, S., eds., Continental Intraplate Earthquakes: Science, Hazard, and Policy Issues: Geological Society of America Special Paper 425, doi: 10.1130/2007.2425(04).

Swanberg, C.A., Mitchell, B.J., Lohse, R.L., and Blackwell, D.D., 1982, Heat flow in the upper Mississippi embayment: U.S. Geological Survey Professional Paper 1236, p. 185–189.

Talwani, P., 1988, The intersection model for intraplate earthquakes: Seismological Research Letters, v. 59, p. 305–310.

Talwani, P., and Acree, S.D., 1984, Pore pressure diffusion and the mechanism of reservoir induced seismicity: Pure and Applied Geophysics, v. 122, p. 947–965.

Van Arsdale, R., 2000, Displacement history and slip rate on the Reelfoot fault of the New Madrid seismic zone: Engineering Geology, v. 55, p. 219–226, doi: 10.1016/S0013-7952(99)00093-9.

Van der Lee, S., and Frederiksen, A., 2005, Surface-wave tomography applied to the North American upper mantle, *in* Nolet, G., and Levander, A., eds., Seismic Earth: Array Analysis of Broadband Seismograms: American Geophysical Union Monograph, v. 157, p. 67-80.

Vlahovic, G., Powell, C., and Lamontagne, M., 2003, A three-dimensional P wave velocity model for the Charlevoix seismic zone, Quebec, Canada: Journal of Geophysical Research, v. 108, doi: 10.1029/2002JB002188.

Wang, K., Mulder, T., Rogers, G.C., and Hyndman, R.D., 1995, Case for a very low coupling stress on the Cascadia subduction fault: Journal of Geophysical Research, v. 100, p. 12,907–12,918, doi: 10.1029/95JB00516.

WGCEP (Working Group on California Earthquake Probabilities), 1995, Seismic hazards in southern California: Probable earthquakes, 1994–2024: Bulletin of the Seismological Society of America, v. 85, p. 379–439.

Zoback, M.D., 1987, New evidence for the state of stress on the San Andreas fault system: Science, v. 238, p. 1105–1111, doi: 10.1126/science.238.4830.1105.

Zoback, M.D., and Townend, J., 2001, Implications of hydrostatic pore pressure and high crustal strength for the deformation of intraplate lithosphere: Tectonophysics, v. 336, p. 19–30, doi: 10.1016/S0040-1951(01)00091-9.

Zoback, M.L., and Richardson, R.M., 1996, Stress perturbation associated with the Amazonas and other ancient continental rifts: Journal of Geophysical Research, v. 101, p. 5459–5475, doi: 10.1029/95JB03256.

Zoback, M.L., and Zoback, M.D., 1980, State of stress in the conterminous United States: Journal of Geophysical Research, v. 85, p. 6113–6156.

MANUSCRIPT ACCEPTED BY THE SOCIETY 29 NOVEMBER 2006

The Geological Society of America
Special Paper 425
2007

Does seismicity delineate zones where future large earthquakes are likely to occur in intraplate environments?

Alan L. Kafka

Weston Observatory, Department of Geology and Geophysics, Boston College, Weston, Massachusetts 02493, USA

ABSTRACT

The spatial distribution of seismicity is often used as one of the indicators of zones where future large earthquakes are likely to occur. This is particularly true for intraplate regions such as the central and eastern United States, where geology is markedly enigmatic for delineating seismically active areas. Although using past seismicity for this purpose may be intuitively appealing, it is only scientifically justified if the tendency for past seismicity to delineate potential locations of future large earthquakes is well-established as a real, measurable, physical phenomenon as opposed to an untested conceptual model. This paper attempts to cast this problem in the form of scientifically testable hypotheses and to test those hypotheses. Ideally, thousands (or even millions) of years of data would be necessary to solve this problem. Lacking such a long-term record of seismicity, I make the "logical leap" of using data from other regions as a proxy for repeated samples of seismicity in intraplate regions. Three decades of global data from the National Earthquake Information Center are used to explore how the tendency for past seismicity to delineate locations of future large earthquakes varies for regions with different tectonic environments. This exploration helps to elucidate this phenomenon for intraplate environments. Applying the results of this exercise to the central and eastern United States, I estimate that future earthquakes in the central and eastern United States (including large and damaging earthquakes) have ~86% probability of occurring within 36 km of past earthquakes, and ~60% probability of occurring within 14 km of past earthquakes.

Keywords: seismicity, earthquakes, intraplate, statistics.

INTRODUCTION

Many seismic hazard studies rely heavily on seismicity as a presumed indicator of zones where future large earthquakes are likely to occur. This is particularly true for intraplate regions, where the cause of earthquakes is largely unknown, and seismically active geological and geophysical features are difficult to delineate. The most recent U.S. National Seismic Hazard Maps, for example, rely heavily on the observed record of seismicity for mapping the hazard in the central and eastern United States (e.g., Frankel, 1995; Frankel et al., 1996; Wheeler and Frankel, 2000).

While this approach may be intuitively appealing, it is only scientifically justified if the tendency for past seismicity to delineate zones where future large earthquakes are likely to occur is well-established as a real, measurable, physical phenomenon as opposed to an untested conceptual model. However, the scientific basis for measuring this tendency has yet to be fully explored, particularly for intraplate regions, and as Lord Kelvin put it: "I often say that when you can measure what you are speaking about, and express it in numbers, you know something about it; but when you cannot measure it, when you cannot express it in numbers, your knowledge is of a meagre and unsatisfac-

Kafka, A.L., 2007, Does seismicity delineate zones where future large earthquakes are likely to occur in intraplate environments?, *in* Stein, S., and Mazzotti, S., ed., Continental Intraplate Earthquakes: Science, Hazard, and Policy Issues: Geological Society of America Special Paper 425, p. 35–48, doi: 10.1130/2007.2425(03).

tory kind ..." (Lord Kelvin, Sir William Thompson, "Electrical Units of Measurement," in *Popular Lectures and Addresses*, v. 1, p. 72–73, 1883). Furthermore, even if we were to establish that this tendency could be measured in principle, there would still remain the questions of: (1) how to actually go about measuring it, (2) whether or not currently available earthquake catalogs provide representative samples of this tendency, and (3) how it varies from one tectonic environment to the next.

The purpose of this paper is to explore these issues, with particular emphasis on intraplate environments, and with the objective of shedding some light on the extent to which seismicity delineates zones where future large earthquakes are likely to occur in the central and eastern United States. The challenge in this type of investigation is to find a way to cast the problem in the form of testable hypotheses and then to use observed earthquake catalogs to test them. Figure 1 illustrates my approach for casting this problem in the form of testable hypotheses: H1 is the hypothesis that future large earthquakes occur only where past earthquakes have occurred; H2 is the hypothesis that future large earthquakes occur only where past earthquakes have not occurred; and H3 represents the hypothesis that future large earthquakes are equally likely anywhere. This study is essentially a test of hypothesis H1 for various tectonic environments, with emphasis on intraplate regions and with a specific focus on the central and eastern United States. I find that, while H1 is not unequivocally verified for the central and eastern United States, it seems as reasonable a conceptual model for the central and eastern United States as it is for many other parts of the world, including some plate-boundary regions.

I began investigating these issues shortly after a workshop on seismic hazard mapping in the northeastern United States

held in 1994 to obtain input for the next generation of the U.S. National Seismic Hazard Maps. The concept of using seismicity to indicate zones where future large earthquakes are likely to occur in the central and eastern United States was presented at the workshop, and this motivated my first investigation of this concept, an attempt to test hypothesis H1 for earthquakes in the northeastern United States. Kafka and Walcott (1998) found that, on average, "future" (i.e., later-occurring) earthquakes in the northeastern United States tended to occur in the vicinity of past earthquakes more frequently than would be expected for a random distribution of future earthquakes, and we were curious to see if the same pattern would be found in other regions. This led us to conduct similar tests of H1 in various regions, including the entire central and eastern United States (Kafka and Levin, 2000). For a variety of regions and tectonic environments, we found that future earthquakes tended to occur in the vicinity of past earthquakes more frequently than would be expected for a random distribution of future earthquakes (Kafka and Levin, 2000; Kafka, 2002). While these results are not particularly surprising, we consider them a preliminary step toward investigating the scientific basis for relying on past seismicity as an indicator of zones where future large earthquakes are likely to occur. A surprising result was that, for the regions and time periods studied, we did not detect any statistically significant difference in the percentage of future earthquakes occurring near past earthquakes for intraplate versus plate-boundary environments. In this paper, I first summarize these previous studies, and then extend these investigations to other parts of the world in an effort to discern how the tendency for past seismicity to delineate zones where future large earthquakes are likely to occur varies with tectonic environment. This provides a foundation for estimating the probability of future large earthquakes occurring near past earthquakes in the central and eastern United States.

METHODS AND REVIEW OF PREVIOUS STUDIES

In addition to the studies associated with the development of the U.S. National Seismic Hazard Maps (e.g., Frankel, 1995; Frankel et al., 1996), other studies have explored the tendency for past seismicity to delineate zones of future large earthquakes. For example, Cao et al. (1996) estimated the seismic hazard in Southern California from background seismicity, using the assumption that future large earthquakes cluster spatially near locations of historical earthquakes of magnitude ≥4.0. Jackson and Kagan (1999) tested forecasts of future earthquakes in the Northwest and Southwest Pacific based on smoothed versions of past seismicity (magnitude ≥5.8) using the Harvard Centroid Moment Tensor (CMT) catalog of earthquakes (e.g., Dziewonski et al., 1999). They found that the actual catalogs for both regions were quite consistent with their forecast model.

In our approach, we use a method that was developed over the course of our past studies of this phenomenon (Kafka and Walcott, 1998; Kafka and Levin, 2000; Kafka, 2002). This

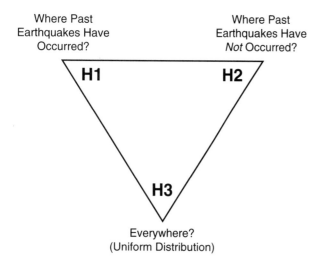

Relative to the Spatial Distribution of Past Earthquakes,

Future Large Earthquakes Occur:

Where Past Earthquakes Have Occurred?

H1

Where Past Earthquakes Have *Not* Occurred?

H2

H3

Everywhere? (Uniform Distribution)

Figure 1. Diagram representing the hypotheses tested in this study (adapted from Kafka and Levin, 2000).

method is "purely statistical" in the sense that no attempt is made here to explain the physical cause of the earthquakes or the physical reasons why the earthquakes occur in some places and don't occur in other places. On a global scale, the locations of future earthquakes will, of course, be dominated by the process of plate tectonics, but the very occurrence of intraplate earthquakes means that delineation of plate boundaries alone is not the sole indicator of where earthquakes occur. While a physical understanding of why intraplate earthquakes occur where they do is an ultimate, fundamental goal, the goal of this study is more modest: to systematically investigate the pattern of the relationship between where intraplate earthquakes occurred in the past versus where they will occur in the future.

Because our method is analogous to the configuration of a cellular phone system, we affectionately refer to it as the "cellular seismology" method. We construct circles of a given radius around each epicenter in an earthquake catalog (the "before" catalog), and investigate the percentage of later-occurring earthquakes (the "after" catalog) that were located within that radius of at least one previous earthquake (Fig. 2). The shaded zones in Figure 2 show the area surrounding the "before" earthquakes, and the filled circles are the "after" earthquakes.[1] The radius is varied so that the shaded circles fill a given percentage of the map area. If a filled circle falls within a shaded zone, we call that a "hit," and the observed proportion of hits is called $\hat{\rho}$ in this paper (see notation below under Statistical Analysis of Percentages of Hits). In the hypothetical case shown in Figure 2, six of the eight filled circles fall within the shaded zones, so $\hat{\rho}$ is 75%. Although this is a rather simple method of characterizing the relationship between past earthquake seismicity and locations of future earthquakes, we tried more complex approaches (including Gaussian smoothing, following the method of Frankel, 1995) and found the results to be quite similar to what we obtained using this simpler cellular method (Kafka and Levin, 2000). Thus, we have adopted the cellular seismology method as a simple and straightforward way of measuring the tendency for past seismicity to delineate zones where future earthquakes are likely to occur.

Figure 3 shows examples of the application of the cellular seismology method to the northeastern United States and Southern California, with radii of circles around the "before" epicenters chosen to fill 33% of the map area. I use the notation "M" in Figure 3, and throughout this paper, to represent magnitude as it was reported in the various earthquake catalogs used in this study. As in Figure 2, the shaded areas in Figure 3 indicate the portions of the map that are near the "before" epicenters, and filled circles are the "after" epicenters. I varied the magnitude cutoffs for "before" and "after" earthquake catalogs so that I could analyze forecasts of earthquakes in the various regions for comparable numbers of events and comparable periods of time.

"Cellullar Seismology" Method

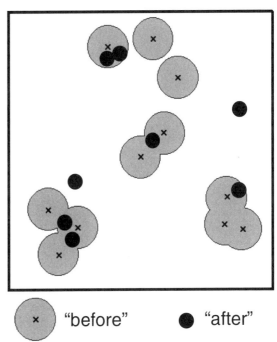

Figure 2. Illustration of the methodology used in this study to measure the extent to which future earthquakes tend to occur near past earthquakes.

ods of time. The values of $\hat{\rho}$ for the northeastern United States and for Southern California are 78% and 79%, respectively.

As another example of the cellular method, Figure 4 shows the application of this method to the entire central and eastern United States for "before" earthquakes, chosen to be events between 1924 and 1987 (M ≥ 3.0), and "after" earthquakes, chosen to be events between 1988 and 2003 (M ≥ 4.5). For this case, with radii of the circles around the "before" epicenters again chosen to fill 33% of the map area, there are 90% hits.

The eventual goal of this type of study is to forecast the locations (if not the times) of future damaging earthquakes. For the U.S. National Seismic Hazard Maps, a minimum magnitude of 5.0 was used in the hazard calculations for the central and eastern United States (Frankel, et al., 1996) because that was considered to be the threshold for an earthquake to cause significant damage. While it would be ideal to limit this study to analyzing "after" earthquakes of about magnitude 5.0 and greater, unfortunately such an approach would mean that I would only be able to analyze very small samples (therefore making any statistical analysis very difficult to carry out). The approach taken in this study is to choose magnitude thresholds for seismicity catalogs based on the data available and the completeness of the catalogs at lower magnitudes. For the "before" catalogs, I chose magnitude thresholds such that I was reasonably confident that the catalogs were complete. For the "after" catalogs, I chose magnitude thresholds such that

---

[1]In this paper, the terms "before" and "after" catalogs refer to the *samples* analyzed, and "past" and "future" earthquakes refer to the *populations* of past and future earthquakes that those samples were selected from.

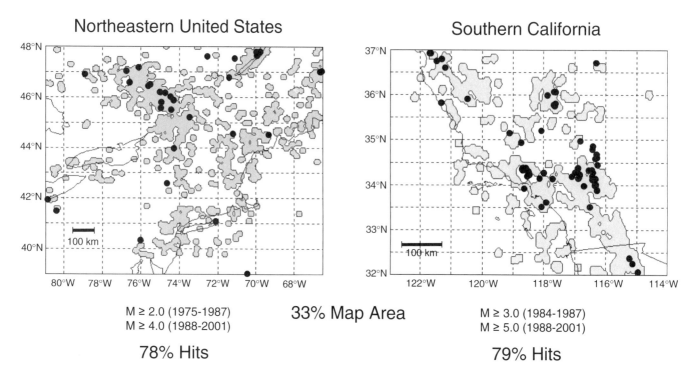

Northeastern United States

M ≥ 2.0 (1975-1987)
M ≥ 4.0 (1988-2001)

78% Hits

33% Map Area

Southern California

M ≥ 3.0 (1984-1987)
M ≥ 5.0 (1988-2001)

79% Hits

Figure 3. Examples of the application of the "cellular seismology" method (described in the text) to the northeastern United States (left) and Southern California (right), adapted from Kafka (2002). Shaded areas are zones surrounding the "before" catalog, and filled circles indicate epicenters of the "after" catalog. For the northeastern United States: "before" corresponds to M ≥ 2.0, 1975–1987, and a radius of 15.5 km, and "after" corresponds to M ≥ 4.0, 1988–2001. For Southern California: "before" corresponds to M ≥ 3.0, 1984–1987, and a radius of 13.2 km, and "after" corresponds to M ≥ 5.0, 1988–2001.

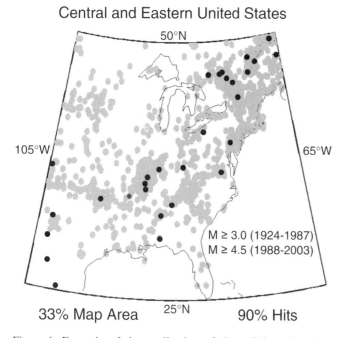

Central and Eastern United States

M ≥ 3.0 (1924-1987)
M ≥ 4.5 (1988-2003)

33% Map Area   **90% Hits**

Figure 4. Example of the application of the cellular seismology method to the entire central and eastern United States. Shaded areas are zones surrounding the "before" catalog, and filled circles indicate epicenters of the "after" catalog. For this case, "before" corresponds to M ≥ 3.0, 1924–1987, and a radius of 36.0 km, and "after" corresponds to M ≥ 4.5, 1988–2003.

the future earthquakes were as large as possible, while still having enough of them to study the values of $\hat{\rho}$ statistically. These magnitude thresholds were then different for different regions analyzed. While this is not an ideal way to investigate the problem, it is an attempt to use the maximum amount of data available. In order to obtain at least a preliminary assessment of the extent to which these results apply to forecasting locations of truly large and damaging earthquakes, I include here an analysis of the effect of magnitude threshold used for the "after" catalogs, and, as will be seen, I found no evidence of an effect of the choice of this magnitude threshold on the results of this study.

The examples shown in Figures 3 and 4 illustrate how this method provides a measure of the tendency for past seismicity to delineate epicenters of future earthquakes. However, there is a fundamental problem in statistical hypothesis testing of these types of results (as well as for seismicity studies in general). Whereas hypotheses should be tested on multiple independent data sets, in earthquake studies of this type, we often have only one observed data set: the observed record of seismicity. (For additional discussion of issues related to hypothesis testing in earthquake studies, see Rhoades and Evison, 1989.) Lacking such multiple independent data sets, I make a "logical leap" of using earthquake catalogs in different regions as a proxy for repeated samples from the central and eastern United States. I then compare how well seismicity "retrodicts" earthquakes in the central and eastern United States with how well seismicity retrodicts earthquakes in other regions (including a variety of tectonic environments).

To make these proxy data sets in various regions as comparable as possible, the essence of the distribution of seismicity must be captured in such a way that the measure of the distribution of seismicity is as independent as possible of the size and shape of the region being investigated. For this purpose, I use the percentage of map area surrounding the past earthquakes (P) as a parameter to characterize the distribution of seismicity for a given region. One might envision that the radius of the circles surrounding the epicenters might be a better (and more fundamental) variable for this purpose because it is more directly related to the physics of the earthquake process. I have found, however, that the value of $\hat{\rho}$ for a given radius is more affected by the characteristics of seismicity specific to a given region than is the value of $\hat{\rho}$ for a given percentage of map area. Thus, I consider the percentage of map area to be a more useful parameter than radius for the purpose of this study, i.e., for comparing the extent to which seismicity delineates zones of future earthquakes from one region to the next. As an illustration, Figure 5 shows the value of $\hat{\rho}$ as a function of radius and P for all regions analyzed by Kafka (2002). Notice that choosing percentage of area rather than radius results in a smaller spread of values of $\hat{\rho}$ for a given mean value of $\hat{\rho}$. Thus, it appears that $\hat{\rho}(P)$ captures the extent to which seismicity delineates zones of future earthquakes in such a way that it minimizes the effect

of the size and shape of the specific region investigated and emphasizes the phenomenon of interest itself, thus justifying the logical leap of using many regions as a proxy for many realizations of earthquake catalogs for the same region.

Figure 6 shows results for the various regions analyzed by Kafka (2002). When the radius is chosen such that 33% of the map area is filled, the average value of $\hat{\rho}$ for all of these data lumped together is 74%. The observed values of $\hat{\rho}$ range from 60% for the southeastern United States to 91% for the entire central and eastern United States. These results show no striking pattern of systematic differences in values of $\hat{\rho}$ for intraplate versus plate-boundary regions. Given the more spatially concentrated seismicity in plate-boundary regions, I had expected (intuitively) to observe a greater tendency for past seismicity to delineate zones of future earthquakes in interplate regions than in plate interiors, but no such pattern was observed.

A simple, straightforward observation from Figure 6 is that the percentage of hits exceeds the percentage of map area in all cases. Although not a surprising result, this provides a measure of the fact that, for these cases, future earthquakes are likely to be more highly concentrated in the vicinity of past earthquakes than would be expected for a random distribution of future earthquakes (i.e., values of $\hat{\rho}$ are greater than P). Without such an explicit comparison of the observations with that expected for a random distribution, there is no empirical basis for arguing that future earthquakes tend to occur in the vicinity of past earthquakes. One could imagine that for some regions $\hat{\rho}$ would be below the line

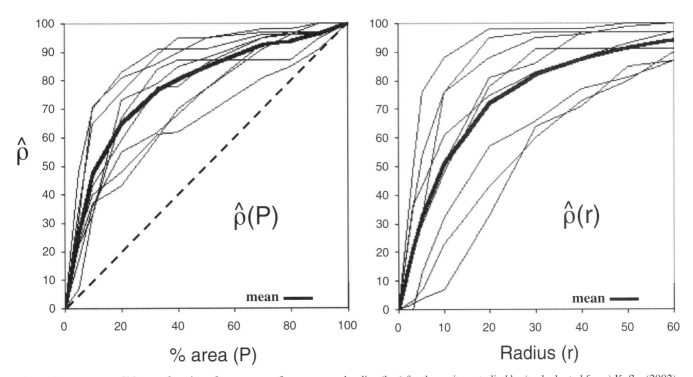

Figure 5. Percentage of hits as a function of percentage of map area and radius (km) for the regions studied by (and adapted from) Kafka (2002). Thin lines denote the individual regions, and thick lines indicate the mean for all regions. Dashed line indicates the locus of points where $\hat{\rho}$ = P, corresponding to hypothesis H3.

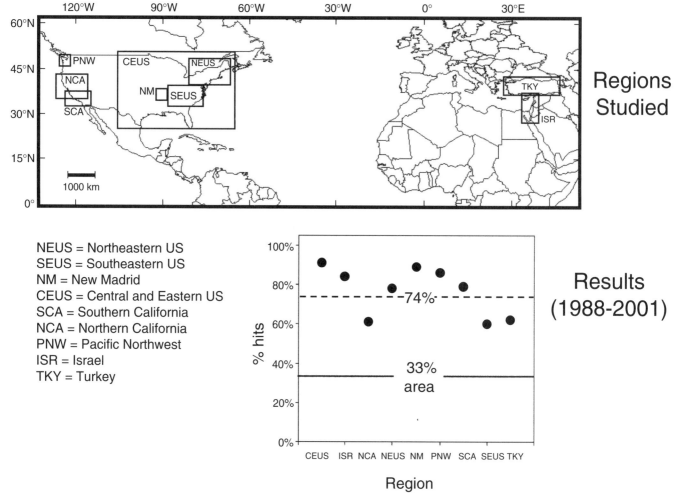

Figure 6. Summary of results for the nine regions analyzed by (and adapted from) Kafka (2002). Radii around the "before" catalog epicenters were chosen such that 33% of the map area was covered. The radii corresponding to 33% map area are: northeastern United States (15.5 km), southeastern United States (31.0 km), New Madrid (13.5 km), central and eastern United States (36.0 km), Southern California (13.2 km), Northern California (10.0 km), Pacific Northwest (9.2 km), Israel (48.0 km), and Turkey (26.0).

representing the 33% map area, which would support hypothesis H2 of Figure 1 for that area. The fact that no such case has been found is encouraging for those who would like to use seismicity as a basis for seismic hazard mapping.

Beyond this simple, straightforward conclusion, however, is the question of how much we can glean from such analyses. To what extent, for example, does the value of $\hat{\rho}$ (for a given P) vary from one region to the next? To what extent does it vary with different magnitude cutoffs for "before" and "after" earthquake catalogs? To what extent does it vary with time after the end of the "before" earthquake catalog? In statistical terms: we can envision an underlying distribution of values of $\hat{\rho}$ for a given value of P, and then attempt to discern how that distribution varies for different regions, different magnitude cutoffs, and different time periods after the end of the "before" earthquake catalog. In the next two sections of this paper, I cast these questions in statistical terms and use that formulation to investigate these variations.

STATISTICAL ANALYSIS OF PERCENTAGES OF HITS

The following statistical formulation of the problem is used for this study. P is the percentage of map area covered by circles of a given radius surrounding "before" earthquake epicenters, $\hat{\rho}$ is the observed percentage of "after" earthquakes that occur "near" (i.e., within that given radius of) at least one of the "before" earthquakes, and ρ is the underlying probability that a future earthquake will occur "near" at least one of the past earthquakes. Our objective, then, is to estimate ρ based on observations of $\hat{\rho}$.

Consider the range of possible distributions of $\hat{\rho}(P)$. At one extreme, we could imagine that $\hat{\rho}$ has equal probability of having any value between 0 and 1, implying that there is no information content in the distribution of past seismicity (or the choice of P) that is relevant to where future earthquakes are likely to occur. At the other extreme, we could imagine that the nature of earthquake processes (and the choice of P) is such that (regardless of the chosen region) there is a very narrow distribution of possible values of $\hat{\rho}$, once the value of P is chosen. Between these two extremes, we can imagine that there is some characteristic (presumably region-dependent) shape to the distribution of $\hat{\rho}$, which provides information about the probability that future earthquakes will be located near past earthquakes for a given value of P.

Figure 7 shows the distribution of values of $\hat{\rho}$ from data analyzed by Kafka (2002), with P chosen to be 33%. The fact that the observed distribution of $\hat{\rho}$ lies consistently to the right of P for all cases studied is evidence that future earthquakes are more likely to occur near past earthquakes than where previous earthquakes have not occurred (regardless of the specific region investigated). If the distribution lay far to the right, it would be evidence that future earthquakes are highly concentrated in zones of past seismicity. A high variance in the distribution would indicate great variation from one region to the next (and presumably one tectonic environment to the next) in the probability of future earthquakes occurring near past ones.

Based on this type of analysis, I conducted systematic hypothesis tests to obtain estimates of $\rho(P)$ based on a given data set of values of $\hat{\rho}$. I treated $\hat{\rho}$ as a random variable, and estimated the probability distribution of $\hat{\rho}$. The process of examining whether a future large earthquake occurs within a given radius of past ones is thus modeled as a binomial experiment with success defined as the event that the future earthquake (filled circle in Fig. 2) occurs within that radius of at least one past earthquake (shaded zones in Fig. 2).

Using this formulation, Kafka (2002) estimated that, at the 95% level of statistical significance, more than 71% of the earthquakes in a region will tend to occur near previous epicenters

(defined by the 33% map area contours). Furthermore, the results of that study suggested that there was no statistically significant difference in the values of $\hat{\rho}$ for intraplate versus plate-boundary regions. While these results are a good first step toward providing statistical support for the notion that past seismicity delineates where future earthquakes are likely to occur, an obvious problem with these studies to date is that they have been based on "convenience sampling" (i.e., choosing data samples because they happen to be available or relatively easy to obtain). The regions studied were chosen primarily because they are areas where high-quality seismic monitoring resulted in earthquake catalogs that were readily available for analysis. Thus, the data were not random samples of seismicity from various tectonic environments, so what seems like statistical significance might be highly biased and merely a coincidence of the particular data sets analyzed. One would instead want to apply these methods to data sets that are more randomly and objectively chosen from different regions and different periods of time.

Another problem with our previous studies was that the sizes and shapes of the regions varied greatly (Fig. 6), and thus the characteristic of "nearness" to previous earthquakes was different for the different regions. Furthermore, in our previous studies the number of earthquakes in the catalogs was not large enough to, in any systematic way, have more than one statistical sample of "future" large earthquakes for a given region. Thus, we didn't really have a basis from which to effectively investigate what the distribution of $\hat{\rho}$ might look like for a given region (i.e., for a given tectonic environment).

The ideal data sets to test these hypotheses are impossible to obtain without another century or more of monitoring. However, the three decades of global earthquake data available from the U.S. Geological Survey (USGS) National Earthquake Information Center (NEIC, Fig. 8) provide an opportunity to test such hypotheses in a more systematic and objective manner than in our previous studies.

IN SEARCH OF A MORE SYSTEMATIC TEST OF H1

Using this NEIC database, the statistical variation of the values of $\hat{\rho}$ was explored by dividing the world into eight subregions (labeled R11 through R24), all with the same shape and same surface area (Fig. 8). The cellular seismology method was then applied uniformly to all of these subregions. In an effort to simulate a series of realizations of hypothetical cases like the situation we face in the central and eastern United States, I envision this as a "thought experiment" in which we have eight regions where earthquakes are occurring, and the earthquakes are caused by some tectonic process (or processes) which I will treat here as unknown. Although we actually are aware of the tectonic processes in these regions, we can still think of them in this manner because the boundaries were chosen independently of tectonic environment. Continuing with this thought experiment, we envision trying to discern from a catalog of earthquakes alone the extent to which past seismicity can be used to delineate zones

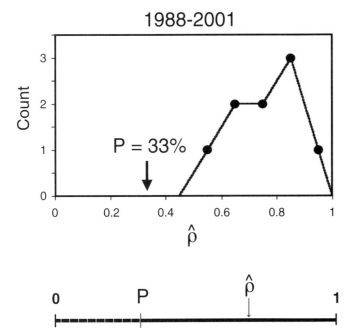

Figure 7. Distribution of $\hat{\rho}$ based on the data analyzed by (and adapted from) Kafka (2002). The $\hat{\rho}$ values shown here are calculated for P = 33%.

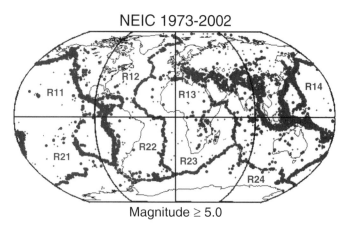

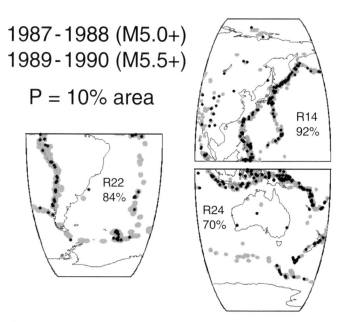

Figure 8. Filled circles indicate shallow (≤70 km) earthquakes M ≥ 5.0 recorded by the National Earthquake Information Center (NEIC) from 1973 to 2002. Size of circles is proportional to magnitude (ranging from 5.0 to 8.8). Also shown are the eight subregions (R11 to R24) of equal surface area and shape for which the cellular seismology method was applied as described in the text.

Figure 9. Examples of the application of the cellular seismology method to the subregions shown in Figure 8. Gray shading shows zones surrounding the "before" catalog, and filled black circles indicate epicenters of the "after" catalog. For this case, "before" corresponds to M ≥ 5.0, 1987–1988, and "after" corresponds to M ≥ 5.5, 1989–1990. Radii around the "before" catalog epicenters were chosen such that 10% of the map area was covered. The radii corresponding to 10% map area are: R22 (168 km), R14 (109 km), and R24 (117 km).

where future large earthquakes are likely to occur. (After this "blind" statistical analysis, I will later speculate on how tectonic environment within these regions actually does contribute to differences in the distributions of $\hat{\rho}$.)

Using the NEIC data shown in Figure 8 (with a lower magnitude cutoff of 5.0), there is a sufficiently large number of events (from a complete earthquake catalog) that the cellular seismology method can be applied to many independent "two-year-before catalogs" forecasting "two-year-after catalogs" (Figs. 9 and 10). Figure 9 shows three examples (regions R14, R22, and R24) where two years of M ≥ 5.0 earthquakes represent past seismicity, two years of M ≥ 5.5 earthquakes represent future earthquakes, and P = 10% of map area. For these cases, the values of $\hat{\rho}$ are: 92% for R14, 84% for R22, and 70% for R24. Given the length of the NEIC catalog, I was able to analyze the data in the manner illustrated in Figure 9 for 14 pairs of two-year-before/two-year-after catalogs for each of the eight regions (Fig. 10).

For simplicity, I use the notation "Mx+" to represent earthquakes of M ≥ x. The top graph in Figure 10 shows results for 14 cases of two years of M5.0+ earthquakes representing the past, and two years of M6.0+ earthquakes representing the future for the eight subregions shown in Figure 8. The bottom graph shows the same M5.0+ "before" earthquake catalogs forecasting locations of M5.5+ earthquakes. For the top graph, the number of "after" earthquakes for the 14 cases ranged from 1 to 109, with a mean of 26 and a standard deviation of 22. For the bottom graph the number of "after" earthquakes ranged from 11 to 312, with a mean of 87 and a standard deviation of 66. Thus, the values of $\hat{\rho}$ shown for each two-year sample for the M5.5+ plot were based on large enough samples to be considered as statistical measures of the underlying value of ρ expected for that region. The results for the M5.5+ case are quite similar to those of the M6.0+ case, except that the variances for the M5.5+ case tended to be lower

than for the M6.0+ case. This observation suggests that the M5.5+ case is measuring the same phenomenon as the M6.0+ case, but is just based on larger sample sizes. Figure 11 shows the mean values for the M6.0+ case for each region.

We can see in Figures 10 and 11 that there is a characteristic mean value of $\hat{\rho}$ for a given subregion, as well as a characteristic variance. For example, region R11 has a high mean (93% for M6.0+ forecasts) and a low variance, while region R23 has a low mean (49%) and a high variance. Below, in the Discussion and Conclusions section, I speculate on possible relationships between the mean and variance of the values of $\hat{\rho}$ and the tectonic environments in the subregions analyzed.

As in the cases analyzed by Kafka (2002), for these eight subregions, the values of $\hat{\rho}$ for a given radius are more affected by the characteristics of the specific region than are the values of $\hat{\rho}$ for a given percentage of map area (Fig. 12). Again, choosing percentage of area rather than radius results in a lower range of values of $\hat{\rho}$ for a given mean value of $\hat{\rho}$. This effect is not as pronounced here as in the cases analyzed by Kafka (2002), and I suspect that this is due to the fact that the eight subregions are all the same size and shape, so that radius and percentage of area are more directly related to each other in this situation. In additional cases investigated next, where I analyze regions with specific tectonic environments and with significant differences in the shape and size of the regions, this effect is again more clearly observed (Figs. 13 and 14).

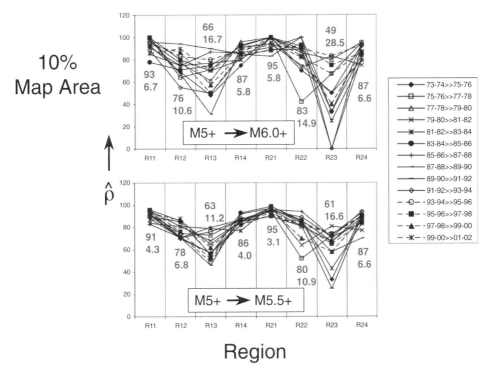

10%
Map Area

Figure 10. Results of the application of the cellular seismology method to the eight subregions shown in Figure 8. Numbers shown in gray are the mean (above) and the standard deviation (below) of $\hat{\rho}$ for each region.

While the analysis of the eight global subregions described here has the advantages of being objective and being based on regions of the same shape and size, a disadvantage of this analysis is that each region arbitrarily includes a variety of tectonic environments. In an effort to discern more specific effects of tectonic environment on the distribution of $\hat{\rho}$, I applied the cellular seismology method to four additional subregions of Earth, each representing a different tectonic environment (Figs. 13 and 14). The tectonic environments chosen for these analyses were: a subduction zone (labeled Subduction in Fig. 13), an oceanic spreading center (labeled Ridge), a region of active continental collision (labeled Continent), and the interior of the North American plate (labeled INAP).

In the Ridge and Subduction regions, the seismicity is (not surprisingly) densely concentrated near the plate boundary. Future earthquakes in these regions have a high probability of occurring near past seismicity, with $\hat{\rho}$ (for P = 33%) equal to 98% for Ridge and 100% for Subduction. The INAP and Continent regions, by contrast, have (also not surprisingly) more diffusely distributed seismicity, and future earthquakes in these regions appear to have

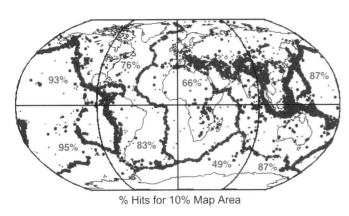

Figure 11. Mean values of $\hat{\rho}$ from the analyses shown in Figure 10 (M6.0+ case).

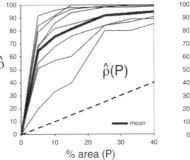
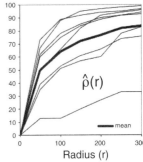

Figure 12. Values of $\hat{\rho}$ as a function of percentage of map area and radius (km) for the eight subregions shown in Figure 8. Thin lines denote the individual regions, and thick lines indicate the mean for all regions. Dashed line indicates the locus of points where $\hat{\rho} = P$, corresponding to hypothesis H3.

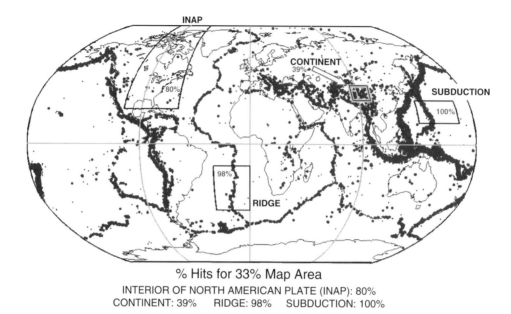

% Hits for 33% Map Area
INTERIOR OF NORTH AMERICAN PLATE (INAP): 80%
CONTINENT: 39%   RIDGE: 98%   SUBDUCTION: 100%

Figure 13. Percentage of hits for 33% map area calculated for regions with different tectonic environments (M5.0+, 1973–1987 forecasting M5.5+, 1988–2002). The radii corresponding to 33% map area are: interior of the North American plate (INAP; 420 km), Continent (63 km), Ridge (320 km), and Subduction (330 km). Percentages of hits for other values of percentage of map area, and for variation in radius, are shown in Figure 14.

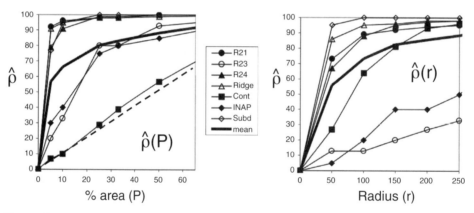

Figure 14. Percentage of hits as a function of percentage of map area and radius (km) for the four tectonic regions shown in Figure 13 (M5.0+, 1973–1987 forecasting M5.5+, 1988–2002), and for regions R21, R23, and R24 of Figure 8 (M5.0+, 1973–1974 forecasting M5.5+, 1975–1976). Thin lines denote the individual regions, and thick black lines indicate the mean for all regions. Dashed line indicates the locus of points where $\hat{\rho}$ = P, corresponding to hypothesis H3. INAP—interior of North American plate, Cont—Continent, and Subd—Subduction.

lower probabilities of occurring near past seismicity, with $\hat{\rho}$ (for P = 33%) equal to 80% for INAP and 39% for Continent. More specific aspects of these results do not have obvious interpretations. For example, why is the value of $\hat{\rho}$ for the Continent region so low (39%), and how might this low value of $\hat{\rho}$ be related to that of R13, which has a relatively low $\hat{\rho}$ for 10% map area and a high variance (see Figs. 10 and 11)? Also, when we compare these results for the Continent region with results for the central and eastern United States, we find that the central and eastern United

States has surprisingly high values of $\hat{\rho}$. For the central and eastern United States case shown in Figure 4, $\hat{\rho}$ is 90% for 33% map area, and $\hat{\rho}$ is 57% for 10% map area.

In an effort to investigate the extent to which these results apply to forecasting locations of truly large and damaging earthquakes, I studied the effect of the minimum magnitude threshold used for the "after" catalogs in these analyses. For nine cases, the minimum magnitude cutoff for "after" earthquakes, M(min), was varied systematically to investigate the effect on

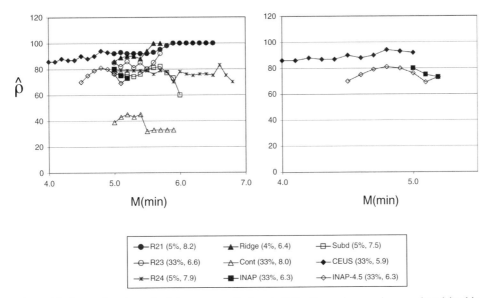

Figure 15. $\hat{\rho}$ as a function of minimum magnitude cutoff for the various regions analyzed in this study. Numbers in parentheses are percentage of map area corresponding to the "before" earthquake catalog and magnitude of largest "after" earthquake for a given region. For this analysis, it was necessary to vary the percentage of map area for the different regions so that the initial value of $\hat{\rho}$ was low enough that the variation in $\hat{\rho}$ could be seen as M(min) increased. INAP—interior of North American plate, Cont—Continent, Subd—Subduction, and CEUS—central and eastern United States.

$\hat{\rho}$ as M(min) increases (Fig. 15). For each region, M(min) was increased by 0.1 unit intervals up to the point where there were at least 10 "after" earthquakes remaining to be analyzed. The minimum magnitude for the "before" earthquake catalogs was 5.0 for all cases shown in Figure 15, except for central and eastern United States (where it was 3.0). Since the INAP region had much fewer earthquakes than the other regions, there are two cases shown for INAP in Figure 15, one in which the "before" and "after" catalogs start at magnitude 5.0, and the other in which they start at magnitude 4.5.

As a simple test of whether there is any systematic effect on $\hat{\rho}$, I counted the number of times that $\hat{\rho}$ increased when M(min) was changed to M(min) + 0.1, the number of times it remained the same, and the number of times it decreased. For example, in the case of central and eastern United States, $\hat{\rho}$ is 86% for M(min) = 4.0, remains the same (86%) for M(min) = 4.1, and then increases to 88% for M(min) = 4.2. If there was, for example, a systematic decrease in $\hat{\rho}$ with increasing M(min), we would expect there to be significantly more decreases than increases. The results were: 30 times for "increase," 27 times for "decrease," and 21 times for "remained the same." The similarity between numbers of increases and numbers of decreases suggests that there is no observed effect of M(min) on $\hat{\rho}$. Using this result as a guide, it appears that the results of this study are independent of M(min), and therefore apply to "large and damaging" earthquakes in the central and eastern United States.

Although somewhat beyond the scope of this study, this investigation would not be complete without at least some mention of the question of the time dependence of the values of

$\hat{\rho}$ for a given region. One would intuitively imagine that the values of $\hat{\rho}$ would tend to decrease for forecasting locations of earthquakes farther into the future. Figure 16 shows the distribution of $\hat{\rho}$ for 10% map area for the eight global regions analyzed (M5.0+ seismicity forecasting locations of M5.5+ earthquakes). The "before" catalog is for 1973–1974, and Figure 16 shows the distribution of values of $\hat{\rho}$ for "after" earthquakes occurring in 1975 and 1976, as well as in 2001 and 2002. What is most striking about the results shown in this figure is the lack of evidence for systematically lower values of $\hat{\rho}$ for the 2001–2002 forecasts than for the 1975–1976 forecasts (even though the 2001–2002 catalog starts 26 yr after the end of the "before" seismicity catalog). For 1975–1976, the mean value of $\hat{\rho}$ is 75.6 and the median is 84, while for 2001–2002, the mean is 79.6 and the median is 82. Thus, at least on the time scale of a few decades, there does not seem to be any evidence of a decrease in the tendency for past seismicity to delineate zones where future earthquakes are likely to occur.

As an additional test of the time dependence of this phenomenon, there were sufficient numbers of M4.0+ earthquakes in the central and eastern United States catalog to divide the results for observed values of $\hat{\rho}$ in the central and eastern United States from 1988 to 2003 into six subsamples of 15 earthquakes each. Each one of the six subsamples is ordered in Figure 17 such that they consist of sequentially later events in time, and for each subsample, $\hat{\rho}$ was calculated. Again as in the analysis described in the previous paragraph, there was no evidence of any decrease in the values of $\hat{\rho}$ as time increased after the end of the past earthquake seismicity catalog. While these issues need to be analyzed

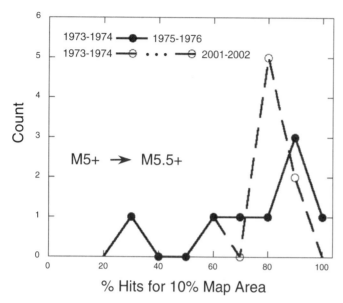

Figure 16. Distribution of percentage of hits for 10% map area for the eight regions shown in Figure 8 (M5.0+ seismicity forecasting locations of M5.5+ earthquakes). Filled circles represent results for 1973–1974 seismicity forecasting locations of earthquakes occurring between 1975 and 1976 (mean percentage of hits = 75.6, median percentage of hits = 84). Open circles are for 1973–1974 seismicity forecasting locations of earthquakes occurring between 2001 and 2002 (mean = 79.6, median = 82).

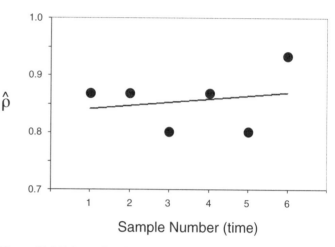

Figure 17. M4.0+ earthquakes in the central and eastern United States catalog (1988–2003) divided into six subsamples of 15 earthquakes each. The six subsamples are ordered such that they consist of sequentially later events in time, and for each subsample, $\hat{\rho}$ was calculated. The solid line is a least squares fit to the data.

I have a large enough sample to apply methods of statistical inference, and we find $\hat{\rho} = 0.86$. Thus, I form a 95% confidence interval, as follows (e.g., Weiss and Hassett, 1982):

$$\rho(CEUS, 0.33) = \hat{\rho} \pm 1.96\sqrt{\frac{\hat{\rho}(1-\hat{\rho})}{n}}, \qquad (1)$$

where $n = 91$ is the number of "after" earthquakes. The 95% confidence interval for this case is $0.79 \leq \rho(CEUS, 0.33) \leq 0.93$, and I estimate that the probability of a given future earthquake occurring in the light shaded zones in Figure 18 is 0.86 ± 0.072. For P = 10% map area (dark shaded zones in Fig. 18), I find that $\hat{\rho} = 0.60 \pm 0.100$, and the 95% confidence interval is $0.50 \leq \rho(CEUS, 0.10) \leq 0.70$.

This estimate of ρ for the central and eastern United States is not inordinately low compared to observed values of $\hat{\rho}$ for regions around the world, but it falls within the lower end of the range of values (Figs. 10 and 13). Stated in terms of radius surrounding the epicenters of past earthquakes, these results suggest that future earthquakes in the central and eastern United States have ~86% probability of occurring within 36 km of past earthquakes, and ~60% probability of occurring within 14 km of past earthquakes.

DISCUSSION AND CONCLUSIONS

Given our less-than-complete understanding of the cause of earthquakes in the central and eastern United States, seismic hazard analysis for this region (and probably most intraplate regions) will likely continue to depend, to a large extent, on seismicity for delineating locations of future large earthquakes. Thus, it is important to understand the scientific basis underlying the tendency for seismicity to delineate zones where future large earthquakes are likely to occur. This will not be an easy task, but if

in greater detail to make any strong conclusions about the time dependence of this phenomenon, I see no evidence of a decrease in the values of $\hat{\rho}$ as time increases.

ESTIMATING ρ FOR THE CENTRAL AND EASTERN UNITED STATES

Having explored the statistical variation of the values of $\hat{\rho}$ for a variety of tectonic environments, magnitude thresholds, and sizes and shapes of regions, it seems clear that the tendency for past seismicity to delineate zones where future large earthquakes are likely to occur is a real, measurable, physical phenomenon. Furthermore, the time scale over which this phenomenon varies appears to be such that it is possible to obtain representative samples of this phenomenon from seismicity maps and to use those samples to estimate the probability of future earthquakes occurring near past earthquakes in a region of interest. Based on these results, I can thus estimate the probability of a future large earthquake occurring in zones delineated by past seismicity in the central and eastern United States, as well as a confidence interval for that estimate.

Let $\rho(CEUS, P)$ represent the probability that a future large earthquake in the central and eastern United States will occur within the zones defined by P% map area, defined as discussed already using the cellular seismology method. Based on M4.0+ earthquakes occurring between 1988 and 2003 (with P = 33%),

Central and Eastern United States

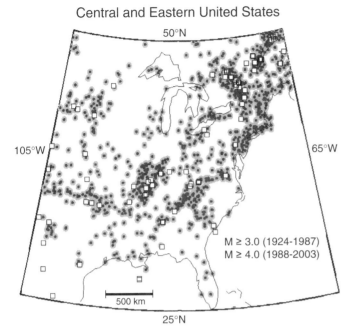

Figure 18. M3.0+ earthquakes in the central and eastern United States (1924–1987) forecasting locations of M4.0+ earthquakes (1988–2003). Lighter shading corresponds to a 36 km radius and 33% map area, and darker shading corresponds to a 14 km radius and 10% map area. M4.0+ earthquake epicenters for 1988–2003 are shown by the open squares.

we do not undertake this task, then one of the major inputs into seismic hazard analysis will be based merely on an untested conceptual model. The analysis presented here supports the notion that the tendency for future large earthquakes to occur in zones delineated by past seismicity is a real, measurable, physical phenomenon that can be investigated scientifically. While this investigation is only at a very early stage of development, it shows some systematic patterns regarding similarities and differences in this tendency for regions characterized by different tectonic processes. The results suggest that using seismicity as an indicator of zones where future large earthquakes are likely to occur is as reasonable a conceptual model for the central and eastern United States as it is for many other parts of the world, including some plate-boundary regions.

The analysis of the eight regional subdivisions of Earth, chosen objectively, shows that a given region, within which some given combination of tectonic processes is occurring, is characterized by a mean value and variance of $\hat{\rho}$. As can be seen in Figures 10 and 11, regions that contain large proportions of subduction zones (such as R11, R14, and R24) have characteristically high mean values of $\hat{\rho}$ and characteristically low variances of $\hat{\rho}$. While this same pattern might have been expected for regions containing large proportions of oceanic spreading centers (such as R23), such regions have characteristically lower mean values of $\hat{\rho}$ and higher variances of $\hat{\rho}$ compared to regions dominated by subduction zones. The variations

in $\hat{\rho}$ seem to be the result of a combination of not only tectonic differences, but also differences in the way in which the NEIC samples the different types of tectonic zones. Although we are just beginning to scratch the surface of how to measure and categorize this phenomenon, it appears that $\hat{\rho}$ is higher when a major subduction zone is included in a region, and it is lower for continental areas and mid-ocean ridges.

The analysis of specific tectonic regions (Fig. 13) supports the (not surprising) result discussed already regarding a strong tendency for future earthquakes to occur near past earthquakes in subduction zones. The very high value of $\hat{\rho}$ for the Ridge region (98%), however, suggests that the relatively low values of $\hat{\rho}$ for region R23 are not merely the result of the high percentage of ridge-type plate boundaries in region R23. Unraveling the effects of tectonic environment on values of $\hat{\rho}$ is, of course, complicated by differences in the sizes and shapes of the regions chosen for analysis (even though use of the percentage of map area does mitigate this problem to some extent).

The value of $\hat{\rho}$ for the Continent region (39% for 33% map area) is among the lowest values of $\hat{\rho}$ for any region analyzed in our studies. In fact, 39% hits for 33% map area is very close to what would be expected for a random spatial distribution of future earthquakes. The reason for this low value is not clear at this point, but it should caution us not to conclude that seismicity will always be a good indicator of where future earthquakes will occur.

The value of $\hat{\rho}$ for the INAP region (80% for 33% map area) is neither particularly high nor particularly low compared to other regions around the world. Thus, while hypothesis H1 is not unequivocally verified for intraplate regions, these results, along with our specific results for the central and eastern United States, suggest that H1 is as reasonable a conceptual model for the central and eastern United States as it is for many parts of the world, including some plate-boundary regions. Our studies, therefore, support the approach of using past seismicity as an indicator of zones where future large earthquakes are likely to occur in the central and eastern United States for the purpose of seismic hazard analysis (at least until the physical cause of earthquakes in this region is better understood).

The analysis of the time dependence of the tendency for future earthquakes to occur in zones delineated by past seismicity yielded no evidence of a decrease in this tendency as time increases after the end of the past seismicity catalog. This result is encouraging for seismic hazard analysis as it suggests that seismic hazard maps developed today may provide meaningful estimates of hazards to facilities that are expected to have very long lifetimes.

The results of these preliminary analyses of time dependence suggest that we are not (yet?) seeing evidence of the "paleoseismicity" model of Ebel et al. (2000). This model hypothesizes that the major earthquakes that have occurred in the central and eastern United States might be long-delayed aftershocks of large early historical or prehistorical earthquakes. If this were the case, we might expect to eventually see a decrease in the number of earthquakes near past historical earthquakes as time increases, and therefore might expect our ability to forecast locations of future earthquakes based on seismicity to decrease as time increases. The time spans covered by the earthquake catalogs used in this study are probably too short, however, to detect such a pattern. Thus, the results of this study cannot rule out the paleoseismicity model.

We find no evidence to suggest that the tendency for seismicity to delineate zones where future large earthquakes are likely to occur is any "less real" for the central and eastern United States than for any other regions. Given what we have been able to discern from this and our previous studies on this topic, we estimate that the probability of a given future earthquake (including large and damaging earthquakes) occurring in the light shaded zones in Figure 18 is 0.86 ± 0.072, and the probability of such an earthquake occurring in the dark shaded zones in that figure is 0.60 ± 0.100.

This study provides evidence that the tendency for seismicity to delineate zones where future large earthquakes are likely to occur is a real, measurable, physical phenomenon. Furthermore, the time scale over which seismicity data are available appears to be representative of the time scale over which this phenomenon occurs. Many years of additional seismicity data will, nonetheless, be required to fully test the ideas discussed in this paper. For this (and many other reasons), there is value in continuing to monitor seismicity in intraplate regions. Continued monitoring of seismicity will provide a better basis for discerning the scientific justification for using seismicity as an indicator of zones where future large earthquakes are likely to occur in the central and eastern United States and elsewhere.

ACKNOWLEDGMENTS

I thank the editors of this volume, Seth Stein and Stéphane Mazzotti, as well as the three anonymous reviewers of this manuscript for their thoughtful critiques, which improved the paper in many ways. This study benefited from many years of valuable discussions with John Ebel on various aspects of intraplate earthquakes in general and "cellular seismology" in particular. I thank Heather Wagner for assisting me with the data processing for the analysis of the eight subregions of Earth.

REFERENCES CITED

Cao, T., Petersen, M.D., and Reichle, M.S., 1996, Seismic hazard estimate from background seismicity in Southern California: Bulletin of the Seismological Society of America, v. 86, no. 5, p. 1372–1381.

Dziewonski, A.M., Ekström, G., and Maternovskaya, N.N., 1999, Centroid-moment tensor solutions for April-June 1998: Physics of the Earth and Planetary Interiors, v. 112, p. 11–19, doi: 10.1016/S0031-9201(98)00157-5.

Ebel, J.E., Bonjer, K.P., and Oncescu, M.C., 2000, Paleoseismicity: Seismicity evidence for past large earthquakes: Seismological Research Letters, v. 71, no. 2, p. 283–294.

Frankel, A., 1995, Mapping seismic hazard in the central and eastern United States: Seismological Research Letters, v. 66, no. 4, p. 8–21.

Frankel, A., Mueller, C., Barnhard, T., Perkins, T.D., Leyendecker, E.V., Dickman, N., Hanson, S., and Hopper, M., 1996, National Seismic Hazard Maps, Documentation, June 1996: U.S. Geological Survey Open-File Report 96-532, 41 p.

Jackson, D.D., and Kagan, Y.Y., 1999, Testable earthquake forecasts for 1999: Seismological Research Letters, v. 70, no. 4, p. 393–403.

Kafka, A.L., 2002, Statistical analysis of the hypothesis that seismicity delineates areas where future large earthquakes are likely to occur in the central and eastern United States: Seismological Research Letters, v. 73, no. 6, p. 990–1001.

Kafka, A.L., and Levin, S.Z., 2000, Does the spatial distribution of smaller earthquakes delineate areas where larger earthquakes are likely to occur?: Bulletin of the Seismological Society of America, v. 90, no. 3, p. 724–738, doi: 10.1785/0119990017.

Kafka, A.L., and Walcott, J.R., 1998, How well does the spatial distribution of smaller earthquakes forecast the locations of larger earthquakes in the northeastern United States?: Seismological Research Letters, v. 69, no. 5, p. 428–439.

Rhoades, D.A., and Evison, F.V., 1989, On the reliability of precursors: Physics of the Earth and Planetary Interiors, v. 58, p. 137–140, doi: 10.1016/0031-9201(89)90049-6.

Weiss, N., and Hassett, M., 1982, Introductory Statistics: Reading, Massachusetts, Addison-Wesley Publishing Co., 651 p.

Wheeler, R.L., and Frankel, A., 2000, Geology in the 1996 USGS Seismic Hazard Maps, central and eastern United States: Seismological Research Letters, v. 71, no. 2, p. 273–282.

Manuscript Accepted by the Society 29 November 2006

The Geological Society of America
Special Paper 425
2007

Limitations of the short earthquake record for seismicity and seismic hazard studies

Laura Swafford
Seth Stein[†]
Department of Earth and Planetary Sciences, Northwestern University, Evanston, Illinois 60208, USA

ABSTRACT

Attempts to study earthquake recurrence in space and time are limited by the short history of instrumental seismology compared to the long and variable recurrence time of large earthquakes. As a result, apparent concentrations and gaps in seismicity and hence seismic hazard within a seismic zone, especially where deformation rates are slow (<10 mm/yr), are likely to simply reflect the short earthquake record. Simple numerical simulations indicate that if seismicity were uniform within a tectonically similar seismic zone, such as the Atlantic coast of Canada, St. Lawrence valley, or the coast of North Africa, thousands of years of record would be needed before apparent concentrations and gaps of seismicity and hazard did not arise. Hence, treating sites of recent seismicity as more hazardous for future large earthquakes is likely to be inappropriate, and it would be preferable to regard the hazard as comparable throughout the seismic zone.

Keywords: Canada, North Africa, seismic hazard, earthquake simulation, earthquake record.

INTRODUCTION

A major theme of this volume is the difficulty in assessing seismic hazards within continental plates and formulating effective mitigation strategies. Two coupled difficulties make these far more difficult than at typical plate boundaries. Unlike the situation at plate boundaries, we have no theoretical basis for predicting deformation rates and hence seismic moment release rates within plate interiors. Hence, we can only rely on the known seismic history. However, because intraplate deformation is typically slower (<1 mm/yr) than at most plate boundaries (Dixon et al., 1996; Newman et al., 1999; Calais et al., 2005; Bada et al., this volume, chapter 16; Camelbeeck et al., this volume, chapter 14; Leonard et al., this volume, chapter 17), the recurrence times for large earthquakes in individual parts of the seismic zone are

longer. As a result, the historic and instrumental seismic record may often give an inaccurate view of the long-term seismicity.

This situation gives rise to a common feature of many seismic hazard maps that predict the maximum shaking expected for given probabilities within specified intervals, namely "bull's-eyes" of high predicted hazard within a tectonically similar seismic zone. For example, bull's-eyes (Fig. 1) appear in the predicted hazard due to intraplate earthquakes along the passive margin of Atlantic Canada and the St. Lawrence valley fault system. However, because the earthquakes are thought to result from reactivation of fossil structures primarily by stresses including those due to postglacial rebound (Stein et al., 1979, 1989; Mazzotti et al., 2005), there is no reason to expect parts of these structures to be more hazardous than others. This situation can also arise at slowly deforming plate boundaries. Hence, although the rate of

[†]Corresponding author e-mail: seth@earth.northwestern.edu.

Swafford, L., and Stein, S., 2007, Limitations of the short earthquake record for seismicity and seismic hazard studies, *in* Stein, S., and Mazzotti, S., ed., Continental Intraplate Earthquakes: Science, Hazard, and Policy Issues: Geological Society of America Special Paper 425, p. 49–58, doi: 10.1130/2007.2425(04). For permission to copy, contact editing@geosociety.org. ©2007 The Geological Society of America. All rights reserved.

Swafford and Stein

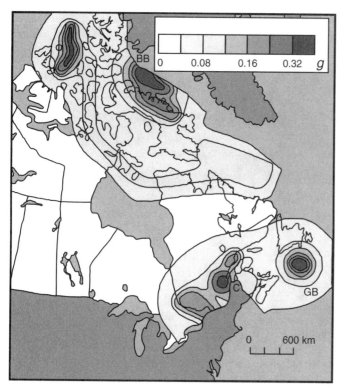

Figure 1. Seismic hazard map of Canada derived in 1985 showing peak ground acceleration expected at 2% probability in 50 yr (Geological Survey of Canada, 1985). This map, based primarily on the recorded and historic seismicity, shows bull's-eyes of high predicted hazard in Baffin Bay (BB), on the Grand Banks (GB), and in the Charlevoix (C) area. More recent maps, also including geological data, reduce the effect of the bull's-eyes and predict more diffuse hazard.

convergence between the Eurasian and Nubian (west African) plates varies smoothly along the coast of North Africa (Argus et al., 1989), the predicted hazard from the resulting earthquakes shows distinct bull's-eyes (Fig. 2).

The bull's-eyes arise because the predicted hazard in these maps depends on the earthquake history. The Canadian coastal maps reflect the 1929 M7.3 Grand Banks and 1933 M7.4 Baffin Bay earthquakes (Stein et al., 1979, 1989; Hasegawa and Kanamori, 1987; Bent, 1995, 2002), the St. Lawrence map reflects five M6 earthquakes in the Charlevoix area since 1663 (Bent, 1992; Schulte and Mooney, 2005), and the bull's-eye on the North Africa map results from the October 1980 M_s 7.3 El Asnam earthquake (Nabelek, 1985).

The utility of this approach is unclear, because in these seismic zones, the deformation rate is slow, so the recurrence time for large earthquakes anywhere in the seismic zone is comparable to the length of the earthquake history. As a result, the spatial and temporal earthquake history may be adequate to show that large earthquakes occur, but is not a good representation of the long-term earthquake distribution. So the question is whether to view the hazard as highest where recent seismicity is concentrated or regard the long-term hazard as essentially uniform within regions

of similar structure. A case can be made for either model. Viewing the presently active areas as most hazardous has the advantage of simplicity, in that such areas are easily identified. These areas may reflect stress concentrations within a seismic zone and thus may be more active than other parts of the seismic zone, even if the other parts have similar structure. Moreover, even if the largest earthquakes migrate within a seismic zone, recent seismicity may still be the best indicator of activity on a hundred year time scale (Kafka, this volume, chapter 3). This is especially likely to be the case if the seismicity is dominated by the smaller aftershocks of large earthquakes (Ebel et al., 2000), as may be the case if intraplate areas have relatively long aftershock sequences owing to the slow stress loading rate (Stein and Newman, 2004).

However, this approach based on historical seismicity, traditionally used in the United States (Frankel et al., 1996) and elsewhere, may lead to an undue focus on the sites of recent earthquakes, when other areas are as or more likely to be the sites of large future earthquakes. This possibility is suggested by the fact that the largest earthquakes to date in North Africa since the El Asnam event have not occurred in regions with the highest predicted hazard (Fig. 2). Hence, a hazard map may reflect more the date when it was made than the actual hazard.

To reduce this difficulty, there is increasing interest in also using paleoseismic and geological data to predict future earthquake locations. For example, Australian paleoseismologists are considering the possibility that seismicity migrates with time, such that faults that seem aseismic from the earthquake record may be the next to generate a damaging earthquake (Clark, 2003; Crone et al., 2003; Leonard et al., this volume, chapter 17). Australian, Hungarian (Toth et al., 2004), and Canadian (Adams et al., 1995; Halchuk and Adams, 1999; Atkinson, this volume chapter 21) researchers are developing maps in which the hazard estimates also reflect geologic structure, and this results in maps with more diffuse hazards (Fig. 3). Although resolving this issue is a formidable task and will take extensive research over time, understanding it will make it possible to adopt the most cost-effective building codes to minimize earthquake damage.

Toward this end, we focus here on a key part of the issue. We use simple numerical simulations to assess how long an earthquake history is needed from a seismic zone in which earthquakes are uniformly distributed to avoid apparent concentrations of seismicity, and, conversely, apparent seismic gaps resulting from a short earthquake history. These simulations are thus a way of exploring the question of whether the portions of a seismic zone in which large earthquakes are known to have occurred are necessarily different from the remainder of the zone, or could simply reflect the short earthquake history.

SIMULATIONS

We chose three seismic zones where hazard maps show distinct bull's-eyes to explore the possibility that these bull's-eyes might be artifacts of the short earthquake history. One is the slowly convergent plate boundary along the coast of North Africa, and the

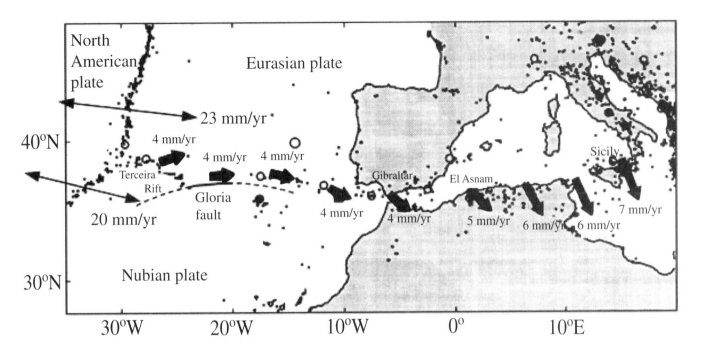

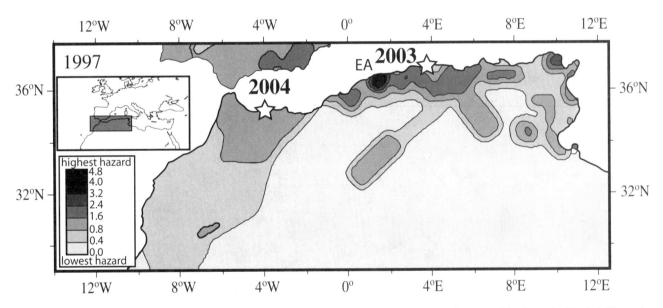

Figure 2. Top: Predicted motion of Eurasia with respect to Africa, showing convergence rate varying smoothly along the North African plate boundary (Argus et al., 1989). Bottom: Portion of Global Seismic Hazard Map (1999) for North Africa, showing peak ground acceleration in m/s² expected at 10% probability in 50 yr. Note prominent bull's-eye at site of the 1980 M$_s$ 7.3 El Asnam (EA) earthquake. The largest subsequent earthquakes to date, the May 2003 M$_s$ 6.8 Algeria and February 2004 M$_s$ 6.4 Morocco events (stars), did not occur in the predicted high hazard regions.

other two are intraplate seismic zones in eastern North America along the passive margin of Atlantic Canada and the St. Lawrence valley fault system. For each, we used a frequency-magnitude (b-value) relation derived from the recorded seismicity. We then computed synthetic space-time histories of earthquakes with M ≥ 7 or 6 (for the St. Lawrence) assuming a Gaussian distribu-tion of recurrence times with standard deviation equal to 0.35 times the mean recurrence time. We assigned each earthquake a loca-tion along the seismic zone (treated as one dimensional) using a uniform random number generator. The results are shown as syn-thetic seismicity maps, with events plotted as 30- or 6- (for the St. Lawrence) km-diameter circles that approximate the fault area.

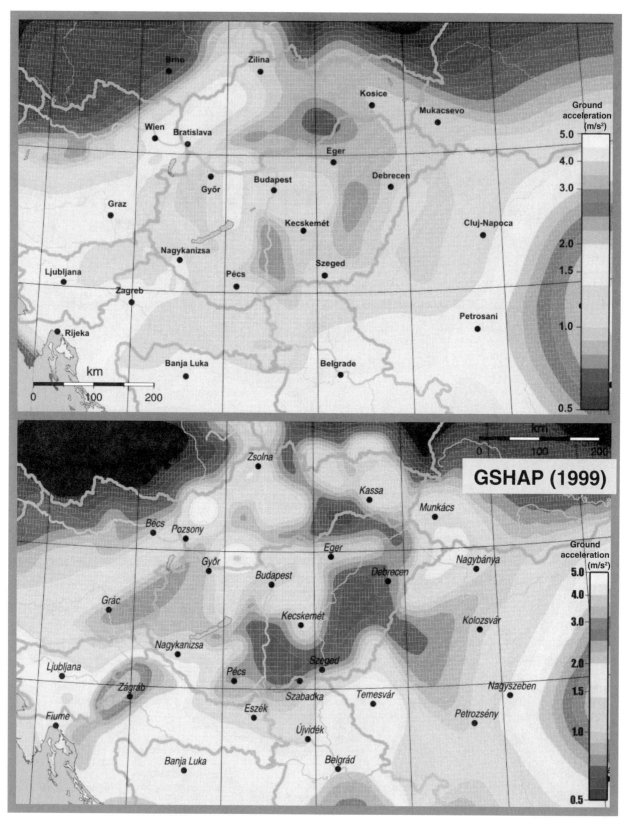

Figure 3. Alternative hazard maps for Hungary and the surrounding area, showing peak ground acceleration in m/s² expected at 10% probability in 50 yr. The Global Seismic Hazard Assessment Program (GSHAP) model, based only on historic and recorded seismicity (bottom), predicts more concentrated hazard near sites of previous earthquakes (e.g., Zagreb), compared to a model that includes geological data (top), which predicts more diffuse hazard (Toth et al., 2004).

North Africa

We simulated the seismicity along the North Africa coast (Fig. 4) from Tunisia to Morocco, which results from the convergence between Eurasia and Nubia at <10 mm/yr (Argus et al., 1989; Sella et al., 2002; Fernandes et al., 2003). We used two frequency-magnitude relations derived from the Advanced National Seismic Systems (ANSS) catalog. The first, derived from 1963 to 2004 seismicity for which magnitudes are presumably well estimated, yields $a = 4.63$ and $b = 0.91$. Hence, an earthquake

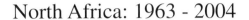

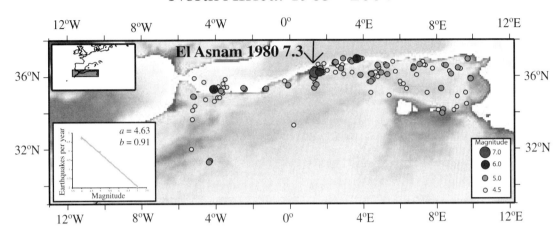

Years	Number of events	Average years between events
100	2	50
500	11	45
1000	20	45
2000	44	45
3000	60	50
4000	78	51
5000	99	50
6000	118	51
7000	136	51
8000	155	52

Figure 4. Seismicity along the North Africa plate boundary for 1963–2004. Simulations using a frequency-magnitude relation derived from these data predict that if seismicity is uniform in the zone, ~8000 yr of record is needed to avoid apparent concentrations and gaps.

with M ≥ 7 occurs somewhere in the seismic zone about every 42 yr on average. Using this relation, apparent concentrations of large earthquakes and gaps without them appear for earthquake records up to thousands of years long. Only after 7000–8000 yr is the uniform nature of the seismicity fully apparent. This effect is even more striking using a frequency-magnitude relation from the 1910–2004 earthquake record, which yields $a = 4.37$ and $b = 0.90$, and thus a 95 yr mean recurrence for M ≥ 7 events. The resulting simulations require 10,000–12,000 yr to show the uniform seismicity (Fig. 5).

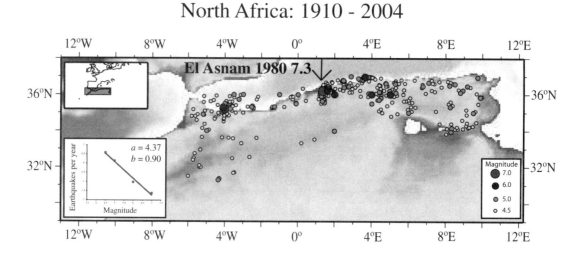

Years	Number of events	Average years between events
100	1	100
500	6	83
1000	12	83
2000	22	90
4000	40	100
6000	59	103
8000	79	101
10,000	101	99
12,000	121	99

Figure 5. Seismicity along the North Africa plate boundary for 1910–2004. Simulations using a frequency-magnitude relation derived from these data predict that if seismicity is uniform in the zone, ~12,000 yr of record is needed to avoid apparent concentrations and gaps.

Atlantic Canada

The eastern coast of Canada shows the most striking example of large earthquakes along a passive continental margin, where ideal plate tectonics predict no relative motion between continental and oceanic portions of the same plate. The 1929 M7.3 Grand Banks and 1933 M7.4 Baffin Bay earthquakes are the best studied examples of this phenomenon, which also occurs on other passive margins (Stein et al., 1989; Schulte and Mooney, 2005). A frequency-magnitude relation derived for events with M ≥ 4.5 from the Geological Survey of Canada catalog from 1925 to 2003 yields $a = 3.38$ and $b = 0.73$, and thus a recurrence time of 47 yr for M ≥ 7. Simulations using this relation yield apparent concentrations of large earthquakes and gaps without them for earthquake records up to thousands of years long (Fig. 6). Approximately 8000–11,000 yr of record is needed to show that the seismicity is uniform.

St. Lawrence Valley

The St. Lawrence valley seismic zone (Fig. 7) (Mazzotti et al., 2005) is generally considered to consist of two seismic zones: the lower St. Lawrence seismic zone, defined by events with M ≤ 5, and the Charlevoix seismic zone, in which larger (M 6) events occur. Our simulations assume that large earthquakes are equally likely throughout the entire zone. A frequency-magnitude relation derived for events with M ≥ 4 from the Geological Survey of Canada catalog from 1925 to 2004 yields $a = 4.52$ and $b = 1.01$ and thus a recurrence time of ~40 yr for M ≥ 7. In the resulting simulations, 12,000–16,000 yr are needed to show the uniform seismicity.

DISCUSSION

The simulations illustrate that if seismicity were uniformly distributed within a seismic zone, apparent concentrations of seismicity and seismic gaps would appear in earthquake records shorter than the time needed for large earthquakes to occur throughout the zone. For the zones used in the simulations, which are deforming more slowly than ~10 mm/yr, the time needed for the uniformity to become clear is thousands of years. Hence, it is plausible that many of the concentrations and gaps seen in real data are artifacts of the short sampling intervals.

Naturally, the simulations simplify real—and poorly understood—aspects of earthquake recurrence. They do not include all

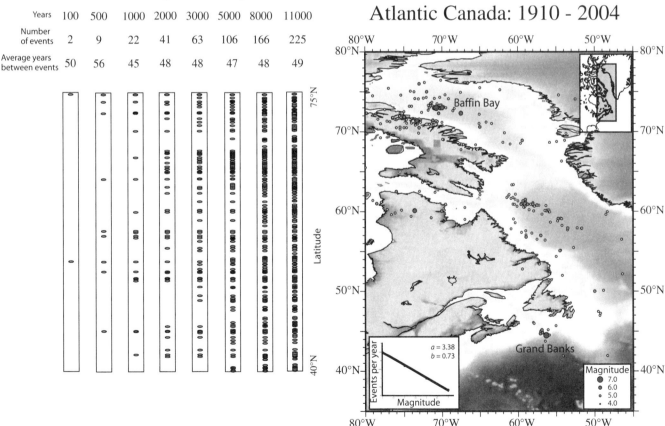

Figure 6. Intraplate seismicity along the eastern coast of Canada. Simulations using a frequency-magnitude relation derived from these data predict that if seismicity is uniform in the zone, ~11,000 yr of record is needed to avoid apparent concentrations and gaps.

St. Lawrence Valley: 1925 - 2004

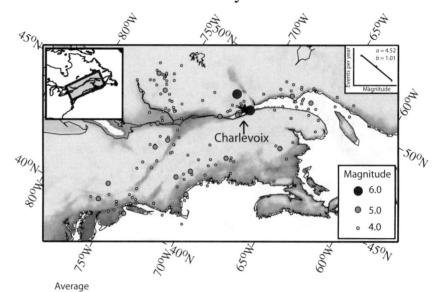

Simulated earthquake history M > 6

Years	Number of events	Average years between events
100	2	50
500	14	35
1000	28	35
2000	61	33
3000	90	33
4000	117	34
6000	173	35
9000	263	34
12,000	349	34
16,000	472	34

75°W   Latitude   60°W

Figure 7. Intraplate seismicity in the St. Lawrence valley. Simulations using a frequency-magnitude relation derived from these data predict that if seismicity is uniform in the zone, ~16,000 yr of record is needed to avoid apparent concentrations and gaps.

possible time- and space-dependant probability effects, including stress transfer, in which stress changes due to earthquakes affect the location of future events (Stein, 1999). Similarly, all earthquakes were assumed to have the same magnitude. Although more complicated simulations are possible, it is tricky to decide which effects to include. At a plate boundary like North Africa, the known convergence rate and direction permit more sophisticated models (Lin and Stein, 2004). Even so, it appears that the size and geometry of large thrust earthquakes are highly variable along a section of a trench (Ando, 1975; Stein et al., 1986; Cisternas et al., 2005), and it is unclear how to relate moment release rate (seismic slip) to plate motion. How to model intraplate areas, where even the fundamental tectonics are unclear, is even more challenging, although models have been made for specific areas such as the New Madrid seismic zone (Mueller et al., 2004; Li et al., this volume, chapter 11). Thus, we used simulations representing the simplest possible behavior, and we believe that the basic results are likely to be robust.

The fact that simulations assuming uniform deformation rates within seismic zones yield bull's-eyes and gaps does not prove that such patterns observed in earthquake records necessarily reflect short sampling. It is also possible that within a tectonically similar seismic zone, local structures or long-lived stress and strain concentrations (Gangopadhyay and Talwani, 2003, and this volume) make some parts more active than others. Of the cases we examined, the Charlevoix concentration looks the most like the latter case. The simulations show that from seismicity data alone—in particular without geodetic data—the two possibilities are hard to distinguish.

Even so, we feel that the results have interesting implications. They bear out the effect that many features of space-time patterns of seismicity may in part reflect short sampling times. For example, modern seismicity maps show essentially no seismicity on the southern segment of the San Andreas where the large 1857 earthquake occurred. Similarly, prior to the great 2004 earthquake, the portion of the Sumatra trench on which it occurred did not appear very active and was not considered particularly dangerous. In both cases, the absence of large earthquakes from the seismological record reflects their long recurrence time. As a result, inferences from the short earthquake history would be misleading. For example, the Sumatra example and data from other trenches suggest that a previously inferred effect—that large trench earthquakes occur only when young lithosphere subducts rapidly—may largely reflect the short earthquake history sampled (Stein and Okal, 2007). Similarly, it seems likely that some cases in which large earthquakes are more frequent (characteristic) or less frequent (uncharacteristic) than expected from the rate of smaller ones may also be a time sampling effect (Stein and Newman, 2004).

The simulations also argue against the practice of treating sites of recent seismicity as more hazardous for future large earthquakes, and in favor of the more recent trend of treating hazard as uniform within a tectonic zone. As noted, the simulations show that apparent concentrations of seismicity are quite plausi-bly sampling artifacts where the earthquake history is short compared to the recurrence time of large earthquakes. In our view, there is no reason to believe these sites are more likely to have future large earthquakes than other sites on the same structures, assuming that they deform at similar rates. Hazard estimates using the seismicity are thus likely to overestimate hazards where earthquakes happen to have occurred recently, and underestimate it elsewhere within the seismic zone, where earthquakes happen to have not occurred. In fact, stress transfer arguments imply that large earthquakes at the other sites may be more likely (e.g., Li et al., this volume, chapter 11). Hence, we favor models that treat hazard as uniform within a tectonic zone. For example, we would regard the seismic hazard as similar all along the eastern Canadian passive margin (Stein et al., 1989; Adams and Basham, 1989), rather than highest near the sites of the Baffin Bay and Grand Banks earthquakes.

ACKNOWLEDGMENTS

We thank Emile Okal and Stéphane Mazzotti for helpful comments.

REFERENCES CITED

Adams, J., and Basham, P., 1989, Seismicity and seismotectonics of Canada's eastern margin and craton, *in* Gregerson, S., and Basham, P., eds., Earthquakes at North Atlantic Passive Margins: Neotectonics and Postglacial Rebound: Dordrecht, Kluwer, p. 355–370.

Adams, J., Basham, P.W., and Halchuk, S., 1995, Northeastern North America earthquake potential—New challenges for seismic hazard mapping, *in* Eastern Canada and National and General Programs (Est du Canada et Programmes Nationaux et Generaux): Geological Survey of Canada, Current Research, v. 1995-D, p. 91–99.

Ando, M., 1975, Source mechanisms and tectonic significance of historical earthquakes along the Nankai Trough, Japan: Tectonophysics, v. 27, p. 119–140, doi: 10.1016/0040-1951(75)90102-X.

Argus, D.F., Gordon, R.G., DeMets, C., and Stein, S., 1989, Closure of the Africa–Eurasia–North America plate motion circuit and tectonics of the Gloria fault: Journal of Geophysical Research, v. 94, p. 5585–5602.

Atkinson, G., 2007, this volume, Challenges in seismic hazard analysis for continental interiors, *in* Stein, S., and Mazzotti, S., eds., Continental Intraplate Earthquakes: Science, Hazard, and Policy Issues: Geological Society of America Special Paper 425, doi: 10.1130/2007.2425(21).

Bada, G., Grenerczy, G., Toth, L., Horvath, F., Stein, S., Windhoffer, G., Fodor, L., Fejes, I., Pinter, N., and Cloetingh, S., 2007, this volume, Motion of Adria and ongoing inversion of the Pannonian Basin: Seismicity, GPS velocities and stress transfer, *in* Stein, S., and Mazzotti, S., eds., Continental Intraplate Earthquakes: Science, Hazard, and Policy Issues: Geological Society of America Special Paper 425, doi: 10.1130/2007.2425(16).

Bent, A., 1992, A re-examination of the 1925 Charlevoix, Quebec, earthquake: Bulletin of the Seismological Society of America, v. 82, p. 2097–2113.

Bent, A., 1995, A complex double-couple source mechanism for the M_s7.2 1929 Grand Banks earthquake: Bulletin of the Seismological Society of America, v. 85, p. 1003–1020.

Bent, A., 2002, The 1933 Ms = 7.3 Baffin Bay earthquake: strike-slip faulting along the northeastern Canadian passive margin: Geophysical Journal International, v. 150, p. 724–736, doi: 10.1046/j.1365-246X.2002.01722.x.

Calais, E., Mattioli, G., DeMets, C., Nocquet, J.-M., Stein, S., Newman, A., and Rydelek, P., 2005, Tectonic strain in plate interiors?: Nature, v. 438, doi: 10.1038/nature04428.

Camelbeeck, T., Vanneste, K., Alexandre, P., Verbeeck, K., Petermans, T., Rosset, P., Everaerts, M., Warnant, R., and Van Camp, M., 2007, this volume, Relevance of active faulting and seismicity studies to assess long-term earthquake activity in northwest Europe, *in* Stein, S., and Mazzotti, S., eds., Continental Intraplate Earthquakes: Science, Hazard, and Policy Issues: Geological Society of America Special Paper 425, doi: 10.1130/2007.2425(14).

Cisternas, M., Atwater, B.F., Torrejon, F., Sawai, Y., Machuca, G., Lagos, M., Eipert, A., Youlton, C., Salgado, I., Kamataki, T., Shishikura, M., Rajendran, C.P., Malik, J.K., Riza, Y., and Husni, M., 2005, Predecessors of the giant 1960 Chile earthquake: Nature, v. 437, p. 404–407, doi: 10.1038/nature03943.

Clark, D., 2003, Earthquakes move on but not without a trace: Ausgeo News, v. 70, p. 30–32.

Crone, A.J., De Martini, P.M., Machette, M.N., Okumura, K., and Prescott, J.R., 2003, Paleoseismicity of two historically quiescent faults in Australia: Implications for fault behavior in stable continental regions: Bulletin of the Seismological Society of America, v. 93, p. 1913–1934.

Dixon, T.H., Mao, A., and Stein, S., 1996, How rigid is the stable interior of the North American plate?: Geophysical Research Letters, v. 23, p. 3035–3038, doi: 10.1029/96GL02820.

Ebel, J.E., Bonjer, K.P., and Oncescu, M.C., 2000, Paleoseismicity: Seismicity evidence for past large earthquakes: Seismological Research Letters, v. 71, p. 283–294.

Fernandes, R., Ambrosius, B., Noomen, R., Bastos, L., Wortel, M., Spakman, W., and Govers, R., 2003, The relative motion between Africa and Eurasia as derived from ITRF2000 and GPS: Geophysical Research Letters, v. 30, no. 16, doi: 10.1029/2003GL017089.

Frankel, A., Mueller, C., Barnhard, T., Perkins, D., Leyendecker, E., Dickman, N., Hanson, S., and Hopper, M., 1996, National Seismic Hazard Maps Documentation: U.S. Geological Survey Open-File Report 96-532, 110 p.

Gangopadhyay, A., and Talwani, P., 2003, Symptomatic features of intraplate earthquakes: Seismological Research Letters, v. 74, p. 863–883.

Global Seismic Hazard Map, 1999, Global Seismic Hazard Map: http://www.seismo.ethz.ch/gshap (November 2004).

Halchuk, S., and Adams, J., 1999, Crossing the border: Assessing the difference between new Canadian and American seismic hazard maps: Proceedings of the 8th Canadian Conference of Earthquake Engineering, p. 77–82.

Hasegawa, H.S., and Kanamori, H., 1987, Source mechanism of the magnitude 7.2 Grand Banks earthquake of November 1929; double couple or submarine landslide?: Bulletin of the Seismological Society of America, v. 77, p. 1984–2004.

Kafka, A., 2007, this volume, Does seismicity delineate zones where future large earthquakes are likely to occur in intraplate environments?, *in* Stein, S., and Mazzotti, S., eds., Continental Intraplate Earthquakes: Science, Hazard, and Policy Issues: Geological Society of America Special Paper 425, doi: 10.1130/2007.2425(03).

Leonard, M., Robinson, D., Allen, T., Schneider, J., Clark, D., Dhu, T., and Burbidge, D., 2007, this volume, Towards a better model for earthquake hazard in Australia, *in* Stein, S., and Mazzotti, S., eds., Continental Intraplate Earthquakes: Science, Hazard, and Policy Issues: Geological Society of America Special Paper 425, doi: 10.1130/2007.2425(17).

Li, Q., Liu, M., Zhang, Q., and Sandoval, E., 2007, this volume, Stress evolution and seismicity in the central-eastern USA: Insights from geodynamic modeling, *in* Stein, S., and Mazzotti, S., eds., Continental Intraplate

Earthquakes: Science, Hazard, and Policy Issues: Geological Society of America Special Paper 425, doi: 10.1130/2007.2425(11).

Lin, J., and Stein, R., 2004, Stress triggering in thrust and subduction earthquakes and stress interaction between the southern San Andreas and nearby thrust and strike-slip faults: Journal of Geophysical Research, v. 109, doi: 10.1029/2003JB002607.

Mazzotti, S., James, T., Henton, J., and Adams, J., 2005, GPS crustal strain, postglacial rebound, and seismic hazard in eastern North America: The Saint Lawrence valley example: Journal of Geophysical Research, v. 110, doi: 10.1029/2004JB003590.

Mueller, K., Hough, S.E., and Bilham, R., 2004, Analyzing the 1811–1812 New Madrid earthquakes with recent instrumentally recorded aftershocks: Nature, v. 429, p. 284–288, doi: 10.1038/nature02557.

Nabelek, J., 1985, Geometry and mechanism of faulting of the 1980 El Asnam, Algeria, earthquake from inversion of teleseismic body waves and comparison with field observations: Journal of Geophysical Research, v. 90, p. 12,713–12,728.

Newman, A., Stein, S., Weber, J., Engeln, J., Mao, A., and Dixon, T., 1999, Slow deformation and lower seismic hazard at the New Madrid seismic zone: Science, v. 284, p. 619–621, doi: 10.1126/science.284.5414.619.

Schulte, S.M., and Mooney, W.D., 2005, An updated global earthquake catalogue for stable continental regions: Reassessing the correlation with ancient rifts: Geophysical Journal International, v. 161, p. 707–721, doi: 10.1111/j.1365-246X.2005.02554.x.

Sella, G.F., Dixon, T.H., and Mao, A., 2002, REVEL: A model for recent plate velocities from space geodesy: Journal of Geophysical Research, v. 107, no. B04, doi: 10.1029/2000JB000033.

Stein, R., 1999, The role of stress transfer in earthquake occurrence: Nature, v. 402, p. 605–609, doi: 10.1038/45144.

Stein, S., and Newman, A., 2004, Characteristic and uncharacteristic earthquakes as possible artifacts: Applications to the New Madrid and Wabash seismic zones: Seismological Research Letters, v. 75, p. 170–184.

Stein, S., and Okal, E.A., 2007, Ultralong period seismic study of the December 2004 Indian Ocean earthquake and implications for regional tectonics and the subduction process: Bulletin of the Seismological Society of America, v. 97, no. 1A, p. S279–S295.

Stein, S., Sleep, N.H., Geller, R.J., Wang, S.C., and Kroeger, G.C., 1979, Earthquakes along the passive margin of eastern Canada: Geophysical Research Letters, v. 6, p. 537–540.

Stein, S., Engeln, J.F., DeMets, C., Gordon, R.G., Woods, D., Lundgren, P., Argus, D., Stein, C., and Wiens, D.A., 1986, The Nazca–South America convergence rate and the recurrence of the great 1960 Chilean earthquake: Geophysical Research Letters, v. 13, p. 713–716.

Stein, S., Cloetingh, S., Sleep, N., and Wortel, R., 1989, Passive margin earthquakes, stresses, and rheology, *in* Gregerson, S., and Basham, P., eds., Earthquakes at North Atlantic Passive Margins: Neotectonics and Postglacial Rebound: Dordrecht, Kluwer, p. 231–260.

Toth, L., Gyori, E., Monus, P., and Zsiros, T., 2004, Seismicity and seismic hazard in the Pannonian Basin, *in* Proceedings of the NATO Adria Research Workshop: The Adria Microplate: Dordrecht, Springer, p. 231–260.

MANUSCRIPT ACCEPTED BY THE SOCIETY 29 NOVEMBER 2006

The Geological Society of America
Special Paper 425
2007

Frequency-size distributions for intraplate earthquakes

Emile A. Okal
Justin R. Sweet
Department of Geological Sciences, Northwestern University, Evanston, Illinois 60208, USA

ABSTRACT

We examine the question of a possible difference in the frequency-size statistics of intraplate earthquakes, as opposed to their more numerous interplate counterparts. We use both the Harvard Centroid Moment Tensor catalogue and the data set of the National Earthquake Information Center. In the former case, we quantify earthquakes through their seismic moment and describe their population distribution through the β-value introduced by Molnar. In the latter case, we use traditional *b*-values computed from both body-wave magnitudes (m_b) and surface-wave magnitudes (M_s). We conclude that both β- and *b*-values for true intraplate earthquakes (i.e., not occurring in areas of broad tectonic deformation) are essentially equivalent to those of interplate earthquakes in similar ranges of moments or magnitudes. This is consistent with a fractal dimension of two for the intraplate seismogenic zones, suggesting that, like along plate boundaries, they consist of two-dimensional faults and not of volumes with greater dimensions. The distribution of earthquakes in deformed regions, principally the Mediterranean-Tethyan belt, follows that of worldwide interplate earthquakes but with a greater value for the critical moment expressing the saturation with depth of the width of the fault at the brittle-ductile transition, suggesting that the latter would take place at greater depths under large-scale orogens.

Keywords: *b*-values, intraplate earthquakes, seismic scaling laws.

INTRODUCTION AND BACKGROUND

The purpose of this paper is to explore the possibility that intraplate earthquakes could feature frequency-size population statistics that differ significantly from those of their counterparts at plate boundaries. We conclude that "true intraplate" earthquakes, i.e., those associated neither with deformed, diffuse plate boundaries, nor with intraplate magmatic centers (hotspots), do not exhibit recognizably different properties in this respect.

It has long been observed in all seismic provinces that there are, simply speaking, more small earthquakes than large ones. Following the introduction by Richter (1935) of the concept of magnitude, Gutenberg and Richter (1954) modeled this

behavior for homogeneous populations of earthquakes (e.g., belonging to a particular seismic area) using a frequency-size relation of the form

$$\log_{10} N = a - bM, \qquad (1)$$

where N is the number of earthquakes in the group with a magnitude equal to or greater than M. The absolute value of the slope of the regression in Equation 1, universally known as the *b*-value of the population, has been found to be remarkably constant and close to unity for a large number of data sets of tectonic earthquakes. It is only when earthquake sources of a different nature are considered that the *b*-value departs significantly from 1, rising, for example, to $b \geq 2$ during volcano-

Okal, E.A., and Sweet, J.R., 2007, Frequency-size distributions for intraplate earthquakes, *in* Stein, S., and Mazzotti, S., ed., Continental Intraplate Earthquakes: Science, Hazard, and Policy Issues: Geological Society of America Special Paper 425, p. 59–71, doi: 10.1130/2007.2425(05). For permission to copy, contact editing@geosociety.org. ©2007 The Geological Society of America. All rights reserved.

seismic swarms (e.g., Mogi, 1963; Minakami, 1974). Such variations have indeed been used to identify volcanic swarms.

Following the generalization of the use of the seismic moment M_0 as a physical measure of the size of an earthquake, Molnar (1979) proposed the symbol β to similarly describe earthquake statistics based on seismic moments, according to

$$\log_{10} N = \alpha - \beta \log_{10} M_0, \qquad (2)$$

where N is now the number of earthquakes with a moment that equals or exceeds M_0. As discussed in detail by Okal and Romanowicz (1994), the relationship between β and b depends on the variation of any particular magnitude scale with seismic moment, which is itself controlled in part by the saturation of magnitude scales as the seismic source grows, which affects first m_b (around 6.3), then M_s (around 8.2) (Geller, 1976).

A considerable amount of literature has been published on the subject of frequency-size distributions, and the reader is referred to, for example, Båth (1981) and Frohlich and Davis (1993) for reviews. Perhaps the most seminal among such papers was Rundle's (1989), in which the author justified a β-value of 2/3 based on the following argument: In a given population of earthquakes, the process of faulting is assumed to be scale-independent (in Rundle's own words "the fault area available to produce events of all sizes is the same" p. 12,338 of Rundle [1989]). This is equivalent to the number of earthquakes of a given size, dN in the notation of Equation 2, being inversely proportional to the area of faulting, S. Earthquake scaling laws (e.g., Geller, 1976) predict that S grows like $M_0^{2/3}$ (at least in a range of source sizes not affected by the intrinsic limits of the dimensions of seismogenic zones), resulting in the theoretical values $\beta = 2/3$ and $b = 1$, as a factor 1.5 has often been introduced between various magnitude scales and $\log_{10} M_0$. Note that the values predicted in Rundle's (1989) derivation of Equations 1 or 2 based on simple physical laws are exact numbers ($\beta = 2/3$ or $b = 1$).

In this respect, it is interesting to note that in the first edition of their classic work, Gutenberg and Richter (1941) described the "rule of tenfold increase," in other words a b-value of exactly 1 as defined by Equation 1. In their more definitive work (Gutenberg and Richter, 1954), they refined this estimate to $b = 0.9$ for shallow earthquakes. Later, Turcotte (1992) related the b-value of a population to the fractal dimension D of its source, assigning $D = 1.8$ on the basis of $b = 0.9$ as reported by Gutenberg and Richter (1954). In this context, Okal and Romanowicz (1994) argued that the exact value $b = 1$ is physically more realistic, since it expresses the geometrical dimension ($D = 2$) of the seismogenic zone, constrained to a two-dimensional *fault*; any small departure from $b = 1$ then illustrates artifacts of the physical saturation of the seismogenic zone, as well as variations with earthquake size in the expected relation between magnitude and moment. Due to the generally small number of large events, there remains some controversy about the actual behavior of β at very large moments—some studies suggest a decrease in β, while others advocate an increase (Romanowicz and Rundle, 1993).

In the case of deep earthquakes, Okal and Kirby (1995) argued that the seismogenic zone may extend over a *volume* of fractal dimension $D = 3$, thus explaining larger β and b-values, as already reported by Gutenberg and Richter (1954). This trend is strongly affected, however, by the finite size of the seismogenic zone inside slabs, which explains the large diversity of observations in various subduction zones (Frohlich and Davis, 1993).

In this general context, we examine here the values of β and b for various groups of intraplate earthquakes, not directly associable with tectonic processes occurring along plate boundaries. We were motivated by the idea that intraplate earthquakes may not necessarily be constrained to a given fault system of fractal dimension $D = 2$. For example, rupture in a seismogenic *volume* of fractal dimension $D = 3$, rather than along a two-dimensional fault, would lead to $\beta = 1$ and similarly to larger values of b (Okal and Kirby, 1995). We also note that Frohlich and Davis (1993) explored the concept of possible regional variations in b-values according to prevailing tectonic regime. Similarly, Bird et al. (2002) presented a formal regionalization of tectonic boundaries and studied the variation of frequency-size relationships among them. While the earlier study by Bergman and Solomon (1980) did address the question of b-values in a few specific oceanic intraplate areas, it lacked a worldwide scope and predated the routine quantification of earthquakes through seismic moment; the other studies mentioned did not consider intraplate earthquakes in their regionalizations.

METHODOLOGY

The two principal catalogues used in the present study were the Harvard Centroid Moment Tensor (CMT) data set (Dziewonski et al., 1983, and subsequent quarterly updates) (1977–March 2003) and the database of the National Earthquake Information Center (NEIC) of the U.S. Geological Survey. The two catalogues were filtered for depth, and only shallow earthquakes ($h < 100$ km) were retained. This threshold may appear exceedingly deep, as most intraplate events are known to occur at lesser depths. However, it has no effect on the final results, as our search algorithms retained insignificant populations at the greater depths (e.g., only 18 earthquakes with $50 < h < 100$ km out of 2737 for the intraplate NEIC population; see following). Such events are generally small and poorly located, making their published depths suspect. The use of a conservative depth threshold (100 km) may indeed guard against inadvertently excluding genuine shallow intraplate events, with once again no tangible effect on our final results.

The intraplate character of an earthquake was assessed by testing the minimum distance from its epicenter to Bird's (2003) discrete set of 12,148 plate-boundary locations. We used a threshold distance of 400 km from the nearest plate boundary to define an earthquake as intraplate. This rather conservative estimate is larger than previously used, for example by Wysession et al. (1991) in the study of Pacific Basin intraplate seismicity. It ensures the elimination of events potentially associated with boundary processes, such as buckling of the plate seaward of a subduction zone. Also, since it is significantly larger than the typical sample spacing in Bird's data set (70 km), it ensures that the identification of an intraplate earthquake is not affected by the discrete nature of the latter.

THE CMT DATA SET

Focusing first on the CMT catalogue, which provides the most homogeneous data set based on the quantitative inversion of a physical model of the seismic source, the above procedure resulted in the retention of only 821 events among the 16,349 shallow CMT solutions (Fig. 1). This data set was further refined by separating those earthquakes that belong to so-called "deformed" provinces, to diffuse plate boundaries, and to identifiable intraplate magmatic centers (hotspots). Deformed provinces are mostly continental, mostly compressional, belts over which intense deformation takes place across widths significantly larger than the 400 km threshold. They include the Tethyan belt, extending from Italy to Burma, including the Tibet-Mongolian system, as well the Rocky Mountain system in North America, and additional pockets of activity in South America and Africa. The 565 events in those provinces are shown as downward-pointing triangles on Figure 1.

An additional 22 events, shown as upward-pointing triangles on Figure 1, were classified as belonging to areas of diffuse plate boundaries, as defined, for example, by Stein and Sella (2002). These earthquakes were located in the eastern Indian Ocean (between the Indian and Australian plates) and near the Macquarie triple junction, the latter of which consisted of the large 1998 earthquake and its aftershocks.

Finally, we classified separately 16 earthquakes evidently associated with activity at hotspots, principally in Hawaii and the Canary Islands; these are shown as squares on Figure 1. This left 218 "truly intraplate" events in the CMT catalogue, shown as circles on Figure 1. Note that this number represents only a very small fraction of the total shallow seismicity of the catalogue (1.3% of the number of earthquakes; 0.3% of the seismic moment released).

β-Values from the CMT Data Set

In order to provide a worldwide reference for the study of intraplate data sets, we first analyzed the frequency-moment relationship for the entire CMT catalogue of 16,349 shallow earthquakes. Individual events were sorted by moment into bins with a width of 0.2 units of $\log_{10} M_0$, with the corresponding populations shown as plus signs on Figure 2A. The larger symbols represent cumulative populations, i.e., for each bin, the number of events (N in Equation 2) with a moment equal to, or greater than, that of the bin. The β-value was obtained by a least-squares regression of N against $\log_{10} M_0$ over specific ranges of moments. The open symbols denote parts of the data set that lacked completeness and thus were ignored from the regressions. The full data set clearly exhibits a change in slope at a critical moment $M_0^c = 10^{27.5}$ dyn-cm, as previously observed by many investiga-

tors (Pacheco et al., 1992; Romanowicz and Rundle, 1993; Okal and Romanowicz, 1994) and interpreted as expressing the saturation of fault width W upon reaching the brittle-ductile transition. Our β-values (0.67 below M_0^c, 1.42 above M_0^c) are in good agreement with Okal and Romanowicz' (1994) values (0.70 and 1.35), which were obtained from a data set roughly half the size of ours. We note however that our critical moment is slightly larger than that observed ($10^{27.2}$) in the previous study. The excellent agreement between the β-value at low moments and its theoretical value (2/3) also confirms the exact fractal dimension $D = 2$ of sources unaffected by saturation.

We then applied the same algorithm to the "deformed" and "true intraplate" populations, of 565 and 218 earthquakes, respectively. (With only 22 and 16 earthquakes, respectively, the diffuse and hotspot populations were too small to provide meaningful results.) As shown on Figure 2B, the "deformed" data set features a slightly lower β = 0.58, with only the hint of a possible elbow at $M_0 = 10^{27.2}$ dyn-cm. In the case of the "intraplate" population (Fig. 2C), β = 0.66 is indistinguishable from its worldwide value and from the theoretical value of 2/3 predicted in Rundle's (1989) model. However, because of the absence of large events in the data set, no critical moment M_0^c can be defined in that case.

b-Values from the CMT Data Set

We next examined the same data set of CMT solutions, but considered its conventional magnitudes. The regression algorithm is similar, but uses bins of 0.2 units of magnitude. We computed b-values from the surface-wave magnitude M_s (measured at 20 s) and from the body-wave magnitude m_b (measured at 1 s).

About three-fourths of the 16,349 events in the global CMT database are assigned M_s values; this data set features a stepwise increase in b-value with magnitude (Fig. 3A) from b = 0.73 in the unsaturated range ($M_s < 6.6$) to around b = 2 for $M_s \geq 8$; these values are in full agreement with Okal and Romanowicz' (1994) results (their Figs. 9b and 9c). The 492 events in "deformed" provinces similarly follow the expected b-value of 2/3 at low magnitudes, for which M_s is directly proportional to $\log_{10} M_0$, but fail to involve a definitive change of slope at larger magnitudes (Fig. 3B). In contrast, the 177 "intraplate" events, which also closely follow b = 2/3 at low magnitudes, appear to involve a different behavior beyond $M_s = 6.5$ (Fig. 3C), but this observation is based on only a handful of events.

When regressed as a function of body-wave magnitude m_b, the full CMT data set features an almost continuous increase in b-value from 1.04 for $m_b < 6.0$ to more than 2.5 around $m_b \approx 7$ (Fig. 3D). These results are, once again, in agreement with Okal and Romanowicz' (1994) (see their Fig. 12). Results for the "deformed" and "intraplate" data sets were essentially equivalent (Figs. 3E and 3F).

In conclusion, the data set of earthquakes inverted as part of the CMT project suggests that intraplate earthquakes follow the same population distributions as their much more numerous counterparts at plate boundaries.

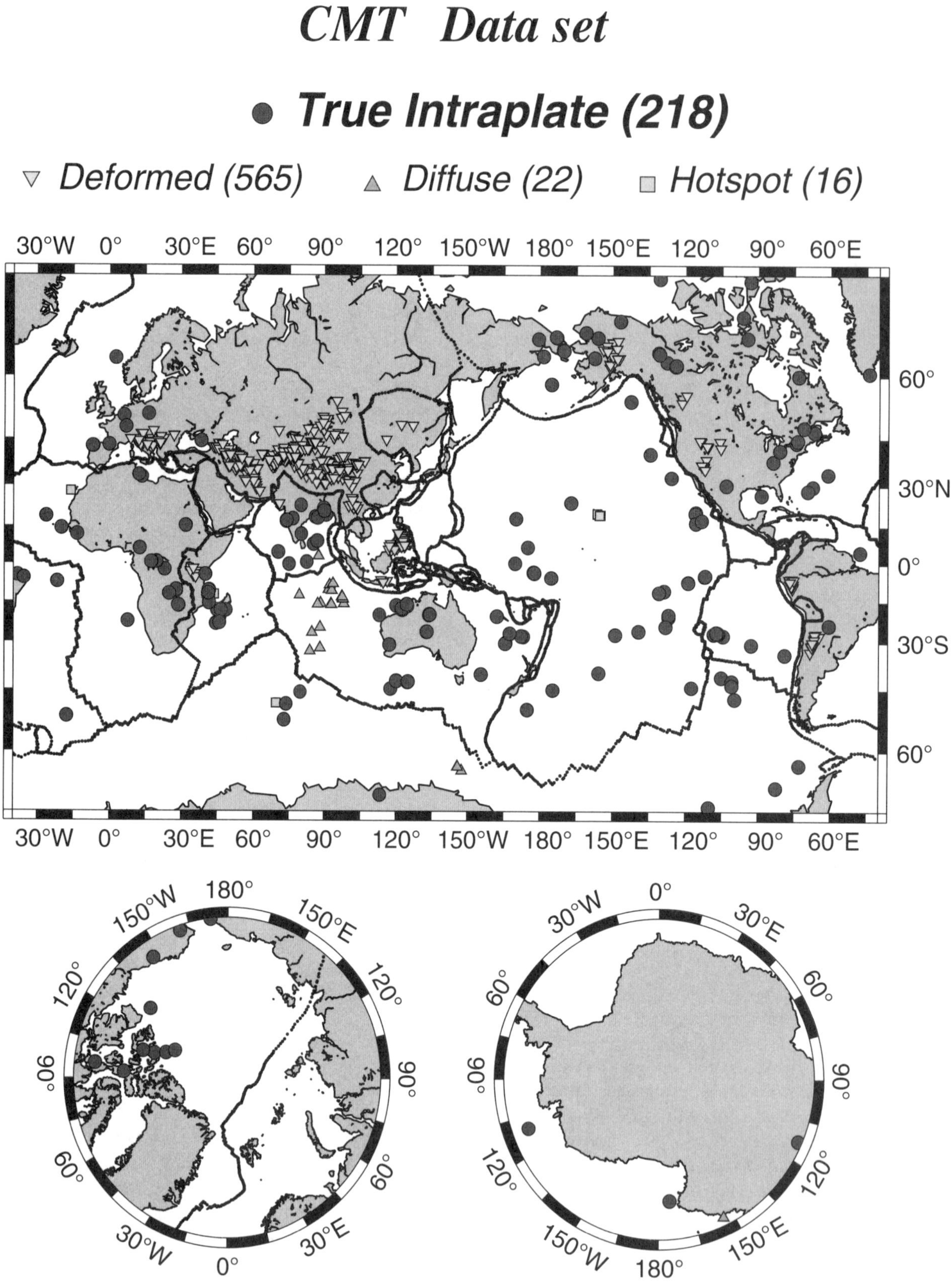

CMT Data set

● True Intraplate (218)

▽ Deformed (565)　　▲ Diffuse (22)　　□ Hotspot (16)

Figure 1. Maps of the Harvard Centroid Moment Tensor (CMT) data set considered in this study. The solid lines (actually composed of 12,148 individual dots) are Bird's (2003) data set of plate-boundary locations. Downward-pointing triangles show events occurring inside deformed areas, upward-pointing triangles events located at diffuse plate boundaries, and squares events associated with intraplate hotspots. The remaining truly intraplate earthquakes are shown as solid circles. Numbers in parentheses show the total populations of the various groups.

THE NEIC DATA SET

While the CMT data set is unique in providing a homogeneous catalogue of earthquake solutions inverted through a common algorithm, it suffers from its youth (no events prior to 1976 and a coarser sampling for that year) and its relatively large size threshold for completeness (in principle, no events below $m_b = 5.0$ are processed into the catalogue, resulting in completeness only for $M_0 \geq 2 \times 10^{24}$ dyn-cm). Accordingly, we consider in this section b-values computed from data sets extracted from the NEIC catalogue, keeping in mind that their interpretation may be more delicate than corresponding β-values due to the progressive saturation of magnitude scales (especially m_b) with earthquake size and its effect on apparent b-values (Okal and Romanowicz, 1994).

For this purpose, we used the NEIC database, extending back to 1963, when systematic reporting started for m_b, as defined by the Prague formula (Váněk et al., 1962). M_s, defined in the same forum, is catalogued starting in 1968. The NEIC database was used up to and including 2002. Out of the 474,203 earthquakes listed in the NEIC database, and applying the same distance threshold as for the CMT database, we identified 2737 truly intraplate shallow earthquakes with $m_b \geq 4$, after eliminating 8614 events belonging to "deformed" regions, 124 to areas of diffuse plate boundaries, and 257 earthquakes presumably correlated with hotspots. Figure 4 shows the geographical repartition of these various data sets. Note that their occasional clustering may reflect the existence of local networks (e.g., in Australia), rather than true regional variations in seismic activity.

Similarly, we identified 2019 events in the NEIC catalogue with a reported M_s, of which 1427 were in deformed areas, 33 were in diffuse regions, and 40 were associated with hotspots, leaving 519 true intraplate earthquakes.

b-Values for the m_b Populations

Figure 5A illustrates a two-segment population for truly intraplate earthquakes, with $b = 0.84$ at lower magnitudes and $b = 2.08$ for $m_b \geq 6.0$. These values are in agreement with those predicted theoretically by Okal and Romanowicz (1994): $b = 2$ in the range $5.7 \leq m_b \leq 6.6$ and b increasing from 2/3 to 1 at lower magnitudes. They are also similar to those obtained by the same authors for a much larger worldwide data set of 90,074 events.

Results are essentially similar for NEIC data sets in the "deformed" regions, which feature an increase from $b = 1.16$ at low magnitudes to $b = 2.06$ for $m_b > 6.0$ (Fig. 5B). Regarding the hotspot and diffuse data sets (Figs. 5C and 5D), the predicted steep trend at large magnitudes ($b = 2$) cannot be resolved on account of their small populations. The increase from $b \approx 2/3$ to $b \approx 1$ at lower magnitudes is well resolved for hotspot events, probably due to the existence of local networks, but not for diffuse regions, where detection capabilities are poorer at low mag-

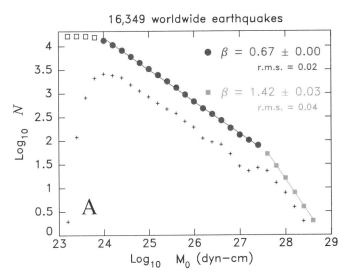

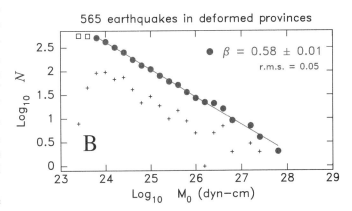

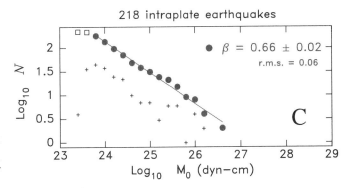

Figure 2. Determination of β-values for the Harvard Centroid Moment Tensor (CMT) data sets (r.m.s. is root mean square). We used bin widths of 0.2 logarithmic units of seismic moment M_0. On each frame, the + signs represent the populations of earthquakes in each bin, and the larger symbols denote the cumulative populations. The open squares at low moments represent bins affected by undersampling due to loss of completeness of the data set. The large solid symbols (circles and squares) show ranges of satisfactory linear regression, with corresponding β-values listed at right. (A) Full CMT data set of shallow earthquakes. (B) CMT solutions in deformed provinces. (C) True intraplate earthquakes. See text for discussion.

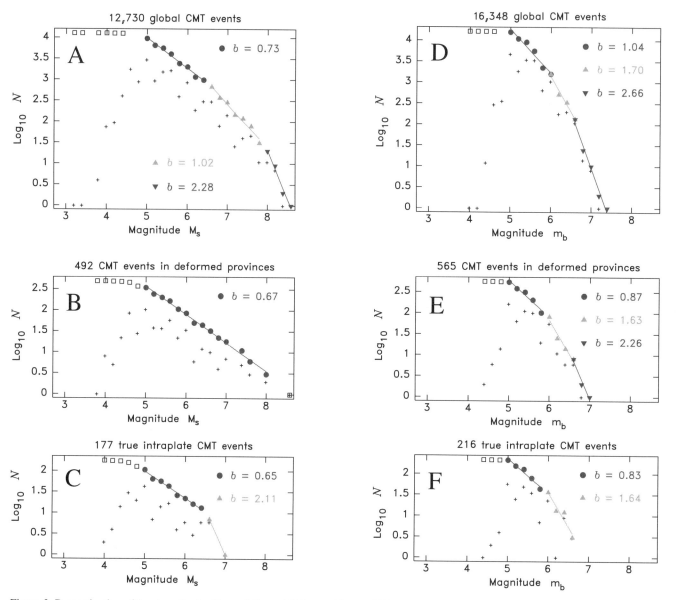

Figure 3. Determination of *b*-values for the Harvard Centroid Moment Tensor (CMT) data sets. The layout of the individual frames is similar to that in Figure 2, except that events are now binned according to their catalogued magnitudes M_s (left) or m_b (right). See text for discussion.

nitudes. In summary, none of the data sets examined exhibits a behavior of the m_b populations recognizably different from that of the global data set, as investigated and justified theoretically by Okal and Romanowicz (1994).

b-Values for the M_s Populations

The largest data set, relative to the deformed regions, features $b = 0.63$ at lower M_s, which is in reasonable agreement with the theoretical value of 2/3 (Okal and Romanowicz, 1994), although the expected transition to higher values of b is not as prominent as for global data sets (Fig. 6B). By contrast, the 519 "true" intraplate earthquakes have $b = 0.78$ for $M_s \leq 6.6$, which is in accept-

able agreement with the predicted $b = 2/3$, and to $b = 2.39$ at higher magnitudes (Fig. 6A). The position of the elbow ($M_s = 6.7$) is remarkably consistent with Okal and Romanowicz' (1994) predictions and observations. The hotspot and diffuse data sets were too small for meaningful regressions.

TARGETED REGIONAL DATA SETS FROM THE NEIC CATALOGUE

In this section, we complement the previous approach with a detailed look at a number of specifically targeted areas, namely the Pacific, North American, and African plates, which regroup most of the intraplate CMT data set, as illustrated on Figure 1.

NEIC Data set

● **True Intraplate (2737)**

▽ *Deformed (8614)*   ▲ *Diffuse (124)*   ▫ *Hotspot (257)*

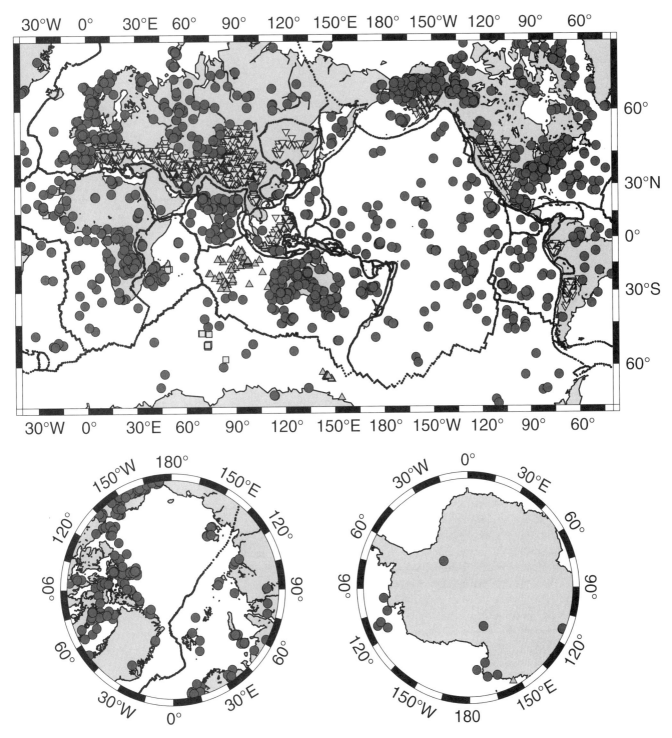

Figure 4. Maps of the National Earthquake Information Center (NEIC) data set. Symbols as in Figure 1.

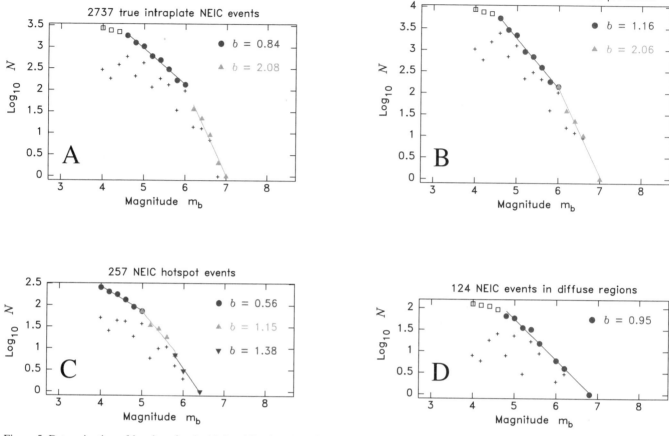

Figure 5. Determination of *b*-values for the National Earthquake Information Center (NEIC) data sets, using the body-wave magnitude m_b and the same procedure as in Figure 3.

The Pacific Data Set

Figure 7A shows the NEIC data set for the Pacific plate, which includes 435 sources. As mentioned by Wysession et al. (1991), a significant portion of the intraplate activity of the basin is expressed through swarms, i.e., episodes of seismicity concentrated in space and time but lacking a clear main shock. The most intense among them was the Gilbert Island swarm of 1981–1984, which included 225 teleseismically recorded events (Lay and Okal, 1983; Okal et al., 1986). The International Seismological Center (ISC) catalogue has reported two additional events in the area in 1997 and 2001, and there have been reports during the 1990s, of tremors felt by residents of the nearby islands. The *b*-value of the Gilbert swarm has previously been reported as 1.35 (Lay and Okal, 1983) and 1.40 (Wysession et al., 1991). Our study (Fig. 7B) confirms these relatively high values and further suggests that a gradual increase in *b*-value (from 1.02 to more than 2) is resolvable around $m_b = 5.4$.

Wiens and Okal (1987) also analyzed a swarm of earthquakes that occurred in the Pacific plate, ~700 km west of the Rivera fracture zone, for which they reported $b = 1.05$.

We reprocessed this data set after including 11 additional earthquakes (for a total of 80), listed by the ISC, but not by the NEIC, and found an equivalent $b = 1.08$ (Fig. 7C). In this respect, the East Pacific swarm does not feature the increase in *b* with size found in the Gilbert swarm. We note that this difference in behavior may also be reflected in the fact that the East Pacific swarm clearly features a main shock (on 02 December 1984), but it remains rather unusual because most of the activity occurred prior to the main shock, with comparatively fewer aftershocks.

After removal of the Gilbert and East Pacific swarms, the Pacific plate NEIC data set is composed of 141 earthquakes, for which *b*-values are presented on Figure 7D. The body-wave coefficient, $b = 1.29$, is larger than expected theoretically ($b = 1$) in the corresponding range of magnitudes, but, on the other hand, it is comparable to the observed worldwide average ($b = 1.35$; Okal and Romanowicz, 1994; their Table 4). Any possible increase in *b*-value with m_b is impossible to resolve because of the narrow band of magnitudes available above the relatively high threshold of completeness in this region ($m_b \geq 5.0$). Similarly, the M_s-derived value ($b = 0.74$) is equivalent to the observed

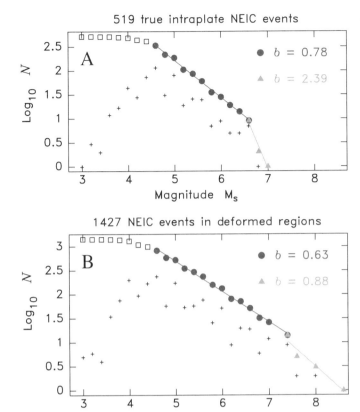

Figure 6. Same as Figure 5, but using the surface-wave magnitude M_s. The "deformed" and "diffuse" data sets are too small for regression.

worldwide average ($b = 0.75$; Okal and Romanowicz, 1994; their Table 3), and it is comparable to the predicted value of 2/3 (Fig. 7E). We note however that the population with reported M_s (43 events) becomes undersampled.

This Pacific data set does include a significant number of earthquakes clustered at locations initially described as regions B and C by Okal et al. (1980), for which these authors proposed very low b-values based on local magnitudes. When further extracted, this data set of only 40 east-central Pacific earthquakes features $b = 1.02$, and the remaining nonclustered data set has $b = 1.31$, both values using m_b. This would generally support the observation by Bergman and Solomon (1980) of a slightly higher b-value inside the Pacific plate but probably not their interpretation that such b-values express a more important contribution of volcanic earthquakes.

The African Plate

We originally extracted from the NEIC database a data set of 285 earthquakes located inside the African plate. Among them, 63 were located in the vicinity of Lake Kariba, on the Zambezi River at the border between Zambia and Zimbabwe (Fig. 8). Gough and Gough (1970a, 1970b) modeled the occur-

rence and evolution of this seismicity as induced by the loading stresses incurred during and after the filling of the reservoir in 1959–1963. They calculated b-values of 1.03 for the full data set spanning 1959–1968, and $b = 1.14$ for the filling stage, culminating with the largest shocks in 1963. They argued that, since these values were computed using local magnitudes M_L, the corresponding m_b-based value should be 1.4 times greater, based on the empirical relationship between m_b and M_L proposed by Gutenberg and Richter (1956). In this context, it seemed warranted to conduct an independent frequency-size study at Kariba based on the NEIC catalogue updated to 2003. Interestingly, we found that activity detected teleseismically continues at Kariba, with a total of 63 events with a reported m_b since 1963, at a rate (~1.5 earthquake per year) essentially equivalent to that during and immediately after filling of the reservoir. Figure 8A shows that this group features an m_b-based b-value of 0.96, which is not significantly different from the M_L-based values reported by Gough and Gough (1970b). In particular, these authors proposed, on their Figure 6, two slightly different regressions for the whole swarm (their open symbols) and the generally smaller events during the filling phase (1959–1963). These expressions lead to $b = 1.02$ for the complementary data set of earthquakes subsequent to filling (1964–1968). The agreement between this value and ours (m_b-based and pertaining to 1963–2003) would suggest first that the seismogenic process subsequent to filling is still going on 35 yr later, and second, that the use of Gutenberg and Richter's (1956) relation to convert M_L into m_b may not be warranted in the context of the Kariba earthquakes. We note in particular that Gutenberg and Richter's work pertained to aftershocks of the 1952 Kern County earthquake, in the magnitude range 5–7, much larger than the Kariba events, and that other relationships were later proposed for M_L versus m_b (Båth, 1981) for various magnitude ranges in various parts of the world, with slopes varying from as low as 0.8 (Chhabra et al., 1975) to as high as 2.04 (Båth, 1978).

After removal of the Kariba seismicity, the remaining 222 NEIC intraplate earthquakes feature a regular $b = 0.92$ between $m_b = 4.6$ and 6, with a lone event ($m_b = 6.4$ in Guinea) suggestive of an increase in b (Fig. 8B). These results are in general agreement with the theoretical models by Okal and Romanowicz (1994), which predict an increase from $b = 2/3$ to $b = 1$ in the range of source sizes covered by the African data set. The data set of events with a reported M_s was too small (39 events) for meaningful processing.

Eastern North America

As presented on Figure 9, we extracted 432 events belonging to the passive (eastern) part of the North American continent. This data set features $b = 1.23$ beyond its limit of completeness ($m_b = 4.6$), with the possible hint of an increase in b for $m_b > 6$. After removal of 57 earthquakes clustered in the New Madrid seismic zone, the b-value is only marginally altered, to $b = 1.19$. The New Madrid events feature a poorly constrained $b \approx 1.25$.

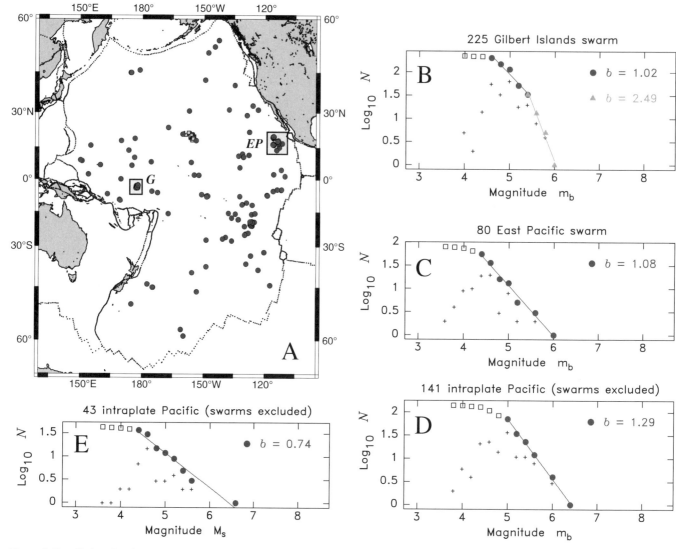

Figure 7. Detailed study of the Pacific plate. (A) The map is a close-up of Figure 4, with the locations of the Gilbert Islands (G) and East Pacific (EP) swarms outlined. The frames at right show the populations of body-wave magnitudes m_b for the two swarms (B, C) and for the remaining 141 true intraplate events (D). (E) The small data set of surface-wave magnitudes M_s.

DISCUSSION AND CONCLUSION

Our examination of global data sets of intraplate earthquakes, extracted both from the CMT and NEIC catalogues, fails to reveal a compelling difference in frequency-size distribution of their populations as compared to their much more numerous interplate counterparts. This suggests a fundamental similarity of the relevant scaling laws governing the populations of intraplate and interplate events. Most importantly, the β-value of 0.66 for the intraplate population (derived from seismic moments, and thus not subject to the saturation affecting the standard magnitude scales) is identical both to its worldwide counterpart (0.67), and to the theoretical value (2/3) predicted at low moments by Romanowicz and Rundle (1993). In the framework of Rundle's (1989) theory, it argues for a fractal

dimension $D = 2$ for the source of intraplate earthquakes, which suggests that they take place on two-dimensional systems of *faults*, rather than inside a seismogenic volume, as do certain deep earthquakes (Okal and Kirby, 1995).

Expectedly, the interpretation of results derived from standard magnitude scales becomes less clear-cut as the characteristic period of the magnitude decreases and the effect of saturation becomes more prevalent for any given size of event. For true intraplate earthquakes, the b-value (0.78) computed at low magnitudes from the 20 s surface-wave magnitude M_s is somewhat higher than predicted theoretically (2/3), but similar to that observed by Okal and Romanowicz (1994) on a global data set (0.75). When derived from the 1 s body-wave magnitude m_b, b-values (0.83 for the CMT data set; 0.84 for the NEIC data set) are lower than reported by Okal and Romanowicz (1994)

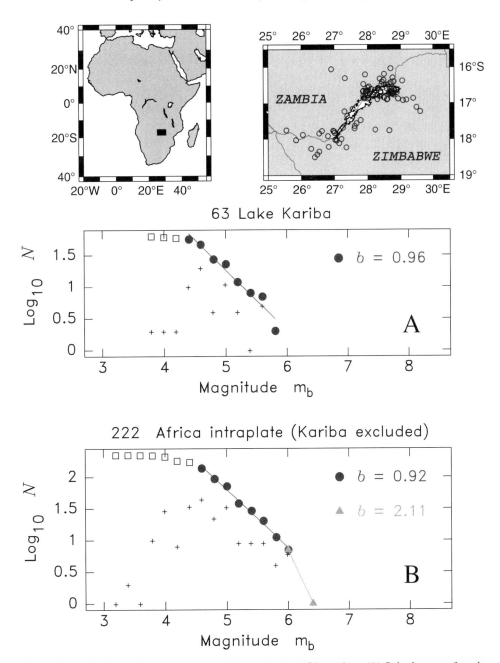

Figure 8. Populations of body-wave magnitudes m_b in the African plate. (A) Sub–data set of earthquakes associated with the filling of Lake Kariba. (B) Remaining true intraplate earthquakes. The maps at top show the distribution of seismicity around Lake Kariba (right) and (left) the location of the study area in Africa (dark rectangle).

for CMT events (1.17) or the full NEIC population (1.35) but still fall in the range (2/3–1) predicted theoretically. We confirm the trend toward larger *b*-values for intraplate events in the Pacific, but only for m_b (1.29), while the M_s-derived *b*-value is the same as in other plates.

On the other hand, events in the so-called "deformed" provinces, consisting mainly of the Mediterranean-Tethyan belt, exhibit slightly reduced values of β and *b* in the low-magnitude

range, but perhaps more interestingly, greater values of the critical earthquake size controlling the elbows in the frequency-size distributions: in the case of the well-sampled M_s distribution, Figure 6B could suggest $M_s^c \approx 7.4$, significantly greater than predicted theoretically (6.7) or observed in true intraplate areas (6.6); in the case of the CMT data set, the absence of any detectable elbow could be explained by a critical moment larger than the global value of $10^{27.5}$ dyn-cm, which would thus fall into a

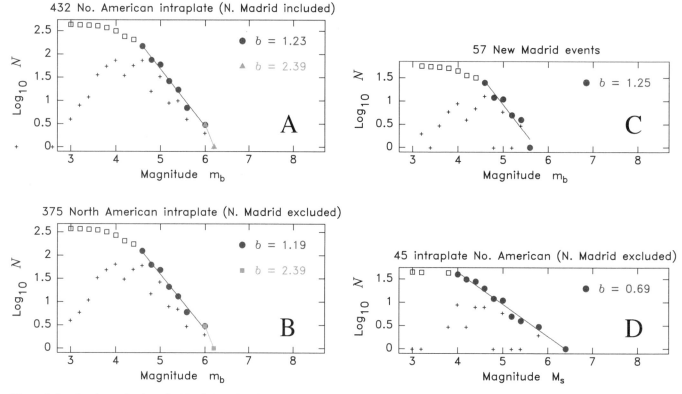

Figure 9. *b*-value determinations for North American intraplate earthquakes. Body-wave *b*-values are shown for the full population (A) and after removal of New Madrid earthquakes (B). The latter are examined in detail (C). (D) The small data set of surface-wave magnitudes M_s.

poorly sampled part of the moment population. Since the size of the critical earthquake reflects the saturation of fault width when it reaches the maximum depth of the seismogenic brittle zone, a larger critical earthquake would argue for a thicker seismogenic zone, which in turn has been both documented and explained in the context of the ongoing deformation and orogeny in the deformed provinces.

Finally, it is interesting to compare our results for the North American data set, which is overwhelmingly continental (Fig. 9A), with those for the Pacific plate, which is exclusively oceanic (Fig. 7D). The similarity between *b*-values (1.23 versus 1.29) suggests a commonality of scaling processes in both provinces. The smaller nature of the Pacific data set precludes the determination of a corner magnitude and, hence, its comparison with the value obtained in North America, which at any rate is only tentative.

ACKNOWLEDGMENTS

We thank Brian Mitchell for originally pointing our attention to the problem of intraplate *b*-values and Giovanni Sella for discussion. The paper was improved through the comments of an anonymous reviewer. Figures were drafted using the GMT software (Wessel and Smith, 1991).

REFERENCES CITED

Båth, M., 1978, Structural and vibrational bedrock properties in Sweden: Journal of Computational Physics, v. 29, p. 344–356, doi: 10.1016/0021-9991(78)90138-9.

Båth, M., 1981, Earthquake magnitude—Recent research and current trends: Earth-Science Reviews, v. 17, p. 315–398, doi: 10.1016/0012-8252(81)90014-3.

Bergman, E.A., and Solomon, S.C., 1980, Oceanic intraplate earthquakes: Implications for local and regional intraplate stress: Journal of Geophysical Research, v. 85, p. 5389–5410.

Bird, P., 2003, An updated digital model of plate boundaries: Geochemistry, Geophysics, Geosystems, v. 4, no. 3, doi: 10.1029/2001GC000252.

Bird, P.G., Kagan, Y.Y., and Jackson, D.D., 2002, Plate tectonics and earthquake potential of spreading ridges and oceanic transform faults: American Geophysical Union Geodynamics Series, v. 30, p. 203–218.

Chhabra, M.P., Chouhan, R.K.S., Srivastava, H.N., and Chaudhury, H.M., 1975, The relationship between body wave and local magnitudes for Himalayan earthquakes: Annali di Geofisica, v. 28, p. 381–391.

Dziewonski, A.M., Friedman, A., Giardini, D., and Woodhouse, J.H., 1983, Global seismicity of 1982: Centroid moment tensor solutions for 308 earthquakes: Physics of the Earth and Planetary Interiors, v. 33, p. 76–90, doi: 10.1016/0031-9201(83)90141-3.

Frohlich, C., and Davis, S.D., 1993, Teleseismic *b*-values; or, much ado about 1.0: Journal of Geophysical Research, v. 98, p. 631–644.

Geller, R.J., 1976, Scaling relations for earthquake source parameters and magnitudes: Bulletin of the Seismological Society of America, v. 66, p. 1501–1523.

Gough, D.I., and Gough, W.I., 1970a, Stress and deflection of the lithosphere near Lake Kariba: Part I: Geophysical Journal of the Royal Astronomical Society, v. 21, p. 65–78.

Gough, D.I., and Gough, W.I., 1970b, Load-induced earthquakes at Lake Kariba: Part II: Geophysical Journal of the Royal Astronomical Society, v. 21, p. 79–101.

Gutenberg, B., and Richter, C.F., 1941, Seismicity of the Earth: Geological Society of America Special Paper 34, 125 p.

Gutenberg, B., and Richter, C.F., 1954, Seismicity of the Earth and Associated Phenomena: Princeton, New Jersey, Princeton University Press, 310 p.

Gutenberg, B., and Richter, C.F., 1956, Earthquake magnitude, intensity, energy and acceleration: Bulletin of the Seismological Society of America, v. 46, p. 105–145.

Lay, T., and Okal, E.A., 1983, The Gilbert Islands (Republic of Kiribati) earthquake swarm of 1981–83: Physics of the Earth and Planetary Interiors, v. 33, p. 284–303, doi: 10.1016/0031-9201(83)90046-8.

Minakami, T., 1974, Seismology of volcanoes in Japan, *in* Civetta, L., Gasparini, P., Luongo, G., and Rapolla, A., eds., Physical Volcanology: Amsterdam, Elsevier, p. 1–27.

Mogi, K., 1963, Some discussions on aftershocks, foreshocks and earthquake swarms—The fracture of a semi-infinite body caused by inner stress origin and its relation to the earthquake phenomena, 3: Bulletin of the Earthquake Research Institute of Tokyo University, v. 41, p. 615–658.

Molnar, P., 1979, Earthquake recurrence intervals and plate tectonics: Bulletin of the Seismological Society of America, v. 69, p. 115–133.

Okal, E.A., and Kirby, S.H., 1995, Frequency-moment distribution of deep earthquakes: Implications for the seismogenic zone at the bottom of slabs: Physics of the Earth and Planetary Interiors, v. 92, p. 169–187, doi: 10.1016/0031-9201(95)03037-8.

Okal, E.A., and Romanowicz, B.A., 1994, On the variation of b-value with earthquake size: Physics of the Earth and Planetary Interiors, v. 87, p. 55–76, doi: 10.1016/0031-9201(94)90021-3.

Okal, E.A., Talandier, J., Sverdrup, K.A., and Jordan, T.H., 1980, Seismicity and tectonic stress in the south-central Pacific: Journal of Geophysical Research, v. 85, p. 6479–6495.

Okal, E.A., Woods, D.F., and Lay, T., 1986, Intraplate deformation in the Samoa–Gilbert–Ralik area: A prelude to a change of plate boundaries in the southwest Pacific?: Tectonophysics, v. 132, p. 69–78, doi: 10.1016/0040-1951(86)90025-9.

Pacheco, J., Scholz, C.H., and Sykes, L.R., 1992, Changes in frequency-size relationships from small to large earthquakes: Nature, v. 355, p. 71–73, doi: 10.1038/355071a0.

Richter, C.F., 1935, An instrumental earthquake magnitude: Bulletin of the Seismological Society of America, v. 25, p. 1–32.

Romanowicz, B.A., and Rundle, J.B., 1993, Scaling relations for large earthquakes: Bulletin of the Seismological Society of America, v. 83, p. 1294–1297.

Rundle, J.B., 1989, Derivation of the complete Gutenberg-Richter magnitude-frequency relation using the principle of scale invariance: Journal of Geophysical Research, v. 94, p. 12,337–12,342.

Stein, S., and Sella, G.F., 2002, Plate boundary zones: Concepts and approaches: American Geophysical Union Geodynamics Series, v. 30, p. 1–26.

Turcotte, D.L., 1992, Fractals and Chaos in Geology and Geophysics: Cambridge, Cambridge University Press, 221 p.

Váněk, J., Zátopek, A., Kárnik, V., Kondorskaya, N.V., Riznichenko, Yu.V., Savarensky, S.F., Solov'ev, S.L., and Shebalin, N.V., 1962, Standardization of magnitude scales: Bulletin of the USSR Academy of Sciences, Geophysics Series, v. 2, p. 108–111.

Wessel, P., and Smith, W.H.F., 1991, Free software helps map and display data: Eos (Transactions, American Geophysical Union), v. 72, p. 441, 445–446.

Wiens, D.A., and Okal, E.A., 1987, Tensional intraplate seismicity in the east-central Pacific: Physics of the Earth and Planetary Interiors, v. 49, p. 264–282, doi: 10.1016/0031-9201(87)90029-X.

Wysession, M.E., Okal, E.A., and Miller, K.L., 1991, Intraplate seismicity of the Pacific Basin, 1913–1988: Pure and Applied Geophysics, v. 135, p. 261–359, doi: 10.1007/BF00880241.

MANUSCRIPT ACCEPTED BY THE SOCIETY 29 NOVEMBER 2006

The Geological Society of America
Special Paper 425
2007

Remotely triggered earthquakes following moderate main shocks

Susan E. Hough

U.S. Geological Survey, 525 S. Wilson Avenue, Pasadena, California 91106, USA

ABSTRACT

Since 1992, remotely triggered earthquakes have been identified following large (M > 7) earthquakes in California as well as in other regions. These events, which occur at much greater distances than classic aftershocks, occur predominantly in active geothermal or volcanic regions, leading to theories that the earthquakes are triggered when passing seismic waves cause disruptions in magmatic or other fluid systems. In this paper, I focus on observations of remotely triggered earthquakes following moderate main shocks in diverse tectonic settings. I summarize evidence that remotely triggered earthquakes occur commonly in mid-continent and collisional zones. This evidence is derived from analysis of both historic earthquake sequences and from instrumentally recorded M5–6 earthquakes in eastern Canada. The latter analysis suggests that, while remotely triggered earthquakes do not occur pervasively following moderate earthquakes in eastern North America, a low level of triggering often does occur at distances beyond conventional aftershock zones. The inferred triggered events occur at the distances at which SmS waves are known to significantly increase ground motions. A similar result was found for 28 recent M5.3–7.1 earthquakes in California. In California, seismicity is found to increase on average to a distance of at least 200 km following moderate main shocks. This supports the conclusion that, even at distances of ~100 km, dynamic stress changes control the occurrence of triggered events. There are two explanations that can account for the occurrence of remotely triggered earthquakes in intraplate settings: (1) they occur at local zones of weakness, or (2) they occur in zones of local stress concentration.

Keywords: earthquake, triggering, aftershock dynamic stress.

INTRODUCTION

In 1992, the Landers earthquake provided unambiguous evidence that the "reach" of a large earthquake can extend far beyond its immediate aftershock zone (e.g., Hill et al., 1993; Bodin and Gomberg, 1994). A similar burst of regional seismicity followed the 16 October, 1999, Mw 7.1 Hector Mine, California earthquake (Gomberg et al., 2001; Glowacka et al., 2002; Hough and Kanamori, 2002). In these and other documented cases, triggered seismicity was observed to occur preferentially, although not exclusively, in active geothermal and volcanic regions, such

as Long Valley Caldera, The Geysers, and the Salton Sea region (e.g., Stark and Davis, 1996; Gomberg and Davis, 1996; Prejean et al., 2005). Triggering has also been observed at geothermal and volcanic sites elsewhere around the world (e.g., Power et al., 2001), leading some to conclude that triggered earthquakes are not observed in other seismotectonic settings (Scholz, 2003).

A number of previous studies have presented compelling evidence that remotely triggered earthquakes are caused by the dynamic stress changes associated with transient seismic waves, typically the high-amplitude S and/or surface-wave arrivals (e.g., Gomberg and Davis, 1996; Kilb et al., 2000). The asso-

Hough, S.E., 2007, Remotely triggered earthquakes following moderate main shocks, *in* Stein, S., and Mazzotti, S., ed., Continental Intraplate Earthquakes: Science, Hazard, and Policy Issues: Geological Society of America Special Paper 425, p. 73–86, doi: 10.1130/2007.2425(06). For permission to copy, contact editing@geosociety.org. ©2007 The Geological Society of America. All rights reserved.

ciation of triggered earthquakes with dynamic stress changes is in contrast to aftershocks, which appear to be caused primarily by local, static stress changes associated with fault movement (e.g., Das and Scholz, 1981; King et al., 1994; Toda and Stein, 2003). (According to convention, aftershocks are generally, albeit vaguely, assumed to be events within 1–2 fault lengths of a main shock.) Recent studies (e.g., Felzer and Brodsky, 2006) suggest that dynamic stress changes might also play an important role in controlling the distribution of aftershocks. While both types of stress change may play a role in aftershock generation, investigations of remotely triggered earthquakes have focused only on dynamic stress changes.

Because almost all of the initial examples of remotely triggered earthquakes were in regions with active volcanic processes or shallow hydrothermal activity—both of which are associated with abundant heat and fluids at shallow depths in Earth's crust—initially proposed triggering mechanisms involved disruption of fluids. Proposed triggering mechanisms involved the effects of seismic waves on bubbles within fluid systems, such as advective overpressure (Linde et al., 1994) and rectified diffusion (Sturtevant et al., 1996; Brodsky et al., 1998). More recently, Brodsky and Prejean (2005) proposed a barrier-clearing model whereby long-period waves generate fluid flow and pore-pressure changes within fault zones.

In this paper, I summarize both previous and new results that provide compelling evidence that remotely triggered earthquakes do occur following even moderate (M5–7) main shocks outside of active geothermal and/or hydrothermal regions, including intraplate regions.

TRIGGERED EARTHQUAKES IN DIVERSE TECTONIC SETTINGS

To facilitate the subsequent discussion of the implications of remotely triggered earthquake results, in this section, I present both new analyses as well as a brief discussion of salient results from previous studies of remotely triggered earthquakes outside of active geothermal or volcanic regions. In addition to the cases listed next, Gomberg et al. (2004) recently concluded that remotely triggered earthquakes occurred in western North America following the 2002 M7.9 Denali earthquake, although at least some of these events appear to have occurred in or near active geothermal regions.

Central and Eastern North America

Investigations of remotely triggered earthquakes in mid-plate settings are inevitably hampered by data limitations. Researchers are typically limited to analysis of macroseismic data from large historic earthquakes or sparse instrumental data from moderate recent earthquakes. Hough (2001) and Hough et al. (2003) presented evidence for remotely triggered earthquakes that occurred both during the 1811–1812 New Madrid earthquake sequence and following the 1886 Charleston, South

Carolina, earthquake. These results suggest that triggering commonly occurs following large earthquakes in the North American mid-continent. There is particularly compelling evidence that moderate earthquakes were triggered in northern Kentucky–southern Ohio during the New Madrid sequence, along or near the Ohio River Valley (Fig. 1). Additionally, Mueller et al. (2004) presented evidence that one of the so-called New Madrid main shocks, conventionally placed in the northern New Madrid seismic zone (e.g., Johnston, 1996; Johnston and Schweig, 1996), may have in fact occurred in the Wabash Valley, ~200 km away from the New Madrid seismic zone.

In retrospect the results of Seeber and Armbruster (1987) also provide evidence that intraplate triggering is common. Although this study talks about "aftershocks" of the 1886 Charleston, South Carolina, earthquake, their inferred locations are distributed over distances of 200–300 km, well outside an expected aftershock zone given the size of the main shock. The triggered events following the Charleston main shock discussed by Hough et al. (2003) were located at even greater distances, for example in the Wabash Valley.

To explore the possibility that remotely triggered earthquakes occur following moderate intraplate earthquakes, I consider recent M4.9–6.1 events in eastern Canada. The Geological Survey of Canada (GSC) operates a network of over 100 seismometers throughout Canada, with especially dense coverage in seismically active areas such as the Charlevoix, Quebec, region of the St. Lawrence Valley. The Canadian National Earthquake Database includes historic earthquakes as far back as 1568, but to focus on earthquakes for which good instrumental data is available, I searched the catalog for M4.9 and greater earthquakes since 1985. The catalog includes 15 such events, the largest two of which are the 1988 Saguenay, Quebec, and the 1989 Ungava earthquakes, both close to M6. Three of the events were in the northernmost United States: since they were recorded in Canada, presumably the network coverage of such events was not ideal. Several earthquakes also occurred in northeastern Canada, where network coverage is presumably also limited. Of the 15 events, 8 occurred in regions where the network should have provided good coverage of small earthquakes (Table 1).

The issue of catalog completeness arises in any seismicity study. In this study, completeness is expected to vary not only with time, but also spatially. However, detecting short-term seismicity fluctuations requires only short-term catalog stability, which can be assumed. One completeness-related issue bears mention, however: completeness invariably degrades in the immediate aftermath of a large regional earthquake. This will hinder the detection of very early triggered earthquakes, in any time or region.

Using catalogs from one month (30 d) before and after each event, I investigated seismicity changes using a standard beta-statistic approach (Matthews and Reasenberg, 1988; Reasenberg and Simpson, 1992). The beta statistic, β, is defined as

$$\beta = N_a - N_e/(v)^{1/2}, \qquad (1)$$

10:40 pm (LT), 7 February 1812

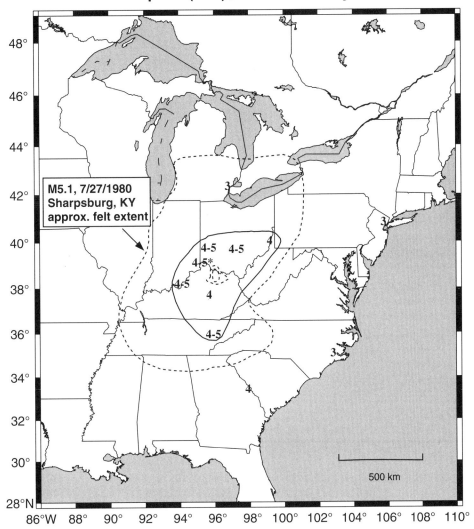

Figure 1. Map showing intensity (MMI) values for an earthquake that occurred at 10:40 p.m. (LT) on 7 February 1812. Within the solid contour, almost all of the accounts describe the event with the word "violent" or "severe." For comparison, the inner dashed lines indicate the MMI V and VI contours for the 1980 M5.1 Sharpsburg, Kentucky, earthquake (Mauk et al., 1982). The outer dashed line indicates the felt area of the Sharpsburg event, which is considerably smaller than that of the 1812 event.

TABLE 1. RECENT MODERATE EARTHQUAKES IN EASTERN CANADA, PLUS THE 2002 AU SABLE FORKS EARTHQUAKE IN NORTHERN NEW YORK STATE

Event	Date	Name	ML	Lat. (°N)	Long. (°W)
1	25 November 1988	Saguenay	6.1	48.12	71.18
2	16 March 1989	Ungava	5.7	60.06	70.06
3	25 December 1989	Ungava(2)	6.1	60.12	73.60
4	18 October 1990	Quebec	5.0	46.47	75.59
5	16 November 1997	St. Lawrence	5.1	46.80	71.42
6	16 March 1999	Quebec(2)	5.1	49.61	66.32
7	1 January 2000	Quebec(3)	5.2	46.84	78.93
8	20 April 2002	Au Sable, NY	5.5	44.53	73.73

Note: Locations and magnitudes (ML) were taken from the Geological Survey of Canada database (http://www.seismo.nrcan.gc.ca/EarthquakesCanada.html).

where N_a is the number of events occurring following an event, N_e is the expected number given the pre–main shock seismicity rates (assuming seismicity is stationary), and v is the variance of N_e. β will be large and positive in regions where seismicity increases. β is correspondingly large and negative in regions where seismicity rate decreases. However, in the "null case," where there are no earthquakes in a given subregion either before or after a main shock, β is not zero. As introduced by Matthews and Reasenberg (1988), N_e represents a probability density function with equal probability over a ±0.5 range bracketing whole numbers. For example, a value of 2 corresponds to an expected range of 1.5–2.5. Because the expected number of earthquakes cannot be negative, if there are zero events in a pre-event window, N_e is set to 0.25. For the analyses in this paper, the baseline value of β is thus not 0 but rather approximately –0.7. β is equal to 0 only if $N_e = N_a$.

Because seismicity levels commonly fluctuate significantly, even a high value of β does not prove that a seismicity increase was caused by a preceding main shock. Other evidence, such as a close temporal correspondence between the main shock timing and the initiation of subsequent events, is needed to establish a causal relationship. This analysis yields no evidence of widespread triggering following any of the events: overall seismicity fluctuations, both positive and negative, appear to be comparable to the usual level of fluctuations observed over time, as illustrated in Figure 2 for a M5.1 event in 1999.

Although the preliminary results are largely negative, several of the beta-statistic maps do reveal a similar feature: an apparent seismicity increase at ~100 km epicentral distance, beyond the presumed aftershock zone for M5–6 earthquakes. To further investigate this result, I calculated the average beta value as a function of epicentral distance from each event. Of the eight earthquakes for which results are shown in Figure 3, five revealed a small increase in β at a distance of ~100 km. None of these small increases was significant by itself, as evidenced by the fact that they are comparable in amplitude to fluctuations seen over the broader region over the same time period. (Also, as noted, even a statistically significant increase would not in itself imply a causal link with the main shock.) What is intriguing, however, is the persistent appearance of a slight increase at a narrow distance range.

The most prominent increase appeared following the mb5.9 1988 Saguenay earthquake. This earthquake was followed within the first month after the main shock by a number of small events along the Charlevoix seismic zone (Fig. 4). The Charlevoix events occurred at relatively shallow depths, whereas the Saguenay source was significantly deeper (Fig. 5). The Charlevoix events were thus clustered in both their epicentral distance from the main shock and their depth distribution. The most straightforward explanation for this clustering is that the events were triggered by postcritical Moho reflections (SmS arrivals), which are known to significantly increase ground motions at a distance of ~100 km. SmS is a body wave, and so SmS-associated triggering would be expected to occur anywhere along the raypath where the wave is of substantial amplitude. If, for example, the Charlevoix events were clustered horizontally but not vertically, this might argue against triggering by body waves. I explore this hypothesis further in a later section.

Northern India

As a second example of triggering outside of geothermal or hydrothermal areas, I summarize recent results from the 1905 Kangra, India, earthquake, for which very early instrumental data are available. The Kangra earthquake has been the subject of

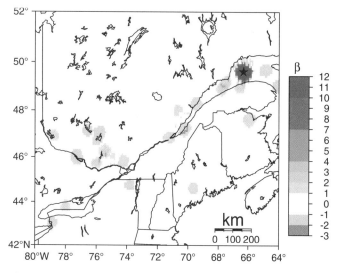

Figure 2. Beta statistic calculated from seismicity during the 30 d following the 16 March 1999 M5.1 earthquake in Quebec compared to 30 d prior to the earthquake. Scale bar indicates shading of beta values between –3 and 12. Within immediate aftershock zones, beta values are much higher. (The same scale is used for all subsequent beta-statistic maps.)

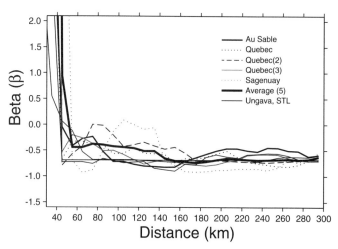

Figure 3. Beta statistic as a function of epicentral distance for the earthquakes listed in Table 1. Five of the curves reveal some hint of a molehill signal. For three events, no post–main shock seismicity was recorded outside the immediate aftershock zone: the two Ungava events and the 16 November 1997 St. Lawrence event (thin dark lines).

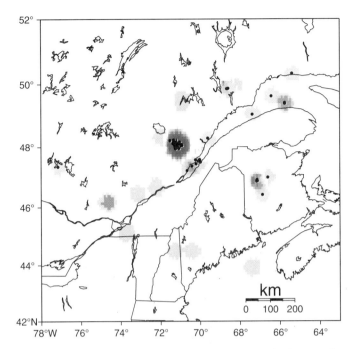

Figure 4. Seismicity following the 1988 Saguenay, Quebec, earthquake. Same color scale as shown in Figure 2.

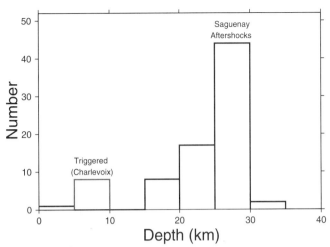

Figure 5. Histogram of depths of Saguenay aftershocks and events along the St. Lawrence shown in Figure 4 (Duberger et al., 1991). Depths (and locations) are from the Canadian National Seismic Network catalog (www.seismo.nrcan.gc.ca/cnsn), which does not report location uncertainties.

debate over the years. Early intensity surveys revealed two separate loci of strong shaking and damage suggestive of two source zones spanning a distance of 400–500 km (e.g., Middlemiss, 1905), and early magnitude estimates suggested an earthquake large enough to connect the two high-intensity regions. However, Molnar (1987) concluded that two distinct loci of high-intensity shaking were resolved by surveys following the earthquake. Moreover, Ambraseys and Bilham (2000) estimated Ms7.8 for the main shock, suggesting that the main shock rupture was not large enough to span the two high-intensity zones. As discussed by Hough et al. (2004), the extensive intensity reevaluation of Ambraseys and Douglas (2004) provided compelling additional evidence for two distinct zones of high intensity. Combined with the geodetic constraints (Bilham, 2001; Wallace et al., 2002), the macroseismic data provided compelling evidence that a substantial second event occurred near the town of Dehra Dun, ~150 km southeast of the inferred terminus of the main shock rupture. The damage near Dehra Dun was not especially severe (Ambraseys and Douglas, 2004), but by modeling predicted shaking from the main shock, Hough et al. (2004) showed that intensities were substantially higher than predicted over a broad region, including many hard rock sites. The intensity pattern was shown to be consistent with a large (M >7) earthquake at a relatively deep (30 km) depth.

As discussed by Hough et al. (2004), a small handful of early instrumental recordings are available for the Kangra earthquake. Of the handful of operating stations included in the U.S. Geological Survey (USGS) microfilm archive, seismograms are either missing for the date of the main shock or else are of poor quality, either because the reproduction quality was poor or because the data

were recorded on undamped instruments that do not reveal clear phase arrivals. However, two records from early Wiechert instruments provide useful information: a recording from Gottingen included in the compilation of Duda (1992) and a recording from Leipzig, Germany (Fig. 6). The latter record is especially clear, and it reveals a sharp initial S-sS arrival followed by presumed S multiples of lower frequency. Approximately 7 min after the first S-sS arrival, a second distinct arrival can be seen on the record, very similar in frequency content and waveform characteristics to the initial S-sS group. This later arrival is clearly distinct from the main shock Love waves and is most obviously interpreted as an S-wave group from a second source. The S-sS separation is moreover larger than that in the initial S-sS group, suggesting that the second source was deeper than the first.

The available instrumental data thus corroborate the conclusions from the macroseismic data analysis and provide compelling evidence that a second substantial earthquake occurred ~7 min after the Kangra main shock, at a distance of ~150 km. Hough et al. (2004) noted that the location of the triggered earthquake coincided with a "halo" of amplified intensity that appears to have been the macroseismic signature of SmS arrivals. While this one correspondence cannot be considered a compelling piece of evidence by itself, the inferred location of the triggered earthquake is consistent with the SmS triggering hypothesis.

SmS TRIGGERING IN CALIFORNIA

The preceding results were derived from small numbers of events and are therefore regarded as intriguing but not conclusive. However, if SmS arrivals increase the likelihood of triggering, such triggering would be expected in any region where a well-defined Moho is present. Thus, to test the hypothesis of

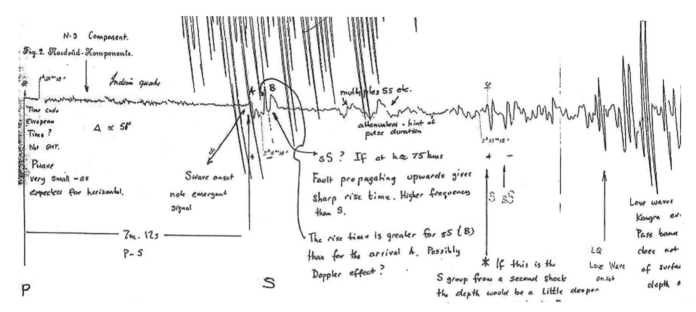

Figure 6. Seismogram of the 1905 Kangra earthquake recorded on an early Wiechert instrument in Leipzig. The record reveals a clear initial S-sS arrival, a longer-period S-multiple arrival, and a second distinct S-sS group prior to the surface waves. The S-sS separation is larger for the second group than for the first.

SmS triggering, and to explore whether remote triggering is in fact pervasive following even moderate main shocks, one can turn to a region where high seismicity rates and good catalogs are available. To investigate whether SmS triggering occurs elsewhere, I considered 27 earthquakes with magnitudes between 4.8 and 6.9 that occurred in California between 1980 and 2004, as well as the Mw7.1 Hector Mine earthquake (Table 2). I did not consider the 1992 Landers earthquake because the extent of the main shock rupture (as well as the occurrence of the Big Bear aftershock) was such that a simple distance metric could not be defined. The list of events was drawn from a compilation of significant earthquakes on the Southern California Earthquake Center (SCEC) web site (http://www.data.scec.org/chrono index/ quakedex.html). Although not complete for events with magnitudes near 5, the compilation does include virtually all significant, independent moderate main shocks in southern California as well as a number of especially well-recorded M4.8–5.0 events. For events larger than M5, the SCEC list omits only aftershocks and a small number of events near the periphery of the network.

Again using the standard beta-statistic approach, I compared seismicity rates during the 30 d after and before each event. The beta statistic was calculated using a grid with 10 km spacing and a smoothing radius of 15 km. The beta-statistic maps again reveal positive and negative seismicity fluctuations outside of the aftershock zone. No widespread, statistically significant triggering is revealed for the events except for Hector Mine, consistent with previous results.

I then calculated the average beta statistic as a function of epicentral distance. Because the greater number of events allowed the possibility of more in-depth statistical analysis than was possible for the eastern Canada events, I treated the "null

case" by calculating averages using only those subregions with at least one earthquake either before or after the main shock. This removed the slight negative bias introduced by the large number of cases for which seismicity rate change effectively cannot be measured, and it allows seismicity rate fluctions to be resolved against a baseline of zero.

For most of the moderate earthquakes in California, as well as the Hector Mine earthquake, β decreases outside of the immediate aftershock zone but increases slightly at a distance of 70–120 km (Fig. 7). The increase is particularly strong for the 1993 Coalinga earthquake (Fig. 8). Figure 7 also reveals a number of large peaks in β at larger distances. Although individual peaks can be large enough to affect the average, none of these peaks is as persistent as that at 70–120 km. The results shown in Figure 7 can be illustrated in map view by shifting all 27 beta-statistic maps to zero latitude/longitude and contouring the aggregate results. (Hector Mine is omitted so that the results are not biased by its relatively large aftershock zone.) Figure 9 clearly reveals that seismicity rates increase on average to a distance of at least 120 km, well beyond the traditional aftershock zone for moderate earthquakes. Seismicity also increases on average, albeit more weakly, to a distance of ~230 km.

In theory, an increase in β can result from a particularly low local standard error of the background rate. However, following earlier studies (e.g., Matthews and Reasenberg, 1988), I assumed seismicity to be Poissonian. The standard error is thus given simply as the square-root of the mean, so the increases in β described in this paper are all associated with seismicity rate increases. In effect, this approach is equivalent to a consideration of absolute seismicity rate fluctuations with an explicit normalization to background rate. In a comparison of Figures 3 and 7, it

TABLE 2. RECENT MODERATE EARTHQUAKES IN SOUTHERN AND CENTRAL CALIFORNIA ANALYZED IN THIS STUDY

Event no.	Date	Event	Mw	Lat. (°N)	Long. (°W)
1	25 February 1980	White Wash	5.5	33.50	116.52
2	26 April 1981	Westmoreland	5.8	33.096	115.625
3	15 June 1982	Anza	4.8	33.548	116.677
4	2 May 1983	Coalinga	6.1	36.228	120.318
5	8 July 1986	North Palm Springs	6.0	33.999	116.608
6	13 July 1986	Oceanside	5.5	32.971	117.874
7	1 October 1987	Whittier	5.9	34.061	118.079
9	24 November 1987	Elmore Ranch	6.2	33.090	115.792
10	26 June 1988	Upland	4.7	34.133	117.708
11	3 December 1988	Pasadena	5.0	34.141	118.133
12	19 January 1989	Malibu	5.0	33.917	118.627
13	28 June 1991	Sierra Madre	5.8	34.270	117.993
14	23 April 1992	Joshua Tree	6.1	33.960	116.317
15	11 July 1992	Mojave	5.7	35.208	118.067
16	28 May 1993	Wheeler Ridge	5.2	35.149	119.104
17	17 January 1994	Northridge	6.7	34.213	118.537
18	17 August 1995	Ridgecrest	5.4	35.776	117.662
19	27 November 1996	Coso	5.3	36.075	117.650
20	18 March 1997	Calico	5.3	34.971	116.819
21	6 March 1998	Coso	5.2	36.067	117.638
22	16 August 1998	San Bernardino	4.8	34.121	116.928
23	27 October 1998	Whiskey Springs	4.8	34.323	116.844
24	16 October 1999	Hector Mine	7.1	34.600	116.270
25	31 October 2001	Anza	5.1	33.508	116.514
26	22 February 2003	Big Bear	5.4	34.319	116.848
27	28 September 2004	Parkfield	6.0	35.819	120.364

Note: Locations and magnitudes are from SCSN/NCSN/CISN online catalogs. SCSN—Southern California Seismic Network, NCSN—Northern California Seismic Network, CISN—California Integrated Seismic Network.

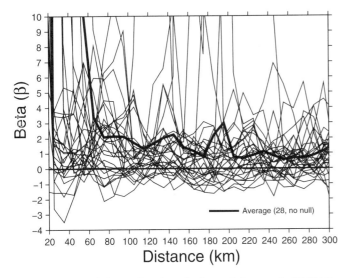

Figure 7. Beta statistic as a function of epicentral distance for 28 M4.8–7.1 earthquakes in California. Average of 28 curves is also shown (dark line). The increase, or "molehill," at 80–100 km is small but robust.

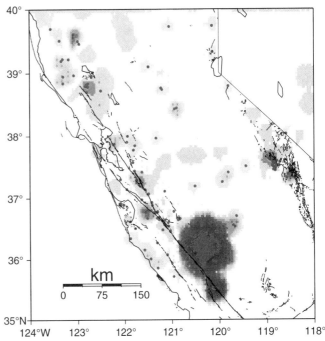

Figure 8. Beta statistic calculated from seismicity during the 30 d following the 1993 Coalinga earthquake compared to 30 d prior to the earthquake. Same color scale as shown in Figure 2.

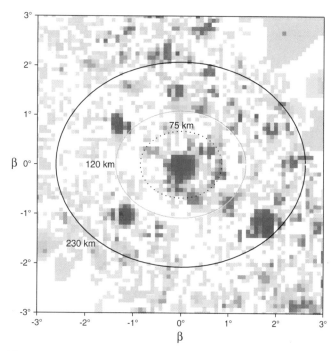

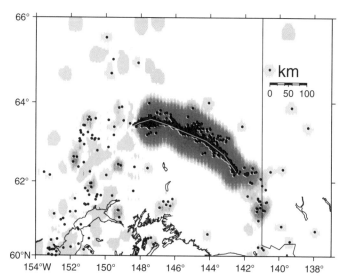

Figure 10. Seismicity fluctuations following the 2002 M7.9 Denali, Alaska, earthquake.

Figure 9. Average seismicity fluctuations, expressed as beta statistic, are shown in map view. To generate this (Mercator projection) map, 27 beta-statistic maps such as that shown in Figure 8 were shifted to zero origin and combined to reveal the average spatial pattern of seismicity fluctuations. The ovals correspond to three radii: (1) 75 km, the distance at which the $\beta(r)$ signal is inferred to peak slightly, (2) 120 km, the distance over which seismicity clearly increases on average, and (3) 230 km, the distance range over which average seismicity rates increase weakly.

is clear that the "molehill" signature, while weak, occurs at a similar range of distances in both California and eastern North America.

Following the 2002 Denali earthquake, seismicity also increased in a region ~100–140 km to the southeast of the main shock rupture (Fig. 10). This location would have also experienced amplified ground motions due to directivity effects, which illustrates an important point: if SmS arrivals do trigger earthquakes, they will be only one of several factors that control the location of triggered events. Previous studies have shown or suggested that triggering also depends on other factors, including directivity (e.g., Kilb et al., 2000) as well as the presence of faults that are susceptible to failure.

In each of the three regions considered, the observed increases of β at 70–140 km were small, and one could not attach statistical significance to the results from any one earthquake. However, the increases were insensitive to the choice of analysis parameters (smoothing distance, etc.), and the persistent appearance of a molehill at a narrow distance range is difficult to dismiss as a fluke.

The molehill signals reflect seismicity increases within a month of the respective main shocks. The temporal sequence characteristics of remotely triggered earthquakes might provide an observational constraint against which one can test theoretical triggering models (e.g., Gomberg, 2001). However, the simplest

explanation for delayed triggering is that transient stress changes cause very early triggered events, either large or small, and these initial triggered events caused local disturbances that generated local sequences (Richter, 1955; Hough and Kanamori, 2002; Hough et al., 2003). In any case, we can explore the timing of the inferred triggered earthquakes identified in this study. Following the 1983 Coalinga earthquake, the molehill was primarily due to a cluster of events to the south-southeast of the main shock (Fig. 8). The earliest recorded event in this cluster occurred ~2.5 d following the main shock. The first recorded event in the lower St. Lawrence followed the 1988 Saguenay, Quebec, main shock by a similar delay (3 d). In the absence of local broadband data, it is impossible to know if triggered earthquakes occurred in these locations immediately after their respective main shocks. However, delays ranging from a few minutes (Kangra) to a few days are consistent with the time delay of remotely triggered earthquakes observed in other regions. The triggered events identified by Hough (2001) occurred ~4 d after the 23 January 1812 main shock and ~16 and 18 h after the 7 February 1812 New Madrid main shock.

To further explore the temporal behavior of the inferred triggered earthquakes, I considered the two earthquakes that had the largest molehills: the 1983 Coalinga and 1999 Hector Mine earthquakes. Considering only the rates of earthquakes at a distance of 70 to 110 km from each main shock, I found that the rates of these events did decrease with time following their respective main shocks (Fig. 11). The time decay of the (inferred) triggered events did not change substantially when considering events between 80 and 110 km. In effect, Figure 11 suggests that the events at this distance range "look like aftershocks" in terms of their sequence statistics. However, as I discuss later, an association with SmS arrivals provides compelling evidence that these events were triggered by dynamic rather than static stress changes. In the following section, I consider the statistical significance of the observations.

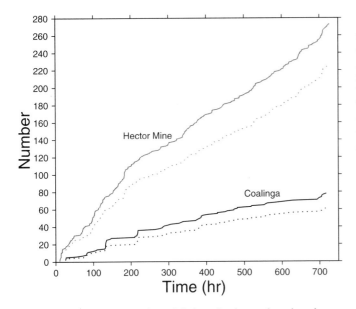

Figure 11. Cumulative number of (inferred) triggered earthquakes as a function of time following the Coalinga earthquake (black lines) and Hector Mine earthquake (gray lines). The solid lines indicate temporal characteristics of earthquakes at 70–110 km distance from the respective epicenters; the dotted lines indicate earthquakes at 80–110 km distance.

STATISTICAL SIGNIFICANCE

As previous studies have pointed out (e.g., Reasenberg and Simpson, 1992), it is difficult to assess the statistical significance of any beta-statistic result. Although one can infer strict confidence levels for different β values, β can increase or decrease substantially because of random seismicity fluctuations.

Typically, the statistical significance of seismicity observations such as those presented in this study can be demonstrated using a Monte Carlo approach whereby the results are compared to results generated with randomized catalogs. A formal Monte Carlo approach is difficult in this case, since it would clearly not be a fair test to compare the results with results from a randomized catalog. If systematic artifacts arise in $\beta(r)$, they will almost certainly be generated by the naturally clustered character of seismicity.

To explore the statistical significance of the results, I conducted the following experiment: First, I calculated beta-statistic maps for several two-month periods that included no conspicuous main shocks or swarms. I then calculated $\beta(r)$ for a series of randomly chosen test epicenters to see how often molehill signals arose. The calculations were done for suites of eight random epicenters. This number is arbitrary, and it was chosen to reflect the numbers of earthquakes analyzed in this study.

Figure 12A presents a beta-statistic map for a two-month period when seismicity did not fluctuate significantly from one month to the next. After calculating $\beta(r)$ curves for a small number of random epicenters, it became clear that there was no significant signal in $\beta(r)$. By the time the curves from the eight events were averaged, the resulting curve was nearly flat (Fig. 13).

Figure 12B presents a beta-statistic map for a two-month period in which seismicity increased modestly in some areas during the second month. Such a signal can result from either a swarm or a modest burst of events after a particularly quiet month. In this case, $\beta(r)$ can reveal peaks with amplitudes similar to those shown in Figures 3 and 7. The question is then, how likely are these signals to survive averaging over a number of randomly chosen epicenters? Again, I used eight random epicenters in each trial. Figure 14 shows the results for eight different sets of random epicenters using the map shown in Figure 12B; in each panel, the individual and average $\beta(r)$ curves are shown.

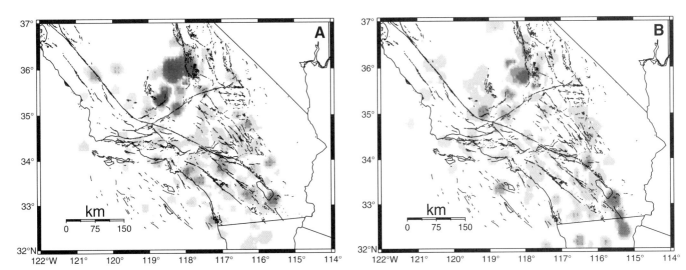

Figure 12. Beta statistic map generated using two different one-month periods during which no notable activity occurred in Southern California: March–April 2001 (A), and May–June 2001 (B). Same color scale as shown in Figure 2.

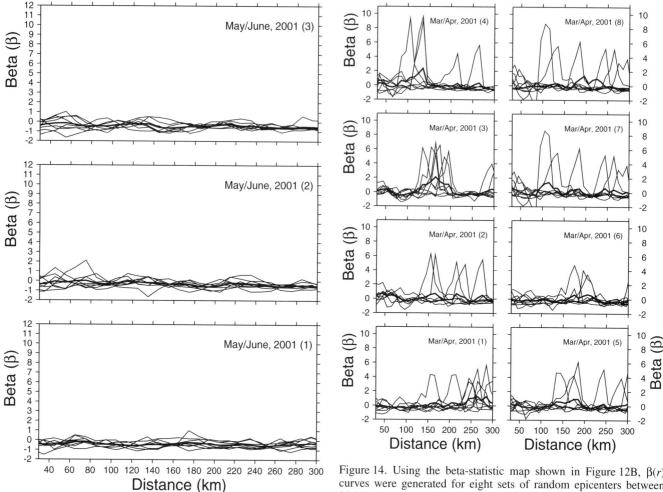

Figure 13. Using the beta-statistic map shown in Figure 12A, $\beta(r)$ curves were generated for three sets of random epicenters between 33.0°N and 35.5°N and 116°W and 119°W. Solid line in each panel indicates average of eight individual $\beta(r)$ curves.

Figure 14. Using the beta-statistic map shown in Figure 12B, $\beta(r)$ curves were generated for eight sets of random epicenters between 33.0°N and 35.5°N and 116°W and 119°W. Solid line in each panel indicates average of eight individual $\beta(r)$ curves. Number in each panel, 1–8, refers to test number.

Figure 14 reveals that some systematic trends in $\beta(r)$ can arise by random chance because of random seismicity fluctuations. In test three, for example, the random epicenters happen to cluster near the center of the region, at a similar distance to the most prominent seismicity increase. However, while a hint of a persistent peak can be seen in some of the averaged curves, it occurs at a range of distances for the different simulations. The individual $\beta(r)$ curves also reveal substantial variability within any one simulation, with equally strong peaks occurring at quite different distances. These suites of curves are qualitatively different from those shown in Figures 3 and 7. The $\beta(r)$ curves in Figure 7 tend to reveal a molehill in the same distance range, or no molehill signal at all. The statistical test assumes that it is a reasonable proxy to use one beta-statistic map with multiple test epicenters instead of the same test epicenter with multiple maps. While this might be open to question, conceptually it seems clear that, if molehill signals following main shocks were simply an

artifact caused by random seismicity fluctuations, those fluctuations would occur at random distances from the main shock, and the statistics would be comparable to those of the test. In fact, while this was not rigorously tested, Figures 3 and 7 provide evidence that strong, random seismicity fluctuations do not occur commonly. That is, whether or not SmS triggering occurs, if fluctuations as strong as those in Figure 13B were common, one would see $\beta(r)$ peaks at a range of distances for any given two-month period, whether or not it was centered at the time of a significant main shock. I thus conclude that, while $\beta(r)$ peaks comparable to the inferred molehills can result as artifacts due to the naturally clustered nature of seismicity, such artifacts are highly unlikely to persist in a certain, narrow distance range.

To test the significance of the inferred seismicity increase at distances over 200 km, I employed a bootstrap approach, calculating average $\beta(r)$ for random subsets of 10 events; i.e., subsets of 10 curves shown in Figure 7. The results confirm the observation that $\beta(r)$ is positive on average out to at least 200 km. Plotting the results on logarithmic axes diminishes the appearance of the molehill signal at 70–120 km but reveals another intriguing result: the suggestion of a slope break at ~50 km (Fig. 15A). At 0–20 km, the shape of the curve will reflect finite-fault effects, as a simple epicentral distance is used. However, the slope from 20 to 50 km is systematically steeper than that at greater distances. These results appear to suggest a transition from an aftershock regime (0–50 km) to a regime in which seismicity increases are caused by triggered earthquakes. Since static stress decays as $1/r^3$, whereas the decay of dynamic stresses is closer to $1/r$, the results are consistent with the conventional interpretation that aftershocks are controlled by static stress changes, whereas remotely triggered earthquakes are controlled by dynamic stress changes. Grouping the events into 0.5-unit-magnitude bins, I found that, as expected, the distance at which the transition occurs scales with the size of the main shock (Fig. 15B).

TRIGGERED EARTHQUAKES AND SmS ARRIVALS

The fundamental result illustrated in Figure 7 is that seismicity increases to a distance of at least 200 km following moderate main shocks in California, a range that is significantly beyond a conventional aftershock zone. Thus, while seismologists have previously regarded earthquakes such as the 8 July 1986 North Palm Springs and 13 July 1986 Oceanside earthquakes as unrelated, the results presented in this study suggest otherwise. The inference of SmS triggering is therefore not surprising: if, as seems nearly certain, the probability of triggering depends on the amplitude of dynamic waves, then anything that increases wave amplitudes will increase the probability of triggering.

Several previous studies have quantified SmS amplitudes in California. Somerville and Yoshimura (1990) showed that post-critical Moho reflections, or SmS arrivals, contributed to damage in the San Francisco Bay area during the 1989 Loma Prieta, California, earthquake. Somerville and Yoshimura showed that SmS arrivals were larger than the direct S arrivals at distances of 50–100 km; later studies (e.g., Mori and Helmberger, 1996) examined recorded waveforms for the 1992 Landers earthquake and found similar results in Southern California. The distance range at which SmS waves appear depends, of course, on Moho depth. In Southern California, SmS arrivals first appear at a distance of ~70 km and can be larger than the direct S wave at distances of 70–170 km (Mori and Helmberger, 1996). Although not always larger than the direct S wave, SmS arrivals are typically of high enough amplitude to increase shaking and damage during large earthquakes (Somerville and Yoshimura, 1990; Hough et al., 2004). (At distances of 70–170 km, a distinct surface-wave group has not generally formed.)

If SmS arrivals do cause triggered earthquakes, one would expect the triggered events to occur at larger epicentral distances in regions where the Moho is deeper. The results presented in this paper are generally consistent with this hypothesis: the triggered earthquakes in eastern North America and India are at somewhat greater distances from their main shocks than the triggered earthquakes in California. However, because SmS arrivals will be of high amplitude over a range of distances, one would expect the signature of Moho depth to be smeared out.

The observations presented and summarized here provide evidence for a correspondence between the locations of remotely triggered earthquakes and the distances at which SmS waves generate large-amplitude arrivals. This correspondence has important implications. First and most fundamentally, it provides additional evidence for the earlier conclusion that triggered earth-

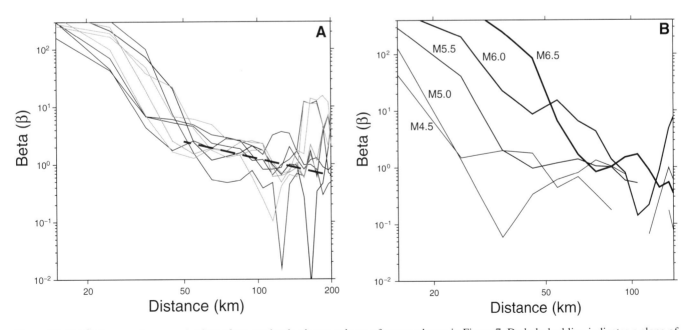

Figure 15. (A) $\beta(r)$ curves for several suites of ten randomly chosen subsets of curves shown in Figure 7. Dark dashed line indicates a slope of −1 on logarithmic axes. (B) $\beta(r)$ curves for California events grouped into magnitude bins, as indicated.

quakes are caused by the dynamic stress changes associated with seismic waves (e.g., Gomberg and Davis, 1996; Kilb et al., 2000; Gomberg et al., 2001), even at relatively short distances, where static stress change is not necessarily negligible.

Further, SmS waves are body waves of relatively high frequency compared to surface waves. This suggests that triggering does not (or does not always) require long-period (>10–15 s) energy, as is apparently the case in Long Valley (Brodsky and Prejean, 2004). However, it is not surprising that the nature of the triggering mechanism might be different in volcanic and nonvolcanic regions.

DISCUSSION

Observational investigations of remotely triggered earthquakes have been limited to a handful of case studies of triggering following recent large earthquakes (e.g., Hill et al., 1993; Gomberg and Davis, 1996; Kilb et al., 2000; Prejean et al., 2005). Investigations of triggering outside of volcanic or geothermal regions are more limited still, as are investigations of the source properties of remotely triggered events.

Before addressing the interpretation of remotely triggered earthquakes, it is useful to consider the implications of remotely triggered earthquake results in intraplate regions. Results to-date suggest that remotely triggered earthquakes outside of geothermal or volcanic regions occur on weak faults. Hough and Kanamori (2002) presented an analysis of remotely triggered earthquakes in the Brawley seismic zone following the 1999 Mw7.1 Hector Mine earthquake. The Brawley seismic zone is generally interpreted as an extensional transform zone in which stress is transferred between the San Andreas and Imperial faults via a zone of oblique extension (e.g., Larsen and Reilinger, 1991). Local extensional forces are associated with geothermal activity: several commercial geothermal power plants are in operation near the southern end of the Salton Sea.

Using an empirical Green's function approach to isolate source properties, Hough and Kanamori (2002) concluded that the 1999 remotely triggered earthquakes had source spectra consistent with expectations for tectonic, brittle-shear-failure earthquakes, with relatively low stress drop values of 0.1–1.0 MPa. That is, although the Brawley seismic zone is an active geothermal area, the radiated spectra of triggered earthquakes in this area do not reveal any evidence of a fluid-controlled source process. These results are not definitive: one could imagine, for example, a fluid-controlled source process that changes pore pressures within a fault zone in such a way that tectonic brittle-shear-failure earthquakes are encouraged. However, high-resolution empirical Green's function analysis can reveal evidence of anomalous source spectra of even very small earthquakes (Hough et al., 2000), and no such evidence can be seen in the triggered Brawley seismic zone events. The results are thus consistent with the simple interpretation that the earthquakes occurred on weak faults. The estimated stress drop values were generally lower than the estimated peak dynamic stress caused by the S-wave–surface-

wave group. If the triggered earthquakes were total stress drop events, the peak dynamic stress exceeded the failure stress.

The inference of triggering on weak faults derives an additional measure of support from a related negative result: even relatively large, relatively close earthquakes do not appear to cause triggered earthquakes on faults such as the San Andreas (Spudich et al., 1995), although triggered slip at shallow depths is fairly common elsewhere along the San Andreas (e.g., Bodin et al., 1994; Rymer, 2000). One might argue that the San Andreas fault is itself weak; however, an interesting new elastodynamic model by Lapusta and Rice (2004) proposes that that the fault is instead brittle (statically strong but dynamically weak.) I suggest that this model provides a cohesive explanation for recent observational results concerned both triggered slip and triggered earthquakes. Triggered slip occurs on such a fault within the shallow, presumably velocity-strengthened regime, as suggested by recent modeling results (Du et al., 2003). Triggered earthquakes do not generally occur on such faults because, as demonstrated by Lapusta and Rice (2004), over most of its extent, a brittle fault will be nowhere near a failure threshold. In the model of Lapusta and Rice (2004), large earthquakes originate in the "defect regions" along a fault where, by virtue of elevated pore pressure or other material properties, the fault is especially weak. I suggest that remotely triggered earthquakes will occur in these same defect regions. By virtue of its high heat flow and extensional tectonic setting, as well as its low stress-drop events, the Brawley seismic zone is an obvious candidate for a defect region abutting the southern terminus of the San Andreas fault.

So how, then, does one explain remotely triggered earthquakes in mid-plate and collisional settings? Two possibilities exist: these events also occur at the defect regions along otherwise strong faults, and/or these events occur where faults are close to failure. As I will discuss shortly, these possibilities are not necessarily mutually exclusive. The latter interpretation was explored by Seeber (2000) and Hough et al. (2004), who showed that in a low-strain-rate environment, a small amount of permanent, aseismic deformation can keep faults close to their failure level for a longer part of the earthquake cycle than faults in high-strain-rate regions.

In the model proposed by Seeber (2000), permanent deformation is assumed to be accommodated by a mechanism such as power-law creep, the key characteristic of which is that the aseismic strain rate depends on stress. Although the real physical processes are likely to be more complex than simple power-law creep, this dependence will, regardless of the details of the mechanism, slow the accumulation of stress available to drive earthquakes. The mechanism thus provides a conceptually simple explanation for the suggestion that intraplate crust is critically stressed (e.g., Townend and Zoback, 2000), which in turn provides a straightforward conceptual explanation for remotely triggered earthquakes.

However, the possibility remains that remotely triggered earthquakes occur where faults are relatively weak. In the most compelling case of triggering in eastern Canada, the events are clustered along the St. Lawrence Seaway, a reactivated failed

Iapetan rift structure that is likely to represent a zone of relative weakness (Roy et al., 1993) in a region where faults are expected to be otherwise strong. Other (inferred) intraplate triggered earthquakes discussed in this paper are located within major river valleys; again, probable zones of (relative) weakness.

Recent results from Gangopadhyay et al. (2004) suggest that in intraplate crust, stress concentrations will develop at pre-existing zones of weakness. Using two-dimensional modeling of discrete crustal blocks, they showed how localized stress concentrations can build around zones with pre-existing intersecting faults, such as the New Madrid seismic zone. If this model is correct, remotely triggered intraplate earthquakes, like their interplate counterparts, would be expected to occur in zones of relative weakness. Unlike the situation in interplate regions, such intraplate zones would also be characterized by long-lived stress concentrations and, therefore, persistent seismicity. Remotely triggered earthquakes may thus serve as beacons of stress concentration in intraplate regions, places where future large earthquakes are possible.

In any region, remotely triggered earthquakes can provide clues into earthquake rupture processes. For example, detailed analysis of source properties could reveal whether or not intraplate triggered earthquakes, like the interplate triggered earthquakes analyzed by Hough and Kanamori (2002), are low-stress-drop events, and whether they are characterized by the more common brittle shear-failure mechanism. Further investigations of remotely triggered earthquakes in intraplate regions will also shed further light on the question of where such events do (and do not) occur.

The results presented in this paper suggest that it is not necessary to wait for rare large intraplate earthquakes to further investigate the properties of intraplate triggered earthquakes. If remotely triggered earthquakes occur more commonly than can be identified with a standard beta-statistic analysis, SmS triggering in particular provides a unique opportunity to stack signals from multiple events—perhaps as small as M5.5—and further explore the prevalence and source properties of remotely triggered earthquakes.

ACKNOWLEDGMENTS

I thank Stephanie Prejean, Jeanne Hardebeck, Karen Felzer, Ned Field, Allan Lindh, and Ross Stein for constructive feedback on the work presented in this paper.

REFERENCES CITED

Ambraseys, N., and Bilham, R., 2000, A note on the Kangra Ms = 7.8 earthquake of 4 April 1905: Current Science, v. 79, p. 101–106.

Ambraseys, N.N., and Douglas, J., 2004, Magnitude calibration of north Indian earthquakes: Geophysical Journal of the Interior, v. 159, p. 165–206, doi: 10.1111/j.1365-246X.2004.02323.x.

Bilham, R., 2001, Slow tilt reversal of the Lesser Himalaya between 1862 and 1992 at 78°E, and bounds to the southeast rupture of the 1905 Kangra earthquake: Geophysical Journal of the Interior, v. 144, p. 1–23, doi: 10.1046/j.0956-540X.2000.01255.x.

Bodin, P., and Gomberg, J., 1994, Triggered seismicity and deformation between the Landers, California, and Little-Skull-Mountain, Nevada, earthquakes: Bulletin of the Seismological Society of America, v. 84, p. 835–843.

Bodin, P., Bilham, R., Behr, J., Gomberg, J., and Hudnut, K.W., 1994, Slip triggered on Southern California faults by the 1992 Joshua Tree, Landers, and Big Bear earthquakes: Bulletin of the Seismological Society of America, v. 84, p. 806–816.

Brodsky, E.E., and Prejean, S.G., 2005, New constraints on mechanisms of remotely triggered seismicity at Long Valley Caldera: Journal of Geophysical Research, v. 110(B4), Article B04302, April.

Brodsky, E.E., Sturtevant, B., and Kanamori, H., 1998, Earthquakes, volcanos, and rectified diffusion: Journal of Geophysical Research, v. 103, p. 23,827–23,838, doi: 10.1029/98JB02130.

Das, S., and Scholz, C.H., 1981, Off-fault aftershock clusters caused by shear-stress increase: Bulletin of the Seismological Society of America, v. 71, p. 1669–1675.

Du, W.X., Sykes, L.R., Shaw, B.E., and Scholz, C.H., 2003, Triggered aseismic slip from nearby earthquakes, static or dynamic effect?: Journal of Geophysical Research, v. 108(B2): Article 2131, February.

Duberger, R., Roy, D.W., Lamontagne, M., Woussen, G., North, R.G., and Wetmiller, R.J., 1991, The Saguenay (Quebec) earthquake of November 25, 1988—Seismological data and geologic setting: Tectonophysics, v. 186, p. 59–74, doi: 10.1016/0040-1951(91)90385-6.

Duda, S.J., 1992, Global earthquakes 1903–1985: U.S. Geological Survey Open-File Report 92-360, 623 p.

Felzer, K.R., Abercrombie, R.E., and Brodsky, E.E., 2005, Testing the stress shadow hypothesis: Journal of Geophysical Research, v. 110, no. B05, S09.

Gangopadhyay, A., Dickerson, J., and Talwani, P., 2004, A two-dimensional numerical model for current seismicity in the New Madrid seismic zone: Seismological Research Letters, v. 75, p. 406–418.

Glowacka, E., Nava, F.A., de Cossio, G.D., Wong, V., and Farfan, F., 2002, Fault slip, seismicity, and deformation in Mexicali Valley, Baja California, Mexico, after the M7.1 1999 Hector Mine earthquake: Bulletin of the Seismological Society of America, v. 92, p. 1290–1299.

Gomberg, J., 2001, The failure of earthquake failure models: Journal of Geophysical Research, v. 106, p. 16,253–16,263, doi: 10.1029/2000JB000003.

Gomberg, J., and Davis, S., 1996, Strain changes and triggered seismicity following the M(w)7.3 Landers, California, earthquake: Journal of Geophysical Research, v. 101, p. 751–764, doi: 10.1029/95JB03251.

Gomberg, J., Reasenberg, P.A., Bodin, P., and Harris, R.A., 2001, Earthquake triggering by seismic waves following the Landers and Hector Mine earthquakes: Nature, v. 411, p. 462–466, doi: 10.1038/35078053.

Gomberg, J., Bodin, P., Larson, K., and Dragert, H., 2004, Earthquake nucleation by transient deformations caused by the M = 7.9 Denali, Alaska, earthquake: Nature, v. 427, p. 621–624, doi: 10.1038/nature02335.

Hill, D.P., Reasenberg, P.A., Michael, A., Arabaz, W.J., Beroza, G., Brunmbaugh, D., Brune, J.N., Castro, R., Davis, S., DePolo, D., Ellsworth, W.L., Gomberg, J., Harmsen, S., House, L., Jackson, S.M., Johnston, M.J.S., Jones, L., Keller, R., Malone, S., Munguia, L., Nava, S., Pechmann, J.-C., Sanford, A., Simpson, R.W., Smith, R.B., Stark, M., Stickney, M., Vidal, A., Walter, S., Wong, V., and Zollweg, J., 1993, Seismicity remotely triggered by the magnitude 7.3 Landers, California, earthquake: Science, v. 260, p. 1617–1623, doi: 10.1126/science.260.5114.1617.

Hough, S.E., 2001, Triggered earthquakes and the 1811–1812 New Madrid, central U.S. earthquake sequence: Bulletin of the Seismological Society of America, v. 91, p. 1574–1581, doi: 10.1785/0120000259.

Hough, S.E., and Kanamori, H., 2002, Source properties of earthquakes near the Salton Sea triggered by the 16 October 1999 M7.1 Hector Mine earthquake: Bulletin of the Seismological Society of America, v. 92, p. 1281–1289, doi: 10.1785/0120000910.

Hough, S.E., Dollar, R., and Johnson, P., 2000, The 1998 earthquake sequence south of Long Valley Caldera, California: Hints of magmatic involvement: Bulletin of the Seismological Society of America, v. 90, p. 752–763, doi: 10.1785/0119990109.

Hough, S.E., Seeber, L., and Armbruster, J.G., 2003, Intraplate triggered earthquakes: Observations and interpretation: Bulletin of the Seismological Society of America, v. 93, p. 2212–2221.

Johnston, A.C., 1996, Seismic moment assessment of earthquakes in stable continental regions. III: New Madrid 1811–1812, Charleston 1886, and Lisbon 1755: Geophysical Journal of the Interior, v. 126, p. 314–344.

Johnston, A.C., and Schweig, E.S., 1996, The enigma of the New Madrid earthquakes of 1811–1812: Annual Reviews of Earth and Planetary Science Letters, v. 24, p. 339–384, doi: 10.1146/annurev.earth.24.1.339.

Kilb, D., Gomberg, J., and Bodin, P., 2000, Triggering of earthquake aftershocks by dynamic stresses: Nature, v. 408, p. 570–574, doi: 10.1038/35046046.

King, G.C.P., Stein, R.S., and Lin, J., 1994, Static stress changes and the triggering of earthquakes: Bulletin of the Seismological Society of America, v. 84(3), p. 935–953.

Larsen, S., and Reilinger, R., 1991, Age constraints for the present fault configuration of the Imperial Valley, California—Evidence for northwestward propagation of the Gulf of California rift system: Journal of Geophysical Research, v. 96, p. 10,339–10,346.

Lapusta, N., and Rice, J.R., 2003, Nucleation and early seismic propagation of small and large events in a crustal earthquake model: Journal of Geophysical Research, v. 108(B4), 2205, doi: 10.1029/2001JB000793.

Linde, A.T., Sacks, I.S., Johnston, M.J.S., Hill, D.P., and Bilham, R.G., 1994, Increased pressure from rising bubbles as a mechanism for remotely triggered seismicity: Nature, v. 371, p. 408–410, doi: 10.1038/371408a0.

Matthews, M.V., and Reasenberg, P.A., 1988, Statistical methods for investigating quiescence and other temporal seismicity patterns: Pure and Applied Geophysics, v. 126, p. 357–372, doi: 10.1007/BF00879003.

Mauk, F., Christensen, D., and Henry, S., 1982, The Sharpsburg, Kentucky, earthquake 27 July 1980: Main shock parameters and isoseismal maps: Bulletin of the Seismological Society of America, v. 72, p. 221–236.

Middlemiss, C.S., 1905, Preliminary account of the Kangra earthquake of 4th April: Memoir of the Geological Society of India 32, part 4: Calcutta, Geological Survey of India, p. 258–294.

Molnar, P., 1987, The distribution of intensity associated with the 1905 Kangra earthquake and bounds on the extent of rupture: Journal of the Geological Society of India, v. 29, p. 221.

Mori, J., and Helmberger, D., 1996, Large-amplitude Moho reflections (SmS) from Landers aftershocks, Southern California: Bulletin of the Seismological Society of America, v. 86, p. 1845–1852.

Mueller, K., Hough, S.E., and Bilham, R., 2004, Analysing the 1811–1812 New Madrid earthquakes with recent instrumentally recorded aftershocks: Nature, v. 429, p. 284–288, doi: 10.1038/nature02557.

Power, J.A., Moran, S.C., McNutt, S.R., Stihler, S.D., and Sanchez, J.J., 2001, Seismic response of the Katmai volcanos to the 6 December 1999 magnitude 7.0 Karluk Lake earthquake, Alaska: Bulletin of the Seismological Society of America, v. 91, p. 57–63, doi: 10.1785/0120000054.

Prejean, S.G., Hill, D.P., Brodsky, E.E., Hough, S.E., Johnston, M.J.S., Malone, S.D., Oppenheimer, D.H., Pitt, A.M., and Richards-Dinger, K.B., 2005, Remotely triggered seismicity on the United States West Coast following the M7.9 Denali fault earthquake: Bulletin of the Seismological Society of America, v. 94, p. S348–S359, doi: 10.1785/0120040610.

Reasenberg, P.A., and Simpson, R.W., 1992, Response of regional seismicity to the static stress change produced by the Loma-Prieta earthquake: Science, v. 255, p. 1687–1690, doi: 10.1126/science.255.5052.1687.

Richter, C.F., 1955, Unpublished notes, Box 7.8, Papers of Charles F. Richter, 1839–1984: Pasadena, California Institute of Technology Archives.

Roy, D.W., Schmitt, L., Woussen, G., and Duberger, R., 1993, Lineaments from airborne SAR images and the 1988 Saguenay earthquake, Quebec, Canada: Photogrammetric Engineering and Remote Sensing, v. 59(8), p. 1299–1305.

Rymer, M.J., 2000, Triggered surface slips in the Coachella Valley area associated with the 1992 Joshua Tree and Landers, California, earthquakes: Bulletin of the Seismological Society of America, v. 90, p. 832–848, doi: 10.1785/0119980130.

Scholz, C.H., 2003, Earthquakes—Good tidings: Nature, v. 425, p. 670–671, doi: 10.1038/425670a.

Seeber, L., 2002, Triggered earthquakes and hazard in stable continental regions: Vicksburg, Mississippi, Report to U.S. Army Corps of Engineers, Waterways Experiment Station, 40 p.

Seeber, L., and Armbruster, J.G., 1987, The 1886–1889 aftershocks of the Charleston, South Carolina, earthquake—A widespread burst of seismicity: Journal of Geophysical Research, v. 92, p. 2663–2696.

Somerville, P., and Yoshimura, J., 1990, The influence of critical Moho reflections on strong ground motions recorded in San Francisco and Oakland during the 1989 Loma Prieta earthquake: Geophysical Research Letters, v. 17, p. 1203–1206.

Spudich, P., Steck, L.K., Hellweg, M., Fletcher, J.B., and Baker, L.M., 1995, Transient stresses at Parkfield, California, produced by the M-7.4 Landers earthquake of June 28, 1992—Observations from the UPSAR dense seismogram array: Journal of Geophysical Research, v. 100, p. 675–690, doi: 10.1029/94JB02477.

Stark, M.A., and Davis, S.D., 1996, Remotely triggered microearthquakes at The Geysers geothermal field, California: Geophysical Research Letters, v. 23, p. 945–948, doi: 10.1029/96GL00011.

Sturtevant, B., Kanamori, H., and Brodsky, E.E., 1996, Seismic triggering by rectified diffusion in geothermal systems: Journal of Geophysical Research, v. 101, p. 25,269–25,282, doi: 10.1029/96JB02654.

Toda, S., and Stein, R.S., 2003, Toggling of seismicity by the 1997 Kagoshima earthquake couplet: A demonstration of time-dependent stress transfer: Journal of Geophysical Research, v. 108(B12): Article 2567, Dec.

Townend, J., and Zoback, M., 2000, How faulting keeps the crust strong: Geology, v. 28, p. 399–402.

Wallace, K., Bilham, R., Blume, F., Gaur, V.K., and Gahalaut, V., 2005, Surface deformation in the region of the 1905 Kangra Mw = 7.8 earthquake in the period 1846–2001: Geophysical Research Letters, v. 32 (15), Article L15307, August.

MANUSCRIPT ACCEPTED BY THE SOCIETY 29 NOVEMBER 2006

The Geological Society of America
Special Paper 425
2007

Two-dimensional numerical modeling suggests preferred geometry of intersecting seismogenic faults

Abhijit Gangopadhyay[†]
Pradeep Talwani
Department of Geological Sciences, University of South Carolina, Columbia, South Carolina 29208, USA

ABSTRACT

We undertook a parametric study, using a two-dimensional distinct element method, to investigate if there is a preferred geometry of intersecting faults that may favor the occurrence of intraplate earthquakes. This model subjects two and three vertical, intersecting faults within a block to a horizontal force across them, representing the maximum horizontal compression (S_{Hmax}). The main fault is oriented at an angle α with respect to S_{Hmax}, and β is the interior angle between the main fault and the intersecting fault. The third fault is oriented parallel to the main fault and is half its length. The distribution of shear stresses is examined along the faults for different values of α and β, and varying lengths of the main and intersecting faults. In all cases, maximum shear stresses are generated at the fault intersections. The modeling results reveal that the magnitudes of the shear stresses depend on the values of α and β, with an optimum range for α between 30° and 60°. In the case where the sign of the shear stress on the intersecting fault is opposite that on the main fault, the largest stresses at the fault intersections are obtained when β is between 65° and 125°. When the stresses on these two faults are of the same sign, the largest stress values at the intersections are obtained when $145° \leq \beta \leq 170°$. The results of the modeling are consistent with the observed geometry of faults in the New Madrid and Middleton Place Summerville seismic zones.

Keywords: intersecting faults, numerical modeling, fault geometry, New Madrid, Charleston.

INTRODUCTION

Intersecting faults have been postulated to play an important role in earthquake mechanics because they interact dynamically and are sometimes responsible for controlling nucleation, dimension, propagation, and termination of earthquake ruptures (e.g.,

Sharp et al., 1982; Harris and Day, 1993; Rousseau and Rosakis, 2003; Spotila and Anderson, 2004). Hence, mechanics of intersecting faults and fault junctions have been extensively studied for decades (e.g., King and Nabelek, 1985; Pollard and Segall, 1987; McCaig, 1988; Andrews, 1989, 1994; Harris and Day, 1993; Robinson and Benites, 1995; Maerten, 2000; Kato, 2001; Fitzenz and Miller, 2001; Crider and Peacock, 2004). However, most of these studies have been directed toward understanding the role of intersecting faults in plate-boundary earthquakes. A spatial association of intersecting faults with continental intraplate earth-

[†]Present address: Institute for Geophysics, University of Texas at Austin, J.J. Pickle Research Campus, Building 196, 10100 Burnet Road, Austin, Texas 78758, USA.

Gangopadhyay, A., and Talwani, P., 2007, Two-dimensional numerical modeling suggests preferred geometry of intersecting seismogenic faults, *in* Stein, S., and Mazzotti, S., ed., Continental Intraplate Earthquakes: Science, Hazard, and Policy Issues: Geological Society of America Special Paper 425, p. 87–99, doi: 10.1130/2007.2425(07).

quakes has also been observed (e.g., Talwani et al., 1979; Illies, 1982; King and Nabelek, 1985; King, 1986; Talwani, 1988), and a causal association has been proposed (Talwani, 1988; Talwani and Rajendran, 1991; Gangopadhyay and Talwani, 2003). Schematic and two-dimensional, linear-elastic and isotropic models that explain the cause of intraplate earthquakes in compressional stress regimes demonstrate that intersecting faults can act as stress concentrators and are spatially correlatable with the locations of observed intraplate seismicity (Talwani, 1988; Jing and Stephansson, 1990; Gangopadhyay et al., 2004; Gangopadhyay and Talwani, 2005). However, geologic data show that intersecting faults are abundant in nature (McCaig, 1988), but only some of them are the locations of intraplate seismicity. This leads us to wonder if there is a preferred geometry for intersecting faults that favors the occurrence of continental intraplate earthquakes. Previous research involving field observations and analytical computations has shown that there is a range of reactivation angles for pre-existing "Andersonian" thrust, normal, and strike-slip faults using the commonly used Coulomb failure criteria and Byerlee-type values of static friction (Sibson, 1985, 1990, 1991). However, the effect of a compressional stress field on sets of two or more intersecting faults in intraplate regions has not yet been considered. Because intersecting faults are abundant, we address this question with a parametric study using two-dimensional (2-D) numerical models. In this paper, we investigate if there are configurations of faults that will optimally respond to plate tectonic stresses, concentrate stresses locally, and thus potentially play a role in the initiation of seismicity within plates. We are not attempting to explain the genesis of faulting over an earthquake cycle or duplicate real situations. We do not address the effect of fluid pressures and variations in frictional properties, which can influence the likelihood of seismicity. This paper is mainly aimed at studying if there are preferred orientations of two or three intersecting faults, with respect to S_{Hmax} and each other, that maximize local concentration of stresses, and if so, determining their optimal geometrical configurations.

NUMERICAL MODELING METHODOLOGY

Two-dimensional distinct element modeling was performed using a program called Universal Distinct Element Code (UDEC) written by Itasca Consulting Group, Inc., Minneapolis, Minnesota (version 3.1, 1999). This 2-D numerical program was first developed by Cundall (1971). The advantage of the Distinct element method over typical continuum-based methods is its ability to change and update joint patterns continuously during the computation process (Jing and Hudson, 2002; Gangopadhyay et al., 2004). The program simulates the response of the discontinuous media to either static or dynamic loading. It models the rock mass as an assemblage of rigid or deformable discrete blocks and the faults as discontinuities. The equations of motion for the blocks are solved by a central difference scheme, and mutual interactions between blocks are included. It uses calculations in the Lagrangian scheme to model large movements and deformations of a system. Several

built-in material behavior models, for both the intact blocks and the faults, permit the simulation of real geologic situations. Displacements are allowed along the faults, which are treated as boundaries between blocks, and the blocks are allowed to move with respect to each other. The individual blocks can be made either rigid or deformable. The deformable blocks are divided into a mesh of triangular constant-strain finite difference zones, and each zone behaves according to a prescribed stress-strain law. In the case of elastic analysis, the formulation of these zones is identical to that of constant-strain finite elements. The relative motions along the discontinuities are constrained by force-displacement relations for movement in both the normal and shear directions. The suitability, efficiency, and adaptability of UDEC in solving two-dimensional, simplified geological problems involving faulted and fractured rocks has already been adequately demonstrated and established (see, e.g., Gangopadhyay et al., 2004, and references therein). For example, it has been used to model tectonic and geologic frameworks of active continental intraplate regions such as New Madrid (Gangopadhyay et al., 2004) and Charleston (Gangopadhyay and Talwani, 2005).

DEVELOPMENT OF THE 2-D MODEL

Model Geometry

Examination of 20 case histories of intraplate regions globally that have been host to 39 earthquakes of M 5.0 or greater revealed that 65% of them included two or three intersecting faults, which acted as stress concentrators and locators of intraplate earthquakes (Gangopadhyay and Talwani, 2003). Thus, in this parametric study, we modeled tectonic frameworks consisting of two and three intersecting faults. Figure 1 shows the block geometries for these two frameworks. They are strictly two-dimensional. The maximum horizontal compressive stress, S_{Hmax}, is oriented E-W in all cases. The block is compressed by a force applied along the x-axis. This force is along the inferred direction of plate motion, and its magnitude is proportional to the plate velocity discussed in detail in the next section. In this case, the main fault, AB, subtends an angle α with the direction of S_{Hmax} and an interior angle β with the intersecting fault, BC (Fig. 1). The blocks and faults are linear-elastic and associated with elastic properties that are detailed in a later section. In the first set of simulations with two intersecting faults, the length of the main fault (AB) was taken as 10 units and that of the intersecting fault (BC) was chosen to be 1, 3, and 5 units (Fig. 1A). A similar set of simulations with two intersecting faults was carried out with length of AB = 10 units, oriented at an angle $(180° - \alpha)$, and length of BC = 3 units (Fig. 1B). A third set of simulations was performed varying the length of AB (1, 3, and 5 units) and keeping BC fixed (10 units) (Fig. 1C). In the case of three intersecting faults (Fig. 1D), the length of the main fault was 10 units, a shorter fault CD was introduced with a length of 5 units, and it was oriented parallel to the main fault. The lengths of faults AB and CD and the orientation of CD

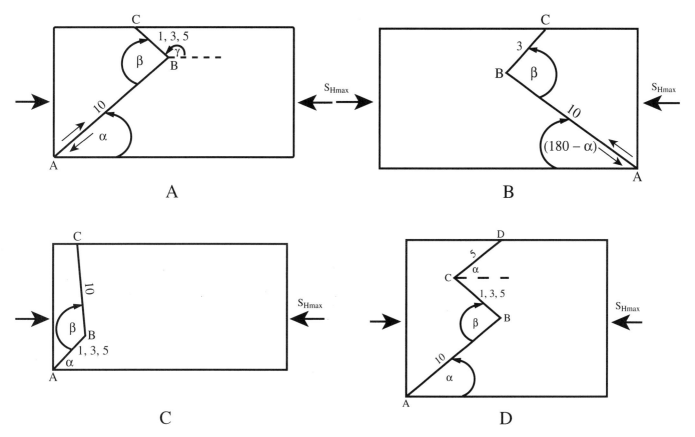

Figure 1. Two-dimensional (2-D) block geometry used to model a set of (A) two intersecting faults with AB = 10 units and BC = 1, 3, and 5 units; (B) two intersecting faults with AB = 10 units, oriented at an angle ($180° − \alpha$) to S_{Hmax}, and BC = 3 units; (C) two intersecting faults with AB = 1, 3, and 5 units and BC = 10 units; and (D) a set of three intersecting faults with AB = 10 units, BC = 1, 3, and 5 units, and CD = 5 units. In all the cases, the main fault AB is oriented at an angle α to S_{Hmax}, and β is the interior angle between AB and the intersecting fault BC. The angle γ is the orientation of BC with respect to S_{Hmax}. In D, the third fault CD is oriented parallel to the main fault AB. The direction of S_{Hmax} is along the *x*-axis. A velocity is applied across the block, decreasing from the right to the left.

were kept fixed. The intersecting fault (BC) connected the two faults AB and CD (Fig. 1D). Simulations of this model were performed with varying lengths of BC (1, 3, and 5 units). In all cases, the block corners were rounded with a circle that was tangential to the two corresponding edges at a specified rounding distance from the corner. In practice, the rounding distance is ~1% of the typical block edge length (UDEC Command Reference Manual, 1999), and the same was utilized in our model simulations. Since our model is two-dimensional, the commonly used plane stress condition is imposed, wherein none of the blocks experience stresses in the vertical direction although they can exhibit strain in that direction.

The computer code divides the deformable blocks into triangular finite difference zones using a built-in automatic mesh generator that decides the size of the zones based on the block lengths, specified rounding length, and the memory availability to perform the computations. All the blocks in our model were deformable and movable with respect to each other. UDEC calculates the amplitude and sign of the shear stress (τ_{xy}) at each node. Shear stress (τ_{xy}) is positive when it tends to rotate the block in

a counterclockwise manner, i.e., by left-lateral strike slip, and it is negative when the block rotation is clockwise. The sign of the shear stress can be used to infer how the block will rotate, and contours show its spatial variation.

Model Parameters

UDEC has seven built-in constitutive models for the blocks and four for the joints that can represent various geologic situations. The simplest constitutive models were utilized for this parametric study. In our model, the blocks conformed to the linearly elastic isotropic model and the faults followed the joint area contact elastic/plastic Coulomb slip failure model. The Linearly Elastic Isotropic Model for the blocks describes the simplest form of material behavior assuming homogeneous and isotropic materials that exhibit linear stress-strain behavior with reversible deformation upon unloading (UDEC Command Reference Manual, 1999). The Joint Area Contact Elastic/Plastic Coulomb Slip Failure Model for the joints is the most commonly used Coulomb slip model that predicts failure or initiation of slip on a fault based on the accumulated shear stress (UDEC Command Reference Manual, 1999). It is represented by the following equations:

$$\Delta\sigma_n = -k_n\Delta u_n \tag{1}$$

and

$$\Delta\tau_s = -k_s\Delta u_s, \tag{2}$$

where, $\Delta\sigma_n$, Δu_n, $\Delta\tau_s$, and Δu_s are the effective normal stress, normal displacement, shear stress, and shear displacement increments, respectively, and k_n and k_s are normal and shear stiffnesses.

The failure criterion for the joints is given by

$$|\tau_s| \geq C + \sigma_n \tan\phi, \tag{3}$$

where τ_s = shear stress, C = cohesive strength of the joint, σ_n = normal stress, and ϕ = friction angle for the joint.

The block assembly was subjected to a horizontal compressive force along the *x*-axis. This was achieved by subjecting the right block boundary to a prescribed displacement resulting from a plate velocity of 5 mm/yr. The left boundary was kept fixed. The plate velocity was determined from geodetic observations in two major active intraplate regions of eastern United States (see Gangopadhyay et al., 2004, for details). The calculated stresses scale linearly with velocity. The applied velocity gradient in our model was larger than those obtained from global positioning system (GPS) measured velocities and was chosen in order to obtain a measurable response with a shorter loading time used in the computer model. The velocity gradient was also assumed to not be a function of depth, and so the whole block was subjected to the same horizontal stress.

Model Properties

Input modeling parameters were based on an earlier study involving the New Madrid seismic zone, and these properties and their derivations have been described in detail in Gangopadhyay et al. (2004). The values are considered to be representative of active intraplate regions. In our computations, we assumed a value of 1.73, 5.63 km/s, and 2690 kg/m^3 for the V_P/V_S ratio, P-wave velocity, and density, respectively, for all the blocks in the model. Utilizing these values, we computed the bulk and shear moduli to be 47.28 GPa and 28.48 GPa, respectively, for the various blocks.

Other input parameters required for modeling the deformation included friction angle, normal and shear stiffnesses, and cohesion of the faults (treated as joints). The chosen values of these parameters were also from the New Madrid seismic zone study (Gangopadhyay et al., 2004): friction angles of 27°, and joint normal and shear stiffnesses of 101 GPa/m and 76 GPa/m, respectively, for all faults. The faults were considered to be cohesionless.

Limitations of the Model

A notable limitation of this model is the fact that it is two-dimensional. Because it is a two-dimensional code, all faults are considered to be vertical, and the effects in the third dimension cannot be observed. The code does not allow us to study isolated faults within a block but requires that faults be extended to block boundaries. However, the computational scheme in UDEC is such that when using a linear, elastic, and isotropic constitutive model, the effect of stress concentration at the boundaries has minimal effects at the fault intersections within the block. In spite of these limitations, by keeping the fault intersections away from the block boundaries, it is possible to study the effects of their geometrical configurations on stress concentration and their possible influence on earthquake generation. Due to memory limitations in our version of UDEC, it was not possible to run the model for a geologically realistic loading time. However, the model was run for different loading times, and the calculated stress was found to be linear with loading time; thus, running the model for shorter times provided relative stress distributions—the objective of this study. In this study, the model simulations were performed for 100,000 cycles (tectonic loading of 1 day).

MODEL SIMULATION RESULTS

Outputs from the model simulations included normal and shear stresses developed in the blocks and along the faults. The shear stresses in response to the applied tectonic loading along the individual fault planes were analyzed. The absolute values of the shear stresses depend on the model parameters and tectonic loading time, and those obtained in this study are representative; however, their relative values are more instructive. The results of the different simulations show that the shear stresses in plan view (Fig. 2) are largest at the intersections of the faults (i.e., at B for faults AB and BC) (Fig. 2A) and near C for fault CD (Fig. 2B). Shear stresses are positive (Fig. 2) when they tend to rotate the block in a counterclockwise manner, i.e., by left-lateral strike slip, and they are negative when the block rotation is clockwise. Shear stresses are also obtained along the faults, and the model simulation results discussed here show the variation of shear stresses at B along AB and BC, and at C along CD for a range of values of α, β, and the length of BC. The results are presented in two sections, first, the case of two intersecting faults, and second, the case of three intersecting faults. For purposes of clarity, the variation of the magnitude of shear stresses for different values of α for each set of simulations is shown in two diagrams.

Case 1: Two Intersecting Faults

The simulations with two intersecting faults were performed using three different block geometries shown in Figures 1A–1C. The main results of all the simulations are presented in Table 1. The first block geometry was composed of a main fault AB of length 10 units and an intersecting fault BC (Fig. 1A). The model

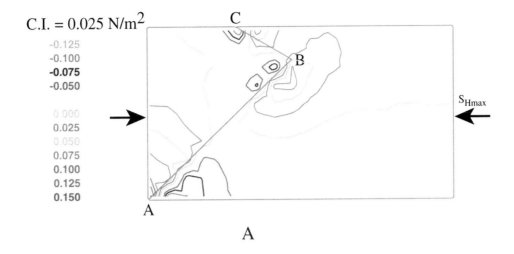

C.I. = 0.025 N/m^2

-0.125
-0.100
-0.075
-0.050

0.000
0.025
0.050
0.075
0.100
0.125
0.150

A

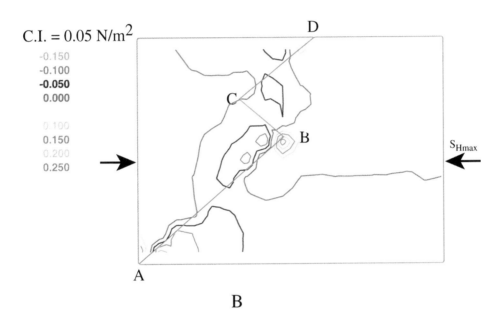

C.I. = 0.05 N/m^2

-0.150
-0.100
-0.050
0.000

0.100
0.150
0.200
0.250

B

Figure 2. Contours of shear stresses are shown in plan view superimposed on the two-dimensional block geometry used to model a set of two intersecting faults (A) and three intersecting faults (B). C.I.—Contour Interval.

was run with α ranging between 20° to 80° and β ranging between 20° to 160° in increments of 20° for each value of α, with an additional value at $\beta = 175°$. Three sets of simulations were carried out, varying the length of BC (1, 3, and 5 units). The shear stresses at the intersection B along the fault planes AB and BC were observed for each set and tabulated (Table 1). They are negative when the motion along the fault is right-lateral (α and γ are acute angles with respect to S_{Hmax}) and positive when the motion along the fault is left-lateral (α and γ are obtuse angles with respect to S_{Hmax}) (Table 1; Figs. 1A and 1B). The data presented in the table are illustrated using one example, where AB is 10 units and BC is 3 units long. Shear stresses were calculated at B along AB (Figs. 3A and 3B) and at B along BC (Figs. 3C and 3D) for vari-

ous values of α (different color curves) as a function of β. For clarity, the results are presented separately for $\alpha < 45°$ (Figs. 3A and 3C) and $\alpha \geq 45°$ (Figs. 3B and 3D). In all cases, along AB, the calculated stresses were negative, implying right-lateral shear. For the first case, the maximum value of shear stress (−1.25 N/m^2) was obtained for $\alpha = 50°$ and $\beta = 80°$ (Fig. 3B). We arbitrarily choose a "favorable range," where the stresses were ≥90% of this maximum value. For this case (assuming linear variation between plotted angles), the favorable ranges were estimated from the curves for both α and β and were found to be 46° to 53°, and 76° to 91°, respectively. These are shown in Table 1. The shear stress at B along BC was found to be both positive and negative, depending on the orientation of BC with

TABLE 1. SUMMARY OF PREFERRED ANGLES OF ORIENTATION (α AND β) FROM MODEL RESULTS

Geometry	Figure	AB (Length units)	BC (Length units)	Observations at B along AB					Observations at B along BC*				
				Maximum			Favorable range (90% max. value)		Maximum			Favorable range (90% max. value)	
				Shear stress (N/m²)	α (°)	β (°)	α (°)	β (°)	Shear stress (N/m²)	α (°)	β (°)	α (°)	β (°)
Two intersecting faults	1A	10	1	−1.03	60	80	45–60	75–100	1.1	45	120	35–45	65–125
									−1.55	40	160	35–40	152–164
			3	−1.25	50	80	46–53	76–91	1.25	45	100	30–45	75–105
									−1.72	40	160	37–41	153–170
			5	−1.58	50	80	48–52	75–89	1.45	30	80	30–37	68–85
									−1.87	40	160	36–42	149–162
Two intersecting faults	1B	10	3	1.32	130	80	127–134	75–90	−1.18	145	80	140–150	70–85
									1.64	140	160	137–143	152–170
Two intersecting faults	1C	1	10	−1.19	45	80	45–60	70–95	1.05	30	80	30–37	70–85
									−1.48	30	160	35–40	150–164
			3	−1.32	45	80	45–50	70–84	1.24	30	80	30–35	72–84
									−1.8	40	160	33–42	151–162
			5	−1.52	45	80	47–52	75–87	1.4	30	80	30–34	67–85
									−1.87	40	160	35–42	150–164

*Positive values for left-lateral region and negative values for right-lateral region.

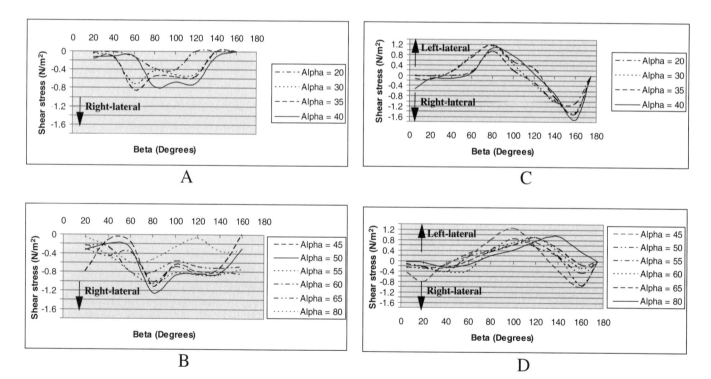

Figure 3. Plot of the magnitude of shear stress at B along fault plane AB (length = 10 units) for (A) $\alpha < 45°$ and (B) $\alpha \geq 45°$, and BC (length = 3 units) for (C) $\alpha < 45°$ and (D) $\alpha \geq 45°$ for a range of $\beta = 20–160°$. The largest shear stress in the right-lateral range occurs for $\alpha = 50°$ at $\beta = 80°$. The largest shear stresses in the left-lateral range occur for $\alpha = 45°$ at $\beta = 100°$.

respect to S_{Hmax} (the angle γ, Fig. 1A). A negative value (indicating right-lateral strike slip along BC) occurred when $\gamma < 90°$, when $\beta >> 90°$, and BC essentially trended in the same direction as AB (Fig. 1A). When $\gamma > 90°$, the shear stress on BC was positive, indicating left-lateral shear. Correspondingly, two peaks in shear stress values, one positive and the other negative, were obtained for various combinations of α and β (Figs. 3C and 3D). These two peaks yielded maximum shear stress values and favorable ranges of α and β for left-lateral (positive shear stress) and right-lateral (negative shear stress) movement. These are listed in Table 1. The maximum shear stress at B along BC associated with the right-lateral motion was found to be -1.72 N/m^2, and it occurred at $\alpha = 40°$ and $\beta = 160°$ (Fig. 3C). The peak shear stress at B along BC (in the left-lateral region, positive shear stress values) was 1.25 N/m^2, and it occurred at $\alpha = 45°$ and $\beta = 100°$ (Table 1; Fig. 3D). The favorable range was found to be $30°$ to $45°$ for α and $75°$ to $105°$ for β, whereas the favorable ranges for right-lateral motion were $37°$ to $41°$ and $153°$ to $170°$ for α and β, respectively (Table 1). The results for the same geometry (Fig. 1A) but for different lengths of BC (1 and 5 units) are presented in Table 1.

In these simulations, the main fault AB was set at an acute angle with S_{Hmax} (α ranged from $20°$ to $80°$). The next simulation was performed for $\alpha > 90°$ (Fig. 1B), with AB = 10 units and BC = 3 units long. The model was run with α ranging between $100°$ to $160°$ (or [$180° - \alpha$] from $80°$ to $20°$) and β ranging between $20°$ to $175°$ for each value of α. The shear stresses at the intersection B along the fault planes AB and BC were observed for each set, and the results are presented in Table 1. We note that the shear stresses at B along AB were positive for this geometry, indicating left-lateral motion along fault AB (Table 1). The largest shear stress at B along AB (1.32 N/m^2) occurred for $\alpha = 130°$ (supplementary to $50°$) at $\beta = 80°$ (Table 1). The estimated favorable ranges for α and β were $127°$–$134°$ and $75°$–$90°$, respectively. Along BC, the shear stresses at B were both positive and negative (depending if γ was $\geq 90°$ or $< 90°$), similar to the geometry in Figure 1A. As before, the positive values (corresponding to left-lateral shear) occurred when $\beta >> 90°$, and negative values occurred for $\gamma < 90°$ (Table 1). Along BC in the right-lateral range, the largest shear stress (-1.18 N/m^2) occurred for $\alpha = 145°$ (supplementary to $35°$) at $\beta = 80°$, and the corresponding favorable ranges were $140°$ to $150°$ for α and $70°$ to $85°$ for β (Table 1). When the motion along BC was left-lateral, the maximum positive shear stress (1.64 N/m^2) occurred for $\alpha = 140°$ at $\beta = 160°$, and the corresponding favorable ranges for α and β were $137°$–$143°$ and $152°$–$170°$, respectively (Table 1). Note that in this simulation, for comparison with the geometry in Figure 1A, we measured the angle α with respect to S_{Hmax} oriented in the negative *x*-direction. The optimum value of α was found to lie between $127°$ and $134°$. When comparing with other models (Figs. 1A, 1C, and 1D), we compared the supplementary angle, i.e., the acute angle $53°$ to $46°$ between the fault and the S_{Hmax} oriented in the positive *x*-direction.

The block geometry shown in Figure 1C was used in the third model run with two intersecting faults. In these runs, the length of fault AB was varied (1, 3, and 5 units), whereas the length of fault BC (10 units) was kept unchanged. The results of all the runs are summarized in Table 1. The maximum value of shear stress at B along AB (-1.32 N/m^2) was obtained for $\alpha = 45°$ and $\beta = 80°$ (Table 1). The favorable ranges were found to be $45°$ to $50°$, and $70°$ to $84°$ for α and β, respectively (Table 1). Similarly, two peaks were found for the shear stresses at B along BC. In the left-lateral range, it was found to be 1.24 N/m^2, and it occurred at $\alpha = 30°$ and $\beta = 80°$ (Table 1), and the corresponding favorable range was found to be $30°$ to $35°$ for α and $72°$ to $84°$ for β (Table 1). In the right-lateral range, the maximum shear stress value was -1.8 N/m^2, which occurred for $\alpha = 40°$ and $\beta = 160°$, and the corresponding favorable ranges were $33°$ to $42°$ and $151°$ to $162°$ for α and β, respectively (Table 1).

Case 2: Three Intersecting Faults

For the case of three intersecting faults (Fig. 1D), CD (5 units long) was oriented parallel to fault AB. In three model runs, the lengths of AB (10 units) and CD (5 units) were kept fixed, while that of BC was varied (1, 3, and 5 units). The angles α and β were varied in the same way as for the case with two intersecting faults, and the shear stresses at the intersections B and C along fault planes AB, BC, and CD were observed. The results of the simulations are presented in Table 2, and results for the case where AB, BC, and CD were 10, 3, and 5 units long, respectively, are shown in Figures 4A–4F and are discussed below. The shear stresses are negative when the motion along the fault is right-lateral (α and γ are acute angles) and positive when the motion along the fault is left-lateral (α and γ are obtuse angles) (Table 2). The largest shear stress at B along AB (-1.42 N/m^2) occurred for $\alpha = 50°$ at $\beta = 80°$ (Fig. 4B). As in the case of two intersecting faults, two peaks of shear stresses were observed at B along BC, one when $\gamma < 90°$ (right-lateral motion along BC) and the other when $\gamma \geq 90°$ (left-lateral motion along BC). Along BC in the left-lateral range, the maximum shear stress (1.32 N/m^2) occurred for $\alpha = 35°$ at $\beta = 80°$ (Fig. 4C), whereas in the right-lateral range, it was -1.80 N/m^2 for $\alpha = 40°$ at $\beta = 160°$ (Table 2). At C along CD, the largest shear stress (-1.39 N/m^2) occurred for $\alpha = 50°$ at $\beta = 80°$ (Fig. 4F) (Table 2). The estimated favorable ranges for α and β, respectively, were $48°$–$53°$ and $69°$–$89°$ when observed along fault AB, $30°$–$38°$ and $68°$–$85°$ when observed along fault BC in the left-lateral range, $34°$–$42°$ and $150°$–$163°$ when observed along fault BC in the right-lateral range, and $48°$–$56°$ and $76°$–$103°$ when observed along fault CD (Table 2).

In summary, all 12 favorable ranges of shear stresses (at B along AB and BC, and at C along CD) occurred only when α was between $30°$ and $60°$ (Table 2). Along AB and CD, the favorable ranges of shear stresses were observed when α was between $45°$–$60°$, whereas along BC, the corresponding range for α was $30°$–$45°$. The favorable ranges of shear stresses along AB, BC (for left-lateral movement), and CD were observed when β was between $65°$ and $105°$, and for right-lateral movement along BC, when β was between $145°$ and $166°$.

TABLE 2. SUMMARY OF PREFERRED ANGLES OF ORIENTATION (A AND B) FROM MODEL RESULTS

Geometry	AB (Length units)	BC (Length units)	CD (Length units)	Observations at B along AB Maximum Shear stress (N/m²)	α (°)	β (°)	Favorable range (90% max. value) α (°)	β (°)	Observations at B along BC* Maximum Shear stress (N/m²)	α (°)	β (°)	Favorable range (90% max. value) α (°)	β (°)	Observations at C along CD Maximum Shear stress (N/m²)	α (°)	β (°)	Favorable range (90% max. value) α (°)	β (°)
Three intersecting faults (Fig. 1D)	10	1	5	−1.19	50	80	45–60	65–95	1.25	35	80	30–40	65–90	−0.15	50	80	45–60	70–105
									−1.595	40	160	35–40	145–166					
		3	5	−1.42	50	80	48–53	69–89	1.32	35	80	30–38	68–85	−1.39	50	80	48–56	76–103
									−1.8	40	160	34–42	150–163					
		5	5	−1.65	50	80	48–53	74–87	1.41	35	80	32–38	70–85	−1.48	50	80	48–52	70–100
									−2.1	40	160	36–42	150–164					

*Positive values for left-lateral region and negative values for right-lateral region.

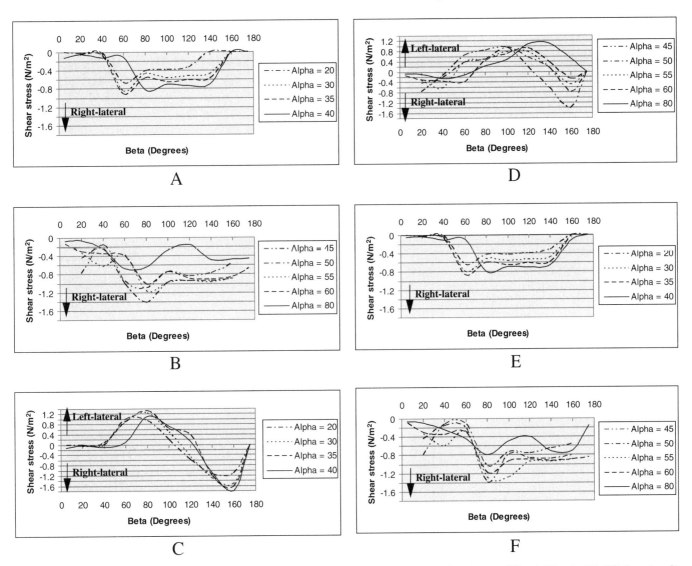

Figure 4. Plot of the magnitude of shear stress at B along fault plane AB (length = 10 units) for (A) $\alpha < 45°$ and (B) $\alpha \geq 45°$; BC (length = 3) for (C) $\alpha < 45°$ and (D) $\alpha \geq 45°$; and at C along fault plane CD (length = 5) for (E) $\alpha < 45°$ and (F) $\alpha \geq 45°$ for a range of $\beta = 20$–$160°$. The largest shear stress in the right-lateral range occurs for $\alpha = 50°$ at $\beta = 80°$. The largest shear stress in the left-lateral range occurs for $\alpha = 35°$ at $\beta = 80°$.

In summary, from the results of model calculations for two and three intersecting faults (Tables 1 and 2), we note the following:

1. The maximum shear stresses are generated at the fault intersections, B and C along the faults.

2. Increasing the length of BC (keeping length of AB fixed) or increasing the length of AB (keeping length of BC fixed) results in an increase in the maximum shear stress at B.

3. The larger shear stress at B is found to be along the shorter of the two intersecting faults AB and BC.

4. Two peaks, one positive and one negative, are obtained for the shear stress along BC at B. The sign of the shear stress is the same as that at B along AB for large values of β ($\gamma << 90°$), i.e., when BC is oriented essentially in the same direction as AB. When γ is $> 90°$, the sign of shear stress at B along BC is opposite that at B along AB.

5. In all cases, the favorable range for orientation of fault AB with respect to S_{Hmax} (α) is found to be ~30°–60°. The largest shear stresses at B along AB are observed when $45° \leq \alpha \leq 60°$ and along BC when $30° \leq \alpha \leq 45°$.

6. Along AB, CD, and BC (when the sign of the shear stress is opposite that along AB), the favorable range of shear stresses is obtained for $\beta = 65°$–$125°$. In the case where the shear stresses along BC and AB are of the same sign, i.e., AB and BC are essentially along the same direction, the favorable range of shear stresses is obtained for $\beta = 145°$–$170°$.

7. The large favorable range of β (65°–125°) for the case AB = 10 and BC = 1 units is probably an artifact of the modeling, as B is close to the horizontal block boundary.

We tested these simple 2-D model results with observations from two active intraplate locations in the eastern United States.

DISCUSSION

The results of the two-dimensional modeling presented here indicate that the magnitude of stress accumulation at fault intersections subjected to plate tectonic forces depends on their orientation with respect to S_{Hmax} and each other. They also suggest that there is an optimal fault geometry for their reactivation. The areal distribution of stress is instructive and can be compared with observed locations of seismicity associated with intersecting faults. However, the comparison is limited to stress concentration in two-dimensions, which may enhance the seismic potential of a region.

We compare our modeling results with observations from New Madrid seismic zone and Middleton Place Summerville seismic zone near Charleston, South Carolina. Both these regions have been extensively studied, and reliable information about their fault geometry is available.

Figure 5A shows the structural framework of New Madrid seismic zone as outlined by Hildenbrand et al. (2001) and the

instrumentally located earthquakes with M 3.0 or greater adapted from the 1974–2002 CERI (The Center for Earthquake Research and Information) Memphis catalog. In the New Madrid seismic zone, within the NE-SW–trending, nearly 400 km long and 100 km wide Reelfoot rift, there are two intersecting fault zones, the ~65-km-long Blytheville fault zone, oriented ~NE-SW, and the ~60-km-long Reelfoot fault zone, oriented NW-SE (Van Arsdale et al., 1995; Johnston and Schweig, 1996). A third fault, the New Madrid North fault, lies outside the edge of the floor of the Reelfoot rift but within its edge (Rhea and Wheeler, 1995). This ~30-km-long NNE-trending fault is considered to be the extension of the aseismic Bootheel lineament (Johnston and Schweig, 1996). The observed seismicity inside the Reelfoot rift is located along the Blytheville fault zone, Reelfoot, and the New Madrid North faults (Fig. 5A), with a clustering of seismicity at and/or near the fault intersections (Fig. 5A). The direction of maximum horizontal stress, S_{Hmax}, in the region is oriented N80°E (Fig. 5; Zoback, 1992). The seismically active Blytheville fault zone and New Madrid North fault are

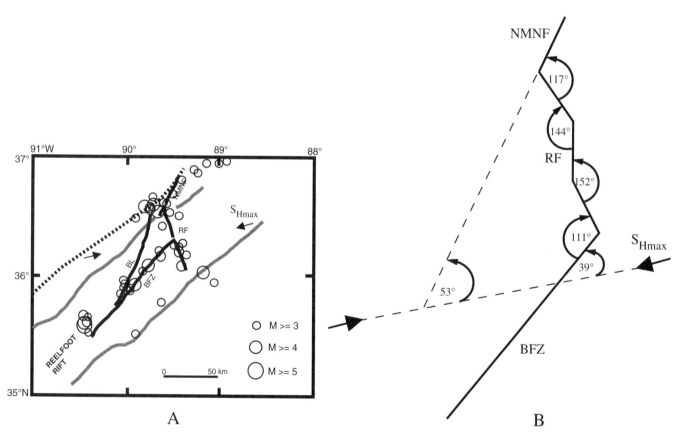

Figure 5. (A) Map showing New Madrid seismic zone. The margins of the Reelfoot rift floor (solid gray lines) and faults (solid black lines) were taken from Hildenbrand et al. (2001). The edge of the western margin of the Reelfoot rift (dotted line) was adopted from Rhea and Wheeler (1995). BFZ—Blytheville fault zone; RF—Reelfoot fault; NMNF—New Madrid North fault; BL—Bootheel Lineament. Open circles represent instrumentally located seismicity of M ≥ 3.0 from 1974 to 2002 from CERI (The Center for Earthquake Research and Information) Memphis catalog. (B) Schematic representation of the seismogenic faults in New Madrid seismic zone (not to scale). Blytheville fault zone and New Madrid North fault are oriented at angles α_1 and α_2 counterclockwise with respect to S_{Hmax} (bold arrows). Based on the analyses of seismicity by Pujol et al. (1997), the Reelfoot fault has been divided into three segments that make angles β_1, β_2, and β_3 with Blytheville fault zone.

oriented at 39° and 53° (α) with respect to S_{Hmax}, respectively (Fig. 5B). Pujol et al. (1997) and Mueller and Pujol (2001) have reanalyzed the seismicity associated with the Reelfoot fault. Based on their analyses, the Reelfoot fault can be divided into three segments that are oriented (from south to north) ~N28°W, ~N-S, and ~N36°W respectively (Fig. 5B). The orientations (internal angles) of the southern segment with respect to the Blytheville fault zone and of the New Madrid North fault with respect to the northern segment of Reelfoot fault are 111° and 117° (β), respectively (Fig. 5B). If we treat the three segments as one Reelfoot fault, the angle between the Blytheville fault zone and Reelfoot fault ranges between ~110° and 120°. These angles (α and β) are within the preferred range obtained from this parametric study (i.e., $30° ≤ α ≤ 60°$, $65° ≤ β ≤ 125°$). The Blytheville fault zone and New Madrid North fault exhibit right-lateral motion, whereas Reelfoot fault (oriented at an obtuse angle with S_{Hmax}) shows some left-lateral motion (Herrmann and Ammon, 1997), consistent with the model results. Due to its two-dimensional nature, this model is incapable of replicating the predominantly reverse movement on Reelfoot fault.

The structural framework of Middleton Place Summerville seismic zone is shown in Figure 6. The ~12-km-long, ~NW-SE Sawmill Branch fault–Ashley River fault system intersects the ~NNE-trending Woodstock fault, which is ~200 km long, dividing it into northern and southern legs (Fig. 6; Dura-Gomez, 2004). The Woodstock fault (north) trends ~N15°E–N28°E, whereas the Woodstock fault (south), which meets at the intersection of Sawmill Branch fault and Ashley River fault, trends ~N30°E (Marple and Talwani, 2000; Dura-Gomez, 2004). The ~6-km-long Sawmill Branch fault is oriented N30°W, whereas the ~6-km-long Ashley River fault trends N60°W (Fig. 6; Dura-Gomez, 2004). Only about a 30 km segment of the Woodstock fault (north) near the intersection is active seismically (dashed circle in Fig. 6). The instrumentally located seismicity (1974–2004) in this region lies mostly along the Sawmill Branch fault and is concentrated near its intersection with Woodstock fault (north) and (south) (Fig. 6). S_{Hmax} in the region is oriented N60°E (Fig. 6; Talwani, 1982; Zoback, 1992). The orientations of the Woodstock fault (north) and (south) with respect to S_{Hmax} ($α_1$ and $α_2$) in Middleton Place Summerville seismic zone are 30°–38° and 30°, respectively (Fig. 6). Sawmill Branch fault subtends an internal angle of 120° (β) with both the Woodstock fault (north) and (south) (Fig. 6). Thus, the range of angles between the seismogenic faults and S_{Hmax}, α, in Middleton Place Summerville seismic zone is 30°–38°, and the interior angle between the intersecting faults, β, is ~120°. These values are consistent with the results of the 2-D modeling (α between 30°–60°, and β between 65°–125°). Similar to the New Madrid seismic zone and matching our model results, in Middleton Place Summerville seismic zone, Woodstock fault (north) and Woodstock fault (south) exhibit right-lateral motion, whereas Sawmill Branch fault (oriented at an obtuse angle with S_{Hmax}) exhibits some left-lateral motion, although the predominant movement on it is reverse. Due to its two-dimensional nature, this model is incapable of replicating any uplift motion.

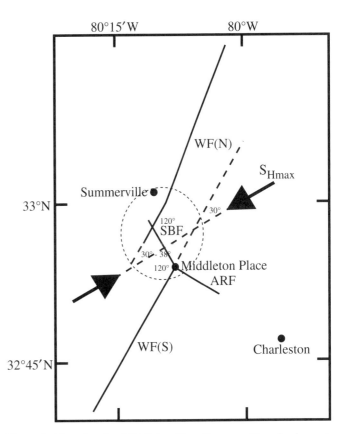

Figure 6. Map showing Middleton Place Summerville seismic zone (MPSSZ). The seismogenic faults (solid black lines) have been taken from Dura-Gomez (2004). WF(N)—Woodstock fault North; SBF—Sawmill Branch Fault; ARF—Ashley River Fault; WF(S)—Woodstock fault South. The dashed circle outlines the area of instrumentally located seismicity. The orientations of WF(N) and WF(S) $α_1$ and $α_2$) with respect to S_{Hmax} (bold arrows) are shown together with the interior angles subtended by Sawmill Branch fault ($β_1$), and Ashley River fault ($β_2$).

CONCLUSIONS

In summary, the results of this parametric study suggest that when subjected to plate tectonic forces, only optimally oriented intersecting faults are reactivated due to a larger shear stress buildup and, thus, may cause intraplate seismicity. The maximum shear stresses are generated at the fault intersection, and increasing the lengths of the intersecting faults results in an increase in the maximum shear stress. The largest shear stress at an intersection is found to be along the shorter of the two intersecting faults. The favorable range for orientation of the main fault with respect to S_{Hmax} (α) is ~30°–60° (Fig. 7). The largest shear stresses at the intersection along the main fault are observed when $30° ≤ α ≤ 45°$ and along the intersecting fault when $45° ≤ α ≤ 60°$. The range of β that yields maximum shear stresses at the intersection is ~65°–125° (Fig. 7) when the motion along the intersecting fault is opposite to that along the main fault. When the two faults have the same sense of motion, β is found to be between 145° and

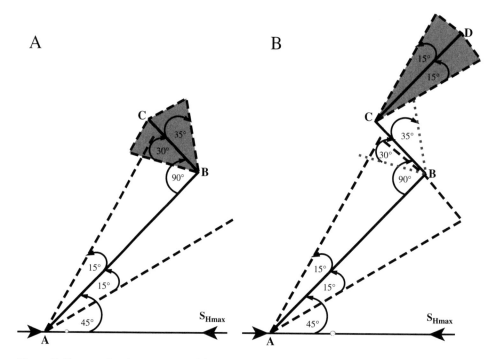

Figure 7. Cartoon showing summary of the modeling results with (A) two and (B) three intersecting faults. The favorable range for orientation of the main fault AB with respect to S_{Hmax} was found to be ~45° ± 15°. A similar range for orientation of fault CD with respect to S_{Hmax} was also obtained (light-gray shaded area in B). The range of β that yielded maximum shear stresses at the intersection was ~90° ± 35°: light-gray shaded area in A and area bounded by light-gray small-dotted line in B when the motion along the intersecting fault BC was opposite that along the main fault AB.

170° for a favorable range of shear stresses. For the case of three intersecting faults, the optimum orientation of CD is the same as that for AB (Fig. 7B).

In spite of the limitations and the two-dimensional nature of the models, by studying the fault geometries of intersecting faults and their orientation with respect to S_{Hmax} and each other, it is possible to identify more likely seismogenic faults in the presence of the ambient stress field. Future studies will include a comparison of model results with orientations of seismogenic faults in other intraplate regions.

ACKNOWLEDGMENTS

We thank the anonymous reviewers and volume editor Stéphane Mazzotti for their helpful comments, which improved the manuscript.

REFERENCES CITED

Andrews, D.J., 1989, Mechanics of fault junctions: Journal of Geophysical Research, v. 94, p. 9389–9397.
Andrews, D.J., 1994, Fault geometry and earthquake mechanics: Annali di Geofisica, v. 37, p. 1341–1348.
Crider, J.G., and Peacock, D.C.P., 2004, Initiation of brittle faults in the upper crust: A review of field observations: Journal of Structural Geology, v. 26, p. 691–707, doi: 10.1016/j.jsg.2003.07.007.
Cundall, P.A., 1971, A computer model for simulating progressive, large scale movement in blocky rock systems, in Proceedings of the Symposium on Rock Fracture: Nancy, International Society of Rock Mechanics, v. I, p. 2–8.
Dura-Gomez, I., 2004, Seismotectonic Framework of Middleton Place Summerville Seismic Zone, Charleston, South Carolina [M.S. thesis]: Columbia, University of South Carolina, 180 p.
Fitzenz, D.D., and Miller, S.A., 2001, A forward model for earthquake generation on interacting faults including tectonics, fluids, and stress transfer: Journal of Geophysical Research, v. 106, p. 26,689–26,706, doi: 10.1029/2000JB000029.
Gangopadhyay, A., and Talwani, P., 2003, Symptomatic features of intraplate earthquakes: Seismological Research Letters, v. 74, p. 863–883.
Gangopadhyay, A., and Talwani, P., 2005, Fault intersections and intraplate seismicity in Charleston, South Carolina: Insights from a 2-D numerical model: Current Science, v. 88, p. 1609–1616.
Gangopadhyay, A., Dickerson, J., and Talwani, P., 2004, A two-dimensional numerical model for the current seismicity in New Madrid seismic zone: Seismological Research Letters, v. 75, p. 406–418.
Harris, R.A., and Day, S.M., 1993, Dynamics of fault interaction: Parallel strike-slip faults: Journal of Geophysical Research, v. 98, p. 4461–4472.
Herrmann, R.B., and Ammon, C.J., 1997, Faulting parameters of earthquakes in the New Madrid, Missouri, region: Engineering Geology, v. 46, p. 299–311, doi: 10.1016/S0013-7952(97)00008-2.
Hildenbrand, T.G., Stuart, W.D., and Talwani, P., 2001, Geologic structures related to New Madrid earthquakes near Memphis, Tennessee, based on gravity and magnetic interpretations: Engineering Geology, v. 62, p. 105–121, doi: 10.1016/S0013-7952(01)00056-4.
Illies, J.H., 1982, Der Hohenzollern Graben und Intraplatten-Seismizitat infolge Vergitterung lamellarer Scherung mit einer Riftstruktur: Oberrheinische Geologische Abhandlungen, v. 31, p. 47–78.
Jing, L., and Hudson, J.A., 2002, Numerical methods in rock mechanics: International Journal of Rock Mechanics and Mining Sciences, v. 39, p. 409–427, doi: 10.1016/S1365-1609(02)00065-5.

Jing, L., and Stephansson, O., 1990, Numerical modeling of intraplate earthquake source by 2-dimensional distinct element method: Gerlands Beitrage zür Geophysik, v. 99, p. 463–472.

Johnston, A.C., and Schweig, E.S., 1996, The enigma of the New Madrid earthquakes of 1811–1812: Annual Review of Earth and Planetary Sciences, v. 24, p. 339–384, doi: 10.1146/annurev.earth.24.1.339.

Kato, N., 2001, Simulation of seismic cycles of buried intersecting reverse faults: Journal of Geophysical Research, v. 106, p. 4221–4232.

King, G., 1986, Speculations on the geometry of the initiation and termination processes of earthquake rupture and its relation to morphological and geological structure: Pure and Applied Geophysics, v. 124, p. 567–585, doi: 10.1007/BF00877216.

King, G., and Nabelek, J., 1985, Role of fault bends in the initiation and termination of earthquake rupture: Science, v. 228, p. 984–987, doi: 10.1126/science.228.4702.984.

Maerten, L., 2000, Variation in slip on intersecting normal faults: Implications for paleostress inversion: Journal of Geophysical Research, v. 105, p. 25,553–25,565, doi: 10.1029/2000JB900264.

Marple, R.T., and Talwani, P., 2000, Evidence for a buried fault system in the coastal plain of the Carolinas and Virginia—Implications for neotectonics in the southeastern United States: Geological Society of America Bulletin, v. 112, p. 200–220, doi: 10.1130/0016-7606(2000)112<0200:EFABFS>2.3.CO;2.

McCaig, A., 1988, Vector analysis of fault bends and intersecting faults: Journal of Structural Geology, v. 10, p. 121–124, doi: 10.1016/0191-8141(88)90134-4.

Mueller, K., and Pujol, J., 2001, Three-dimensional geometry of the Reelfoot blind thrust: Implications for moment release and earthquake magnitude in the New Madrid seismic zone: Bulletin of the Seismological Society of America, v. 91, p. 1563–1573.

Pollard, D.D., and Segall, P., 1987, Theoretical displacements and stresses near fractures in rock: With applications to faults, joints, veins, dikes, and solution surfaces, *in* Atkinson, B.K., ed., Fracture Mechanics of Rock: London, Academic Press, p. 277–349.

Pujol, J., Johnston, A., Chiu, J.M., and Yang, Y.T., 1997, Refinement of thrust faulting models for the central New Madrid seismic zone: Engineering Geology, v. 46, p. 281–298, doi: 10.1016/S0013-7952(97)00007-0.

Rhea, S., and Wheeler, R.L., 1995, Map showing synopsis of seismotectonic features in the vicinity of New Madrid, Missouri: U.S. Geological Survey Miscellaneous Investigations Series, I-2521, scale 1:250,000.

Robinson, R., and Benites, R., 1995, Synthetic seismicity models of multiple interacting faults: Journal of Geophysical Research, v. 100, p. 18,229–18,238, doi: 10.1029/95JB01569.

Rousseau, C.E., and Rosakis, A.J., 2003, On the influence of fault bends on the growth of sub-Rayleigh and intersonic dynamic shear ruptures: Journal of Geophysical Research, v. 108, no. B9, p. 2411, doi: 10.1029/2002JB002310.

Sharp, R.V., Lienkaemper, J.J., Bonilla, M.G., Burke, D.B., Fox, B.F., Herd, D.G., Miller, D.M., Morton, D.M., Ponti, D.J., Rymer, M.J., Tinsley, J.C., Yount, J.C., Kahle, J.E., and Hart, E.W., 1982, Surface faulting in the central Imperial valley, *in* The Imperial Valley, California, Earthquake of October 15, 1979: U.S. Geological Survey Professional Paper 1254, p. 119–144.

Sibson, R.H., 1985, A note on fault reactivation: Journal of Structural Geology, v. 7, p. 751–754, doi: 10.1016/0191-8141(85)90150-6.

Sibson, R.H., 1990, Rupture nucleation on unfavorably oriented faults: Bulletin of the Seismological Society of America, v. 80, p. 1580–1604.

Sibson, R.H., 1991, Loading of faults to failure: Bulletin of the Seismological Society of America, v. 81, p. 2493–2497.

Spotila, J.A., and Anderson, K.B., 2004, Fault interaction at the junction of the Transverse Ranges and Eastern California shear zone: A case study of intersecting faults: Tectonophysics, v. 379, p. 43–60, doi: 10.1016/j.tecto.2003.09.016.

Talwani, P., 1982, Internally consistent pattern of seismicity near Charleston, South Carolina: Geology, v. 10, p. 654–658, doi: 10.1130/0091-7613(1982)10<654:ICPOSN>2.0.CO;2.

Talwani, P., 1988, The intersection model for intraplate earthquakes: Seismological Research Letters, v. 59, p. 305–310.

Talwani, P., and Rajendran, K., 1991, Some seismological and geometric features of intraplate earthquakes: Tectonophysics, v. 186, p. 19–41, doi: 10.1016/0040-1951(91)90383-4.

Talwani, P., Amick, D.C., and Logan, R., 1979, A model to explain the intraplate seismicity in the South Carolina Coastal Plain: Eos (Transactions, American Geophysical Union), v. 60, p. 311.

Universal Distinct Element Code (UDEC), 1999, Universal Distinct Element Code, version 3.1: Minneapolis, Minnesota, ITASCA Corporation (CD-ROM).

Van Arsdale, R.B., Kelson, K.I., and Lumsden, C.H., 1995, Northern extension of the Tennessee Reelfoot scarp into Kentucky and Missouri: Seismological Research Letters, v. 66, p. 57–62.

Zoback, M.L., 1992, Stress field constraints on intraplate seismicity in eastern North America: Journal of Geophysical Research, v. 97, p. 11,761–11,782.

MANUSCRIPT ACCEPTED BY THE SOCIETY 29 NOVEMBER 2006

The Geological Society of America
Special Paper 425
2007

Integrated geologic and geophysical studies of North American continental intraplate seismicity

Xavier van Lanen[†]
Walter D. Mooney[‡]

U.S. Geological Survey, MS 977, 345 Middlefield Road, Menlo Park, California 94025, USA

ABSTRACT

The origin of earthquakes within stable continental regions has been the subject of debate over the past thirty years. Here, we examine the correlation of North American stable continental region earthquakes using five geologic and geophysical data sets: (1) a newly compiled age-province map; (2) Bouguer gravity data; (3) aeromagnetic anomalies; (4) the tectonic stress field; and (5) crustal structure as revealed by deep seismic-reflection profiles. We find that: (1) Archean-age (3.8–2.5 Ga) North American crust is essentially aseismic, whereas post-Archean (less than 2.5 Ga) crust shows no clear correlation of crustal age and earthquake frequency or moment release; (2) seismicity is correlated with continental paleorifts; and (3) seismicity is correlated with the NE-SW structural grain of the crust of eastern North America, which in turn reflects the opening and closing of the proto– and modern Atlantic Ocean. This structural grain can be discerned as clear NE-SW lineaments in the Bouguer gravity and aeromagnetic anomaly maps. Stable continental region seismicity either: (1) follows the NE-SW lineaments; (2) is aligned at right angles to these lineaments; or (3) forms clusters at what have been termed stress concentrators (e.g., igneous intrusions and intersecting faults). Seismicity levels are very low to the west of the Grenville Front (i.e., in the Archean Superior craton). The correlation of seismicity with NE-SW–oriented lineaments implies that some stable continental region seismicity is related to the accretion and rifting processes that have formed the North American continental crust during the past 2 b.y. We further evaluate this hypothesis by correlating stable continental region seismicity with recently obtained deep seismic-reflection images of the Appalachian and Grenville crust of southern Canada. These images show numerous faults that penetrate deep (40 km) into the crust. An analysis of hypocentral depths for stable continental region earthquakes shows that the frequency and moment magnitude of events are nearly uniform for the entire 0–35 km depths over which crustal earthquakes extend. This is in contradiction with the hypothesis that larger events have deeper focal depths. We conclude that the deep structure of the crust, in particular the existence of deeply penetrating faults, is the controlling parameter, rather than lateral variations in temperature, rheology, or

[†]E-mail: xavier_van_lanen@yahoo.co.uk.
[‡]E-mail: mooney@usgs.gov.

van Lanen, X., and Mooney, W.D., 2007, Integrated geologic and geophysical studies of North American continental intraplate seismicity, *in* Stein, S., and Mazzotti, S., ed., Continental Intraplate Earthquakes: Science, Hazard, and Policy Issues: Geological Society of America Special Paper 425, p. 101–112, doi: 10.1130/2007.2425(08). For permission to copy, contact editing@geosociety.org. ©2007 The Geological Society of America. All rights reserved.

high pore pressure. The distribution of stable continental region earthquakes in eastern North America is consistent with the existence of deeply penetrating crustal faults that have been reactivated in the present stress field. We infer that future earthquakes may occur anywhere along the geophysical lineations that we have identified. This implies that seismic hazard is more widespread in central and eastern North America than indicated by the limited known historical distribution of seismicity.

Keywords: intraplate earthquakes, seismic refraction, Bouger gravity, aeromagnetics, North America.

INTRODUCTION

The large impact of moderate and large earthquakes in stable continental regions on human society provides a strong motivation to study the systematics of their occurrence. Unfortunately, limited historical information and the infrequent occurrence of surface ruptures (Johnston et al., 1994) complicate efforts to better understand these intraplate events (Stein et al., 1989; Newman et al., 1999; Li et al., this volume, chapter 11; McKenna et al., this volume, chapter 12).

Several authors have presented models to explain the occurrence of stable continental region earthquakes. These models require conditions, or a combination of conditions, that result in either a zone of weakness or a zone of stress concentration in the crust. Such conditions include intersecting fault systems, igneous intrusions, glacial unloading, crustal strength due to high pore pressure or high heat flow, and other factors that increase stress (Gangopadhyay and Talwani, 2003; Campbell, 1978; Liu and Zoback, 1997; Kenner and Segall, 2000; Long, 1988; Stein et al., 1979; Talwani, 1988, 1999; Talwani and Rajendran, 1991; Vinnik, 1989; Zoback and Richardson, 1996; Stein and Newman, 2004; Mazzotti et al., 2005). In principle, each of these models assumes that seismicity results from the reactivation of preexisting zones of weakness (Sykes, 1978).

Based on these models, a spatial correlation of stable continental region earthquakes with extended and/or rifted regions is expected. However, studies of the stable continental region historical seismicity record (Johnston et al., 1994; Schulte and Mooney, 2005) show that only ~48%–62% of all events can be related to ancient extended or rifted regions. Furthermore, many continental paleorifts show little or no seismic activity. Thus, 38%–52% of stable continental region seismicity does not fit the rifted-crust model, and other mechanisms are required to explain this seismic activity.

The historical record of stable continental region earthquakes covers a relatively short time interval (~400 yr), especially considering the long recurrence intervals (500–1000 yr or more) of major stable continental region events (Johnston and Kanter, 1990). This makes it difficult to obtain a true picture of where past moderate-to-large stable continental region earthquakes might have occurred. This is an important point, since most seismic hazard assessments have relied on the main assumption that future moderate and large earthquakes will occur in the same region as historical events. As a result, the seismic hazard in several historically aseismic stable continental region areas could be underestimated (Kafka, this volume, chapter 3; Leonard et al., this volume, chapter 17; Swafford and Stein, this volume, chapter 4).

Paleoseismic studies reported by Crone et al. (1997, 2003) further strengthen this argument by showing that faults in stable continental regions have a typical long-term behavior characterized by episodically clustered activity separated by large quiescent intervals. This clustering hypothesis is consistent with the historical record of North American seismicity.

Spatial clustering of seismicity has also been observed. An analysis of the stable continental region earthquakes by Schulte and Mooney (2005) showed that for rift-related stable continental region earthquakes, only nine ancient interior rifts account for 76% of the events and 99% of the total seismic moment release. Similar numbers apply to continental (passive) margins. However, it is hard to determine whether stable continental region seismic events cluster as a rule due to the limited historical record of these events. Thus, seismically quiescent areas may still have high seismic potential if the crust contains zones of weakness (e.g., faults and intrusions) that have favorable orientations in the current stress field.

In this paper, we focus on stable continental region earthquakes within central and eastern North America (Fig. 1). These earthquakes occur primarily in crust of Proterozoic age and extend from the southern Grenville and Appalachian provinces to Baffin Island and Baffin Bay (Fig. 1). We use a variety of geologic and geophysical data to assess the spatial characteristics of these stable continental region earthquakes, and we address the questions: (1) Are there geologic or geophysical trends that correlate with North American stable continental region seismicity? (2) Can geophysical data provide insights into the physical properties of the zones of weakness that promote stable continental region seismic activity?

METHODS

To investigate the spatial distribution of historical stable continental region earthquakes and their correlation with geophysical lineaments, we used standard data processing methods to enhance the expression of geophysical lineaments (Sreedhar-Murthy, 2002). We defined geophysical lineaments as clear geometric trends that can be observed in gravity and/or magnetic anomaly data. We also considered geological trends, such as terrane boundaries.

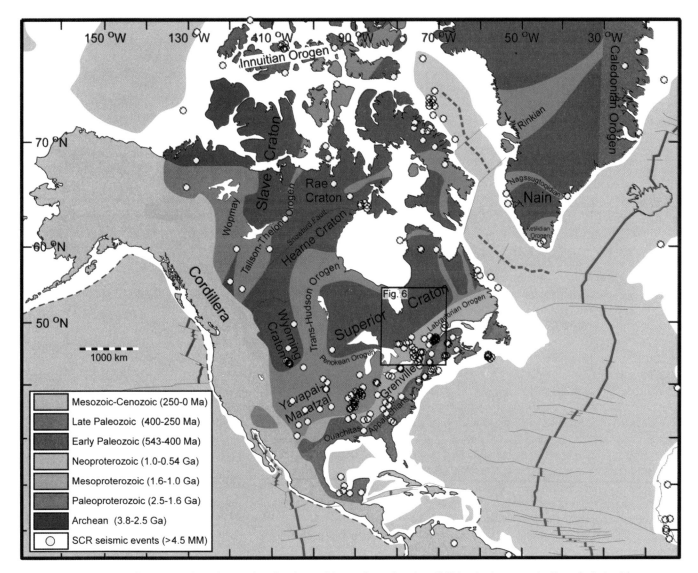

Figure 1. Basement province map of North America showing stable continental region (SCR) seismic events (yellow circles) with a moment magnitude (MM) ≥ 4.5 (after Schulte and Mooney, 2005). The box outlines the region under study in Figure 6.

Several other researchers have analyzed geologic and geophysical lineaments to investigate the occurrence of stable continental region earthquakes. Wallach et al. (1998) conducted geologic field work along two regional geophysical lineaments near western Lake Ontario to search for evidence of historical brittle faulting. Positive evidence was found, and the identification of those faults, considered as potential seismogenic sources, yielded an improved assessment of the seismic hazards in the area (Wallach et al., 1998). McBride et al. (2002) discussed the possible earthquake source region within the Wabash Valley seismic zone (Illinois, Indiana, and Kentucky) based on an interpretation of reprocessed seismic-reflection and gravity and magnetic potential field data. They concluded that the most likely source zone in the Wabash Valley is a reverse fault that was identified as a dipping seismic reflector, which is coincident with the

Commerce geophysical lineament, as identified in potential field data (Langenheim and Hildenbrand, 1997). Using geophysical, geological, and seismic-reflection data, Lamontagne et al. (2003) defined major Precambrian basement lineaments in the Lower St. Lawrence seismic zone and examined their correlation with the local earthquakes. These regional studies all demonstrate a correlation of seismic activity with geophysical lineaments, but these correlations may be expected if one accepts the general assumption that stable continental region earthquakes occur near ancient zones of weakness, as outlined already.

In this study, we examined the correlation between seismicity and geophysical anomalies on a continental scale for eastern North America. Geophysical anomalies, such as regional (short-wavelength) gravity and magnetic anomalies, are caused by lateral variations in crustal density and magnetizations, and therefore provide insight into the crustal structure. In many regions, geophysical anomaly maps are complemented by multichannel seismic-reflection profiles. Such profiles provide the highest resolution image of the structure of the crust. The study of the continental crust with seismic-reflection profiles has evolved rapidly since the earliest measurements were made in the mid-1970s. In North America, two research consortiums have been responsible for much of the data collected so far. In the United States, Consortium for Continental Reflection Profiling (COCORP) was founded in 1975, and it evolved to a nearly continuously operating field crew. In Canada, the LITHOPROBE program was initiated in 1984, and it coordinates the collection of several thousands of kilometers of deep crustal reflection profiles.

We used seismic-reflection profiles from regions with stable continental region earthquakes in our analysis. We also considered stress indicators that define tectonic stress orientation (Zoback and Zoback, 1989). Several hundred measurements are available from eastern North America, and these provide a clear picture of the tectonic stress orientation.

DATA

Stable Continental Region Earthquake Catalog

We used the global stable continental region earthquake catalog of Schulte and Mooney (2005), which contains a record of events with moment magnitudes ≥4.5. This catalog uses the definition for stable continental regions described in Johnston et al. (1994), which defined them as regions of continental crust that have not experienced any major tectonism, magmatism, basement metamorphism, or anorogenic intrusions since the Early Cretaceous and no rifting, major extension, or transtension since the Paleogene.

The earliest seismic events described in the database for eastern and central North America were mainly located in a few locations, such as the New Madrid, Missouri, seismic events of 1811–1812. The historical description of these and other seismic events is dependent on the written observations of settlers. Consequently, the historical seismic record of eastern and central North America covers different time intervals for each region. The portion of North America up to Montréal, Canada, has the longest historical record (~300–400 yr), whereas the Midwestern USA and south-central Canada for the most part have less than 200 yr of historical seismic records. Instrumental recording, which has greatly increased the documentation of earthquakes, began after 1910. As a result of these facts, there is a relatively short historical seismic record for North America.

Much of the historically recorded seismicity of eastern North America has occurred in several geographic clusters (Fig. 1), including the well-investigated New Madrid seismic zone, the Wabash Valley seismic zone (Illinois, Indiana, and Kentucky), St. Lawrence seismic zone (SE Canada and NE USA), Baffin Bay, and Grand Banks (both Canada).

Potential Fields Data

The potential fields data consist of Bouguer gravity and aeromagnetic anomalies, were compiled as part of the Geological Society of America Decade of North American Geology (DNAG) program, and were taken from the Geosoft DAP data server. These data have been used by others to investigate crustal structure and infer geologic and tectonic processes (Kane and Godson, 1989). These authors divided the conterminous United States into a series of zones based on both geophysical anomaly maps. We used their observations and zoning for guidance in our study. Hinze (1985) presented additional examples of the use of gravity and magnetic anomaly maps to define large- and small-scale crustal features.

Stress Field

The tectonic stress map for central and eastern North America was taken from the World Stress Map (WSM) project (Reinecker et al., 2005). These authors provided a quality ranking of each data point, which consisted of four types: (1) earthquake focal mechanisms; (2) well-bore breakouts and drilling-induced fractures; (3) in situ stress measurements (overcoring, hydraulic fracturing, borehole slotter); and (4) young geologic data (fault-slip analysis and volcanic vent alignments).

Multichannel Seismic-Reflection Profiles

To better understand the relation between crustal structure and seismicity, we selected a detailed crustal cross section based on high-resolution seismic-reflection data, the LITHOPROBE Abitibi-Grenville transect (Ludden and Hynes, 2000). This transect is ideally located due to its coincidence with active seismicity. The transect is composed of four profiles that each cross the Grenville Front.

INTERPRETATION

Crustal Age

North American stable continental region seismicity shows a strong correlation with crustal age (Fig. 1). With few exceptions, the Archean (≥2.5 Ga) cratons, such as the Superior craton, are nearly aseismic. Seismicity associated with Archean cratons is concentrated at the margins of these cratons. This observation is consistent with other evidence that indicates that Archean crust is commonly underlain by a thick (≥200 km) lithospheric keel that provides a high degree of stability to the overriding crust (Pollack, 1986; Jordan, 1988). For crust that is post-Archean in age, North American seismicity data do not show a strong correlation with crustal age. Rather, seismicity occurs in clusters within all age provinces, from the Paleoproterozoic (2.5–1.6 Ga) to the late Paleozoic (400–250 Ma; Fig. 1). This observation contradicts the hypothesis that Phanerozoic stable continental region crust is uniformly more seismically active than Precambrian crust, with the exception of Archean crust, which is nearly aseismic.

We note that crustal age, as depicted in Figure 1, may not reflect later tectonic events that have modified the crust. Thus, the New Madrid earthquakes are located within crust of a Paleoproterozoic (2.5–1.6 Ga) age that was modified by late Neoproterozoic (1.0–0.54 Ga) rifting. The prevalence of tectonic events affecting older crust may explain the lack of correlation of seismicity with crustal age for post-Archean crust. Conversely, much of the Archean-age crust has not been affected by post-

Archean tectonic events, such as rifting, and therefore it remains largely aseismic. The Archean Wyoming craton of the western United States, which has undergone significant post-Archean tectonic activity, is a prominent exception to this rule.

Bouguer Gravity and Aeromagnetic Anomaly Data

Figure 2 shows the Bouguer gravity and aeromagnetic anomaly maps for central and eastern North America with the seismicity overlain. In general, the Bouguer gravity anomalies

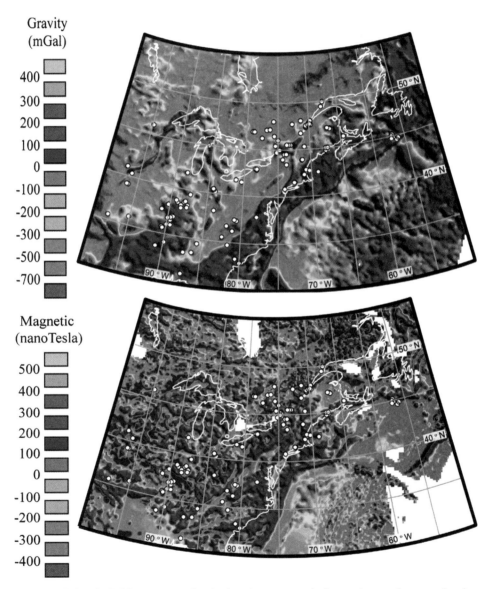

Figure 2. Color-shaded Bouguer gravity (top) and aeromagnetic (bottom) anomaly maps showing the stable continental region seismic events (white circles) with a moment magnitude ≥ 4.5 (after Schulte and Mooney, 2005). The geophysical data were compiled as part of the Geological Society of America Decade of North American Geology (DNAG) program and were taken from the Geosoft DAP data server. The program was a joint effort by the Geological Society of America, Geological Survey of Canada (GSC), U.S. Geological Survey (USGS), and Consejo de Recursos Minerales of Mexico (CRM). An interpretation of the correlation between geophysical lineations and epicentral locations is given in Figure 3.

have longer wavelengths than the aeromagnetic anomalies, and both anomaly maps show well-defined lineations (cf. Kane and Godson, 1989). The most easily defined trends can be identified where gravity and magnetic trends terminate abruptly against trends with a different azimuthal orientation. In several cases, a trend can only be identified in one anomaly map. This indicates that the trend is purely due to either a density or magnetization anomaly.

The gravity anomaly map shows a clear coast-parallel lineation, whereas the aeromagnetic anomaly map is more complex and contains both NE-SW– and NW-SE–oriented lineations. The NE-SW aeromagnetic lineation is offshore, roughly parallel with the ocean-continent transition, and it is the strongest anomaly located east of the coastline. Despite the prominence of this ocean-continent geophysical anomaly, seismicity within the region is limited; the largest event was the 1929 Mw = 7.2 Grand Banks events located at 45°N, 56.5°W (Bent, 1995). This earthquake had a complex source mechanism dominated by NW-SE–oriented strike-slip motion and a well-constrained hypocentral depth of 20 ± 2 km (Bent, 1995). Another locus of seismic activity that correlates with the ocean-continent tran-

sition is at Baffin Bay in northernmost Canada, as discussed by Stein et al. (1979, 1989) and Swafford and Stein (this volume, chapter 4). It is worth emphasizing that earthquakes that occur on offshore faults may pose a hazard from strong ground motion and from the possible generation of a tsunami, such as the one that devastated Lisbon, Portugal, in 1755.

The major lineations identified in the Bouguer gravity and aeromagnetic maps are summarized in Figure 3. There is a clear correlation between epicentral locations and the lineations, indicating that lateral variations in crustal properties have an association with seismicity in eastern North America.

Crustal Stress

Several hundred determinations of crustal stress are available for central and eastern North America (Reinecker et al., 2005). Among the stress indicators, normal faulting earthquake focal mechanisms are rare. The stress data indicate an average N60°E maximum compressive stress orientation (Fig. 4), approximately parallel to the direction of absolute plate motion (Demets et al., 1990). This orientation aligns the stress field at 0–30° with respect

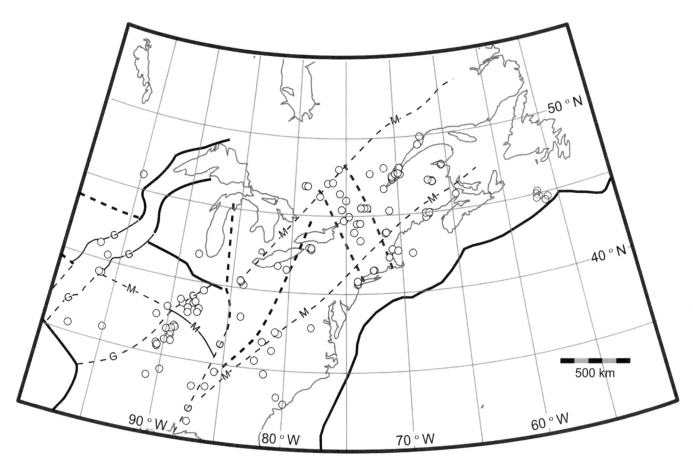

Figure 3. Interpretive map showing the major observed gravimetric and magnetic lineaments (following Kane and Godson, 1989). Lineaments are indicated with solid or dashed lines representing well-defined or less well-defined regions, respectively. When the lineament line contains either the letter G or M, it was defined only by gravity or magnetic data, respectively. The stable continental region seismic events with a moment magnitude ≥ 4.5 (after Schulte and Mooney, 2005) are indicated by white circles.

to the NE-SW lineations identified in Figure 3. The implication is that the crustal stress field is favorably oriented to activate faulting on NE-SW–oriented lineations, either in strike-slip or reverse mechanisms, which are indeed the predominate source mechanisms in eastern North America. The stress map (Fig. 4) also indicates numerous perturbations to the average stress orientation. As has previously been suggested, these perturbations are likely due to local zones of weakness and/or stress concentrations.

Hypocentral Depths and Moment Magnitudes

Thus far, we have considered only earthquake epicentral locations and correlated these with geologic and geophysical trends. Stable continental region earthquake hypocentral depths and magnitudes provide additional insights into the conditions that give rise to this seismicity. Depending on the available seismic network station coverage, hypocentral depths of stable continental region earthquakes may have a large uncertainty. However,

in many cases, a special effort has been made to constrain depth by detailed waveform modeling (e.g., Bent, 1995). Keeping these uncertainties in mind, we consider hypocentral depth estimates from both the entire global stable continental region catalogue (Fig. 5A) and North America alone (Fig. 5B). The global data show a maximum number of events within the 5–10-km-depth bin (255 events). The number of events decreases to <20% of this number in both the 20–25-km-depth and 25–30-km-depth bins (~45 events within each depth bin). The global catalogue shows a second strong peak for all events greater than 30 km depth, which is likely an artifact of a 33 km starting depth. This artifact has been removed from the North American catalog due to the relatively dense station coverage, which permits more accurate depth estimates than is possible globally.

We also consider minimum, average, and maximum moment magnitude versus depth (Fig. 5A). The minimum moment is bounded by the seismic detection capability. Significantly, the maximal moment magnitude gradually increases from the 0–5-km-

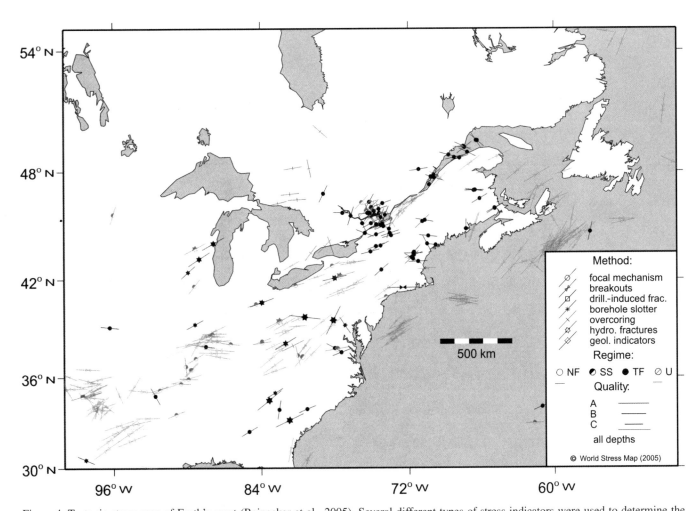

Figure 4. Tectonic stress map of Earth's crust (Reinecker et al., 2005). Several different types of stress indicators were used to determine the tectonic stress orientation. They were usually grouped into four categories: (1) earthquake focal mechanisms; (2) well-bore breakouts and drilling-induced fractures; (3) in situ stress measurements (overcoring, hydraulic fracturing, borehole slotter); and (4) geological indicators, such as fault-slip analysis and dike alignments. NF—normal fault; SS—strike slip fault; TF—transform fault; U—unknown.

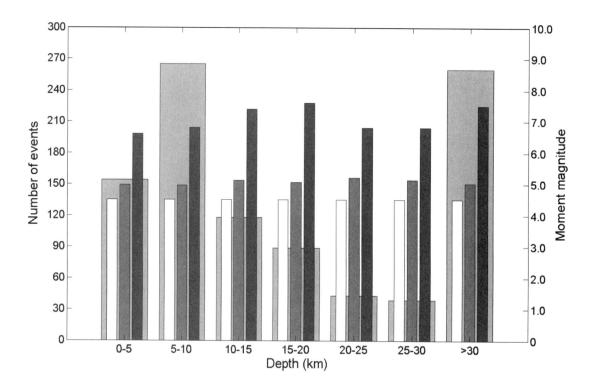

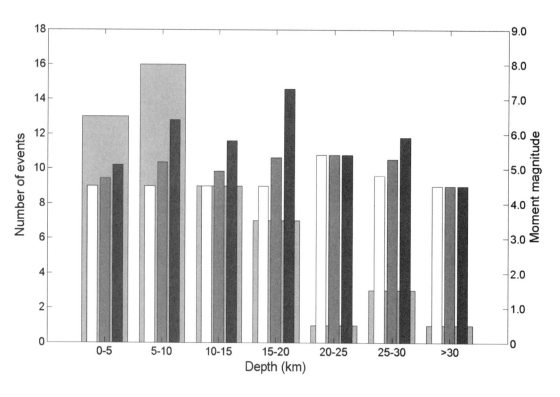

Figure 5. Distribution of stable continental region (SCR) seismic events (after Schulte and Mooney, 2005) as function of depth (*x*-axis) and the corresponding minimal, average, and maximal moment magnitudes (MM; *y*-axis). The global (top) and eastern North America (bottom) data sets are both plotted.

depth bin to reach a peak of Mw = 7.7 within the 15–20-km-depth bin. A plausible interpretation of this observation is that the largest stable continental region earthquakes initiate at ~20 km depth and rupture up to the surface. A second peak in maximal moment magnitude is found for the bin at ≥30 km depth.

A similar pattern can be seen in the North American catalog, with 88% of all events reported within the upper 20 km of the crust (Fig. 5B). Of the 49 earthquakes with a moment magnitude greater than 4.5, 44 earthquakes are above 20 km depth and only 5 earthquakes are located below 20 km depth. The results from North America are more reliable than the global compilation due to the relatively good seismic station coverage in North America. A commonly cited interpretation of the dominance of shallow (≤20 km) hypocentral depths is the existence of a brittle-ductile transition at this depth. In this model, deformation occurs by creep below 20 km depth and by brittle faulting above. We compare these observations with results obtained from deep seismic-reflection profiles in the following section.

Crustal Structure from Deep Seismic-Reflection Profiles

Deep seismic-reflection profiles provide the most detailed geophysical images of the structure of the crust to seismogenic depths. We consider four interpretations of recently recorded profiles across the Grenville Province of Ontario, Canada (Fig. 6). Three of these profiles (A, B, and C; Fig. 6) are located immediately north of Lake Ontario, and the fourth profile (D; Fig. 6) crosses the Grenville Front at 52.5°N. This region of mid-Proterozoic (1.1 Ga) crust has a relatively high level of seismicity (Fig. 6, top). The seismic-reflection profiles reveal that the crust is made up of NW-vergent, thick-skinned thrust sheets (Fig. 6, bottom). Each sheet consists of a 50–150-km-wide assemblage of rocks that is composed of a distinct accreted terrane. While we lack sufficient hypocentral accuracy to correlate individual earthquakes with specific fault planes in the seismic-reflection data, two observations are clear: (1) most seismicity is located SE of the Grenville Front, which is consistent with the SE-dipping structure interpreted in Figure 6B; and (2) the seismicity is widely distributed in map view (Fig. 6, top), which is consistent with faulting occurring on a series of low-angle fault planes.

These observations, together with the histograms of hypocentral depths (Fig. 5), show that seismicity is concentrated above 20 km depth and imply that the observed seismicity in SE Canada (Fig. 6) occurs on the low-angle thrust sheets that are clearly imaged in the seismic-reflection profiles. The regional stress field in SE Canada has a NE-SW orientation (Fig. 4), and fault slip occurs predominantly as thrust motions on preexisting thrust faults.

DISCUSSION AND CONCLUSIONS

Earthquakes in stable continental regions are commonly attributed to the reactivation of upper crustal zones of weakness that are favorably orientated with respect to the regional stress field. The most commonly cited examples of zones of weakness are faults associated with ancient continental interior rifts and paleorifted margins. However, a previous study determined that these two tectonic environments only provide an explanation of 48%–62% of stable continental region earthquakes (Schulte and Mooney, 2005). In this paper, we: (1) further evaluated the interior rift–rifted margin model and (2) tried to find an explanation for the remaining 38%–52% of stable continental region earthquakes that are not explained by this model.

North American stable continental region earthquakes are rare in Archean crust, which we attribute to the cold and thick lithospheric keel that provides stability to the overlying crust (Pollack, 1986; Jordan, 1988; Artemieva and Mooney, 2001). Post-Archean crust does not show a correlation between crustal age and seismic activity, perhaps because more recent tectonic events can modify and weaken all crust that is post-Archean in age.

The geologic structure of eastern North America has a NE-SW strike, which corresponds to the lateral accretion of the Grenville and Appalachian provinces in the mid-Proterozoic and Paleozoic. Geophysical anomalies, as expressed in Bouguer gravity and aeromagnetic anomaly maps, follow this NE-SW trend. Aeromagnetic anomalies also define some perpendicular NW-SE trends. There is a remarkable correlation between stable continental region earthquakes and the trends defined by geophysical data. However, it is notable that the ocean-continent transition, which lies 100–300 km offshore, has been nearly aseismic for the past 200+ yr, despite a clear demarcation in the aeromagnetic anomaly map. Two prominent exceptions are the seismicity at (1) the Grand Banks and (2) Baffin Bay, Canada. These exceptions raise the possibility that other portions of the North American ocean-continent transition may have a significant seismic potential (Stein et al., 1979, 1989; Swafford and Stein, this volume, chapter 4).

The hypocentral depths of stable continental region earthquakes are less accurate than their epicentral locations. We used a statistical approach and plotted all reported depths in 5 km bins. More than 90% of all hypocentral depths for North American stable continental region earthquakes are at or above 20 km, whereas the crust has an average thickness of 38 km. Thus, the seismicity is concentrated in the upper crust. Deep seismic-reflection profiles reveal low-to-moderate angle, east-dipping faults within the upper crust that accommodated thrusting during the accretion of the Grenville and Appalachian provinces. We lack sufficient hypocentral accuracy to correlate individual earthquakes with specific faults imaged by the seismic-reflection data. Nevertheless, the thrust faulting source mechanisms of the earthquakes in SE Canada are in excellent agreement with the geometry of the faults imaged in the seismic-reflection profiles. We conclude that there is a correlation between seismicity and deeply penetrating crustal faults, which are pervasive in the accreted Grenville and Appalachian provinces. This means that seismic hazards are far more widespread in central and eastern North America than is indicated by the limited known historical distribution of seismicity (Swafford and Stein, this volume, chapter 4).

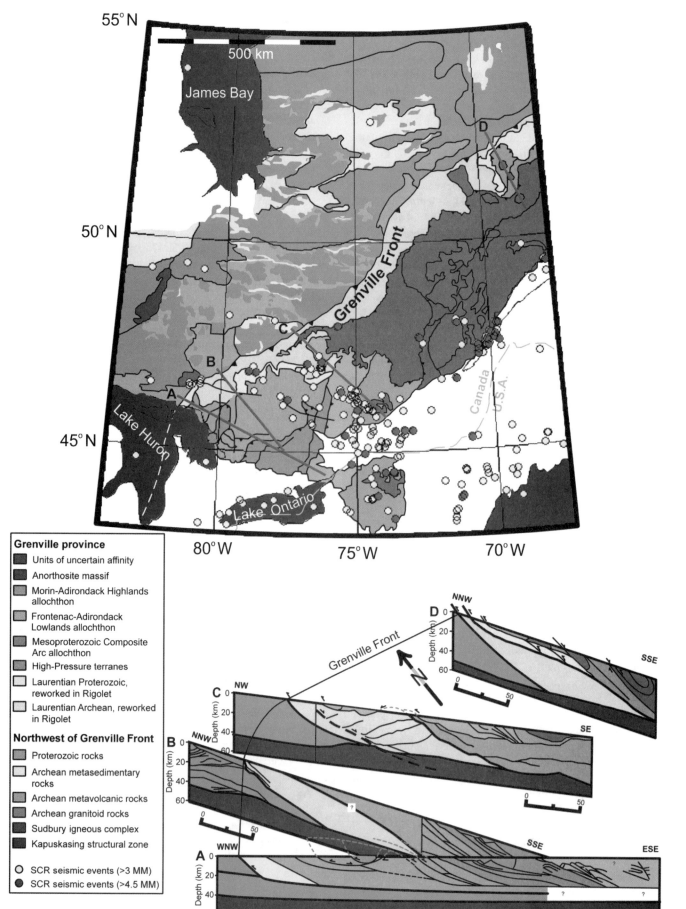

Grenville province

- Units of uncertain affinity
- Anorthosite massif
- Morin-Adirondack Highlands allochthon
- Frontenac-Adirondack Lowlands allochthon
- Mesoproterozoic Composite Arc allochthon
- High-Pressure terranes
- Laurentian Proterozoic, reworked in Rigolet
- Laurentian Archean, reworked in Rigolet

Northwest of Grenville Front

- Proterozoic rocks
- Archean metasedimentary rocks
- Archean metavolcanic rocks
- Archean granitoid rocks
- Sudbury igneous complex
- Kapuskasing structural zone

○ SCR seismic events (>3 MM)
● SCR seismic events (>4.5 MM)

Figure 6. Stable continental region (SCR) earthquakes. Deep structure of the Grenville Province, eastern Canada, from mulit-channel seismic reflection data. (Top) Geological sketch map and location of four seismic reflection profiles (solid red lines) (after Ludden and Hynes, 2000). SCR seismic events superposed on the geological sketch map show events with a moment magnitude (MM) ≥ 4.5 (Schulte and Mooney, 2005) and ≥ 3.0 (Armbruster and Seeber, 1992) are indicated by red and yellow circles, respectively. (Bottom) Interpretive cartoons of the crustal structure of the Grenville province for profiles A through D. Low-to-moderate angle thrust faults bound a series of terranes that were accreted to the east coast of cratonic North America. Seismicity is associated with the upper 20 km of these faults.

Numerous previous studies have established that earthquake hazards are high in paleorifts and passive continental margins (Stein et al., 1979, 1989; Schulte and Mooney, 2005, and references therein). Based on the observations summarized here, we propose the "suture model" of earthquake genesis, whereby many stable continental region earthquakes occur on low-to-moderate angle faults that mark crustal sutures. The most important sutures are associated with the major terrane boundaries of eastern North America that formed during the mid-Proterozoic Grenville orogeny and the Paleozoic Appalachian orogeny. The trends of these sutures can be observed using geophysical anomaly maps (summarized in Fig. 3), and we infer that seismic hazards are higher than previously recognized along these trends.

ACKNOWLEDGMENTS

It is a pleasure to acknowledge the Vrije University Amsterdam, especially C. Biermann, for their support for the first author during a research internship at the U.S. Geological Survey. We also would like to thank S. Detweiler for his help with data compilation, technical suggestions, and advice on constructing the figures and obtaining potential fields data. We are grateful to M. Barazangi, K. Bergen, S. Detweiler, R. England, J. Gomberg, S. Murthy, E. Schwieg, S. Stein, S.T. White, and an anonymous reviewer for discussions and/or critical reviews and suggestions.

REFERENCES CITED

Armbruster, J., and Seeber, L., 1992, NCEER-91 Earthquake Catalog for the United States: Buffalo, National Center for Earthquake Engineering Research, State University of New York: http://folkworm.ceri.memphis.edu/catalogs/html/cat_nceer.html (last accessed 11 July 2005).

Artemieva, I.M., and Mooney, W.D., 2001, Thermal structure and evolution of Precambrian lithosphere: A global study: Journal of Geophysical Research, v. 106, p. 16,387–16,414, doi: 10.1029/2000JB900439.

Bent, A.L., 1995, A complex double-couple source mechanism for the Ms 7.2 1929 Grand Banks earthquake: Bulletin of the Seismological Society of America, v. 85, p. 1003–1020.

Campbell, D.L., 1978, Investigation of the stress-concentration mechanism for intraplate earthquakes: Geophysical Research Letters, v. 5, no. 6, p. 477–479.

Crone, A.J., Machette, M.N., and Bowman, J.R., 1997, Episodic nature of earthquake activity in stable continental regions revealed by palaeoseismicity studies of Australian and North American Quaternary faults: Australian Journal of Earth Sciences, v. 44, p. 203–214.

Crone, A.J., De Martini, P.M., Machette, M.N., Okumura, K., and Prescott, J.R., 2003, Paleoseismicity of two historically quiescent faults in Australia: Implications for fault behavior in stable continental regions: Bulletin of the Seismological Society of America, v. 93, no. 5, p. 1913–1934, doi: 10.1785/0120000094.

Demets, C., Gordon, R.G., Argus, D.F., and Stein, S., 1990, Current plate motions: Geophysical Journal International, v. 101, p. 425–478.

Gangopadhyay, A., and Talwani, P., 2003, Symptomatic features of intraplate earthquakes: Seismological Research Letters, v. 74, p. 863–883.

Hinze, W.J., ed., 1985, The Utility of Regional Gravity and Magnetic Anomaly Maps: Tulsa, Oklahoma, Society of Exploration Geophysicists, 454 p.

Johnston, A.C., and Kanter, L.R., 1990, Earthquakes in stable continental regions: Scientific American, v. 262, no. 3, p. 68–75.

Johnston, A.C., Coppersmith, K.J., Kanter, L.R., and Cornell, C.A., 1994, The earthquakes of stable continental regions: Palo Alto, California, Electric Power Research Institute (EPRI), Report TR-102261.

Jordan, T.H., 1988, Structure and formation of the continental tectosphere: Journal of Petrology, v. 29, p. 11–37.

Kafka, A., 2007, this volume, Does seismicity delineate zones where future large earthquakes are likely to occur in intraplate environments?, in Stein, S., and Mazzotti, S., ed., Continental Intraplate Earthquakes: Science, Hazard, and Policy Issues: Geological Society of America Special Paper 425, doi: 10.1130/2007.2425(03).

Kane, M.F., and Godson, R.H., 1989, A crust/mantle structural framework of the conterminous United States based on gravity and magnetic trends, in Pakiser, L.C., and Mooney, W.D., eds., Geophysical Framework of the Continental United States: Geological Society of America Memoir 172, p. 383–403.

Kenner, S.J., and Segall, P., 2000, A mechanical model for intraplate earthquakes: Application to the New Madrid seismic zone: Science, v. 289, p. 2329–2332, doi: 10.1126/science.289.5488.2329.

Lamontagne, M., Keating, P., and Perreault, S., 2003, Seismotectonic characteristics of the Lower St. Lawrence seismic zone, Quebec: Insights from geology, magnetics, gravity, and seismics: Canadian Journal of Earth Sciences, v. 40, p. 317–336, doi: 10.1139/e02-104.

Langenheim, V.E., and Hildenbrand, T.G., 1997, Commerce geophysical lineament—Its source, geometry, and relation to the Reelfoot rift and New Madrid seismic zone: Geological Society of America Bulletin, v. 109, no. 5, p. 580–595, doi: 10.1130/0016-7606(1997)109<0580:CGLISG>2.3.CO;2.

Leonard, M., Robinson, D., Allen, T., Schneider, J., Clark, D., Dhu, T., and Burbidge, D., 2007, this volume, Toward a better model for earthquake hazards in Australia, in Stein, S., and Mazzotti, S., ed., Continental Intraplate Earthquakes: Science, Hazard, and Policy Issues: Geological Society of America Special Paper 425, doi: 10.1130/2007.2425(17).

Li, Q., Liu, M., Zhang, Q., and Sandovol, E., 2007, this volume, Stress evolution and seismicity in the central-eastern USA: Insights from geodynamic modeling, in Stein, S., and Mazzotti, S., ed., Continental Intraplate Earthquakes: Science, Hazard, and Policy Issues: Geological Society of America Special Paper 425, doi: 10.1130/2007.2425(19).

Liu, L., and Zoback, M.D., 1997, Lithospheric strength and intraplate seismicity in the New Madrid seismic zone: Tectonics, v. 16, no. 4, p. 585–595, doi: 10.1029/97TC01467.

Long, L.T., 1988, A model for major intraplate continental earthquakes: Seismological Research Letters, v. 59, no. 4, p. 273–278.

Ludden, J., and Hynes, A., 2000, The Lithoprobe Abitibi-Grenville transect: Two billion years of crust formation and recycling in the Precambrian Shield of Canada: Canadian Journal of Earth Sciences, v. 37, p. 459–476, doi: 10.1139/cjes-37-2-3-459.

Mazzotti, S., James, T.S., Henton, J., and Adams, J., 2005, GPS crustal strain, postglacial rebound, and seismic hazard in eastern North America: The Saint Lawrence valley example: Journal of Geophysical Research, v. 110, p. 1–16, doi: 10.1029/2004JB003590.

McBride, J.H., Hildenbrand, T.G., Stephenson, W.J., and Potter, C.J., 2002, Interpreting the earthquake source of the Wabash Valley seismic zone (Illinois, Indiana, and Kentucky) from seismic-reflection, gravity and magnetic-intensity data: Seismological Research Letters, v. 73, no. 5, p. 660–686.

McKenna, J., Stein, S., and Stein, C., 2007, this volume, Is the New Madrid seismic zone hotter and weaker than its surroundings?, in Stein, S., and Mazzotti, S., ed., Continental Intraplate Earthquakes: Science, Hazard, and Policy Issues: Geological Society of America Special Paper 425, doi: 10.1130/2007.2425(12).

Newman, A., Stein, S., Weber, J., Engeln, K., Mao, A., and Dixon, T., 1999, Slow deformation and lower seismic hazards at the New Madrid seismic zone: Science, v. 284, p. 619–621, doi: 10.1126/science.284.5414.619.

Pollack, H.N., 1986, Cratonization and thermal evolution of the mantle: Earth and Planetary Science Letters, v. 80, p. 175–182, doi: 10.1016/0012-821X(86)90031-2.

Reinecker, J., Heidbach, O., Tingay, M., Sperner, B., and Müller, B., 2005, The 2005 Release of the World Stress Map: www.world-stress-map.org (last accessed 11 July 2005).

Schulte, S., and Mooney, W.D., 2005, An updated global earthquake catalogue for stable continental regions: Reassessing the correlation with ancient rifts: Geophysical Journal International, v. 161, p. 707–721, doi: 10.1111/j.1365-246X.2005.02554.x.

Sreedhar-Murthy, Y., 2002, On the correlation of seismicity with geophysical lineaments over the Indian subcontinent: Current Science, v. 83, no. 6, p. 760–766.

Stein, S., and Newman, A., 2004, Characteristic and uncharacteristic earthquakes as possible artifacts: Applications to the New Madrid and Wabash seismic zones: Seismological Research Letters, v. 75, p. 173–187.

Stein, S., Sleep, N., Gellar, R.J., Wang, S.C., and Kroeger, G.C., 1979, Earthquakes along the passive margin of eastern Canada: Geophysical Research Letters, v. 6, no. 7, p. 537–540.

Stein, S., Cloetingh, S., Sleep, N., and Wortel, R., 1989, Passive margin earthquakes, stresses, and rheology, *in* Gregerson, S., and Basham, P., eds., Earthquakes at North Atlantic Passive Margins: Neotectonics and Postglacial Rebound: Dordrecht, Kluwer, p. 231–260.

Swafford, L., and Stein, S., 2007, this volume, Limitations of the short earthquake record for seismicity and seismic hazard studies, *in* Stein, S., and

Mazzotti, S., ed., Continental Intraplate Earthquakes: Science, Hazard, and Policy Issues: Geological Society of America Special Paper 425, doi: 10.1130/2007.2425(04).

Sykes, L.R., 1978, Intraplate seismicity, reactivation of preexisting zones of weakness, alkaline magmatism, and other tectonism postdating continental fragmentation: Reviews of Geophysics and Space Physics, v. 16, no. 4, p. 621–688.

Talwani, P., 1988, The intersection model for intraplate earthquakes: Seismological Research Letters, v. 59, no. 4, p. 305–310.

Talwani, P., 1999, Fault geometry and earthquakes in continental interiors: Tectonophysics, v. 305, p. 371–379, doi: 10.1016/S0040-1951(99)00024-4.

Talwani, P., and Rajendran, K., 1991, Some seismological and geometric features of intraplate earthquakes: Tectonophysics, v. 186, p. 19–41, doi: 10.1016/0040-1951(91)90383-4.

Vinnik, L.P., 1989, The origin of strong intraplate earthquakes: Translated from O prirode sil'nykh vnutrilplitovykh zemletyaseniy: Doklady Akademii Nauk SSSR, v. 309, no. 4, p. 824–827.

Wallach, J.L., Mohajer, A.A., and Thomas, R.L., 1998, Linear zones, seismicity, and the possibility of a major earthquake in the intraplate western Lake Ontario area of eastern North America: Canadian Journal of Earth Sciences, v. 35, p. 762–786, doi: 10.1139/cjes-35-7-762.

Zoback, M.L., and Richardson, R.M., 1996, Stress perturbation associated with the Amazonas and other ancient continental rifts: Journal of Geophysical Research, v. 97, p. 11,761–11,782.

Zoback, M.L., and Zoback, M.D., 1989, Tectonic Stress Field of the Continental United States: Geological Society of America Memoir 172, p. 523–539.

MANUSCRIPT ACCEPTED BY THE SOCIETY 29 NOVEMBER 2006

The Geological Society of America
Special Paper 425
2007

Effects of a lithospheric weak zone on postglacial seismotectonics in eastern Canada and the northeastern United States

Patrick Wu

Department of Geology and Geophysics, University of Calgary, Calgary, AB T2N 1N4, Canada

Stéphane Mazzotti

Geological Survey of Canada, Pacific Geoscience Centre, 9860 West Saanich Road, Sidney, BC V8L 4B2, Canada

ABSTRACT

At postglacial rebound time scales, the intraplate continental lithosphere typically behaves as an elastic solid. However, under exceptional conditions, the effective viscosity of the lower crust and lithospheric mantle may be as low as ~10^{20} Pa s, leading to ductile behavior at postglacial rebound time scales. We studied the effects of a lithospheric ductile zone on postglacial rebound–induced seismicity and deformation in eastern Canada and the northeastern United States using three types of models: (1) a reference model with no lithospheric ductile layer; (2) a model with a uniform, 25-km-thick, ductile layer embedded in the middle of the lithospheric column; and (3) a model with a dike-like vertical ductile zone, extending from mid-crust level down to the bottom of the lithosphere, along the Precambrian rift structure of the St. Lawrence Valley. Based on geothermal and rock physics data, the viscosity of the ductile zone is set to either 10^{20} or 10^{21} Pa s. We found that a narrow ductile zone cutting vertically through the lithosphere has larger effects than the uniformly thick horizontal ductile layer. Effects of a lithospheric weak zone on uplift rates may be large enough to be detected by global positioning system (GPS) measurements, especially for low viscosities. While the effect on fault stability is also large, the impact on the onset time of instability is small for sites within the ice margin. The impact on the onset time is more significant for sites outside the ice margin. Effects of a lithospheric weak zone are also significant on present-day horizontal velocities and strain rates and are at the limit of resolution for GPS measurements.

Keywords: postglacial rebound, seismotectonics, North America, lithospheric ductile zone.

INTRODUCTION

Eastern North America is affected by large intraplate earthquakes, with magnitude as large as M7–8. One possible explanation for this intraplate seismicity is related to stress perturbations produced by glacial loading and unloading events during the last Ice Age. According to Walcott (1970), the stress difference induced by ice-sheet building and melting cycles may be large enough to exceed the strength of crustal rocks. In the past 30 yr, several studies have addressed the impact of glacial and postglacial stress perturbations on the seismicity of eastern North America and other glaciated intraplate regions. A general result

Wu, P., and Mazzotti, S., 2007, Effects of a lithospheric weak zone on postglacial seismotectonics in eastern Canada and the northeastern United States, *in* Stein, S., and Mazzotti, S., ed., Continental Intraplate Earthquakes: Science, Hazard, and Policy Issues: Geological Society of America Special Paper 425, p. 113–128, doi: 10.1130/2007.2425(09). For permission to copy, contact editing@geosociety.org. ©2007 The Geological Society of America. All rights reserved.

of these models is that ice-sheet loading tends to prohibit seismicity underneath the load, whereas melting tends to promote earthquakes within and outside of the formerly glaciated region (e.g., Johnston, 1987; Wu and Hasegawa, 1996a, 1996b). The onset time of seismicity and the amplitude of the stress and strain perturbations are sensitive to the model details and, in particular, to the assumed rheology of the lithosphere and asthenosphere (e.g., Wu, 1997, 1998b).

The purpose of this study was to investigate the effect of heterogeneity in the lithosphere strength, expressed as an effective viscosity, on crustal motion, fault stability, and strain rates in eastern Canada and the northeastern United States. In typical postglacial rebound models, the lithosphere is described as a purely elastic solid sitting on a ductile viscoelastic upper mantle. A few postglacial rebound studies have considered the impact of a weak ductile zone in the lithosphere. Grollimund and Zoback (2001) showed that a small local weak zone in the New Madrid area, central United States, can produce an increase in the predicted crustal strain rates of two to three orders of magnitude. Other results have suggest that a ductile layer embedded in the elastic lithosphere can only produce significant effects if it is associated with very low viscosity (Wu, 1997; Klemann and Wolf, 1999; Di Donato et al., 2000). This idea of a lithospheric "jelly sandwich" was originally proposed for active tectonic regions (e.g., Chen and Molnar, 1983; Jackson, 2002), but it seems unreasonable for stable intraplate environments. In order to test the impact of lithospheric strength heterogeneities on fault stability and seismicity in eastern Canada and the northeastern United States, we considered two models: a horizontal uniform ductile layer embedded halfway in the lithosphere and a localized ductile weak zone cutting vertically through the lower crust and lithospheric mantle.

After a brief discussion of previous studies on the relation between fault stability and postglacial rebound, we review the evidence pointing to a limited range of possible viscosity values for the lower crust and lithospheric upper mantle. We used this range of viscosity contrast to constrain the postglacial rebound models. We also compared the predicted crustal velocities and strain rates to global positioning system (GPS) measurements to assess whether the relatively small variations due to postglacial rebound signals could be detected.

EARTHQUAKES, FAULT STABILITY, AND POSTGLACIAL REBOUND

The relationship between intraplate earthquakes and postglacial rebound stress perturbations has been considered since the late 1970s. Stein et al. (1979, 1989) showed that flexural stresses due to glacial loading can produce normal faulting in the deglaciated region and thrust faulting farther away. This is consistent with the observations in Baffin Island and Baffin Bay but not with the observations in other parts of eastern Canada, where thrust faulting is the dominant mechanism. Quinlan (1984) studied the effects of background stresses on fault stability and concluded that background stresses are as important as the post-

glacial rebound stresses in triggering earthquakes in eastern Canada. Johnston (1987, 1989) showed that large continental ice sheets can suppress earthquakes, but it is not clear if the removal of the earthquake-suppressing stresses results in a reactivation of faults and initiation of seismicity.

In these early studies, only the flexural stress in the lithosphere due to simple ice loads was considered, and the relaxation in the mantle was neglected. James (1991) and Spada et al. (1991) used a full glacial rebound model, including mantle relaxation, to compute rebound stress differences but did not use the Coulomb-Mohr failure criterion to relate stress and fault instability. Wu and Hasegawa (1996a, 1996b) used full glacial rebound models with both simple and realistic ice models to compute the evolution of postglacial rebound stresses in eastern Canada. By extending the idea of a fault stability margin for Coulomb-Mohr failure (Johnston 1987, 1989), they calculated the fault stability when both ambient tectonic stresses and overburden pressure are included. Temporal changes in fault stability predicted a pulse of earthquake activity with timing and mode of failure consistent with observations (Shilts et al., 1992; Obermeier et al., 1991; Adams, 1996). Their model further showed that although postglacial rebound stresses have decayed since early postglacial time and are not large enough to cause fracture today, they can trigger current earthquakes by reactivating fault zones that have been brought close to failure by tectonic stress.

Data from the World Stress Map Project (Adams, 1989; Adams and Bell, 1991; Zoback, 1992a, 1992b) show that the observed orientation of the present-day maximum horizontal stress S_{Hmax} does not appear to be influenced by the effects of past glaciation. Rather, S_{Hmax} in eastern Canada and some parts of the northeastern United States is mainly aligned in a NE direction, which can be explained by the ridge-push forces at the Mid-Atlantic Ridge (Richardson and Reding, 1991). On the other hand, postglacial thrust faults indicate that the S_{Hmax} orientation of the paleostress field in southeastern Canada during early postglacial time was oriented NW-SE, consistent with the direction of ice retreat (Adams 1989, 1995). Thus, S_{Hmax} in southeastern Canada has rotated by ~90° since postglacial times. Since the viscosity of the mantle controls the rate of decay of rebound stresses, it should also affect this change in stress orientation. Wu (1996) showed that this rotation in stress orientation could be explained only if the viscosity of the lower mantle is lower than 10^{22} Pa s.

This raises an important question: how does mantle viscosity affect postglacial seismicity? Wu (1997) showed that viscosity structure strongly affects the onset timing and the magnitude of stress at sites outside the ice margin. On the other hand, the viscosity structure of the mantle does not significantly affect the mode of failure nor the onset timing of fault instability for sites within the ice margin, which are dominated by the ice history (Wu et al., 1999). In general, the viscosity structure controls the rate of change of fault instability and is thus important for considering seismic hazards to nuclear plants and storage facilities for toxic waste (Wu, 1998a).

All the models discussed heretofore are flat-Earth models except those in James (1991) and Spada et al. (1991). Wu and Johnston (2000) showed that the effects of sphericity and self-gravity of a solid Earth and the oceans are small on the timing of instability within the ice margin.

Recently, the effects of nonlinear rheology (Wu, 2002) and lateral heterogeneity in mantle viscosity (Kaufmann and Wu, 2002; Wu, 2005) on postglacial rebound–induced seismicity and crustal motion have been investigated. Again, their effects on onset time are significant for sites outside the ice margin but small for sites within. The effect of lateral heterogeneity in the mantle (both upper and lower mantle) and ice history on fault stability in Antarctica has also been studied. Kaufmann et al. (2005) found that inclusion of lateral heterogeneities, together with a long glaciation phase, explains the Balleny Island earthquake, which occurred just outside the ice margin in Antarctica.

In summary, previous studies have shown that Pleistocene deglaciation can reactivate preexisting faults and trigger seismicity in the zones brought close to failure in eastern Canada. Mantle viscosity can affect the onset timing of fault instabilities outside the ice margin, and the rotation of stress orientation within the ice margin, but not the mode of failure.

In all these studies, the continental lithosphere was treated as a purely elastic solid encompassing the crust and uppermost mantle down to 70–200 km depth. However, under certain circumstances, the mechanical behavior of the lower crust and lithospheric mantle can be represented as steady-state plastic flow. Long-term plastic deformation of the lower crust has been proposed in active orogens such as Tibet (e.g., Clark and Royden, 2000) or the northern Canadian Cordillera (Mazzotti and Hyndman, 2002). Ductile behavior of the lower crust is also involved in modulating postseismic strains (e.g., Rydelek and Pollitz, 1994). Wu (1997) showed that the effect of a 25-km-thick ductile layer (with viscosity of 10^{22} Pa s) within the lithosphere does not significantly affect the fault stability analysis. Klemann and Wolf (1999) found that the effect becomes significant if the viscosity of the ductile layer is reduced to 10^{17} Pa s. The presence of such low viscosity in a 10-km-thick ductile layer shifts the onset timing earlier by 1000 yr for sites within the ice margin and also significantly affects the magnitude of the horizontal velocity. Di Donato et al. (2000) used 10^{18} Pa s viscosity in a 15-km-thick layer and found that it had a significant effect on crustal velocities along the U.S. East Coast. Using similar values, Kendall et al. (2003) found that it affected the inference of lithospheric thickness in Australia.

The viscosities of the ductile layer used in most of these studies were based on estimates derived in active tectonic settings (e.g., Kaufman and Royden, 1994), which are associated with a hot geotherm and possibly significant amounts of water in the crust and upper mantle (Dixon et al., 2004). In the next section, we show that the viscosity of the lower crust and lithospheric mantle in eastern Canada and the northeastern United States is not likely to be lower than $\sim 10^{20}$ Pa s.

DUCTILE LAYER IN THE LITHOSPHERE

Plastic flow in the crust and lithospheric mantle is commonly associated with dislocation creep and is represented using a power-law relation between the flow strain rate $\dot{\varepsilon}$ and the differential stress σ (e.g., review in Evans and Kohlstedt, 1995):

$$\dot{\varepsilon} = A\sigma^n \exp\left(-\frac{Q+PV}{RT}\right), \tag{1}$$

where n, A, and Q are the power-law exponent, the activation energy, and a material parameter, respectively. R is the gas constant and T is the absolute temperature. The pressure term PV is commonly regarded as second order and can be omitted. Under constant strain rate and differential stress, power-law creep can be associated with an effective viscosity η_{eff} defined by the stress-strain rate relation:

$$\eta_{eff} = \frac{1}{2}\frac{\sigma}{\dot{\varepsilon}}. \tag{2}$$

If Equations 1 and 2 are combined, the effective viscosity relates to the strain rate and differential stress as:

$$\eta_{eff} = \frac{1}{2}\dot{\varepsilon}^{\frac{1-n}{n}} A^{-\frac{1}{n}} \exp\left(\frac{Q}{nRT}\right) \tag{3a}$$

and

$$\eta_{eff} = \frac{1}{2}\sigma^{1-n} A^{-1} \exp\left(\frac{Q}{RT}\right). \tag{3b}$$

Thus, the primary factors controlling the effective viscosity are the material parameters (n, A, Q) and the temperature (T).

In order to assess the range of reasonable viscosity values for the lower crust and lithospheric mantle under typical continental intraplate conditions, we applied Equation 3a to different rock rheologies with a range of temperatures and strain rates. We used four types of possible lower-crustal rheologies (from weak to strong): wet granite (Hansen and Carter, 1982), felsic granulite (Wilks and Carter, 1990), mafic granulite (Wilks and Carter, 1990), and dry diabase (Mackwell et al., 1998), and two types of mantle rheologies: dry and wet dunite (Hirth and Kohlstedt, 1996).

Eastern Canada is characterized by a cold geotherm with surface heat-flow values between 30 and 60 mW/m². On average, lower-crustal temperatures are low (400–500 °C), especially in and near the Precambrian cratons (Mareschal et al., 2000). Small-scale variations in surface heat flow are related to local perturbations in upper-crustal heat production (Mareschal et al., 2000) or possibly to locally higher mantle heat flow, although the source of such an anomaly remains unclear (cf. discussion in Pollitz et al., 2001). Typical intraplate strain rates are commonly estimated to be smaller than $\sim 10^{-10}$ yr^{-1} ($\sim 10^{-17}$ s^{-1}; cf. discussion in Mazzotti, this volume, chapter 2). Plate tectonic reconstructions put an upper bound to intraplate strain rates at ~10^{-11} yr^{-1} (e.g., Gordon, 1998). Average seismic strain rates derived from earthquake catalogues are within 10^{-11}–10^{-12} yr^{-1} (i.e., about one M6–7 earthquake per 1000 km per 10,000 yr), although, locally, seismic strain rates may be as high as 10^{-9}–10^{-10} yr^{-1} (Anderson, 1986; Mazzotti and Adams, 2005). Postglacial rebound strain rates in eastern Canada are ~10^{-9}–10^{-11} yr^{-1} (e.g., Wu, 1998b).

Under typical intraplate conditions (400–600 °C and 10^{-10}–10^{-12} yr^{-1}), effective viscosity values for lower-crustal rocks (granulites and diabase, Fig. 1A) and upper-mantle rocks (Fig. 1B) are very high, ~10^{24}–10^{28} Pa s. For a viscoelastic lower crust and upper mantle, these high viscosity values mean that millions to billions of years pass before the rock can transit from elastic to viscous behavior. In other words, the crust and upper mantle behave purely elastically at postglacial rebound time scales of 10,000 yr. Viscous flow behavior under postglacial rebound time scales would require effective viscosities smaller than ~10^{22} Pa s. As shown in Figure 1, such low viscosity values may be reached in the lower crust and uppermost mantle under a hot geotherm ($T > 600$ °C) and high strain rates (more than ~10^{-15} s^{-1}). The presence of water and/or weak material would also promote a lower viscosity.

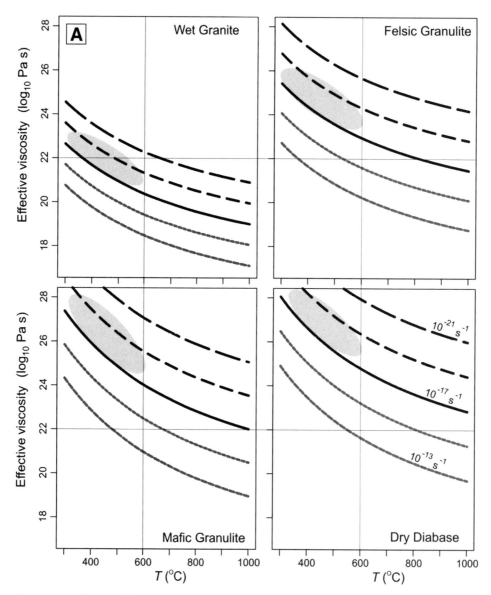

Figure 1 (*on this and following page*). Effective viscosity of lower crust (A) and upper mantle (B). The effective viscosity is shown as a function of temperature for six different rock rheologies (lower crust from weak to strong: wet granite, felsic granulite, mafic granulite, dry diabase) at typical tectonic and intraplate strain rates (from 10^{-13} to 10^{-21} s^{-1} every two orders of magnitude). The light-gray shading shows the area of typical intraplate thermal and strain-rate conditions.

These estimates are made for a steady-state stress-strain approximation of the power-law creep. Within a given weak zone, the effective viscosity values would likely evolve with time because transient perturbations of the stress field, such as postglacial rebound, can produce local strain concentration and result in an apparent drop of the effective viscosity. In other words, viscosity values associated with higher strain rates may be more appropriate as an input to our Newtonian viscoelastic modeling approach.

Thus, we believe that the lowest reasonable viscosity for a ductile layer in eastern Canada is ~10^{20} Pa s. Such low viscosity may be reached locally in paleo–tectonic zones, where the crust and upper mantle may be weakened by the presence of fluids, felsic intrusions, or weak materials (serpentinite, talc, etc.). The St. Lawrence valley, which lies along the Late Precambrian Iapetan rift margin, is the most likely candidate for a low viscosity zone in eastern Canada (Fig. 2; cf. discussion in Mazzotti, this volume, chapter 2). On the other hand, models with a spatially uniform, very low-viscosity layer embedded in the lithosphere (e.g., Klemann and Wolf, 1999; Di Donato et al., 2000) appear to be unlikely because they violate typical thermal and strain conditions in intraplate regions.

For viscosities as high as 10^{22} Pa s, a spatially uniform ductile layer has no effect on seismotectonics in eastern Canada (Wu, 1997). Thus, in this paper, we focus on ductile layers with viscosities in the range 10^{20} to 10^{21} Pa s.

MODELS

We used a finite-element model to compute the spatial and temporal variation of glacial-induced surface motions, strain rates, stresses, and fault stability. The model consisted of two inputs: the loading history and the viscosity structure of Earth. The ice model contained two sawtooth glacial cycles that had a slow buildup time of 90 k.y. The first cycle had a deglaciation time of 10 k.y. The ICE-3G model of Tushingham and Peltier (1991) was used to describe the last deglaciation phase. Increasing the number of glacial cycles had little effect on our results. Ocean loading was included and was given by eustatic sea levels.

Five Earth models are discussed in this paper. The reference model RF contained a 125-km-thick lithosphere overlying a mantle with uniform viscosity of 10^{21} Pa s. Model RF contained no ductile layer within the elastic lithosphere. For the other models, there was a ductile layer in the lithosphere, but the viscosity in the mantle was the same as that of the reference model RF. In models UD1 and UD2, there was a uniform, 20-km-thick ductile layer at a depth of 20–40 km within the lithosphere (lower crust level). In model UD1, the viscosity of the ductile layer was 10^{21} Pa s, while in UD2, the viscosity was 10^{20} Pa s. In models SLV1 and SLV2, a vertical ductile zone was modeled along the St. Lawrence Valley weak zone, shown in Figure 2, and it extended vertically from 20 km depth to the bottom of the lithosphere (125 km depth). The width of the zone varied from 100 to 300 km. The viscosity of the ductile zone was 10^{21} Pa s in model SLV1 and 10^{20} Pa s in SLV2. Other laterally heterogeneous models with uniformly thick ductile layers under the continents or oceans were also considered, but their effects were similar to those of models UD1 and UD2, except that the effects were restricted to where the ductile layer existed. Thus, the results of these models are not presented here.

The total stress field is composed of the postglacial rebound stress, tectonic stress, and overburden stress. The orientation of the first-order tectonic maximum horizontal principal stress was taken to be in the N60°E direction. The magnitude of the

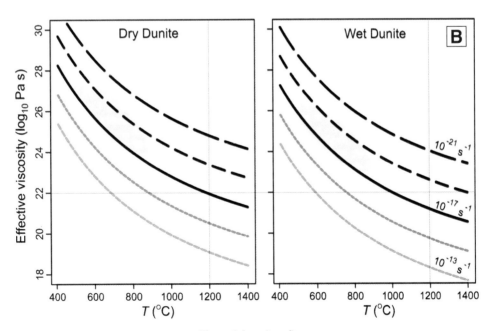

Figure 1 (*continued*).

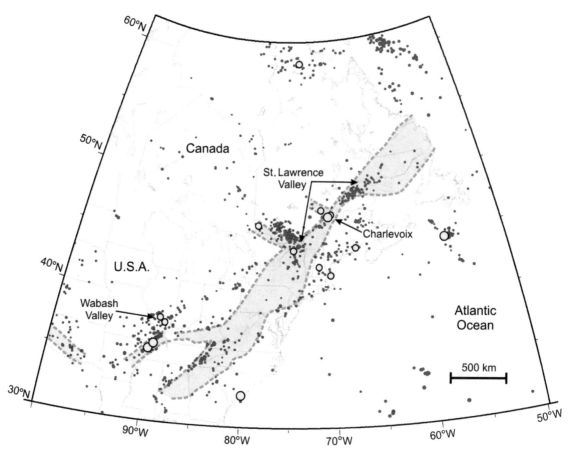

Figure 2. Location of a possible ductile weak zone along the St. Lawrence valley. The red and yellow circles represent background seismicity (M ≥ 3 since 1973) and large (M ≥ 6) historical earthquakes, respectively. The green shaded area shows the Iapetan rifted margin and aulacogens (cf. Mazzotti, this volume, chapter 2).

maximum and minimum horizontal tectonic stresses is largely unknown. However, tectonic stress magnitudes have little effect on fault stability (Wu and Hasegawa, 1996a, 1996b), and only their difference affects total stress orientation. In this paper, the difference between maximum and minimum horizontal tectonic stresses was taken to be 5 MPa. Tectonic stresses were included in all models.

Changes in the stability of faults (δFSM) are related to changes in the state of stress by (Wu and Hasegawa, 1996a):

$$\delta\mathrm{FSM}(t) = \left\{ \left[\sigma_1(t_0) - \sigma_3(t_0) \right] - \left[\sigma_1(t) - \sigma_3(t) \right] \right\}/2 \\ + \mu\beta\left\{ \left[\sigma_1(t) + \sigma_3(t) \right] - \left[\sigma_1(t_0) + \sigma_3(t_0) \right] \right\}, \tag{4}$$

where the spatial dependence has been suppressed. Here, $\beta = \sin(\arctan[\mu])/2\mu$, and μ is the coefficient of friction, taken to be 0.6. Also, t is the time; t_0 is the initial time (before the onset of glaciation); and σ_1, σ_2, σ_3 are the maximum, intermediate, and minimum (compressive) principal stresses, respectively. Ivins et al. (2003) have studied the difference between

the Mogi-von Mises failure criteria and the Coulomb criteria that is used here. They found that both give similar predictions. δFSM is our measure of fault-stability potential: a negative value enhances the likelihood of faulting for optimally oriented faults, whereas a positive value promotes fault stability. When the faults are initially close to failure and optimally oriented, a small negative value of δFSM may be sufficient to trigger an earthquake. In other cases, the initial value of δFSM is positive (i.e., all faults in a region are more than marginally stable). In that case, even optimally oriented faults require large and negative values of δFSM to overcome the initial stability and trigger earthquakes. For optimally oriented faults, the mode of failure depends on which of the principal stresses is closest to the vertical. If σ_1 is nearly vertical, the mode of failure is normal; if σ_3 is close to the vertical, thrusting occurs; otherwise, the mode of failure is strike slip.

In eastern Canada, most postglacial thrust faults are steeply dipping (50° or more) and are not optimally oriented. To reactivate these high-angle preexisting faults, the Mohr circle has to rise above the line of failure (see Figure 2c in Wu, 1998a); thus, a large negative δFSM is required.

RESULTS

Fault Stability Margin

The spatial variation of δFSM in eastern Canada at the present time predicted by the reference model RF is shown in Figure 3A. Dashed contours are for negative values of δFSM, indicating that fault instability is promoted. The stress level available for triggering earthquakes is ~1 MPa at present. This level of stress is able to activate optimally oriented preexisting faults at near-failure equilibrium but is too small to cause rock fracture. The results are similar to those of model L1 in Wu (1997).

The predictions from the other Earth models are shown in Figures 3B, 3C, and 3D. The effect of a uniformly thick ductile layer is quite significant on the spatial distribution of δFSM. The main effect is to reduce the values offshore along the Labrador coast by more than 1 MPa. On land, δFSM is barely reduced in all the area surrounding Hudson Bay. The most significant effect is for model UD1, where an area with fault stability (positive values of δFSM) is produced in Labrador (Fig. 3B). If the ductile zone is restricted along the St. Lawrence valley weak zone (Fig. 3D), the changes in δFSM mainly lie along the weak zone, with variations as large as 1.5 MPa. Results for model SLV2 are similar to those of model SLV1 (with slightly larger amplitudes) and are not shown.

With the large differences in the spatial distribution of δFSM, one might expect that the presence of the ductile zones would also significantly affect the onset time of paleo-

Present-Day δFSM

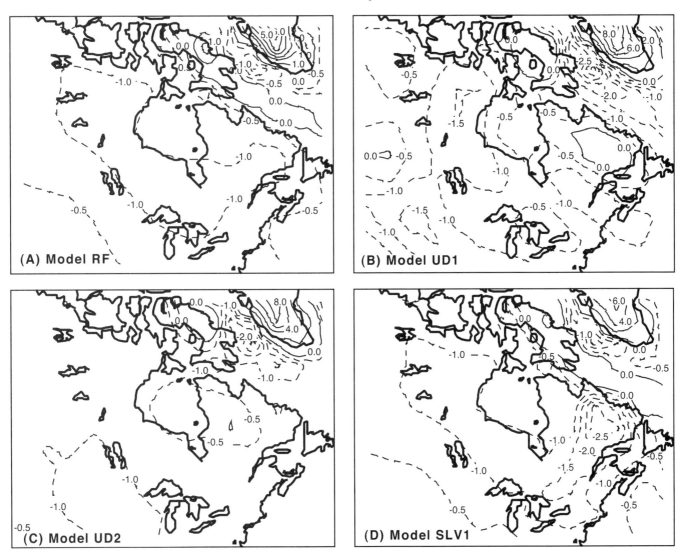

Figure 3. Spatial variation of changes in Fault Stability Margin (δFSM) at 5 km depth in eastern Canada and northeastern United States at present time predicted by models (A) RF, (B) UD1, (C) UD2, and (D) SLV1. Contour interval is 0.5 MPa. Dashed contours are for negative values.

earthquakes. However, the temporal history of δFSM shows that the effect on onset timing is only significant for sites outside the ice margin; the only exception within the ice margin is the small area in Labrador (Fig. 3B). This is illustrated in Figure 4, which shows the evolution of δFSM at two sites where observational data about the timing of paleo-earthquakes are available. The first site is within the ice margin in Charlevoix, Quebec, where the observed timing deduced from earthquake-triggered mud slumping events in nearby Lac Temiscouata is 9 ± 1 ka (Shilts et al., 1992). The predicted onset timing of all models agrees quite well with the observed values, indicating that the presence of a ductile layer has very little effect. This is different from the finding of Klemann and Wolf (1999), who used a lower-viscosity ductile layer. Our finding agrees with other studies (Wu et al., 1999; Wu, 2002; Kaufmann and Wu, 2002) that have

shown that mantle viscosity variations have very small effect on the timing of instability inside the ice margin. The reason is that the onset time there is mainly controlled by the ice deglaciation history. Figure 4A also shows that the effect of the ductile layer (its viscosity and its lateral variation) on the amplitude of δFSM is also small until the last 7 ka.

For sites outside the ice margin, the effect of the ductile layer is more significant. This is demonstrated in Figure 4B where the evolution history in the Wabash Valley, Indiana, is shown. Model SLV1 predicts a very early onset time, around 14 ka, 6 k.y. earlier than predicted by the reference model RF. However, the instability for model SLV1 vanishes around 10–9 ka. Model UD2 also predicts an onset time that is 4 k.y. earlier than that for model RF. For other models, the shift in onset time is very small. The timing of the Wabash Valley seismicity has been dated by paleo-liquefaction data and most likely ranges between 8 and 1 ka (Obermeier et al., 1991); thus the early onset times predicted by models UD2 and SLV1 fall too far outside the error bar to be considered likely.

The predicted mode of failure is thrust faulting for all models. For sites within the ice margin, this is consistent with most of the observations in eastern Canada, except in Baffin Island and Baffin Bay. For sites outside the ice margin, pure thrust-fault earthquakes have been observed in the northeastern United States (Herrmann, 1979; Nabelek and Suarez, 1989); however, the current predominant mode of failure is strike slip. Since postglacial rebound stresses diminish rapidly south of the Wisconsin ice margin (Wu and Johnston, 2000), earthquakes in the northeastern United States are dominated by tectonic stresses rather than postglacial rebound stresses.

In summary, the effects of ductile zones on the magnitude of δFSM are significant, but the effects on the timing of the onset of instability or the mode of failure are small for sites within the ice margin. An important question is whether the effect of a ductile layer is also small for surface motion within the ice margin.

Uplift Rates

Figure 5A shows the predicted land uplift rate for the reference model. The pattern of uplift is dominated by the deglaciation history, while the magnitudes are more dependent on mantle viscosity. The introduction of ductile layers only affects the predicted uplift rates in a subtle way. To facilitate the comparison, we plot the difference between the models with ductile layer and the reference model. Figures 5B and 5C show that changes associated with the thin ductile layers are generally small: peak values for the differences are 0.6 mm/yr for model UD1 and 1.2 mm/yr for model UD2. If the ductile layer is localized and cuts through to the bottom of the lithosphere, the differences are limited to the weak zone area, with peak values reaching 4 mm/yr for model SLV1 and 3 mm/yr for SLV2 (the plot for SLV2 is similar to Fig. 5D, except for a slightly smaller peak, so it is not shown).

The large amplitudes of vertical velocities in eastern Canada make a good measurement target for geodetic experiments. Grav-

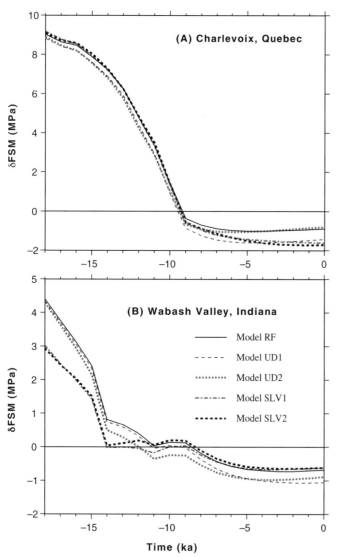

Figure 4. Temporal evolution of stability of faults (δFSM) at (top) Charlevoix, Quebec, and (bottom) Wabash Valley, Indiana, predicted by the various models.

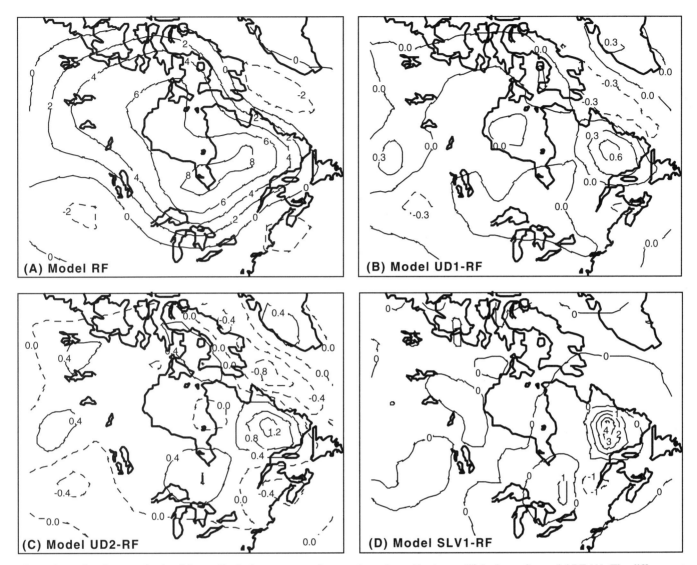

Figure 5. Predicted current land uplift rate. Dashed contours are for negative values. Absolute uplift is shown for model RF (A). The differences between the predicted current uplift rate for models UD1, UD2, and SLV1 and the reference model RF are given in B, C, and D, respectively.

ity, leveling, and, more recently, global positioning system (GPS) data have shown that, to a first order, surface uplift in central and eastern Canada follows the predicted postglacial rebound pattern of large uplift around Hudson Bay and little to no uplift near the former ice margin (e.g., Larson and van Dam, 2000; Sella et al., 2007). The more subtle variations in uplift rates between the reference model and the ductile-layer models are just at the resolution level of GPS measurements. Uncertainties on GPS-derived vertical velocities depend mostly on the sampling rate and length of the time series. Typically, resolution levels of ~1–3 mm/yr (standard deviation) can be achieved with yearly campaign data over 5–10 yr or permanent data over 3–5 yr (e.g., Mao et al., 1999; Langbein et al., 2002).

As an example, Figure 6 shows a compilation of vertical GPS data for eastern Canada and the northeastern United States. Campaign station velocities along the lower St. Lawrence valley

are based on 6–9 yr of yearly measurements and have a typical uncertainty of ~1.5 mm/yr (Mazzotti et al., 2005). Permanent station velocities are based on continuous 3 yr time series, and their uncertainty is ~1.5–3 mm/yr. The hinge line between uplift and subsidence runs along New Brunswick, Maine, New York State, and southern Ontario, in agreement with the predicted postglacial rebound pattern (Fig. 5A). Small variations as predicted by the ductile-layer models are not apparent in this data set and may be below the current resolution level.

Horizontal Velocities

Figure 7A shows that the horizontal velocities predicted by the reference model diverge from the center of rebound, with higher velocity along the Atlantic coast. This is consistent with the finding of Peltier (1998). With the inclusion of a ductile layer,

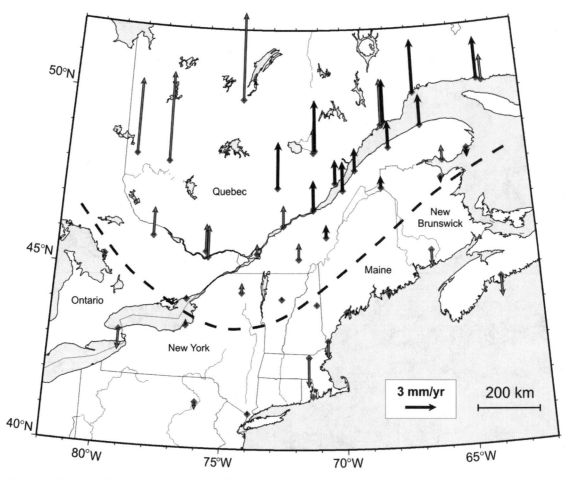

Figure 6. Global positioning system (GPS) uplift rates in southeastern Canada and northeastern United States. Black arrows show the vertical velocities (in ITRF2000 reference frame) for campaign sites along the lower St. Lawrence valley based on 6–9 yr of yearly surveys (Mazzotti et al., 2005). Gray arrows show the vertical velocities (in ITRF2000 reference frame) for permanent sites in Quebec, Ontario, and New England, based on 3 yr of continuous data. The dashed line shows the approximate location of the uplift-subsidence hinge line.

the horizontal motion remains mainly divergent, and the effect is best revealed by plotting the difference in velocity.

The differences between models UD1 and RF, shown in Figure 7B, are very small and indicate a mainly convergent motion toward the center of rebound for sites outside the ice margin. Within the ice margin, the difference has a peak near 0.4 mm/yr. The difference in velocity between models UD2 and RF, shown in Figure 7C, is mainly divergent outside the ice margin. For sites within, the difference is a convergent motion toward several centers near the ice margin. However, the largest difference, with magnitude as large as 0.9 mm/yr, is found along the Labrador coast.

A large difference in velocity between model SLV1 (or SLV2) and the reference model results from the presence of the ductile zone that cuts through the lower crust and lithospheric mantle. The effect is to turn the velocities toward the ductile zone. The difference in horizontal velocity for model SLV1 reaches a peak magnitude of 1.2 mm/yr (Fig. 7D), while that for model SLV2 is 1.0 mm/yr.

Typical uncertainty levels of GPS horizontal velocities are ~0.5–1.5 mm/yr for 5–10-yr-long campaign data or 3–5-yr-long continuous data (Mao et al., 1999; Langbein et al., 2002). Thus, the first-order pattern of postglacial rebound horizontal velocities (divergence away from the center of load) is detectable by GPS measurements. Horizontal velocities along the lower St. Lawrence valley measured by campaign GPS indicate a general southeastward motion of ~0.7 mm/yr relative to North America, in agreement with postglacial rebound patterns (Mazzotti et al., 2005). These rates are smaller than the predictions of our reference model (Fig. 7A) and would be consistent with the presence of a weak ductile zone along the St. Lawrence valley (models UD2 and SLV2, Figs. 7C and 7D). However, this level of detail is at the limit of resolution of the GPS measurements. More accurate results in eastern Canada and the northeastern United States could possibly resolve local velocity variations as predicted in our weak-layer models.

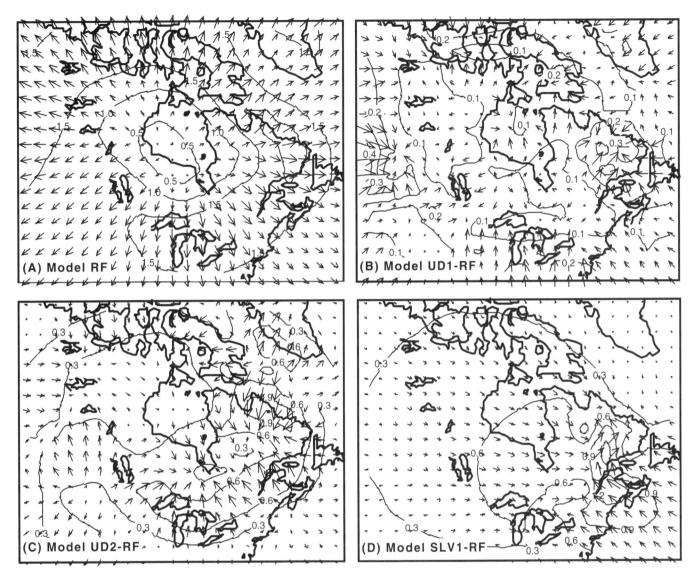

Figure 7. (A) Predicted current horizontal velocity for model RF. Contours give the magnitude of the velocities in mm/yr. The differences between the predicted current uplift rate for models UD1, UD2, and SLV1 and the reference model RF are shown in B, C, and D, respectively.

Strain Rates

The differences in horizontal and vertical motions give rise to differences in strain rates. Two complementary forms of strain estimates are the dilatational strain rate and the effective strain rate. The former is related to deformation that causes a fractional change in volume and is given by:

$$\dot{\theta} = \frac{1}{3}\dot{\varepsilon}_{rr} = \frac{1}{3}\left(\dot{\varepsilon}_{11} + \dot{\varepsilon}_{22} + \dot{\varepsilon}_{33}\right). \quad (5)$$

The effective strain rate is the second invariant of the strain deviator $\dot{\varepsilon}'_{ij} = \dot{\varepsilon}_{ij} - \frac{1}{3}\dot{\varepsilon}_{rr}\delta_{ij}$ and is given by:

$$\dot{\varepsilon}_{\text{effective}} = \sqrt{\frac{1}{2}\dot{\varepsilon}'_{ij}\dot{\varepsilon}'_{ij}} = \sqrt{\frac{1}{2}\left(\dot{\varepsilon}'^2_{11} + \dot{\varepsilon}'^2_{22} + \dot{\varepsilon}'^2_{33}\right) + \dot{\varepsilon}'^2_{12} + \dot{\varepsilon}'^2_{13} + \dot{\varepsilon}'^2_{23}}. \quad (6)$$

The strain rates are calculated at 5 km depth, in the upper part of the crust. The magnitude of the strain rates is small, on the order of 10^{-10} yr^{-1} (strain-rate values have been multiplied by 10^{10} before being contour plotted in Figs. 8, 9, and 10). These magnitudes are in general agreement with the finding of James and Bent (1994), who showed that strain rates due to glacial unloading are ~1–3 orders of magnitude larger than average seismic strain rates.

Figure 8 shows the present-day dilatational rates. The contours are positive where the land, which suffered compression during glacial loading, is now decompressing. By comparing Figures 8B, 8C, and 8D with Figure 8A, one can see that ductile

Present-Day Rate of Dilatation x 10^{10} (1/yr)

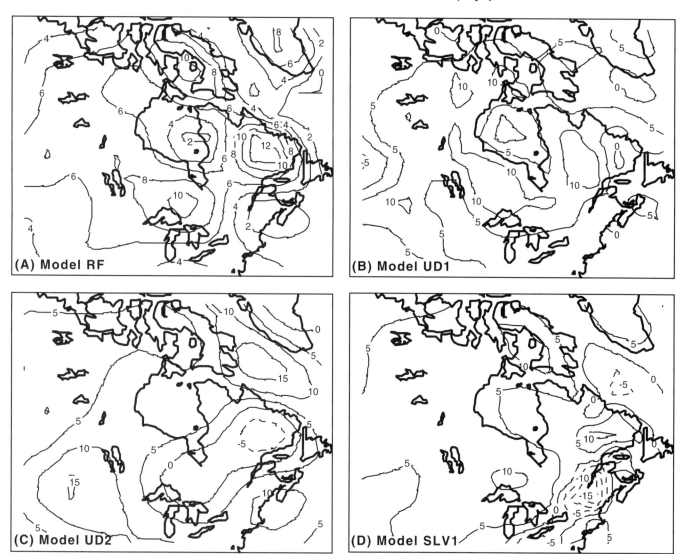

Figure 8. Contour plots of the present-day rate of dilatation multiplied by 10^{10}. The models used are (A) RF, (B) UD1, (C) UD2, and (D) SLV1. Contour interval is 5 units except for A. Dashed contours are for negative values.

layers have large effects on the dilatational rate. A lower viscosity in the uniform ductile layer causes a significant decrease in the dilatational rate in southern Quebec and Labrador (Figs. 7B and 7C). Similarly, the ductile zone along the St. Lawrence valley (model SLV2) gives a peak rate of 20×10^{-10} yr^{-1} in the east Labrador coast compared to 10×10^{-10} yr^{-1} predicted by SLV1. The weak St. Lawrence valley model also predicts large contraction rates between Quebec City and Montreal (Fig. 7D), where all other models predict dilatation. These variations, compared to the reference model, are mainly along the weak zone (Fig. 7D).

Figure 9 shows that the effective strain rates are significantly larger than the dilatational rates. The low viscosity in the ductile layer in model UD1 gives slightly larger present-day effective

strain rates than the reference model RF (Fig. 9B). In contrast, as viscosity decreases to 10^{20} Pa s in UD2, the effective strain rates decrease slightly. With a ductile zone along the St. Lawrence valley, the effective strain rate becomes much more concentrated near the weak zone, and the magnitude increases to the peak value of 60×10^{-10} yr^{-1} (Fig. 9D). The effective strain rates for SLV2 are similar to those shown in SLV1 except for peak values of 70×10^{-10} yr^{-1}.

Figure 10 shows the horizontal shear strain rates, $\dot{\varepsilon}_{12}$, at the present time. The effect of a uniformly thick ductile layer is significant along the coastal area from Baffin Island to Labrador. Comparisons of Figures 10B and 10C show that a reduction in viscosity to 10^{20} Pa s causes a large reduction in the strain rate.

Present-Day Effective Strain Rate x 10^{10} (1/yr)

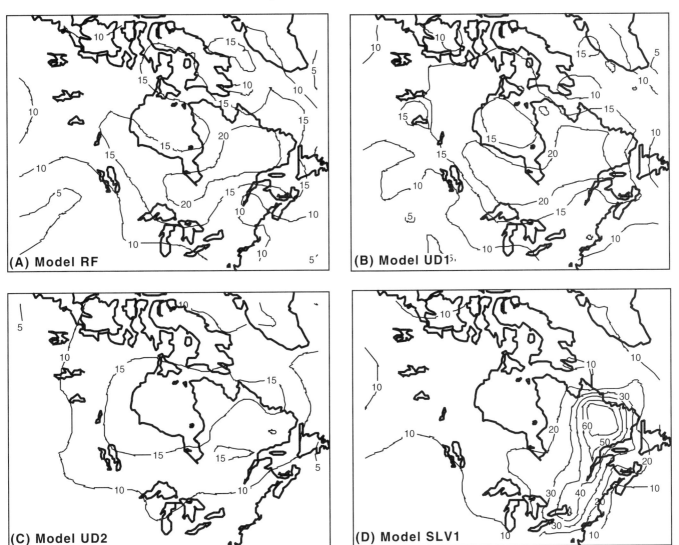

Figure 9. Similar to Figure 7 except for present-day effective (shear) strain rate multiplied by 10^{10}. Contour intervals are 5 units, except for D, where interval is 10 units.

A ductile zone along the St. Lawrence valley causes the strain rates to be concentrated near the weak zone and the magnitude to increase by a factor of 2–10, to more than 40×10^{-10} yr^{-1}. However, as viscosity decreases further as in SLV2, the change decreases. For sites near Charlevoix, the increase in strain rate is a factor of 8 compared with the reference model and a factor of 5–6 compared to models UD1 or UD2.

These large variations in strain rates might be detectable by local, high-quality GPS measurements. However, GPS strain rate estimates are typically noisier than velocity estimates. Campaign GPS data along the lower St. Lawrence valley indicate that the horizontal strain rates are essentially uniaxial E-W shortening at ~20–40 × 10^{-10} yr^{-1} (±10–40 × 10^{-10} yr^{-1}) (Mazzotti et al., 2005).

These measurements are at the limit of resolution of the GPS data, and thus they cannot discriminate between the different models discussed here.

CONCLUSIONS

The effect of a lithospheric ductile layer with uniform thickness generally increases as its viscosity decreases from 10^{21} to 10^{20} Pa s; the only exceptions are for fault stability, δFSM, and strain rate, $\dot{\varepsilon}_{12}$. A narrow ductile zone that cuts vertically through the lower crust and lithospheric mantle generally has a larger effect on crustal motion and strain rates than a horizontal, uniform, 25-km-thick ductile layer. A smaller viscosity in the narrow

Present-Day $\dot{\varepsilon}_{12}$ Strain Rate x 10^{10} (1/yr)

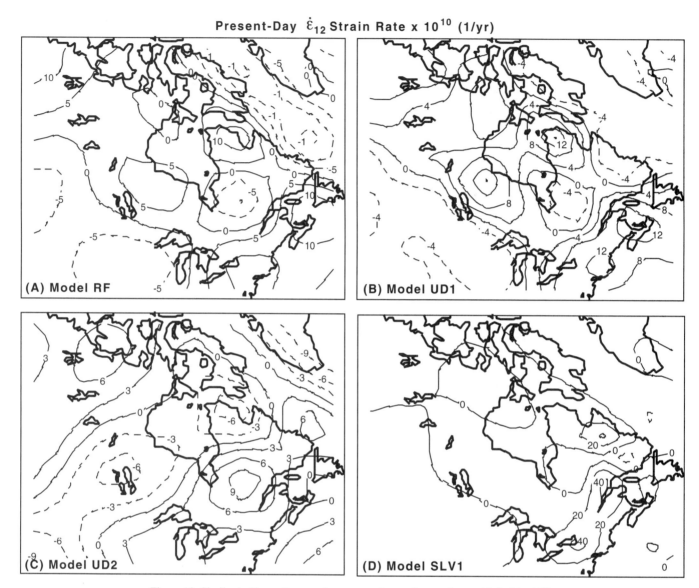

Figure 10. Similar to Figure 8 except for present-day strain rate, $\dot{\varepsilon}_{12}$, multiplied by 10^{10}.

vertical ductile zone generally has similar effects to the higher-viscosity model, except for slightly smaller amplitudes.

For sites within the ice margin, the effect of lithospheric weak zones on δFSM is ~0.5–0.6 MPa, but the effect on the timing of instability onset is insignificant. For sites outside the ice margin, the onset timing of instability is more sensitive to the presence of the lithospheric weak zone. Provided that ICE-3G is an accurate description of the deglaciation history, the models with the narrow ductile zone along the St. Lawrence valley have the best chance of explaining the timing of the Wabash Valley earthquakes (among all the models considered).

Among the models considered, the models with the narrow ductile zone along St. Lawrence Valley have the largest effects on uplift rate, horizontal velocity, and present-day strain rate. The deviations from the reference model are large enough to be resolvable by high-resolution GPS measurements. As the accuracies of GPS measurements improve in the coming years, the effects of possible ductile zones can be resolved, and this will give us a better understanding of the dynamics of the lithosphere and the implications for seismotectonics in eastern Canada.

ACKNOWLEDGMENTS

We would like to thank Seth Stein (editor), Detlef Wolf, and an anonymous reviewer for their useful comments and suggestions. Calculations were performed with the ABAQUS package from Hibbitt, Karlsson, and Sorensen, Inc. This research was supported by a discovery grant from Natural Sciences and Engineering Research Council of Canada. This is Geological Survey of Canada contribution 2005802.

REFERENCES CITED

Adams, J., 1989, Postglacial faulting in eastern Canada: Nature, origin and seismic hazard implications: Tectonophysics, v. 163, p. 323–331, doi: 10.1016/0040-1951(89)90267-9.

Adams, J., 1995, The Canadian Crustal Stress Database—A compilation to 1994: Part I: Geological Survey of Canada Open-File Report 31223122, 38 p.

Adams, J., 1996, Paleoseismology in Canada: A dozen years of progress: Journal of Geophysical Research, v. 101, p. 6193–6207, doi: 10.1029/95JB01817.

Adams, J., and Bell, J.S., 1991, Crustal stresses in Canada, *in* Slemmons, D.B., et al., eds., Neotectonics of North America: Boulder, Colorado, Geological Society of America, Decade of North American Geology, Decade Map, v. 1, p. 367–386.

Anderson, J.G., 1986, Seismic strain rates in central and eastern United States: Bulletin of the Seismological Society of America, v. 76, p. 273–290.

Chen, W.P., and Molnar, P., 1983, Focal depths of intracontinental and intraplate earthquakes and their implications for the thermal and mechanical properties of the lithosphere: Journal of Geophysical Research, v. 88, p. 4183–4215.

Clark, M.K., and Royden, L.H., 2000, Topographic ooze: Building the eastern margin of Tibet by lower crustal flow: Geology, v. 28, p. 703–706, doi: 10.1130/0091-7613(2000)28<703:TOBTEM>2.0.CO;2.

Di Donato, G., Mitrovica, J.X., Sabadini, R., and Vermeersen, L.L.A., 2000, The influence of a ductile crustal zone on glacial isostatic adjustment: Geodetic observables along the U.S. East Coast: Geophysical Research Letters, v. 27, p. 3017–3020, doi: 10.1029/2000GL011390.

Dixon, J.E., Dixon, T.H., Bell, D.R., and Malservisi, R., 2004, Lateral variations in upper mantle viscosity: Role of water: Earth and Planetary Science Letters, v. 222, p. 451–467, doi: 10.1016/j.epsl.2004.03.022.

Evans, B., and Kohlstedt, D.L., 1995, Rheology of rocks, *in* Ahrens, T.J., ed., Rock Physics and Phase Relations: A Handbook of Physical Constants: Washington, D.C., American Geophysical Union, Reference Shelf Volume 3, p. 148–165.

Gordon, R.G., 1998, The plate tectonic approximation: Plate non-rigidity, diffuse plate boundaries, and global plate reconstructions: Annual Review of Earth and Planetary Sciences, v. 26, p. 615–642, doi: 10.1146/annurev.earth.26.1.615.

Grollimund, B., and Zoback, M.D., 2001, Did deglaciation trigger New Madrid seismicity?: Geology, v. 29, p. 175–178, doi: 10.1130/0091-7613(2001)029<0175:DDTISI>2.0.CO;2.

Hansen, F.D., and Carter, N.L., 1982, Creep of selected crustal rocks at 1000 MPa: Eos (Transactions, American Geophysical Union), v. 63, p. 437.

Herrmann, R.B., 1979, Surface wave focal mechanisms for eastern North American earthquakes with tectonic implications: Journal of Geophysical Research, v. 84, p. 3543–3552.

Hirth, G., and Kohlstedt, D.L., 1996, Water in the oceanic upper mantle; implications for rheology, melt extraction and the evolution of the lithosphere: Earth and Planetary Science Letters, v. 144, p. 93–108, doi: 10.1016/0012-821X(96)00154-9.

Ivins, E.R., James, T.S., and Klemann, V., 2003, Glacial isostatic stress shadowing by the Antarctic ice sheet: Journals of Geophysical Research, v. 108, no. B12, 2560, doi: 10.1029/2002JB002182.

Jackson, J., 2002, Strength of the continental lithosphere: Time to abandon the jelly sandwich?: GSA Today, v. 12, no. 9, p. 4–10, doi: 10.1130/1052-5173(2002)012<0004:SOTCLT>2.0.CO;2.

James, T.S., 1991, Post-glacial Deformation [Ph.D. thesis]: Princeton, New Jersey, Princeton University Press, 190 p.

James, T.S., and Bent, A.L., 1994, A comparison of eastern North American seismic strain-rates to glacial rebound strain-rates: Geophysical Research Letters, v. 21, p. 2127–2130, doi: 10.1029/94GL01854.

Johnston, A.C., 1987, Suppression of earthquakes by large continental ice sheets: Nature, v. 330, p. 467–469, doi: 10.1038/330467a0.

Johnston, A.C., 1989, The effect of large ice sheets on earthquake genesis, *in* Gregersen, S., and Basham, P.W., eds., Earthquakes at North Atlantic Passive Margins: Neotectonics and Postglacial Rebound: Dordrecht, Kluwer Academic, p. 581–599.

Kaufman, P.S., and Royden, L.H., 1994, Lower crustal flow in an extensional setting: Constraints from the Halloran Hill region, eastern Mojave Desert: Journal of Geophysical Research, v. 99, p. 15,723–15,739, doi: 10.1029/94JB00727.

Kaufmann, G., and Wu, P., 2002, Glacial isostatic adjustment on a three-dimensional laterally heterogeneous Earth: Examples from Fennoscandia and the Barents Sea, *in* Mitrovica, J.X., and Vermeersen, L.L.A., eds., Ice Sheets, Sea Level and the Dynamic Earth: American Geophysical Union Geodynamics Series 29, p. 293–309.

Kaufmann, G., Wu, P., and Ivins, E.R., 2005, Lateral viscosity variations beneath Antarctica and their implications on regional rebound motions and seismotectonics: Journal of Geodynamics, v. 39, no. 2, p. 165–181, doi: 10.1016/j.jog.2004.08.009.

Kendall, R., Mitrovica, J.X., and Sabadini, R., 2003, Lithospheric thickness inferred from Australian post-glacial sea-level change: The influence of a ductile crustal zone: Geophysical Research Letters, v. 30, no. 9, p. 1461, doi: 10.1029/2003GL017022.

Klemann, V., and Wolf, D., 1999, Implications of a ductile crustal layer for the deformation caused by the Fennoscandian ice sheet: Geophysical Journal of the Interior, v. 139, p. 216–226, doi: 10.1046/j.1365-246X.1999.00936.x.

Langbein, J., Heflin, M., Hurst, K., Kedar, S., Herring, T., King, N.E., and Prawirodirdjo, L., 2002, Noise level in Southern California Integrated GPS Network (SCIGN) data; Preliminary results: Eos (Transactions, American Geophysical Union), v. 83, no. 47, p. F377.

Larson, K.M., and van Dam, T., 2000, Measuring postglacial rebound with GPS and absolute gravity: Geophysical Research Letters, v. 27, p. 3925–3928, doi: 10.1029/2000GL011946.

Mackwell, S.J., Zimmerman, M.E., and Kohlstedt, D.L., 1998, High-temperature deformation of dry diabase with application to tectonics on Venus: Journal of Geophysical Research, v. 103, p. 975–984, doi: 10.1029/97JB02671.

Mao, A.C., Harrisin, G.A., and Dixon, T.H., 1999, Noise in GPS coordinate time series: Journal of Geophysical Research, v. 104, p. 2797–2816, doi: 10.1029/1998JB900033.

Mareschal, J.C., Jaupart, C., Gariepy, C., Cheng, L.Z., Guillou-Frotier, L., Bienfait, G., and Lapointe, R., 2000, Heat flow and deep thermal structure near the southeastern edge of the Canadian Shield: Canadian Journal of Earth Sciences, v. 37, p. 399–414, doi: 10.1139/cjes-37-2-3-399.

Mazzotti, S., 2007, Geodynamic paradigms for earthquake studies in intraplate regions, *in* Stein, S., and Mazzotti, S., ed., Continental Intraplate Earthquakes: Science, Hazard, and Policy Issues: Geological Society of America Special Paper 425, doi: 10.1130/2007.2425(02).

Mazzotti, S., and Adams, J., 2005, Rates and uncertainties on seismic moment and deformation in eastern Canada: Journal of Geophysical Research, v. 110, p. B09301, doi: 10.1029/2004JB003510.

Mazzotti, S., and Hyndman, R.D., 2002, Yakutat collision and strain transfer across the northern Canadian Cordillera: Geology, v. 30, p. 495–498, doi: 10.1130/0091-7613(2002)030<0495:YCASTA>2.0.CO;2.

Mazzotti, S., James, T.S., Henton, J., and Adams, J., 2005, GPS crustal strain, postglacial rebound, and seismic hazard in eastern North America: The St. Lawrence valley example: Journal of Geophysical Research, v. 110, p. B11301, doi: 10.1029/2004JB003590.

Nabelek, J., and Suarez, G., 1989, The 1983 Goodnow earthquake in central Adirondacks, New York: Rupture of a simple, circular crack: Bulletin of the Seismological Society of America, v. 79, p. 1762–1777.

Obermeier, S.F., Bleuer, N.R., Munson, C.A., Munson, P.J., Martin, W.S., McWilliams, K.M., Tabaczynski, D.A., Odum, J.K., Rubin, M., and Eggert, D.L., 1991, Evidence of strong earthquake shaking in the lower Wabash Valley from prehistoric liquefaction features: Science, v. 251, p. 1061–1063, doi: 10.1126/science.251.4997.1061.

Peltier, W.R., 1998, Postglacial variations in the level of the sea: Implications for climate dynamics and solid-earth geophysics: Reviews of Geophysics, v. 36, p. 603–689, doi: 10.1029/98RG02638.

Pollitz, F.F., Kellogg, L., and Burgmann, R., 2001, Sinking mafic body in a reactivated lower crust: A mechanism for stress concentration at the New Madrid seismic zone: Bulletin of the Seismological Society of America, v. 91, p. 1882–1897, doi: 10.1785/0120000277.

Quinlan, G., 1984, Postglacial rebound and the focal mechanisms of eastern Canadian earthquakes: Canadian Journal of Earth Sciences, v. 21, p. 1018–1023.

Richardson, R.M., and Reding, L.M., 1991, North American plate dynamics: Journal of Geophysical Research, v. 96, p. 12,201–12,223.

Rydelek, P.A., and Pollitz, F.F., 1994, Fossil strain from the 1811–1812 New Madrid earthquakes: Geophysical Research Letters, v. 21, p. 2303–2306, doi: 10.1029/94GL02057.

Sella, G.F., Stein, S., Dixon, T.H., Craymer, M., James, T.S., Mazzotti, S., and Dokka, R.K., 2007, Observation of glacial isostatic adjustment in "stable" North America with GPS: Geophysical Research Letters, v. 34, L02306, doi: 10.1029/2006GL027081.

Shilts, W.W., Rappol, M., and Blais, A., 1992, Evidence of late and postglacial seismic activity in the Temiscouata-Madawaska Valley, Quebec—New Brunswick, Canada: Canadian Journal of Earth Sciences, v. 29, p. 1043–1059.

Spada, G., Yuen, D.A., Sabadini, R., and Boschi, E., 1991, Lower-mantle viscosity constrained by seismicity around deglaciated regions: Nature, v. 351, p. 53–55, doi: 10.1038/351053a0.

Stein, S., Sleep, N.H., Geller, R.J., Wang, S.C., and Kroeger, G.C., 1979, Earthquakes along the passive margin of eastern Canada: Geophysical Research Letters, v. 6, p. 537–540.

Stein, S., Cloetingh, S., Sleep, N.H., and Wortel, R., 1989, Passive margin earthquakes, stresses and rheology, *in* Gregersen, S., and Basham, P.W., eds., Earthquakes at North Atlantic Passive Margins: Neotectonics and Postglacial Rebound: Dordrecht, Kluwer Academic Publications, p. 231–259.

Tushingham, A.M., and Peltier, W.R., 1991, Ice-3G: A new global model of late Pleistocene deglaciation based upon geophysical predictions of postglacial relative sea level change: Journal of Geophysical Research, v. 96, p. 4497–4523.

Walcott, R.I., 1970, Isostatic response to loading of the crust in Canada: Canadian Journal of Earth Sciences, v. 7, p. 716–727.

Wilks, K.R., and Carter, N.L., 1990, Rheology of some continental lower crustal rocks: Tectonophysics, v. 182, p. 57–77, doi: 10.1016/0040-1951(90)90342-6.

Wu, P., 1996, Changes in orientation of near-surface stress field as constraints to lower-mantle viscosity and horizontal principal tectonic stress difference in eastern Canada: Geophysical Research Letters, v. 23, p. 2263–2266, doi: 10.1029/96GL02149.

Wu, P., 1997, Effect of viscosity structure of fault potential and stress orientations in eastern Canada: Geophysical Journal of the Interior, v. 130, p. 365–382.

Wu, P., 1998a, Will earthquake activity in eastern Canada increase in the next few thousand years?: Canadian Journal of Earth Sciences, v. 35, p. 562–568, doi: 10.1139/cjes-35-5-562.

Wu, P., 1998b, Intraplate earthquakes and postglacial rebound in eastern Canada and northern Europe, *in* We, P., ed., Dynamics of the Ice Age Earth: A Modern Perspective: Uetikon-Zuerich, Switzerland, Trans Tech Publications, p. 603–628.

Wu, P., 2002, Effects of mantle flow law stress exponent on postglacial induced surface motion and gravity in Laurentia: Geophysical Journal of the Interior, v. 148, p. 676–686, doi: 10.1046/j.1365-246X.2002.01620.x.

Wu, P., 2005, Effects of lateral variations in lithospheric thickness and mantle viscosity on glacially induced surface motion in Laurentia: Earth and Planetary Science Letters, v. 235, p. 549–563, doi: 10.1016/j.epsl.2005.04.038.

Wu, P., and Hasegawa, H., 1996a, Induced stresses and fault potential in eastern Canada due to a disc load: A preliminary analysis: Geophysical Journal of the Interior, v. 125, p. 415–430.

Wu, P., and Hasegawa, H., 1996b, Induced stresses and fault potential in eastern Canada due to a realistic load: A preliminary analysis: Geophysical Journal of the Interior, v. 127, p. 215–229.

Wu, P., and Johnston, P., 2000, Can deglaciation trigger earthquakes in N. America?: Geophysical Research Letters, v. 27, p. 1323–1326, doi: 10.1029/1999GL011070.

Wu, P., Johnston, P., and Lambeck, K., 1999, Postglacial rebound and fault instability in Fennoscandia: Geophysical Journal of the Interior, v. 139, p. 657–670, doi: 10.1046/j.1365-246x.1999.00963.x.

Zoback, M.L., 1992a, First- and second-order patterns of stress in the lithosphere: The World Stress Map Project: Journal of Geophysical Research, v. 97, p. 11,703–11,728.

Zoback, M.L., 1992b, Stress field constraints on intraplate seismicity in eastern North America: Journal of Geophysical Research, v. 97, p. 11,761–11,782.

MANUSCRIPT ACCEPTED BY THE SOCIETY 29 NOVEMBER 2006

The Geological Society of America
Special Paper 425
2007

Popup field in Lake Ontario south of Toronto, Canada: Indicators of late glacial and postglacial strain

Robert D. Jacobi[†]

UB Rock Fracture Group, Department of Geology, 876 NSC, University at Buffalo, Buffalo, New York 14068, USA

C.F. Michael Lewis[‡]

Geological Survey of Canada Atlantic, Natural Resources Canada, Box 1006, Dartmouth, Nova Scotia B2Y 4A2, Canada, and
Graduate School of Oceanography, University of Rhode Island, Narragansett, Rhode Island 02882, USA

Derek K. Armstrong[§]

Ontario Geological Survey, Ministry of Northern Development and Mines,
993 Ramsey Lake Road, Sudbury, Ontario P3E 6B5, Canada

Stephan M. Blasco[#]

Geological Survey of Canada Atlantic, Natural Resources Canada, Box 1006, Dartmouth, Nova Scotia B2Y 4A2, Canada

ABSTRACT

A field of stress-release bedrock structural features occurs on the floor of western Lake Ontario south of Toronto, Canada. These features were investigated using side-scan and multibeam sonars, high-resolution seismic profiling, and submersible dive observations. The study region was mostly stripped of its glacial drift in late glacial time, and the region has since accumulated only a relatively thin, discontinuous cover (1–2 m) of lacustrine sediment. The stress-release features affect the flat to gently dipping interbedded shales and calcareous siltstones of the Upper Ordovician Georgian Bay Formation. The features consist of sub-lakefloor buckles, about 50–100 m wide with structural relief of 5+ m, and surface bedrock popups, 10–15 m wide with a general relief of 1–2 m. Deeper bedrock faults are possibly associated with some of the sub-lakefloor buckles. Trends of the popups and buckles can be grouped into six modes from 7.5° to 347.5°. Abutting and sediment onlap relationships suggest that the popups formed throughout late and postglacial time following the Last Glacial Maximum ~20,000 yr ago. The earliest set of popups is estimated to have formed before 9500 B.P.; they trend WNW, collinear with isobases of glacial rebound, and do not parallel major geophysical or structural linear zones in the region. These and other factors suggest that this set developed in response to glacial rebound-induced stress. Later

[†]E-mail: rdjacobi@geology.buffalo.edu.
[‡]E-mail: miklewis@nrcan.gc.ca.
[§]E-mail: derek.armstrong@ndm.gov.on.ca.
[#]E-mail: sblasco@nrcan.gc.ca.

Jacobi, R.D., Lewis, C.F.M., Armstrong, D.K., and Blasco, S.M., 2007, Popup field in Lake Ontario south of Toronto, Canada: Indicators of late glacial and postglacial strain, *in* Stein, S., and Mazzotti, S., ed., Continental Intraplate Earthquakes: Science, Hazard, and Policy Issues: Geological Society of America Special Paper 425, p. 129–147, doi: 10.1130/2007.2425(10). For permission to copy, contact editing@geosociety.org. ©2007 The Geological Society of America. All rights reserved.

popups form an irregular pattern with several orientations of axes, suggesting that the horizontal principal stress vectors were of similar magnitude. The decrease of rebound strain with time and clockwise rotation of modern contours of basin tilting relative to glacial lake isobases suggest that popups today are likely a response to reduced glacial stress combined with far-field tectonic stress.

Keywords: popups, Lake Ontario, faults, glacial rebound, seismicity.

INTRODUCTION

Popups are elongate anticlinal structures of meter-scale amplitude that form in surface rocks under horizontal compressive stress. These rocks usually have horizontal planes of weakness that allow the upper buckling units to decouple from the lower unaffected units (Roorda et al., 1982; Wallach et al., 1993). Horizontal rock strata in eastern North America commonly exhibit popups (also called buckles, pressure ridges, or stream anticlines) where overburden is thin (e.g., Wallach et al., 1993). In quarries, where the overburden or surface bedrock load has been removed, popups may form rapidly on quarry floors (e.g., Adams, 1982). Some geologists have suggested that popups may be an indicator of regions that are vulnerable to moderate- to large-magnitude earthquakes (e.g., Thomas et al., 1989a, 1989b; Wallach et al., 1993).

Individual popups and groups of a few popups have been extensively documented in the regions surrounding Lake Ontario (Sbar and Sykes, 1973; Fakundiny et al., 1978; Karrow, 1987; Wallach et al., 1993; McFall, 1993; Rutty and Cruden, 1993; Karrow and White, 2002). However, extensive fields of popups, and lakefloor popups, were unknown until the late 1980s, when R.L. Thomas and others first recognized a field of popups on sidescan sonar records that imaged a swath about 500 m wide on the floor of Lake Ontario south of Toronto (Thomas et al., 1989a, 1989b). The popup field was confirmed by systematic sidescan sonar and seismic-reflection surveys in western Lake Ontario (Lewis et al., 1992). These surveys revealed an extensive lakefloor area approximately 16 km (east-west) by 6 km (north-south) of thin sediment cover and bedrock outcrop with popups at a spacing of about 2–3 per km of survey line (Lewis et al., 1995).

Thomas et al. (1993) described the popups in Lake Ontario as beds dipping abruptly away from a broken axial zone (i.e., angular buckles), with dimensions reaching 7 m wide, 1.5–2 m high, and 1.5 km long. Thomas et al. (1993, p. 332) believed that the popups "pierce through the overlying late- and postglacial sedimentary cover, making a geologically young age a distinct possibility." It was thought that the young age might indicate neotectonic activity close to Toronto (e.g., Thomas, et al., 1993; Wallach et al., 1993). The proximity of the popup field to known relatively elevated levels of seismicity in the western Lake Ontario basin, such as the Toronto-Hamilton seismicity zone (e.g., Mohajer, 1993; Bowlby et al., 1996; Wallach et al., 1998; Boyce and Morris, 2002; Jacobi, 2002; see Fig. 1), supported the concern that these popups might indicate unstable near-surface crustal conditions. The possibility that these popups might indicate a broader earthquake source zone in close proximity to Toronto warranted further investigation.

A 1 km × 5 km portion of the popup field approximately 4 km south of the Toronto Islands (Fig. 1B) has been studied extensively over the past 11 years by the authors and their associates using high-resolution boomer (seismic-reflection profiler), sidescan and multibeam sonars, and submersible dives (e.g., Armstrong et al., 1996). The current report presents results of the ongoing studies, including characterization of popups, as well as popup age relationships derived from abutting relationships of popup trends and sediment onlaps onto up-tilted limbs of the popups. These results are based on images of 228 popups (although some are continuous from ship track to ship track).

METHODS

We used Klein sidescan and Simrad 3000 multibeam sonar records to map the structural features in the surficial bedrock. From the sidescan sonar records, we also measured the length and width of features and described the geometry of the features and the structural intersections (e.g., abutting relationships). Relief and subsurface structure were obtained from the high-resolution boomer seismic profiles (Huntec deep tow, 0.2–0.3 m resolution). The profiles of sediment-covered popup limbs were scrutinized for relative age information, including whether unconsolidated sediment layers lapped onto the structure (i.e., sediment younger than the structure) or unwarped over the structure (i.e., sediment older than the structure). Thicknesses and depths in the profiles were based on assumed sound velocities of 1500 m s^{-1} in unlithified sediment and 4725 m s^{-1} in bedrock. The bedrock velocity was adopted from Hobson's (1960) determination of P-wave velocities from reflection and refraction surveys for the Meaford-Dundas (now Georgian Bay) Formation (which crops out in the survey area), and it is consistent with velocities deduced from synthetic seismograms by Beinkafner (1983) for correlative units in western New York State. The velocity of the Georgian Bay Formation observed on a sonic log conducted in a hole ~20 km west of our survey site in Mississauga, Ontario (Johnson, 1983) is lower than that determined from seismic surveys.

Visual observations of some structural features were made from the Canadian Armed Forces submersible SDL-1 (Armstrong et al., 1998). Video and standard camera images of the lakebed were obtained during transects over features. Customized instruments were used to make measurements and take

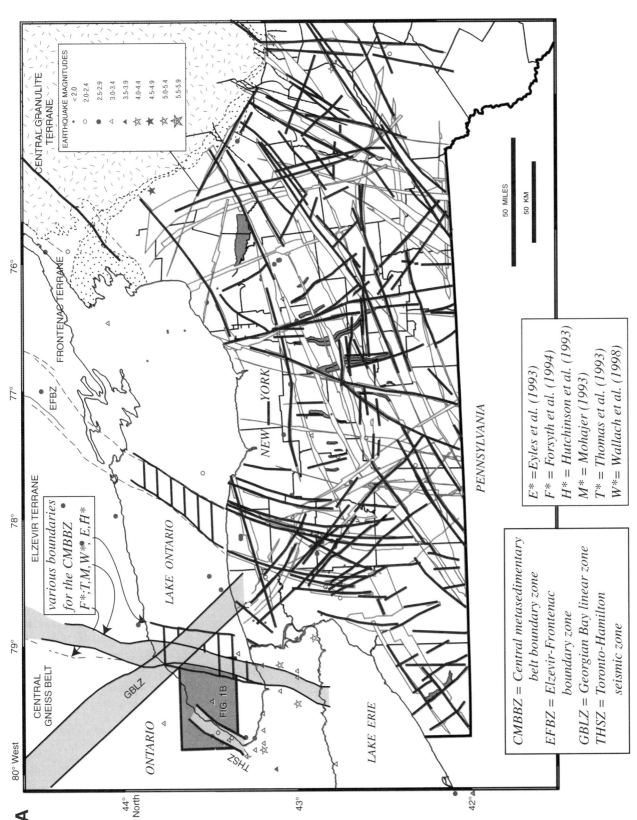

Figure 1 (*on this and following page*). Study location maps. (A) Generalized fault map displaying selected major fault systems, lineament trends, and location of seismic events (after Jacobi, 2002). Location of B is indicated by the labeled green area. Yellow bands indicate potential extent of fault systems, based on EarthSat (1997) lineament bundles. Brown lines indicate potential and known faults based on integrated studies (see Jacobi, 2002). (B) Regional bathymetry of Lake Ontario surrounding the survey area and location of the study area (bathymetry in meters; after Virden et al., 1999).

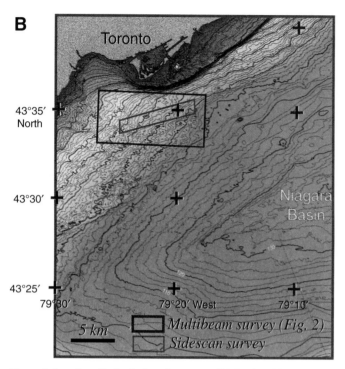

B

Toronto

43°35′
North

43°30′

43°25′

79°30′   79°20′ West   79°10′

5 km   Multibeam survey (Fig. 2)
          Sidescan survey

Niagara
Basin

Figure 1 (*continued*). Study location maps. (B) Regional bathymetry of Lake Ontario surrounding the survey area and location of the study area (bathymetry in meters; after Virden et al., 1999).

samples. We laid scaled metal squares, 1 m × 1 m and 0.5 m × 0.5 m, on the outcrops for image analysis of bedrock fracture patterns. A 0.5-m-long, scaled pipe was employed to probe the sediment cover and as a scale for photographs. Push-core tubes (0.4 m long) were utilized to sample the unconsolidated sediment cover. The submersible's manipulator arms and claws were used to obtain bedrock samples.

RESULTS

Bathymetry and Surficial Geology

The imaged portion of the popup field is located in water depths between ~55 m and ~95 m on the multibeam data (Fig. 2). Bedrock here and to the west consists of flat-lying to gently dipping interbedded shales and calcareous siltstones of the Upper Ordovician Georgian Bay Formation (e.g., Johnson 1983). In most of the multibeam region, the bedrock surface is nearly exposed at the lake bottom and regionally slopes southeastward with the lakebed as shown in the bathymetry of Figure 1B. The southern flank of the Laurentide ice sheet covered and extended ~150 km south of Lake Ontario (Dyke and Prest, 1987). Overlying sediments in the field of popups, interpreted from surface samples and boomer seismic profiles, include sporadic thin (<1 m) patches of glacial till, as well as thin discontinuous occurrences of glacial lacustrine and postglacial clastic, silt-clay sediment. The region of thin unconsolidated sediment is bounded to

the north and south by relatively thick sequences (up to 40 m and more) of continuous glacial, proglacial, and postglacial sediment over the bedrock surface (Lewis et al., 1995). This widely distributed thick sequence is thought to have been removed from the region of visible bedrock popups during and shortly after deglaciation, between about 14,000 and 9000 ^{14}C yr B.P. (about 16,700 and 10,100 cal. yr B.P.), possibly first by concentrated subglacial flows of meltwater (Lewis et al., 1999), and later by shoreface erosion during low stages of early Lake Ontario (Anderson and Lewis, 1985).

The postglacial lacustrine silty clay cover is generally <2 m thick, and therefore represents average sedimentation rates of about 0.1–0.2 mm/yr if sediment accumulation was continuous after removal of the glacial and proglacial sediment sequences. The lacustrine sediment, though relatively thin, has not been studied in detail. However, pollen grains in short cores (<30 cm long) recovered during submersible dives are differentially preserved, possibly as a result of episodic exposure to oxidizing conditions (Sunderland, 1998). These conditions suggest that the upper sediments have been episodically resuspended and redeposited during and following episodes of strong bottom currents throughout deposition of the lacustrine sediment sequence. These conditions are consistent with visual observations of the lakefloor, which is veneered with an easily disturbed "fluff" composed of loose particles of fine-grained sediment and organic detritus that would likely be removed during periods of accelerated bottom-water movement. This dynamic environment apparently results in a relatively low net rate of sediment accumulation compared with offshore basins, where accumulation rates over the past one to two centuries are known to be in the range of 0.5 to 1.0 mm/yr (Kemp and Harper, 1976; Farmer, 1978). Thus, the lacustrine sediments in the popup area may have accumulated at rates up to 1 mm/yr for short periods, but they likely have been winnowed and reworked intermittently by bottom currents, resulting in a long-term net accumulation rate of about 0.1–0.2 mm/yr.

Popup Geometry

The bathymetric relief across the axial zone of the popups is generally about 1–2 m, based on the record from the downward-looking beam on the Klein sidescan sonar instrument. However, a submersible dive measured a relief approaching 3 m on a popup in the northeastern part of the study area (dive 1040, location shown on Fig. 2). The detectable width of the popups is variable, but it is usually <10–15 m (based on the sidescan sonar records).

Popup Trends

The popups can be grouped into sets based on the orientations of 228 popups observed on the sidescan records. The popup sets strike primarily NNE, NE, ENE, WNW, NW, and NNW (Fig. 3), with frequency modes at 7.5°, 42.5°, 67.5°/77.5°, 277.5°/287.5°, 322.5°, and 347.5°, respectively, as determined from the sidescan records and confirmed in the multibeam image. The longest linear

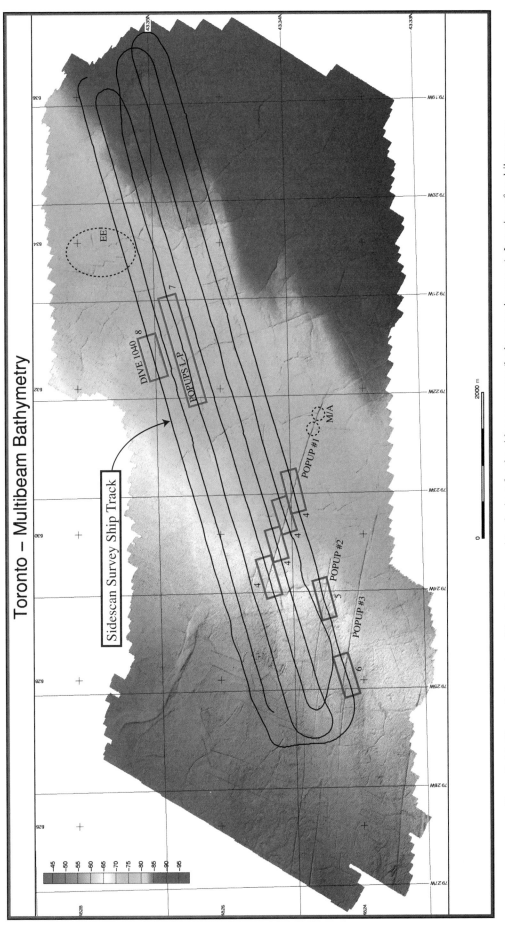

Figure 2. Multibeam bathymetry of the study area and track chart for the sidescan survey (bathymetry in meters). Location of multibeam survey is shown in Figure 1B. Red rectangles indicate locations of sidescan sonar images displayed in other figures; number adjacent to rectangle denotes the figure number in which the feature is portrayed. Labeled popups are discussed in the text and featured in other figures. The two circled popup intersections along popup #1 (labeled "M/A") illustrate master and abutting relationships: WNW-striking popup #1 is master to both of the northerly striking popups; i.e., the northerly popups abut popup #1 in the circled areas. The development of popups in area EE area may have been controlled by fractures/riedel shears above a subsurface NNW-striking wrench fault.

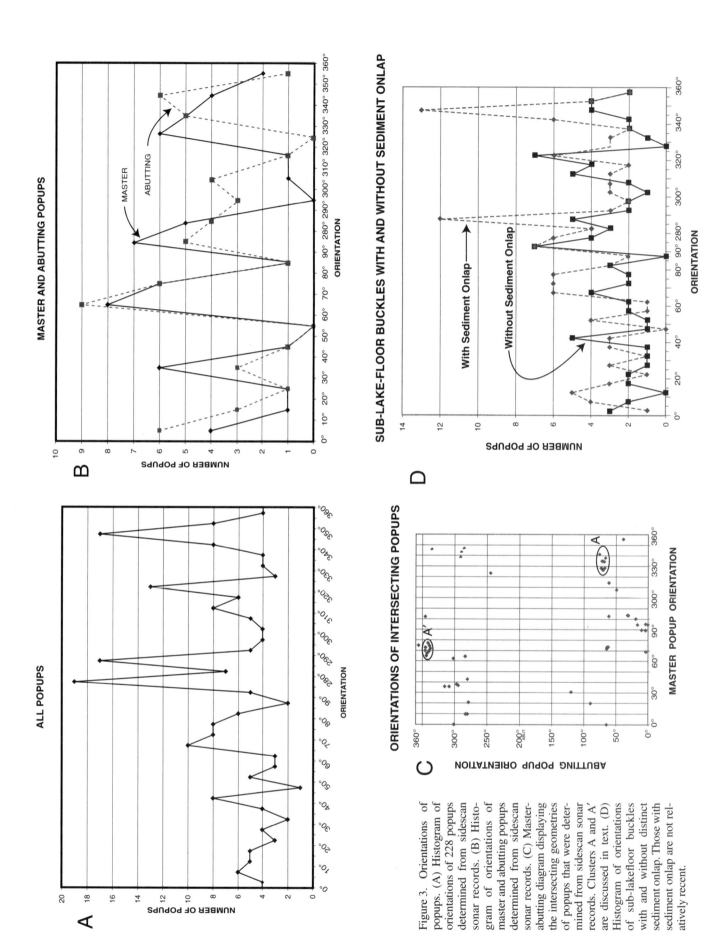

Figure 3. Orientations of popups. (A) Histogram of orientations of 228 popups determined from sidescan sonar records. (B) Histogram of orientations of master and abutting popups determined from sidescan sonar records. (C) Master-abutting diagram displaying the intersecting geometries of popups that were determined from sidescan sonar records. Clusters A and A′ are discussed in text. (D) Histogram of orientations of sub-lakefloor buckles with and without distinct sediment onlap. Those with sediment onlap are not relatively recent.

popups trend WNW and reach lengths of 2–3 km in the southwestern and central part of the multibeam survey area (Fig. 2). These results support Thomas et al.'s (1993) contention that the set of popups with the cumulative longest length strike WNW (280°). The multibeam image also shows that popups with other trends are significantly shorter, and some have curvilinear to "ragged" axes. The popups generally describe an irregular pattern across the survey region, compared to either the WNW-striking popups or to a "brick" pattern common to fractured bedrock in regions surrounding Lake Ontario (e.g., to the south; Jacobi, 2002). The pattern of NNE-striking popups in area EE on Figure 2 has an appearance similar to en-echelon fractures or riedel shears above an (inferred) subsurface NNW-striking, dextral wrench fault. If the contemporary maximum principal compressive stress trends ENE, collinear with that measured onshore, then the proposed fault is approximately orthogonal to the contemporary maximum principal compressive stress; the fault therefore should experience primarily dip-slip motion, not wrench motion. We suggest, therefore, that the riedal shears/fractures developed in an older stress field, and that the riedel shears/fractures influenced the orientation of the developing more recent popups. Although the NNW-trend is accentuated in the multibeam data, the sidescan survey data indicate that NW-striking popups are nearly as prevalent as the NNW-striking features (Fig. 3A). Northwesterly (WNW, NW, and NNW) striking popups account for 60% of the popups observed on the sidescan records, similar to the fraction of northwesterly trending popups in eastern North America (two thirds of 337 popups), as found by Wallach et al. (1993).

Bedrock Sub-Lakefloor Buckles and Faults

High-resolution boomer records reveal that many of the narrow axial zones of the popups at the lakefloor are located at the edges of deeper structural blocks that have undergone rigid body rotations and have structural relief of 5+ m. The widths of the tilted blocks are significantly wider than the surficial popups. For example, the WNW-striking popup structure #1 (location in Fig. 2) has a narrow axial popup (<10 m wide, as displayed on Fig. 4A), but the boomer record indicates that the upwarped sub-lakefloor[1] bedrock forms a buckle (anticline or popup) ~50 m wide on line 5 in Figure 4B (correcting for ship's track orientation). Similar relationships can be observed for structural features #2 (Fig. 5) and #3 (Fig. 6). For structural feature #3 (Fig. 6), the sub-lakefloor buckle is ~100 m, whereas the surface popup is ~20 m wide.

Many of the sub-lakefloor buckles are asymmetrical. For example, the southwest limb of structural feature #1 (Fig. 4) is twice as long as, and dips more gently than, the northeastern limb on line 5. The asymmetry of the sub-lakefloor buckles is not consistent across the map area. For example, popup #3 (Fig. 6) has the long limb southwest of the axis, but the adjacent, collinear

popup to the south (feature MM on Fig. 6) has the long limb on the northeast side of the axis.

There is an apparent gradation between significantly asymmetric sub-lakefloor buckles and what appear to be faults (as defined by reflector offsets across a narrow zone). Part of the apparent gradation is a function of resolution in the boomer records. For example, popup #3 (Fig. 6) may be located on a fault that separates essentially flat-lying beds to the north from upwarped bedrock dipping homoclinally south. However, the resolution of the boomer record does not allow definitive recognition of the fault in the records, and in fact, the "fault" could be a very short, steeply dipping limb chiefly hidden in the hyperbolic reflectors at the popup. Our interpretation is that this discontinuity may be a fault, but the lack of deeper penetration in the boomer records does not allow us to determine if the sedimentary layers are offset below the obvious upper 5–10 m of the sub-lakefloor buckle. The nondefinitive nature of this interpretation is common for many of the faults we have tentatively identified (Table 1). For comparison, a structural feature we are confident is a fault is shown in Figure 7.

An additional factor that can lead to equivocal interpretations of faults involves apparent offsets of the lakefloor—these interpretations are important because such an offset would imply modern tectonism. However, in the boomer records we examined, careful inspection of the "lakefloor offsets" showed that most of the offsets resulted from sediment ponding at a popup. For example, in Figure 8, recent sediment onlaps the east-dipping limb of a sub-lakefloor buckle, which leads to the question: did this popup form a dam that allowed the sediment to pond at a higher level on the east, or was the clearly ponded sediment later offset along a fault at the western edge of the sub-lakefloor buckle? In our interpretation, the thicker interval between reflectors on the ponded side suggests that the popup acted as a dam, and significant faulting of these sediments is considered less likely. The equivocal nature of this interpretation is obvious.

Another factor that can contribute to equivocal interpretations of lakefloor faults are cuestas on dipping beds. For example, an apparent fault at "FFF" near popup #2 (Fig. 5) appears to offset the lakefloor. However, the apparent lakefloor offset may rather represent a cuesta on gently west-dipping layers. Note the subtly west-dipping reflectors immediately east of the feature (in the apparent trough), and the truncation of these reflectors at the lakefloor. Additionally, the strong sub-bottom reflectors may be continuous across the apparent fault zone. Minor sediment ponding may also play a minor role in the apparent lakefloor offset.

Table 1 presents features we conservatively judge are likely to be faults. We also included monoclines in the fault category, since the monoclines offset the bedrock units across the structure (unlike the usual surficial popup). An example of the probable faults noted in Table 1 is displayed in Figure 7, where feature "M" offsets all the subbottom reflectors by 2.5 m. The small number of faults in Table 1 makes any general conclusions drawn from the data tenuous.

---

[1]Hereafter, "sub–lakefloor" will be used to distinguish the wider buckles that involve units beneath the lakefloor from the narrow, surficial popups.

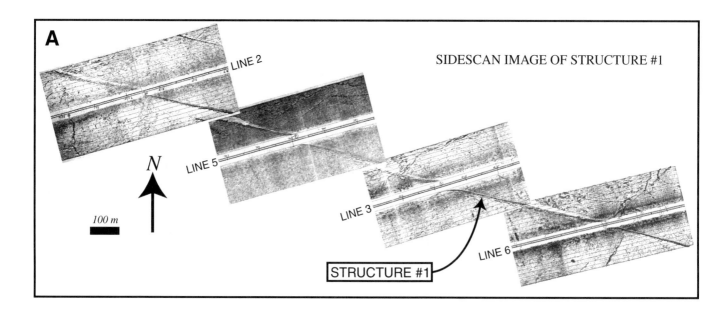

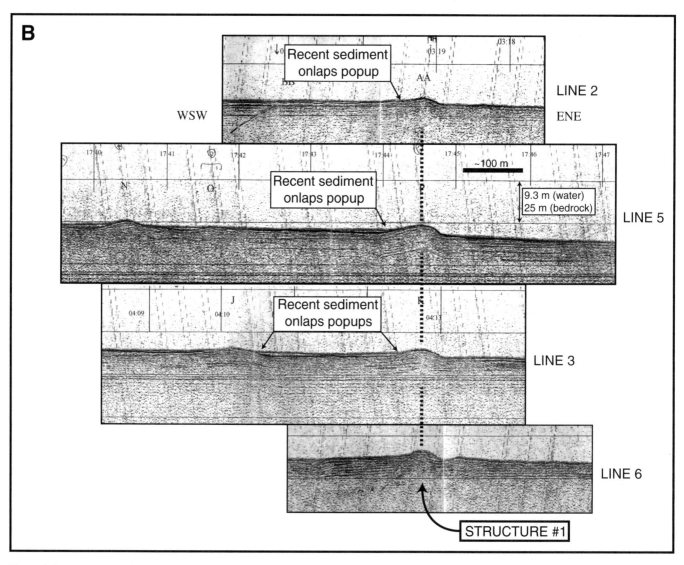

Figure 4. Structure #1 (a sub-lakefloor buckle). (A) Sidescan image of structure 1. (B) High-resolution "boomer" records (from a Huntec DTS [Deep Tow System] unit) across structure 1. Note that structure 1 is onlapped by sediment; the thickness of sediment suggests that the sub-lakefloor buckle has not been active for >9500 yr, based on a conservative sedimentation rate of 0.1–0.2 mm/yr. For location of the images, see Figure 2.

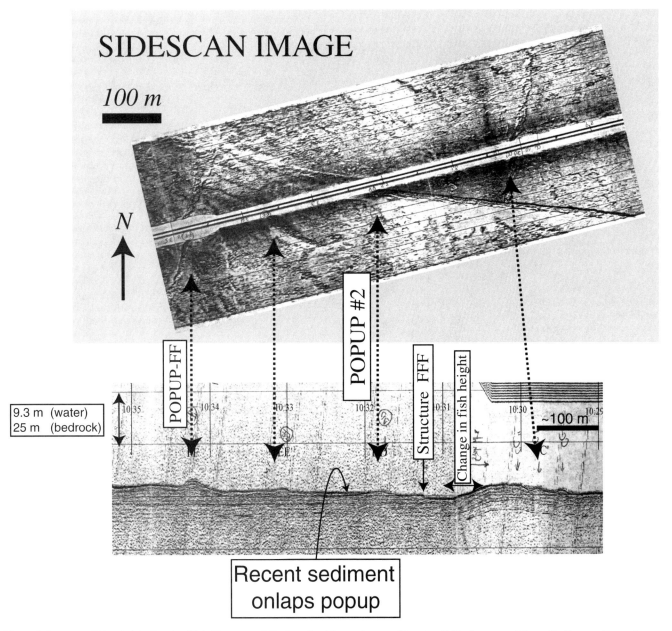

Figure 5. Structure #2 (a sub-lakefloor buckle). Upper panel displays the sidescan image of structure 2. The lower panel shows the "boomer" record across structure 2. Note that structure 2 is onlapped by sediment, suggesting that the sub-lakefloor buckle has not been active for at least 8000–19,000 yr, based on a conservative sedimentation rate of 0.1–0.2 mm/yr. The feature "FFF" demonstrates the equivocal nature of apparent "fault" interpretations in the boomer records. Feature FFF appears to offset the lakefloor (and is therefore a recent fault), but the apparent lakefloor offset may rather represent a cuesta on gently west-dipping layers (see text for further discussion). For location of the images, see Figure 2.

We identified faults that trend NS, NNE, NE, EW to WNW, and NW (Table 1). About 64% of the faults trend EW to WNW. The apparent sense of offset is variable for all fault trends except the WNW-striking faults, which consistently show a down-on-the-SSW sense of offset, based on reflector offsets (Table 1). A fault that strikes NE (48.5°) locally displays a down-to-the-NW offset, but more generally exhibits a down-to-the-SE offset. The latter offset is consistent with the multibeam image, which exhibits sharp drops in bathymetry along ENE trends in the eastern part of the survey (Fig. 2). Perhaps these sharp bathymetric drops represent additional NE-striking faults or monoclines.

Popup Age I: Relative Age from Abutting Relationships

The relative age of the initial development of popups can be determined from the intersection geometries of their axial trends, following accepted principles in fracture genera-

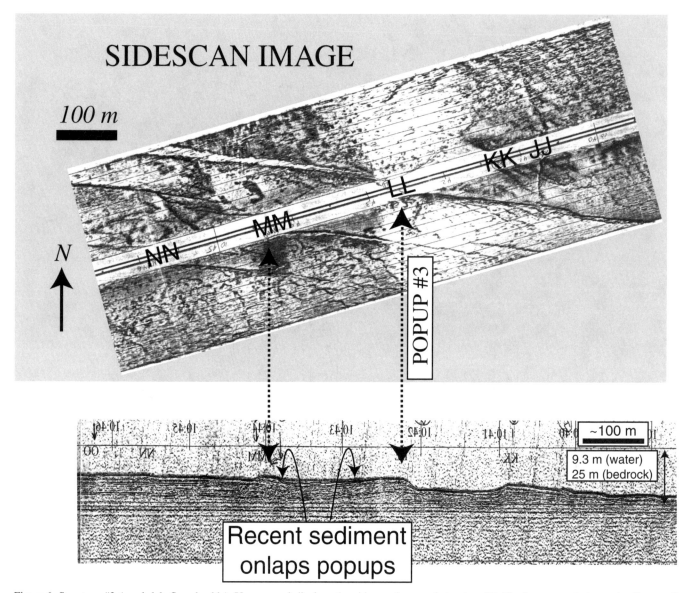

Figure 6. Structure #3 (a sub-lakefloor buckle). Upper panel displays the sidescan image of structure #3. The lower panel shows the "boomer" record across structure 3. Note that structure 3 is onlapped by sediment, suggesting that the sub-lakefloor buckle has not been active for at least 8000–19,000 yr, based on a conservative sedimentation rate of 0.1–0.2 mm/yr. For location of the images, see Figure 2.

TABLE 1. POSSIBLE FAULTS OF DEEPER BEDROCK AND/OR SURFACE UNITS

Line	Feature	Orientation	Affects surface	Sedimentary cover?	Apparent offset down to:	Notes
Line 6		350.5	?	Y	E	Asymmetrical buckle with long limb sloping down to W, "fault" on E side
Line 6	F	2.5	?	N	W	Monocline/fault
Line 2	Z	19.5	Y	N	W	Possibly surface sediment offset
Line 4	N	48.5	?	Y	NW +SE	Down on west locally, and down on the east regionally
Line 3	B	271	Y	N	E	Anticline/monocline
Line 6	NN	284	?	Y	NNE	
Line 6	S	289	N	N	NNE	
Line 1	Z	306.5	Y	N	NNE	Monocline
Line 1	M	307	Y	N	WSW	
Line 4	I	317.5	N	?	SSW	Thin sediment drape is faulted

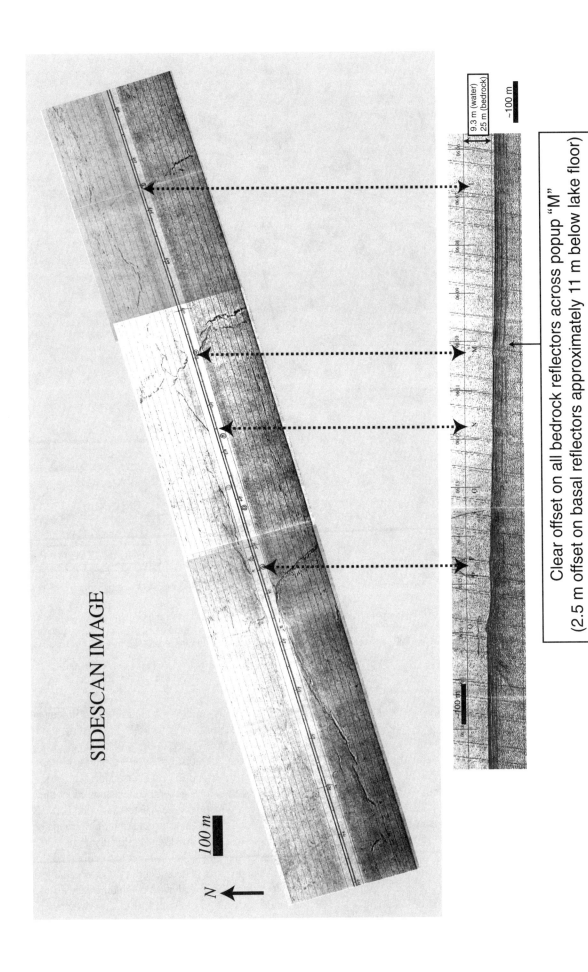

Figure 7. Line #4 structures (sub-lakefloor buckles). Upper panel displays the sidescan image of structures imaged on ship track #4. The lower panel shows the "boomer" record across the same structures. For location of the images, see Figure 2.

SIDESCAN IMAGE

N

100 m

Clear offset on all bedrock reflectors across popup "M"
(2.5 m offset on basal reflectors approximately 11 m below lake floor)

9.3 m (water)
25 m (bedrock)

~100 m

~100 m

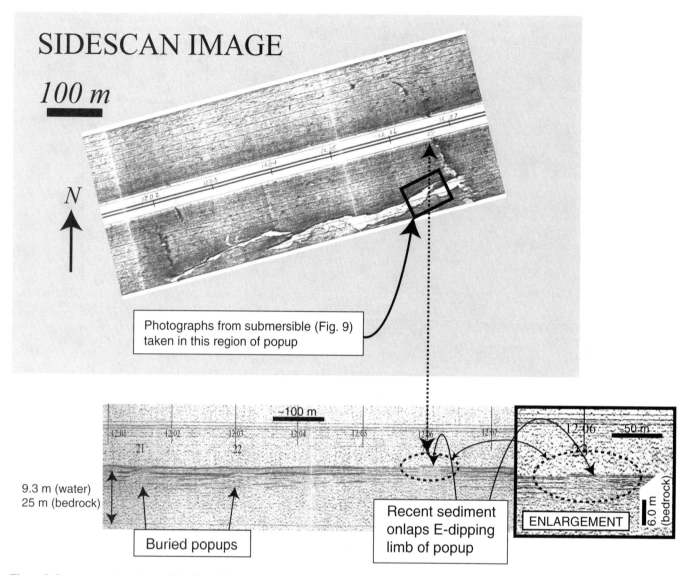

SIDESCAN IMAGE

100 m

N

Photographs from submersible (Fig. 9)
taken in this region of popup

~100 m

9.3 m (water)
25 m (bedrock)

Buried popups

Recent sediment
onlaps E-dipping
limb of popup

ENLARGEMENT

Figure 8. Structures at the submersible dive 1040 site. Upper panel displays the sidescan image of structures imaged in the region of the submersible dive. The lower panel shows the "boomer" record across the same structures. Note the buried structures #21, #22, and #23, suggesting that these sub-lakefloor buckles have not been active for >9500 yr, based on a conservative sedimentation rate of 0.1–0.2 mm/yr. For location of the images, see Figure 2.

tion (Pollard and Aydin, 1988). In the simplest case, the later popup abuts the pre-existing master (though-going) popup. For example, in Figure 2, the circled popup intersections at "M/A" show that popup #1 is master to the northerly trending popups (which abut popup #1); popup #1 is thus thought to be older than the northerly trending popups. Examination of the multibeam bathymetric map (Fig. 2) shows that the distinctly longest four popups, the WNW-striking popups in the southwestern and central parts of the survey area, are master to all other popup trends that intersect these four popups east of longitude 79°25′W. These master popups therefore predate the intersecting features. For example, near the southeastern end of popup #1 (near 79°22′W and at the circles "M/A" on Fig. 2), popup #1

is master to several N- and NNW-trending curvilinear popups. Similarly, popup #2 is master to a NNW-trending curvilinear popup (near 79°24′W; Fig. 2).

Except for the four distinct WNW-trending popups, most popup sets do not display consistent intersection geometries across the mapped area; therefore, general conclusions cannot be drawn concerning which popup sets are master and which are abutting. Figure 3B shows that both master and abutting popups occur at most orientation frequency modes. Additionally, Figure 3C, which displays master/abutting relationships for individual intersections among popup pairs (where identified), shows that groups of popup intersections generally have a complementary inverse intersection field. For example, group A (Fig. 3C), which represents

325°–345°–striking popups that are master to 65°–80°–striking popups, has an inverse field, group A', wherein 65°–80°–striking popups are master to 335°–350°–striking popups.

The apparent contradiction of popups with similar orientations displaying both abutting and master relationships with popups of another orientation (Fig. 3) suggests that most of the sets developed over a similar time period. The horizontal stress vectors controlling popup orientation may have switched from time to time to promote inconsistent master/abutting popup relationships across the area of interest; modeling shows that fractures can develop such inconsistent relationships across an area during times of "fracture-filling" without resorting to regional stress changes (Bai et al., 2002). The inconsistent master/abutting relationships across the survey area are also reflected in the irregular pattern of popups evident in the multibeam survey data.

Exceptions to the observation that popup sets have inconsistent intersection relationships across the map area are:

(1) the long WNW-trending popups (striking ~275°–290°) are master to NNW-striking (~340°–355°) popups (Fig. 2);

(2) popups striking 290°–308° abut other fracture sets; they are rarely masters (Figs. 3B and 3C);

(3) popups striking northeasterly (30°–40°) are master to popups striking northwesterly (290°–320°) (Fig. 3C); and

(4) popups striking 310°–315° are consistently master to popups striking 50°–60° (although we only have two intersections) (Fig. 3C).

These exceptions suggest that the long WNW-striking popups (and perhaps those that strike 310°–315°) may have developed relatively early and the 290°–308°–striking popups may have developed relatively late. The northeasterly (30°–40°) striking popups predate the 290°–308°–striking popups.

Popup Age II: Sediment Onlap Considerations

Popups are generally assumed to be younger than the last glacial cover, based on several considerations (see review in Wallach et al., 1993; Rutty and Cruden, 1993; Karrow and White, 2002). These lines of evidence include (1) upwarp of glacial and postglacial sediments over the developing popup, (2) short exposure time of popup "hogbacks" since weathering has not rounded the edges of the bedrock, (3) occurrence of popups in the thalwegs of postglacial stream beds, where the overburden has been removed by postglacial stream erosion, (4) occurrence of popups in quarries and other excavations where overburden has been removed anthropogenically, and (5) the assumption that popups would not occur under the normal loading of glacial ice, or if they did, they could not survive shear of warm-based ice across them (Wallach et al., 1993). In the case of the Lake Ontario popups, Thomas et al. (1993) believed that the sidescan sonar records indicated that the popups "pierced" postglacial sediments, from which they inferred that the popups were modern features. Earlier submersible dives on the popups also suggested a fairly modern age, since the edges

on popup "hogbacks" appeared to be angular (lacking rounded, weathered corners), consistent with the interpretation that the popups were modern. However, the "fresh-looking" (i.e., angular) appearance of the rock edges may be sustained over relatively long periods of time due to the relatively slow rate of mechanical weathering in this environment (Armstrong et al., 1998). Furthermore, grab samples from the dive site show extensive weathering "rinds" that are centimeters thick, indicating significant geochemical weathering, which is not consistent with a modern age of exposure.

The boomer data show that postglacial sediment clearly onlaps the limbs of more than half (60%) of the sub-lakefloor buckles (Fig. 3D), although regions with minimal (or no) recent sediments make this determination impractical in those areas. Buckles with significant sediment onlap imply that more than half of these are not modern. Especially notable are the NNW- and WNW-striking sets of sub-lakefloor buckles; in these sets, buckles with onlapped sediment are three times as prevalent as those without sediment onlap. Similarly, ENE-striking sub-lakefloor buckles with sediment onlap are twice as common as those without sediment onlap. In contrast, distinctly more NW-striking (300°–320°) sub-lakefloor buckles are without sediment onlap than with onlap (Fig. 3D).

All the sub-lakefloor buckle sets have a relatively long history of development that may continue into modern times, since each sub-lakefloor buckle set consists of some individual buckles with onlapped sediment and other sub-lakefloor buckles without onlapped sediment. However, the variations in the relative number of sediment-onlapped buckles in the different sub-lakefloor buckle sets suggest that, in general, some of the sets developed at different times. The NNW- and WNW-striking sets generally developed earliest, followed by the ENE-striking set, and the NW-striking sub-lakefloor buckles are the most recent. The orientation of these recent buckles is consistent with having formed in the contemporary stress field. Similarly, the suggestion that the WNW-trending sub-lakefloor buckles are oldest, based on sediment onlap relationships, is consistent with the abutting relationships discussed earlier.

All the long sub-lakefloor buckles along popups #1, #2, and #3 (Fig. 2) and collinear "MM" (Fig. 6), a subset of the WNW-striking sub-lakefloor buckles, are partly buried by lacustrine sediments (Figs. 4B, 5, and 6). Significant motion on any of these structures is not likely to have occurred recently. At least 1.9 m to 2.3 m of sediment onlaps these structures; thus, each of the WNW-striking sub-lakefloor buckles last experienced significant motion more than 9500 yr ago, using a long-term sedimentation rate of 0.1–0.2 mm/yr. Examples of the other set we believe to be relatively old, the NNW-striking set, are shown in Figure 8, where NNW-striking sub-lakefloor buckles #21 and #22 are totally buried by 3.0 m of sediments. These sub-lakefloor buckles therefore could be of an age similar to the WNW-striking set, or even older, up to 15,000 yr old.

Submersible SD-1 dove on a surficial, ENE-striking popup (dive 1040, Fig. 8) and found that the bedrock fractures had angular edges (Fig. 9), but the dive collected a 0.3 m push-core

Jacobi et al.

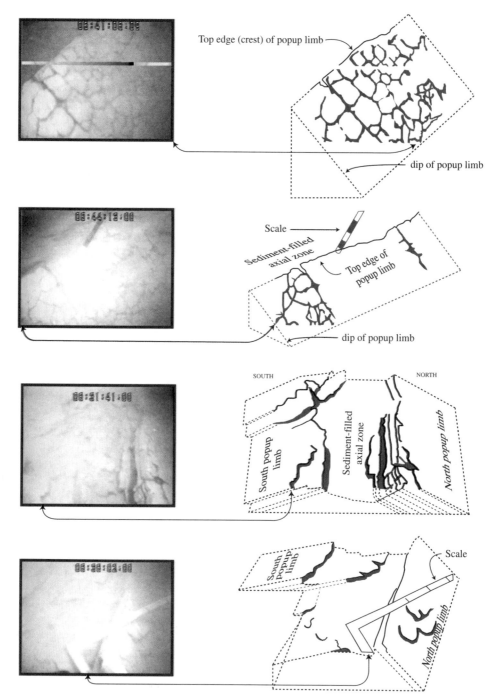

Figure 9. Photographs and interpretations from submersible of popup at dive site 1040. Upper two panels display the relatively steeply dipping, fractured north limb of popup. Lower two panels display the axial zone of popup (note opposing dips on opposite limbs and angular corners on limbs). Scale divisions are 10 cm in both panels. For location of the images taken at dive site 1040, see Figure 2.

of sediment in the axis of the popup. Thus, the axis of this popup is locally covered by at least 0.3 m of sediment, which could represent 1500–3000 yr given the range of inferred sedimentation rates. The maximum thickness of sediment in the axis is unknown, and so the true minimum age of the popup is also unknown.

We found four faults that may offset the lakefloor, and if so, they are modern (within the limits of resolution of the boomer records; Table 1): three trend WNW and one trends NNE. In addition to the problems discussed already in interpreting lakefloor offsets, sediment drape over a popup or sub-lakefloor buckle can have the appearance of a lakefloor offset on a fault. Note that the deeper reflectors do not deflect across the zone where at depth the fault should exist. Good-quality boomer records across the four faults indicate that we can rule out the sediment drape option for these four faults within the resolution of the boomer records.

ORIGIN AND DISCUSSION OF POPUPS

The longest and probably some of the oldest popups (older than about 9500 yr B.P.), the WNW-trending set, are collinear with the glacial rebound isobases that are based on warping of glacial Lake Iroquois beaches (Fig. 10). The collinearity between rebound isobases and the WNW-trending popups might indicate a causal relationship, as has been proposed by Adams (1989) for popups oriented parallel to the Laurentide ice front.

In the simplest model, the depression of the crust from glacial loading forms a chord across the former arc of Earth's surface and, therefore, results in shortening and compression. This compression could have resulted in popups, but these pop-

ups may not have survived basal glacial shear stress. However, a more realistic model involving plate flexure and dissipation of the peripheral bulge during glacio-isostatic rebound also results in a zone of compression. For example, in James and Bent's (1994) model of a spherical polar ice cap, rebound produces a present-day zone (~1000 km wide) in which horizontal compression is oriented radially from the center of the former ice sheet. The radial horizontal strain in this zone reaches a maximum of about -2×10^{-9} per year and has a northerly orientation in the popup area. The maximum strain is centered on the maximum extent of the former ice sheet (which was ~150 km south of the popup area). By considering both the vertical and horizontal stresses, James and Bent (1994) found that thrust faulting would be promoted in a zone about 400–500 km wide immediately north of the glacial maximum extent. These thrusts would strike parallel to the ice margin (and the rebound isobases) in the absence of relatively strong tectonic, far-field plate stresses. The popup area is within this zone of potential thrusting, and the oldest popup set, which strikes WNW (e.g., popups #1, #2, #3), is collinear with the thrust direction proposed from the modeling. As a cautionary note, however, stress calculations from a different rebound

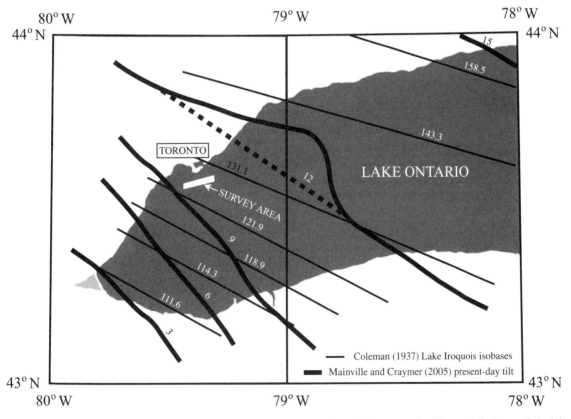

Figure 10. Map of western Lake Ontario showing the popup survey area (parallelogram south of Toronto), isobases of glacial Lake Iroquois (thin straight lines marked in elevations of meters above sea level [asl] from 111.6 m to 158.5 m), and contours of present basin tilting based on water-level gauge data (thick lines marked in values of tilt rate from 3 to 15 cm/100 yr). The 12 cm/100 yr contour has been smoothed (straightened, dashed section) for this analysis. Curves drawn through the survey area normal to the isobases and contours illustrate the clockwise rotation (15°–20°) of postglacial strain to a more southwesterly direction between the time of Lake Iroquois (11,700 ^{14}C yr B.P.; 13,550 cal. yr B.P.) and the present, respectively.

model, the ICE-3G (Tom James, 2005, personal commun.), indicate that S_H was directed easterly in times more recently than 12,000 yr B.P. (with a generally negligible S_h) (where S_H = maximum principal horizontal stress and S_h = minimum principal horizontal stress). However, from 16,000 to 12,000 yr. B.P., the ICE-3G model suggests that S_H and S_h were almost equal in magnitude and would have promoted thrusting. This time-frame is consistent with the NNW-striking sub-lakefloor buckles that could be as old as 15,000 years B.P., and the WNW-trending popups that are older than about 9500 yr B.P.).

By combining vertical strain rates with tangential and radial horizontal strain rates, James and Bent (1994) calculated that the total annual strain rate from rebound (10^{-9} per year) is presently two orders of magnitude greater than that which Anderson (1986) estimated for seismic strain rates in the eastern United States, and three orders of magnitude greater than the historical model of Mazzotti and Adams (2005), in which they calculated a present strain rate of 10^{-12} per year for the region that includes the popup area.

The presently relatively high modeled strain rate from rebound would have been even higher during initial deglaciation. Observed tilt rates for western Lake Ontario based on data from paleobeach elevations show that uplift and tilt rates have decayed exponentially since at least the time of glacial Lake Iroquois (11,700 ^{14}C B.P.; 13,550 cal. yr B.P.) (see Appendix; Lewis et al., 2007). Thus, strain rates calculated from a simplified glacial loading model and inferred from observed glacio-isostatic rebound suggest that the strain rates would have been relatively high during early deglaciation and that the orientation of thrusting during this high strain rate regime would have been strongly influenced by the rebound. The early age (> ca. 9500 yr B.P.) and orientation (WNW-striking) of the longest, early popups observed in the study area are thus consistent with a rebound stress mechanism. Further, this WNW trend is not parallel to proposed NNE- and NNW-striking tectonic zones in the Toronto region (e.g., Wallach et al., 1998).

The significant decrease in assumed radial horizontal strain based on rebound curves suggests that stress from far-field tectonic plate stresses would play an increasingly more important role as the rebound rate decreased. Thus, the development of the popups, as a result of the combined plate tectonic and rebound stresses, should increasingly reflect far-field plate tectonic stresses, rather than the stronger effect of the initially high rebound rates. The locally variable trends of the younger popup axes, and the wide range of popup set orientations, are consistent with this change in stress. Additionally, the clockwise rotation from the oldest popup set, which strikes WNW (about 275°–290°), to one of the youngest sets, which strikes 320°, is reflected in clockwise rotation of the basin tilt, ranging from about 016°–026° (NNE) for the Lake Iroquois isobases (11,700 ^{14}C yr B.P.; 13,550 cal. yr B.P.) to about 032°–048° (NE) at present for the contours of lake-level gauge data (Fig. 10). This rotation of the direction of horizontal strain, expressed by the empirical evidence of tilting

of the western part of the Lake Ontario basin, suggests that the popup area was strongly affected by glacial rebound in the millennia following the Last Glacial Maximum, but, as time progressed, the area was proportionately more influenced by tectonic stresses in the North American plate, which are known to exert maximum horizontal stress at present in an ENE-WSW direction (Zoback, 1992).

The contemporary horizontal compressive stress in the Toronto region is relatively high. The maximum horizontal compressive stress in the Paleozoic cover sequence is oriented generally ENE in the near-surface and has values of 6–13 MPa (see reviews in Wallach et al., 1993; Karrow and White, 2002). However, the minimum horizontal compressive stress is also relatively high, and varies between 4 and 9 MPa. Such values are not sufficient to cause failure in competent limestones; for example, the uniaxial strength of the calcareous beds in the Georgian Bay Formation varies from 140 MPa to 206 MPa (Lo et al., 1987; Lo and Cooke, 1989). However, Byerlee's law (1978) indicates that prefractured calcareous beds, typical of this area, could sustain further failure under the present stress conditions. Additionally, the interbedded shales have low uniaxial strengths of 11 MPa to 20 MPa (Lo et al., 1987; Lo and Cooke, 1989).

It is probably not fortuitous that some of the modes of popup orientations match the modes of fractures that Andjelkovic et al. (1996, 1997) measured north of Lake Ontario. For example, their ENE-striking fracture set at 81° is similar to the Lake Ontario ENE-striking set (modes at 67.5°/77.5°), and their set at 120° (300°) is comparable to our mode at 315°/322.5°, although the orientation in the lake appears to have rotated slightly clockwise. The ENE-striking set also occurs along the southern shore of Lake Ontario, where abutting relationships and intersection patterns suggest that this set has multiple times of generation, including pre- and post-Alleghanian development (Harper and Jacobi, 2000; Harper, 2001). The similarity in trend of these popup and fracture sets suggests that these popup sets probably utilized pre-existing joint sets, a conclusion also reached by Rutty and Cruden (1993). The fractures that are collinear with the popup axis evident in the photographs from the submersible dive are consistent with this interpretation (Fig. 9). The dip of these fractures (observed at the edge of the upturned beds) is orthogonal to the present dip on the popup limb, indicating that the fractures were rotated along with the bed, and therefore they predate the popup (Fig. 9).

Other popup sets do not have clear fracture equivalents on land north of Lake Ontario: the popup modes at 347.5°, 277.5°/287.5°, and 42.5° are distinctly underrepresented in the fracture data of Andjelkovic et al. (1996, 1997). However, these trends do occur across the lake along the southern shore of Lake Ontario (Harper and Jacobi, 2000; Harper, 2001). It is possible, therefore, that most of the major popup sets utilized preexisting fracture sets. In that case, the contemporary stress field, with both S_H and S_h relatively high, could have induced popups along these older fracture trends.

In summary, during initial glacio-isostatic rebound, the stress field associated with rebound near the Last Glacial Maximum was of such magnitude that it generated popups aligned along the rebound isobases. As the glacial rebound rate decreased, and the associated stresses dropped, the tectonic plate stresses became relatively more important. The result was popups with variable trends. The contemporary stress field in the Toronto area is sufficient to generate popups along pre-existing fractures in the interbedded calcareous and shale beds, but not in unfractured calcareous interbeds. From modeling, it might appear that the observed relatively high S_H and S_h values are partly a result of the rebound, and thus, the fact that popups, rather than near-surface strike-slip faults, occur in this region at present is a result of the vector addition of the rebound and tectonic stresses.

Whether these popups are a definitive warning indicator for deeper, moderate to large seismic events is beyond the scope of this report. However, the popups confirm that the present stress field is capable of rupturing the prefractured surface bedrock, and they suggest that stresses related to glacial rebound contributed to the formation of several of the popups. Because some earthquakes in eastern North America have been ascribed to glacial rebound (e.g., Stein et al., 1979; Mazzotti et al., 2005), the rebound-induced popups might indicate that the region has been susceptible to earthquakes resulting from a combination of rebound and tectonic stresses.

CONCLUSIONS

Our study of a field of popups in interbedded shales and calcareous siltstones of the Upper Ordovician Georgian Bay Formation on the Lake Ontario lakefloor south of Toronto Harbor resulted in the following conclusions:

- 228 popups (stress-relief features) observed by sidescan and multibeam sonars strike primarily in modes at 7.5°, 42.5°, 67.5°/77.5°, 277.5°/287.5°, 322.5°, and 347.5°.
- Popup relief is generally 1–2 m with a local maximum of 3 m.

- Two scales of compressive stress-release features were recognized:
  (1) Sub-lakefloor buckles, about 50–100 m wide, manifested as tilted blocks with structural relief of ~5+ m, were detected in high-resolution seismic profiles. Asymmetric buckles appear to grade to bedrock faults, two-thirds of which trend EW to WNW.
  (2) Surface bedrock popups, 10–15 m wide and with an axial fracture, commonly occur at the edges of the rotated structural blocks.
- Sediment onlap and abutting relationships suggest the popups formed throughout late and postglacial time following the Last Glacial Maximum about 20,000 yr ago. The abutting relationships indicate that buckles striking WNW formed relatively early, and popups striking 290°–320° developed relatively late.
- Abutting, younger popups form an irregular pattern of fractures throughout the popup field, suggesting that horizontal principal stress vectors were of similar magnitude, and probably locally switched episodically through postglacial time.
- The earliest popups (formed before 9500 yr B.P.) are master to most other popup sets, are the longest popups, are collinear with isobases of glacial rebound, and are not parallel to major geophysical or structural zones in the western Lake Ontario region.
- The clockwise rotation from earliest popups (striking ~275°–290°) to more recent popups (striking 290°–320°) is mirrored in the clockwise rotation of glacial lake rebound isobases to modern basin tilting.
- Previously published radial strain calculations for present glacial rebound and decreasing glacial rebound through time suggest that early popup formation was greatly influenced by stress induced by rebound. The decrease of rebound strain with time and clockwise rotation of modern contours of basin tilting relative to the glacial lake isobases suggest that popups today are likely the combined response to reduced glacial stress and far-field tectonic stress.

Jacobi et al.

APPENDIX

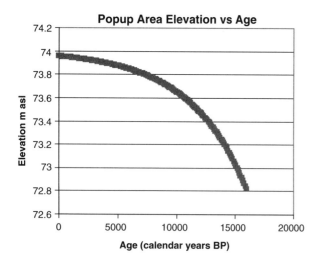

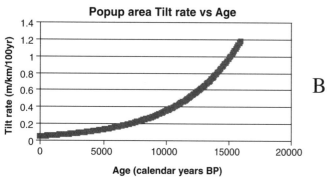

Figure A1. Effects of postglacial rebound in western Lake Ontario. (A) Change of elevation with time of the NNE end of a 1-km transect oriented in the direction of maximum postglacial rebound where the SSW end is fixed at 74 m a.s.l., in the popup area in western Lake Ontario. Since basin tilting is active at present, the initial position of the NNE end of the transect is selected so that this end reaches 74 m a.s.l. and is level with the SSW end of the transect about 4 periods of exponential relaxation, or 19,760 years in the future. (B) Change in tilt rate for a crustal segment downtilted to the NNW, in the popup area in western Lake Ontario. Since basin tilting is active at present, the segment becomes level about 4 periods of exponential relaxation or 19,760 years in the future. After Lewis et al. (2007).

ACKNOWLEDGMENTS

Drs. Owen White and Paul Karrow kindly took time to show onshore popups north of Lake Ontario to some of the authors in preparation for our examination of similar features on the floor of Lake Ontario. We thank the officers and crews of CCGS *Samuel Risley* and HMCS *Cormorant* for their enthusiastic interest and assistance in obtaining geophysical survey data and submersible data, respectively, from the popup study area. We are indebted to John Bowlby and Arsalan Mohajer, who assisted with many aspects of the submersible dive operations. We acknowledge the spirited discussions with Joe Wallach. The multibeam survey data were provided by the Canadian Hydrographic Service, Burlington, Ontario. We thank Darren Keyes, who constructed a mosaic of the sidescan records, and Robert Colville, who processed the multibeam data. Many other colleagues have contributed assistance in related research as well as stimulating discussions about the origin and significance of the Lake Ontario popups. We thank Paul Karrow, Alexander Cruden, and the editors for thoughtfully reviewing the manuscript.

REFERENCES CITED

Adams, J., 1982, Stress-relief buckles in the McFarland Quarry, Ottawa: Canadian Journal of Earth Sciences, v. 19, p. 1883–1887.

Adams, J., 1989, Postglacial faulting in eastern Canada: Nature, origin, and seismic hazard implications, *in* Morner, N.-A., and Adams, J., eds., Paleoseismicity and Neotectonics: Tectonophysics, v. 163, p. 321–331.

Anderson, J.G., 1986, Seismic strain rates in the central and eastern United States: Bulletin of the Seismological Society of America, v. 76, p. 273–290.

Anderson, T.W., and Lewis, C.F.M., 1985, Postglacial water-level history of the Lake Ontario basin, *in* Karrow, P.F., and Calkin, P.E., eds., Quaternary Evolution of the Great Lakes: Geological Association of Canada Special Paper 30, p. 231–253.

Andjelkovic, A.R., Cruden, A.R., and Armstrong, D.K., 1996, Structural geology of south-central Ontario: Preliminary results of joint mapping studies, *in* Baker, C.L., Thurston, P.C., Gerow, M.C., Kaszycki, C.A., Laderoute, D.G., Merlino, G., Newsome, J.A., Owsiacki, L., Spiers, G., and Fyar, J.A., eds., Summary of Field Work and Other Activities: Ontario Geological Survey Miscellaneous Paper 166, p. 103–107.

Andjelkovic, A.R., Cruden, A.R., and Armstrong, D.K., 1997, Joint orientation trajectories in south-central Ontario, *in* Ayer, J.A., Baker, C.L., Laderoute, D.G., and Thurston, P.C., eds., Summary of Field Work and Other Activities: Ontario Geological Survey Miscellaneous Paper 168, p. 127–133.

Armstrong, D.K., Lewis, C.F.M., Bowlby, J.R., Jacobi, R.D., and Mohajer, A.A., 1996, Submersible investigation of linear bedrock features on the floor of western Lake Ontario: A progress report, *in* Baker, C.L., Thurston, P.C., Gerow, M.C., Kaszycki, C.A., Laderoute, D.G., Merlino, G., Newsome, J.A., Owsiacki, L., Spiers, G., and Fyar, J.A., eds., Summary of Field Work and Other Activities: Ontario Geological Survey Miscellaneous Paper 166, p. 108–109.

Armstrong, D.K., Lewis, C.F.M., Jacobi, R.D., Bowlby, J.R., Mohajer, A.A., and White, O.L., 1998, A bedrock pop-up field beneath Lake Ontario, south of Toronto, Ontario: Geological Society of America Abstracts with Programs, v. 30, no. 7, p. A-295.

Bai, T., Maerten, L., Gross, M.R., and Aydin, A., 2002, Orthogonal cross joints: Do they imply a regional stress rotation?: Journal of Structural Geology, v. 24, p. 77–88, doi: 10.1016/S0191-8141(01)00050-5.

Beinkafner, K.J., 1983, Deformation of the Subsurface Silurian and Devonian Rocks of the Southern Tier of New York State [Ph.D. thesis]: Syracuse, New York, Syracuse University, 332 p.

Bowlby, J.R., Thomas, R.L., Jacobi, R.D., McMillan, R.K., and Lewis, C.F.M., 1996, Submersible observations of neotectonic structures and their implications for geologic hazards in the western Lake Ontario area: Geological Society of America Abstracts with Programs, v. 28, no. 3, p. 41.

Boyce, J.I., and Morris, W.A., 2002, Basement-controlled faulting of Paleozoic strata in southern Ontario, Canada: New evidence from geophysical lineament mapping, *in* Fakundiny, R.H., Jacobi, R.D., and Lewis, C.F.M., eds., Neotectonics and Seismicity in the Eastern Great Lakes Basin: Tectonophysics, Special Issue, v. 353, p. 151–171.

Byerlee, J.D., 1978, Friction of rocks: Pure and Applied Geophysics, v. 116, p. 615–626, doi: 10.1007/BF00876528.

Coleman, A.P., 1937, Lake Iroquois, *in* Forty-Fifth Annual Report of the Ontario Department of Mines, part 7, and Map No. 45f, scale: 1:316,800, p. 1–36.

Dyke, A.S., and Prest, V.K., 1987, Late Wisconsinan and Holocene Retreat of the Laurentide Ice Sheet: Geological Survey of Canada Map 1702A, scale 1:5,000,000.

EARTHSAT (Earth Satellite Corporation), 1997, Remote sensing and fracture analysis for petroleum exploration of Ordovician to Devonian fractures reservoirs in New York State: Albany, New York, New York State Energy Research and Development Authority, 35 p.

Eyles, N., Boyce, J., and Mohajer, A.A., 1993, The bedrock surface of the western Lake Ontario region: Evidence of reactivated basement structures?, *in* Wallach, J.L., and Heginbottom, J., eds., Neotectonics of the Great Lakes Area: Géographie physique et Quaternaire, v. 47, p. 269–283.

Fakundiny, R.H., Pomeroy, P.W., Pferd, J.W., and Nowak, T.A., 1978, Structural instability features in the vicinity of the Clarendon-Linden fault system, western New York and Lake Ontario, *in* Symposium on Advances in Analysis of Geotechnical Instabilities: Waterloo, Ontario, Canada, Study 13, p. 121–178.

Farmer, J.G., 1978, The determination of sedimentation rates in Lake Ontario using the ^{210}Pb dating method: Canadian Journal of Earth Sciences, v. 15, p. 431–437.

Forsyth, D.A., Milkereit, B., Zelt, C.A., White, D.J., Easton, R.M., and Hutchinson, D.R., 1994, Deep structure beneath Lake Ontario; crustal-scale Grenville subdivisions: Canadian Journal of Earth Sciences, v. 31, p. 255–270.

Harper, A., 2001, Fracture Analysis in Five 7.5′ Quadrangles on the Southwest Shore of Lake Ontario: Implications for an Extension of the St. Lawrence Rift System through Lake Ontario [M.A. thesis]: Buffalo, New York, University at Buffalo, 214 p.

Harper, A., and Jacobi, R.D., 2000, Fracture analysis along the southwest shores of Lake Ontario: Implications for extension of the St. Lawrence rift system through Lake Ontario: Geological Society of America Abstracts with Programs, v. 32, no. 1, p. A23.

Hobson, G.D., 1960, A reconnaissance seismic refraction and reflection survey in southwestern Ontario: Canadian Mining Journal, v. 81, no. 4, p. 83–87.

Hutchinson, D.R., Lewis, C.F.M., and Hund, G.E., 1993, Regional stratigraphic framework of surficial sediments and bedrock beneath Lake Ontario, *in* Wallach, J.L., and Heginbottom, J., eds., Neotectonics of the Great Lakes Area: Géographie physique et Quaternaire, v. 47, p. 337–352.

Jacobi, R.D., 2002, Basement faults and seismicity in the Appalachian Basin of New York State, *in* Fakundiny, R.H., Jacobi, R.D., and Lewis, C.F.M., eds., Neotectonics and Seismicity in the Eastern Great Lakes Basin: Tectonophysics, v. 353, p. 75–113.

James, T.S., and Bent, A.L., 1994, A comparison of eastern North American seismic strain-rates to glacial rebound strain-rates: Geophysical Research Letters, v. 21, p. 2127–2130, doi: 10.1029/94GL01854.

Johnson, M.D., 1983, Oil shale assessment project, deep drilling results 1982/83, Toronto region: Ontario Geological Survey Open-File Report 5477, 17 p.

Karrow, P.F., 1987, Quaternary geology of the Hamilton-Cambridge area, southern Ontario: Ontario Geological Survey Report 225, 94 p.

Karrow, P.F., and White, O.L., 2002, A history of neotectonic studies in Ontario, *in* Fakundiny, R.H., Jacobi, R.D., and Lewis, C.F.M., eds., Neotectonics and Seismicity in the Eastern Great Lakes Basin: Tectonophysics, v. 353, p. 3–15.

Kemp, A.L.W., and Harper, N.S., 1976, Sedimentation rates and sediment budget for Lake Ontario: Journal of Great Lakes Research, v. 2, p. 324–340.

Lewis, C.F.M., Sherin, A.G., Atkinson, A.S., Harmes, R.H., Jodrey, F.D., Nielsen, J.A., Parrott, D.R., and Todd, B.J., 1992, Report of Cruise 92–800, CCGS *Griffon*, Lake Ontario: Geological Survey of Canada Open-File Report 2678, 69 p.

Lewis, C.F.M., Cameron, G.D.M., King, E.L., Todd, B.J., and Blasco, S.M., 1995, Structural contour, isopach and feature maps of Quaternary sediments in western Lake Ontario: Atomic Energy Control Board (Canada) Report INFO-0555, 67 p.

Lewis, C.F.M., Barnett, P.J., Blasco, S.M., and Cameron, G.D.M., 1999, Subglacial erosion of diamicton beneath the Erie lobe of the Laurentide ice sheet about 13.5 ka: Canadian Quaternary Association Meeting Program and Abstracts, August 20–30, 1999, Calgary, Alberta, p. 41.

Lewis, C.F.M., Blasco, S.M., and Gareau, P.L., 2007, Glacial isostatic adjustment of the Laurentian Great Lakes basin: Using the empirical record of strandline deformation for reconstruction of early Holocene paleo-lakes and discovery of a hydrologically closed phase: Géographie physique et Quaternaire (2005), v. 59, p. 187–210.

Lo, K.T., and Cooke, B.H., 1989, Foundation design for the Skydome Stadium, Toronto: Canadian Geotechnical Journal, v. 26, p. 22–33.

Lo, K.T., Cooke, B.H., and Dunbar, D.D., 1987, Design of buried structures in squeezing rock in Toronto, Canada: Canadian Geotechnical Journal, v. 24, p. 232–241.

Mainville, A., and Craymer, M.R., 2005, Present-day tilting of the Great Lakes region based on water level gauges: Geological Society of America Bulletin, v. 117, p. 1070–1080, doi: 10.1130/B25392.1.

Mazzotti, S., and Adams, J., 2005, Rates and uncertainties on seismic moment and deformation rates in eastern Canada: Journal of Geophysical Research, v. 110, p. B09301, doi: 10.1029/2004JB003510.

Mazzotti, S., Thomas, J.S., Henton, J., and Adams, J., 2005, GPS crustal strain, postglacial rebound, and seismicity in eastern North America: The St. Lawrence Valley example: Journal of Geophysical Research, v. 110, p. B11301, doi: 10.1029/2004JB003590.

McFall, G.H., 1993, Structural elements and neotectonics of Prince Edward County, southern Ontario, *in* Wallach, J.L., and Heginbottom, J. eds., Neotectonics of the Great Lakes Area: Géographie physique et Quaternaire, v. 47, p. 303–312.

Mohajer, A.A., 1993, Seismicity and seismotectonics of the western Lake Ontario region, *in* Wallach, J.L., and Heginbottom, J., eds., Neotectonics of the Great Lakes Area: Géographie physique et Quaternaire, v. 47, p. 353–362.

Pollard, D.D., and Aydin, A., 1988, Progress in understanding jointing over the past century: Geological Society of America Bulletin, v. 100, p. 1181–1204, doi: 10.1130/0016-7606(1988)100<1181:PIUJOT>2.3.CO;2.

Roorda, J., Thompson, J.C., and White, O.L., 1982, The analysis and prediction of lateral stability in highly stressed, near-surface rock strata: Canadian Geotechnical Journal, v. 19, p. 451–462.

Rutty, A.L., and Cruden, A.R., 1993, Pop-up structures and the fracture pattern in the Balsam Lake area, southern Ontario, *in* Wallach, J.L., and Heginbottom, J., eds., Neotectonics of the Great Lakes Area: Géographie physique et Quaternaire, v. 47, p. 379–388.

Sbar, M.L., and Sykes, L.R., 1973, Contemporary compressive stress and seismicity in eastern North America: An example of intra-plate tectonics: Journal of Geophysical Research, v. 84, p. 1861–1882.

Stein, S., Sleep, N., Geller, R., Wang, S., and Kroeger, G., 1979, Earthquakes along the passive margin of eastern Canada: Geophysical Research Letters, v. 6, p. 537–540.

Sunderland, B.P., 1998, The Paleoseismicity of the Georgian Bay Linear Zone (GBLZ) Determined by Stratigraphic, Geophysical and Paleoenvironmental Indicators [B.S. thesis]: St. Catharines, Ontario, Department of Earth Sciences, Brock University, 121 p.

Thomas, R.L., McMillan, R.K., Keyes, D.L., and Mohajer, A.A., 1989a, Surface sedimentary signatures of neotectonism as determined by sidescan sonar in western Lake Ontario: Geological Association of Canada Program with Abstracts, v. 14, p. A128.

Thomas, R.L., McMillan, R.K., and Keyes, D.L., 1989b, Acoustic surveys: Implications to the geoscience discipline. Lighthouse: Journal of the Canadian Hydrographic Association, v. 40, p. 37–42.

Thomas, R.L., Wallach, J.L., McMillan, R.K., Bowlby, J.R., Frape, S., Keyes, D., and Mohajer, A.A., 1993, Recent deformation in the bottom sediments of western and southeastern Lake Ontario and its association with major structures and seismicity, *in* Wallach, J.L., and Heginbottom, J., eds., Neotectonics of the Great Lakes Area: Géographie physique et Quaternaire, v. 47, p. 325–336.

Virden, W.T., Warren, J.S., Holcombe, T.L., Reid, D.F., and Berggren, T.L., 1999, Bathymetry of Lake Ontario: Boulder, Colorado, National Geophysical Data Center, World Data Center for Marine Geology and Geophysics, Report MGG-15, scale 1:275,000.

Wallach, J.L., Mohajer, A.A., McFall, G.H., Bowlby, J.R., Pearce, M., and McKay, D.A., 1993, Popups as geological indicators of earthquake-prone areas in intraplate eastern North America: Quaternary Proceedings, v. 3, p. 67–83.

Wallach, J.L., Mohajer, A.A., and Thomas, R.L., 1998, Linear zones, seismicity, and the possibility of a major earthquake in the intraplate western Lake Ontario area of eastern North America: Canadian Journal of Earth Sciences, v. 35, p. 762–786, doi: 10.1139/cjes-35-7-762.

Zoback, M.L., 1992, First- and second-degree patterns of stress in the lithosphere—The World Stress Project: Journal of Geophysical Research, v. 97, p. 11,708–11,728.

Manuscript Accepted by the Society 29 November 2006

The Geological Society of America
Special Paper 425
2007

Stress evolution and seismicity in the central-eastern United States: Insights from geodynamic modeling

Qingsong Li
Mian Liu
Qie Zhang
Eric Sandvol
Department of Geological Sciences, University of Missouri, Columbia, Missouri 65211, USA

ABSTRACT

Although the central and eastern United States is in the interior of the presumably stable North American plate, seismicity there is widespread, and its causes remain uncertain. Here, we explore the evolution of stress and strain energy in intraplate seismic zones and contrast it with that in interplate seismic zones using simple viscoelastic finite-element models. We find that large intraplate earthquakes can significantly increase Coulomb stress and strain energy in the surrounding crust. The inherited strain energy may dominate the local strain energy budget for thousands of years following main shocks, in contrast to interplate seismic zones, where strain energy is dominated by tectonic loading. We show that strain energy buildup from the 1811–1812 large events in the New Madrid seismic zone may explain some of the moderate-sized earthquakes in this region since 1812 and that the inherited strain energy is capable of producing some damaging earthquakes (M >6) today in southern Illinois and eastern Arkansas, even in the absence of local loading. Without local loading, however, the New Madrid seismic zone would have remained in a stress shadow where stress has not been fully restored from the 1811–1812 events. We also derived a Pn velocity map of the central and eastern United States using available seismic data; the results do not support the New Madrid seismic zone being a zone of thermal weakening. We simulated the long-term Coulomb stress in the central and eastern United States. The predicted high Coulomb stress concentrates near the margins of the North American tectosphere, correlating spatially with most seismicity in the central and eastern United States.

Keywords: New Madrid seismic zone, intraplate earthquakes, stress, seismicity, modeling.

INTRODUCTION

Plate tectonic theory provides a successful geodynamic framework for understanding the majority of earthquakes that occur along plate-boundary zones. However, it offers no ready explanation for earthquakes in the presumably rigid plate interiors. One such region is the central-eastern United States, defined broadly as the region of the continental United States east of the Rocky Mountains. Although the central and eastern United States is in the middle of the North America plate where Cenozoic crustal deformation is minimal, both historic earthquakes and instrumentally recorded earthquakes are abundant. Major seismic zones in this area include the following (Dewey et al., 1989) (Fig. 1):

Li, Q., Liu, M., Zhang, Q., and Sandvol, E., 2007, Stress evolution and seismicity in the central-eastern United States: Insights from geodynamic modeling, *in* Stein, S., and Mazzotti, S., ed., Continental Intraplate Earthquakes: Science, Hazard, and Policy Issues: Geological Society of America Special Paper 425, p. 149–166, doi: 10.1130/2007.2425(11). For permission to copy, contact editing@geosociety.org. ©2007 The Geological Society of America. All rights reserved.

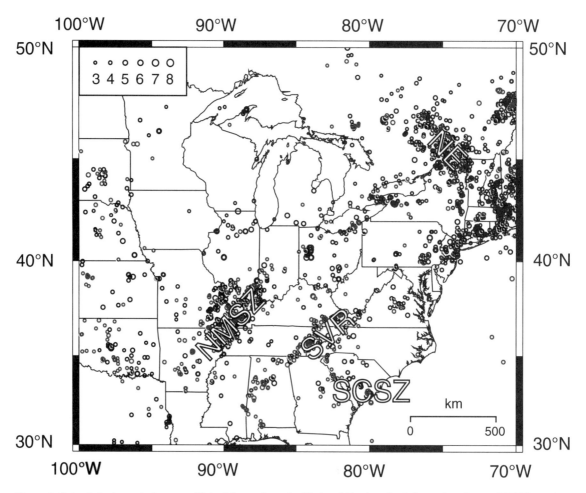

Figure 1. Seismicity in central-eastern United States from the National Earthquake Information Center (NEIC) catalog. Black circles: historic events (1800–1973); red circles: modern events (1973–2004). Major seismic zones include the New Madrid seismic zone (NMSZ); the Southern Valley and Ridge (SVR); the South Carolina seismic zone (SCSZ); and the New England and the St. Lawrence River valley seismic zone (NE).

1. The New Madrid seismic zone and Mississippi Embayment, which was the site of the famous 1811–1812 large earthquakes. The magnitudes of the largest three events were Mw 7–7.5 (Hough et al., 2000). Paleoseismic results indicate that major earthquakes occurred ca. 900 and 1400 A.D. (Kelson et al., 1996; Tuttle et al., 2002). Modern instrumentation has recorded thousands of events since 1977.

2. Southern Valley and Ridge, also referred to as the eastern Tennessee seismic zone, where modern seismicity is concentrated beneath the Valley and Ridge Province, near the western edge of the Appalachians. The largest historical event here was the M 5.8 Giles County, Virginia, earthquake of 31 May 1897 (Nuttli et al., 1979).

3. South Carolina seismic zone, where the best-known event was a destructive (M ~6.5–7.0) event near Charleston, South Carolina, on 31 August 1886 (Nuttli et al., 1979). Paleoseismic studies indicate at least two prehistoric earthquakes in the past 3000 yr (Obermeier et al., 1985; Talwani and Cox, 1985).

4. New England and the St. Lawrence River valley: Earthquake epicenters in central New England, upstate New York, and adjacent Canada form a northwest-trending belt of seismicity, sometimes called the Boston-Ottawa zone (Diment et al., 1972; Sbar and Sykes, 1973). The largest historic earthquake in the U.S. part of this zone was probably the M ~6 Cape Ann, Massachusetts, earthquake of 1755 (Street and Lacroix, 1979). Further north in the St. Lawrence River valley, numerous events with M 6–7 have been recorded.

Despite intensive studies, the mechanics of earthquakes in the central and eastern United States remain poorly understood. Some workers have suggested that these seismic zones occur within ancient rifts, thus proposing crustal weakness as the main cause of these earthquakes (e.g., Johnston and Kanter, 1990; Johnston, 1996). Others have suggested stress concentration by various factors, including regional and local crustal structures, as the main cause (e.g., Grana and Richardson, 1996; Stuart et al., 1997; Kenner and Segall, 2000; Grollimund and Zoback,

2001; Pollitz et al., 2001a). Although intraplate earthquakes are commonly believed to differ fundamentally from interplate earthquakes, their differences in dynamics are not clear. In this study, we first explore the basic mechanics of intraplate seismic zones and compare them to the mechanics of interplate seismic zones. We then apply the results to investigate seismicity in the New Madrid seismic zone. Finally, we present a regional geodynamic model of the central and eastern United States constrained by the lithospheric structure based on seismic studies by others and our Pn tomography.

THE MECHANICS OF INTRAPLATE VERSUS INTERPLATE SEISMIC ZONES

We developed three-dimensional viscoelastic models to explore the differences in stress evolution between intraplate and interplate seismic zones. To illustrate the basic physics, we kept the models relatively simple. We considered two contrasting properties of these zones: (1) intraplate seismic zones are of finite length, surrounded by strong ambient crust, whereas interplate seismic zones are effectively infinitely long; and (2) tectonic loading for intraplate seismic zones is applied at far-field plate boundaries and typically produce low strain rates, whereas interplate seismic zones are loaded directly by relative motions of tectonic plates, causing relatively high strain rates (Fig. 2). The model rheology is a viscoelastic (Maxwellian) medium. The model domain is 500 km × 500 km. Both models include a 20-km-thick stiff upper crust and a ductile lower crust. The viscosity for the upper crust is 8.0×10^{23} Pa s, which makes it essentially elastic for the time scales considered here (thousands of years). For the lower crust, a range of viscosity values (1.0×10^{19} to 1.0×10^{21} Pa s) was explored for the effects of postseismic relaxation. For the intraplate model (Fig. 2A), a 150-km-long fault zone was used to simulate a finite seismic zone. The boundary conditions included

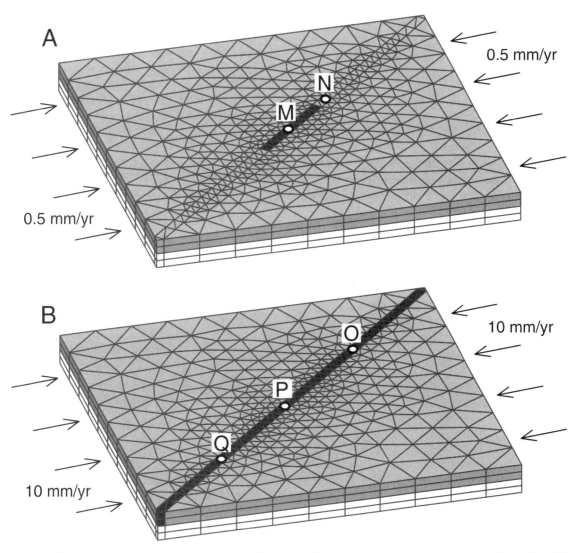

Figure 2. Finite-element models for intraplate (A) and interplate (B) seismic fault zones (in dark shading). Points M and N are where stress evolution is shown in Figure 4, and points O, P, and Q are where stress evolution is shown in Figure 7.

0.5 mm/yr compression imposed on the two sides of the model domain. The resulting strain rate is ~2.0 × 10⁻⁹ yr⁻¹, which is close to the upper bound for the central and eastern United States (Newman et al., 1999; Gan and Prescott, 2001). Young's modulus and the Poisson's ratio are 8.75×10^{10} Pa and 0.25, respectively, for the entire crust (Turcotte and Schubert, 1982). For the interplate model, the fault zone cuts across the entire model domain (Fig. 2B), and the boundary condition is 10 mm/yr on both sides, causing an average slip rate of ~28 mm/yr along the fault zone in the model, which is similar to that on the San Andreas fault (Bennett et al., 2004; Becker et al., 2005). In both models, the fault zones are represented by special elements, on which we simulate earthquakes by using instant plastic strain to lower the stress to below the yield strength (Li et al., 2005).

Intraplate Seismic Zones

Many seismic zones in the central and eastern United States are marked by past large earthquakes, the initial triggering mechanisms of which are uncertain. Here, we focus on stress evolution in the seismic zones following a large earthquake.

Figure 3 shows the calculated evolution of the Coulomb stress following a large intraplate earthquake. The Coulomb stress on a plane is defined as

$$\sigma_f = \tau_\beta - \mu\sigma_\beta, \qquad (1)$$

where τ_β is the shear stress on the plane, σ_β is the normal stress, and μ is the effective coefficient of friction (King et al., 1994). Outside the main fault zone, we calculated the optimal Coulomb stress, which is the stress on planes optimally orientated for failure

(King et al., 1994). We assumed that, initially, the stress in the upper crust is close to the yield strength, a condition that might be applicable to many continental interiors (Townend and Zoback, 2000; Zoback et al., 2002) and is consistent with the widely scattered seismicity in and around the seismic zones in the central and eastern United States (Fig. 1). The model started with a large earthquake, simulated by a 7.5 m sudden slip across the entire fault plane. This event is equivalent to an M_w ~8.0 earthquake, which caused ~5 MPa stress drop within the fault zone. Coseismic stress release from the fault zone migrates to the tip regions of the fault zone and loads the lower crust below the fault zone. Postseismic viscous relaxation in the lower crust then causes the stress to reaccumulate within the upper crust, mainly near the tips of the fault zone. Similar results have been reported in previous viscoelastic models (Freed and Lin, 2001; Pollitz et al., 2001b). Note that 200 yr after the main earthquake, the fault zone remains in a stress shadow where the stress relieved during the earthquake has not been fully restored. This is mainly due to slow tectonic loading, the effects of which are insignificant over 200 yr (compare Figures 3B and 3C). In this model, we assumed a complete healing of the fault zone, such that the yield strength returned to the original level immediately following the main shock. If the fault zone were unhealed or partially healed, stress reaccumulation within the fault zone would be even slower.

In addition to the rate of tectonic loading, postseismic stress evolution depends on the rheology of the lithosphere, especially the lower crust. Figure 4 shows the effects of lower-crustal viscosity on the modeled stress evolution. A less viscous lower crust causes more rapid viscous relaxation and stress reloading in the upper crust. However, without fast tectonic loading from the far field, the total amount of stress restoration within the fault zone is

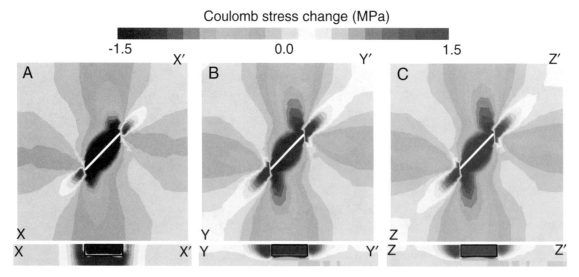

Figure 3. Predicted Coulomb stress change following a large earthquake in the intraplate seismic zone. (A) Coseismic stress change. (B) The sum of coseismic and postseismic (200 yr) stress change. (C) Same as B but without boundary loading. The bottom panels are depth sections, with 200% vertical exaggeration. The white lines (map view) and black frames (depth section) show the ruptured fault zone. Values were calculated assuming a viscosity of 10^{19} Pa s for the lower crust.

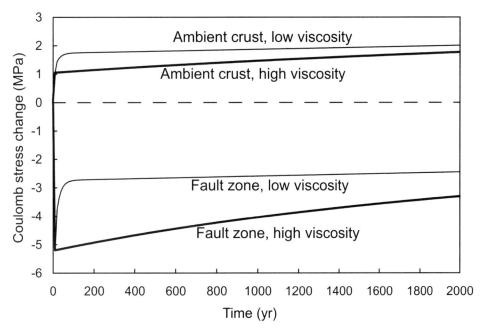

Figure 4. Predicted Coulomb stress evolution in the fault zone (point M in Fig. 2A) and the ambient crust near the fault tips (point N in Fig. 2A) using two values for the lower crust viscosity: high (10^{21} Pa s) and low (10^{19} Pa s). The Coulomb stress in the fault zone drops instantly during an earthquake. The initial stress restoration is accelerated by viscous relaxation in the lower crust at a rate that is sensitive to the viscosity. Further stress restoration is mainly controlled by tectonic loading.

largely determined by the stress relieved from the earthquake. For the viscosity range typical of the lower crust (10^{19}–10^{21} Pa s), viscous relaxation and far-field loading cannot fully restore stress in the fault zone thousands of years after the major earthquake. This may be a fundamental difference between intraplate and interplate seismic zones; the latter are directly loaded by plate motions at high rates, and a ruptured fault segment can also be influenced by earthquakes on nearby fault segments (see next section).

There is a migration and accumulation of strain energy associated with the stress evolution, defined as

$$E = \frac{1}{2}\sigma'_{ij}\varepsilon'_{ij}, \qquad (2)$$

where σ'_{ij} and ε'_{ij} are the deviatoric stress and strain tensors, respectively, using the Einstein summation convention for indexes i and j. Figure 5 shows the coseismic and postseismic changes of strain energy. Similar to the stress change (Fig. 3), most increase of strain energy is near the tips of the fault zone. Because of the slow tectonic loading, much of the strain energy is inherited from the main shock. It would take thousands of years for the far-field tectonic loading to accumulate a comparable amount of strain energy.

Interplate Seismic Zones

We may better appreciate the stress and strain energy evolution in intraplate seismic zones by contrasting them with interplate seismic zones (Fig. 2B). Some of the processes are similar.

An interplate earthquake relieves stress to the lower crust and the tips of the ruptured fault segment. Viscous relaxation in the lower crust further loads the upper curst, similar to what occurs in intraplate seismic zones. However, the high strain rates associated with plate motions restore stress in the ruptured segment more quickly than in intraplate fault zones. The essentially infinitely long fault zone also confines earthquakes to largely be within, and migrate along, the fault zone (Fig. 6A). Usually, other segments rupture before an earthquake repeats on the same segment (Fig. 6B). Thus, for each segment of the ruptured fault zone, postseismic stress recovery may be affected by three major factors: tectonic loading, viscous relaxation, and stress migration from nearby earthquakes. Figure 7 illustrates such stress evolution at three neighboring points in the fault zone. An earthquake at one of these points causes an instant stress drop. Postseismic stress restoration at the ruptured segment is fast within the first few tens of years because of both tectonic loading and viscous relaxation in the lower crust, transferring stress to the upper crust. A period of roughly steady-state stress buildup follows, resulting from tectonic loading. Sudden stress jumps may occur when a nearby segment ruptures, which may trigger a new earthquake. Such dynamic behavior has been reported in many interplate seismic zones, including the San Andreas fault (Stein et al., 1997; Rydelek and Sacks, 2001; Lin and Stein, 2004).

We have shown that, in an intraplate seismic zone, strain energy released from a large earthquake will migrate to the surrounding regions and dominate the local strain energy budget for thousands of years (Fig. 5). This is generally not true for

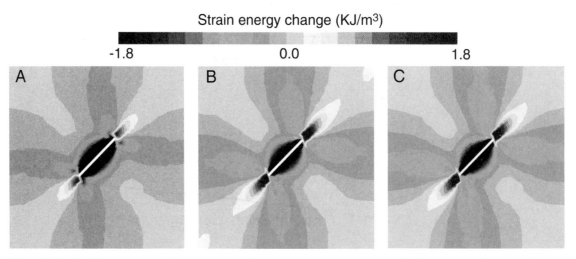

Figure 5. Predicted strain energy change in an intraplate seismic zone in map view. (A) Coseismic strain energy change. (B) Total strain energy change 200 yr after the main shock. (C) Same as B but without tectonic loading. The similarity between B and C shows the dominance of the inherited strain energy. White lines show the fault zone that ruptured.

interplate seismic zones, where tectonic loading dominates the strain energy budget. Figure 8 shows one selected episode of the interplate model experiment. Here, postseismic energy evolution is influenced by strain energy migration from the ruptured segment, viscoelastic reloading, and tectonic loading. Figures 8C and 8D compare strain energy in the model crust 200 yr after the earthquake, with and without tectonic loading. Clearly, tectonic loading dominates the strain energy evolution. To show the cumulative strain energy produced from tectonic loading, we artificially prohibited earthquakes for the period shown here. In reality, strain energy will be modulated by ruptures of other segments of the fault zone (Figs. 6 and 7).

STRESS EVOLUTION AND SEISMICITY IN THE NEW MADRID SEISMIC ZONE

In this section, we apply the model results to the New Madrid seismic zone, perhaps the best known seismic zone in the central and eastern United States. At least three large earthquakes occurred here within three months in the winter of 1811–1812. The magnitudes of these events have been estimated to be 7–7.5 (Hough et al., 2000). Since then, a dozen or so major events (M 5–6) have occurred in the New Madrid seismic zone and surrounding regions, and modern instruments have recorded thousands of small events in the past few decades (Fig. 9).

The New Madrid seismic zone fault zone is generally delineated by seismicity. Only one segment of the fault system, the NW-trending Reelfoot fault, has clear surface expression. Other parts of the New Madrid fault zone, including the southwestern segments (Cottonwood Grove fault) and the northeastern segment (New Madrid North fault), are inferred mainly from reflection seismic and aeromagnetic data, and seismicity (Hildenbrand and Hendricks, 1995; Johnston and Schweig, 1996). The Reelfoot fault is a reverse fault, whereas the southwestern and northeastern

segments are inferred to be right-lateral faults from morphologic and geologic features (Gomberg, 1993). These faults are believed to be within a failed rift system formed in Late Proterozoic to Early Cambrian times (Ervin and McGinnis, 1975). Herein, we use "New Madrid seismic zone" when referring to the geographic region of concentrated seismicity, and "New Madrid fault zone" when referring to these fault structures.

The results in the previous section indicated that following the 1811–1812 large earthquakes, the New Madrid fault zone would still remain in a stress shadow with Coulomb stress lower than the pre–1811–1812 level; this condition would be unfavorable for repetition of large earthquakes. This has important implications for assessing earthquake hazards in the New Madrid seismic zone. Here, we further explore this issue with a more realistic model (Fig. 9). The New Madrid fault zones are represented by two vertical strike-slip branches that approximate the northeastern and southwestern fault segments, connected by a NW-trending reverse fault that dips 45°SW, based on inferred geometry of the Reelfoot fault (Chiu et al., 1992; Mueller and Pujol, 2001). The compressive stresses across the North American plate were simulated by applying a 0.5 mm/yr velocity boundary condition on the eastern and western edges of the model domain (Fig. 9). This produced a strain rate of ~2 × 10^{-9} yr^{-1} within the model domain, which is likely the upper bound of internal deformation rate within the North American plate based on global positioning system (GPS) and seismological data (Anderson, 1986; Newman et al., 1999; Zoback et al., 2002). Other model parameters, including the initial conditions and rheological structures, are similar to those in Figure 2A.

Previous work has concluded that three large earthquakes occurred on the Reelfoot fault and the southwestern branch of the New Madrid seismic zone (the Cottonwood Grove fault) between December 1881 and February 1882 (Johnston, 1996; Johnston and Schweig, 1996). Some recent studies have suggested that

Coulomb stress change (MPa)

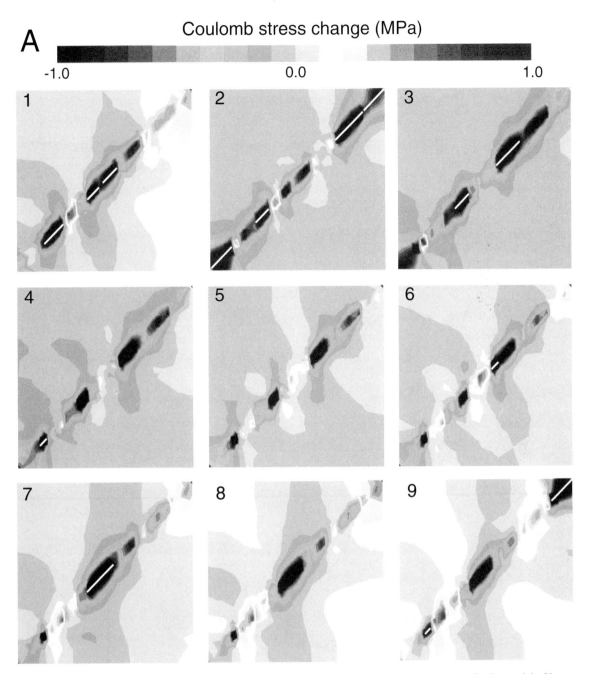

Figure 6 (*on this and following page*). (A) Snapshots of the predicted stress evolution (map view) for the model of interplate seismic zone (Fig. 2B). The time interval between successive panels is 20 yr. The sequence shows a selected period of the simulations of stress changes with ruptures of the high-stress segments.

there may have been four large events in the 1811–1812 sequence of events (Hough et al., 2000), and at least one of the main shocks may have been outside the New Madrid seismic zone (Mueller et al., 2004; Hough et al., 2005). Because our focus here is the long-term effects of the large 1811–1812 events, we ignore the detailed rupture sequences and treat these large events as having occurred simultaneously along the entire fault zones. This was simulated with ~5 m instant slip along the model fault zones,

resulting in a Coulomb stress drop of 5 MPa within the fault zones, as estimated by Hough et al. (2000). Figure 10 shows the calculated Coulomb stress evolution following the 1811–1812 events. In the upper crust, the increases in maximum stress occur near the NE and SW ends of the New Madrid fault zones. Conversely, stress decreases within the New Madrid fault zones and along a broad zone extending roughly NNW-SSE across the New Madrid seismic zone. This general pattern is similar to that in Figure 3.

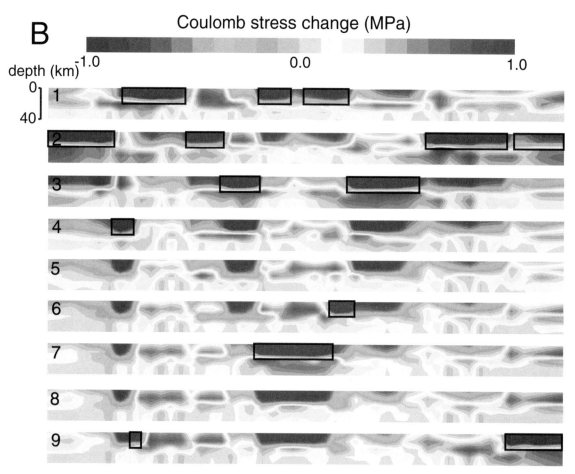

Figure 6 (*continued*). (B) Depth sections of the predicted stress evolution shown in A. The labels of the panels correspond to the map-view panels in A. The white lines (map view) and black frames (depth sections) mark the fault segments that have ruptured during the 20 yr time interval.

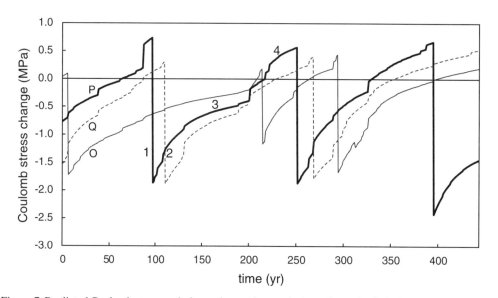

Figure 7. Predicted Coulomb stress evolution at three points on the interplate seismic fault zone (see Fig. 2B). At each point, the cycle of stress evolution includes four stages: (1) stress drop during an earthquake; (2) accelerated stress restoration because of viscous relaxation of the lower crust; (3) steady stress increase mainly from tectonic loading; and (4) stress jump due to triggering effect of nearby earthquakes.

Strain energy change (KJ/m³)

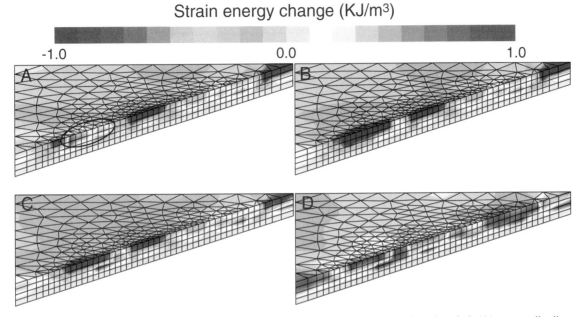

-1.0 0.0 1.0

Figure 8. Predicted evolution of strain energy in an interplate seismic zone for a selected period: (A) energy distribution before an earthquake (ellipse marks the ruptured area); (B) energy distribution immediately after the earthquake; (C) energy distribution after 200 yr without tectonic boundary loading; and (D) energy distribution after 200 yr with tectonic boundary loading.

Each of the large 1811–1812 events was followed by numerous large aftershocks (M >6.0) (Johnston and Schweig, 1996), and since 1812, a dozen or so moderate-sized events (M >5) have occurred in the New Madrid seismic zone and surrounding regions (Fig. 9). Although not all these events were included in the calculation, their effects have likely been minor in terms of energy release. This can be seen from the stress perturbation by two of the largest earthquakes in the New Madrid seismic zone region since 1812: the 1895 Charleston, Missouri, earthquake (M = 5.9) and the 1843 Marked Tree, Arkansas, earthquake (M = 6.0) (Stover and Coffman, 1993) (Fig. 10). The results show some local stress changes near the epicenters of these events, but the general stress pattern remains dominated by the 1811–1812 large events, leaving the New Madrid seismic zone in a stress shadow where stress has not reached the pre–1811–1812 level. The largest Coulomb stress increases are in southern Illinois and eastern Arkansas. Interestingly, these are located where many of the major earthquakes (M >5) since 1812 have occurred (Fig. 10). The spatial correlation is not perfect because seismicity is controlled by both stress and crustal strength, but the lateral variations of strength in the ambient crust are not included in the model. Thus, the clustering of moderate earthquakes in southern Illinois and western Indiana may be attributable to both the increased Coulomb stress and the relatively weak crust in the Wabash Valley seismic zone, and seismicity near the Missouri-Illinois border, where the Coulomb stress actually decreased following the 1811–1812 large events, may indicate weakness in both crust and the uppermost mantle (see following). If one of the 1811–1812 main shocks occurred outside of the New Madrid

seismic zone, as suggested by Mueller et al. (2004) and Hough et al. (2005), the predicted stress field may differ somewhat from that in Figure 10.

The predicted stress evolution is consistent with seismic energy release in the New Madrid seismic zone and surrounding regions following the 1811–1812 large events. Figure 11A shows the calculated seismic energy release based on historic and modern earthquake data from the National Earthquake Information Center (NEIC) catalog (http://neic.usgs.gov/neis/epic/epic.html). We used the Gutenberg-Richter formula (Lay and Wallace, 1995), and approximated all magnitudes as M_s. The spatial pattern is dominated by a dozen moderate-sized events (M >5) since 1812, especially the two M ~6 events near the NE and SW tips of the New Madrid seismic zone (Fig. 9). Figure 11B shows the excess strain energy, calculated by assigning a strain change in each element, if needed, to bring the deviatoric stress below the yield strength of the crust during a time step. A vertical integration of the product of such strain changes and stress gives the total excess strain energy accumulated over a single time step at a given place. The spatial pattern of the calculated excess strain energy is consistent with the seismic energy release in the past two centuries (Fig. 11A), but the magnitude of the excess strain energy is two to three orders higher, presumably because not all energy has been released via earthquakes. The relation between strain energy before the large earthquakes, the energy released during them, and the fraction of energy radiated as seismic waves remains unclear (Kanamori, 1978). It is possible that the fraction of the energy release that is radiated as seismic waves (seismic efficiency) is

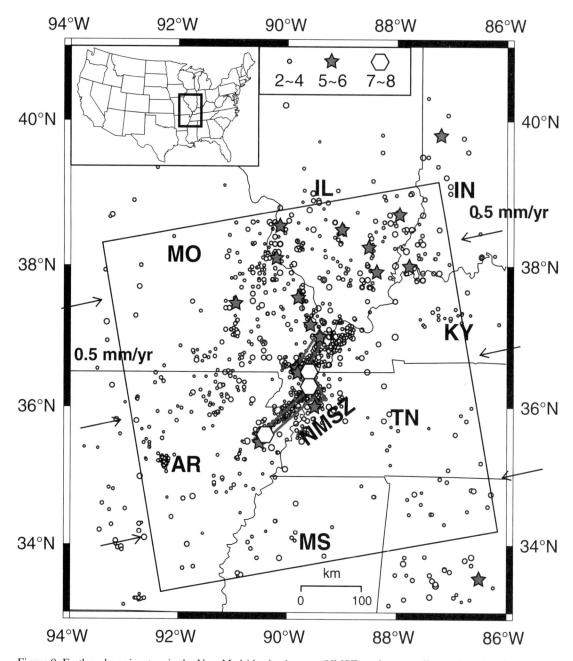

Figure 9. Earthquake epicenters in the New Madrid seismic zone (NMSZ) and surrounding regions (the inset shows the location). Modern earthquake data (for events with M >2 since 1974, circles) are from the National Earthquake Information Center (NEIC) and Center for Earthquake Research and Information (CERI) catalog (1974–2003); pre–1974 and historic earthquake data (M >5, stars) are from Stover and Coffman (1993). Hexagons show the large 1811–1812 events (Stover and Coffman, 1993). The New Madrid seismic zone is delineated by thick gray lines. The frame and arrows show the model domain and boundary conditions. State abbreviations: IL—Illinois, IN—Indiana, KY—Kentucky, TN—Tennessee, MS—Mississippi, AR—Arkansas, MO—Missouri.

only ~10% (Lockner and Okubo, 1983). Multiplication of the estimated seismic energy release (Fig. 11A) by a factor of 10 provides an estimate of total energy released by earthquakes. Subtracting it from the excess strain energy in Figure 11B gives the residual strain energy, some of which may be released by future earthquakes. The partition between seismic and aseismic

energy release is uncertain—estimations range from 2% to 80% (Ward, 1998). Figure 11C shows the estimated seismic energy in the New Madrid seismic zone region assuming that 10% of the total excess strain energy will be released in future earthquakes. This energy is capable of producing a number of Mw 6–7 earthquakes in southern Illinois and eastern Arkansas.

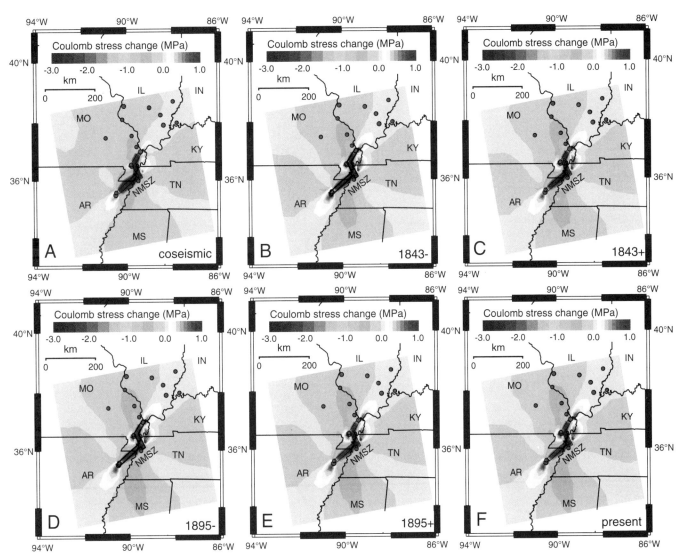

Figure 10. Predicted Coulomb stress evolution in the New Madrid seismic zone (NMSZ) and surrounding regions following the 1811–1812 large events: (A) coseismic; (B) before the 1843 Marked Tree, Arkansas, earthquake; (C) after the 1843 Marked Tree event; (D) before the 1895 Charleston, Missouri, earthquake; (E) after the 1895 Charleston event; and (F) at present. The red dots are the epicenters of the major events (M >5) since 1812 (Stover and Coffman, 1993). State abbreviations: IL—Illinois, IN—Indiana, KY—Kentucky, TN—Tennessee, MS—Mississippi, AR—Arkansas, MO—Missouri.

The basic mechanics illustrated by the simple model of intraplate seismic zones (Fig. 2A) thus appear to apply to the New Madrid seismic zone. Without some kind of local loading, the New Madrid fault zone is expected to remain in a stress shadow today, and the repetition of large earthquakes within the New Madrid fault zones would be unlikely in the next few hundred years. On the other hand, much of the strain energy released by the 1811–1812 events has migrated to southern Illinois and eastern Arkansas, where a number of moderate earthquakes have occurred since 1812. Based on this model, the residual strain energy in these regions, even without additional contribution from local loading, is capable of producing damaging earthquakes.

LITHOSPHERIC STRUCTURE AND SEISMICITY IN THE CENTRAL AND EASTERN UNITED STATES

So far our discussion of intraplate earthquakes has focused on postseismic evolution after a large earthquake. Given the low strain rates in the central and eastern United States and most other stable continents, it remains unclear what caused these large earthquakes in the first place. It has been suggested that most intraplate earthquakes, especially the large events (Mw >6.0), occur in ancient rift zones (Johnston and Kanter, 1990). This is true for the New Madrid seismic zone, which is within the Mesozoic Reelfoot rift system (Ervin and McGinnis, 1975). Most hypotheses of local loading mechanisms responsible for the large earthquakes

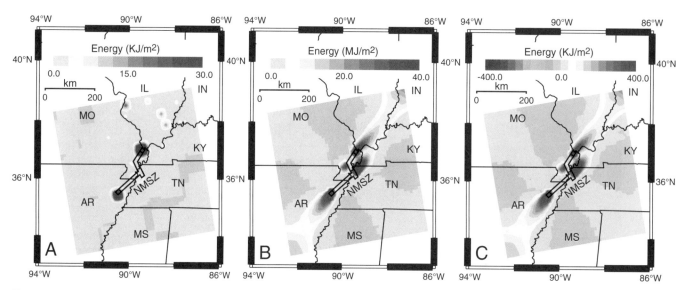

Figure 11. (A) Estimated seismic energy release in the New Madrid seismic zone (NMSZ) and surrounding regions since 1812. (B) Predicted total excess strain energy since the 1811–1812 events. (C) Predicted seismic strain energy in the crust today available for producing earthquakes, assuming 10% of the total excess strain energy will be released in future earthquakes. Here all energies are vertically integrated values. State abbreviations: IL—Illinois, IN—Indiana, KY—Kentucky, TN—Tennessee, MS—Mississippi, AR—Arkansas, MO—Missouri.

in the New Madrid seismic zone are based on inferred properties of the rift, including the sinking of an intrusive mafic body in the rift (Grana and Richardson, 1996; Pollitz et al., 2001a), detachment faulting at the base of the rift (Stuart et al., 1997), and an unspecified sudden weakening of the lower crust (Kenner and Segall, 2000). However, Figure 12 shows that not all seismic zones in the central and eastern United States are associated with rifts, and not all rifts are seismically active. One notable example is the Mid-Continent Rift, one of the most prominent rift systems in the central and eastern United States that has been essentially aseismic in historic times. On the other hand, earthquakes in the central and eastern United States appear to concentrate along the margins of the seismologically inferred North American craton, or the "tectosphere" (Jordan, 1979), which is defined by abnormally thick lithosphere.

Stress Field in the Central and Eastern United States

Could the lithosphere-tectosphere transition zone concentrate stresses and thus contribute to seismicity in the central and eastern United States? To address this question, we developed a finite element model for the central and eastern United States (Fig. 13). To simulate the long-term stress pattern, we treated the lithosphere as a power-law fluid continuum with a relatively high viscosity (10^{24} Pa s), underlain by a viscous asthenosphere with a lower viscosity (10^{21} Pa s). The thickness of the model lithosphere was based on seismologically derived thermal lithosphere thickness (Goes and van der Lee, 2002). The bottom of the model domain was a free slip boundary. The model domain was loaded on both sides by a 30 MPa compressive stress oriented N60°E, the direction of maximum tectonic compression for the central and

eastern United States (Zoback and Zoback, 1989). The calculated Coulomb stress was concentrated in the zones of relatively thin lithosphere, around the margin of the North American tectosphere and under the Mississippi Embayment (Fig. 14). The regions of high Coulomb stress showed a strong spatial correlation with seismic zones in the central and eastern United States, suggesting that the lateral heterogeneity of lithospheric structures is an important factor for seismicity in the central and eastern United States.

Pn Tomography of the Central and Eastern United States

The calculated high Coulomb stress in the Mississippi Embayment results from relatively thin lithosphere inferred from low Vs velocities (Goes and van der Lee, 2002) (Fig. 12), which relate to heat-flow anomalies in the New Madrid seismic zone region (Liu and Zoback, 1997). To refine the uppermost mantle velocity structure beneath this area, we derived a preliminary Pn velocity map (Fig. 15). Pn is a leaky mode guided wave that travels primarily through the uppermost mantle and is therefore most sensitive to seismic velocity fluctuations there. Pn tomography has become a common method of exploring the lithospheric mantle velocity structure (Hearn et al., 1994). This method commonly uses a least-squares algorithm (Paige and Saunders, 1982) to iteratively solve for all event-station pairs to obtain slowness, anisotropy, and station and event delays. The method includes damping parameters on both velocity and anisotropy to regularize the solution and reduce noise artifacts. P-wave traveltime residuals (<8 s) from sources at 2° to 14° are inverted for uppermost mantle velocity. A straight-line fit for the initial traveltime residuals versus distance gives an apparent Pn velocity of 8.1 km/s for the study area.

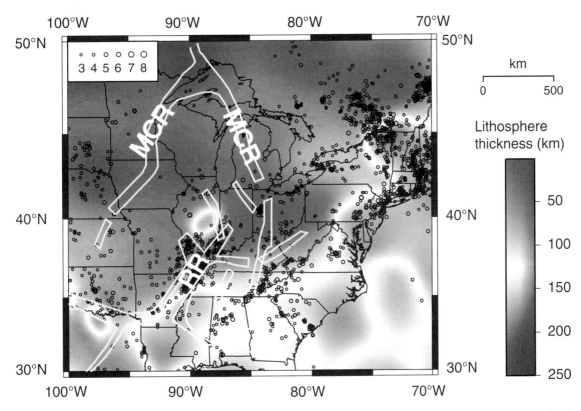

Figure 12. Thermal lithospheric thickness (Goes and van der Lee, 2002) and seismicity (1800–2004) in the central and eastern United States. Yellow lines show the rift zones. MCR—Mid-Continent Rift; RR—Reelfoot rift.

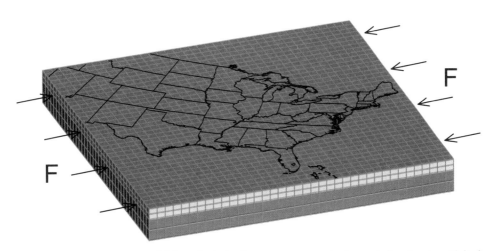

Figure 13. Finite-element model for calculating long-term stresses in the central and eastern United States. See text for detail.

To map the Pn velocity structure in the central and eastern United States, we collected ~10,000 Pn traveltimes from International Seismological Centre (ISC), NEIC, and 750 handpicked arrivals from both permanent and temporary stations throughout the central and eastern United States. To compensate for the relatively small numbers of ray paths, we used a relatively large cell size in our model parameterization (0.5° × 0.5°). Overall, we have a relatively high density of ray paths within the active seismic zones in the central and eastern United States and lower ray coverage to the west of the Great Lakes and along the southern coastline of the United States (Fig. 15A).

We found a first-order agreement between the NA00 model (Goes and van der Lee, 2002) and our Pn tomographic velocity model (Fig. 15B). However, we also observed interesting small-

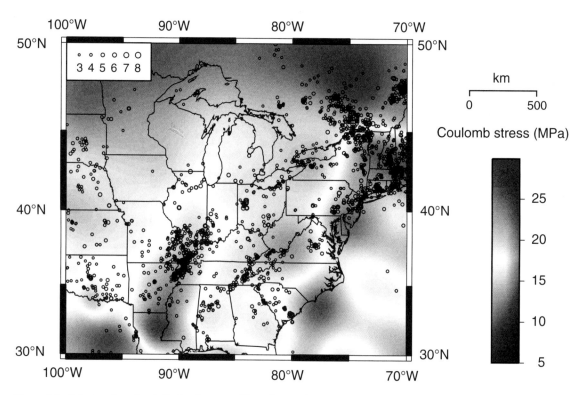

Figure 14. Calculated optimal Coulomb stress. Note the spatial correlation between seismicity and regions of high Coulomb stress in the central and eastern United States.

scale heterogeneity, such as the high velocities (~8.25 km/s) beneath both the New Madrid Seismic Zone and the Eastern Tennessee Seismic Zone. A low velocity zone (~8.0 km/s) is found in the western Ohio as well. The lithospheric mantle velocities within the North American shield are consistent with the high S-wave velocities measured at 100 km depth. Our results also show relatively low velocities (~8.05 km/s) in the northern and southern Appalachians.

The primary difference between our P-wave velocity measurements and the surface wave velocities (Goes and van der Lee, 2002) are in the Eastern Tennessee seismic zone, where we found a region of relatively high velocity that is not apparent in the NA00 model. A viscosity contrast and hence a change of lithospheric mantle properties here may concentrate stress and thus help to explain the Eastern Tennessee seismic zone seismicity.

DISCUSSION

One major result from this study is that the strain energy inherited from large intraplate earthquakes may dominate the local strain energy budget for hundreds to thousands of years. This result is expected, given the generally low strain rates in stable continents, including the North American plate interior (Dixon et al., 1996; Gan and Prescott, 2001). Applied to the New Madrid seismic zone, we have shown that the predicted spatial pattern and values of the stress and strain energy buildup following the 1811–1812 large

events may explain the occurrence of many moderate-sized earthquakes in areas surrounding the New Madrid seismic zone since 1812. Some of these events may be viewed as aftershocks, as slow loading usually causes a long duration of aftershocks (Stein and Newman, 2004). Many of these events occurred outside the New Madrid seismic zone and were triggered or even directly produced (in terms of energy source) by the main events. Furthermore, we have shown that, after large earthquakes, intraplate seismic zones tend to stay in a stress shadow where full stress restoration may take thousands of years, longer than predictions based solely on regional strain rate estimates. This is because seismic zones within a stable continent are of finite length and are surrounded by relatively strong crust. As long as deviatoric stresses are supported by the ambient crust, little stress is available to reload the fault zones. This result is consistent with geodetic measurements in the New Madrid seismic zone and surrounding regions, which show that the current strain rates are very slow (0 ± 2 mm/yr) (Newman et al., 1999; Gan and Prescott, 2001), rather than 5–8 mm/yr reported earlier (Liu et al., 1992). More recent GPS data have confirmed the low strain rate around the New Madrid seismic zone (Smalley et al., 2005a); whether or not higher strain rates within the fault zone can be detected from present GPS data is debatable (Calais et al., 2005; Smalley et al., 2005b).

These results do not contradict the present rate of seismicity in the New Madrid seismic zone. Although thousands of events have been recorded in the New Madrid seismic zone in recent

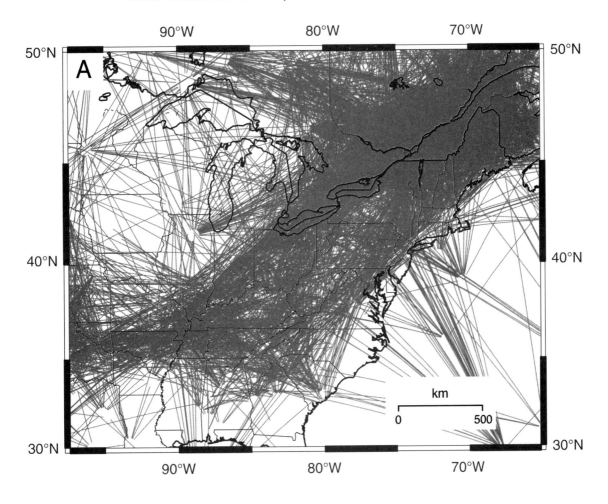

Figure 15 (*on this and following page*). (A) Ray coverage for Pn paths between 2° and 14° distance in the central and eastern United States. Approximately 10,000 raypaths are used in the model.

decades, most are small (M <4) and thus release little energy. No major (M >5) events have occurred within the New Madrid fault zone since 1812, and the two largest events in the past two centuries, the 1895 Charleston, Missouri, earthquake (M = 5.9) and the 1843 Marked Tree, Arkansas, earthquake (M = 6.0), occurred near the tip of the inferred New Madrid fault zones, which is consistent with the model prediction.

However, the model results are inconsistent with paleoseismic data indicating that at least two more events similar to the 1811–1812 large events occurred in the New Madrid seismic zone, around A.D. 900 and 1400 (Kelson et al., 1996; Tuttle et al., 2002). Given the difficulties in determining the size and location of paleoearthquakes from liquefaction data, it is not surprising that some conclusions drawn from paleoseismic data are questionable. Newman et al. (1999), for instance, argued that the size of these paleoevents may have been overestimated: these may have been M ~7, rather than M ~8, events, which would be more in line with the new estimates for the 1811–1812 events (Hough et al., 2000). However, our model shows that repetition of even M ~7 events every few hundred years would be difficult in the

New Madrid fault zone. Thus, explanations of paleoseismic data require local loading. Various local loading mechanisms have been proposed, including sinking of a "mafic pillow" within the Reelfoot rift (Grana and Richardson, 1996; Pollitz et al., 2001a). We have avoided including these models in our calculations because of the large uncertainties of these models. Because seismic activity in the New Madrid seismic zone likely started in the Holocene (Pratt, 1994; Schweig and Ellis, 1994; Van Arsdale, 2000), any local loading mechanism must also explain why it started in the Holocene. Stress triggering associated with glacial isostatic adjustment (GIA) provides some interesting possible causes. James and Bent (1994) and Wu and Johnston (2000) concluded that GIA may be significant for seismicity in the St. Lawrence valley but not for the more distant New Madrid seismic zone, because GIA predicts predominately thrust faulting in the New Madrid seismic zone, not strike-slip faulting as expected, and the stress change would be too small (0.01 MPa). Hough et al. (2000) suggested that the main mechanism in the New Madrid seismic zone is actually reverse faulting. Grollimund and Zoback (2001) showed that GIA could have caused three orders of seis-

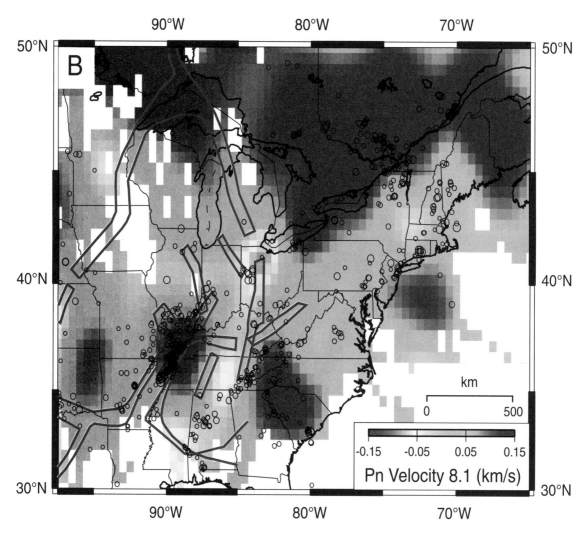

Figure 15 (*continued*). (B) Preliminary Pn tomographic map for the central and eastern United States. The minimum hit count for plotting is 3. Black circles are the same earthquake epicenters shown in Figure 9.

mic strain rate increase in the vicinity of the New Madrid seismic zone if there were a weak zone there. Refined imaging of crustal and lithospheric structures under the New Madrid seismic zone and other seismic zones in the central and eastern United States would help to test potential local loading mechanisms.

The stress field in the central and eastern United States is characterized by a nearly horizontal, NE- to E-striking axis of maximum compressive stress (Sbar and Sykes, 1973; Herrmann, 1979; Zoback and Zoback, 1989). The uniformity of stress-tensor orientation over a broad area of the central and eastern United States suggests that the stress field arises from forces that drive or resist plate motions (Richardson and Solomon, 1979; Zoback and Zoback, 1989). Given the rather uniform far-field stresses and the stability of the plate interior, crustal weakness, often found in ancient rift zones, is commonly related to intraplate earthquakes (Johnston and Kanter, 1990; Johnston and Schweig, 1996). This seems true in the central United States, especially in the Mississippi Embayment (Fig. 12), but not in the eastern United States, where seismic zones

seem to be spatially associated with ancient faults that developed when the eastern United States was near plate boundaries (Dewey et al., 1989), or faults that may be related to transform fracture zones in the Atlantic Ocean floor (Sykes, 1978). Whereas these seismic zones may be associated with different structural causes, we suggest that there may be a common and deep cause for most of the seismicity in the central and eastern United States: the transition zone between the thick North American tectosphere and the surrounding lithosphere. Our calculations show that such lateral heterogeneity of lithospheric structure could concentrate stress near the margins of the tectosphere, and the predicted regions of high stresses have a strong correlation with seismicity in the central and eastern United States. However, our regional model does not include lateral heterogeneity of crustal structure, which would further affect stress distribution and seismicity. Further testing of the causal relationship between lithospheric structures and seismicity must await more detailed data about crustal and lithospheric structures of the central and eastern United States.

CONCLUSIONS

1. Intraplate seismic zones tend to remain in a Coulomb stress shadow for thousands of years following large earthquakes. The slow far-field tectonic loading rates and the relatively strong ambient crust make stress reaccumulation within intraplate fault zones difficult, unless there are some local loading mechanisms. On the other hand, a significant amount of the stress relieved from large intraplate earthquakes, and the associated strain energy, may migrate to and be trapped within the neighboring crust, mainly near the tips of the fault zones. Such inherited strain energy may dominate the strain energy budget in the intraplate fault zone and surrounding regions for hundreds to thousands of years, and it can produce aftershocks hundreds of years after the main shocks. These behaviors are fundamentally different from interplate seismic zones, which are constantly loaded by plate motions.

2. The 1811–1812 large earthquakes in the New Madrid seismic zone caused significant buildup of Coulomb stress and strain energy in the surrounding regions, mainly southern Illinois and eastern Arkansas. Many of the moderate-sized earthquakes (M >5) in these regions since 1812 may have been triggered or facilitated by stress and strain energy inherited from the 1811–1812 large events. The residual strain energy from the 1811–1812 main shocks is capable of producing some damaging (M >6) earthquakes in areas surrounding the New Madrid seismic zone today, even in the absence of local loading. Conversely, the New Madrid fault zones should remain in a stress shadow, and thousands of years may be needed to restore the stress to the pre–1811–1812 level. Thus, some local loading mechanism must be active if numerous large events similar to the 1811–1812 events have occurred in the fault zones during the Holocene, as suggested by paleoseismic data. Although a number of local loading mechanisms have been proposed, more studies, including refined imaging of the crustal and lithospheric structures in the New Madrid seismic zone region, are needed to test these hypotheses.

3. Seismicity in the central and eastern United States shows a strong spatial correlation with the margins of the North American tectosphere, consistent with our model prediction of high Coulomb stress in the tectosphere-lithosphere transition zones. In the New Madrid seismic zone, the seismicity seems to be related to an abnormally thin lithosphere under the Mississippi Embayment. Again, further imaging of the crustal and lithospheric structure will help to address the cause of seismicity in the New Madrid seismic zone and other seismic zones in the central and eastern United States.

ACKNOWLEDGMENTS

We have benefited from helpful discussions with Seth Stein, Jian Lin, and Andy Newman. Careful reviews by Susan Hough, Alan Kafka, and Seth Stein improved this paper. This work was supported by U.S. Geological Survey NEHRP (National Earthquake Hazards Reduction Program) grant 04HQGR0046.

REFERENCES CITED

Anderson, J.G., 1986, Seismic strain rates in the central and eastern United States: Seismological Society of America Bulletin, v. 76, p. 273–290.

Becker, T.W., Hardebeck, J.L., and Anderson, G., 2005, Constraints on fault slip rates of the southern California plate boundary from GPS velocity and stress inversions: Geophysical Journal International, v. 160, p. 634–650, doi: 10.1111/j.1365-246X.2004.02528.x.

Bennett, R.A., Friedrich, A.M., and Furlong, K.P., 2004, Codependent histories of the San Andreas and San Jacinto fault zones from inversion of fault displacement rates: Geology, v. 32, p. 961–964, doi: 10.1130/G20806.1.

Calais, E., Mattioli, G., Demets, C., Nocquet, J.M., Stein, S., Newman, A., and Rydelek, P., 2005, Tectonic strain in plate interiors?: Nature, v. 438, p. E9–E10, doi: 10.1038/nature04428.

Chiu, J.M., Johnston, A.C., and Yang, Y.T., 1992, Imaging the active faults of the central New Madrid seismic zone using PANDA array data: Seismological Research Letters, v. 63, p. 375–393.

Dewey, J.W., Hill, D.P., Ellsworth, W.L., and Engdahl, E.R., 1989, Earthquakes, faults, and the seismotectonic framework of the contiguous United States, *in* Pakiser, L.C., and Mooney, W.D., eds., Geophysical Framework of the Continental United States: Boulder, Colorado, Geological Society of America Memoir 172, p. 541–576.

Diment, W.H., Urban, T.C., and Revetta, F.A., 1972, Some geophysical anomalies in the eastern United States, *in* Robertson, E.C., ed., The Nature of the Solid Earth: New York, McGraw-Hill Book Co., p. 544–572.

Dixon, T.H., Mao, A., and Stein, S., 1996, How rigid is the stable interior of the North American plate?: Geophysical Research Letters, v. 23, p. 3035–3038, doi: 10.1029/96GL02820.

Ervin, C.P., and McGinnis, L.D., 1975, Reelfoot rift; reactivated precursor to the Mississippi Embayment: Geological Society of America Bulletin, v. 86, p. 1287–1295, doi: 10.1130/0016-7606(1975)86<1287:RRRPTT>2.0.CO;2.

Freed, A.M., and Lin, J., 2001, Delayed triggering of the 1999 Hector Mine earthquake by viscoelastic stress transfer: Nature, v. 411, p. 180–183, doi: 10.1038/35075548.

Gan, W., and Prescott, W.H., 2001, Crustal deformation rates in central and eastern U.S. inferred from GPS: Geophysical Research Letters, v. 28, p. 3733–3736, doi: 10.1029/2001GL013266.

Goes, S., and van der Lee, S., 2002, Thermal structure of the North American uppermost mantle inferred from seismic tomography: Journal of Geophysical Research, v. 107, p. 2050, doi: 10.1029/2000JB000049.

Gomberg, J.S., 1993, Tectonic deformation in the New Madrid seismic zone; inferences from map view and cross-sectional boundary element models: Journal of Geophysical Research, ser. B, Solid Earth and Planets, v. 98, p. 6639–6664.

Grana, J.P., and Richardson, R.M., 1996, Tectonic stress within the New Madrid seismic zone: Journal of Geophysical Research, v. 101, p. 5445–5458, doi: 10.1029/95JB03255.

Grollimund, B., and Zoback, M.D., 2001, Did deglaciation trigger intraplate seismicity in the New Madrid seismic zone?: Geology, v. 29, p. 175–178, doi: 10.1130/0091-7613(2001)029<0175:DDTISI>2.0.CO;2.

Hearn, T.M., Rosca, A.C., and Fehler, M.C., 1994, Pn tomography beneath the southern Great Basin: Geophysical Research Letters, v. 21, p. 2187–2190, doi: 10.1029/94GL02054.

Herrmann, R.B., 1979, Surface wave focal mechanisms for eastern North American earthquakes with tectonic implications: Journal of Geophysical Research, v. 84, p. 3543–3552.

Hildenbrand, T.G., and Hendricks, J.D., 1995, Geophysical setting of the Reelfoot rift and relations between rift structures and the New Madrid seismic zone: U.S. Geological Survey Professional Paper 1538-E, p. E1–E30.

Hough, S.E., Armbruster, J.G., Seeber, L., and Hough, J.F., 2000, On the modified Mercalli intensities and magnitudes of the 1811–1812 New Madrid earthquakes: Journal of Geophysical Research, ser. B, Solid Earth and Planets, v. 105, p. 23,839–23,864, doi: 10.1029/2000JB900110.

Hough, S.E., Bilham, R., Mueller, K., Stephenson, W., Williams, R., and Odum, J., 2005, Wagon loads of sand blows in White County, Illinois: Seismological Research Letters, v. 76, p. 373–386.

James, T.S., and Bent, A.L., 1994, A comparison of eastern North American seismic strain-rates to glacial rebound strain-rates: Geophysical Research Letters, v. 21, p. 2127–2130, doi: 10.1029/94GL01854.

Johnston, A.C., 1996, Seismic moment assessment of earthquakes in stable continental regions; III. New Madrid 1811–1812, Charleston 1886 and Lisbon 1755: Geophysical Journal International, v. 126, p. 314–344.

Johnston, A.C., and Kanter, L.R., 1990, Earthquakes in stable continental crust: Scientific American, v. 262, no. 3, p. 68–75.

Johnston, A.C., and Schweig, E.S., 1996, The enigma of the New Madrid earthquakes of 1811–1812: Annual Review of Earth and Planetary Sciences, v. 24, p. 339–384.

Jordan, T.H., 1979, Mineralogies, densities and seismic velocities of garnet lherzolites and their geophysical implications, *in* Boyd, F.R., and Meyer, H.O.A., eds., The mantle sample: Inclusions in kimberlites and other volcanics: Washington, D.C., American Geophysical Union, p. 1–14..

Kanamori, H., 1978, Quantification of earthquakes: Nature, v. 271, p. 411–414, doi: 10.1038/271411a0.

Kelson, K.I., Simpson, G.D., Van Arsdale, R.B., Haraden, C.C., and Lettis, W.R., 1996, Multiple late Holocene earthquakes along the Reelfoot fault, central New Madrid seismic zone: Journal of Geophysical Research, v. 101, p. 6151–6170, doi: 10.1029/95JB01815.

Kenner, S.J., and Segall, P., 2000, A mechanical model for intraplate earthquakes; application to the New Madrid seismic zone: Science, v. 289, p. 2329–2332, doi: 10.1126/science.289.5488.2329.

King, G.C.P., Stein, R.S., and Lin, J., 1994, Static stress changes and the triggering of earthquakes: Bulletin of the Seismological Society of America, v. 84, p. 935–953.

Lay, T., and Wallace, T.C., 1995, Modern Global Seismology: San Diego, California, Academic Press, 521 p.

Li, Q., Liu, M., and Sandvol Eric, A., 2005, Stress evolution following the 1811–1812 large earthquakes in the New Madrid seismic zone: Geophysical Research Letters, v. 32, doi: 10.1029/2004GL022133.

Lin, J., and Stein, R.S., 2004, Stress triggering in thrust and subduction earthquakes, and stress interaction between the southern San Andreas and nearby thrust and strike-slip faults: Journal of Geophysical Research, v. 109, doi: 10.1029/2003JB002607.

Liu, L., and Zoback, M.D., 1997, Lithospheric strength and intraplate seismicity in the New Madrid seismic zone: Tectonics, v. 16, p. 585–595, doi: 10.1029/97TC01467.

Liu, L., Zoback, M.D., and Segall, P., 1992, Rapid intraplate strain accumulation in the New Madrid seismic zone: Science, v. 257, p. 1666–1669, doi: 10.1126/science.257.5077.1666.

Lockner, D.A., and Okubo, P.G., 1983, Measurements of frictional heating in granite: Journal of Geophysical Research, v. 88, p. 4313–4320.

Mueller, K., and Pujol, J., 2001, Three-dimensional geometry of the Reelfoot blind thrust: Implications for moment release and earthquake magnitude in the New Madrid seismic zone: Bulletin of the Seismological Society of America, v. 91, p. 1563–1573, doi: 10.1785/0120000276.

Mueller, K., Hough, S.E., and Bilham, R., 2004, Analysing the 1811–1812 New Madrid earthquakes with recent instrumentally recorded aftershocks: Nature, v. 429, p. 284–288, doi: 10.1038/nature02557.

Newman, A., Stein, S., Weber, J., Engeln, J., Mao, A., and Dixon, T., 1999, Slow deformation and lower seismic hazard at the New Madrid seismic zone: Science, v. 284, p. 619–621, doi: 10.1126/science.284.5414.619.

Nuttli, O.W., Bollinger, G.A., and Griffiths, D.W., 1979, On the relation between modified Mercalli intensity and body-wave magnitude: Bulletin of the Seismological Society of America, v. 69, p. 893–909.

Obermeier, S.F., Gohn, G.S., Weems, R.E., Gelinas, R.L., and Rubin, M., 1985, Geologic evidence for recurrent moderate to large earthquakes near Charleston, South Carolina: Science, v. 227, p. 408–411, doi: 10.1126/science.227.4685.408.

Paige, C.C., and Saunders, M.A., 1982, LSQR: An algorithm for sparse linear equations and sparse least squares: ACM Transactions on Mathematical Software, v. 8, p. 43–71, doi: 10.1145/355984.355989.

Pollitz, F.F., Kellogg, L., and Buergmann, R., 2001a, Sinking mafic body in a reactivated lower crust; a mechanism for stress concentration at the New Madrid seismic zone: Bulletin of the Seismological Society of America, v. 91, p. 1882–1897, doi: 10.1785/0120000277.

Pollitz, F.F., Wicks, C., and Thatcher, W., 2001b, Mantle flow beneath a continental strike-slip fault; postseismic deformation after the 1999 Hector Mine earthquake: Science, v. 293, p. 1814–1818, doi: 10.1126/science.1061361.

Pratt, T.L., 1994, How old is the New Madrid seismic zone?: Seismological Research Letters, v. 65, p. 172–179.

Richardson, R.M., and Solomon, S.C., 1979, Tectonic stress in the plates: Reviews of Geophysics and Space Physics, v. 17, p. 981–1019.

Rydelek, P.A., and Sacks, I.S., 2001, Migration of large earthquakes along the San Jacinto fault; stress diffusion from 1857 Fort Tejon earthquake: Geophysical Research Letters, v. 28, p. 3079–3082, doi: 10.1029/2001GL013005.

Sbar, M.L., and Sykes, L.R., 1973, Contemporary compressive stress and seismicity in eastern North America; an example of intra-plate tectonics: Geological Society of America Bulletin, v. 84, p. 1861–1881, doi: 10.1130/0016-7606(1973)84<1861:CCSASI>2.0.CO;2.

Schweig, E.S., and Ellis, M.A., 1994, Reconciling short recurrence intervals with minor deformation in the New Madrid seismic zone: Science, v. 264, p. 1308–1311, doi: 10.1126/science.264.5163.1308.

Smalley, R., Ellis, M.A., Paul, J., and Van Arsdale, R.B., 2005a, Space geodetic evidence for rapid strain rates in the New Madrid seismic zone of central USA: Nature, v. 435, p. 1088–1090, doi: 10.1038/nature03642.

Smalley, R., Ellis, M.A., Paul, J., and Van Arsdale, R.B., 2005b, Tectonic strain in plate interiors? Reply: Nature, v. 438, p. E10, doi: 10.1038/nature04429.

Stein, R.S., Barka, A.A., and Dieterich, J.H., 1997, Progressive failure on the North Anatolian fault since 1939 by earthquake stress triggering: Geophysical Journal International, v. 128, p. 594–604.

Stein, S., and Newman, A., 2004, Characteristic and uncharacteristic earthquakes as possible artifacts: Applications to the New Madrid and Wabash seismic zones: Seismological Research Letters, v. 75, p. 173–187.

Stover, C.W., and Coffman, J.L., 1993, Seismicity of the United States, 1568–1989 (revised): U.S. Geological Survey Professional Paper 1527, 418 p.

Street, R., and Lacroix, A., 1979, An empirical study of New England seismicity: 1727–1977: Bulletin of the Seismological Society of America, v. 69, p. 159–175.

Stuart, W.D., Hildenbrand, T.G., and Simpson, R.W., 1997, Stressing of the New Madrid seismic zone by a lower crust detachment fault: Journal of Geophysical Research, v. 102, p. 27,623–27,633, doi: 10.1029/97JB02716.

Sykes, L.R., 1978, Intraplate seismicity, reactivation of preexisting zones of weakness, alkaline magmatism, and other tectonism postdating continental fragmentation: Reviews of Geophysics and Space Physics, v. 16, p. 621–688.

Talwani, P., and Cox, J., 1985, Paleoseismic evidence for recurrence of earthquakes near Charleston, South Carolina: Science, v. 229, p. 379–381, doi: 10.1126/science.229.4711.379.

Townend, J., and Zoback, M.D., 2000, How faulting keeps the crust strong: Geology, v. 28, p. 399–402, doi: 10.1130/0091-7613(2000)28<399:HFKTCS>2.0.CO;2.

Turcotte, D.L., and Schubert, G., 1982, Geodynamics: Applications of Continuum Physics to Geological Problems: New York, John Wiley & Sons, 450 p.

Tuttle, M.P., Schweig, E.S., Sims, J.D., Lafferty, R.H., Wolf, L.W., and Haynes, M.L., 2002, The earthquake potential of the New Madrid seismic zone: Bulletin of the Seismological Society of America, v. 92, p. 2080–2089, doi: 10.1785/0120010227.

Van Arsdale, R., 2000, Displacement history and slip rate on the Reelfoot fault of the New Madrid seismic zone: Engineering Geology, v. 55, p. 219–226, doi: 10.1016/S0013-7952(99)00093-9.

Ward, S.N., 1998, On the consistency of earthquake moment rates, geological fault data, and space geodetic strain; the United States: Geophysical Journal International, v. 134, p. 172–186, doi: 10.1046/j.1365-246x.1998.00556.x.

Wu, P., and Johnston, P., 2000, Can deglaciation trigger earthquakes in N. America?: Geophysical Research Letters, v. 27, p. 1323–1326, doi: 10.1029/1999GL011070.

Zoback, M.D., Townend, J., and Grollimund, B., 2002, Steady-state failure equilibrium and deformation of intraplate lithosphere: International Geology Review, v. 44, p. 383–401.

Zoback, M.L., and Zoback, M.D., 1989, Tectonic stress field of the continental United States, *in* Pakiser, L.C., and Mooney, W.D., eds., Geophysical Framework of the Continental United States: Boulder, Colorado, Geological Society of America Memoir 172, p. 523–539.

MANUSCRIPT ACCEPTED BY THE SOCIETY 29 NOVEMBER 2006

The Geological Society of America
Special Paper 425
2007

Is the New Madrid seismic zone hotter and weaker than its surroundings?

Jason McKenna

U.S. Army Engineer Research and Development Center, 3909 Halls Ferry Road, Vicksburg, Mississippi 39180, USA

Seth Stein[†]

Department of Earth and Planetary Sciences, Northwestern University, Evanston, Illinois 60208, USA

Carol A. Stein

*Department of Earth and Environmental Sciences, University of Illinois at Chicago,
m/c 186, 845 West Taylor St., Chicago, Illinois 60607-7059, USA*

ABSTRACT

A fundamental question about continental intraplate earthquakes is why they are where they are. For example, why are the New Madrid seismic zone earthquakes concentrated on the Reelfoot rift when the continent contains many fossil structures that would seem equally likely candidates for concentrated seismicity? A key to answering this question is understanding of the thermal-mechanical structure of the seismic zone. If it is hotter and thus weaker than surrounding regions, it is likely to be a long-lived weak zone on which intraplate strain release concentrates. Alternatively, if it is not significantly hotter and weaker than its surroundings, the seismicity is likely to be a transient phenomenon that migrates among many similar fossil weak zones. These different models have important implications for the mechanics of the seismic zone, stress evolution after and between large earthquakes, and seismic hazard assessment.

The sparse heat-flow data in the New Madrid area can be interpreted as supporting either hypothesis. There is a possible small elevation of heat flow in the area compared to its surroundings, depending on the New Madrid and regional averages chosen. The inferred high heat flow has been interpreted as indicating that the crust and upper mantle are significantly hotter and thus significantly weaker than surrounding areas of the central and eastern United States. In this model, the weak lower crust and mantle concentrate stress and seismicity in the upper crust. However, reanalysis of the heat flow indicates that the anomaly is either absent or much smaller (3 ± 23 rather than mW m^{-2}) than assumed in the previous analyses, leading to much smaller (~90%) temperature anomalies and essentially the same lithospheric strength. Moreover, if a small heat-flow anomaly exists, it may result from groundwater flow in the rift's fractured upper crust, rather than higher temperatures. The latter interpretation seems more consistent with studies that find low seismic velocities only in parts of the seismic zone and at shallow depths. Hence, although the question

[†]E-mail: seth@earth.northwestern.edu.

McKenna, J., Stein, S., and Stein, C.A., 2007, Is the New Madrid seismic zone hotter and weaker than its surroundings?, *in* Stein, S., and Mazzotti, S., ed., Continental Intraplate Earthquakes: Science, Hazard, and Policy Issues: Geological Society of America Special Paper 425, p. 167–175, doi: 10.1130/2007.2425(12).

cannot be resolved without additional heat-flow data, we find no compelling case for assuming that the New Madrid seismic zone is significantly hotter and weaker than its surroundings. This result is consistent with migrating seismicity and the further possibility that the New Madrid seismic zone is shutting down, which is suggested by the small or zero motion observed geodetically. If so, the present seismicity are aftershocks of the large earthquakes of 1811 and 1812, and such large earthquakes will not recur there for a very long time.

Keywords: New Madrid earthquakes, thermal structure of faults, intraplate earthquakes.

INTRODUCTION

One of the biggest challenges in understanding the tectonics of continental interiors and the hazard posed by earthquakes within them, such as those in the New Madrid seismic zone, is that we do not understand whether the present seismic zone is fundamentally different from similar structures that appear less seismically active. North American intracontinental earthquakes appear to be concentrated in a number of seismic zones. Some, such as New Madrid, seem related to failed rift zones, whereas other seismicity is not. Conversely, other prominent structures, such as the Mid-Continent Rift, have little seismicity. Hence, it is unclear why, at present and within the past few thousand years, earthquakes are concentrated on the Reelfoot rift when the continent contains many fossil structures that seem equally likely candidates for concentrated seismicity. As discussed by several papers in this volume, this issue is fundamental to assessing seismic hazards and hence mitigating risk in the central and eastern United States or other continental interiors.

Insight into this issue can come from many approaches, including assessment of the thermal-mechanical structure of the seismic zones. If they are hotter and weaker than surrounding regions, they are likely to be long-lived weak zones on which intraplate strain release concentrates. Alternatively, if they are not significantly hotter and weaker than their surroundings, the seismicity is likely to be a transient phenomenon that migrates among many fossil weak zones. The latter possibility is suggested by an increasing body of data showing that continental intraplate faults tend to have episodic seismicity separated by quiescent periods (Crone et al., 2003). The different models (Fig. 1) have important implications both for the long-term mechanics of seismic zones and for stress evolution after and between large earthquakes.

NEW MADRID HEAT FLOW

Assessments of whether, and if so, how, geotherms and hence strength profiles differ between the New Madrid seismic zone and its surroundings depend on two key questions. First, what heat-flow values inside and outside the New Madrid seismic zone should be compared? Second, if the New Madrid seismic zone heat flow is higher than for the surroundings, is the difference primarily an effect of higher conductive heat transfer and thus temperatures, or does it reflect hydrothermal heat transport in the rift zone?

LONG-TERM SEISMICITY IN WEAK ZONE

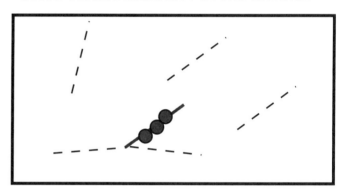

SEISMICITY MIGRATES BETWEEN ZONES OF SIMILAR STRENGTH

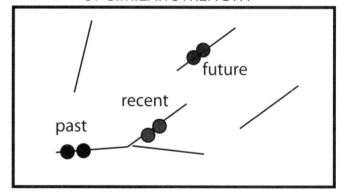

Figure 1. Schematic illustration of alternative models for continental intraplate seismicity.

The few heat-flow data in the New Madrid area can be interpreted as showing a possible small elevation of heat flow in the area compared to its surroundings, depending on how the New Madrid and regional averages are chosen. The most recent compilation (Blackwell and Richards, 2004) shows seven heat-flow measurements within the Reelfoot rift (Fig. 2). These values (44, 50, 55, 55, 58, 60, and 65 mW m^{-2}) yield a mean value of 55 ± 7 mW m^{-2}. Whether this value is anomalous, and if so, by how much, depends on the region used for comparison. The New Madrid seismic zone average is slightly higher than, although not statistically different from, the mean central and eastern U.S.

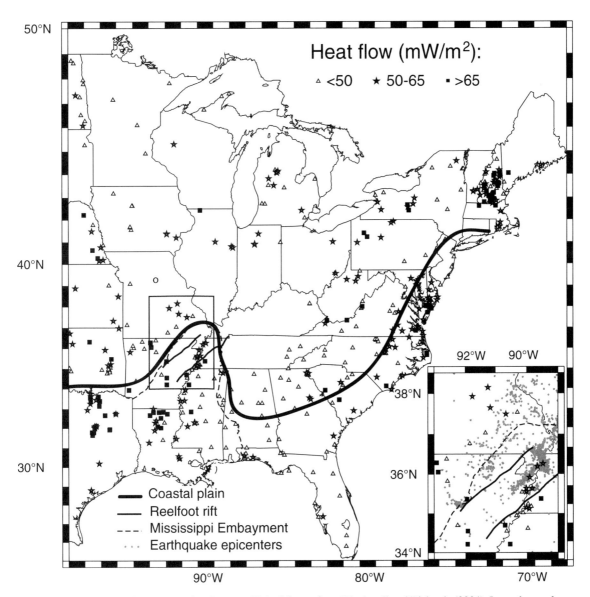

Figure 2. Heat-flow data in the central and eastern United States from Blackwell and Richards (2004). Inset shows close-up of the New Madrid seismic zone and surroundings with heat-flow sites and earthquake epicenters. Solid line shows northern boundary of coastal plain heat-flow province (Morgan and Gosnold, 1989).

heat flow of 52 ± 22 mW m^{-2}, which emerges from Blackwell and Richards' (2004) data (Fig. 2), or that of 51 ± 20 mW m^{-2} from an earlier data set (Morgan and Gosnold, 1989).

Figures 2 and 3 illustrate how New Madrid seismic zone heat flow compares to that in its surroundings. It is higher than observed to the southeast, and comparable to that observed in the other three quadrants. However, the sparse data have considerable scatter, owing to uncertainties of measurement and variations in crustal thickness, which controls radiogenic heat production and fluid flow. Hence, given the uncertainties in estimating mean heat flow both in the New Madrid seismic zone and outside it, the New Madrid seismic zone heat flow may or may not be slightly higher than some of its surroundings but is

well within the range of the observed values (Sass et al., 1976; Blackwell and Richards, 2004).

Assuming that a New Madrid seismic zone heat-flow anomaly exists, two interpretations have been made. In one (Fig. 4), Liu and Zoback (1997) argued that the New Madrid seismic zone heat flow is significantly (15 mW m^{-2}) higher than in the surroundings, and they interpreted it as indicating lower crust and upper mantle that is several hundred degrees hotter. This approach assumes that the heat flow observed reflects a geothermal gradient unperturbed by groundwater flow, which is extrapolated downward by incorporating the effects of crustal heat production. Hence, the lower crust and upper mantle in the New Madrid seismic zone would be significantly weaker than in surrounding

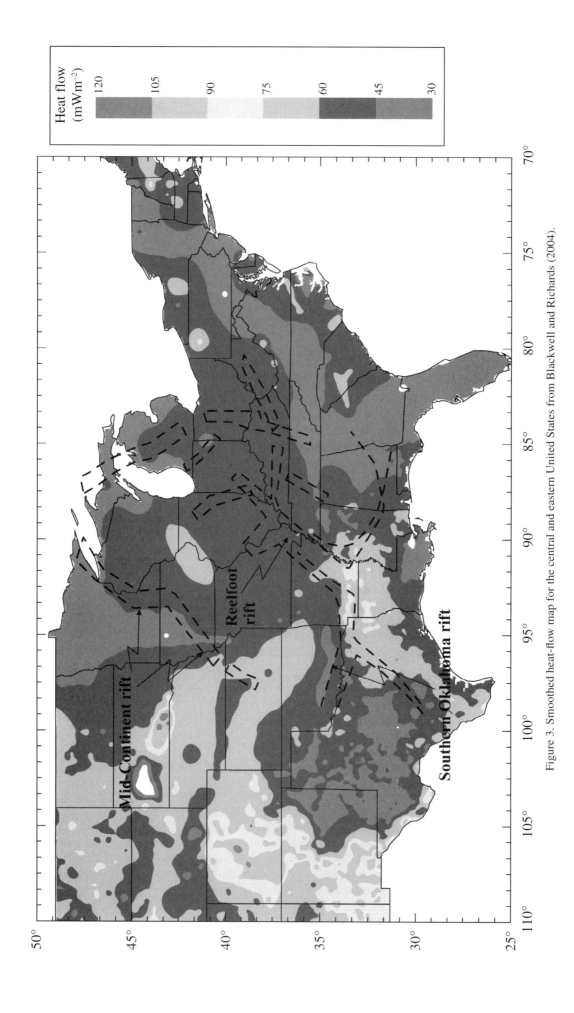

Figure 3. Smoothed heat-flow map for the central and eastern United States from Blackwell and Richards (2004).

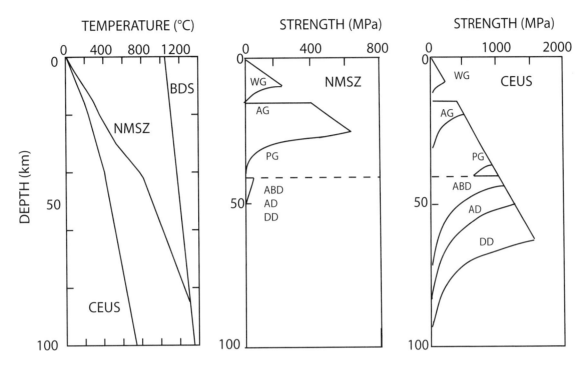

Figure 4. Thermal and mechanical structure beneath the New Madrid seismic zone (NMSZ), assuming higher temperatures and hence significant weakening relative to typical central and eastern United States (CEUS) values. Ductile-flow portions are shown for various flow laws: Westerly granite (WG), Adirondack and Pikwitonei granulite (AG and PG), Anita Bay, Aheim, and dry dunites (ABD, AD, and DD). New Madrid seismic zone curves are for strain rate of 10^{-15} s^{-1}; central and eastern United States curves are for strain rates of 10^{-16} s^{-1}; BDS is basalt dry solidus. (After Liu and Zoback, 1997.)

areas, such that plate-driving forces would be concentrated in the upper crust, causing deformation and seismicity.

Such long-term weakness of the lower crust and mantle would have a significant effect on the seismic cycle in the area and on stress transfer following large earthquakes such as those in 1811–1812. For example, Kenner and Segall (2000) proposed that a weak zone under the New Madrid seismic zone has recently relaxed, such that for a few earthquake cycles, strains can be released faster than they are observed to accumulate at present by geodesy. A limitation of this hypothesis is that there is no evidence for such a weak zone and no obvious reason for why the proposed weakening occurred. Elevated temperatures at depth are also assumed by models in which the seismicity results from sinking of a high-density mafic body (Grana and Richardson, 1996; Stuart et al., 1997) due to recent weakening of the lower crust in the past 9 k.y. (Pollitz et al., 2001), or by Grollimund and Zoback's (2001) model in which deglaciation stresses act on a weak lower crust.

Alternatively, the inferred high heat flow has been interpreted as resulting from groundwater flow in the fractured upper crust, such that the New Madrid seismic zone is not necessarily hotter than its surroundings. In such cases, the measured high heat flow includes convective heat transfer by upward water flow, so temperatures at depth will be overestimated unless this effect is included. This effect is illustrated schematically in Figure 5

for a simple one-dimensional model of heat transfer by fluid flow in a porous medium (Bredehoeft and Papadopulos, 1965), which is often used to analyze heat-flow data (Anderson et al., 1979; Langseth and Herman, 1981). Relative to heat transfer by conduction alone, upward fluid flow increases the surface heat flow and decreases temperature at depth. In a realistic geometry, upward flow would be expected along the rift-bounding faults and above the subsurface faults associated with the earthquakes, redistributing heat laterally and causing a pattern of higher and lower heat-flow values.

Swanberg et al. (1982) favored such an interpretation, noting that their four heat-flow measurements in the Reelfoot rift were made in wells that failed to penetrate the Paleozoic basement and, thus, seemed likely to be affected by groundwater flow within the Cretaceous sands and underlying fractured basement rocks. They also favored this interpretation for the larger number of bottom-hole temperatures, which offer better spatial coverage. Because only a few of the wells within the most seismically active part of the New Madrid seismic zone have unusually high bottom-hole temperatures, they favored the hypothesis that these data reflected groundwater flow associated with the subsurface faults. This interpretation seems more consistent with studies that find low seismic velocities only in parts of the seismic zone and at shallow depths (Al-Shukri and Mitchell, 1987; Vlahovic et al., 2000; Vlahovic and Powell, 2001).

REANALYSIS

To explore this issue, we reexamined Liu and Zoback's (1997) estimates. We found that their inferred large temperature differences between the New Madrid seismic zone and the average central and eastern United States resulted from two effects. First, their analysis assumed a much larger heat-flow anomaly than shown by recent data. Second, plotting errors in their paper increased the difference in geotherms even further.

The anomaly inferred by Liu and Zoback (1997) assumed average New Madrid seismic zone and central and eastern United States heat-flow values of 60 and 45 mW m^{-2}. Their New Madrid seismic zone value came from a combination of five heat-flow measurements within the Reelfoot rift (Fig. 2) (Swanberg et al., 1982; McCartan and Gettings, 1991), with a mean value of 55 ± 9 mW m^{-2} and a value of 75 mW m^{-2} just outside the rift, for a mean value of 58 ± 12 mW m^{-2}. This value is plausible, though slightly higher than given by the recent data. The more important issue is the choice of a central and eastern United States value to characterize the surroundings. Their central and eastern United States value was inferred from a 42 mW m^{-2} average for the coastal plain given by Morgan and Gosnold (1989), which is significantly lower than the 55 ± 21 mW m^{-2} average calculated from the later Blackwell and Richards (2004) compilation. Moreover, as Figures 2 and 3 show, heat flow in the coastal plain and surroundings is highly variable. Parts of the coastal plain east of 90°W, and the area immediately to the north, show lower heat flow than the New Madrid seismic zone. However, the coastal plain west of 90°W

shows average heat flow of 67 ± 17 mW m^{-2}, higher than the New Madrid seismic zone, and heat flow to the north of the New Madrid seismic zone is comparable to that within it.

Figure 6 illustrates the resulting geotherms. Liu and Zoback (1997) showed geotherms predicting that, relative to the central and eastern United States, the New Madrid seismic zone is ~100 °C hotter at 20 km, near the deepest earthquakes, 400 °C hotter at 42 km, an approximate Moho depth, and ~650 °C hotter at 80 km. However, when we calculated the geotherms using their values for surface heat flow (their Fig. 4), thermal conductivity, and heat production (Table 1), we found a central and eastern United States geotherm in the lower crust and mantle that is 50 °C hotter than their plotted one, and an New Madrid seismic zone geotherm in the lower crust and mantle significantly (~110 °C) cooler than their plotted one. Hence, the difference between geotherms shown in their Figure 3 (our Fig. 4) is ~160 °C greater than predicted by their model. Part of the difference is an apparent error in which their New Madrid seismic zone geotherm increases from 28 to 42 km. This "kink" is

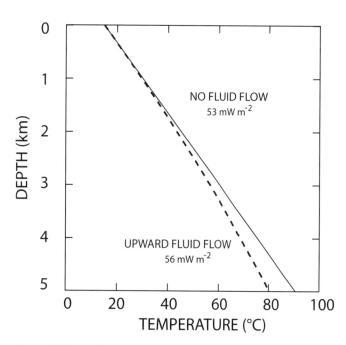

Figure 5. Schematic comparison of geotherms, illustrating how upward fluid flow corresponds to higher heat flow and lower temperatures at depth.

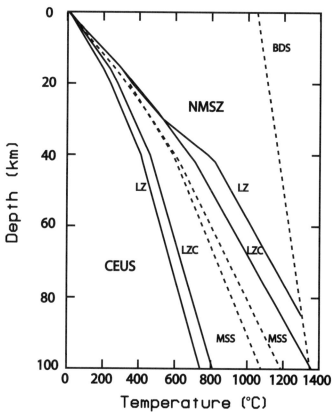

Figure 6. Alternative thermal models for the New Madrid seismic zone (NMSZ) and central and eastern United States (CEUS). LZ denotes geotherms plotted in Liu and Zoback (1997), LZC denotes geotherms computed from Liu and Zoback (1997) values. MSS denotes geotherms for models in this paper, showing much smaller differences between the New Madrid seismic zone and central and eastern United States, due to the much smaller assumed heat-flow difference. BDS is basalt dry solidus.

TABLE 1. GEOTHERMAL PARAMETERS FOR THE NEW MADRID SEISMIC ZONE (NMSZ) AND CENTRAL AND EASTERN UNITED STATES (CEUS)

Region	Layer	Thickness (km)	Conductivity (W m^{-1} K^{-1})	Heat production (μW m^{-3})
NMSZ	Sediments	3.0	3.5	1.50
	Low-velocity zone	2.0	3.0	1.20
	Upper crust	11	2.5	1.10
	Lower crust	12	2.4	0.20
	Altered lower crust	14	2.6	0.02
	Upper mantle	58	3.4	0.01
CEUS	Upper crust	16	2.5	1.17
	Lower crust	24	2.5	0.26
	Upper mantle	60	3.4	0.01

Note: From Liu and Zoback (1997).

implausible because heat flow decreases with depth since some of the heat was generated above that depth. Hence, temperature gradient decreases unless the conductivity decreases so dramatically that it offsets the lower heat flow.

For comparison, we calculated geotherms for the same thermal conductivity and heat production versus depth used by Liu and Zoback (1997), but different surface heat flow. For the New Madrid seismic zone, we used the average from the recent compilation, 55 mW m^{-2}, slightly lower than the Liu and Zoback (1997) value. For the central and eastern United States, we used the central and eastern U.S. average of 52 mW m^{-2}, which is significantly higher than the 45 mW m^{-2} value they used. For this much smaller—and statistically insignificant—heat-flow difference (3 ± 23 mWm-2), the corresponding geotherms are very similar.

The resulting geotherms (Fig. 6) predict that the New Madrid seismic zone is cooler, and the central and eastern United States is hotter, than in Liu and Zoback's (1997) model. As a result, the inferred temperature anomaly is much lower. We predict much smaller differences: ~10 °C versus 100 °C at 20 km, ~20 °C versus 400 °C at 42 km, and ~80 °C versus 650 °C at 80 km. So, in our model, temperature differences are trivial in the seismogenic crust and small in the mantle.

We thus also predict much smaller differences in strength between the New Madrid seismic zone and central and eastern United States. Figure 7 shows strength envelopes for our thermal model. Upper-crustal strength first increases with depth in the brittle region according to Byerlee's law (Brace and Kohlstedt, 1980), computed assuming hydrostatic pore pressure and using the vertical stress as the least compressive principal stress. At depth, strength decreases due to increasing temperature according to ductile-flow law. For comparison with Liu and Zoback (1997), we used the same flow laws. The upper crust is modeled as Westerly granite, and lower crustal strengths are bounded by flow laws for Adirondack and Pikwitonei granulite. A range of upper-mantle strengths are modeled using Anita Bay, Aheim, and dry dunites.

We illustrate the comparison assuming a strain rate within the New Madrid seismic zone of 10^{-16} s^{-1}, approximately corresponding to the geodetically estimated 1 mm/yr across 100 km (Newman et al., 1999; Calais et al., 2005). If the central and eastern United States had the same strain rate, the temperature differences would yield essentially the same strength profile (Fig. 7, center). Assuming a lower central and eastern United States strain rate of 10^{-18} s^{-1}, consistent with the average seismic moment release rate (Anderson, 1986), weaker ductile behavior

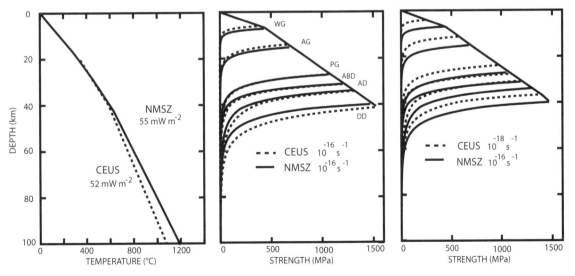

Figure 7. Strength envelopes for our thermal model. Ductile-flow portions are shown for various flow laws: Westerly granite (WG), Adirondack and Pikwitonei granulite (AG and PG), Anita Bay, Aheim, and dry dunites (ABD, AD, and DD). New Madrid seismic zone curves (NMSZ) are for strain rate of 10^{-16} s^{-1}; central and eastern United States (CEUS) curves are for strain rates of 10^{-16} s^{-1} (center) and 10^{-18} s^{-1} (right). The two regions have essentially the same strength profile.

is predicted (Fig. 7, right). This weakening more than offsets the fact that the central and eastern United States is slightly cooler, making the New Madrid seismic zone slightly stronger.

Our results are quite different from those of Liu and Zoback (1997). They assumed that the New Madrid seismic zone is much hotter than its surroundings, so the New Madrid seismic zone lower crust is much weaker than central and eastern United States crust, and the New Madrid seismic zone mantle has essentially no strength. For our model with a much smaller temperature contrast, the New Madrid seismic zone and central and eastern United States have essentially the same strength. Hence, there would be no tendency for upper-crustal stresses to be concentrated in the New Madrid seismic zone.

DISCUSSION

The temperature structure under the New Madrid seismic zone and its surroundings will remain poorly known until additional heat-flow data become available. Moreover, even if the thermal structure were better known, its implications depend on the area to which the New Madrid seismic zone is compared. The data can be interpreted as showing that the New Madrid seismic zone has higher heat flow and is thus hotter than areas to the southeast. Alternatively, we can view New Madrid seismic zone heat flow as essentially the same as for most of the central and eastern United States. It is worth noting that Li et al. (this volume) do not find low Pn velocity under the New Madrid seismic zone, which would be expected if it were hot and weak.

We view the latter interpretation—that any thermal differences are minor—as more useful. If so, then strength differences between the New Madrid seismic zone and its surroundings are small. Although the specific strength envelopes depend on the assumed thermal structure, rheology, and strain rate, we think it is hard to make a strong case that the New Madrid seismic zone is thermally weaker than its surroundings. It is worth noting that Liu and Zoback (1997) used a strain rate within the New Madrid seismic zone of 10^{-15} s^{-1}, somewhat higher than the value that can be inferred geodetically. Adopting this value would make the New Madrid seismic zone even stronger than in our model. Moreover, we suspect that much of the small heat-flow anomaly is due to groundwater flow associated with the subsurface faults. If so, the temperature and strength differences would be even less. There is also the possibility that the heat-flow anomaly results from differences in radiogenic heat production, which is not well known.

The interpretation that the New Madrid seismic zone is not significantly hotter and weaker than its surroundings argues against models in which platewide stresses are concentrated there for long times. Instead, it favors models in which the New Madrid seismic zone became active with the past few thousand years (Schweig and Ellis, 1994), perhaps in the most recent cluster of large earthquakes (Holbrook et al., 2006), and will shut down at some point, perhaps for a long time. In this case, the locus of large earthquakes may migrate. Moreover, this shutdown may be occurring at present (Newman et al., 1999). In this case, the recent small earthquakes

are aftershocks of the large earthquakes of 1811–1812 (Ebel et al., 2000; Stein and Newman, 2004). The possibility that the New Madrid seismic zone is shutting down is suggested by geodetic observations, which show little or none of the expected interseismic motion expected before a future large earthquake (Newman et al., 1999; Calais et al., 2005). If geodetic data continue to show essentially no motion as their uncertainties decrease due to longer spans of observations, the idea of the New Madrid seismic zone shutting down will seem increasingly plausible.

ACKNOWLEDGMENTS

We thank Stéphane Mazzotti, Douglas Wiens, and John Holbrook for helpful comments.

REFERENCES CITED

Al-Shukri, H.J., and Mitchell, B., 1987, Three-dimensional velocity variations and their relation to the structure and tectonic evolution of the New Madrid seismic zone: Journal of Geophysical Research, v. 92, p. 6377–6390.

Anderson, J.G., 1986, Seismic strain rates in the central and eastern United States: Seismological Society of America Bulletin, v. 76, p. 273–290.

Anderson, R.N., Langseth, M.G., and Hobart, M.A., 1979, Geothermal convection through oceanic crust and sediments in the Indian Ocean: Science, v. 204, p. 828–832, doi: 10.1126/science.204.4395.828.

Blackwell, D.D., and Richards, M., 2004, Geothermal Map of North America: American Association of Petroleum Geologists, scale 1:6,500,000. (Data at http://www.smu.edu/geothermal/georesou/usa.htm.)

Brace, W.F., and Kohlstedt, D.L., 1980, Limits on lithospheric stress imposed by laboratory experiments: Journal of Geophysical Research, v. 85, p. 6248–6252.

Bredehoeft, J.D., and Papadopulos, I.S., 1965, Rates of vertical groundwater movement estimated from the Earth's thermal profile: Water Resources Research, v. 2, p. 325–328.

Calais, E., Mattioli, G., DeMets, C., Nocquet, J.-M., Stein, S., Newman, A., and Rydelek, P., 2005, Tectonic strain in plate interiors?: Nature, v. 438, doi: 10.1038/nature04428.

Crone, A.J., De Martini, P.M., Machette, M.N., Okumura, K., and Prescott, J.R., 2003, Paleoseismicity of two historically quiescent faults in Australia: Implications for fault behavior in stable continental regions: Bulletin of the Seismological Society of America, v. 93, p. 1913–1934.

Ebel, J.E., Bonjer, K.P., and Oncescu, M.C., 2000, Paleoseismicity: Seismicity evidence for past large earthquakes: Seismological Research Letters, v. 71, p. 283–294.

Grana, J.P., and Richardson, R.M., 1996, Tectonic stress within the New Madrid seismic zone: Journal of Geophysical Research, v. 101, p. 5445–5458, doi: 10.1029/95JB03255.

Grollimund, B., and Zoback, M.D., 2001, Did deglaciation trigger intraplate seismicity in the New Madrid seismic zone?: Geology, v. 29, p. 175–178, doi: 10.1130/0091-7613(2001)029<0175:DDTISI>2.0.CO;2.

Holbrook, J., Autin, W.J., Rittenour, T.M., Marshak, S., and Goble, R.J., 2006, Stratigraphic evidence for millennial-scale temporal clustering of earthquakes on a continental-interior fault: Holocene Mississippi River floodplain deposits, New Madrid seismic zone, USA: Tectonophysics, v. 420, p. 431–454.

Kenner, S.J., and Segall, P., 2000, A mechanical model for intraplate earthquakes: Application to the New Madrid seismic zone: Science, v. 289, p. 2329–2332, doi: 10.1126/science.289.5488.2329.

Langseth, M.G., and Herman, B.M., 1981, Heat transfer in the oceanic crust of the Brazil basin: Journal of Geophysical Research, v. 86, p. 10,805–10,819.

Li, Q., Liu, M., Zhang, Q., and Sandvol, E., 2007, Stress evolution and seismicity in the central-eastern United States: Insights from geodynamic modeling, in Stein, S., and Mazzotti, S., ed., Continental Intraplate Earthquakes: Science, Hazard, and Policy Issues: Geological Society of America Special Paper 425, doi: 10.1130/2007.2425(11).

Liu, L., and Zoback, M.D., 1997, Lithospheric strength and intraplate seismicity in the New Madrid seismic zone: Tectonics, v. 16, p. 585–595, doi: 10.1029/97TC01467.

McCartan, L., and Gettings, M.E., 1991, Possible relation between seismicity and warm intrusive bodies in the Charleston, South Carolina, and New Madrid, Missouri, area: U.S. Geological Survey Bulletin, v. 1953, p. 1–18.

Morgan, P., and Gosnold, W.D., 1989, Heat flow and thermal regimes in the continental United States, *in* Pakiser, L.C., and Mooney, W.D., eds., Geophysical Framework of the Continental United States: Geological Society of America Memoir 172, p. 493–522.

Newman, A., Stein, S., Weber, J., Engeln, J., Mao, A., and Dixon, T., 1999, Slow deformation and lower seismic hazard at the New Madrid seismic zone: Science, v. 284, p. 619–621, doi: 10.1126/science.284.5414.619.

Pollitz, F., Kellogg, L., and Burgmann, R., 2001, Sinking mafic body in a reactivated lower crust: A mechanism for stress concentration in the New Madrid seismic zone: Bulletin of the Seismological Society of America, v. 91, p. 1882–1897, doi: 10.1785/0120000277.

Sass, J., Diment, W., Lachenbruch, A., Marshall, B., Monroe, R., Moses, T., and Urban, T., 1976, A new heat flow map of the conterminous United States: U.S. Geological Survey Open-File Report 76-0756, p. 76–756.

Schweig, E.S., and Ellis, M.A., 1994, Reconciling short recurrence intervals with minor deformation in the New Madrid seismic zone: Science, v. 264, p. 1308–1311, doi: 10.1126/science.264.5163.1308.

Stein, S., and Newman, A., 2004, Characteristic and uncharacteristic earthquakes as possible artifacts: Applications to the New Madrid and Wabash seismic zones: Seismological Research Letters, v. 75, p. 170–184.

Stuart, W.D., Hildenbrand, T.G., and Simpson, R.W., 1997, Stressing of the New Madrid seismic zone by a lower crust detachment fault: Journal of Geophysical Research, v. 102, p. 27,623–27,633, doi: 10.1029/97JB02716.

Swanberg, C.A., Mitchell, B.J., Lohse, R.L., and Blackwell, D.D., 1982, Heat flow in the upper Mississippi Embayment: U.S. Geological Survey Professional Paper 1236, p. 185–189.

Vlahovic, G., and Powell, C.A., 2001, Three-dimensional S wave velocity structure and Vp/Vs ratios in the New Madrid seismic zone: Journal of Geophysical Research, v. 106, p. 13,501–13,512, doi: 10.1029/2000JB900466.

Vlahovic, G., Powell, C.A., and Chiu, J.M., 2000, Three-dimensional P wave velocity structure in the New Madrid seismic zone: Journal of Geophysical Research, v. 105, p. 7999–8011, doi: 10.1029/1999JB900272.

MANUSCRIPT ACCEPTED BY THE SOCIETY 29 NOVEMBER 2006

The Geological Society of America
Special Paper 425
2007

Upland Complex of the central Mississippi River valley: Its origin, denudation, and possible role in reactivation of the New Madrid seismic zone

Roy Van Arsdale
Ryan Bresnahan
Natasha McCallister
Department of Earth Sciences, University of Memphis, Memphis, Tennessee 38152, USA

Brian Waldron
Department of Civil Engineering, University of Memphis, Memphis, Tennessee 38152, USA

ABSTRACT

Approximately 8000 lignite exploration cores, each 91 m (300 ft) deep, were used to map the gravel facies of the Upland Complex (Lafayette gravel) preserved on drainage divides in western Kentucky and Tennessee and on Crowley's Ridge in southeastern Missouri and eastern Arkansas. The Upland Complex is interpreted to be the remnant of a high-level terrace of the ancestral Mississippi-Ohio River system. The longitudinal profile of the Upland Complex and its projection on sea-level curves suggest that this alluvial deposit is early Pliocene in age (5.5–4.5 Ma). Sea level during the early Pliocene was +100 m, and the Upland Complex is interpreted to have been an ~100-m-thick floodplain when initially deposited. Sea-level decline to −20 m at 4 Ma resulted in incision through the Pliocene floodplain, which formed the high-level terrace. Incision through the floodplain occurred in the Western and Eastern Lowlands of eastern Arkansas and their tributary valleys. The upper silt and sand facies of the terrace (~60 m) were eroded, leaving the basal gravel-rich Upland Complex preserved on drainage divides.

The New Madrid seismic zone lies beneath the Eastern Lowlands. There has been up to 100 m of denudation above the seismic zone in the past 4 m.y., and the most recent denudation occurred in the Holocene due to the confluence of the Mississippi and Ohio Rivers stepping north to Thebes Gap, Missouri. The late Wisconsin and Holocene denudation may have perturbed the local stress field and reactivated the New Madrid seismic zone.

Keywords: Upland Complex, ancestral Ohio River, New Madrid seismic zone, Lafayette gravel.

Van Arsdale, R., Bresnahan, R., McCallister, N., and Waldron, B., 2007, Upland Complex of the central Mississippi River valley: Its origin, denudation, and possible role in reactivation of the New Madrid seismic zone, *in* Stein, S., and Mazzotti, S., ed., Continental Intraplate Earthquakes: Science, Hazard, and Policy Issues: Geological Society of America Special Paper 425, p. 177–192, doi: 10.1130/2007.2425(13). For permission to copy, contact editing@geosociety.org. ©2007 The Geological Society of America. All rights reserved.

INTRODUCTION

The Upland Complex is a high-level fluvial deposit that is discontinuously preserved on drainage divides from western Kentucky to Louisiana along the eastern side of the Mississippi River valley and on Crowley's Ridge from southern Illinois into eastern Arkansas (Fig. 1) (Autin et al., 1991; Saucier, 1994). This high-level fluvial deposit is also called the Lafayette gravel in western Kentucky and northern Tennessee (Potter, 1955), Grover gravel and Mounds gravel in southern Illinois (Willman and Frye, 1970; Harrison, et al., 1999), Citronelle Formation in Louisiana (Doering, 1958), and pre-loess sand and gravel in Mississippi (Dockery, 1996). The depositional environment of the topographically highest and oldest portion of the Upland Complex has been debated for many years, and the principal theories state it was: (1) a south-flowing ancestral Mississippi-Ohio River system (Fisk, 1944; Russell, 1987; Autin et al., 1991), (2) a compound alluvial fan complex at the head of the Mississippi Embayment fed by the ancestral Mississippi, Ohio-Cumberland, and Tennessee Rivers (Potter, 1955), and (3) an alluvial-fan deposit spread westward from the Nashville dome (Self, 1993; Saucier, 1994). Over most of its extent, the Upland Complex is disconformably buried beneath Pleistocene loess, and it disconformably overlies Eocene Jackson, Claiborne, and Wilcox Formations (Saucier, 1994) (Figs. 1 and 2). The age of the Upland Complex is poorly constrained, and proposed ages are Miocene (May, 1981), Pliocene (Potter, 1955; Stringfield and LaMoreaux, 1957; Anthony and Granger, 2006), early Pleistocene (Fisk, 1944; Doering, 1958), and Pliocene-Pleistocene (Autin et al., 1991).

The Upland Complex ranges in thickness from 0 to 100 m and consists of well-oxidized fluvial chert gravel fine- to coarse-grained quartz sand, silt, and clay (Autin et al., 1991). Approximately 86% of the gravel consists of limonite coated, well-rounded chert gravel (Potter, 1955; Guccione et al., 1990) with grain diameters generally less than 2 cm, but individual clasts up to 60 cm in diameter are present (Russell, 1987). The remaining 14% of the gravel consists of well-rounded quartz and quartzite clasts (Potter, 1955; Guccione et al., 1990).

According to Saucier (1994), the Upland Complex was preserved as an erosional remnant caused by entrenchment of the ancestral Mississippi and Ohio Rivers ~3 m.y., commensurate with the onset of continental glaciation. During the late Pliocene through the Illinoian, the ancestral Mississippi and Ohio rivers flowed down the Western and Eastern Lowlands, respectively, and merged south of Helena, Arkansas (Saucier, 1994) (Fig. 1). Radiometric dates reveal that the ancestral Mississippi River formed a new course through Crowley's Ridge at the Bell City–Oran Gap (Fig. 3B) during isotope stages 4–3 (ca. 60,000 yr B.P.) (Blum et al., 2000). This new Mississippi River course flowed down the west side of Sikeston Ridge and apparently merged with the ancestral Ohio River in the southern portion of the Eastern Lowlands (Saucier, 1994). Between 11,500 and 10,500 yr B.P. (Fig. 3D), the ancestral Mississippi River initi-

ated a new course through Thebes Gap, which flowed parallel to the ancestral Ohio River before merging with the ancestral Ohio again, presumably in the southern portion of the Eastern Lowlands. At 9500 yr B.P. (Fig. 3F), the ancestral Mississippi River joined with the ancestral Ohio River at Cairo, Illinois, to form the Holocene river system we have today. Denudation (lowering of the landscape by erosion) of the Eastern Lowlands within the Wisconsinan (last ~60,000 yr) thus appears to have been in part due to the northern stepping of the confluence of the ancestral Mississippi and Ohio Rivers.

The New Madrid seismic zone is located within the Precambrian to Cambrian Reelfoot rift beneath the Eastern Lowlands (Fig. 4). This seismic zone is interpreted to be the result of a right-lateral strike-slip fault system with a compressional left step-over (Schweig and Ellis, 1994; Purser and Van Arsdale, 1998) within an east-northeast regional compressive stress field generated by ridge push (Zoback and Zoback, 1989). Faults within the New Madrid seismic zone have a long history of small amounts of displacement occurring intermittently since the Late Cretaceous (Van Arsdale, 2000). The left step-over zone is the Reelfoot thrust fault, which has accumulated 73 m of slip since the Late Cretaceous. Of that 73 m, there has been 16 m of Holocene displacement, 5.4 m of which has occurred since A.D. 900. Thus, either the faults of the New Madrid seismic zone are starting an active phase of unknown duration (Schweig and Ellis, 1994), or they experienced a temporary burst of activity that is now terminated. In either case, the questions remain: what is driving the New Madrid seismic zone, why has it recently turned on, and should we expect more large earthquakes?

The origin of the stresses driving active deformation in this intraplate seismic zone is not known; however, a number of models have been proposed. A rift pillow underlies the Reelfoot rift (Mooney et al., 1983; Stuart et al., 1997), and Grana and Richardson (1996) proposed that this rift pillow causes local stress concentration. Stuart et al. (1997) postulated that a weak subhorizontal detachment fault exists in the lower crust above the rift pillow that causes local stress concentration. Liu and Zoback (1997) believed that high local heat flow creates high ductile strain rates in the upper mantle and lower crust, thus causing seismicity in the upper crust. More recently, Grollimund and Zoback (2001) proposed that glacial unloading north of the New Madrid seismic zone at the close of the Wisconsinan increased seismic strain rates in the New Madrid seismic zone and initiated the Holocene seismicity. Kenner and Segall (2000) modeled the area of the New Madrid seismic zone as having a weak lower-crustal zone within an elastic lithosphere. In their model, some local or regional perturbation of the stress field, pore pressure, or thermal state is responsible for triggering weak-zone relaxation. Kenner and Segall (2000) stated that a strong candidate for this perturbation is recession of the Laurentian ice sheet ca. 14 ka.

In this study, we map the distribution of the gravel facies of the Upland Complex along the crests of the highest drainage divides in western Kentucky and Tennessee and on Crowley's

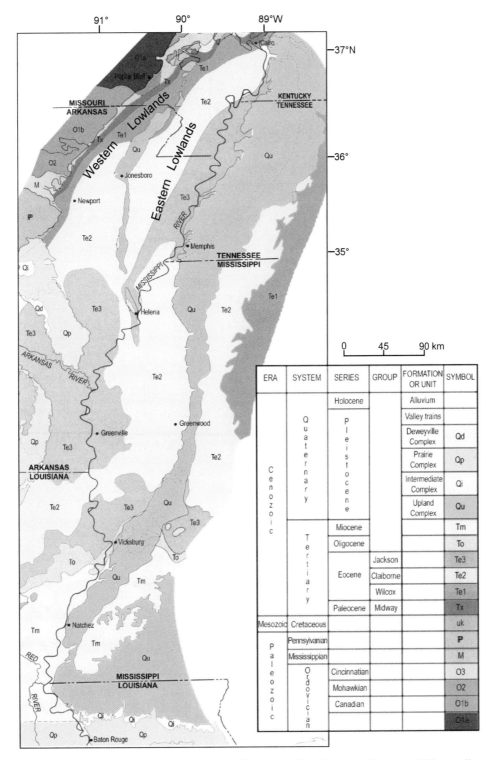

ERA	SYSTEM	SERIES	GROUP	FORMATION OR UNIT	SYMBOL
Cenozoic	Quaternary	Holocene	Pleistocene	Alluvium	
				Valley trains	
				Deweyville Complex	Qd
				Prairie Complex	Qp
				Intermediate Complex	Qi
				Upland Complex	Qu
	Tertiary	Miocene			Tm
		Oligocene			To
		Eocene	Jackson		Te3
			Claiborne		Te2
			Wilcox		Te1
		Paleocene	Midway		Tx
Mesozoic	Cretaceous				uk
Paleozoic	Pennsylvanian				P
	Mississippian				M
	Ordovician	Cincinnatian			O3
		Mohawkian			O2
		Canadian			O1b
					O1a

Figure 1. Geologic outcrop and subcrop (pre-Wisconsinan) of the lower Mississippi River valley (from Saucier, 1994).

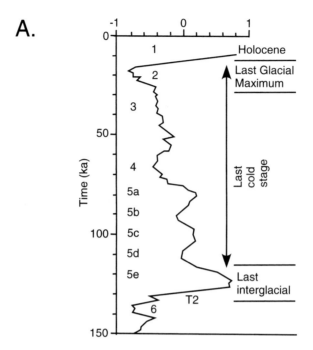

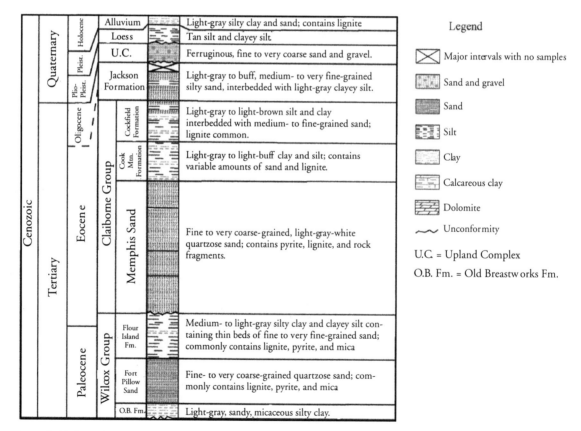

Figure 2. (A) Late Quaternary oxygen isotope stages 1–6 (from Martinson et al., 1987). (B) Cenozoic stratigraphic column of the northern Mississippi Embayment (modified from Van Arsdale and TenBrink, 2000).

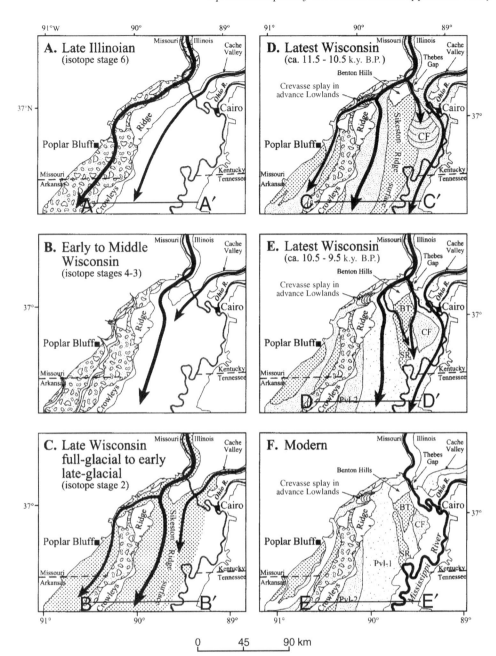

Figure 3. Blum et al. (2000) interpretation of the evolution of the lower Mississippi River valley from the late Illinoian to present day. Cross-section lines A–A′ to E–E′ correspond to Figures 13A–13E. CF—Charleston fan, SR—Sikeston Ridge, BT—Blodgett Terrace, PvL—Late Wisconsin Valley Train.

Ridge in southeastern Missouri and eastern Arkansas from lignite exploration borings (Fig. 5). For clarity of presentation, we hereafter use the term Upland gravel instead of the more cumbersome gravel facies of the Upland Complex. The purposes of our investigation were to map the gravel's current distribution, to estimate the gravel's original distribution, to test models proposed for the gravel's depositional environment, to establish the age of the gravel, to document the denudation of this unit and the associated denudational history of the Eastern Lowlands of the Mississippi River valley, and, lastly, to explore the possible relationship between denudation of the Eastern Lowlands and seismicity within the New Madrid seismic zone.

METHODOLOGY

Extensive lignite exploration conducted by Philips Coal Company (now owned by North American Coal Corporation) during the 1970s resulted in thousands of 300-ft-deep (91.4 m) borings in the lower Mississippi River valley (Fig. 5). Lithologic logs were made of these borings by Philips Coal Company geologists. We picked the top and bottom of the gravel facies of the Upland Complex from the borings drilled on the upland drainage divides of western Kentucky and Tennessee and on Crowley's Ridge in southeastern Missouri and eastern Arkansas (Fig. 5). The gravel facies was mapped because it was

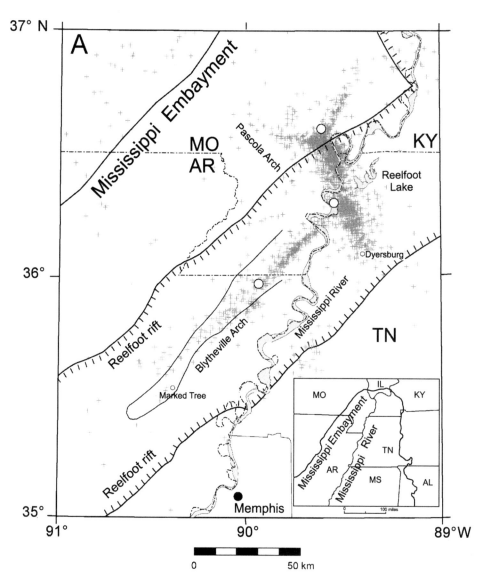

Figure 4. Map of the Reelfoot rift, the New Madrid seismic zone, and the Mississippi Embayment (inset); sediments reflect a Late Cretaceous and early Tertiary embayment of the Gulf of Mexico. Crosses represent microseismicity, and open circles are the three large earthquakes of the 1811–1812 sequence (from Johnston and Schweig, 1996). MO—Missouri, KY—Kentucky, AR—Arkansas, TN—Tennessee.

often impossible to differentiate between Upland Complex sand and underlying Eocene sands from the lithologic descriptions. Only drainage divide borings were selected to avoid including lower and younger terrace deposits or landslides.

Structure contour maps, cross sections, isopachous maps, and planar trend surfaces were made of the Upland gravel to determine its distribution, thickness, and contact slopes using Golden Software's Surfer and Grapher programs (McCallister, 2004; Bresnahan, 2004). Maps and cross sections were made of the gravel's distribution east of the Mississippi River valley, and a second set of maps and cross sections was made of the gravel's distribution along Crowley's Ridge. Cross sections were made from the structure contour maps and planar trend surfaces. Subsequently, all the gravel data were combined in a third set of maps and cross sections to evaluate the gravel's original regional distribution and geometry.

Elevations of the base of alluvium for the Mississippi River valley from Cairo, Illinois, to Baton Rouge, Louisiana, were

determined from floodplain cores by the U.S. Army Corps of Engineers. These 5965 cores allowed us to construct cross sections of the Mississippi valley Quaternary alluvium, to make a planar trend surface of the base of the alluvium, and to compare the Quaternary alluvial section with the Upland gravel.

Elevations of the fluvial surfaces in Figure 3 (Blum et al., 2000) were determined from topographic maps and used to illustrate denudation of the Eastern Lowlands.

RESULTS

Upland Gravel in Western Kentucky and Tennessee

Lignite exploration borings analyzed in western Kentucky and Tennessee reveal that the gravel facies is discontinuous (Fig. 5). Of the 7382 drainage divide logs in this area, only 405 contain Upland gravel. The borings also reveal that the Upland

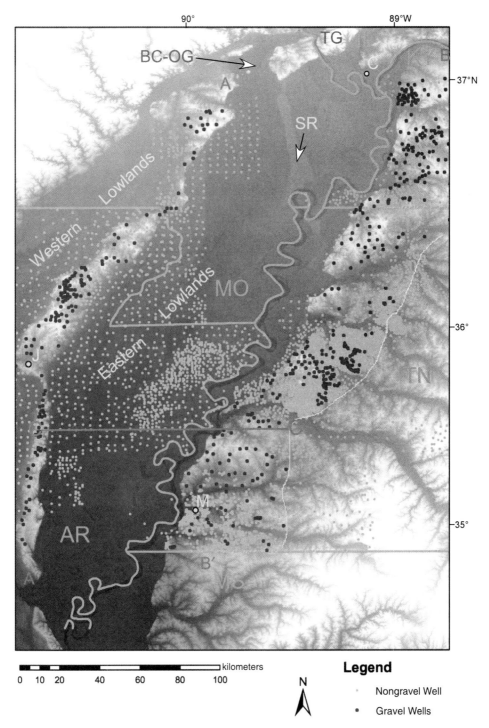

Figure 5. Digital elevation model of the upper Mississippi Embayment and the distribution of lignite wells (red contain gravel and blue do not), the eastern margin of the Upland gravel belt in Tennessee (yellow line) from McCallister (2004), and the cross-section locations of Figures 7 and 9. M—Memphis, J—Jonesboro, SR—Sikeston Ridge, C—Cairo, TG—Thebes Gap, BC-OG—Bell City–Oran Gap.

gravel varies in thickness from 0 to 35 m, with an average thickness of 9 m. A planar trend surface of gravel thickness illustrates that the Upland gravel thickens toward the present Mississippi River at ~0.2 m/km (Fig. 6).

Figure 5 illustrates the western margin of the Upland gravel at the Mississippi River bluff line and also an eastern gravel margin in Tennessee. Regression lines fit to the north-south cross section reveal that the top and bottom of the gravel slopes southerly at 0.25 m/km and 0.30 m/km respectively (Fig. 7). The southerly slope is characteristic of the entire unit as revealed in the slopes of the planar trend surfaces of the top (128°, 0.3 m/km) and bottom (132°, 0.27 m/km) of the gravel (Fig. 8).

East-west cross sections were constructed in western Tennessee to illustrate the position of the Upland gravel in the landscape and to better understand the eastern margin of the gravel belt (McCallister, 2004). The cross sections show an abrupt eastern margin to the Upland gravel and a ground surface that rises east of the gravel margin, thus revealing that this margin is erosionally inset into the underlying Eocene section (Fig. 9).

Upland Gravel in Crowley's Ridge of Southeastern Missouri and Eastern Arkansas

Crowley's Ridge is a topographic divide between the Eastern and Western Lowlands of the Mississippi River valley. This divide was formed primarily as a consequence of Pleistocene incision of the ancestral Ohio River in the Eastern Lowlands and the ancestral Mississippi River in the Western Lowlands (Saucier, 1994; Blum et al., 2000), along with minor late Quaternary tectonic uplift (Cox, 1988; Van Arsdale et al., 1995). Regression lines fit to a north-south cross section reveal that the top and bottom of the gravel slopes south at 0.30 m/km and 0.21 m/km, respectively, and that the gravel has an average thickness of 10 m (Fig. 7). Planar trend surfaces of the top (128°, 0.21 m/km) and bottom (131°, 0.18 m/km) also reveal southerly slopes (Fig. 8).

Interpolated Regional Distribution of the Upland Gravel

Figure 5 illustrates the distribution of preserved Upland gravel in western Kentucky and Tennessee and eastern Missouri and Arkansas. We combined all the gravel data to make planar trend surfaces of the gravel's top and bottom across the Eastern Lowlands (Fig. 8). Figure 8 reveals southerly slopes of the interpolated top (151°, 0.25 m/km) and bottom (153°, 0.27 m/km) of the gravel.

Eastern Lowlands Alluvium

Lignite borings were used to make a cross section of the Quaternary alluvium near Memphis (Fig. 9), and U.S. Army Corps of Engineers boring data of the elevation of the base of the Quaternary alluvium (Saucier, 1994) within the modern Mississippi River valley were used to make a planar trend surface of the base of the alluvium, and a north-south cross section from Cairo, Illinois, to 75 km

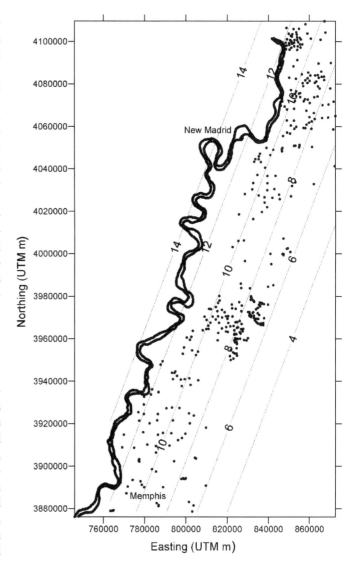

Figure 6. Planar trend surface of the Upland gravel thickness in meters in western Kentucky and Tennessee.

south of Baton Rouge, Louisiana (Figs. 10 and 11). The longitudinal profile (Fig. 11) illustrates that the base of the Quaternary alluvium slopes south at 0.17 m/km, and the trend surface (Fig. 10) reveals a southeast (156°) slope of 0.19 m/km. It is also evident from Figure 11 that the top of the Mississippi River floodplain is at sea level 65 km south of Baton Rouge. At this same location, the base of the Mississippi River's Quaternary alluvium is 65 m below sea level, and the base of the modern Mississippi River channel is 35 m below sea level (U.S. Army Corps of Engineers, 1990).

The east-west cross section near Memphis (Fig. 9) illustrates that the base of the Upland gravel in Crowley's Ridge and western Tennessee lies ~80 m above the base of the modern Mississippi River floodplain gravel. This cross section also illustrates the Mississippi River's Quaternary alluvium beneath the Eastern Lowlands with its basal gravel facies.

Cross Section of the Upland Gravel along Crowley's Ridge

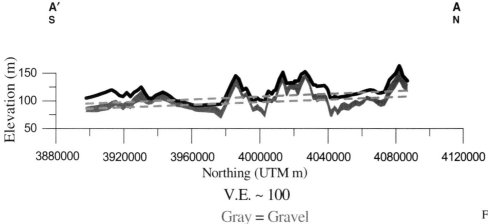

V.E. ~ 100

Gray = Gravel
Black = Surface

Cross Section of the Upland Gravel in Western Tennesse and Kentucky

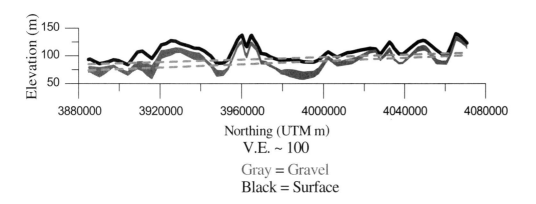

V.E. ~ 100

Gray = Gravel
Black = Surface

Figure 7. Geologic cross sections illustrating the Upland gravel on Crowley's Ridge (A–A′) and western Tennessee–Kentucky (B–B′) with dashed regression lines. UTM—Universal Transverse Mercator projection. Cross section end points are shown in Figure 5. V.E.—vertical exaggeration.

Denudation of the Eastern Lowlands

Denudation of the Eastern Lowlands since deposition of the Upland Complex can be calculated by subtracting the projected upper planar trend surface of the Upland gravel from the landscape topography (Fig. 12). Of course, there have been multiple cycles of degradation and aggradation through the Pleistocene, with sea level, discharge, and sediment load changes from continental ice-sheet growth and decay. Thus, we are presenting a calculation of the total denudation that has occurred since Pliocene time. As revealed in Figure 12, this calculation results in a local maximum of 35 m of denudation. However, as discussed next, the amount of denudation was probably greater.

Blum et al. (2000) illustrated the geomorphic evolution of the Mississippi-Ohio River system in the upper Mississippi Embayment (Fig. 3). By assigning elevations to their geomorphic surfaces, one can see the Illinoian through Holocene denudation of the New Madrid seismic zone within the Eastern Lowlands in some detail (Fig. 13). These Figure 13 cross sections illustrate the progressive denudation of the Eastern Lowlands, which appears to be related to changes in the courses of the ancestral Mississippi and Ohio Rivers through the relevant time intervals.

DISCUSSION

The distribution of the Upland gravel presented herein allows us to test whether the westward prograding alluvial-fan depositional model proposed by Self (1993) and Saucier (1994) is reasonable for western Tennessee. Although the Upland gravel is discontinuous in western Tennessee, there is a distinct eastern margin to the gravel (Fig. 5) (McCallister, 2004).

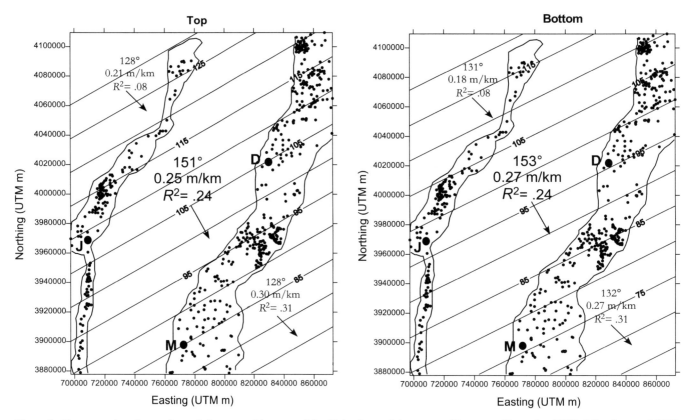

Figure 8. Planar trend surface values of the top and bottom of the Upland gravel in western Tennessee–Kentucky (128°, 0.3 m/km, and 132°, 0.27 m/km), Crowley's Ridge (128°, 0.21 m/km, and 131°, 0.18 m/km), and surface (contour lines) combining the data from western Kentucky–Tennessee and Crowley's Ridge (151°, 0.25 m/km and 153°, 0.27 m/km). Contour interval = 5 m. M—Memphis, J—Jonesboro, D—Dyersburg.

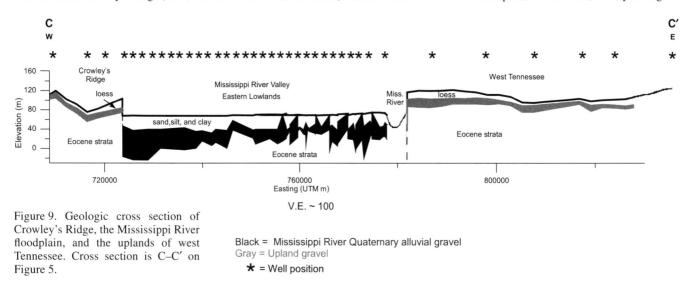

Figure 9. Geologic cross section of Crowley's Ridge, the Mississippi River floodplain, and the uplands of west Tennessee. Cross section is C–C' on Figure 5.

Black = Mississippi River Quaternary alluvial gravel
Gray = Upland gravel
★ = Well position

If indeed the Upland Complex was derived from the Nashville dome and prograded westward as an alluvial fan across western Tennessee to its present location, as proposed by Self (1993), then one would expect gravel to exist east of its current extent. Furthermore, the north-south cross section (Fig. 7) and base-of-gravel trend surface for western Tennessee (Fig. 8) reveal that the Upland gravel slopes southerly, not west, as an eastern source would require. Lastly, the isopachous planar trend surface map (Fig. 6) reveals that the Upland gravel thickens westerly in Tennessee, and not eastward as one would expect if the source were the Nashville dome. Thus, the Upland gravel distribution in western Tennessee does not support the interpretation that the gravel is an alluvial fan complex spread westward from the Nashville dome.

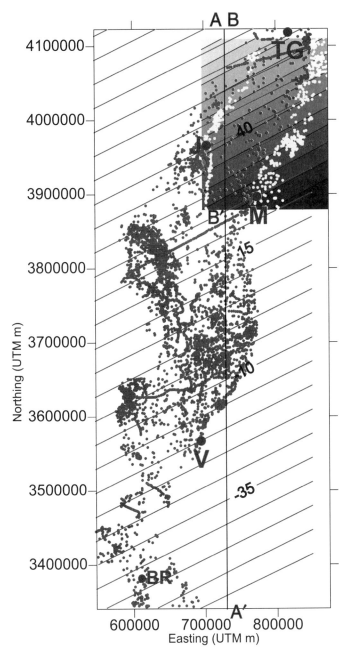

Figure 10. Planar trend surface of the elevation of the base of the modern Mississippi River alluvium (circles—wells), which slopes to the southeast (156°) at 0.19 m/km and base of the Upland gravel (inset of Fig. 8). Contour interval = 5 m. TG—Thebes Gap, J—Jonesboro, M—Memphis, V—Vicksburg, BR—Baton Rouge. Cross sections A–A′ and B–B′ are illustrated in Figure 11.

Our data better support the interpretation that the gravel is a terrace of a south-flowing ancestral Mississippi-Ohio River system. The sinuous eastern margin of the Upland gravel belt appears to be erosionally inset into the landscape, thus suggesting the eastern margin of a meandering river valley. Furthermore, the land surface rises to the east of the gravel belt margin, as one

would expect if a valley wall had been at this location when the Upland Complex river was present (Fig. 9). The southerly slope of the base of the Upland gravel in western Kentucky and Tennessee strongly suggests a south-flowing river system.

The north-south cross section (Fig. 7) and trend surface maps of Crowley's Ridge (Fig. 8) show that the gravel slopes southerly, that the elevation of the base of the gravel is very similar to the basal elevation of the gravel in western Kentucky and Tennessee, and that the thickness of the gravel is similar to the gravel thickness in western Kentucky and Tennessee. Our data agree with Potter (1955) and Guccione et al. (1990) that the Upland gravel sections of western Kentucky, Tennessee, and Crowley's Ridge are all parts of a once-continuous gravel sheet. Based on this interpretation, we combined all the gravel data into one data set. The trend surface maps (Fig. 8) of this combined data set reveal a southerly sloping Upland Complex river floodplain that we reason used to extend from western Kentucky and Tennessee over the current Eastern Lowlands, across Crowley's Ridge, and possibly into the Western Lowlands.

Our data discussed thus far support the interpretation by Autin et al. (1991) that the Upland Complex of western Tennessee and eastern Arkansas is a high-level terrace of an ancestral Ohio-Mississippi River system. Potter (1955) interpreted the Lafayette gravel (Upland Complex) to have been deposited in a shallow braided-river system. However, we believe that the Upland gravel in Tennessee and Arkansas may represent the remains of a basal gravel of a deep river. To be more specific, the Upland Complex was the basal sand and gravel facies of a fluvial section that may have been at least as thick as the Quaternary Mississippi River alluvial section (100 m thick near Memphis in Fig. 9) (Fig. 11). Our justification for this interpretation is derived from a number of observations. (1) The upper contact of the Upland Complex is erosional, and thus more alluvium clearly used to overlie the preserved Upland gravel and Upland Complex. (2) At some locations, the Upland Complex is 100 m thick (Autin et al., 1991). (3) The Upland gravel belt is 110 km wide near Memphis, and thus it is similar to the 135-km-wide Quaternary Mississippi River floodplain (Eastern and Western Lowlands) at the same location. If indeed the Upland gravel belt originally extended west of Crowley's Ridge, then 110 km is a minimum width value. (4) The geomorphology of the Upland Complex river valley was probably very similar to today's Mississippi River valley since both rivers flowed through the same Eocene sediments. (5) Both Potter (1955) and Guccione et al. (1990) believed that the climate during deposition of the Upland Complex was humid temperate with forest vegetation, much like the present-day climate. Thus, assuming similar drainage-basin sizes, the Upland Complex river should have had a discharge comparable to the Holocene Mississippi River discharge. (6) The base of the Upland gravel and base of the Quaternary Mississippi River alluvium have similar southerly slopes from Cairo, Illinois, to Memphis (Fig. 11), thus indicating that the Upland Complex river had a similar discharge to the Quaternary Mississippi River. (7) A high Upland Complex river discharge is indicated by the large size of gravel clasts

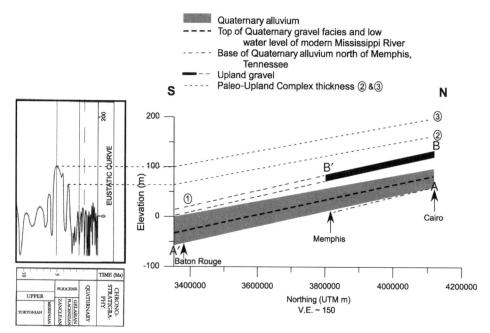

Figure 11. Geologic cross section of the Quaternary Mississippi River alluvial package, the Upland gravel, paleo–Upland Complex thickness estimates 2 and 3, and the sea-level fluctuations from the Pliocene to Holocene (Abreu and Anderson, 1998). The apparent convergence of the Upland gravel and the Mississippi River floodplain is an artifact of fitting a regression line to the upstream portion of the Upland gravel and the full length of the Mississippi floodplain. The dash-dot line beneath the Mississippi River floodplain is the regression line for the upper portion of the Mississippi floodplain, and, as can be seen, it parallels the slope of the Upland gravel. Cross sections are located on Figure 10.

Figure 12. Isopachous map exhibiting the denudation of the Eastern Lowlands since Upland Complex time using the current thickness of the Upland gravel (min) and two estimated thicknesses (int. and max) from Figure 11. Black lines labeled NMSZ represent trace of New Madrid seismic zone. NM—New Madrid, D—Dyersburg, J—Jonesboro, M—Memphis, H—Helena.

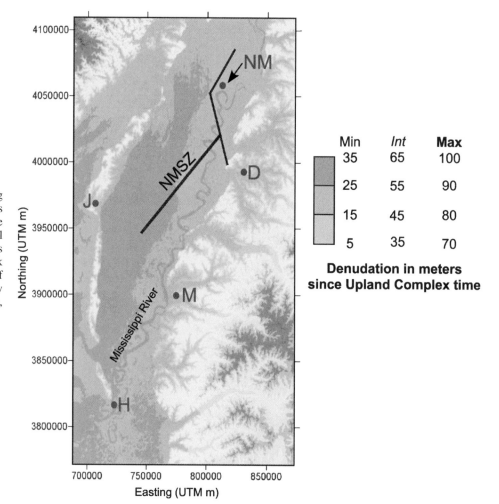

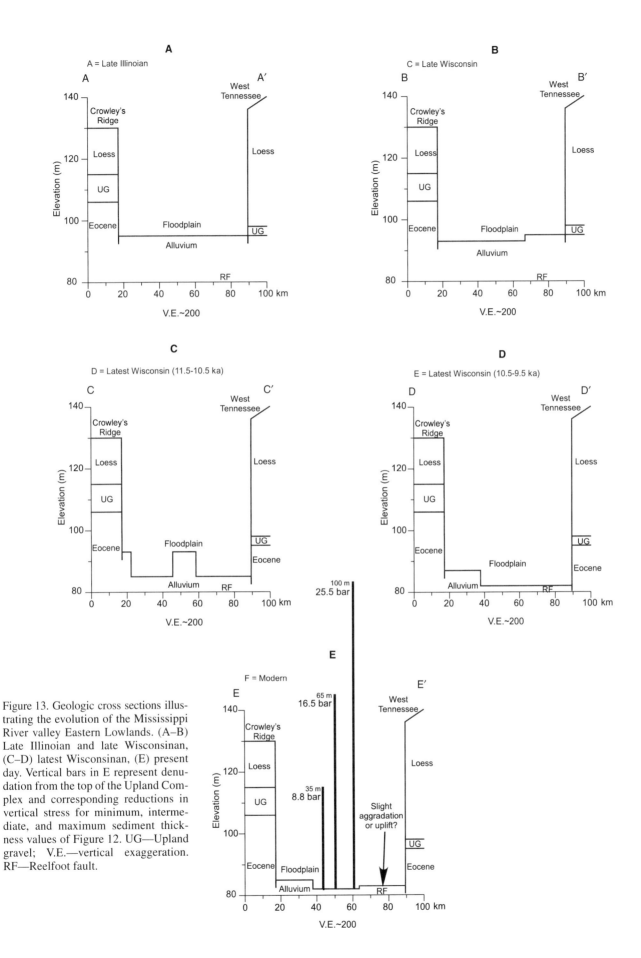

Figure 13. Geologic cross sections illustrating the evolution of the Mississippi River valley Eastern Lowlands. (A–B) Late Illinoian and late Wisconsinan, (C–D) latest Wisconsinan, (E) present day. Vertical bars in E represent denudation from the top of the Upland Complex and corresponding reductions in vertical stress for minimum, intermediate, and maximum sediment thickness values of Figure 12. UG—Upland gravel; V.E.—vertical exaggeration. RF—Reelfoot fault.

in the Upland gravel. There is no evidence of high relief in the Pliocene-Pleistocene central United States, and therefore the best way to transport the large gravel clasts would be by having a deep river that would provide an adequately high basal shear stress (Leopold et al., 1964). Additional support for our interpretation of a major Upland Complex river system comes from the following comparison of the Upland gravel with the Quaternary Mississippi River alluvium and Pliocene eustatic sea-level changes.

The Mississippi River floodplain surface is at sea level 65 km south of Baton Rouge, Louisiana (Fig. 11). At this same location, the base of the Mississippi River Quaternary alluvium is 65 m below sea level. Saucier (1994) proposed that the basal alluvium beneath the Holocene Mississippi meander belt may be Pleistocene or Holocene in age. We reason that most of the alluvium within the Holocene floodplain of the Mississippi River is probably Holocene in age. Support for this interpretation comes from Winkley (1994), wherein he identified ongoing bedrock scour along the base of the Mississippi River in a reach near Greenville, Mississippi. Best and Ashworth (1997) noted that scour at channel bends can be up to five times deeper than the mean channel depth. Based on these profile and flood scour considerations, we believe that the Holocene Mississippi River floodplain is on average 50 m thick (Fig. 11).

Since our data support the interpretation that the Upland gravel reflects an ancestral Mississippi–Ohio River system, then the elevation of the Upland Complex river system would have been controlled by sea level. Thus, the surface of the Upland Complex river floodplain would grade to sea level at the time of its deposition, and the base of the floodplain would have been at a sub–sea-level depth at its ocean shoreline, like the modern Mississippi River 65 km south of Baton Rouge (Fig. 11). We believe that the Upland Complex river had a discharge at least as large as the Holocene Mississippi River, and, if so, it would have had a floodplain averaging at least 50 m thick.

Based on the eustatic sea-level curve by Abreu and Anderson (1998), sea level was one hundred meters higher (+100 m) than today during much of the early Pliocene (5.5–4.5 Ma) (Fig. 11). At ca. 4 Ma, sea level dropped to −20 m, at 3.5 Ma, it rose to +60 m, and then at 3 Ma, it dropped again to −70 m. Based on these arguments and Figure 11, sea level during Upland Complex time had to be greater than 20 m above modern sea level (base of the Upland gravel is +20 m). At first glance, it would appear that the Upland Complex floodplain surface may have been graded to the 3.5 Ma middle Pliocene sea level (Fig. 11, dotted line 2). However, we think this is not likely. The ancestral Mississippi River system experienced a sea-level drop to −20 m ca. 4 Ma, which would have resulted in a floodplain basal elevation lower than −20 m. Pursuing this possible history, this would mean that the Mississippi River then aggraded the Upland Complex to +60 m at 3.5 Ma (Fig. 11, dotted line 2). This is unlikely for three reasons: (1) the base of the Upland gravel is +20 m near Baton Rouge (not below −20 m), (2) any river valley aggradation due to sea-level rise at 3.5 Ma would extend a relatively short distance up the ancestral Mississippi River valley to perhaps Vicksburg, Mississippi (Blum

and Tornqvist, 2000), not over 1000 km to Cairo, Illinois, and, (3) the broad expanse of the Upland Complex terrace suggests a long period of stability, not a relatively short sea-level excursion. Thus, we believe it is more likely that the Upland Complex is an early Pliocene 5.5–4.5 Ma deposit. This early Pliocene age interpretation is consistent with Potter's (1955) conclusion, based on the absence of glacial sediments, that the Lafayette gravel (Upland Complex) is Pliocene. Based on the oxygen isotope record in the Gulf of Mexico, the first glacial meltwater entered the gulf ca. 2.3 Ma (Joyce et al., 1993). Since the Upland Complex is not of glacial origin and the first North American continental ice sheet has an age of ca. 2.3 Ma, then the Upland Complex must be Pliocene in age and older than 2.3 Ma.

If our interpretation is correct, then the Upland Complex was erosionally removed from the Eastern and Western Lowlands ca. 4 Ma during the −20 m lowstand. The Upland Complex, preserved on the drainage divides of the Mississippi River valley, has therefore been subject to denudation between periods of loess burial over the past 4 m.y.

We propose that the Upland Complex was originally at least 50 m thick at Memphis, Tennessee, that it could have been 80 m thick, and that the upper portion of Upland Complex river alluvium, which probably consisted primarily of sand overlain by silt and clay like the modern Mississippi River alluvium (Fig. 9), has been stripped off of the Upland gravel over the past 4 m.y. If indeed the Upland Complex was originally 80 m thick, then a maximum of 100 m of sediment (top of restored Upland Complex to top of Holocene Mississippi River floodplain) has been eroded from the Eastern Lowlands in the last 4 m.y. (Figs. 12 and 13).

Blum et al. (2000) interpreted the late Illinoian through Holocene drainage history of the upper Mississippi Embayment (Fig. 3). Figure 3 illustrates the river course history and northward jumping of the confluence of the ancestral Ohio and Mississippi Rivers during this time period. Denudation of the Eastern Lowlands appears to have occurred as a consequence of these river course changes and confluence jumps. The Mississippi-Ohio River confluence was located south of Helena, Arkansas, during the late Pliocene through late Illinoian, jumped to the southern portion of the Eastern Lowlands at ca. 60 ka, and finally jumped to Cairo, Illinois, 9500 yr ago. Thus, the full discharge of the Mississippi River over the New Madrid seismic zone is unique to the Holocene. By assigning elevations to the late Illinoian, Wisconsinan, and Holocene landforms of Figure 3, the late Illinoian to Holocene denudation of the upper Mississippi Embayment is demonstrated (Fig. 13) (Bresnahan, 2004).

Holocene Reactivation of the New Madrid Seismic Zone

As discussed already, a number of the most recent studies attribute recession of the Laurentian ice sheet as the initiation of the New Madrid seismic zone. A problem with initiation by Laurentian ice-sheet retreat is the nonuniqueness of the last glacial retreat. There have been numerous continental ice sheets in the northern United States (Easterbrook, 1999), yet only evidence of Holocene faulting within the Quaternary. We speculate that denudation of the northern portion of the Eastern Lowlands may have initiated Holocene fault reactivation within the New Madrid seismic zone by providing the perturbation necessary in the Kenner and Segall (2000) model.

Johnston (1987, 1989) and Sauber and Molnia (2004) argued that melting of the Fennoscandian continental ice sheet and resultant vertical stress reduction of 200 bars caused reverse faulting. Specifically, ice removal reduces the vertical compressive stress, and where crustal rock is near failure by reverse faulting due to the local tectonic stress field, the ice-melt stress reduction is sufficient for fault reactivation. This vertical stress reduction process has also been documented to initiate seismicity in quarry excavations (Cook, 1976; Pomeroy et al., 1976; Gibowitz, 1982; Simpson, 1986). It is apparent from Figures 12 and 13 that the Eastern Lowlands and the New Madrid seismic zone has experienced at least 35 m (8.8 bars) to perhaps as much as 100 m (25.5 bars) of denudation and resultant reduced vertical compressive stress since the early Pliocene, with the most recent denudation occurring 9500 yr ago with the formation of the confluence of the Mississippi and Ohio Rivers at Cairo, Illinois. Since 9500 yr B.P., there has been minor aggradation along the Mississippi River, but degradation has occurred across the southern arm of the New Madrid seismic zone. We speculate that denudation of the Eastern Lowlands may have caused the crossing of a stress threshold, which resulted in Holocene reactivation of the New Madrid seismic zone.

CONCLUSIONS

We interpret the Upland Complex in western Kentucky, Tennessee, and on Crowley's Ridge in southeastern Missouri and eastern Arkansas to be a high-level terrace of the ancestral Pliocene Mississippi and Ohio Rivers that used to extend across the Eastern Lowlands and Crowley's Ridge, perhaps into the Western Lowlands (Potter, 1955; Guccione et al., 1990). This terrace is thus the erosional remnant of a large river floodplain near Memphis, Tennessee, that was up to 80 m thick and had a minimum east-west width of 110 km. This large Upland Complex valley width implies longevity. We speculate that this great river valley existed through the sea-level highstand of the early Pliocene from ca. 5.5 to 4.5 Ma (Fig. 11). A major eustatic sea-level decline to −20 m at 4 Ma resulted in incision through the Upland Complex sediments, formation of the Upland Complex terrace, and initial denudation of the Western and Eastern Lowlands.

A 120 m eustatic sea-level decline at 4 Ma would probably have resulted in global river entrenchment, and thus Pliocene high-level river terraces should be found throughout the world's coastal rivers. Indeed, there are Pliocene high-level river terrace deposits along the eastern coast of North America, which Colquhoun et al. (1991) attributed to Pliocene sea-level decline.

Denudation of the Western and Eastern Lowlands appears to have initiated 4 m.y. with the incision of the ancestral Mississippi and Ohio Rivers below the base of the Upland Complex and into the underlying Eocene section. Subsequently, the easily eroded sand and silt upper facies of the Upland Complex were eroded off the basal gravel facies, thus leaving the erosionally more resistant basal gravel (Upland gravel) on high Mississippi River valley drainage divides.

Denudation of the Eastern Lowlands occurred during the Illinoian and into the Holocene with the course changes of the ancestral Mississippi and Ohio Rivers. Specifically, with each jump of the confluence of these two rivers, the Eastern Lowlands became lower and expanded northward. Denudation atop the New Madrid seismic zone appears to have accelerated in the late Wisconsinan, with the most recent denudation occurring after the modern course of the Mississippi River formed 9500 yr ago. This late Quaternary denudation, perhaps in combination with the retreating Laurentide forebulge, may have been sufficient to initiate Holocene seismicity by causing a perturbation in the local stress field (Kenner and Segall, 2000). We close by presenting this denudational history for consideration in future New Madrid seismic zone kinematic model assessments.

ACKNOWLEDGMENTS

This research was supported by the Earthquake Engineering Research Centers Program of the National Science Foundation under award number EEC-9701785. We want to thank North American Coal Company for providing the geologic boring logs, Paul Potter and Randy Cox for their informal reviews, and Ryan Csontos for his help with the figures.

REFERENCES CITED

Abreu, V.S., and Anderson, J.B., 1998, Glacial eustasy during the Cenozoic: Sequence stratigraphic implications: American Association of Petroleum Geologists Bulletin, v. 82, p. 1385–1400.

Anthony, D.M., and Granger, D.E., 2006, Five million years of Appalachian landscape evolution preserved in cave sediments, *in* Harmon, R.S., and Wicks, C., eds., Perspectives on Karst Geomorphology, Hydrology, and Geochemistry—A Tribute Volume to Derek C. Ford and William B. White: Geological Society of America Special Paper 404, p. 39–50.

Autin, W.J., Burns, S.F., Miller, B.J., Saucier, R.T., and Snead, J.I., 1991, Quaternary geology of the Lower Mississippi valley, *in* Morrison, R.B., ed., Quaternary Nonglacial Geology: Conterminous U.S.: Boulder, Colorado, Geological Society of America, Geology of North America, v. K-2, p. 547–582.

Best, J.L., and Ashworth, P.J., 1997, Scour in large braided rivers and the recognition of sequence stratigraphic boundaries: Nature, v. 387, p. 275–277, doi: 10.1038/387275a0.

Blum, M.D., and Tornqvist, T.E., 2000, Fluvial responses to climate and sea-level change: A review and look forward: Sedimentology, v. 47, p. 2–48, doi: 10.1046/j.1365-3091.2000.00008.x.

Blum, M.D., Guccione, M.J., Wysocki, D.A., Robnett, P.C., and Rutledge, E.M., 2000, Late Pleistocene evolution of the lower Mississippi River valley, southern Missouri to Arkansas: Geological Society of America Bulletin, v. 112, p. 221–235, doi: 10.1130/0016-7606(2000)112<0221: LPEOTL>2.3.CO;2.

Bresnahan, R., 2004, The Origin and Denudation of the Upland Gravel in the Upper Mississippi Embayment and its Possible Tectonic Implications [M.S. thesis]: Memphis, University of Memphis, 56 p.

Colquhoun, D.J., Johnson, G.H., Peebles, P.C., Huddlestun, P.F., and Scott, T., 1991, Quaternary geology of the Atlantic coastal plain, *in* Quaternary Nonglacial Geology: Conterminous U.S.: Boulder, Colorado, Geological Society of America, Geology of North America, v. K-2, p. 629–650.

Cook, N.G.W., 1976, Seismicity associated with mining: Engineering Geology, v. 10, p. 99–122, doi: 10.1016/0013-7952(76)90015-6.

Cox, R.T., 1988, Evidence of Quaternary ground tilting associated with the Reelfoot rift zone, northeast Arkansas: Southeastern Geology, v. 28, p. 211–224.

Dockery, D.T., 1996, Toward a revision of the generalized stratigraphic column of Mississippi: Mississippi Geology, v. 17, no. 1, 8 p.

Doering, J.A., 1958, Citronelle age problem: American Association of Petroleum Geologists Bulletin, v. 42, p. 764–786.

Easterbrook, D.J., 1999, Surface Processes and Landforms: Upper Saddle River, New Jersey, Prentice Hall, 546 p.

Fisk, H.N., 1944, Geological investigation of the alluvial valley of the lower Mississippi River: Vicksburg, Mississippi, U.S. Army Corps of Engineers, Mississippi River Commission, 78 p.

Gibowitz, S.J., 1982, The mechanism of large mining tremors in Poland, *in* Gay, N.C., and Wainwright, E.H., eds., Proceedings of the International Congress on Rockbursts and Seismicity: Mines (1st edition): Johannesburg, South African Institute of Mining and Metallurgy, p. 17–28.

Grana, J.P., and Richardson, R.M., 1996, Tectonic stress within the New Madrid seismic zone: Journal of Geophysical Research, v. 101, p. 5445–5458, doi: 10.1029/95JB03255.

Grollimund, B., and Zoback, M.D., 2001, Did deglaciation trigger intraplate seismicity in the New Madrid seismic zone?: Geology, v. 29, p. 175–178, doi: 10.1130/0091-7613(2001)029<0175:DDTISI>2.0.CO;2.

Guccione, M.J., Prior, W.L., and Rutledge, E.M., 1990, The Tertiary and early Quaternary geology of Crowley's Ridge, *in* Guccione, M., and Rutledge, E., eds., Field Guide to the Mississippi Alluvial Valley Northeast Arkansas and Southeast Missouri: Fayetteville, Arkansas, Friends of the Pleistocene, South Central Cell, p. 23–44.

Harrison, R.W., Hoffman, D., Vaughn, J.D., Palmer, J.R., Wiscombe, C.L., McGeehin, J.P., Stephenson, W.J., Odum, J.K., Williams, R.A., and Forman, S.L., 1999, An example of neotectonism in a continental interior—Thebes Gap, Midcontinent, United States: Tectonophysics, v. 305, p. 399–417, doi: 10.1016/S0040-1951(99)00010-4.

Johnston, A.C., 1987, Suppression of earthquakes by large continental ice sheets: Nature, v. 330, p. 467–469, doi: 10.1038/330467a0.

Johnston, A.C., 1989, The effect of large ice sheets on earthquake genesis, *in* Gregersen, S., and Basham, P.W., eds., Earthquakes at North Atlantic Passive Margins: Neotectonics and Postglacial Rebound: Dordrecht, The Netherlands, Kluwer Academic Publishers, p. 581–599.

Johnston, A.C., and Schweig, E.S., 1996, The enigma of the New Madrid earthquakes of 1811–1812: Annual Review of Earth and Planetary Sciences, v. 24, p. 339–384, doi: 10.1146/annurev.earth.24.1.339.

Joyce, J.E., Tjalsma, L.R.C., and Prutzman, J.M., 1993, North American glacial meltwater history for the past 2.3 m.y.: Oxygen isotope evidence from the Gulf of Mexico: Geology, v. 21, p. 483–486, doi: 10.1130/0091-7613(1993)021<0483:NAGMHF>2.3.CO;2.

Kenner, S.J., and Segall, P., 2000, A mechanical model for intraplate earthquakes: Application to the New Madrid seismic zone: Science, v. 289, p. 2329–2332, doi: 10.1126/science.289.5488.2329.

Leopold, L.B., Wolman, M.G., and Miller, J.P., 1964, Fluvial Processes in Geomorphology: San Francisco, W.H. Freeman and Company, 522 p.

Liu, L., and Zoback, M.D., 1997, Lithospheric strength and intraplate seismicity in the New Madrid seismic zone: Tectonics, v. 16, p. 585–595, doi: 10.1029/97TC01467.

Martinson, D.G., Pisias, N.G., Hays, J.D., Imbrie, J., Moore, T.C., and Shackleton, N.J., 1987, Age dating and the orbital theory of ice ages: Development of a high resolution 0–300,000 year chronostratigraphy: Quaternary Research, v. 27, p. 1–29, doi: 10.1016/0033-5894(87)90046-9.

May, J.H., 1981, The updip limit of Miocene sediments in Mississippi: Geological Society of America Abstracts with Programs, v. 13, no. 1, p. 29.

McCallister, N., 2004, The Distribution and Possible Fault Displacement of the Upland Gravel of West Tennessee and Mississippi [M.S. thesis]: Memphis, University of Memphis, 75 p.

Mooney, W.D., Andrews, M.C., Ginzburg, A., Peters, D.A., and Hamilton, R.M., 1983, Crustal structure of the northern Mississippi Embayment and a comparison with other continental rift zones: Tectonophysics, v. 94, p. 327–348, doi: 10.1016/0040-1951(83)90023-9.

Pomeroy, P.W., Simpson, D.W., and Sbar, M.L., 1976, Earthquakes triggered by surface quarrying—Wappingers Falls, New York sequence of June, 1974: Bulletin of the Seismological Society of America, v. 66, p. 685–700.

Potter, E.P., 1955, The petrology and origin of the Lafayette gravel: The Journal of Geology, v. 63, no. 1, p. 1–38, no. 2, p. 115–132.

Purser, J.L., and Van Arsdale, R.B., 1998, Structure of the Lake County uplift: New Madrid seismic zone: Seismological Society of America Bulletin, v. 88, p. 1204–1211.

Russell, E.E., 1987, Gravel aggregate in Mississippi—Its origin and distribution: Mississippi Geology, v. 7, no. 3, 7 p.

Sauber, J.M., and Molnia, B.F., 2004, Glacier ice mass fluctuations and fault instability in tectonically active southern Alaska: Global and Planetary Change, v. 42, p. 279–293, doi: 10.1016/j.gloplacha.2003.11.012.

Saucier, R.T., 1994, Geomorphology and Quaternary geologic history of the lower Mississippi valley: Vicksburg, Mississippi, U.S. Army Engineer Waterways Experiment Station, Volume 1, 364 p.

Schweig, E.S., and Ellis, M.A., 1994, Reconciling short recurrence intervals with minor deformation in the New Madrid seismic zone: Science, v. 264, p. 1308–1311, doi: 10.1126/science.264.5163.1308.

Self, R.P., 1993, Late Tertiary to early Quaternary sedimentation in the Gulf Coastal Plain and lower Mississippi valley: Southeastern Geology, v. 33, p. 99–110.

Simpson, D.W., 1986, Triggered earthquakes: Annual Review of Earth and Planetary Sciences, v. 14, p. 21–42, doi: 10.1146/annurev.ea.14.050186.000321.

Stringfield, V.T., and LaMoreaux, P.E., 1957, Age of Citronelle Formation in Gulf Coastal Plain: American Association of Petroleum Geologists Bulletin, v. 41, p. 742–757.

Stuart, W.D., Hildenbrand, T.G., and Simpson, R.W., 1997, Stressing of the New Madrid seismic zone by a lower crust detachment fault: Journal of Geophysical Research, v. 102, p. 27,623–27,633, doi: 10.1029/97JB02716.

U.S. Army Corps of Engineers, 1990, Mississippi River hydrographic survey: Vicksburg, Mississippi, U.S. Army Corps of Engineers Waterways Experiment Station, 208 p.

Van Arsdale, R.B., 2000, Displacement history and slip rate on the Reelfoot fault of the New Madrid seismic zone: Engineering Geology, v. 55, p. 219–226, doi: 10.1016/S0013-7952(99)00093-9.

Van Arsdale, R.B., and TenBrink, R.K., 2000, Late Cretaceous and Cenozoic geology of the New Madrid seismic zone: Bulletin of the Seismological Society of America, v. 90, p. 345–356, doi: 10.1785/0119990088.

Van Arsdale, R.B., Williams, R.A., Schweig, E.S., Shedlock, K.M., Odum, J.K., and King, K.W., 1995, The origin of Crowley's Ridge, northeastern Arkansas: Erosional remnant or tectonic uplift?: Seismological Society of America Bulletin, v. 85, p. 963–985.

Willman, H.B., and Frye, J.C., 1970, Pleistocene stratigraphy of Illinois: Illinois State Geological Survey Bulletin, v. 94, 204 p.

Winkley, B.R., 1994, Response of the lower Mississippi River to flood control and navigation, *in* Schumm, S.A., and Winkley, B.R., eds., The Variability of Large Alluvial Rivers: New York, American Society of Civil Engineers, p. 45–74.

Zoback, M.L., and Zoback, M.D., 1989, Tectonic stress field of the continental United States, *in* Pakiser, L.C., and Mooney, W.D., eds., Geophysical Framework of the Continental United States: Geological Society of America Memoir 172, p. 523–539.

MANUSCRIPT ACCEPTED BY THE SOCIETY 29 NOVEMBER 2006

The Geological Society of America
Special Paper 425
2007

Relevance of active faulting and seismicity studies to assessments of long-term earthquake activity and maximum magnitude in intraplate northwest Europe, between the Lower Rhine Embayment and the North Sea

Thierry Camelbeeck[†]
Kris Vanneste
Pierre Alexandre
Koen Verbeeck
Toon Petermans
Philippe Rosset
Michel Everaerts
René Warnant
Michel Van Camp
Royal Observatory of Belgium, Avenue Circulaire, 3 BE-1180 Brussels, Belgium

ABSTRACT

We provide a synthesis of the long-term earthquake activity in the region of northwest Europe between the Lower Rhine Embayment and the southern North Sea. Reevaluated historical earthquake and present-day seismological data indicate that much of the known seismic activity is concentrated in the Roer graben. Nevertheless, the three strongest known earthquakes with estimated magnitude ≥ 6.0 occurred outside of this active structure, in the northern Ardenne, the southern North Sea, and the Strait of Dover. During the past 700 yr, destructive earthquakes generally have occurred at different locations, indicating a migration of seismicity with time. Because in plate interiors the present seismic activity does not necessarily reflect past and future activity, we discuss the necessity to use the geologic record to infer long-term earthquake activity. Thus, we synthesize and discuss paleoseismic investigations in the Roer graben that provide evidence that large earthquakes with magnitude up to 7.0 have occurred since the late Pleistocene. We also show that tectonic deformation is close to or below the accuracy of current geodetic techniques. Thus, it is necessary to have longer periods of observation to compare present geodetic deformation rates with the observed seismic moment release and the geologic strain rates. Based on these results, we present methods to define seismic zoning and evaluate the maximum credible earthquake and its magnitude relevant for seismic hazard assessment.

Keywords: seismicity, active faults, earthquakes, intraplate deformation, northwest Europe.

[†]E-mail: thierry.camelbeeck@oma.be.

Camelbeeck, T., Vanneste, K., Alexandre, P., Verbeeck, K., Petermans, T., Rosset, P., Everaerts, M., Warnant, R., and Van Camp, M., 2007, Relevance of active faulting and seismicity studies to assessments of long-term earthquake activity and maximum magnitude in intraplate northwest Europe, between the Lower Rhine Embayment and the North Sea, *in* Stein, S., and Mazzotti, S., ed., Continental Intraplate Earthquakes: Science, Hazard, and Policy Issues: Geological Society of America Special Paper 425, p. 193–224, doi: 10.1130/2007.2425(14). For permission to copy, contact editing@geosociety.org. ©2007 The Geological Society of America. All rights reserved.

INTRODUCTION

In regions where lithospheric deformation is associated with plate boundaries, the relationship among earthquake activity, landscape morphology, and geologic record is generally well recognized. In the interior of plates, the fingerprint of seismic activity in the landscape is often weak or invisible because the governing tectonic processes are slow, and large earthquakes that affect the whole seismogenic layer are rare (Crone et al., 1997). In regions where large earthquakes are absent from the historical seismicity catalogue, the apparent lack of evidence for tectonic activity in the morphology is often considered as a proof of the absence of large earthquakes during recent geological times. In fact, the traces of past activity in the morphology can be masked by depositional and (or) erosional processes (McCalpin, 1996) if these processes are more rapid than the tectonic deformation. Thus, in these areas, present earthquake activity does not neces-

sarily reflect their potential long-term activity. As in any region of the world, the only way to study long-term tectonic activity is to investigate the geologic record.

Compared to other stable continental regions, seismicity in the region of northwest Europe from the Lower Rhine Embayment to the southern North Sea (Fig. 1) appears to be significant, with three and fourteen earthquakes, respectively, with magnitude greater than or equal to 6.0 and 5.0 since 1350 (Table 1). The geological structure is also better known than in many other intraplate regions because geological investigations have been done continuously since the end of the nineteenth century, mainly in relationship with mining and quarrying industries. These characteristics explain why we use this part of Europe to study the relationship between tectonic deformations in the geological record and present-day deformation measured by geodesy and earthquake activity. The importance of studying seismic activity here is also linked to the increasing risk from

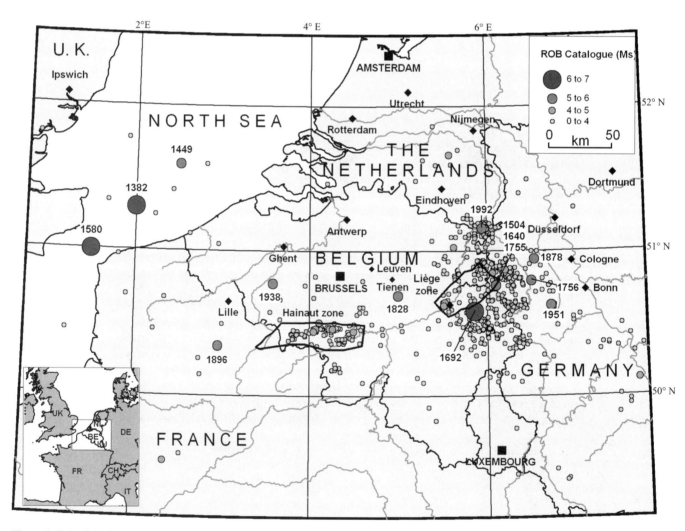

Figure 1. Seismicity between the Lower Rhine Embayment and the southern North Sea. Data come from the earthquake catalogue of the Royal Observatory of Belgium (ROB). The strong historical earthquakes listed in Table 1 are reported by their year of occurrence. The Hainaut and Liège seismic zones are also indicated.

TABLE 1. STRONG AND LARGE EARTHQUAKES IN INTRAPLATE NORTHWEST EUROPE,
BETWEEN THE LOWER RHINE EMBAYMENT AND THE NORTH SEA

Year	Mon	Day	H	Min	Lat. (°N)	Long (°E)	Magnitude M_S (ROB)	Magnitude M (EPRI)	Location
1382	5	21	15		51.30	2.00	6	5.4	Offshore Kent
1449	4	23	3		51.60	2.50	5 2/4	No info	Southern North Sea
1504	8	23	22		50.77	6.10	5	No info	Aachen
1580	4	6	18		51.00	1.50	6	6.7	Strait of Dover
1640	4	4	3	30	50.77	6.10	5 2/4	No info	Aachen
1692	9	18	14	30	50.59	5.86	6 1/4	5.1	Verviers
1755	12	27	0	0	50.77	6.10	5 1/4	5.5	Aachen
1756	2	18	8		50.80	6.50	5 3/4	5.3	Düren
1828	2	23	8	15	50.70	5.00	5	No info	Hesbaye
1878	8	26	8	55	50.95	6.53	5 2/4	4.9	Tolhausen
1896	9	2	21	15	50.35	2.96	5	4.9	Lens-Arras
1938	6	11	10	57	50.78	3.58	5.0	5.3	Nukerke
1951	3	14	9	46	50.63	6.72	5.3	5.5	Euskirchen
1992	4	13	1	20	51.16	5.95	5.4	–	Roermond

Note: The list, based on the catalogue of the Royal Observatory of Belgium (ROB), includes all the earthquakes with $M_S \geq 5.0$ that occurred in the studied area since 1350. The origin time (columns 1–5), location (columns 6 and 7), and M_S magnitude (column 8) come from the ROB catalogue. For earthquakes before 1910, M_S is given by steps of 1/4 (in place of 0.25) of value, because the uncertainty on magnitudes is estimated around 0.5. The magnitudes given by EPRI (1994) are also reported in column 9. The indication «No info» means that the earthquake was not reported in the EPRI (Electric Power Research Institute) catalogue.

earthquakes in these densely populated and highly industrialized areas. The damaging earthquakes of Liège (Belgium) on 8 November 1983 ($M_S = 4.7$) and Roermond (The Netherlands) on 13 April 1992 ($M_S = 5.4$) demonstrated the vulnerability of the area to even small or moderate earthquakes and thus the necessity to better understand the seismicity and the potential for large earthquakes.

As in all intraplate regions, it is challenging to infer the location, magnitude, and average return period of future large earthquakes. During the last ten years, we have developed a multidisciplinary approach to answering these questions.

In the first part of the paper, we present the characteristics of the seismic activity and the general geological structure of the region. We then discuss geodetic data and the preliminary results of experiments we have conducted to evaluate present-day crustal movements. A synthesis of the seismotectonic framework of the region is also given.

The second part of this work discusses our recent investigations suggesting that large earthquakes are more likely than once thought and assessing their locations. In the first and second sections, we discuss our studies on strong historical earthquakes and on the Quaternary faults in the Lower Rhine Embayment. Two regional examples, the north of France and the Roer graben, explain how we define seismogenic sources and capable faults. In section 2.3, we discuss the arguments supporting the occurrence of large earthquakes as causes of the observed surface faulting on the border faults of the Roer graben. This problem is important because it has a strong impact on long-term seismic activity assessment, and it contradicts the theory of Ahorner (1968, 1975, 1996), which explained the surface fault movements as the result of continuous aseismic deformation.

The third and last part of the paper discusses how to evaluate the maximum credible earthquake in the different seismic sources of the area considered. We do not consider estimating the return period of these earthquakes.

REVIEW OF SEISMIC ACTIVITY, SEISMOTECTONICS, AND GEOLOGICAL STRUCTURE OF NORTHWEST EUROPE

Seismic Activity in the Region Extending from the Lower Rhine Embayment to the North Sea

In this paper, precise definitions of "large" and "strong" earthquakes are used. A large earthquake is an event with an inferred rupture dimension that equals or exceeds the width of the seismogenic layer (Scholz, 1990). Hence, it produces a measurable deformation of the ground surface. A strong earthquake is an event that causes damages without any consideration of its magnitude.

The basic information we use to identify the location and the parameters of future large and strong earthquakes is the known seismicity. The earthquake epicenters from 1350 to 2004 are plotted on Figure 1. The reliability of the information in the earthquake catalogue depends strongly on the epoch considered (Alexandre and Vogt, 1994). The seismic history of the region begins around 700 A.D. Until the fourteenth century, the rare historical sources allow the establishment of a list of the strongest earthquakes, but few can be reliably assessed in terms of magnitude and location. Since the fourteenth century, the number of different sources (chronicles, annotations, parish registers, account registers, etc.) has increased significantly. These give more details on local effects and allow more reliable estimation of damage and felt areas of the earthquakes. It is possible to estimate their probable epicentral areas and also their magnitudes by comparison with recent earthquakes for which the magnitude was instrumentally determined (e.g., Ambraseys, 1985b; Johnston, 1996).

Even when instrumental recording of earthquakes began in the last century, the low density of seismic stations made the reliability of the calculated earthquake parameters unsatisfactory (Camelbeeck, 1993) at least until 1965 for the epicenters and 1980 for focal depths. Before 1960, except for larger shocks, macroseismic reports often provided better epicenters than those calculated from arrival times on the seismograms.

The analysis of historical sources, especially those written before the nineteenth century, requires the expertise of historians. For example, to judge if information concerning an earthquake can be used, it should be established when, where, and why the account was written. Unfortunately, many earthquake catalogues do not provide this information, and the reported information duplicates that of previous catalogues based on accounts dating in some cases centuries after the event's supposed occurrence. It is also important for this basic source of information to be made available, so that later researchers can add new findings or reinterpret them. This reappraisal of historical seismicity in Europe was undertaken in the 1980s, and it yielded studies, involving historians and seismologists, that can be used to assess seismic activity. Our earthquake catalogue is based on such sources. We use the work of Alexandre (1990) for continental earthquakes before 1259, Alexandre's list of events in Camelbeeck (1993) for earthquakes felt in Belgium before 1910, and sources in Ambraseys and Melville (1983) for the earthquakes in England before 1800. Earthquakes that occurred since 1910 were studied by Camelbeeck (1993). The magnitudes of the strongest earthquakes in northwest Europe since 1800 were provided by Ambraseys (1985a, 1985b). Lambert and Levret-Albaret (1996) tabulated seismicity data since 1800 in the French part of the study area.

An important problem concerning earthquake catalogues in northwest and central Europe involves evaluating the magnitudes of earthquakes during the last 20–40 yr. Despite the enormous work done in the past (Karnik, 1969; Ambraseys, 1985a), which has provided methods to evaluate M_S in Europe, the magnitude furnished by local and regional seismological centers is always M_L or pseudo-M_L (magnitude labeled as M_L, but determined with a procedure similar to that used for m_b evaluation). Thus, magnitudes given by different networks are not always consistent and can differ greatly from magnitude based on the inferred earthquake source dimension. In our catalogue, for earthquakes for which instrumental data are available, M_S was determined using the methods proposed by Ambraseys (1985a). To evaluate the magnitude of historical events, the relationships established by Ambraseys (1985b) between M_S and macroseismic information were used. For earthquakes before 1910, M_S is given by step of ¼ of value, because the uncertainty on magnitudes is estimated around 0.5. In parallel, the catalogue of Reamer and Hinzen (2004) for the northern Rhine area for 1975–2002 provides

moment magnitudes obtained by a relationship between M_W and M_L. Hinzen and Oemisch (2001) also used published intensity data to evaluate the location and magnitude of recent and historic earthquakes in the same region.

Based on our catalogue as of July 2004, we evaluated the seismic activity rate in the region in Figure 1. For a given magnitude, the annual occurrence rate for the entire region is estimated for the period of time for which the catalogue can be considered complete. Figure 2 plots the cumulative number of earthquakes, for magnitude ranges 1.8–3.0, 3.0–4.0, 4.0–5.0, and 5.0+, versus the start date of the catalogue. The completeness for a given magnitude range is assumed back to the oldest date for which a relatively constant activity rate similar to the most recent one is observed. On each diagram in Figure 2, the date for which it is not possible to fit the data with a coherent linear regression is used as the completeness date for the given magnitude range. The choice is often subjective because the catalogue is not homogeneous, and the seismic activity is sporadic.

For magnitudes below 3.0, the date of the installation of the modern Belgian seismic network in 1985 is used. For historical events with magnitude greater than 5.0, we consider the catalogue to be almost complete since 1350. For events of magnitude between 4.0 and 5.0, it is not easy to fix a date because the catalogue provides only 20 events, but we believe it is complete since 1925. For magnitudes ranging between 3.0 and 4.0, we assume the catalog is complete since 1960, when denser seismic networks began to be installed in northern Europe.

Assuming the completeness of the catalogue for these four time periods and the corresponding magnitude ranges, we estimated the seismic activity rate. The cumulative annual magnitude-frequency relationship was calculated both using the maximum likelihood method (Weichert, 1980) and least-square fitting (Fig. 3). If only $M_S \geq 3.0$ is considered in the maximum likelihood method, the results are similar for the two methods, giving

$$\log N(M_S) = 3.46(\pm 0.15) - 0.98(\pm 0.05)M_S,$$

where $N(M_S)$ is the cumulative number of events per year with magnitude greater than or equal to M_S.

This relationship gives a first-order estimation of the seismicity for Belgium and the surrounding areas. The average return periods for earthquakes of magnitude greater than or equal to 6.0, 5.0, and 4.0 are 260, 30, and 3 yr, respectively.

The seismicity (Fig. 1) is mainly concentrated in the border region between Belgium, the Netherlands, and Germany. The Roer graben area appears to be the most active area, with at least six earthquakes with $M \geq 5.0$, the strongest being the 1756 Düren (Germany) event with a M_S magnitude estimated around 5 3/4. The most important recent earthquakes are the 1951 Euskirchen (Germany) $M_S = 5.3$ and 1992 Roermond (The Netherlands) $M_S = 5.4$ (van Eck and Davenport, 1994) events. To the west of the Roer graben, notable activity exists in the Belgian Ardenne and Eifel Mountains. The strongest known seismic event in this entire region occurred on 18 September 1692, and it had a magnitude

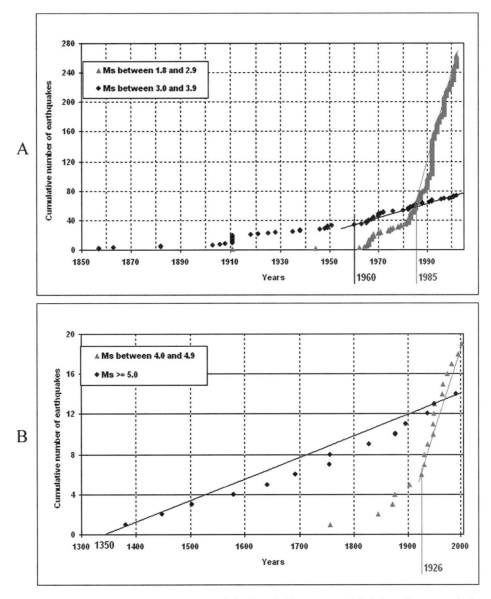

Figure 2. Completeness of the catalogue of the Royal Observatory of Belgium. Four magnitude ranges are considered: (A) M_S ranging from 1.8 to 2.9 and 3.0 to 3.9. (B) M_S ranging from 4.0 to 4.9 and greater than or equal to 5.0. The completeness dates for the four magnitude ranges are indicated by a vertical bar crossing the corresponding curve, and the corresponding year is written below the time axis.

estimated around 6 1/4 (Camelbeeck et al., 1999). Descriptions of the resulting destruction can be found in Alexandre (1997) and Camelbeeck et al. (1999). Interestingly, this earthquake was one of the most strongly felt in England, although this country was at the periphery of its felt zone (Morse, 1983). Near Liège, seismicity is less, but on 8 November 1983, a $M_S = 4.7$ earthquake caused damages (Melchior, 1985). There is also a concentration of earthquakes in the Hainaut region (Fig. 1), but up to now, their magnitudes have not exceeded $M_S = 4.5$. In other parts of the study area, earthquake activity is more diffuse and sporadic with time. The best example is the southern North Sea and the Strait

of Dover, where activity has been weak since the seventeenth century, but two earthquakes have occurred, in 1382 and 1580 (Melville et al., 1996), with an estimated magnitude greater than or equal to 6.

We list the earthquakes with $M_S > 5.0$ since 1350 on Figure 1 and Table 1. The contribution of the recent investigations is evidenced by comparisons between these earthquakes' magnitude value as compiled by the Electric Power Research Institute (EPRI, 1994) and our values (Table 1). The possible different interpretations of the isoseismals suggest that the uncertainty in the magnitude is on the order of 0.5.

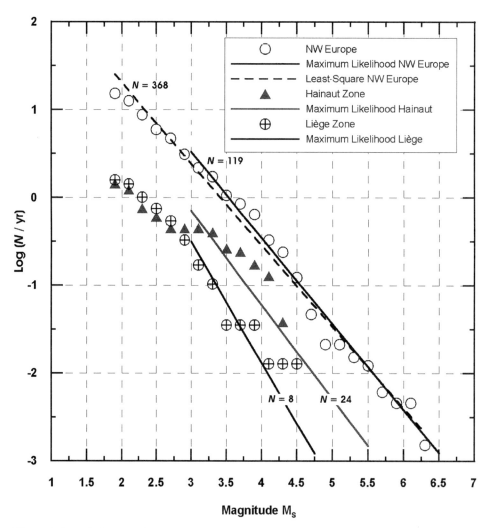

Figure 3. Cumulative annual magnitude-frequency relationships for the study area and the Hainaut and Liège seismic zones. Computations were based on the catalogue of the Royal Observatory of Belgium for earthquakes since 1350. Maximum likelihood (for $M_S \geq 3.0$) and least square (for $M_S \geq 1.8$) methods were applied for northwest Europe. Maximum likelihood method was used with $M_S \geq 3.0$ data for the Hainaut and Liège seismic zones. The number of events (N) is indicated for each curve.

We also reevaluated the earthquake locations. The most significant modification concerns the 18 September 1692 earthquake, reported in Tienen (Brabant Massif) by EPRI (1994), based on earlier catalogues. Vogt (1984) suggested an epicenter 100 km more to the east, near Liège, Aachen, and Maastricht. A new study using new contemporary historical documents (Alexandre, 1997; Camelbeeck et al., 1999) confirmed that this earthquake originated in the northern part of the Belgian Ardennes.

Regional Geology

The geology of this part of northwestern Europe is shown in Figure 4. Its basement consists of rocks that have undergone two major orogenies, the Lower Paleozoic (570–400 Ma)

Caledonian/Acadian phase and the Upper Paleozoic (400–250 Ma) Variscan/Hercynian phase. The present structural pattern results from post-Paleozoic intraplate deformation. The most important of these recent structures is the rift that developed from late Eocene to Holocene times across the Rhenish Shield and Lower Rhine Embayment to the North Sea. It is a part of the Cenozoic rift system of western and central Europe (Illies, 1981; Ziegler, 1994). At the time of this rift development, other parts of the region were subjected to compressive stresses, as evidenced by the inversion of the Weald–Artois Mesozoic tensional basins (Auffret and Colbeaux, 1977).

The Variscan Front, which splits the region, consists of a zone of southward-dipping thrust faults and folded foreland basins. The northernmost of these fault zones is the Variscan Midi-Eifel thrust (Fig. 4).

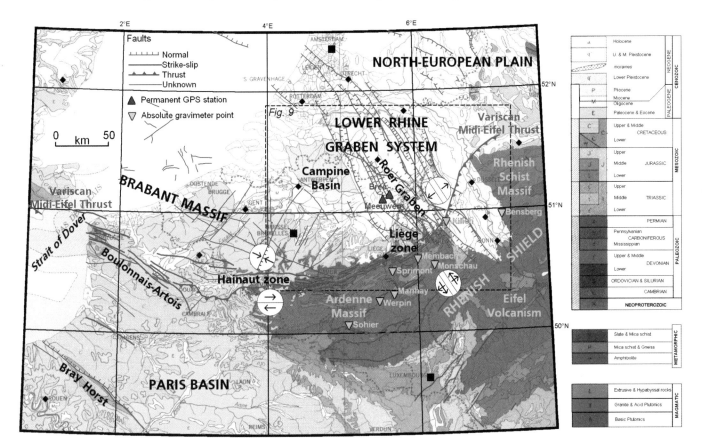

Figure 4. Geology of the studied area. The figure was modified and completed from the *Atlas of Belgium* (de Béthune and Bouckaert, 1968). The locations of the two global positioning system (GPS) sites of Bree and Meeuwen and of the stations of the absolute gravity profile across the Ardenne and the Roer graben are indicated.

North of the Variscan Front and from west to east, three provinces can be distinguished in the European Precambrian Platform: the Anglo-Brabant Massif, the North Sea and Lower Rhine graben system, and the North European Plain.

The Anglo-Brabant Massif is a stable province that consists of Late Proterozoic to Silurian deposits that have been deformed by the Acadian orogeny and partly covered by Variscan over-thrusting (Debacker et al., 2002; Verniers et al., 2002). It is overlain by tilted post-Paleozoic deposits (Cretaceous chalk, Tertiary sands and clays, and Quaternary loess) that thicken strongly toward the northeast, with a maximum of more than 1000 m above the Campine Basin. During the Devonian and Carboniferous, this thickening was caused by NW-SE normal faults that delimit the Campine Basin (Demyttenaere, 1989; Wouters and Vandenberghe, 1994). The Lower Paleozoic rocks of the Anglo-Brabant Massif only outcrop in valleys where the Cenozoic cover has been eroded. Not much is known about recent tectonic deformation. The incised rivers in the southern part suggest some relative uplift, while the northern part is subsiding. The recent northward tilting appears to be gradual and not associated with faulting. The rivers follow a NW-SE direction, parallel to one of the structural trends in the basement

corresponding to the so-called transverse faults (De Vos et al., 1993; Everaerts et al., 1996; Legrand, 1968). Several historical earthquakes are located in the Brabant Massif (Fig. 1), such as those of 21 May 1382 ($M_S = 6$) and 11 June 1938 ($M_S = 5.0$).

The North Sea and Lower Rhine graben system represent a continental extension province that developed during the Cenozoic (Ziegler, 1994). The Roer graben, which is a part of the Lower Rhine Embayment, is the southern part of this graben system. It is filled by up to 2000 m of Upper Oligocene to Quaternary sediments (Geluk et al., 1994). It is bounded by two NNW-SSE–trending normal fault systems that have been active during the Quaternary, the Peelrand fault to the northeast and the Feldbiss fault zone to the southwest. The morphology reflects recent fault movements, and several moderate historical earthquakes have occurred here (Camelbeeck and van Eck, 1994). The 1756 M_S 5 3/4, Düren earthquake was the largest and most destructive event in this zone. Our paleoseismic investigations (Camelbeeck and Meghraoui, 1996, 1998; Vanneste et al., 1999, 2001; Vanneste and Verbeeck, 2001) reveal that some of these faults are active and capable of producing earthquakes of magnitude up to 7.0.

The province east of this graben is the western part of the North European Plain. The post-Paleozoic deposits of this region are gradually tilted toward the north. Seismicity in this region is low, and some is induced by gas extraction (Dost and Haak, 2007).

South of the Variscan Front, the basement consists of rocks that have been folded and faulted during the Variscan orogeny. Three major units are distinguished: the prolongation of the Lower Rhine graben system, the Rhenish Shield, and the Paris Basin.

The Lower Rhine graben system extends south of the Variscan Front and divides the Rhenish Shield into the Rhenish Schist Massif to the east and the Ardenne Massif and Eifel Mountains to the west. Toward the southeast, the graben structure is gradually less developed and narrower, and both the morphological evidences of Quaternary tectonic activity and the historical seismicity diminish. The Rhenish Shield forms the northern part of the Rhenohercynian belt and consists of Upper Paleozoic (Devonian-Carboniferous) calcareous and siliciclastic formations that were deformed during the Variscan orogeny. Some remnants of Permian, Cretaceous, and Tertiary deposits and Quaternary loam cover the basement of the Rhenish Shield. The important Quaternary uplift and the corresponding erosion and deep incision of the rivers have removed the thin post-Paleozoic cover. The Ardenne Massif has been uplifted more than the Rhenish Schist Massif (Demoulin, 1995; Fuchs et al., 1983). Seismicity in the eastern Ardenne is higher than in the rest of the Rhenish Shield but lower than in the bordering Lower Rhine graben system. Several recent earthquakes with magnitudes up to 4.4 have been recorded, some of which seem to be clustered around NNW-SSE–oriented zones, parallel to the faults of the Lower Rhine graben system. The largest known earthquake is the 1692 M_S 6 1/4, Verviers earthquake. South of the Lower Rhine graben structure, there is the Quaternary Eifel volcanic zone (Fuchs et al., 1983). Following earlier Mesozoic (Cretaceous) and Tertiary volcanism, ~300 small eruptions occurred between 700,000 and 10,800 yr B.P. The total erupted volume of <15 km^3 represents an average magma flux that is several thousand times smaller than that of the Hawaiian hotspot. The volcanic activity in the Eifel has been accompanied by up to 250 m of uplift in the last 600,000 yr. Analyzing teleseismic waveforms, Ritter et al. (2001) detected a mantle plume that may be the source for the Eifel volcanism. This plume extends from the uppermost mantle at least to the top of the transition zone (400 km depth). The Eifel volcanic zone seismicity is characterized by small earthquake swarms (Ahorner, 1983).

The basement of the northern part of the Paris Basin has the same basement architecture as the Rhenish Shield, but the post-Paleozoic development of this basin is mainly characterized by subsidence and deposition. The sediments thicken and are tilted toward the center of the basin near Paris. During basin development, two main horst structures developed: the Artois-Boulonnais horst along the Variscan Front and the Bray horst along the limit between the Saxothuringian and Rhenohercynian zone. The post-Paleozoic tectonic activity in the Artois-Boulonnais is briefly described in the section: Large and strong historical earthquakes and their relationship with geological structure. The 1580 Strait of Dover earthquake is located in this region.

Present-Day Crustal Movements

Until recently, the only available information on present-day crustal movement in northwest Europe came from analysis of the differences among first-order levelings conducted periodically in Belgium (Pissart et Lambot, 1989), the Netherlands, France (Fourniguet, 1987), and Germany (Quitzow and Vahlensieck,1955; Waalewijn,1961; Mälzer et al., 1983) since the end of the nineteenth century. The measured vertical geodetic rates (van den Berg et al., 1994) in the Netherlands suggest a relative movement of the Roer graben compared to the South Limburg block (Fig. 5) of 0.8 ± 0.5 mm yr^{-1} (1σ confidence level). Measurements in the southern part of the Roer graben, at the border with the Rhenish Massif (Mälzer et al., 1983) show an even higher vertical relative movement rate of ~1–2 mm yr^{-1}. These movements are an order of magnitude higher than the vertical relative movement rates inferred geologically for the Quaternary (Ahorner, 1975, 1983). Thus, if these estimates of movement rate are reliable, there is a discrepancy between geodetic vertical relative movement rates and inferred geological movements. The causes of these present-day movements, if real, can be varied. They may be linked to active faults in the Ardenne and bordering the Roer valley graben, to the possible Eifel plume (Ritter et al., 2001), or other causes, such as Fennoscandian postglacial rebound. In a recent paper, Milne et al. (2001) examined postglacial rebound models that fit the global positioning system (GPS) data in Fennoscandia. The models predict Belgium to be on the peripheral bulge of the deformation, with a subsidence rate of −0.9 mm/yr, which does not indicate differential movement across a fault, but could indicate long-wavelength movements.

Ahorner Hypothesis on the Fault Movements in the Lower Rhine Graben Based on Levelings

Ahorner (1968, 1996) proposed that aseismic creep plays a prominent role in the neotectonic deformation of the Lower Rhine Embayment. He based his interpretation on old high-precision leveling across the Erft fault system near Cologne and the Peelrand fault near Roermond (Fig. 5), which show movements of the order of 1–2 mm/yr (Fig. 6A); this is much faster than values implied by the total seismic moment of historical seismicity. On the other hand, based on his geological investigations, he proposed that the average geological rates of ~0.1 mm/yr were due to bursts of aseismic motion with vertical deformation rates similar to those observed today. He considered our epoch to be a similar burst of aseismic motion.

Present-Day Crustal Movement in Northwest Europe by the Comparison of Levelings—Are They Reliable?

A simple analysis of the stochastic errors of the leveling data used by Ahorner, without even considering possible systematic errors in the surveys, indicates that the calculated movements are not statistically significant.

Figure 6A shows the relative height changes across the Erft fault system between 1933 and 1952 (Ahorner, 1996). It shows a

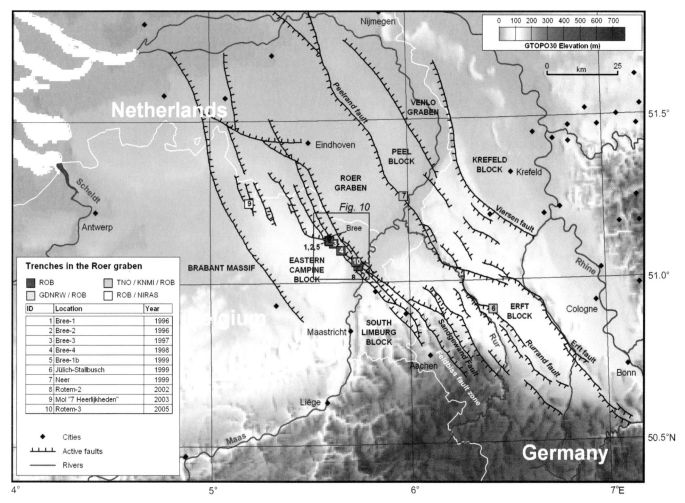

Figure 5. Limits of the crustal blocks inside the Lower Rhine Embayment showing the location of the Quaternary faults in the region and trenches excavated across the border faults of the Roer graben that we used in the analysis. The different involved institutions are: the Royal Observatory of Belgium (ROB), The Geological Survey of North-Rhine Westphalia (GDNRW), the Geological Survey of the Netherlands (TNO), The Royal Netherlands Meteorological Institute (KNMI), and the Belgian agency for radioactive waste and enriched fissile materials (NIRAS).

step across the fault zone, suggesting elevation change across the fault. A first question is whether this step is a real movement or noise due to the accuracy of the measurements. Without the actual leveling data, it is difficult to infer the real stochastic errors, but since the error in a leveling profile scales with the length of the profile, a better data representation is the difference in movements between successive points of the profile (Fig. 6B). This provides a homogeneous representation of the errors along the profile if the section lengths are similar. This diagram suggests that the difference of height change across the Erft fault zone is not significant compared to that between other pairs of points, which are supposed to be stable. Along the profile, the average movement between pairs of points is 0.53 cm. This represents the uncertainty in the measurements if we assume an absence of real movements within the two blocks separated by the fault. Thus, the maximum value in the profile, which corresponds to the leveling section across the Erft fault, is 1.1 cm and is only twice the average value. We assert that such a movement is not significant.

Our analysis is supported by the results of Mälzer et al. (1983) and van den Berg et al. (1994), which suggest that relative height differences calculated from the comparison of first-order leveling surveys in northwest Europe have to be considered with a critical eye. Generally, they are not significant enough (at 2σ or 3σ confidence) to measure vertical ground displacements. We note also the van den Berg et al.'s (1994) results, which incorporate more than 100 yr of leveling conducted in the Netherlands in the Maas valley, contradict the results of Waalewijn (1961), which showed movement in the Maas valley and which were used by Ahorner (1968, 1998) to infer relative movements across the Peelrand fault.

Moreover, even if some movements are significant, they need not represent tectonic movements. From an example in the Belgian Ardennes, Camelbeeck et al. (2002) concluded that the observed vertical movement in poorly consolidated sediments, as are generally also found in the Roer graben, can be related to the local geology rather than tectonic crustal movements. The sedi-

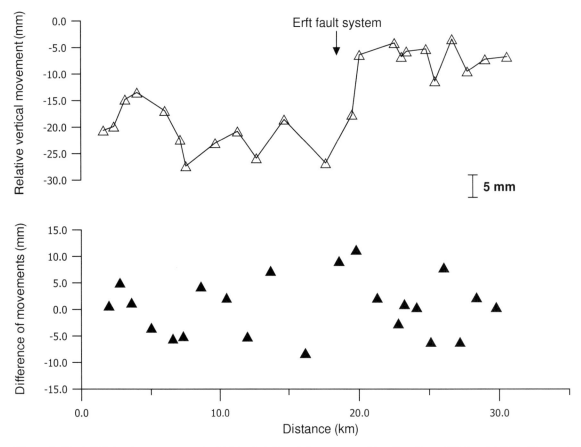

Figure 6. (Top) Height changes across the Erft fault inferred from repeated leveling lines (1932 and 1953). (Bottom) Difference of height changes between successive points on the leveling line.

ments seem to decrease the quality of the measurements along the sections and to induce movement of the benchmarks, likely linked to hydrological conditions.

One example of such nontectonic movements is subsidence induced by water pumping. Since ca. 1960, most of the German part of the Roer graben has been affected by rapid (up to 3 m) subsidence induced by groundwater lowering necessary for large open-pit mines (Schaefer, 1999). Due to the differences in thickness of compacting clay layers, differential subsidence occurs across Quaternary faults. The Rurrand (Fig. 5) fault shows an aseismic slip of ~1 cm/yr on its upper few hundred meters, unrelated to any tectonic process. This subsidence also extends to the Dutch part of the Roer graben. Thus, in a part of the Roer graben, mainly in Germany, any tectonic movements since 1960 would be masked by man-made subsidence. The profiles used by Ahorner (1968) did not suffer from that effect, however.

Although Ahorner's conclusions have been accepted in analyses of earthquake activity in this part of Europe during the past 30 yr, we find that the leveling data across the border faults of the Roer graben do not need tectonic vertical motion to be explained. This conclusion is also supported (see section on the Relevance of paleoseismic data in the Roer Graben:seismic

or aseismic movement) by paleoseismic investigations indicating the absence of any fault displacement that could be correlated with this present deformation inferred from leveling data, except for the differential subsidence induced by the groundwater lowering for the large open-pit mines at the Rurrand fault (Vanneste and Verbeeck, 2001).

Present-Day Crustal Movements in Northwest Europe Using GPS and Absolute Gravimetry

For the past 10 yr, geodesy has been furnishing interesting data on crustal movements in Europe and elsewhere. Ward (1994, 1998) calculated geodetic strain rates from site velocities in Europe. His data set, including Very Long Baseline Interferometry, Satellite Laser Ranging, and GPS stations, yielded local site velocities in northwest Europe with less than 2σ uncertainties. Using GPS networks, Nocquet and Calais (2003, 2004) showed that central Europe (east of the Rhine graben system, north of the Alps and Carpathians, south of Fennoscandia) behaves rigidly at a 0.4 mm/yr level and defines a stable Europe reference frame. In particular, they found no significant relative motion between sites west of the Rhine graben and central Europe. Due to the more limited number of stations, this sets an upper bound of 0.6 mm/yr on horizontal motion across the Rhine graben.

Considering the importance of measuring the present-day crustal deformation to interpret the slippage along active faults in the Lower Rhine Embayment, the Royal Observatory undertook two long-term (more than 10 yr) experiments:

1. The first study used permanent GPS stations on either side of the Feldbiss fault. Permanent stations 7.5 km apart were installed in the villages of Bree and Meeuwen, in the Roer valley graben and on the Campine Plateau (Figs. 4 and 5). Five years of continuous measurements since summer 1997 show no observable relative movement. The repeatability of the measurements is such that a vertical or horizontal relative movement between the two sites would have been observed if it were larger than 0.5 mm/yr and 0.3 mm/yr, respectively. These values are already smaller than those obtained by the repeated levelings and agree with the results of Nocquet and Calais (2003).

2. The second study was a profile of absolute gravity measurements across the Ardenne and the Roer graben to infer vertical crustal movements (Van Camp et al., 2002). This eight-station profile is 140 km long, and measurements are performed twice a year. It should allow detection of the spatial extent of the uplift in the Ardenne expected from the tectonic activity and discrimination between it and long-wavelength phenomena like postglacial rebound (Francis et al., 2004). However, there are still significant uncertainties in the estimated rates of postglacial rebound, which should be better constrained by the profile.

The first results of the profile already show no detectable movement corresponding to gravity changes greater than 13 nm s^{-2} yr^{-1} at a 2σ level. This is equivalent to a vertical movement of 6.5 mm/yr because uplift or subsidence would modify the gravity by ~1 µGal for 5 mm, taking into account free air and Bouguer corrections (Williams et al., 2001). Van Camp et al. (2005) estimated that repeated absolute gravity campaigns should constrain gravity rates of change with an uncertainty of 1 nm/s² (or 0.5 mm of vertical movement) after 15–25 yr, depending on the noise affecting measurements.

An attractive feature of absolute gravity measurements is that, unlike any relative geodetic technique, an absolute gravimeter could go back to any undisturbed gravity point even after 100 yr and make a relevant measurement.

Hence, intraplate deformation in active tectonic structures in northwest Europe such as the Roer graben is close to or below the accuracy of current space and ground-based geodetic techniques. It will be necessary to have longer periods of observation before interpreting the GPS and absolute gravity data in terms of tectonic rates of deformation and hence furnishing arguments in favor of one (or more) of the hypotheses concerning the origin of the crustal movement.

Fault-Plane Solutions: Seismotectonic Context

Fault-plane solutions for the earthquakes since 1980 in the region are presented in Figure 7 and Table 2. We refer to them in the text and in Figure 7 according to their number in Table 2. Solutions for those earthquakes that occurred after July 1992 are

presented in Figure 8. Since the beginning of the 1980s, the density of seismic stations has increased significantly in the region, allowing more reliable determination of source parameters for earthquakes down to magnitude 1.0. For most of the earthquakes, focal mechanisms have been obtained from P-wave first motions. For eight of them (9, 10, 11, 12, 16, 17, 18, and 19), a method combining P-wave first motion sense with apparent seismic moment for P, SV, or SH waves has been used (Camelbeeck, 1993). The focal mechanisms are maximum likelihood solutions found by a grid search over the strike, dip, and rake of the mechanism. The quality of the solution is evaluated by comparing the 99% confidence regions for the pole positions representing the slip vectors of the two nodal planes. A good solution is often characterized by small confidence regions, and a poor one, by large regions.

Figure 4 shows a simplified sketch indicating the main pattern of seismic deformation inferred from our analysis of fault-plane solutions in the Roer graben, Eifel–Hautes-Fagnes, Hainaut and western Ardenne, and south of the Brabant Massif. Arrows inside the white circles indicate the direction and sense of slip along faults represented by a line in their strike direction.

The classic seismotectonic model for northwest Europe, mainly developed by Ahorner (1975, 1985), assumes a NW–SE–directed maximum horizontal compressive stress. This induces tensional forces in the Roer graben perpendicular to the graben axis, which explain the extensional crustal deformation. Focal mechanisms in the Roer graben show mostly dip-slip normal faulting earthquakes along NW-SE–striking faults, in agreement with this model. The two largest earthquakes for which a reliable mechanism (25 and 44) has been calculated, those of Roermond (M = 5.4) on 13 April 1992 and Alsdorf (M = 4.8) on 22 July 2002, are of this type.

In the western part of the Rhenish Massif, Ahorner (1983) interpreted part of the seismic activity as associated with the Midi-Eifel thrust (Fig. 4). He based his assumption on the focal mechanism of the 10 June 1978 Roetgen M_L = 3.0 earthquake and a composite solution from six events (not identified) during 1975–1978. We recalculated the focal mechanism of the Roetgen earthquake with the eleven P-wave first motion data provided by Ahorner (1983). For our best solution (Fig. 8A), the confidence regions for the two poles have relatively small dimensions, but if they give a reliable solution for strike and dip of the quasi-vertical nodal plane, the azimuth of the relatively flat nodal plane is completely undetermined. On the other hand, the best solution disagrees with two of eleven first motions and constrains, except for one datum, the information very close to the nodal planes. Clearly, we lack enough information to find a good solution. In addition, there is no geologic indication of post-Variscan movement on the WSW-ENE–trending thrust faults, like the Midi Eifel thrust, whereas these faults have been displaced by more recent movements on some of the NNW-SSE–trending faults in the region. For these reasons, we consider the interpretation of Ahorner (1983) questionable. The quality of fault-plane solutions for earthquakes in this region improved after the installation of the new Belgian seismic network in 1985. For these more recent

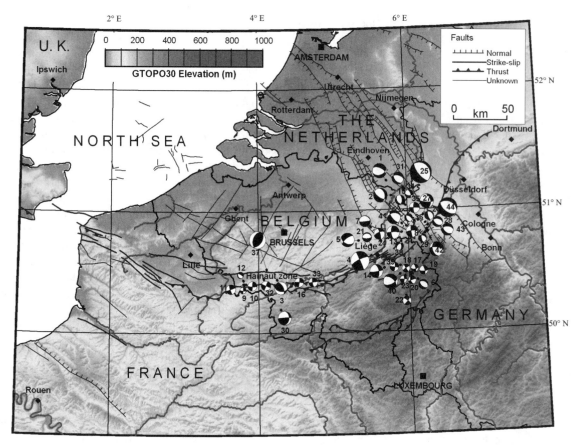

Figure 7. Earthquake focal mechanisms in the region from the Lower Rhine Embayment to the North Sea. The size of the focal mechanisms is proportional to the earthquake magnitude. They are numbered as in Table 2.

events, the mechanisms range from pure normal faulting (6 and 40) on NW-SE–striking faults to quasi-pure strike-slip faulting on NNW-SSE or ENE-WSW faults (14, 22, 38, and 39). Part of the activity is linked to the NNW-SSE–striking Hockai fault zone, already defined by Ahorner (1983). The earthquake spatial distribution and the fault-plane solutions (17, 18, 19, and 20) during a seismic sequence along this fault zone in 1989–1990 suggest that activity is related mainly to sinistral strike-slip mechanisms (Camelbeeck, 1993).

Camelbeeck and van Eck (1994) argued from earthquake focal mechanisms for a complex tectonic deformation field in the small region extending from the Roer graben to the city of Liège, including the eastern part of the Brabant Massif. This will not be discussed here.

Ahorner (1975) defined the whole region extending from the Liège area and the North Sea as the Belgian strike-slip zone. This hypothesis was mainly based on interpretations of the elongation of the isoseismals during the 11 June 1938 Brabant Massif earthquake ($M_S = 5.0$) that indicated a right-lateral strike slip along a WNW-ENE–trending shear zone extending from the Liège area to the North Sea. Up to now, no reliable fault-plane solution has been published for this earthquake due to the

lack of data. On the other hand, it has been possible to link the intensity distribution with the soft-sediment thickness, which suggests that the regional geology and the resulting site effects play a role in the damage distribution (Nguyen et al., 2004). Unfortunately, the region has little present-day seismic activity and only one well-constrained focal mechanism has been determined, that of a $M_S = 4.3$ earthquake in 1995 close to the border of the Hainaut zone. Its mechanism (37) is reverse on a NNE-SSW–striking fault, parallel to the structural trend indicated by river courses in the Brabant Massif. We cannot base a seismotectonic interpretation on a single fault-plane solution, but it demonstrates that Ahorner's hypothesis of a Belgian strike-slip zone still lacks evidence in the Brabant Massif.

On the other hand, Ahorner's hypothesis is supported south of the Brabant Massif by geomorphologic investigations in the north of France (Colbeaux et al., 1977). These authors defined the Zone de Cisaillement Nord-Artois, a presumed WNW-ESE dextral strike-slip zone corresponding more or less to the Variscan Front from west of the city of Liège to the Strait of Dover. Although the existence of this broad zone of deformation has not yet been confirmed by geologic investigations, earthquake focal mechanisms in Hainaut and the western Ardennes agree with this hypothesis. For

TABLE 2. EARTHQUAKE FOCAL MECHANISMS BETWEEN THE LOWER RHINE EMBAYMENT AND THE SOUTHERN NORTH SEA

No.[†]	Ref[‡]	Date[§]			Time[#]	Longitude[††]	Latitude[‡‡]	Depth[§§]	M_S[##]	Strike1[†††]	Dip1[†††]	Rake1[†††]	Strike2[‡‡‡]	Dip2[‡‡‡]	Rake2[‡‡‡]
		Year	Month	Day	(h:min:s)	(°E)	(°N)	(km)		(°)	(°)	(°)	(°)	(°)	(°)
1	(1)	1980	06	05	12:11:39	5.74	51.23	14	3.6	298	57	−62	73	42	−126
2	(1)	1982	05	22	6:00:02	5.80	51.11	14	3.5	125	62	−114	349	36	−52
3	(1)	1982	09	14	19:24:34	4.25	50.45	2	3.2	334	57	126	100	47	48
4	(1)	1983	11	08	0:49:34	5.51	50.63	6	4.7	336	88	14	245	76	178
5	(1)	1984	07	09	23:19:01	5.35	50.75	5	3.4	117	43	−34	233	68	−128
6	(1)	1985	05	12	21:47:51	6.10	50.43	11	2.5	295	36	−104	132	55	−80
7	(1)	1985	07	16	5:33:49	5.51	50.85	11	2.9	85	84	−96	309	9	−47
8	(1)	1985	12	07	23:09:25	5.97	50.87	2	3.0	174	82	−12	266	78	−172
9	(2)	1987	03	21	0:57:37	3.81	50.41	7	1.7	5	84	348	96	78	−174
10	(2)	1987	03	21	14:47:24	3.82	50.41	7	2.4	93	81	179	183	89	9
11	(2)	1987	03	22	21:05:35	3.82	50.41	7	2.5	162	78	1	71	89	168
12	(2)	1987	04	19	0:22:31	3.80	50.42	5	1.0	300	72	256	159	22	−54
13	(1)	1988	10	17	19:39:54	5.93	50.81	22	3.2	270	83	−154	177	64	−7
14	(1)	1988	12	27	11:53:12	5.69	50.52	19	3.3	84	84	−148	350	58	−7
15	(1)	1989	01	26	10:09:00	5.99	51.03	14	2.5	2	55	−68	146	40	−119
16	(1)	1989	12	02	21:59:03	4.44	50.41	1	2.3	278	86	178	8	88	3
17	(1)	1990	01	29	2:09:09	6.02	50.47	8	1.9	346	70	−26	85	65	−158
18	(1)	1990	02	05	3:33:02	6.02	50.47	8	2.0	328	89	6	238	84	179
19	(1)	1990	02	07	2:43:38	6.02	50.47	8	2.2	86	69	−160	349	71	−22
20	(1)	1990	02	21	12:33:35	6.02	50.46	8	2.0	254	85	−148	161	58	−6
21	(1)	1990	04	19	5:35:18	5.51	50.70	17	2.4	276	21	−84	89	69	−92
22	(1)	1991	03	27	23:00:49	5.95	50.30	11	2.2	307	76	−12	40	78	−166
23	(1)	1991	05	18	9:52:03	5.95	50.46	19	1.8	270	81	162	3	72	9
24	(1)	1991	08	10	14:55:31	5.64	50.73	11	2.8	250	65	166	346	77	25
25	(3)	1992	04	13	1:20:02	5.95	51.16	17	5.4	314	22	−98	143	68	−87
26	(3)	1992	04	13	4:32:47	6.26	50.85	14	2.4	166	70	−72	302	26	−131
27	(3)	1992	04	14	1:06:46	6.21	50.94	17	3.5	351	42	−18	94	78	−131
28	(3)	1992	04	14	1:36:23	6.23	50.83	15	2.7	130	48	−70	281	46	−111
29	(3)	1992	04	20	16:50:08	6.24	50.81	15	2.1	174	68	−52	290	43	−147
30	(4)	1992	08	29	9:22:25	4.29	50.20	10	3.4	267	83	124	6	35	12
31	(4)	1993	06	03	12:57:38	5.96	51.17	12	3.1	143	44	−60	284	53	−116
32	(4)	1994	07	20	18:01:56	4.07	50.45	8	2.5	202	90	0	112	90	180
33	(4)	1994	10	12	10:34:14	4.82	50.43	9	2.3	172	74	356	263	86	−164
34	(4)	1994	12	17	5:34:42	5.95	51.16	10	2.1	341	56	−44	99	55	−137
35	(4)	1995	01	16	22:20:18	6.14	50.90	12	2.7	273	86	−162	182	72	−5
36	(4)	1995	03	10	22:59:00	6.11	50.90	12	2.1	299	56	−118	162	43	−55
37	(4)	1995	06	20	1:54:47	4.11	50.51	24	4.3	37	47	102	200	44	78
38	(4)	1995	07	18	12:51:23	5.90	50.74	24	2.1	355	53	−132	231	53	−48
39	(4)	1995	10	30	23:16:47	5.90	50.54	22	2.3	8	79	−4	98	86	−169
40	(4)	1996	07	23	22:30:21	5.89	50.48	16	3.6	310	55	−42	67	57	−137
41	(4)	2001	06	23	1:40:00	5.92	50.88	10	3.4	144	70	−56	260	39	−147
42	(4)	2002	03	17	14:46:30	6.29	50.76	13	3.6	164	43	−32	279	69	−128
43	(4)	2002	04	22	21:17:00	6.25	50.82	12	3.1	273	51	−118	133	47	−60
44	(4)	2002	07	22	5:45:05	6.20	50.87	13	4.8	130	55	−80	293	36	−104

[†]Column 1: number in the list that is also referenced on Figures 7 and 8.

[‡]Column 2: number referring to the publication where the solution has been presented: (1) Camelbeeck (1993); (2) Englert (1995); (3) Camelbeeck and van Eck (1994); (4) present publication.

[§]Column 3: earthquake date.

[#]Column 4: earthquake origin time.

[††]Column 5: epicenter longitude.

[‡‡]Column 6: epicenter latitude.

[§§]Column 7: focal depth.

[##]Column 8: magnitude (M_S).

[†††]Columns 9, 10, and 11: strike, dip, and rake of the first nodal plane.

[‡‡‡]Columns 12, 13, and 14: strike, dip, and rake of the second nodal plane.

example, in the western Ardennes, the two available mechanisms (30 and 34) suggest strike-slip faulting, and the preferred nodal solution is dextral slip on a fault striking approximately E-W.

In the Hainaut region, relatively well-determined fault-plane solutions are available only for recent small earthquakes. They generally show strike-slip mechanisms (9, 10, 11, 16, and 33)

similar to the two mechanisms in the Ardennes, consistent with the existence of a dextral strike-slip zone. Mechanisms for the larger events during 1965–1976 (Camelbeeck, 1993) are not reliable enough because they are poorly constrained and are based on data from stations at distances greater than 200 km, with only a few at shorter distances.

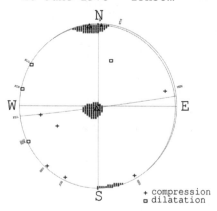

10 June 1978 - 13h58m

+ compression
□ dilatation

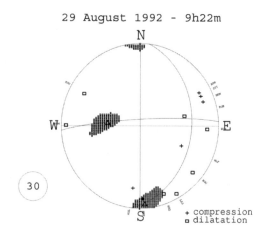

29 August 1992 - 9h22m

30

+ compression
□ dilatation

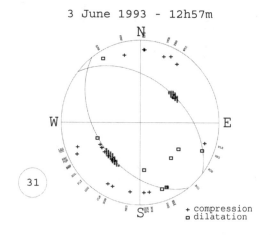

3 June 1993 - 12h57m

31

+ compression
□ dilatation

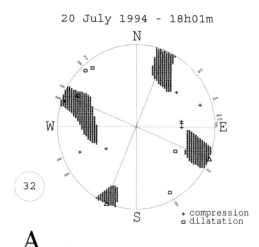

20 July 1994 - 18h01m

32

+ compression
□ dilatation

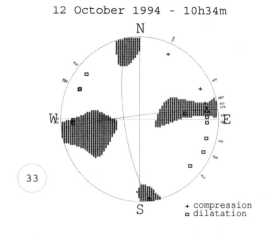

12 October 1994 - 10h34m

33

+ compression
□ dilatation

A

Figure 8 (*on this and following two pages*). (A–C) Fault-plane solution for the Roetgen earthquake (10 June 1978) and the earthquakes between July 1992 and 2005. The maximum likelihood solution of the mechanisms and the 99% confidence region for their poles are plotted for each mechanism. With the exception of the first diagram, which corresponds to the Roetgen earthquake, which is discussed in the text, the other mechanisms are numbered as in Table 2.

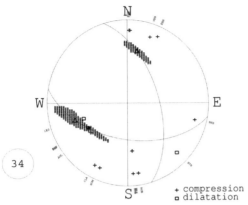

17 December 1994 - 5h34m

16 January 1995 - 22h20m

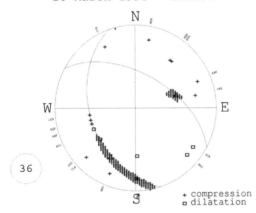

10 March 1995 - 22h59m

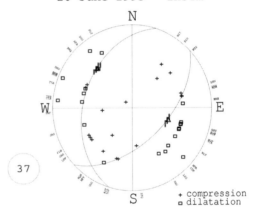

20 June 1995 - 1h54m

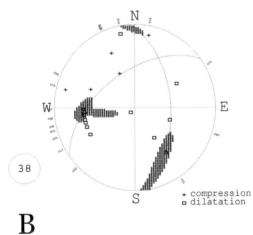

18 July 1995 - 12h51m

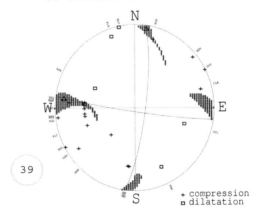

30 October 1995 - 23h16m

B

Figure 8 (*continued*).

23 July 1996 - 22h30m

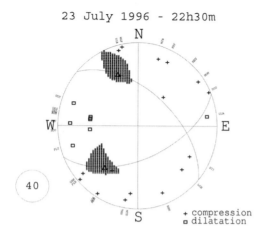

23 June 2001 - 1h40m

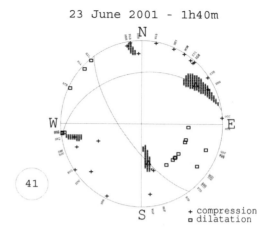

17 March 2002 - 14h46m

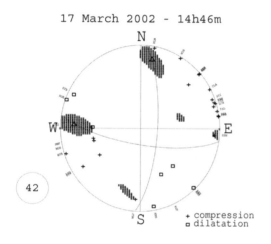

22 April 2002 - 21h17m

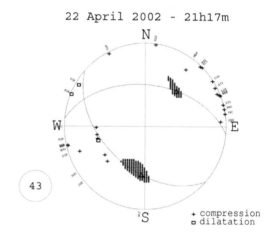

22 July 2002 5h45m

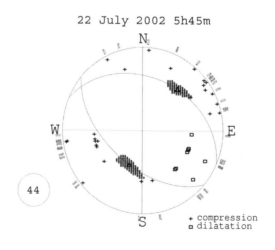

C

Figure 8 (*continued*).

In the north of France, we see tenuous evidence of recent tectonic activity (see section on large and strong historical earthquakes and their relationship with geological structure). The present-day seismicity is weak, and the data do not yield reliable fault-plane solutions. Nicolas et al. (1990) published two mechanisms, but we do not consider them reliable enough because they are constrained by first motions at too large of distances. A better understanding of the seismotectonics of the Boulonnais-Artois region will require more intensive work.

LARGE EARTHQUAKE POTENTIAL IN NORTHWEST EUROPE

Based on the synthesis and analysis in the first part of this paper, we now discuss methods of evaluating the potential for large earthquakes in northwest Europe.

We already presented the investigations on historical earthquakes suggesting that at least three of them were large. Our paleoseismic investigations in the Roer graben (Camelbeeck and Meghraoui, 1996, 1998; Vanneste et al., 1999, 2001; Meghraoui et al., 2000; Vanneste and Verbeeck, 2001) also suggest that coseismic surface ruptures have occurred from earthquakes with magnitude greater than 6.0 during the Holocene and late Pleistocene. A synthesis of these results will be presented in this part of the paper. Thus, there is no doubt that large earthquakes occurred in the past and will occur in the future between the Lower Rhine Embayment and the North Sea. The main issues are to estimate the possible locations, magnitudes, and return periods of future events. The earthquake catalogue shows that the stronger historical earthquakes generally occurred at different locations. Thus, it is very likely that the next one could occur in an unexpected site. Hence, we focus on methods of evaluating the probable location of seismogenic sources capable of generating large earthquakes.

In the fault database presently in development at the Royal Observatory of Belgium, we adapted the concepts and definitions from the database of potential sources for large earthquakes in Italy (Valensise and Pantosti, 2001). The zones defining the known sources of large earthquakes in a region are also seismogenic sources. In addition, we defined capable faults as ones that have sufficient dimension to generate large earthquakes but for which there is presently no evidence of very recent geological activity. Two main sources of information define these seismogenic sources: geology and seismic activity, mainly strong or large historical earthquakes. We now discuss how we use those two types of data to evaluate the location of seismogenic sources. In the first section of this chapter, taking the Boulonnais-Artois region as an example, we present the method used to define seismogenic sources of strong historical earthquakes and how knowledge of these events can help to define capable faults. The seismogenic sources defined from paleoseismic studies are presented in the second section, in which we focus on the Roer graben. The relevance of these data to evaluate long-term seismicity will be discussed in the last section of the chapter.

Large and Strong Historical Earthquakes and Relationship with Geological Structure

Definition of Seismogenic Source Associated with Historical Earthquakes of M >5.0

In regions of moderate seismicity, like northwest Europe, earthquake activity is the basic information for defining some of the seismogenic sources. During historical times, several strong earthquakes have occurred in the area investigated. As for the Italian database, our objective is to define the location and importance of seismogenic sources by evaluation of earthquake source parameters using recorded intensity patterns.

Gasperini et al. (1999) developed a method, called BOXER, that estimates the location, physical dimensions, and the orientation of large historical earthquakes using intensity data. This method estimates the macroseismic epicenter, epicentral intensity, magnitude, and also the azimuth, length, and width (downdip) of the fault that generated the earthquake. The fault length and width are calculated using Wells and Coppersmith's (1994) empirical relationships and default parameters estimated for subsurface rupture length of "all types of fault." To illustrate this on a map, the program computes "box vertices," the corners of a rectangle, centered on the epicenter, which is supposed to represent the surface projection of the seismogenic fault assuming it dips 45°. Application of the method to recent earthquakes in Italy gave stable results that are generally in agreement with the analyses of instrumental data. This program has been used as an input to the database of potential sources for earthquakes larger than M 5.5 in Italy (Valensise and Pantosti, 2001).

We tested the applicability of BOXER to seismicity in northwest Europe. Because the attenuation of intensity with distance, magnitude-epicentral intensity, and magnitude-macroseismic radii relationships differ in northwestern Europe from those in Italy, we modified these relations in BOXER. For historical earthquakes, for which the density and the quality of information are far from uniform, errors and uncertainties can be very important, mainly for source orientation evaluation. This is particularly true in northwest Europe, where earthquakes of sufficient size to apply the method are rare. It is also important not to forget that strong ground motion results from the convolution of source, path, and site effects. Thus, in some cases, the damage and intensity distribution are more related to the local (or regional) geologic structure than to the orientation of the seismogenic source. In general, we consider it difficult to reliably infer source geometry using intensity data. Nevertheless, the BOXER model for the earthquake source is related to the orientation of the area of high intensity, and we believe it gives a good first estimate of the source dimension. Thus, despite its limitations, we consider BOXER to be a useful tool to map earthquake sources and use it for our database for earthquakes that have magnitudes greater than or equal to 5.0.

We chose this minimum magnitude because we considered the source dimensions of such earthquakes, a fault surface of

10–15 km² and slip of a few decimeters, to be significant and representative of the regional deformation, even if they do not correspond to a large earthquake rupturing the whole seismogenic layer. Two other reasons also favor this choice. First, this magnitude corresponds to moderate earthquakes that can be destructive, so mapping their sources is of interest for seismic hazard studies.

Second, considering only large earthquakes would reduce the number of usable events.

As a methodological example, we estimate source properties of the most severe earthquakes that have affected northern France (Fig. 9B): those of 6 April 1580 (M = 6.0) and 2 September 1896 (M = 5.0).

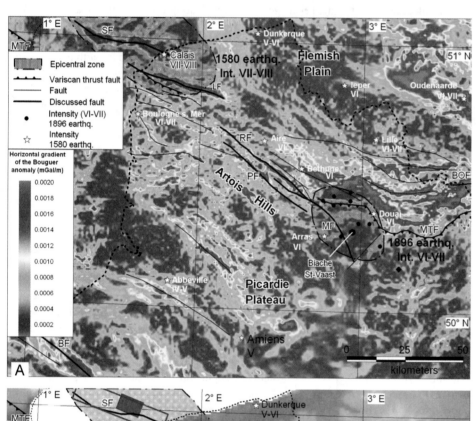

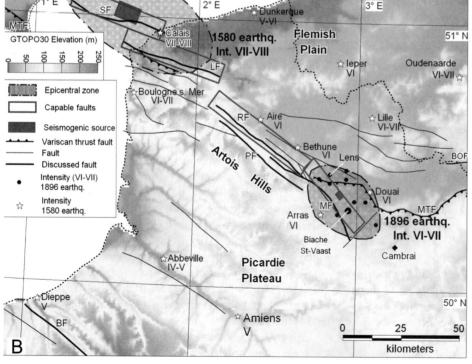

Figure 9. Seismogenic sources associated with historical earthquakes and capable faults in the Boulonnais-Artois region. (A) Fault identification by the horizontal gradient of the Bouguer gravity anomaly. BF—Bray fault, BOF—Border fault, LF—Landrethun fault, MTF—Midi thrust fault, SF—Sangatte fault, RF—Ruitz fault, PF—Pernes fault, and MF—Marqueffles fault. The Bouguer anomaly was calculated using the gravity database of the Royal Observatory of Belgium (ROB) (Everaerts, 2000). For France and the United Kingdom, the data were provided by the French Geological Survey and British Geological Survey, respectively. All gravity data are referenced to the gravity datum of Uccle (1976) (IGSN71–0.048mGal). The density reduction used for the calculation of the Bouguer anomaly on land was 2.67. Above the sea, the free-air anomaly was used. The theoretical gravity was computed using the 1980 geodetic reference system formula. (B) Digital elevation model—the 1580 and 1896 earthquakes seismogenic source—capable faults. The seismogenic sources of the 1580 and 1896 earthquakes are the red rectangles. Their location and orientation are based on the BOXER program, but theoretical scaling relationships between source dimension and magnitude have been used to calculate the rectangle dimensions. The capable faults are represented by blue rectangles, with lengths corresponding to the fault length and widths without any signification.

The 6 April 1580 earthquake (Melville et al., 1996) shook a large area of northwest Europe, including parts of France, England, Belgium, the Netherlands, and Germany. The most affected regions were Flanders, Kent, Artois, and in particular Calais.

The earthquake of 2 September 1896 was felt over a smaller area (Lancaster, 1896). The farthest city where it was felt, Leuven in Belgium, was less than 150 km from the epicentral region. The earthquake caused damages interpreted as intensity VII (MSK scale) in its epicentral area, between the cities of Lens and Arras (Fig. 9B).

Identification of Capable Faults Based on the Knowledge of Strong Historical Earthquakes

In contrast to active regions where potential seismogenic sources are easily identified, only a few can generally be identified within plates. Therefore, we sought to relate locations of M ≥5.0 earthquakes to geologic structures, so as to delineate other geologic structures in the same region that may be capable faults. In a second step, it will be necessary to establish that these structures are active faults and thus real seismogenic sources.

As an example, we investigated the possible relationship between the seismogenic source of the 1580 and 1896 earthquakes in the north of France and known faults.

The region where these two events occurred separates the Flemish plain to the north and the chalky Picardie Plateau (Fig. 9B) to the south. It corresponds to a NW-SE alignment of hills from Boulogne to Cambrai. As in the Weald, Wessex, and southern North Sea basins, the Artois corresponds to a small anticlinal flexure cut by fractures resulting from the early Tertiary tectonic inversion, during which faults slipped in a reverse sense to that during the Paleozoic (Auffret and Colbeaux, 1977).

Interpretation of gravity data provides information on the main geological structures in the region. Figure 9A shows the geographic variation of the horizontal gradient of the Bouguer gravity anomaly, which illustrates abrupt lateral changes of density. In this region, the sharp contrasts mainly indicate faults (Everaerts, 2000; Everaerts and Mansy, 2001) and so illustrate faults that are not visible at the surface and allow us to better understand their geometrical properties and relationship. From this analysis, Everaerts (2000) concluded that the faults have lengths ranging from 15 to 40 km and are arranged en echelon in longer fault zones. Those faults, reactivated during the Tertiary, have slightly different orientations in the Boulonnais (WNW-ESE) and in the Artois (NW-SE).

The faults have influenced the river development, independently of lithological contacts, suggesting possible Quaternary tectonic activity (Auffret and Colbeaux, 1977; Colbeaux et al., 1993, 1979; Hennebert, 1998; Mansy et al., 2003; Sommé, 1967; Van Vliet-Lanoë et al., 2002). Quaternary activity of the Marqueffles fault has been suggested in the archaeological site of Biache-Saint-Vaast (Colbeaux et al., 1981). There, fluvial deposits and the loess covering it are affected by a small dextral horizontal displacement that occurred along WNW-ESE and NNE-SSW faults.

Figure 9B suggests that the 1580, although its location is not well constrained, and 1896 earthquakes could be related to the Sangatte and Marqueffles faults bounding the Artois hills. The weak evidence of recent geological activity makes it reasonable to consider these and the other faults of the same system as capable faults. One of our future objectives is to study the most recent geological activity of these faults in more detail with methods developed in studying the border faults of the Roer graben.

Paleoseismic Investigations in the Roer Graben

In seismically active regions, geology can be the best way to assess seismogenic sources because it can identify and locate active faults. Within plates, unfortunately, geologic information is often neglected for reasons explained in the introduction. However, where the seismic activity is unknown, evaluation of the earthquake potential must be based on geological data. If geological evidence of recent (Quaternary) fault activity exists, characteristics of the fault zones can provide basic information on possible seismogenic sources. Thus, paleoseismic investigations of faults can provide valuable information. Without evidence of which faults are active, a first step is to consider that each existing fault with sufficient dimensions at depth and at the surface could be a capable fault. This is of course very improbable, but depending on the kind of hazard evaluation (e.g., siting a dangerous industry), it cannot necessarily be excluded.

In the Lower Rhine Embayment, Ahorner (1975) provided a comprehensive seismotectonic study establishing the relationship between the seismic activity and normal faults offsetting Quaternary deposits by up to 175 m. He did not consider the possibility of coseismic surface rupturing and inferred that the largest earthquakes should be similar in size to the largest historical earthquakes, corresponding to M_S around 5.5.

Since 1995, we have been conducting a program to identify and study active faults in Belgium along the western border fault of the Roer graben, the Feldbiss fault zone. The Feldbiss fault zone is a wide zone that shows a left-stepping pattern and includes, amongst others, the Sandgewand (Fig. 5), Feldbiss, Geleen, and Heerlerheide faults (Fig. 10). Southeast of the town of Bree, the Geleen fault appears as a NW-SE–oriented escarpment that is easily identified on topographical maps and aerial photographs. This escarpment has a vertical offset of 15–20 m, and it defines the morphological border between the elevated Campine Plateau, covered by the middle Pleistocene main terrace of the Maas River, and the subsiding Roer graben, where the main terrace is buried below late Pleistocene terraces and eolian sediments. The total vertical offset of the base of the main terrace (the age of which is poorly constrained between ca. 350,000 and 770,000 yr B.P.) is ~40 m. The escarpment is quasi-rectilinear over a distance of 11.5 km, and it was named the "Bree fault scarp" (Camelbeeck and Meghraoui, 1996). The scarp is compound, and a small scarplet at the base of the main escarpment offsets young deposits and alluvial terraces in the flat valley.

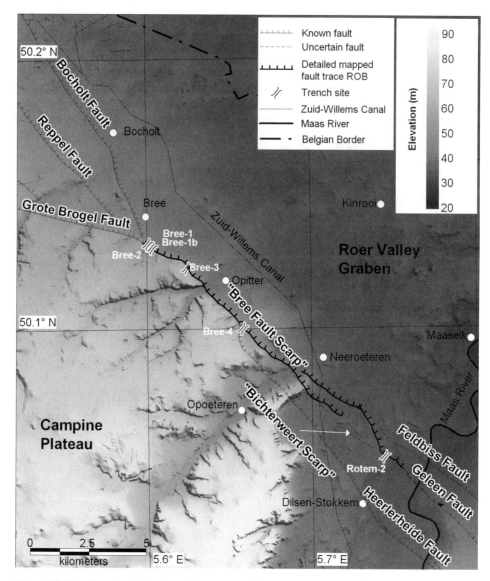

Figure 10. Sites investigated and trenches excavated by the Royal Observatory of Belgium along the western border fault of the Roer graben in Belgium.

Between 1995 and 2000, five paleoseismic trenches (Figs. 5 and 10) were excavated across the Bree fault scarp. A detailed geological description of the trench logs has been provided for trenches 1–4 by Camelbeeck and Meghraoui (1998), Vanneste et al. (1999), Meghraoui et al. (2000), and Vanneste et al. (2001). Trench 1bis was excavated at the same location as trench 1 to collect samples for Infrared Stimulated Luminescence (IRSL) dating. This trench is shown in Figure 11. In each trench, the scarp coincides with tectonic faults affecting late Pleistocene deposits and Holocene soils, indicating that it is the morphologic expression of the most recent earthquakes (Camelbeeck and Meghraoui, 1996, 1998; Vanneste et al., 2001; Meghraoui et al., 2000). We identified up to five surface-rupturing earthquakes since 101.4 ± 9.6 k.y. B.P. (Vanneste et al., 2001). Table 3 shows the correlation with vertical offsets and age constraints for the three youngest events in the different trenches.

The most recent large earthquake has been identified in all of the trenches. In trenches 1 (Camelbeeck and Meghraoui, 1998), 1bis (Fig. 11), and 4 (Vanneste et al., 2001), this event is evidenced by displacement of 0.5 m (1.0 m including flexure), 0.65 m, and 0.12 m (0.71 m including flexure), respectively, of a thin gravel bed, the Beuningen desert pavement, dating back to the Last Glacial Maximum (ca. 14–19 k.y. B.P.; Kolstrup, 1980), and of overlying eolian sands of late glacial age (ca. 10–13 k.y. B.P.; Kolstrup, 1980). In trenches 2 and 3 (Meghraoui et al., 2000), the fault juxtaposes the middle Pleistocene main terrace of the Maas River against late Pleistocene and early Holocene deposits. In trench 2, the most recent displacement is evidenced by a well-defined colluvial wedge, allowing estimation of the slip as ~0.7 m. In trench 3, unfortunately, strong scarp erosion due to a steep surface slope and intense farming

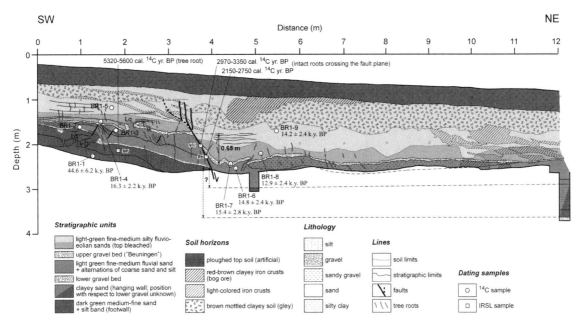

Figure 11. Log of northwest wall of trench 1bis across the Bree fault scarp. One faulting event is evidenced by 0.65 m of vertical offset of a gravel bed that correlates with the regionally known "Beuningen" horizon dating back to the Last Glacial Maximum. Overlying late glacial fluvial-eolian sands, as well as the Holocene soil, are displaced by the same amount. The event horizon could not be identified, and it has most likely been destroyed by the artificial plough zone. Two plant roots crosscutting the fault plane and sealing fault movement were radiocarbon dated between 2150 and 3350 cal. yr B.P. A second event may be present just below the Beuningen horizon, but correlation between footwall and hanging wall is ambiguous and could not be verified because of the shallow water table. Infrared Stimulated Luminescence (IRSL) ages are from Frechen et al. (2001). Lq—liquefaction.

activities made it impossible to measure the most recent displacement. Initial radiocarbon dating in trenches 1 and 2 constrained the event's age between ca. 600 and 900 A.D. However, new IRSL dating suggests that these radiocarbon ages are most likely too young. In trench 1bis (Fig. 11), we obtained a larger age bracket between 12.9 ± 2.4 k.y. B.P. (IRSL age of displaced eolian sands above Beuningen level) and 1020–1400 cal. ^{14}C yr B.C. (radiocarbon age of a plant root that penetrated the fault plane but was not displaced).

Soils that are also affected by faulting do not allow quantitative dating but indicate that the real age is probably close to the upper end of the age bracket. In trenches 1 and 1bis (Fig. 11), Holocene soil that developed in the youngest eolian sands is clearly displaced and eroded from the footwall, while in trench 4 (Fig. 12), a thin, possibly historical, illuvial horizon resulting from the dissolution and precipitation at depth of clay and iron material that developed above the Holocene banded Bt-soil is faulted as well (Vanneste et al., 2001).

The penultimate large earthquake could only be observed in the trenches situated in an interfluvial setting. In trench 1, in the center of a small alluvial valley crossing the fault scarp, all deposits postdate this event (Camelbeeck and Meghraoui, 1998). The analysis of trench 1bis (Fig. 11) suggests that a second event may be present just below the Beuningen horizon, but the evidence is ambiguous. In trenches 3 and 4, the penultimate large earthquake is identified by vertical offsets of 0.5 m and

0.33 m, respectively, of the Older Coversand I unit, an eolian sand unit with a reported age of 22–27 k.y. B.P. (Kolstrup, 1980) located below the Beuningen gravel bed (Meghraoui et al., 2000; Vanneste et al., 2001). We were not able to identify the colluvial wedge associated with this event, most likely because it was removed during the extensive erosional period leading to the development of the Beuningen desert pavement. In trench 2, the penultimate event could not be directly observed due to the complexity of the fault zone. However, soft-sediment deformations were observed in trenches 2 and 4, which we attribute to this event and discuss in the next section. The age of the penultimate event is constrained between 9.6 and 13.6 k.y. B.P.

A third large earthquake can be characterized by the analysis of trenches 3 and 4. In trench 3, this event is evidenced by a vertical displacement of 1.2 m of an eolian sand unit older than the Older Coversand I unit (Meghraoui et al., 2000). In trench 4, it is recognized by an elongated wedge of material derived from the footwall in between eolian sands, which are generally restricted to the hanging wall. This is interpreted as a colluvial wedge generated under periglacial conditions (Vanneste et al., 2001). The thickness of this wedge suggests a total displacement in excess of 1 m. The age of the third event is constrained between 22.2 ± 2.5 k.y. B.P. (IRSL dating in trench 4) and 44.79 ± 1.1 k.y. B.P. (radiocarbon age from trench 3). Soft-sediment deformations in trench 2 that are older than the radiocarbon age of 28.08–34.80 k.y. B.P. possibly correlate with this event.

TABLE 3. VERTICAL DISPLACEMENTS AND DATING OF PALEOEARTHQUAKES ALONG THE BREE FAULT SCARP

Paleoearthquake	Trench 1	Trench 2	Trench 3	Trench 4	Correlation
Latest = event 1	0.5 m (fault slip) + 0.5 m (modeled flexure)	0.7 m (fault slip)	0.9 m (fault slip)	0.12 m (fault slip) +0.7 m (modeled flexure)	0.55 m average displacement (0.85 m incl. flexure)
	>1002–1272 yr B.P. (^{14}C)[†] <1181–1337 yr B.P. (^{14}C)[†] In Trench 1bis: >2970–3350 yr B.P. (^{14}C)[§§] <12.9 ± 2.4 k.y. B.P. (IRSL)[‡‡]	>907–986 yr B.P. (^{14}C)[†] <2151–2342 yr B.P. (^{14}C)[†]	<Older Coversand II (ca. 13–10 k.y. B.P.)[##]	>480–550 yr B.P. (^{14}C)[#] <10.11–10.40 k.y. B.P. (^{14}C)[#] <6.8 ± 1.2 k.y. B.P. (IRSL)[††] <2nd-generation Bt-soil fiber (historical?)[#]	(1337–1002 yr B.P.)[†] 2970–8000 yr B.P.
Penultimate = event 2		Lateral spreading, folding	0.5 m (fault slip)	0.33 m (fault slip)	0.4 m average displacement
		>Beuningen horizon (ca. 19–14 k.y. B.P.) <21.33–26.43 k.y. B.P. (^{14}C)[‡]	≥ Beuningen horizon (ca. 19–14 k.y. B.P.)[##] ≤ Older Coversand I (~27–22 k.y. B.P.)[##]	>Beuningen horizon >11.4 ± 1.8 k.y. B.P. (IRSL)[††] <12.0 ± 1.6 k.y. B.P. (IRSL)[††]	9.6–13.6 k.y. B.P.
Antepenultimate = event 3		Soft-sediment deformation (folding)	1.2 m (fault slip)	0.85 + 0.40 m (fault slip)	1.2 m average displacement
		>28.08–34.80 k.y. B.P. (^{14}C)[‡]	>Older Coversand I (ca. 27–22 k.y. B.P.)[##] <44.79 ± 1.1 k.y. B.P. (^{14}C)[§]	>22.2 ± 2.5 k.y. B.P. (IRSL)[††] <69.0 ± 9.0 k.y. B.P. (IRSL)[††]	19.7 (28.08?)–45.9 k.y. B.P.

Note: The information results from the analysis of four trenches excavated across the Bree fault scarp (Camelbeeck and Meghraoui, 1998; Meghraoui et al., 2000; Vanneste et al., 2001). To facilitate comparison with luminescence ages, radiocarbon ages have been converted from calibrated calendar years into yr B.P. (with respect to 1950 A.D.) where applicable.
[†]Camelbeeck and Meghraoui (1996), radiocarbon samples possibly contaminated.
[‡]Range of two radiocarbon dates from the same stratigraphic horizon (Vanneste et al., 1999).
[§]Meghraoui et al. (2000).
[#]Vanneste et al. (2001).
[††]Vanneste et al. (2001), revised by Frechen et al. (2001).
[‡‡]Frechen et al. (2001).
[§§]Previously unpublished.
[##]Ages from stratigraphic correlation with Kolstrup (1980), given here for reference but not used to determine age brackets.

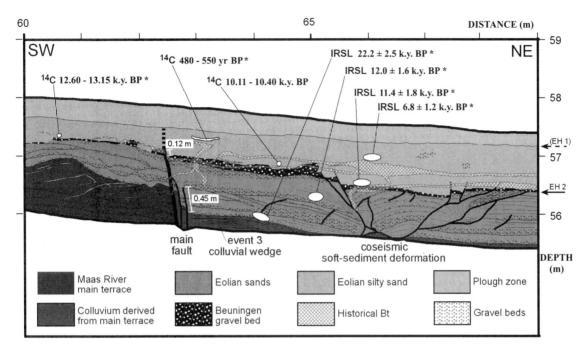

Figure 12. Coseismically induced soft-sediment deformations, involving both faulting and folding, observed on the southeast wall of trench 4 along the Bree fault scarp. These deformations occurred in two phases, correlating with the coseismic offsets of 0.12 m and 0.33 m visible at the fault. Note that a thin, second-generation illuviation horizon (Bt), which is likely historical in age, has been faulted in the last event as well. Infrared Stimulated Luminescence (IRSL) ages are from Frechen et al. (2001), and radiocarbon ages are from Vanneste et al. (1999). Position of samples marked with * is projected from the northwest wall.

Based on all these data, we can estimate the fault slip rate and return period for large earthquakes on the Bree fault scarp (Fig. 13). If we consider the two most recent complete earthquake cycles (between events 3 and 1), which are best constrained in time and can be correlated across the entire scarp, we obtain an average return period of 13.7 ± 7.8 k.y. The average fault slip rate for the same interval, averaging the displacements of events 1 and 3, is 0.050 ± 0.036 mm/yr. Using the longer faulting record from trench 4, we can make the same calculations for the last 100 k.y. These values differ from those of Vanneste et al. (2001), because the IRSL ages have been revised by Frechen et al. (2001). Considering that five paleoearthquakes are recorded in trench 4 since 101.4 ± 9.6 k.y. B.P., corresponding to four or five complete earthquake cycles, we calculate an average return period of 22.7 ± 4.3 k.y. The corresponding average fault slip rate is 0.031 ± 0.012 mm/yr, which is in good agreement with the values for the two last earthquake cycles. Two large earthquakes have occurred during the last 20 k.y., whereas four other similar events have been identified during the previous 80 k.y. However, it is statistically difficult to interpret this observation in terms of variable slip rates or return periods.

Since 2001, we also investigated the SE extension of the Bree fault scarp. In contrast to the Bree fault scarp, the Geleen fault (Fig. 10) has little topographic expression as it traverses the Belgian Maas valley, where the late Weichselian Maas River terrace lies at the surface. Based on scattered boreholes and

electric soundings, previous investigators mapped a displacement of the base of the gravel deposits, which they named the buried Bichterweert scarp, and concluded that the faults in the Belgian Maas valley show no post–Upper Pleniglacial (Paulissen et al., 1985) or post-Weichselian (Beerten et al., 1999) activity. However, using a combination of geomorphic and geophysical analysis, we were able to identify and map the surface trace of an active fault in this area, in some places coinciding with, but in other places significantly diverging from, the previously mapped Bichterweert scarp (Vanneste et al., 2002). Ground-penetrating radar and two-dimensional (2-D) resistivity profiles demonstrate that this fault displaces the top of the late Weichselian Maas River terrace. At a few sites, this fault is also associated with a subtle (<1 m) topographic scarp. In most places, however, there is no geomorphic expression at all, but we suspect that it has been erased by human modification of the landscape. A trench excavated near Rotem (Fig. 14), just outside the Holocene alluvial plain, confirmed that the late Weichselian terrace and the overlying late glacial eolian sands and silts are vertically displaced by 0.75–1 m, and that the Bichterweert scarp, similar to the Bree fault scarp farther north, has experienced one surface-rupturing earthquake during the Holocene. Complete results of this trench are not yet available and will be published elsewhere.

Investigations of other faults in the Roer graben have been conducted in the framework of the EC-project PALEOSIS (trenches 6 and 7 in Fig. 5) in cooperation with the Geological

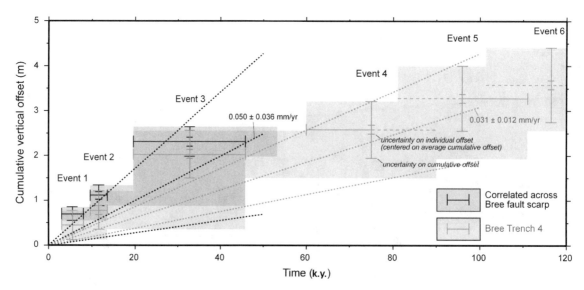

Figure 13. Time-displacement diagram for paleoearthquakes along the Bree fault scarp. The diagram shows the envelope of time–cumulative offset solutions (incorporating their uncertainty) inferred from a correlation of the three most recent events (event 1 is the most recent) across all trenches (darker shade) and for the last six events identified in trench 4 (lighter shade). Superposed are average fault-slip rates (dotted lines) calculated from the last two and five complete earthquake cycles, respectively. Although the plot seems to suggest an increase of fault activity with time, the longer-term fault-slip rate falls entirely within the uncertainty limits of the shorter-term slip rate.

Survey of the Netherlands and the Netherlands Royal Meteorological Institute, for trench 7 excavated across the Peelrand fault, and with the Geologische Dienst Nordrhein-Westfalen, for trench 6 excavated across the Rurrand fault.

In The Netherlands, investigations on the Peelrand fault, the eastern border fault of the Roer graben, recovered faulting history for the last 25 k.y. (van den Berg et al., 2002). Two large earthquakes occurred in a relatively short time span around 15 k.y. ago. A third occurred in the Holocene. Vertical surface offset (Camelbeeck et al., 2001) is 1.3 m along the Peelrand fault, and this corresponds to the tectonic deformation since the Last Glacial Maximum. By comparison, the offset is 1 m along the Bree and Bichterweert fault scarps in Belgium.

Vanneste and Verbeeck (2001) provided a detailed paleoseismic analysis of a trench across the Rurrand fault (Fig. 5), the prolongation of the Peelrand fault in Germany. The Rurrand fault appears to be a complex fault zone with at least five separate, SW-dipping, normal fault strands that displace an early Pleistocene terrace of the Rhine River. Most of the deformation occurred during or after deposition of a stratified loess unit of Weichselian (ca. 117–10 k.y. B.P.) to possibly Saalian age (ca. 185–130 k.y. B.P.). The faulting is clearly episodic, with different fault stands active at different times. Growth faulting was not observed, and two layers of colluvial loess immediately downslope of the fault zone can be interpreted as colluvial wedges. The fault displacement during the last 40 k.y. is at least 3.1 m and possibly up to 6.7 m. Excluding the 0.4 m of modern fault motion induced by manmade groundwater lowering, this yields a long-term fault slip rate of 0.05–0.20 mm/yr.

Paleoseismic investigations thus indicate that Quaternary faults in the Lower Rhine Embayment have to be included in our catalogue of seismogenic sources. The main problem is to define the fault segments that have to be considered as single sources. This is discussed in the section on the maximum credible earthquake based on paleoseismic data.

Relevance of Paleoseismic Data in the Roer Graben: Seismic or Aseismic Movement

There has been (Ahorner, 1996) and continues to be (Houtgast et al., 2003, 2005) debate about whether the fault displacements observed in our and other paleoseismic trenches in the Roer graben are evidence of coseismic surface rupturing, or whether they could also be explained by (episodic) aseismic fault slip.

Paleoseismic studies of low-slip-rate faults in densely populated and relatively high-latitude areas like northwestern Europe face many problems. First, tectonic deformation processes do not outpace other processes modifying the landscape. Tectonic landforms are therefore difficult to recognize in areas of active deposition such as alluvial valleys and are easily destroyed by human activities such as farming. These environments also tend to be wet and difficult to excavate. Outside alluvial valleys, surface deposits are older, and fault scarp morphology is more pronounced. These places, however, generally lack Holocene sediments and, thus, stratigraphic resolution. Because surface offsets are relatively small, on the order of 1 m or less, the event horizon and colluvial wedge, particularly of the most recent large earthquake, are situated close below the ground surface. Since the upper 30 cm

Figure 14. Photograph of northwestern wall of trench Rotem-2 in the Belgian Maas valley (location in Fig. 10), showing the Geleen fault vertically displacing the top of the late Weichselian terrace and overlying eolian sands and silts by 0.75–1 m.

of the soil are usually destroyed by plowing, and stratification is often wiped out down to 1 m due to soil leaching, these features are difficult to identify. The same is also true for older events, since recurrence intervals on a given fault segment may exceed 10 k.y., which implies a long period of soil development that may overprint all stratigraphic evidence, as well as considerable climatic variation. The penultimate and antepenultimate events along the Bree fault scarp likely occurred during glacial periods, when the region was under the influence of permafrost and soil and slope processes were much more active than today (e.g., Vandenberghe, 1985).

Another major problem is dating the deposits affected by faulting, especially for the most recent paleoearthquake. The shallow stratigraphic position of this event increases the likelihood that radiocarbon samples are contaminated, displaced by bioturbation, or disturbed by human activity. Also, the sediments predating the last event are generally much older (late glacial to early Holocene), making it impossible to provide a narrow age bracket. Older events are also difficult to date precisely, because glacial sediments are generally devoid of organic matter. Many samples appear to be reworked. We therefore have to rely on luminescence dating techniques, which are more reliable in this context, but less precise. Another consequence of the insufficient stratigraphic resolution and the wide age brackets is that it is impossible to positively discriminate or correlate paleoearthquakes across possible segment boundaries, which in turn has implications for the estimation of the maximum magnitude (see section on the evaluation of the magnitude of the maximum credible earthquake).

We have encountered similar problems in paleoseismic studies in southern Bulgaria, another low-slip-rate area, where large surface-rupturing earthquakes occurred on 14 and 18 April 1928 with magnitudes of 6.8 and 7.2 (Vanneste et al., 2006).

Despite these limitations, our efforts have yielded strong arguments in favor of a coseismic origin of the deformation observed in the morphology and trench stratigraphy:

First, the geomorphic signatures of the Bree fault scarp and other tectonic escarpments in the Roer graben are similar to those of other active normal faults, only smaller in amplitude. The ~1 m frontal escarpment along the Bree fault scarp and that, for instance, along the Peelrand fault in the Netherlands are clear in the landscape. Trench observations correlate the scarp offsets well with the amount of late Holocene deformation that is visible as a single step in the stratigraphy (Camelbeeck et al., 2001). To have withstood centuries of intense agricultural activity, these scarps likely resulted from sudden fault movement. Up to now, no fault displacement has been observed that could be correlated with the gradual present deformation inferred from leveling data. There is only one place, in the trench across the Rurrand fault (Fig. 5) at Jülich-Stallbusch, where a zone of distributed cracks was observed that affected the plowed layer and extended to the ground surface, which we attribute to 40 cm of differential subsidence induced by regional groundwater lowering for open-cast mining (Vanneste and Verbeeck, 2001).

All the trenches excavated so far indicate episodic faulting events separated by relatively long intervals in which no fault slip occurs. Growth faulting in the form of a gradual increase of fault displacement without discernible intervening stratigraphic breaks, which would indicate aseismic slip, has never been observed. In a few cases, for example, the most recent event in trench 2, we have been able to identify a colluvial wedge (Camelbeeck and Meghraoui, 1996), which is primary evidence for coseismic faulting (McCalpin, 1996). For four older events in trench 4, we observed elongated wedge-shaped bodies directly downslope of the fault, which we interpret as colluvial wedges that originated in a periglacial environment (Vanneste et al., 2001). In trench 1bis, we discovered a plant root radiocarbon dated at 1020–1400 cal. yr. B.C. penetrating through but not displaced by the fault plane, indicating that this fault has not experienced slip for 3000 yr.

In addition to this primary evidence, we have observed, in nearly all trenches, various examples of soft-sediment deformation not previously reported. Admittedly, these features are not diagnostic for coseismic deformation and are sometimes difficult to distinguish from deformation by periglacial processes (cryoturbation), which was a widespread phenomenon in the area during glacial periods (Vandenberghe, 1985). However, the close association with a fault and their appearance in almost all trenches and in different stratigraphic intervals, including sediments that were deposited during the waning stages of glaciation and that show no other glacigenic features (such as ice-wedge casts or frost wedges), make an origin by coseismic liquefaction more than likely. Most deformations were observed in fine, silty eolian sands, which have a grain-size distribution that is prone to liquefaction (Vanneste et al., 1999). An example occurs in the hanging wall of trench 2, where a 1-m-thick stratified sand layer is involved in a series of flow folds, coupled with a drastic thickness decrease. At centimetric scale, this layer shows abundant water-escape features. We interpreted this as lateral spreading and flow failure of a liquefied sand layer (Vanneste et al., 1999). In trench 4, we observed a deformation feature several meters wide directly downslope of the fault zone, which contained both folds and small-scale faults (Fig. 12). This particular combination of quasi-simultaneous plastic and brittle deformation is again not untypical of liquefaction (Owen, 1987). A glacigenic origin is not likely, because we should expect the deformation to affect a wide area instead of being limited to the immediate vicinity of the fault zone. Retrodeformation indicates that the deformation took place in two distinct phases, which most likely correlate with the two different faulting events that we could identify on the nearby fault (Vanneste et al., 2001).

Recent paleoseismic studies in the Dutch part of the Roer graben did not find evidence of coseismic surface faulting (Houtgast et al., 2003, 2005). We feel, however, that these studies lacked a general understanding of earthquake geology practice, because the presence of wedge-shaped deposits in the hanging wall was ignored, and haphazard slope failure not related to faulting was invoked to explain deposits directly downslope of the excavated fault.

We already discussed in a previous section that present-day tectonic deformation in northwest Europe is below the resolution of current space and ground-based geodetic techniques and

that it will be necessary to have longer periods of observation before interpreting data to determine seismic or aseismic deformation. The only possible comparison is between geologic rates and observed seismic moment. In the Roer graben, the observed cumulative seismic moment (Camelbeeck and Meghraoui, 1998) is not enough to explain the geological rates. This is why it is so important to provide evidence of the coseismic surface-rupturing character of the deformation identified in the morphology and the geologic record. In the other parts of the region between the Lower Rhine Embayment and the North Sea, Quaternary geologic movements across known faults are so low that presently they are not known. Thus, the three historical earthquakes that occurred there with magnitude around or greater than 6.0 are the evidence that most of the deformation is seismic.

On the other hand, a relationship between postglacial rebound and seismic activity in some intraplate regions of the world has been suggested by many authors (e.g., Gregersen and Basham, 1989). The most spectacular example is the 155-km-long Pärvie fault in northern Fennoscandia, where the corresponding fault scarp has been interpreted as the result of an earthquake with an estimated magnitude of 7.9 that occurred 9000 yr B.P. within the latest phase of deglaciation (Muir-Wood, 1989). It could also have resulted from a succession of smaller-magnitude earthquakes. Coseismic deformations observed in the Roer graben trenches occurred independently of the large climatic variations and were not restricted to the end or just after the end of the glacial periods. This observation suggests that postglacial rebound should not have a major influence on the long-term seismic activity in the Roer graben.

EVALUATION OF THE MAGNITUDE OF THE MAXIMUM CREDIBLE EARTHQUAKE

The maximum credible earthquake can be defined as the largest earthquake that appears capable of occurring along a fault or in a specific region considering the existing tectonic stress environment. Its magnitude is defined as Mmax.

Even if we don't intend to discuss the evaluation of the return period of large earthquakes, it is interesting to note that our paleoseismic data in the Roer Graben do not allow to favor one of the models of large earthquake recurrence—variable slip, uniform slip, or characteristic earthquake (Scholz, 1990). There is no data elsewhere in northwest Europe.

Following our definition of seismogenic sources, the maximum credible earthquake should correspond to the earthquake associated with the largest seismogenic source in the region considered. This is relatively easy to assess in well-studied seismically active regions where most of the active faults have been identified and historical seismicity includes a major event that can be considered as the maximum credible earthquake. In intraplate tectonic environments, the major problem is identification of the possible seismogenic sources and evaluation of their dimensions. We have discussed two cases, in the north of France and the Roer graben, for which the location of seismogenic sources was obtained using historical seismicity and active fault studies. We discuss now estimating Mmax for these and other regions lacking evidence to define seismogenic sources.

Maximum Credible Earthquake Based on Historical Information

In the north of France, the seismogenic sources corresponding to the $M_S = 6$ 1580 and $M_S = 5$ 1896 earthquakes were defined from the intensity distribution. We already briefly discussed the fault system affecting Tertiary (Eocene) sediments, along which one local geological observation and possible morphological evidence of Quaternary fault activity exist. We classified all the faults of the system as capable (Fig. 9B). This approach is used to define potential sources of large earthquakes for seismic hazard assessment. Depending on the purpose of the evaluation of the maximum credible earthquake, the capable faults may be included in the analysis or not, based on expert judgment. If the capable faults are included in the evaluation, they increase Mmax because some reach lengths of 45 km as evidenced in the section on Large and Strong Historical Earthquakes and Relationship with Geological Structure, implying a magnitude around 7.0 if they are completely ruptured by an earthquake. If only the seismogenic sources defined by the strongest historical earthquakes are taken into account, the strongest one defines Mmax. In this case, it is the 1580 earthquake with an estimated magnitude of 6.

In any case, the historical seismic activity along a fault or in a region provides a minimum value for Mmax. In some cases, it is reasonable to suppose that this corresponds to the true Mmax.

The validity of this hypothesis should be assessed by evaluation of the possible completeness of the seismic catalogue in terms of incorporating the maximum credible earthquake. This can be done in seismically active regions by comparing the total seismic moment release by the earthquakes of the catalogue with that deduced from geologic and present-day geodetic strain-rate estimates. If these rate estimates are similar, it can be concluded that the catalogue is complete, that it incorporates the maximum earthquake, and also that most of the deformation is caused by seismic slip. If earthquake activity is less and cannot explain the geologic and (or) the geodetic rates, two different assumptions can be taken. The first one assumes no aseismic slip, and the maximum credible earthquake should be greater than the largest one in the catalogue. The second considers that the catalogue includes the maximum credible earthquake and that aseismic slip occurs and could be the dominant mode of deformation.

We already discussed the present limitations of geodetic data in the northwestern European tectonic context. If we consider the Roer graben, the only region of northwest Europe where recent geological deformation rates have been estimated, the largest known historical earthquake has a magnitude of 5 3/4. If the observed deformation at the geologic scale (late Pleistocene–Holocene) across the border faults of the Roer graben resulted mostly from earthquake activity, as it is discussed in the section Relevance of Paleoseismic Data in the Roer Graben, and

assuming that large earthquakes follow the frequency-magnitude relationship established from the known seismic activity, Camelbeeck and Meghraoui (1998) calculated that earthquakes with a maximum magnitude around 6.8 should occur with a 3000 yr average return period to explain the observed geological rates. These computations give only a rough evaluation but suggest that the historical catalogue is not complete and does not incorporate the largest earthquakes. Thus, in the Roer graben, paleoseismology should provide a better estimate of Mmax. On the other hand, there is no widespread evidence of Quaternary deformations in the north of France, or in other regions like the Brabant Massif and the Hautes-Fagnes–Ardennes regions, where $M_S \geq 6.0$ historical earthquakes have occurred. Thus, these earthquakes explain the very low (unknown) geologic strain rates and could be considered as maximum credible earthquakes. Up to now, the largest seismogenic source in these regions has been defined by the large historical earthquake, but definition of the seismogenic sources is based on incomplete information. Future work assessing the most recent activity of capable faults could change the maximum credible earthquake.

Maximum Credible Earthquake Based on Paleoseismic Data

When paleoseismic information is available, estimates of Mmax along a given fault can be based on the relationship between earthquake magnitude and fault rupture length and (or) the relationship between earthquake magnitude and amount of fault displacement.

The Roer graben is the only seismic source for which the results from detailed paleoseismic investigations can be used to define Mmax. First, we discuss evaluating Mmax along the Belgian part of the Roer graben (Fig. 10), which is the best-studied area. Up to now, our investigations have not allowed us to determine the length of the fault section ruptured during a single earthquake. We assume that it is at least 11.5 km from field observations of the Bree fault scarp. This minimum value seems compatible (Wells and Coppersmith, 1994) with the observed vertical offsets for the more recent and penultimate large earthquakes (see section on Paleoseismic Investigations in the Roer Graben).

North of Bree (Fig. 10), the fault splits into at least three separate faults, the Grote Brogel, Reppel, and Bocholt faults, at a point that we consider to be a segment boundary. South of the Bree fault scarp, in the Maas valley, the Geleen fault transects much younger (Saalian and late Weichselian) terraces of the Maas River, and its geomorphic expression (Bichterweert scarp) is consequently reduced. Using geophysical techniques (two-dimensional resistivity imaging, ground-penetrating radar, and seismic reflection), we traced the fault through the Belgian part of the Maas valley, where it displaces the top of the late Weichselian terrace ±1 m, as confirmed in a trench at Rotem. This study also revealed major fault complexity: a seismic-reflection profile in the Zuid-Willemsvaart canal crossing the SE end of the Bree fault scarp showed a very wide fault zone with many anti- and synthetic faults northeast of the main fault (Vanneste et al., 1997). Just southeast of this site, our geophysical profiles show a change in fault trend and a considerable left step of ~500 m. Further southeast, ~11 km from the Belgian-Dutch border, offsets on the Geleen fault disappear and seem to be transferred to the Feldbiss fault (Houtgast et al., 2002), at a point which corresponds to a left step on the order of 2 km, which is likely a segment boundary.

The question is whether the step between the Bree fault scarp and the Bichterweert fault scarp is a segment boundary. In general, steps of only a few hundreds of meters are not enough to stop rupture propagation. Also, the step cannot be resolved on the gravimetric map. We do not at present have enough data to decide if this is a segment boundary, so two hypotheses remain:

1. The Bree fault scarp (11.5 km) and the Bichterweert fault scarp (16.5 km) are different segments of the Geleen fault with a different history of rupturing.

2. The Bree fault scarp and the Bichterweert fault scarp form a single segment, with a length of ~28 km.

It may also be possible that most ruptures only involve one fault scarp, whereas only the largest earthquakes propagate beyond the step and rupture the entire length of the Geleen fault. In view of the metric vertical displacement suggested for the antepenultimate earthquake, it could be that this event was larger and ruptured a greater fault length than the Bree fault scarp.

For the two hypotheses on the fault length, the empirical relationships (Wells and Coppersmith, 1994) for normal-faulting earthquakes yield different moment magnitude values.

1. A surface rupture length of 11.5 km yields M = 6.3 ± 0.3, which would be a minimum, as discussed above.

2. Alternatively, a surface rupture length of 28 km yields M = 6.7 ± 0.3.

The vertical displacements evaluated by the analysis of the trenches excavated along the Bree fault scarp are summarized in Table 3. The larger inferred displacements for the antepenultimate earthquake suggest that this event was larger than the two most recent ones and ruptured a greater fault length.

The relationships of Wells and Coppersmith (1994) for normal-faulting earthquakes yield magnitude values for the two most recent earthquakes. An average slip of 0.4 m yields M = 6.5 ± 0.3, whereas a maximal slip of 0.7 m gives M = 6.5 ± 0.3. If the antepenultimate earthquake had an average slip of 1 m, M = 6.8 ± 0.3, whereas a maximal slip of 1.2 m corresponds to M = 6.7 ± 0.3. Inferences from the fault length and estimated slips are consistent and suggest a value of Mmax around 6.7 along the fault segment including the Bree fault and Bichterweert scarps.

Single sites along the Rurrand and Peelrand faults in the Roer graben also have been investigated. The Hambach trench across the Rurrand fault in Germany (Vanneste and Verbeeck, 2001) suggests that a number of large earthquakes causing 1.5–2.5 m slip could have occurred in the very recent past. Hence, along the Rurrand fault, Mmax could approach 7.0, higher than that estimated for the Bree fault. The Peelrand fault scarp and the trench near the city of Neer provide evidence of three large earthquakes during the last 15 k.y. (van den Berg et al., 2002). From the old-

est to the youngest, they correspond to normal slip movements of 0.55 m, 0.30 m, and 0.05–0.10 m, which can be correlated to earthquakes with magnitudes ranging from 6.0 to 6.5. These estimates should be improved by the study of these faults at other sites, as has been done along the Bree fault scarp. We note that the fault segments seem longer and long-term slip rates appear higher along the eastern border of the graben than along the western border, which is in accord with long-term geology.

Maximum Credible Earthquake in Regions Where Information Is Lacking

Many intraplate regions lack information to define the maximum credible earthquake. This is not only the case where seismic activity is unknown or weak and diffuse but also where seismic activity is noticeable but has characteristics that do not clearly delineate seismogenic sources. The Hainaut and Liège zones (Fig. 1) are two typical examples in northwestern Europe. These seismic zones have been defined from specific characteristics that appear different from the earthquake activity in the neighboring regions. The known seismicity is characterized by relatively shallow hypocenters and magnitudes not exceeding 4.5 and 4.7, respectively. It is important to note that all the listed activity has occurred since the beginning of the twentieth century, whereas no known historical earthquakes have occurred there. Extensive coal mining activity during the past 200 yr ended during the 1970–1980s. Earthquakes occur at depths greater than that of the mining activity. The seismic activity is purely tectonic but could be partly due to local stress modification caused by the extensive rock volume removed underground by mining. Hence, it is unclear how to evaluate the maximum credible earthquake.

The catalogue contains 43 and 35 earthquakes with magnitude greater than 1.8 in the Hainaut and Liège areas, respectively. Earthquake frequency-magnitude relations have been evaluated using the maximum likelihood method (Weichert, 1980) for earthquakes with $M_S \geq 3.0$. The annual rate of earthquakes with magnitude greater than or equal to M_S is Log $N(M_S) = a - b\, M_S$, with a = 3.06 (± 0.37) and b = 1.07 (± 0.12) for Hainaut, and a = 3.64 (± 0.95) and b = 1.38 (± 0.32) for Liège region.

Due to the limited data, the uncertainties on the *a* and *b* values are significant. In these regions, recent earthquakes have been destructive (at least locally) at a magnitude around 4.0, indicating the high vulnerability of many buildings, in part due to the low standards of maintenance and construction design. Hence, a stronger earthquake could be more destructive, indicating the necessity for a realistic evaluation of the maximum credible earthquake. Unfortunately, scientific evidence is lacking to estimate Mmax.

In similar cases, the Electric Power Research Institute (1994) recommended using expert experience and judgment to evaluate Mmax. They discussed six methods for determining the maximum credible earthquake for a given source: (1) addition of an increment to the largest historical earthquake; (2) extrapolation of the magnitude-frequency relationship; (3) use of the dimension of the possible seismogenic sources; (4) statistical approaches (application of extreme value theory and maximum likelihood techniques); (5) strain rate or moment release rate methods; and (6) reference to a global database.

Because the level of seismic activity is sufficient to determine the frequency-magnitude relationship, we calculated the magnitude (method 2) corresponding to a return period of 475 yr (the return period corresponding to the hazard requirement in the EUROCODE-8 building code). In Hainaut, M_S (475 yr) = 5.4 ± 0.9, whereas in Liège, it is M_S = 4.6 ± 1.2. The large uncertainties on these results are due to the large uncertainty in the *a* and *b* values.

Method 1, addition of a magnitude increment, generally 0.5 or 1.0, artificially increases the return period by a factor of 3.4–11.7 in Hainaut and 4.9–24 in Liège. This increment gives Mmax values of 5.2–5.7 in Liège and 5.0–5.5 in Hainaut.

Method 3 requires a hypothesis on the seismogenic sources. Considering the maximal depth of 8 km in Hainaut and 6 km in Liège determined from well-located earthquakes, one hypothesis is to suppose that the seismogenic layer extends from the surface to these depths. We also consider that earthquakes rupturing the whole seismogenic layer extend at the surface along faults with a minimal length equivalent to the maximal depths. The empirical relationship between magnitude and surface rupture length (Wells and Coppersmith, 1994), yields magnitudes of 6.0 and 5.9 for Hainaut and Liège, respectively.

Up to now, detailed statistical analysis (method 4) of the available database has not been done because the data are too sparse. Method 5 cannot be applied in our context because tectonic strain rates are unknown and earthquake moment release is insignificant. We consider also that using a global database (method 6) is not pertinent for considering zones with such small dimensions (40 km × 15 km).

Mmax values obtained using three methods ranged between 5.0 and 6.0 in Hainaut and 4.6 and 5.9 in Liège. Earthquakes of that size would have serious consequences. At present, there are no prevention measures against earthquakes in Belgium. Thus, adopting appropriate prevention measures to diminish the impact of destructive earthquakes would be a step in the right direction. Therefore, in these regions, it seems appropriate to consider a definition of the maximum credible earthquake depending on the purpose of the evaluation. Mmax could be defined as the magnitude of the largest possible earthquake expected with a specified probability during a specified exposure time, or of the largest earthquake considered likely to occur in a "reasonable" amount of time (e.g., lifetime of facility involved). A similar debate exists in the eastern United States concerning the seismic hazard evaluation in the New Madrid seismic zone (Stein, 2005).

DISCUSSION AND CONCLUSIONS

The Electric Power Research Institute (EPRI) provided in 1994 the first comprehensive compilation and analysis of earthquakes in stable continental regions worldwide. Since that time, our knowledge and understanding of long-term seismic activity in northwest Europe has increased significantly. We have focused on collection of historical data (Melville et al., 1996; Alexandre, 1997) about past earthquakes and investigations of the most recent activity of the Quaternary faults in the Lower Rhine Embayment and their relationship with paleoearthquakes. This reassessment of historical earthquakes demonstrates that earthquakes with magnitude equal to or greater than 6.0 have occurred since the fourteenth century in the region extending from the Lower Rhine Embayment to the southern North Sea. The most significant advance is evidence that large surface-rupturing earthquakes have occurred along the border faults of the Roer graben during the Holocene and late Pleistocene. These results illustrate the relevance of active faulting studies in assessments of long-term seismic activity in this part of northwest Europe. In this paper, we discussed field observations that support the validity of our assumption. Evidence of strong past earthquakes has also been found recently by Hinzen and Schütte (2002) and Hinzen (2005) in Roman archaeological vestiges in Cologne and Tolbiacum in Germany.

These results imply that long-term seismic activity in northwest Europe can be clarified. Tools to identify seismogenic sources exist, but some seismogenic sources will escape from our catalogue. Therefore, seismic hazard assessment in intraplate regions will always depend on expert judgment based on a largely incomplete scientific knowledge base. Better understanding of the present tectonic deformation and its relationship with seismic activity requires long-term geodetic measurements and more appropriate geological investigations. Up to now, geodetic techniques are at the limit of their resolution to identify the low strain rates in the Lower Rhine Embayment, but it can be expected that GPS and absolute gravity measurements will provide conclusive results in the next two decades. After investigating the Roer graben during the last ten years, we also started investigations in other parts of northwest Europe to identify possible seismogenic active faults.

ACKNOWLEDGMENTS

This study resulted from scientific investigations conducted in the framework of two European Commission projects, PALEOSIS (ENV4-CT97-0578) and SAFE (EVG1-CT-2000-00023), and three projects of the Belgian Federal Public Planning Service Science Policy, MO/33/006, MO/33/010, and MO/33/011. We are grateful to Mustapha Meghraoui for the stimulating discussions we had on earthquake geology and the fruitful field work along the Bree fault scarp from 1995 to 1997. We thank Simon D.P. Williams for applying the maximum likelihood estimation technique to constrain the rate of absolute gravity changes. Many thanks are due to our colleagues B. Bukasa, S. Castelein, F. Collin, F. De Vos, M. Hendrickx, H. Martin, G. Rapagnani, W. Vandeputte, and the late M. Snissaert for their continuous help during our research work. We thank also Seth Stein and the two anonymous reviewers for helping us to improve our text.

REFERENCES CITED

Ahorner, L., 1968, Erdbeben und jüngste Tektonik im Braunkohlenrevier der Niederrheinischen Bucht: Zeitschrift der Deutschen Geologischen Gesellschaft, v. 118, p. 150–160.

Ahorner, L., 1975, Present-day stress field and seismotectonic block movements along major fault zones in central Europe: Tectonophysics, v. 29, p. 233–249, doi: 10.1016/0040-1951(75)90148-1.

Ahorner, L., 1983, Historical seismicity and present-day activity of the Rhenish Massif, central Europe, *in* Fuchs, K., von Gehlen, K., Mälzer, H., Murawski, H., and Semmel, A., eds., Plateau Uplift, the Rhenish Shield—A Case History: Berlin, Springer-Verlag, p. 198–221.

Ahorner, L., 1985, The general pattern of seismotectonic dislocations in central Europe as the background for the Liège earthquake on November 8, 1983: Seismic activity, *in* Melchior, P., ed., Western Europe: Dordrecht, D. Reidel Publishing Company, p. 41–56.

Ahorner, L., 1996, How reliable are speculations about large paleo-earthquakes at the western border fault of the Roer valley graben near Bree, *in* Bonatz, M., ed., Comptes-Rendus des 81ièmes: Journées Luxembourgeoises de Géodynamique, Walferdange, Grand Duchy of Luxemburg, p. 39–57.

Ahorner, L., 1998, Möglichkeiten und Grenzen paläoseismologischer Forschung in mitteleuropäischen Erdbebengebieten, *in* Savidis, S.A., ed., Publication of the Deutschen Gesellschaft für Erdbeben-Ingenieurwesen und Baudynamik (DGEB), 9, Paläoseismologie, Eurocode 8 und Schwingungsisolierung: Berlin, DGEB, p. 9–42.

Alexandre, P., 1990, Les séismes en Europe Occidentale de 324 à 1259: Nouveau catalogue critique: Bruxelles, Série Géophysique, Observatoire Royal de Belgique, n° hors série, 267 p.

Alexandre, P., 1997, Le tremblement de terre de 1692: Feuillets de la Cathédrale de Liège, v. 28, no. (3)2, p. 3–19.

Alexandre, P., and Vogt, J., 1994, La crise séismique de 1755–1762 en Europe du nord-ouest, *in* Albini, P., and Moroni, A., eds., Historical Investigation of European Earthquakes, Materials of the CEC Project: Review of Historical Seismicity in Europe: Milano, Consiglio Nazionale delle Ricerche, v. 2, p. 37–76.

Ambraseys, N.N., 1985a, Magnitude assessment of northwestern European earthquakes: Earthquake Engineering & Structural Dynamics, v. 13, p. 307–320.

Ambraseys, N.N., 1985b, Intensity-attenuation and magnitude-intensity relationships for northwest European earthquakes: Earthquake Engineering & Structural Dynamics, v. 13, p. 733–778.

Ambraseys, N.N., and Melville, C., 1983, Seismicity of the British Isles and the North Sea: London, Marine Technology Centre, 132 p.

Auffret, J.-P., and Colbeaux, J.-P., 1977, Étude structurale du Boulonnais et son prolongement sous-marin en manche orientale: Bulletin de la Société Géologique de France, v. 19, p. 1047–1055.

Beerten, K., Brabers, P., Bosch, P., and Gullentops, F., 1999, The passage of the Feldbiss Bundle through the Maas Valley: Aardkundige Mededelingen, v. 9, p. 153–158.

Camelbeeck, T., 1993, Mécanisme au foyer des tremblements de terre et contraintes tectoniques: Le cas de la zone intraplaque Belge [Ph.D. thesis]: Louvain-la-Neuve, Université Catholique de Louvain, 295 p.

Camelbeeck, T., and Meghraoui, M., 1996, Large earthquakes in northern Europe more likely than once thought: Eos (Transactions, American Geophysical Union), v. 77, p. 405–409, doi: 10.1029/96EO00274.

Camelbeeck, T., and Meghraoui, M., 1998, Geological and geophysical evidence for large palaeoearthquakes with surface faulting in the Roer graben (northwestern Europe): Geophysical Journal International, v. 132, p. 347–362, doi: 10.1046/j.1365-246x.1998.00428.x.

Camelbeeck, T., and van Eck, T., 1994, The Roer valley graben earthquake in northern Europe of 13 April 1992 and its seismotectonic setting: Terra Nova, v. 6, p. 291–300.

Camelbeeck, T., Vanneste, K., and Alexandre, P., 1999, L'Europe Occidentale n'est pas à l'abri d'un grand tremblement de terre: Ciel et Terre, v. 115, p. 13–23.

Camelbeeck, T., Martin, H., Vanneste, K., Verbeeck, K., and Meghraoui, M., 2001, Morphometric analysis of active normal faulting in slow-deformation areas: Examples in the Lower Rhine Embayment: Netherlands Journal of Geosciences, v. 80, p. 95–107.

Camelbeeck, T., Van Camp, M., Jongmans, D., Francis, O., and van Dam, T., 2002, Nature of the recent vertical ground movements inferred from high-precision levelling data in an intraplate setting: NE Ardenne, Belgium:

Comment: Journal of Geophysical Research, v. 107, p. 2281–2287, doi: 10.1029/2001JB000397.

Colbeaux, J.-P., Beugnies, A., Dupuis, C., Robaszynski, F., and Sommé, J., 1977, Tectonique de blocs dans le sud de la Belgique et le nord de la France: Annales de la Société Géologique du Nord, v. XCVII, p. 191–222.

Colbeaux, J.-P., Leplat, J., Paepe, R., and Sommé, J., 1979, Tectonique récente dans le nord de la France et le sud de la Belgique: Exemple de la plaine de la Lys (Feuille d'Hazebrouck à 1/50000): Annales de la Société Géologique du Nord, v. 97, p. 179–188.

Colbeaux, J.-P., Sommé, J., and Tuffreau, A., 1981, Tectonique Quaternaire dans le nord de la France: L'apport du gisement paléolithique de Biache-Saint-Vaast: Bulletin de l'Association Française pour l'Étude du Quaternaire, v. 3, p. 183–192.

Colbeaux, J.-P., Amedro, F., Bergerat, F., Bracq, P., Crampon, N., Delay, F., Dupuis, C., Lamouroux, C., Robaszynski, F., Sommé, J., Vandycke, S., and Vidier, J.-P., 1993, Un enregistreur des épisodes tectoniques dans le bassin de Paris: Le Boulonnais: Bulletin de la Société Géologique de France, v. 164, p. 63–102.

Crone, A.J., Machette, M.N., and Bowman, J.R., 1997, The episodic nature of earthquakes in the stable interior of continents as revealed by paleoseismicity studies of Australian and North American Quaternary faults: Australian Journal of Earth Sciences, v. 44, no. 1, p. 203–214.

Debacker, T.N., Sintubin, M., and Verniers, J., 2002, Timing and duration of the progressive deformation of the Brabant Massif (Belgium): Aardkundige Mededelingen, v. 12, p. 73–76.

de Béthune, P., and Bouckaert, L., 1968, Geological Map of Belgium and Neighbouring Countries: Brussels, Geological survey of Belgium Miscellaneous Geological Maps, scale 1:2,000,000, 1 sheet.

Demoulin, A., ed., 1995, L'Ardenne-Essai de Géographie Physique: Liège, Publication de Département de Géographie Physique et Quaternaire, Université de Liège, 138 p.

Demyttenaere, R., 1989, The post-Paleozoic geological history of north-eastern Belgium: Mededelingen van de Koninklijke Acadademie voor Wetenschappen, Letteren en Schone Kunsten van België: Klasse der Wetenschappen, v. 51, p. 51–80.

De Vos, W., Verniers, J., Herbosch, A., and Vanguestaine, M., 1993, A new geological map of the Brabant Massif, Belgium: Geological Magazine, v. 130, p. 605–611.

Dost, B., and Haak, H.W., 2007, Natural and induced seismicity, *in* Wong, Th.E., Batjes, D.A.J., and De Jager, J., eds., Geology of the Netherlands: Amsterdam, Royal Netherlands Academy of Arts and Sciences, p. 219–235.

Electric Power Research Institute (EPRI), 1994, The Earthquakes of Stable Continental Regions. Volume 1: Assessment of Large Earthquake Potential: Electric Power Research Institute Open-File Report TR-102261s-V1-V5, p. 362.

Englert, S., 1995, Mécanisme au Foyer des Tremblements de Terre de la Sequence de Dour (1987): Apport à la Séismotectonique de la Zone du Hainaut [mémoire de fin d'études]: Bruxelles, Université Libre de Bruxelles, Département des Sciences de la Terre et de l'Environnement, 63 p.

Everaerts, M., 2000, L'interprétation Structurale de la Manche au Rhin: Apport Structural des Champs de Potentiel [Ph.D. thesis]: Louvain-la-Neuve, Université Catholique de Louvain, 167 p.

Everaerts, M., and Mansy, J.-L., 2001, Le filtrage des anomalies gravimétriques; une clé pour la compréhension des structures tectoniques du Boulonnais et de l'Artois (France): Bulletin de la Société Géologique de France, v. 172, p. 267–274, doi: 10.2113/172.3.267.

Everaerts, M., Poitevin, C., De Vos, W., and Sterpin, M., 1996, Integrated geophysical/geological modelling of the western Brabant Massif and structural implications: Bulletin de la Société Belge de Geologie, v. 105, no. 1–2, p. 41–59.

Fourniguet, J., 1987, Géodynamique Actuelle dans le Nord et le Nord-Est de la France: Bureau de Recherches Géologiques et Minières Open-File Report 127, 160 p.

Francis, O., Van Camp, M., van Dam, T., Warnant, R., and Hendrickx, M., 2004, Indication of the uplift of the Ardenne in long term gravity variations in Membach (Belgium): Geophysical Journal International, v. 158, p. 346–352, doi: 10.1111/j.1365-246X.2004.02310.x.

Frechen, M., Vanneste, K., Verbeeck, K., Paulissen, E., and Camelbeeck, T., 2001, The deposition history of the coversands along the Bree fault escarpment, NE Belgium: Netherlands Journal of Geosciences: Geologie en Mijnbouw, v. 80, no. 3–4, p. 171–185.

Fuchs, K., von Gehlen, K., Mälzer, H., Murawski, H., and Semmel, A., 1983, Plateau Uplift, the Rhenish Shield—A Case History: Berlin, Springer-Verlag, 411 p.

Gasperini, P., Bernardini, F., Valensisi, G., and Boschi, E., 1999, Defining seismogenic sources from historical earthquake felt reports: Geological Society of America Bulletin, v. 89, no. 1, p. 94–110.

Geluk, M.C., Duin, E.J.T.H., Dusar, M., Rijkers, M.H.B., van den Berg, M.W., and van Rooijen, P., 1994, Stratigraphy and tectonics of the Roer valley graben: Geologie en Mijnbouw, v. 73, p. 129–141.

Gregersen, S., and Basham, P.-W., eds., 1989, Earthquakes at North Atlantic Passive Margins: Neotectonics and Postglacial Rebound: Dordrecht, Kluwer Academic Publishers, 716 p.

Hennebert, M., 1998, L'anticlinal faillé du Mélantois-Tournaisis fait partie d'une "structure en fleur positive" tardi-varisque: Annales de la Société Géologique du Nord, v. 6, p. 65–78.

Hinzen, K.-G., 2005, The use of engineering seismological models to interpret archaeoseismological findings in Tolbiacum, Germany, a case study: Bulletin of the Seismological Society of America, v. 95, p. 521–539, doi: 10.1785/0120040068.

Hinzen, K.-G., and Oemisch, M., 2001, Location and magnitude from seismic intensity data of recent and historic earthquakes in the northern Rhine area, central Europe: Bulletin of the Seismological Society of America, v. 91, p. 40–56, doi: 10.1785/0120000036.

Hinzen, K.-G., and Schütte, S., 2002, Evidence for earthquake damage on Roman buildings in Cologne, Germany: Seismological Research Letters, v. 74, p. 121–137.

Houtgast, R.F., Balen, R.T.V., Brouwer, L.M., Brand, G.B.M., and Brijker, J.M., 2002, Late Quaternary activity of the Feldbiss fault zone, Roer valley rift system, the Netherlands, based on displaced fluvial terrace fragments: Tectonophysics, v. 352, p. 295–315, doi: 10.1016/S0040-1951(02)00219-6.

Houtgast, R.F., Balen, R.T.V., Kasse, K., and Vandenberghe, J., 2003, Late Quaternary tectonic evolution and postseismic near surface fault displacements along the Geleen fault (Feldbiss fault zone–Roer valley rift system, the Netherlands), based on trenching: Geologie en Mijnbouw, v. 82, no. 2, p. 177–196.

Houtgast, R.F., Balen, R.T.V., and Kasse, C., 2005, Late Quaternary evolution of the Feldbiss fault (Roer valley rift system, the Netherlands) based on trenching, and its potential relation to glacial unloading: Quaternary Science Reviews, v. 24, no. 3–4, p. 489–508, doi: 10.1016/j.quascirev.2004.01.012.

Illies, J.H., 1981, Mechanism of graben formation: Tectonophysics, v. 73, p. 249–266, doi: 10.1016/0040-1951(81)90186-4.

Johnston, A.C., 1996, Seismic moment assessment of earthquakes in stable continental regions: II. Historical seismicity: Geophysical Journal International, v. 125, p. 639–678.

Karnik, V., 1969, Seismicity of the European Area (2 volumes): Dordrecht, Reidel Publishing Company, 364 p.

Kolstrup, E., 1980, Climate and stratigraphy in northwestern Europe between 30,000 B.P. and 13,000 B.P., with special reference to the Netherlands: Mededelingen Rijks Geologische Dienst, v. 32, no. 15, p. 181–253.

Lambert, J., and Levret-Albaret, A., 1996, Mille ans de Séismes en France: Catalogue d'Épicentres—Paramètres et Références: Nantes, Ouest Editions, Presses Académiques, 80 p.

Lancaster, M.A., 1896, Le tremblement de terre du 2 Septembre 1896: Ciel et Terre, v. 14, p. 411–422.

Legrand, L., 1968, Le Massif du Brabant: Geological Survey of Belgium Open-File Report 9, 148 p.

Mälzer, H., Hein, G., and Zippelt, K., 1983, Height changes in the Rhenish Massif: Determination and analysis, in Fuchs, K., et al., eds., Plateau Uplift: Berlin, Springer-Verlag, p. 164–176.

Mansy, J.-L., Manby, G.M., Averbuch, O., Everaerts, M., Bergerat, F., Van Vliet-Lanoë, B., Lamarche, J., and Vandycke, S., 2003, Dynamics and inversion of the Mesozoic Basin of the Weald-Boulonnais area: Role of basement reactivation: Tectonophysics, v. 373, p. 161–179, doi: 10.1016/S0040-1951(03)00289-0.

McCalpin, J.P., ed., 1996, Paleoseismology: San Diego, Academic Press, 583 p.

Meghraoui, M., Camelbeeck, T., Vanneste, K., Brondeel, M., and Jongmans, D., 2000, Active faulting and paleoseismology along the Bree fault zone, Lower Rhine graben (Belgium): Journal of Geophysical Research, v. 105, p. 13,809–13,841, doi: 10.1029/1999JB900236.

Melchior, P., ed., 1985, Seismic Activity in Western Europe with Particular Consideration of the Liège Earthquake of November 8, 1983: Dordrecht, D. Reidel Publishing, NATO ASI Series, v. 144, 448 p.

Melville, C., Levret, A., Alexandre, P., Lambert, J., and Vogt, J., 1996, Historical seismicity of the Strait of Dover–Pas de Calais: Terra Nova, v. 8, p. 626–647.

Milne, G.A., Davis, J.L., Mitrovica, J.X., Scherneck, H.-G., Johansson, J.M., Vermeer, M., and Koivula, H., 2001, Space-geodetic constraints on glacial isostatic adjustment in Fennoscandia: Science, v. 291, p. 2381–2385, doi: 10.1126/science.1057022.

Morse, T., 1983, "How near we were to ruine": The effects in England of the earthquake of 8th September 1692: Disasters, v. 7, p. 272–282.

Muir Wood, R., 1989, Extra-ordinary deglaciation reverse faulting in northern Fennoscandia, in Gregersen, S., and Basham, P.-W., eds., Earthquakes at North Atlantic Passive Margins: Neotectonics and Postglacial Rebound: Dordrecht, Kluwer Academic Publishers, Series C: Mathematical and Physical Sciences, v. 266, p. 141–173.

Nguyen, F., Teerlynck, H., Van Rompaey, G., Van Camp, M., Jongmans, D., and Camelbeeck, T., 2004, Use of microtremor measurement for assessing site effects in northern Belgium—Interpretation of the observed intensity during the Ms = 5.0 June 11, 1938 earthquake: Journal of Seismology, v. 8(1), p. 41–56.

Nicolas, M., Santoire, J.-P., and Delpech, P.-Y., 1990, Intraplate seismicity: New seismotectonic data in Western Europe: Tectonophysics, v. 179, p. 27–53, doi: 10.1016/0040-1951(90)90354-B.

Nocquet, J.-M., and Calais, E., 2003, Crustal velocity field of Western Europe from permanent GPS array solutions 1996–2001: Geophysical Journal International, v. 154, p. 72–88, doi: 10.1046/j.1365-246X.2003.01935.x.

Nocquet, J.-M., and Calais, E., 2004, Geodetic measurements of crustal deformation in the Western Mediterranean and Europe: Pure and Applied Geophysics, v. 161, p. 661–681, doi: 10.1007/s00024-003-2468-z.

Owen, G., 1987, Deformation processes in unconsolidated sands, in Jones, M.E., and Preston, R.M.F., eds., Deformation of Sediments and Sedimentary Rocks: Geological Society [London] Special Publication 29, p. 11–24.

Paulissen, E., Vandenberghe, J., and Gullentops, F., 1985, The Feldbiss fault in the Maas valley bottom (Limburg, Belgium): Geologie en Mijnbouw, v. 64, p. 79–87.

Pissart, A., and Lambot, J., 1989, Les mouvements actuels du sol en Belgique; comparaisons de deux nivellements IGN (1946–1948 et 1976–1980): Annales de la Société Géologique de Belgique, v. 112, no. 2, p. 495–504.

Quitzow, H.W., and Vahlensieck, O., 1955, Über Pleistozäne Gebirgsbildung und rezente Krustenbewegungen in der Niederrheinischen Bucht: Geologische Rundschau, v. 43, p. 56–67, doi: 10.1007/BF01764085.

Reamer, S.K., and Hinzen, K.-G., 2004, An earthquake catalog for the northern Rhine area, central Europe (1975–2002): Seismological Research Letters, v. 75, p. 713–725.

Ritter, J.R.R., Jordan, M., Christensen, U.R., and Achauer, U., 2001, A mantle plume below the Eifel volcanic fields, Germany: Earth and Planetary Science Letters, v. 186, p. 7–14, doi: 10.1016/S0012-821X(01)00226-6.

Schaefer, W., 1999, Bodenbewegungen und Bergschadensregulierung im Rhenischen Braunkohlenrevier, in Proceedings of the 42, Deutscher Markscheider-Verein-Tagung: Cottbus, September 1999, p. 1–10.

Scholz, C.H., 1990, The Mechanics of Earthquakes and Faulting: Cambridge, Cambridge University Press, 471 p.

Sommé, J., 1967, Tectonique récente dans la région de Lille: Pays de Weppes et Mélantois occidental: Revue de Géomorphologie Dynamique, v. 17, p. 55–65.

Stein, S., 2005, Comment on "How Can Seismic Hazard in the New Madrid Seismic Zone Be Similar to that in California": Seismological Research Letters, v. 76, no. 3, p. 364–365.

Valensise, G., and Pantosti, D., eds., 2001, Database of potential sources for earthquakes larger than M 5.5 in Italy, version 2.0: Annali di Geofisica, v. 44, no. 3, p. 797–964.

Van Camp, M., Camelbeeck, T., and Francis, O., 2002, An experiment to evaluate crustal motions across the Ardenne and the Roer graben (northwestern Europe) using absolute gravity measurements: Metrologia, v. 39, p. 503–508, doi: 10.1088/0026-1394/39/5/12.

Van Camp, M., Williams, S.D.P., and Francis, O., 2005, Precision and accuracy of absolute gravity measurements: Journal of Geophysical Research, v. 110, B05406, doi: 10.1029/2004JB003497.

van den Berg, M.W., Groenewoud, W., Lorenz, G.K., Lubbers, P.J., Brus, D.J., and Kroonenberg, S.B., 1994, Patterns and velocities of recent crustal movements in the Dutch part of the Roer valley rift system: Geologie en Mijnbouw, v. 73, no. 2–4, p. 157–168.

van den Berg, M.W., Vanneste, K., Dost, B., Lokhorst, A., Eijk, M.V., and Verbeeck, K., 2002, Paleoseismic investigations along the Peelrand fault: Geological setting, site selection and trenching results: Geologie en Mijnbouw, v. 81, no. 1, p. 39–60.

Vandenberghe, J., 1985, Paleoenvironment and stratigraphy during the last glacial in the Belgian-Dutch border region: Quaternary Research, v. 24, p. 23–38, doi: 10.1016/0033-5894(85)90081-X.

van Eck, T., and Davenport, C.A., eds., 1994, Seismotectonics and seismic hazard in the Roer valley graben; with emphasis on the Roermond earthquake of April 13, 1992: Geologie en Mijnbouw, Special Issue, v. 73, no. 2–4, 441 p.

Vanneste, K., and Verbeeck, K., 2001, Paleoseismological analysis of the Rurrand fault near Jülich, Roer valley graben, Germany: Geologie en Mijnbouw, v. 80, p. 155–169.

Vanneste, K., De Batist, M., Demanet, D., Versteeg, W., Wislez, W., Jongmans, D., and Camelbeeck, T., 1997, High-resolution seismic reflection research of active faulting: Aardkundige Mededelingen, v. 8, p. 197–199.

Vanneste, K., Meghraoui, M., and Camelbeeck, T., 1999, Late Quaternary earthquake-related soft-sediment deformation along the Belgian portion of the Feldbiss fault, Lower Rhine graben system: Tectonophysics, v. 309, p. 57–79, doi: 10.1016/S0040-1951(99)00132-8.

Vanneste, K., Verbeeck, K., Camelbeeck, T., Renardy, F., Meghraoui, M., Jongmans, D., Paulissen, E., and Frechen, M., 2001, Surface rupturing history of the Bree fault escarpment, Roer valley graben: New trench evidence for at least six successive events during the last 150 to 185 kyr: Journal of Seismology, v. 5, p. 329–359, doi: 10.1023/A:1011419408419.

Vanneste, K., Verbeeck, K., and Camelbeeck, T., 2002, Exploring the Belgian Maas valley between Neeroeteren and Bichterweert for evidence of active faulting: Aardkundige Mededelingen, v. 12, p. 5–8.

Vanneste, K., Radulov, A., De Martini, P., Nikolov, G., Petermans, T., Verbeeck, K., Camelbeeck, T., Pantosti, D., Dimitrov, D., and Shanov, S., 2006, Paleoseismologic investigation of the fault that ruptured in the April 14, 1928, Chirpan earthquake (M 6.8), southern Bulgaria: Geophysical Journal International, v. 111, p. B01303.

Van Vliet-Lanoë, B., Vandenberghe, N., Laurent, M., Laignel, B., Lauriat-Rage, A., Louwye, S., Mansy, J.-L., Mercier, D., Hallégouët, B., Laga,

P., Laquement, F., Meilliez, F., Michel, Y., Moguedet, G., and Vidier, J.-P., 2002, Palaeogeographic evolution of northwestern Europe during the Upper Cenozoic: Geodiversitas, v. 24, p. 511–541.

Verniers, J., Pharaoh, T., André, L., Debacker, T.N., De Vos, W., Everaerts, M., Herbosch, A., Samuelsson, J., Sintubin, M., and Vecoli, M., 2002, The Cambrian to mid Devonian basin development and deformation history of Eastern Avalonia, east of the Midlands microcraton: New data and a review, *in* Winchester, J.A., Pharaoh, T.C., and Verniers, J., eds., Paleozoic Amalgamation of Central Europe: Geological Society [London] Special Publication 201, p. 47–93.

Vogt, J., 1984, Révisions des deux séismes majeurs de la région d'Aix-la-Chapelle-Verviers-Liège ressentis en France: 1504 1692, *in* Tremblements de Terre: Histoire et Archéologie: Valbonne, Actes du Colloque d'Antibes 2-4/11/1983, p. 13–21.

Waalewijn, A., 1961, Crustal movements in the Netherlands: Bulletin Geodesique, v. 62, p. 369–371.

Ward, S.N., 1994, Constraints on the seismotectonics of the central Mediterranean from very long baseline interferometry: Geophysical Journal International, v. 117, p. 441–452.

Ward, S.N., 1998, On the consistency of earthquake moment release and space geodetic strain rates: Europe: Geophysical Journal International, v. 135, p. 1011–1018, doi: 10.1046/j.1365-246X.1998.t01-2-00658.x.

Weichert, D.H., 1980, Estimation of the earthquake recurrence parameters for unequal observation periods for different magnitudes: Bulletin of the Seismological Society of America, v. 70, no. 4, p. 1337–1346.

Wells, D.L., and Coppersmith, K.J., 1994, Empirical relationships among magnitude, rupture length, rupture area and surface displacement: Bulletin of the Seismological Society of America, v. 84, p. 974–1002.

Williams, S.D.P., Baker, T.F., and Jeffries, G., 2001, Absolute gravity measurements at UK tide gauges: Geophysical Research Letters, v. 28, no. 12, p. 2317–2320, doi: 10.1029/2000GL012438.

Wouters, L., and Vandenberghe, N., 1994, Geologie van de Kempen: Een Synthese: Brussels, Belgian agency for radioactive waste and enriched fissile materials, 208 p.

Ziegler, P.A., 1994, Cenozoic rift system of western and central Europe: An overview: Geologie en Mijnbouw, v. 73, no. 2–4, p. 99–127.

MANUSCRIPT ACCEPTED BY THE SOCIETY 29 NOVEMBER 2006

The Geological Society of America
Special Paper 425
2007

Seismicity, seismotectonics, and seismic hazard in the northern Rhine area

Klaus-G. Hinzen[†]
Sharon K. Reamer
*Earthquake Observatory Bensberg, Department of Earthquake Geology, University of Cologne,
Vinzenz-Pallotti-Strasse 26, D-51429 Bergisch Gladbach, Germany*

ABSTRACT

The northern Rhine area covers an area of more than 40,000 km² and is one of the most important areas of earthquake recurrence in Europe north of the Alps. The Lower Rhine Embayment, a part of the northern Rhine area that extends over parts of western Germany, eastern Belgium, and the southern Netherlands, displays basin-like subsidence and, along with the Roer valley graben, has been the source of most of the historical and recent earthquake activity, particularly along the western border faults. Other important earthquake-prone areas include the Stavelot-Venn Massif and the Neuwied Basin, the latter of which is an area of periodically intensive microseismicity. Seismic instrumentation in the area has accumulated steadily since the early 1950s and presently consists of ~50 stations. Although the strongest instrumentally recorded earthquake in the region occurred in 1992 near Roermond, with a local magnitude of 6.0, studies of historical earthquakes, such as Vervier (1692) and Düren (1756), estimate macroseismic magnitudes of 6.8 and 6.4, respectively. Paleoseismic studies indicate that even stronger, surface-rupturing earthquakes have occurred during the Holocene. The recent M_L 4.9 earthquake in Alsdorf in 2002 caused some structural damage and was preceded by several small earthquakes and was followed by several aftershocks. Seismic hazard evaluation is hampered by the sparse earthquake record. A hybridized instrumental and historical earthquake catalog compiled from events over the past 300 yr combined with seismotectonic aspects indicates a maximum magnitude of 7.0. A site-intensity map based on macroseismic intensities from the hybridized catalog identifies concentrations of recent activities in the western part of the Lower Rhine Embayment, east of the city of Aachen.

Keywords: Lower Rhine Embayment, northern Rhine area, seismicity, seismotectonic.

INTRODUCTION

The term "Rhineland" is usually associated with the area along the rivers Rhine and Mosel north of the city of Frankfurt am Main to the border between Germany and the Netherlands.

Officially, however, "Rhineland" denotes the former Prussian province. The area of interest for the purpose of this discussion is shown in Figure 1 and will be referred to as the northern Rhine area.

Since the earliest days of scientific interest in earthquakes, the intraplate seismicity of the northern Rhine area has been researched. Astronomer Johann F.J. Schmidt developed an inter-

[†]E-mail: hinzen@uni-koeln.de.

Hinzen, K.-G., and Reamer, S.K., 2007, Seismicity, seismotectonics, and seismic hazard in the northern Rhine area, *in* Stein, S., and Mazzotti, S., ed., Continental Intraplate Earthquakes: Science, Hazard, and Policy Issues: Geological Society of America Special Paper 425, p. 225–242, doi: 10.1130/2007.2425(15). For permission to copy, contact editing@geosociety.org. ©2007 The Geological Society of America. All rights reserved.

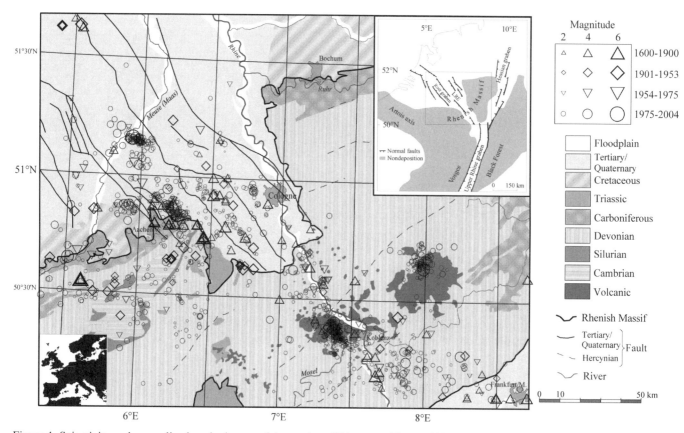

Figure 1. Seismicity and generalized geologic map of the northern Rhine area. The small inset at bottom left shows the location in Europe. The inset in the upper-right corner indicates the main tectonic elements of the northern part of the Rhine-Rhone rift system. Different symbols for the epicenters relate to the time period of the earthquakes; size of the symbol varies with magnitude (see legend). LRE—Lower Rhine Embayment.

est in earthquakes and used a damaging earthquake near St. Goar in 1846 (Meidow, 1995) to calculate seismic velocities. His monograph "Studien über Erdbeben" (Studies about Earthquakes) introduced the term epicenter (Schmidt, 1879; Yeats et al., 1997). As early as 1906, a seismograph station was established in a mining school in Aachen (Fig. 2). Former station AAC was equipped with a 1.3 t Wiechert pendulum, which remained in operation until WWII, with recording gaps during WWI. Nearly a decade after the war, catalyzed by the 1951 M_L 5.1 earthquake near the town of Euskirchen (Fig. 2), a new measuring system was adopted by M. Schwarzbach, founder of the Bensberg seismographic station (BNS) of Cologne University (Robel, 1959). In the 1970s, the station gradually evolved into a short period network. Currently, more than 50 stations, short period and broadband, are operated by universities, governmental agencies, and observatories in Belgium, Germany, Luxemburg, and the Netherlands (Fig. 2).

In this paper, we summarize the instrumental and historical seismicity in the northern Rhine area, discuss the seismotectonic framework, analyze selected significant earthquakes and earthquake sequences, and present a preliminary overview of the seismic hazard in this region.

GEOLOGY AND TECTONIC SETTING OF THE NORTHERN RHINE AREA

The northern Rhine area is part of the Rhine-Rhone rift system (Ziegler, 1992, 1994, 2001; Van den Berg, 1994). Figure 1 shows the main tectonic features of the rift system north of the Alps: the Upper Rhine graben in the south between the Vosges Mountains and the Black Forest, the Hessian graben, and the Lower Rhine Embayment, with the Roer valley graben in the northern Rhine area. The current period of tectonic movements in the Lower Rhine Embayment can be closely correlated with late Tertiary graben structures and was initiated by small but widely distributed fault displacements in the late Miocene. During the Pliocene, faulting was more intense and cumulated in the late Pliocene and early Pleistocene. Considerable synsedimentary and intersedimentary crustal displacements occurred in the Quaternary when the older and younger main terraces of the Rhine and Meuse (Maas) Rivers were accumulated (Ahorner, 1962a). The Upper Rhine graben and the Lower Rhine Embayment have shown moderate seismicity in present and historic times, whereas the Hessian graben is mostly aseismic.

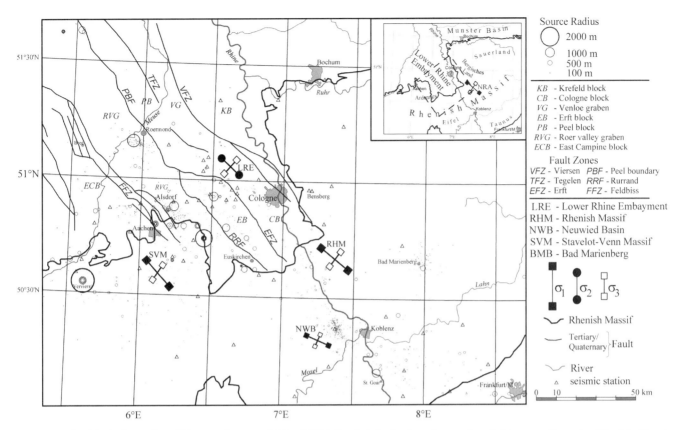

Figure 2. Seismicity in the northern Rhine area between 1600 and 2004. Symbol size for the epicenters is scaled to source dimension (Brudzinski and Chen, 2003); however, it is different from the map scale as shown in the legend. Seismic stations that were in operation between 1975 and present are shown as triangles. The trends of (sub-)horizontal components of the principal stress from Hinzen (2003a) are shown for the whole northern Rhine area in the inset, and for the subregions Lower Rhine Embayment (LRE), Rhenish Massif (RHM), Neuwied Basin (NWB), and Stavelot-Venn Massif (SVM). The dashed line in the inset indicates the location of the block shown in Figure 3, and gray circles show smaller cities.

The present geological structural activity in the Lower Rhine Embayment is characterized by regional flexures, tilted blocks, basin-like subsidence (Fig. 3), and frequent normal faulting occurring subvertically, at least at shallow depth. The main Lower Rhine Embayment faults, with a cumulative length of more than 400 km, generally strike NW-SE (Fig. 2). Some faults active in the Quaternary express a maximum vertical displacement of 174 m. The structural features are expressions of a regional crustal tension in a NE-SW direction, confirmed by stress inversions of fault-plane solutions (Hinzen, 2003a). The total crustal extension in Quaternary times is estimated at 90–180 m (Ahorner, 1962a). Additionally, the structural shape of the northern Rhine area as a whole indicates an uplift and tilting of the northwestern part of Germany toward the northwest. The volcanic Siebengebirge (Fig. 3) is located in the southeast corner of the Lower Rhine Embayment.

Just south of the Lower Rhine Embayment, there lies the Eifel mountain volcanic regime, located in the middle of the European plate. The two parallel Quaternary volcanic zones, the East and West Eifel link the Lower Rhine Embayment to the Upper Rhine graben and the Alpine collisional belt. In this narrow seismoactive zone, the presence of ejected xenoliths indicates upper-mantle

shearing (Regenauer-Lieb, 1999). Important CO_2-dominated mantle degassing observed in mineral springs, lakes, or dry degassing suggests a currently active flushing of brines (Regenauer-Lieb, 1999). Uplift of the central Eifel mountain region is estimated at 200–250 cm over the past 800,000 yr (Meyer and Stets, 1998) and reportedly exceeded a rate of 1 mm/yr in Pleistocene to recent times (Meyer and Stets, 2002). In the west of the Rhenish Massif/ Ardennen, the uplift has been on the order of 15–35 cm during the past 800,000 yr (Van Balen et al., 2000). Garcia-Castellanos et al. (2000) showed that the large-scale pattern of the late Quaternary uplift of the Ardennes–Rhenish Massif region can be explained in terms of a flexural response of the lithosphere to a buoyant sub-crustal load. Ritter et al. (2001) inferred the existence of a plume structure in the upper mantle between ~40 and 400 km depth by traveltime and Q-tomography.

The Roer Valley Rift System

Present-day seismicity in the Lower Rhine Embayment is mainly concentrated in the Roer valley graben, which includes parts of the southern Netherlands, eastern Belgium, and western Germany. The Lower Rhine Embayment is flanked on the east

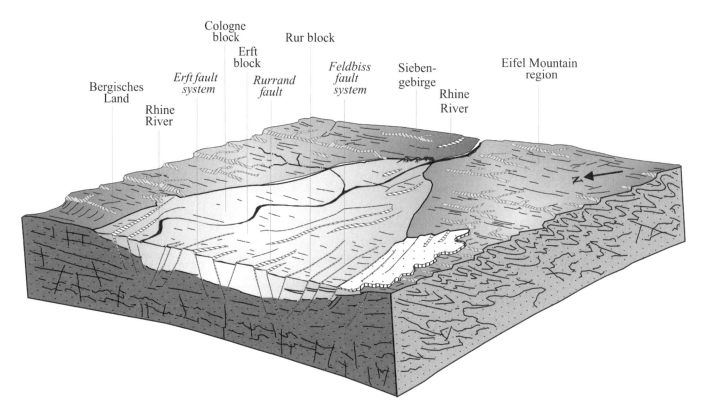

Figure 3. Perspective view of the Lower Rhine Embayment and basic geologic structure. Dark-gray colors indicate the folded Devonian base rock, lighter colors indicate Tertiary and Quaternary sediments, and the light dotted section is Triassic. East-west extension is roughly 90 km, and scale varies with perspective. The location of the block is indicated in the inset of Figure 2. The blocks of the basin are indicated only schematically because the depth extension (not to scale) of the faults is not known in detail (after Schwarzbach, 1951).

by the Bergisches Land (Fig. 2) as part of the Hercynic Rhenish Massif, a series of predominantly Devonian shales, sandstones, and limestones (Figs. 1 and 2). These deposits were strongly folded during the Hercynian orogeny and have undergone uplift of ~300 m since the Pleistocene (Illies et al., 1979).

The eastern tectonic elements of the Lower Rhine Embayment are the Krefeld block in the north and Cologne block in the south. The Venlo graben lies to the west of the Krefeld block and is separated from it by the Viersen fault zone (Fig. 2). The Tegelen fault zone separates the Peel block (also named Peel horst) from the Venlo graben. In the southern part of the Lower Rhine Embayment, the Erft fault zone separates the Cologne block from the Erft block. The Rurrand fault, which passes northward into the Peel boundary fault zone, is the northeastern border fault of the Roer valley graben. While this fault system is relatively narrow at shallow depths, the southwestern border is composed of the much wider and more complex Feldbiss fault zone. To the north, the rift system extends into the Central and West Netherlands Basins, which comprise the southern end of the North Sea Basin.

Recent studies in the Netherlands (Cohen et al., 2002) have shown that fluvial deposits of late Weichselian and Holocene Rhine and Meuse tributaries are vertically displaced along the northern shoulder of the Roer valley graben system. Displace-

ments in the top of the pleniglacial terrace along the Peel boundary fault extend upward to 1.4 m. The maximum displacement between the Peel horst and the Roer valley graben is 2.3 m, and the relative tectonic movement rates averaged over the last 15,000 yr are 0.09 to 0.15 mm/yr (Cohen et al., 2002). The Erft fault zone shows a cumulative vertical movement since the late Tertiary of 900–1000 m, with an average rate between 0.05 and 0.1 mm/yr (Ahorner, 2001). Precision tilt measurements (Kümpel at al., 2001) show episodic step-like tilt anomalies with amplitudes up to 22 mrad close to the Erft fault zone. These observations could reflect creep-like postseismic movements on the active fault. Vanneste and Verbeeck (2001) estimated an average displacement rate of 0.05–0.2 mm/yr on the Rurrand fault over the past 43,000 yr from analysis of a paleoseismic trench.

SEISMICITY OF THE NORTHERN RHINE AREA

Most of the stronger historic and instrumental earthquakes with magnitudes above 4 are connected to recent movements on the border faults of the Roer valley graben. The earliest published studies of earthquakes in the northern Rhine area date from the early nineteenth century. Nöggerath (1828, 1847, 1870) discussed earthquakes in 1828, 1846, 1868, and 1869, three of which caused damage. Lersch (1897), a spa doctor in Aachen, composed a 7000 page handwritten earthquake chronicle spanning 2362 B.C. to 1897 A.D. A comprehensive catalog of historic earthquakes including the northern Rhine area was first compiled by Sieberg (1940), originally from Aachen, and was later extended by Sponheuer (1952). Early catalogs for Belgium and the Netherlands were compiled by Charlier (1951) and van Rummelen (1945). Ahorner (1968, 1970, 1975) included data from early instrumental records and later summarized the seismicity of the Rhenish Massif (Ahorner, 1983a), and Ahorner and Pelzing (1983) studied source parameters of events between 1980 and 1981.

Intensely felt and damaging earthquakes of the northern Rhine area have been extensively studied (i.e., Gutenberg and Landsberg, 1930; Robel, 1958; Robel and Ahorner, 1958; Ahorner, 1962b, 1963; Ahorner and Pelzing, 1985; Camelbeeck et al., 1994; Ahorner, 1994; Pelzing, 1994; Scherbaum, 1994; Hinzen, 2003b). A contemporary catalog of instrumental seismicity since introduction of seismic networks in the northern Rhine area is given by Reamer and Hinzen (2004). The historical earthquakes in this study (from the year 1600) are taken from the historic catalog of Cologne University, which is to a large extent similar to the German earthquake catalogue from Leydecker (2004) (starting with the year 800). Figure 1 shows a current map of the 44,000 km^2 area of seismicity in the northern Rhine area from 1600 to 2004. Different symbols for the epicenters indicate the time period (and database) for the events.

The stress field of the northern Rhine area has been the target of several studies. Ahorner (1975) discussed the stress field and block movements along major faults zones in central Europe. From the P axes of earthquake focal mechanisms, he deduced an average direction of N142°E for the maximum horizontal stress for central Europe. A recent study (Hinzen, 2003a) presented stress tensor inversions for the northern Rhine area as a whole and for several subregions based on focal mechanisms of 110 earthquakes with magnitudes from $1 \leq M_L \leq 6$. Inversion of data from the whole northern Rhine area shows some ambiguity between a strike-slip and extensional stress regime, with a vertical axis for the medium principal stress and a trend of N305°E and N35°E for the largest and smallest principal stress directions, respectively. Results of the inversion of earthquake subsets with epicenters within the Lower Rhine Embayment, Hercynic Rhenish Massif, Neuwied Basin and vicinity, and Stavelot-Venn Massif (SVM) are shown in Figure 2.

Figure 4 shows the cumulative number of earthquakes in the northern Rhine area per year based on the duration and completeness of the catalog as shown in Figure 5 and the total area of 73,700 km^2 (see caption of Fig. 4). A least-squares fit using a Gutenberg-Richter model for the magnitude range from $1.8 \leq M_W \leq 5.2$ yields:

$$\log(N/\text{yr}) = -1.083 M_W + 3.434, \qquad (1)$$

where *N* is the cumulative number of earthquakes. However, paleoseismic studies at or near the border faults of the Roer valley graben

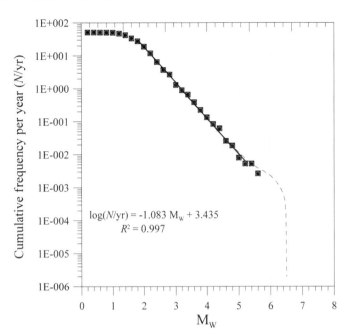

Figure 4. Cumulative frequency of earthquakes per year in the northern Rhine area. Levels of completeness of the catalog are assumed as shown in Figure 5. The circles represent values assuming a continuous record; the filled squares account for recording gaps during the First and Second World Wars. The Gutenberg Richter line is a least-squares fit for $1.8 \leq M_W \leq 5.2$. The correlation coefficient is 0.997. The thin line is a truncated Gutenberg Richter model assuming a maximum magnitude of M_W 7.0. The dashed line is a characteristic earthquake model (Youngs and Coppersmith, 1985) with a recurrence interval for the characteristic event (M_W 6.5) of 5000 yr. Epicenters were selected within a polygon used by Reamer and Hinzen (2004) with the corner points at 49.5°N, 4.0°E; 52°N, 4°E; 52°N, 5.7°E; 51.25°N, 6.4°E; 51.25°N, 8.5°E; and 49.5°N, 8.5°E.

show that the instrumentally and historically recorded earthquakes from the past 300 yr do not include the largest events possible. Hence, the seismotectonic potential of the Lower Rhine Embayment inferred from these earthquakes is probably underestimated and may be biased toward the low to moderate end. Camelbeeck and Meghraoui (1998) found evidence for at least three surface-rupturing paleoearthquakes on the Breé section of the Feldbiss fault during the Holocene with an estimated average recurrence interval of 3500–5000 yr. These earthquakes most probably attained M_W 6.3. Studies along other sections of the Breé fault (Vanneste et al., 2001) have shown evidence for six surface-rupturing events since the late Pleistocene. Preliminary results from paleoseismic trenches at the Viersen fault system, the Erft fault system and the southern part of the Feldbiss fault also show signs of possible coseismic movements. Techmer et al. (2005) described a 15–20 cm structural offset at the Viersen fault, which might have been produced by a single earthquake. Paleoseismic trench observations at the Rurrand fault (Vanneste and Verbeeck, 2001) also allow an interpretation of surface-rupturing earthquakes with M_W 6.9 within the past 2 k.y. Several authors have estimated a maximum magnitude for earthquakes in the Lower Rhine Embayment of M_W 7.0 (i.e., Ahorner,

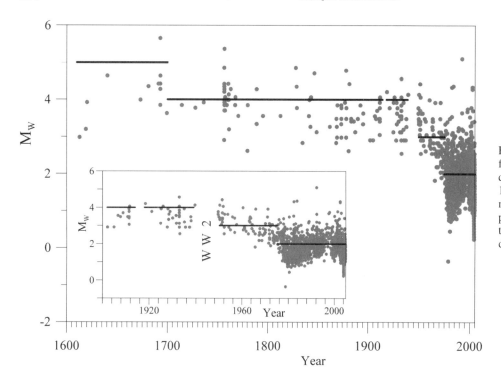

Figure 5. Moment magnitudes M_W as a function of occurrence time for earthquakes in the northern Rhine area since 1600 A.D. Black lines indicate the estimated lower magnitude bound of completeness. The inset shows details since the early 1900s. WW2 marks the gap due to World War II (1939–1945).

2001; Pelzing, 2002; Hinzen, 2005a). Schmedes et al. (2005) estimated the deformation rate in the Lower Rhine Embayment using an integrated approach, systematically incorporating the range of uncertainties. This method resulted in models of the seismic moment release from synthetic catalogs with short catalog length, as for the historic and instrumental catalog, that systematically underestimate the long-term moment rate and indicate an almost complete seismic coupling for the northern Rhine area.

As pointed out by Brudzinski and Chen (2003), plotting seismicity with equal-size symbols on geologic or tectonic maps may obscure important features in seismicity or create misleading visual effects. Symbols scaled by magnitudes (Fig. 1), which are logarithmic measures, preserve the relative earthquake size, albeit in a highly compressed manner. Therefore, the symbol size in Figure 2 was scaled to source dimension, derived through the empirical relation between the measured M_L and seismic moment M_0 calibrated for the northern Rhine area (Reamer and Hinzen, 2004):

$$\log_{10} M_0 = 1.083 M_L + 10.215, \tag{2}$$

where M_0 is measured in Nm, and the relation between M_0 and the radius r of a circular rupture area (Aki, 1967) is:

$$r = \left(\frac{7}{16} \frac{M_0}{\Delta\sigma} \right)^{\frac{1}{3}}, \tag{3}$$

assuming a constant stress drop $\Delta\sigma$ of 100 bar. Magnitudes for the historically recorded earthquakes in the catalog of Cologne University are linked to macroseismic intensities observed at a distance of 10 km, I_{10km}, through the empirical relation (Ahorner, 1983b):

$$I_{10\ km} = 1.5\ M_L - 1.0(\pm 0.6). \tag{4}$$

Within the Hercynic Rhenish Massif, seismicity can be characterized as diffuse, particularly for the Taunus region in the southeast, Hunsrück, and Eifel regions in the south, and Stavelot-Venn Massif in the northwest. An isolated source around 8°E and 50°40′N (Fig. 2) is the M_L 5.0 1982 Bad Marienberg earthquake, with its fore- and aftershock sequence.

Neuwied Basin and Vicinity

A second distinguishable source area within the Hercynic Rhenish Massif is the tectonic depression, the Neuwied Basin (NWB), particularly the eastern edge (Figs. 2 and 6). Although the largest event of this area reaches only M_W 3.6, the area shows significant microseismic activity (Ahorner, 1983a). The Neuwied Basin includes parts of the city of Koblenz and is bounded to the south and the east by the confluence of the Mosel and Rhine Rivers. Nearby, there are the Quaternary volcanic areas of the East Eifel, including the Laacher See volcano (Fig. 6). The Laacher See eruption is the only large explosive eruption known to have occurred in central Europe during late Quaternary times (e.g., Schmincke, 1970; Schmincke et al., 1999; Wörner et al., 1985). The epicenters in the Neuwied Basin cluster in a NW-SE direction starting just south of the Laacher See and extending across the Mosel River valley to the south (Fig. 6). The proximity to the volcanic field and the shallow depth of seismicity (~3–12 km) in the Neuwied Basin with respect to the rest of the northern Rhine area (Hinzen, 2003a; Reamer and Hinzen, 2004) indicate a possible connection to hydraulic processes in the subsurface.

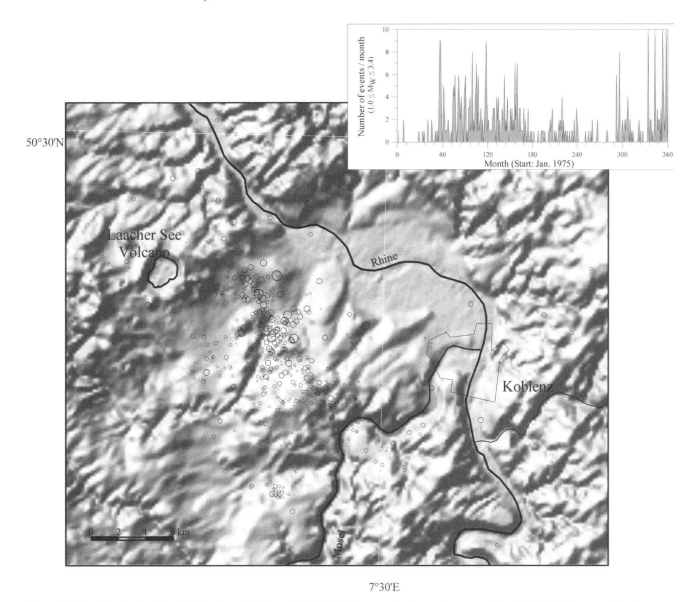

Figure 6. Seismicity in the Neuwied Basin and vicinity (1976–2004). The magnitudes are between 1.0 and 3.4 (M_W). The graph in the upper right corner shows the number of events per month for a 360-month time period.

As a function of time, seismicity in the Neuwied Basin exhibits periods of increased activity. In active months, up to 10 microearthquakes occur, whereas, periods of several months pass with no recorded events. Increased microseismic activity was observed from the end of 1979 through 1989, followed by a period of relative quiescence. Since the end of 2002, microseismic activity has again increased. A clear correlation between microseismic activity with a particular season has not been found; however, the cumulative number of events per month over the past 30 yr shows a minimum of 21 for the month of January and a maximum of 61 for the month of October. The stress tensor inverted from events from the Neuwied Basin and vicinity (Hinzen, 2003a) strongly indicates a strike-slip regime, whereas the surrounding middle Rhine area without the Neuwied Basin tends toward a normal-faulting regime.

Lower Rhine Embayment and Stavelot-Venn Massif

Though it is hard to define a clear border between earthquakes assigned to the Lower Rhine Embayment and the Stavelot-Venn Massif, the inversion of the stress field from fault-plane solutions (Hinzen, 2003a) indicates a clear transition. Whereas Stavelot-Venn Massif events show a strike-slip regime, the Lower Rhine Embayment earthquakes clearly indicate a normal-faulting regime. The epicenters of the historical and relocated earthquakes (Fig. 2) show some alignment with the major fault systems. An unambiguous correlation between epicenters and/or hypocenters and major tectonic features is hampered by a lack of knowledge about the depth extension of the major faults. The border faults exhibit some surface expressions, often clearly visible in the morphology with undulated steps of 10–12 m, but the structure in the seismogenic zone between 5 km and 20 km can only be conjectured. For example, as shown in Figure 2, a band of epicenters in the Erft block is roughly parallel to the Erft fault zone ~6–12 km southwest of the surface expression of the fault system. This projection would geometrically agree with an average southwestward dip of the faults of ~60°. However, the dip of the faults in the shallower region above the seismogenic region is generally steeper. Therefore, an argument can be made for a lystric fault near the bottom of the seismogenic layer. For example, the fault mechanism of the well-observed Alsdorf (M_L 4.9) earthquake of 2002 correlates better with movement on the west-dipping Rurrand fault (or a parallel subfault) than with the east-dipping Feldbiss fault.

The problem with a linear downward extrapolation of the trend and dip of major surface faults was evident with the Roermond 1992 (M_L 6.0) earthquake, the strongest instrumentally observed earthquake in the northern Rhine area. Camelbeeck at al. (1994) and Ahorner (1994) found an ~20° difference between the general strike of the Peel boundary fault, with which this event is associated, and the favored nodal plane from the fault-plane solution. They concluded that a more complicated fault structure persists at 15–20 km depth than would be inferred by a flat plane depth extrapolation of the surface fault. However, a moment tensor solution by Dziewonski et al. (1993) showed a strike consistent with the surface trend.

Figure 7 shows the cumulative number of earthquakes for six subregions of the northern Rhine area between 1600 and the present. Polygon corner points of the selected subregions, their size, total time span, and *a* and *b* values obtained from least-squares regression of the data to a Gutenberg and Richter model are listed in Table 1. The earthquakes from the Bad Marienberg area resemble the 1982 M_L 5.0 earthquake and its aftershocks, with a total of 87 located earthquakes. The ~500 events in the Neuwied Basin and vicinity represent the most prolific subgroup, even though there are only nine recorded events prior to 1975, and the largest event has a M_L of only 4.0. The Erft block, Rur block, Cologne block, and Peel block subregions (Fig. 2) display different seismicity patterns. The least active subregion is the Cologne block, with only 27 events. Seismicity in the Peel block in the northern part of the Roer valley graben is dominated by the Roermond 1992 aftershock sequence. The Rur block in the southern part of the Roer valley graben is the subregion in the Lower Rhine Embayment with the most earthquakes (197). While the total number of known events is almost the same in the Erft block as in the Peel block, the cumulative frequency curve does not exhibit a significant aftershock character. The bend in the cumulative curves of the Erft block and Rur block at M_L 3.5 reflects the different degrees of completeness over the 400 yr time period represented in the hybridized catalog.

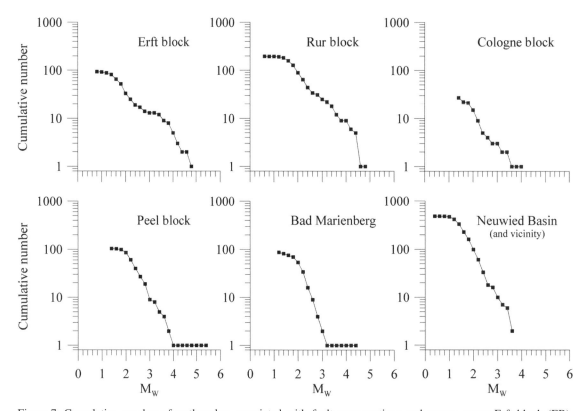

Figure 7. Cumulative number of earthquakes associated with fault zone sections and source areas: Erft block (EB), Rur block (Rurrand fault), Cologne block, Peel block (Peel boundary fault), Bad Marienberg source area, Neuwied Basin and vicinity.

TABLE 1. GEOGRAPHICAL COORDINATES OF POLYGON CORNER POINTS OUTLINING SELECTED SUBREGIONS OF SEISMICITY

Point number	Erft block Latitude (°N)	Longitude (°E)	Rur block Latitude (°N)	Longitude (°E)	Cologne block Latitude (°N)	Longitude (°E)	Peel block Latitude (°N)	Longitude (°E)	Bad Marienberg Latitude (°N)	Longitude (°E)	Neuwied Basin (and vicinity) Latitude (°N)	Longitude (°E)
1	50.628	6.903	50.627	6.905	50.969	6.676	51.270	5.554	50.599	8.154	50.420	7.148
2	50.676	6.803	50.604	6.742	50.995	6.762	51.358	5.853	50.590	7.865	50.438	7.277
3	50.722	6.704	50.633	6.604	50.994	6.822	51.312	5.893	50.720	7.839	50.451	7.376
4	50.764	6.624	50.698	6.451	50.999	6.881	51.256	5.959	50.771	8.024	50.425	7.493
5	50.819	6.561	50.772	6.311	50.954	6.976	51.175	6.088	50.692	8.162	50.378	7.597
6	50.861	6.505	50.813	6.243	50.897	7.075	51.068	6.171			50.312	7.544
7	50.896	6.460	50.861	6.188	50.840	6.954	51.058	6.200			50.269	7.508
8	50.929	6.399	50.920	6.119	50.802	6.879	50.996	6.081			50.245	7.445
9	50.963	6.314	50.949	6.065			50.962	5.948			50.243	7.364
10	51.049	6.282	50.960	6.040			51.020	5.838			50.261	7.270
11	51.036	6.391	51.008	6.169			51.125	5.688			50.313	7.170
12	51.020	6.575	51.017	6.293							50.378	7.107
13	51.018	6.609	50.963	6.312								
14	50.921	6.727	50.930	6.395								
15	50.833	6.845	50.894	6.462								
16	50.719	6.961	50.830	6.546								
17	50.650	7.064	50.768	6.621								
18			50.688	6.777								
G&R *a* and *b* value	2.54	0.48	3.08	0.58	2.17	0.57	3.73	0.89	2.93	0.76	4.74	1.48
Time span (yr)	364.6		248.8		158.9		305.4		14.6		169.9	
Area (km²)	896		902		320		1004		351		552	

Note: The coordinates of the polygons, the corresponding *a* and *b* values of Gutenberg and Richter (G&R) relationships derived by least-squares fits and the time spans are listed.

SELECTED SIGNIFICANT EARTHQUAKES AND SEQUENCES

The cumulative seismic moment for the northern Rhine area (Fig. 8) was calculated from moments estimated from the empirical relation in Equation 2. The strongest earthquake in the catalog starting in 1600 is the 18 September 1692 Verviers event. While some of the significant earthquakes are single events with a typical aftershock sequence (e.g., Tollhausen 1878, Roermond 1992), others appear as doublets. Time intervals of 2, 44, and 12 mo separate the two events for the Düren (1755, 1756), Herzogenrath (1873, 1877), and Euskirchen (1950, 1951) earthquakes, respectively.

Verviers (1692)

The 18 September 1692 earthquake is sometimes named the Brabant earthquake because early macroseismic investigations (Sieberg, 1940) placed the epicenter in Brabant, even though the exact location was not definable. Studies of original historical sources by Alexandre and Kupper (1997) and an analysis by Camelbeeck et al. (2000) locate the major damage in the Liége-Vervier area, although damage was widespread from Kent (England) to the Rhenan region in Champagne (France). Application of the Bakun and Wentworth (1997) method for locating the intensity center and estimating a magnitude for

historic earthquakes places this earthquake south of the city of Liége with an M_L-equivalent magnitude of 6.8 (Hinzen and Oemisch, 2001). This result also defines Vervier as the strongest historic earthquake in the northern Rhine area.

Düren (1755–1756)

Only eight weeks after the famous Lisbon earthquake of 1 November 1755, a sequence of earthquakes began in the southwestern part of the Lower Rhine Embayment and lasted for over a year. This sequence of more than 70 reported earthquakes, analyzed in detail by Meidow (1995), culminated in two earthquakes on 27 December 1755 and 18 February 1756 with maximum intensities of VII and VIII, respectively. By empirically deriving the macroseismic magnitudes of 5.7 and 6.4 for the main events, the total moment of this doublet is estimated at 1.6×10^{17} Nm.

Roermond (1992)

The 13 April 1992 Roermond earthquake is the strongest instrumentally recorded event in the northern Rhine area (M_L 6.0; Reamer and Hinzen, 2004). The source parameters, aftershock sequence, and macroseismic effects have been exhaustively studied (i.e., Camelbeeck et al., 1994; Ahorner, 1994; Braunmiller et al., 1994; Prinz et al., 1994; Haak et al., 1994). The Roermond earthquake was followed by an aftershock sequence typical for

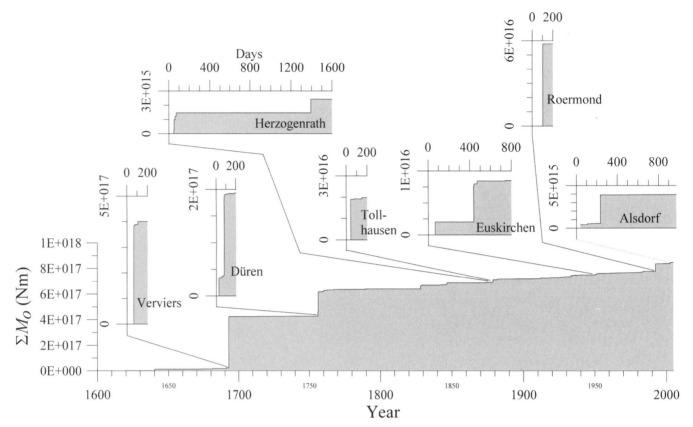

Figure 8. Cumulative moment of earthquakes in the northern Rhine area from 1600 to 2004. The small diagrams highlight individual earthquakes or earthquake sequences. The time axis for all insets is in days.

its size (Camelbeeck et al., 1994). However, the map in Figure 9 indicates two distinct clusters of aftershocks: one close to the main shock and the second ~30–40 km roughly southeast. Figure 10 shows a time series of 113 relocated aftershocks in the two years following the main shock. The distances as a function of time between the main shock and the aftershocks also clearly separate the latter into two clusters. While events from one cluster are less than 15 km from the main event, the second cluster of aftershocks has hypocenters between 25 km and more than 40 km distant from the main shock.

Analysis of the complete instrumental catalog indicates a probable causal connection between the main shock and the more remote aftershock locations. The hypocenter of the strongest Roermond aftershock, which occurred 24 h after and 30 km from the main shock, is located only 8 km from the hypocenter of the 22 July 2002 Alsdorf earthquake. This M_L 4.9 earthquake is the strongest in the northern Rhine area since the Roermond earthquake. The Alsdorf earthquake epicenter lies very close to the centroid of the more distant aftershock cluster of the Roermond earthquake. This result indicates the presence of a zone of critical stress, probably on the depth extension of the northern section of the Rurrand fault (Fig. 2). As noted by Camelbeeck et al. (1994), this phenomenon could not be explained by simple postseismic stress variation caused by the main shock because these effects

would not be noticeably different than the variations caused by tidal movements. However, recent work on stress and tidal triggering (i.e., Gomberg et al., 1998; Stein, 2003, 2004; Hough, 2005; Toda et al., 2005) elevates the effectiveness of remote-triggering mechanisms. Parsons (2005) presented a hypothesis for delayed dynamic triggering on faults following a rate-state friction law. Hence, the doublets in the Lower Rhine Embayment earthquake history, the remote aftershock activity of the Roermond earthquake, and the closeness of the later Alsdorf earthquake to this aftershock zone are being examined using stress-triggering models of Lower Rhine Embayment seismicity.

Figure 11 shows the depth (and time) distribution of the Roermond earthquake aftershocks on vertical planes roughly parallel and perpendicular to the trend of the aftershock zone from the relocated catalog data from Reamer and Hinzen (2004). Although the sequence in the distant cluster does not possess a clear structure, the sequence in the near aftershock cluster shows an identifiable decrease in depth as a function of time. A dashed line in the left part of the NW-SE profile in Figure 11 encloses most of the aftershocks in the nearby cluster. These aftershocks are located outside the rupture zone, which has an estimated diameter of ~3–4 km (Camelbeeck et al., 1994), and migrate upward and slightly southeast with time. The SW-NE cross section through the close aftershock cluster places the relocated

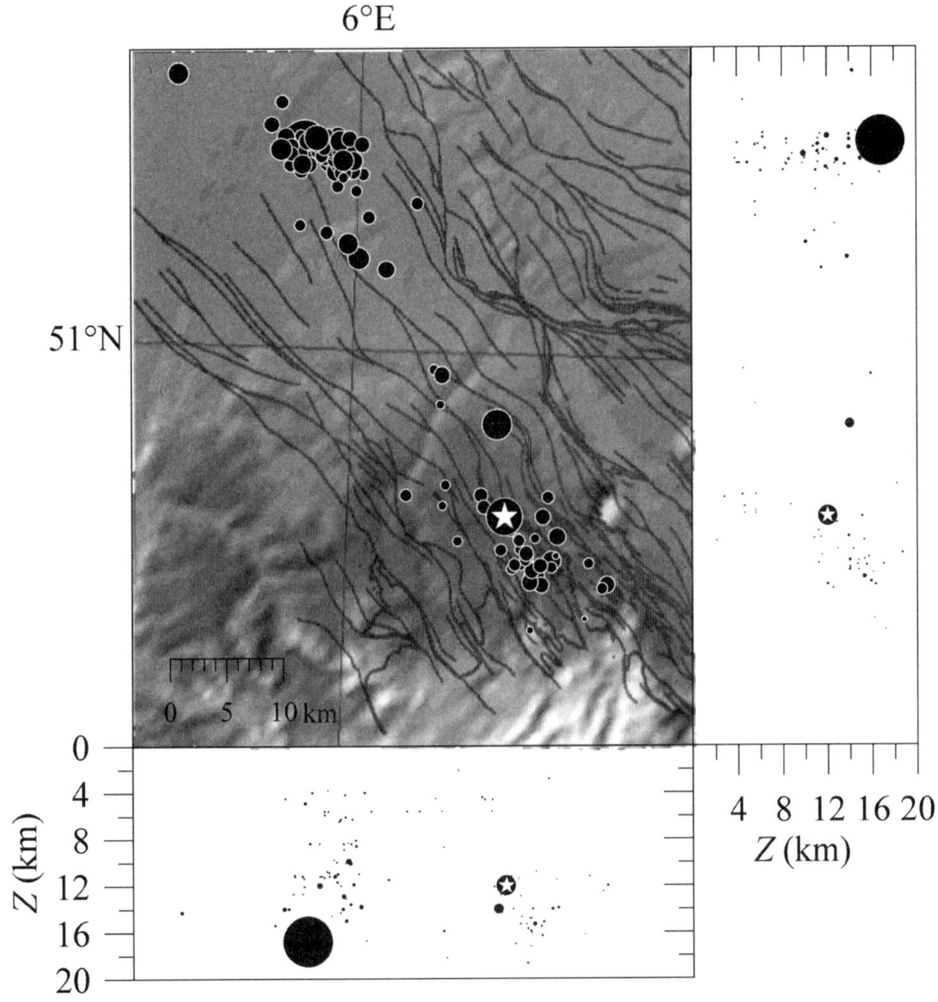

Figure 9. Location of the 13 April 1992 Roermond (M_L 6.0) earthquake and its aftershocks, relocated with a one-dimensional velocity model for the northern Rhine area (Reamer and Hinzen, 2004). The main map shows the epicenters and faults in the source area. Symbol size varies with magnitude. In the north-south and east-west cross sections, the depth distribution is shown with symbols scaled linearly to source dimension. The white stars indicate the location of the 22 July 2002 Alsdorf (M_L 4.9) earthquake, which was the strongest in the northern Rhine area since the Roermond earthquake.

hypocenters in a vertical column above the main event rather than on the assumed trace of the Peel boundary fault. Hence, they can be interpreted as resulting from sagging in the hanging wall rather than a downdip movement along the fault itself.

Alsdorf Earthquake

At 05:45 UTC on 22 July 2002, an earthquake shook the Lower Rhine Embayment and large parts of the surrounding areas in Germany, Belgium, the Netherlands, Luxembourg, and France. The epicenter was close to the city of Alsdorf, ~14 km northeast of Aachen and 54 km east of Cologne. The earthquake was the first with slight damage (intensity VI MSK) in the German part of the Lower Rhine Embayment since the 1992 Roermond earthquake. The seismicity in the vicinity of the Alsdorf earthquake is shown in the map in Figure 12. A series of small earthquakes occurred prior to the main event. The small map on the right bottom of Figure 12 shows the results of double-difference relocation (Waldhauser and Ellsworth, 2000) of eight events during seven months before the main shock with magni-

tudes from M_L 1.3 to 3.6, as well as the main event (M_L 4.9) and 12 of its aftershocks with M_L between 1.6 and 2.6. The depth of the main earthquake is 15.8 km. While the preceding earthquakes occurred from 6 to 12 km primarily southeast of the main event, most of the aftershocks occurred on or above the fault plane of the main event. The mechanism was normal faulting, with a small strike-slip component. Figure 13 shows the P-wave first-motion fault-plane solutions from a grid test procedure (Hinzen, 2003a) using a 5° sampling of the parameter space of the angles of orientation. Moment tensor determination from regional surface-wave analysis (Braunmiller, 2002) shows the same mechanism. The fault-plane solutions of three aftershocks from the same day with M_L >2 have almost identical mechanisms as the main shock (Fig. 13). The dips of the two fault planes (55°W and 35°E) are indicated in the cross section through the hypocenter of the event in Figure 12. As previously discussed, knowledge of the depth extension of major faults in the Lower Rhine Embayment is sparse. While in some areas, the near-surface expression of the faults has been well explored due to the extensive brown-coal-mining activities in the vicinity, the trend and dip of

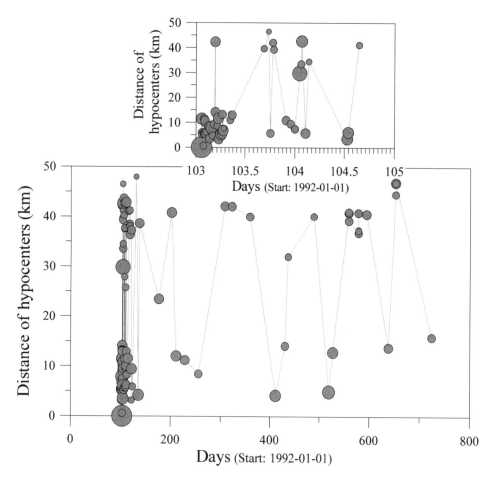

Figure 10. Time series of the Roermond earthquake (1992) and its aftershock sequence. The distance is measured between the hypocenters of the main shock and the relocated aftershocks. Main plot shows the event sequence up to 2 yr after the main shock; inset highlights the first 48 h after the main shock. Symbol size varies with magnitude ($0.8 \leq M_L \leq 6$).

faults within the seismogenic zone between 5 and 20 km remain largely unknown. Though the epicenter is close to the eastern border fault of the Alsdorf horst (Fig. 12), the source geometry precludes movement along the sides of the horst structure. The 55°W-dipping fault plane may coincide with the depth extension of the Rurrand fault or a parallel fault. A downdip movement on the 35°E-dipping fault plane seems unlikely; however, this feature could be interpreted as movement on a fault from the western border of the Roer valley graben (i.e., the Feldbiss fault) with a lystric depth extension in the form of a detachment fault.

SEISMIC HAZARD

Seismic hazard in the northern Rhine area has been addressed in several publications based on a deterministic approach (Ahorner et al., 1970) and a probabilistic approach (i.e., Ahorner and Rosenhauer, 1975; Rosenhauer und Ahorner, 1994; Grünthal and Bosse, 1996). Many detailed site evaluations have been performed, especially as part of seismic risk evaluations prior to construction of nuclear facilities, dams, and chemical plants. These reports, however, are not usually made available for public inspection. Even though a complete seismic hazard analysis for the northern Rhine area is beyond the scope of this paper, a few general observations can be made.

In Figure 4, a bounded Gutenberg-Richter model with a maximum magnitude of M_W 7.0 and a characteristic earthquake model (Youngs and Coppersmith, 1985) with a recurrence interval of 5000 yr for the characteristic earthquake of M_W 6.5 have been fit to the observed data. The Gutenberg-Richter model predicts a recurrence interval for a M_W 6.0 earthquake of 1200 yr, whereas the interval decreases to 440 yr for the characteristic earthquake model. Based on the simple assumption that the intensity of shaking at a site depends only on the strength of the earthquake and the distance of the site from the earthquake, we estimated the numbers of felt (intensity III and above) and possibly slightly damaging (intensity V–VI and above) earthquakes for the past 300 yr from the historic and instrumental earthquake catalog of Cologne University as shown in Figure 14. During this time, more than 120 earthquakes were felt in the southern and western part of the Lower Rhine Embayment. The numbers of felt events were over 110 for the city of Aachen, 100 for Cologne, and 70 for Koblenz (see Fig. 2 for locations). The area with the highest occurrence of intensities V–VI and above was a small area east of Aachen, where 11 possibly damaging events occurred. Outside this "hotspot," the number decreases rapidly. For Cologne, the number predicted from this simple model is five events. This area is roughly congruent with a range in the hazard map of Ahorner and Rosenhauer (1975) with an

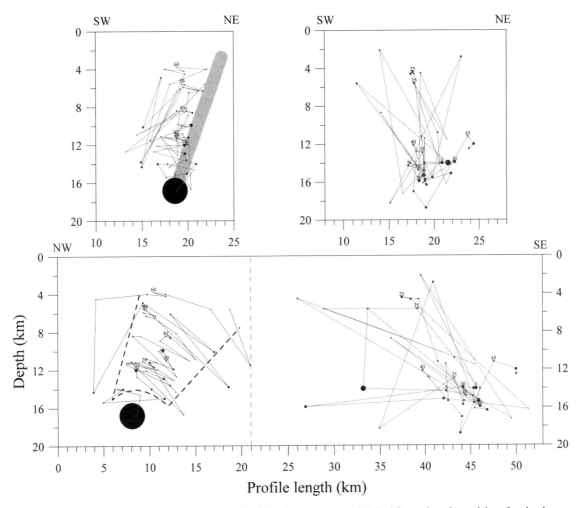

Figure 11. Depth profiles showing projected foci of the 1992 Roermond (M_L 6.0) earthquake and its aftershocks on NW-SE– and SW-NE–trending vertical planes. The former is roughly parallel to and the latter perpendicular to the trend of the aftershock zone. Diagrams at the top are cross sections of the closer and more distant aftershocks. The line of separation between the two clusters is indicated on the bottom profile by a dashed line. Dots are scaled to source dimension. Lines between the dots connect the events in a chronological order. Small labels are chronologically sequenced numbers in the two clusters. The thick gray line in the top left diagram indicates the assumed dip of the Peel boundary fault.

average recurrence interval for peak accelerations of 1.0 and 3.0 m/s² of 250 and 7000 yr, respectively, and probabilistic intensities (MSK) of 7.25 and 8 for 1000 and 10,000 yr, respectively (Rosenhauer and Ahorner, 1994).

From a purely tectonic perspective, the western and eastern border faults of the Lower Rhine Embayment do not differ significantly in their seismotectonic potential. However, both historic and instrumentally recorded events occur with significantly higher frequency on the western faults (i.e., Rurrand, Peel, Feldbiss) than on the eastern faults (i.e., Erft, Viersen). For example, the Erft fault zone is ~50 km long and has a maximum vertical movement of 900–1000 m (Ahorner, 1962a). The average rate of vertical movement is in the range from 0.05 to 0.1 mm/yr; however, geodetic measurements show values on some fault segments of up to 0.9 mm/yr (Ahorner, 2001). This variation in rate may be an indication of increased postseismic

movement. For strong but not surface-rupturing earthquakes, movements at depth in the seismogenic layer reach the surface as a signal smeared over both the temporal and spatial domains. Ahorner (2001) derived a recurrence interval of 2200 yr for a M_w 6.0 earthquake on the Erft fault system and 18,000 yr for M_w 6.7 based on seismotectonic balancing. With such recurrence intervals, a possible hypothesis for the higher activity in the western part of the Lower Rhine Embayment observed in the past 300 yr (Fig. 14) is long-term aftershock activity resulting from an earlier strong earthquake(s), as suggested by Ebel et al. (2000) for the Breé area. Stein and Newman (2004) studied the relation between recurrence time and aftershock sequence duration. They suggested that rate-state friction would predict very long intraplate aftershock sequences. The rates and sizes of the largest observed earthquakes in such a region may differ significantly from their true long-term values (Stein and Newman,

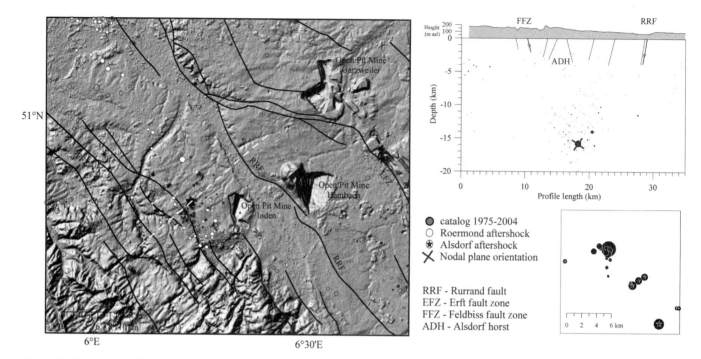

Figure 12. Main map: Seismicity (1975–2004) in the vicinity of the 2002 (M_L 4.9) Alsdorf earthquake (star). Circle size varies with magnitude. The white filled circles are epicenters of earthquakes occurring after the 1992 (M_L 6.0) Roermond earthquake. The star-filled circles are epicenters of earthquakes in 2002 after the Alsdorf earthquake. Major surface fault lines are shown with the Rurrand fault (RRF), Erft fault system (EFZ), and the Feldbiss fault zone (FFZ). The terrain data are based on SRTM (Shuttle Radar Topography Mission) (National Aeronautics and Space Administration, http://srtm.usgs.gov/index.html) measurements, which show surface mining activity in the area. The diagram at the top right shows a vertical cross section along the white dashed line (asl—above sea level). Symbol size represents the equivalent size of a circular rupture area. The lines at the location of the main event indicate the dip of the fault planes from Figure 13 of 55°W and 35°E, respectively. The topographic profile at the top has been exaggerated vertically by a factor of 10. The position and dip of major faults in the upper 4 km were taken from geologic maps (von Kamp, 1986). The small map on the bottom right shows the epicenters of foreshocks (stars) and aftershocks (filled circles) in the source area from double difference localization (Waldhauser and Ellsworth, 2000).

2004). As a consequence of this hypothesis, the seismotectonic potential of all known faults in the Lower Rhine Embayment should be reevaluated in the form of a complete seismic hazard analysis when defining source areas or single fault sources. Ongoing work will combine such models with detailed amplification responses for the soft-rock layers of the sedimentary basin of the Lower Rhine Embayment and the Neuwied Basin.

SUMMARY

Tectonically active intracontinental (intraplate) rifts, such as the Rhine-Rhone rift, East African rift zone, and Shanxi rift of China, correspond to important earthquake and volcanic hazard zones (Ziegler, 2001). Intracontinental earthquake zones may pose an insidious threat because of the lack of public and political awareness of the potential seismic hazard due to the rareness of significant and damaging earthquakes. The northern Rhine area of the central European intraplate rift system presently displays low to moderate seismicity. However, over the past 300 yr, more than 20 damaging earthquakes have occurred in this region. Damages have ranged from minor to the 125 million Euros of insured damage during the 1992 Roermond earthquake series (Bertz, 1994).

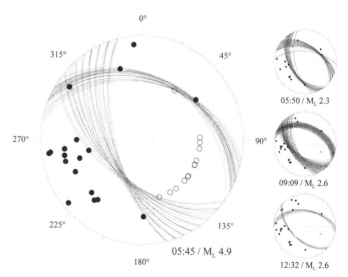

Figure 13. Fault-plane solution from P-wave polarities of the 2002 Alsdorf earthquake and three of its aftershocks. Compressional and dilatational first-motion readings are indicated by filled and open circles, respectively. Source time and magnitude of the events are given below the projections of the lower focal sphere.

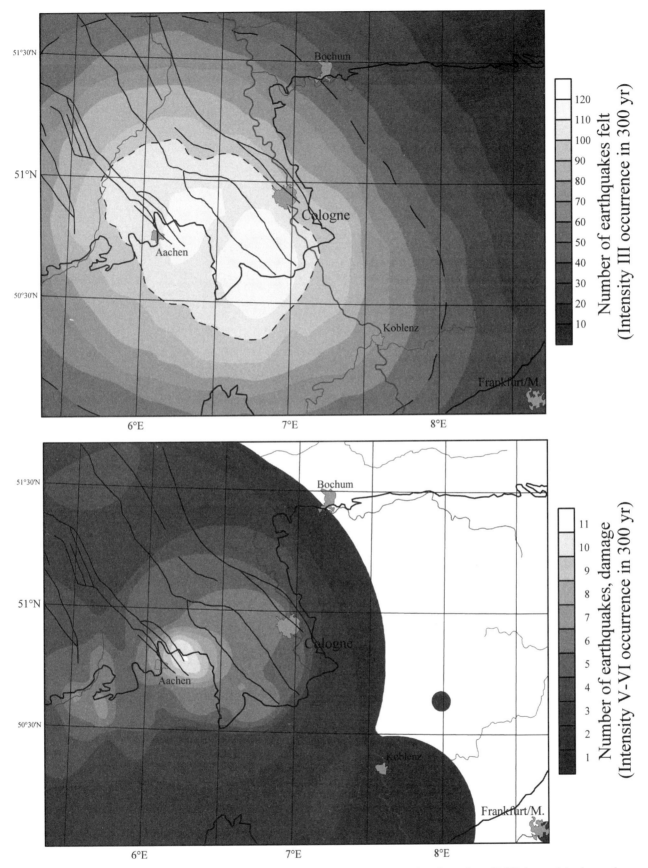

Figure 14. Frequency of felt (intensity ≥ III, top) and potentially damaging earthquakes (intensity ≥ V–VI, bottom) in the northern Rhine area from the hybrid catalog data from 1600 to 2004. Frequencies are based on a simple intensity distance relation (Hinzen and Oemisch, 2001), neglecting possible site effects.

The area between the cities of Bonn in the south, Aachen in the west, and Dortmund in the east is home to approx. 11.5 million people, more than greater Paris and almost as much as greater London. Several recent paleoseismological studies (i.e., Camelbeeck and Meghraoui, 1998; Vanneste et al., 2001; Vanneste and Verbeeck, 2001; Techmer et al., 2005) have shown that at least some of the more than 400 km of active faults have produced surface-rupturing earthquakes in the Holocene. Although the recurrence interval of such events is on the order of several thousand years, and a strong earthquake with moment magnitudes of 7 may have an equivalent probability of occurrence as a new active phase of Eifel volcanism (Ahorner, 2001), thorough evaluation of hazard and risk mitigation is an important scientific and public task and a matter for intensified cross-border preparedness. As discussed in this paper, the seismicity in the northern Rhine area over the past 300 yr is reasonably well documented. However, hazard studies could be greatly improved from the compilation of a unified catalog including subsurface models and utilizing both instrumental and historic seismicity from pooled data from Belgium, Germany, Luxemburg, and the Netherlands.

The Lower Rhine Embayment with the Roer valley graben represents an ideal area for the development and application of advanced modeling techniques for seismic hazard analysis (i.e., Hinzen et al., 2000; Pelzing, 2002; Grünthal and Wahlström, 2004), soil amplification studies (i.e., Scherbaum et al., 2003; Parolai et al., 2005; Hinzen et al., 2004), 3D wave-propagation studies (Ewald et al., 2006), earthquake engineering (i.e., Berger, 1994; Dost et al., 2004; Hinzen, 2005b), paleoseismology (i.e., Camelbeeck and Meghraoui, 1998; Camelbeeck et al., 2000; Vanneste et al., 2001), and archaeoseismological studies (Hinzen and Schütte, 2003; Hinzen, 2005a). Additional strong motion stations in the German part of the Lower Rhine Embayment supplementing the existing networks in Belgium and the Netherlands will improve the database for future hazard and risk studies.

ACKNOWLEDGMENTS

We thank Günter Leydecker, Frank Scherbaum, and Seth Stein for helpful criticism and suggestions to improve the manuscript.

REFERENCES CITED

Ahorner, L., 1962a, Untersuchungen zur quartären Bruchtektonik in der Niederrheinischen Bucht: Eiszeitalter und Gegenwart, v. 13, p. 24–105.

Ahorner, L., 1962b, Das Erdbeben im Saar-Nahe-Becken vom 17 August 1960: Sonderveröffentlichungen des Geologischen Institutes der Universität zu Köln: Rheinische Erdbeben, v. II, p. 3–25.

Ahorner, L., 1963, Das Erdbeben vom 25 Juni 1960 im Belgisch-Niederländischen Grenzgebiet: Sonderveröffentlichungen des Geologischen Institutes der Universität zu Köln, 9: Rheinische Erdbeben, v. III, p. 5–28.

Ahorner, L., 1968, Erdbeben und jüngste Tektonik im Braunkohlenrevier der Niederrheinischen Bucht: Zeitschrift der Deutschen Geologischen Gesellschaft, v. 118, p. 150–160.

Ahorner, L., 1970, Seismotectonic relations between the graben zones of the Upper and Lower Rhine valley, in Illies, H.J., and Müller, St., eds., Graben Problems: Stuttgart, Schweizerbart, p. 155–166.

Ahorner, L., 1975, Present-day stress field and seismotectonic block movements along major fault zones in central Europe, in Pavoni, N., and Green, R., eds., Recent Crustal Movements: Tectonophysics, v. 29, p. 233–249.

Ahorner, L., 1983a, Historical seismicity and present-day microearthquake activity of the Rhenish Massif, central Europe, in Fuchs, K., et al., eds., Plateau Uplift: Berlin, Springer-Verlag, p. 198–221.

Ahorner, L., 1983b, Seismicity and neotectonic structural activity of the Rhine graben system in central Europe, in Ritsema, A.R., and Gürpinar, A., eds., Seismicity and Seismic Risk in the Offshore North Sea Area: Dordrecht, D. Reidel, p. 101–111.

Ahorner, L., 1994, Fault-plane solutions and source parameters of the 1992 Roermond, the Netherlands, mainshock and its stronger aftershocks from regional seismic data: Geologie en Mijnbouw, v. 73, p. 199–214.

Ahorner, L., 2001, Abschätzung der statistischen Wiederkehrperiode von starken Erdbeben im Gebiet von Köln auf Grund von geologisch-tektonischen Beobachtungen an aktiven Störungen: Mitteilungen der Deutschen Geophysikalischen Gesellschaft, v. 2, p. 2–9.

Ahorner, L., and Pelzing, R., 1983, Seismotektonische Herdparameter von digital registrierten Erdbeben der Jahre 1981 und 1982 in der westlichen Niederrheinischen Bucht: Geologisches Jahrbuch, v. E26, p. 35–63.

Ahorner, L., and Pelzing, R., 1985, The source characteristics of the Liège earthquake on November 8, 1983, from digital recordings in West Germany, in Melchior, P., ed., Seismic Activity in Western Europe: Dordrecht, D. Reidel Publishing, p. 263–289.

Ahorner, L., and Rosenhauer, W., 1975, Probability distribution of earthquake accelerations with applications to sites in the northern Rhine area, central Europe: Journal of Geophysics, v. 41, p. 581–594.

Ahorner, L., Murawski, H., and Schneider, G., 1970, Die Verbreitung von schadenverursachenden Erdbeben auf dem Gebiet der Bundesrepublik Deutschland: Zeitschrift für Geophysik, v. 36, p. 313–343.

Aki, K., 1967, Scaling law of seismic spectrum: Journal of Geophysical Research, v. 72, p. 1217–1231.

Alexandre, P., and Kupper, J.-L., 1997, Le tremblement de terre de 1692: Feuillets de la Chathedrale de Liège, v. 28–32, p. 3–19.

Bakun, W.H., and Wentworth, C.M., 1997, Estimating earthquake location and magnitude from seismic intensity data: Bulletin of the Seismological Society of America, v. 87, p. 1502–1521.

Berger, N., 1994, Attenuation of seismic ground motion due to the 1992 Roermond earthquake, the Netherlands: Geologie en Mijnbouw, v. 73, p. 309–313.

Bertz, G., 1994, Assessment of the loss caused by the 1992 Roermond earthquake, the Netherlands (extended abstract): Geologie en Mijnbouw, v. 73, p. 281.

Braunmiller, J., 2002, Swiss moment tensor solutions—2002, http://www.seismo.ethz.ch/moment_tensor/2002/ (last accessed 16 March 2005).

Braunmiller, J., Dahm, T., and Bonjer, K.-P., 1994, Source mechanism of the 1992 Roermond, the Netherlands, earthquake from inversion of regional surface waves: Geologie en Mijnbouw, v. 73, p. 225–227.

Brudzinski, M.R., and Chen, W.-P., 2003, Visualization of seismicity along subduction zones: Towards a physical basis: Seismological Research Letters, v. 74, p. 731–738.

Camelbeeck, T., and Meghraoui, M., 1998, Geological and geophysical evidence for large paleoearthquakes with surface faulting in the Roer graben (northwest Europe): Geophysical Journal International, v. 132, p. 347–362, doi: 10.1046/j.1365-246x.1998.00428.x.

Camelbeeck, T., vanEck, T., Pelzing, R., Ahorner, L., Loohuis, L., Haak, H.W., Hoang-Trong, P., and Hollnack, D., 1994, The 1992 Roermond earthquake, the Netherlands, and its aftershocks: Geologie en Mijnbouw, v. 73, p. 181–197.

Camelbeeck, T., Alexandre, P., Vanneste, K., and Meghraoui, M., 2000, Long-term seismicity in regions of present day low seismic activity: The example of western Europe: Soil Dynamics and Earthquake Engineering, v. 20, p. 405–414, doi: 10.1016/S0267-7261(00)00080-4.

Charlier, Ch., 1951, Etude systématique des tremblements de terre Belges récents (1900–1950): 4e Partie: La Séismicité de la Belgique: Publication du Service Séismologique et Gravimétrique de l'Observatoire Royal de Belgique, v. 10, p. 60.

Cohen, K.M., Stouthamer, E., and Berendsen, H.J.A., 2002, Fluvial deposits as a record for late Quaternary neotectonic activity in the Rhine-Meuse delta, the Netherlands: Netherlands Journal of Geosciences: Geologie en Mijnbouw, v. 81, p. 389–405.

Dost, B., van Eck, T., and Haak, H., 2004, Scaling of peak ground acceleration and peak ground velocity recorded in the Netherlands: Bollettino di Geofisica Teorica ed Applicata, v. 45, p. 153–168.

Dziewonski, A.M., Ekström, G., and Salganik, M., 1993, Centroid moment tensor solutions for April–June 1992: Physics of the Earth and Planetary Interiors, v. 77, p. 151–163, doi: 10.1016/0031-9201(93)90095-Q.

Ebel, J.E., Bonjer, K.-P., and Oncescu, M.C., 2000, Paleoseismicity: Seismicity evidence for past large earthquakes: Seismological Research Letters, v. 71, p. 283–294.

Ewald, M., Igel, H., Hinzen, K.-G., and Scherbaum, F., 2006, Basin-related effects on ground motion for earthquake scenarios in the Cologne basin, Germany: Geophysical Journal International, v. 166, p. 197–212.

Garcia-Castellanos, D., Cloetingh, S., and Van Balen, R., 2000, Modelling the middle Pleistocene uplift in the Ardennes–Rhenish Massif: Thermomechanical weakening under the Eifel?: Global and Planetary Change, v. 27, p. 39–52, doi: 10.1016/S0921-8181(01)00058-3.

Gomberg, J., Beeler, N.M., Blanpied, M.L., and Bodin, P., 1998, Earthquake triggering by transient and static deformations: Journal of Geophysical Research, v. 103, p. 24,411–24,426, doi: 10.1029/98JB01125.

Grünthal, G., and Bosse, Ch., 1996, Probabilistische Karte der Erdbebengefährdung der Bundesrepublik Deutschland—Erdbebenzonierungskarte für das nationale Anwendungsdokument zum Eurocode 8: Potsdam, Forschungsbericht, STR96/10, GeoForschungsZentrum, 24 p.

Grünthal, G., and Wahlström, R., 2004, New generation of probabilistic seismic hazard assessment for the area Cologne/Aachen considering the uncertainties of input data: Natural Hazards Special Issue: German Research Network Natural Disasters, v. 38, nos. 1–2, p. 159–176.

Gutenberg, B., and Landsberg, H., 1930, Das Taunusbeben vom 22 (Januar 1930): Natur und Museum, Senkenbergische Naturforschende Gesellschaft, v. 4, p. 1–6.

Haak, H.W., Bodegraven, J.A., Dleeman, R., Verbeiren, R., Ahorner, L., Meidow, H., Grunthal, G., Hoang-Trong, P., Musson, R.M.W., Henni, P., Schenková, Z., and Zimová, R., 1994, The macroseismic map of the 1992 Roermond earthquake, the Netherlands: Geologie en Mijnbouw, v. 73, p. 265–270.

Hinzen, K.-G., 2003a, Stress field in the northern Rhine area, central Europe, from earthquake fault plane solutions: Tectonophysics, v. 377, no. 3–4, p. 325–356, doi: 10.1016/j.tecto.2003.10.004.

Hinzen, K.-G., 2003b, Source parameters of the M_L 3.8 earthquake on January 20, 2000, near Meckenheim, Germany: Journal of Seismology, v. 7, p. 347–357, doi: 10.1023/A:1024510932145.

Hinzen, K.-G., 2005a, The use of engineering seismological models to interpret archaeoseismological findings in Tolbiacum, Germany, a case study: Bulletin of the Seismological Society of America, v. 95, p. 521–539, doi: 10.1785/0120040068.

Hinzen, K.-G., 2005b, Ground motion parameters of the 22 July 2002 M_L 4.9 Alsdorf (Germany) earthquake: Bolletino di Geofisica, v. 46, p. 303–318.

Hinzen, K.-G., and Oemisch, M., 2001, Location and magnitude from seismic intensity data of recent and historic earthquakes in the northern Rhine area, central Europe: Bulletin of the Seismological Society of America, v. 91, p. 40–56, doi: 10.1785/0120000036.

Hinzen, K.-G., and Schütte, S., 2003, Evidence for earthquake damage on Roman buildings in Cologne, Germany: Seismological Research Letters, v. 74, p. 124–139.

Hinzen, K.-G., Pelzing, R., Reamer, S.K., and Mackedanz, J., 2000, Seismic risk evaluation for the northern Rhine area based on the seismotectonic potential of active faults [abs.]: Eos (Transactions, American Geophysical Union), v. 81(48), Fall Meeting Supplement, abstract S52A-25.

Hinzen, K.-G., Scherbaum, F., and Weber, B., 2004, Study of the lateral resolution of H/V measurements across a normal fault in the Lower Rhine Embayment, Germany: Journal of Earthquake Engineering, v. 8, p. 909–926, doi: 10.1142/S136324690400178X.

Hough, S., 2005, Remotely triggered earthquakes following moderate mainshocks (or, why California is not falling into the ocean): Seismological Research Letters, v. 76, p. 58–66.

Illies, J.H., Prodehl, C., Schmincke, H.U., and Semmel, A., 1979, The Quaternary uplift of the Rhenish Shield in Germany, *in* McGechtin, T.R., and Merril, R.B., eds., Plateau Uplift: Mode and Mechanism: Tectonophysics, v. 61, p. 197–225.

Kümpel, H.-J., Lehmann, K., Fabian, M., and Mentes, G., 2001, Point stability at shallow depths: Experience from tilt measurements in the Lower Rhine Embayment, Germany, and implications for high-resolution GPS and gravity recordings: Geophysical Journal International, v. 146, p. 699–713, doi: 10.1046/j.1365-246X.2001.00494.x.

Lersch, B.M., 1897, Erdbeben-Chronik für die Zeit von 2362 v. Chr. bis 1897: Aachen, Neuzehnbändige Handschrift.

Leydecker, G., 2004, Erdbebenkatalog für die Bundesrepublik Deutschland mit Randgebieten für die Jahre 800–2003: Datenfile, Bundesanstalt für Geowissenschaften und Rohstoffe Hannover.

Meidow, H., 1995, Rekonstruktion und Reinterpretation von historischen Erdbeben in den nördlichen Rheinlanden unter Berücksichtigung der Erfahrungen bei dem Erdbeben von Roermond am 13 April 1992 [Inaugural-Dissertation]: Köln, Mathematisch-Naturwissenschaftlichen Fakultät der Universität zu Köln, 305 p.

Meyer, W., and Stets, J., 1998, Junge Tektonik im Rheinischen Schiefergebirge und ihre Quantifizierung: Zeitschrift der Deutschen Geologischen Gesellschaft, v. 149, p. 359–379.

Meyer, W., and Stets, J., 2002, Pleistocene to recent tectonics in the Rhenish Massif (Germany): Netherlands Journal of Geosciences, v. 81, p. 217–222.

Nöggerath, J.J., 1828, Das Erdbeben vom 23ten Februar 1828 im Königreich der Niederlande und in den Königl: Preussisch Rheinisch-Westfälische Provinzen. Jahrbuch der Chemie und Physik, v. 23, p. 1–61.

Nöggerath, J.J., 1847, Das Erdbeben vom 29: Juli 1846 im Rheingebiet und den benachbarten Ländern: Bonn, 60 p.

Nöggerath, J.J., 1870, Die Erdbeben im Rheingebiet in den Jahren 1868, 1869 und 1870, Verhandlungen des Naturhistorischen Vereins der preussischen Rheinlande und Westphalens, v. 27, 132 p.

Parolai, S., Picozzi, M., Richwalski, S.M., and Milkereit, B., 2005, Joint inversion of phase velocity dispersion and H/V ratio curves from seismic noise recordings using a generic algorithm, considering higher modes: Geophysical Research Letters, v. 32, L01303, doi: 10.1029/2004GL021115.

Parsons, T., 2005, A hypothesis for delayed dynamic earthquake triggering: Geophysical Research Letters, v. 32, p. L04302, doi: 10.1029/2004GL021811.

Pelzing, R., 1994, Source parameters of the 1992 Roermond earthquake, the Netherlands, and some of its aftershocks recorded at the stations of the Geological Survey of Northrhine-Westphalia: Geologie en Mijnbouw, v. 73, p. 215–223.

Pelzing, R., 2002, Seismizität und Erdbebengefährdung in der Niederrheinischen Bucht: Jahrestagung Deutsche Geophysikalische Gesellschaft in Hannover, http://www.dggonline.de/tagungen/dgg2002/abstracts/SO/Pelzing_P.htm (last accessed 14 April 2005).

Prinz, D., Hollnack, D., and Wohlenberg, J., 1994, The seismic activity near Aachen following the 1992 Roermond earthquake, the Netherlands: Geologie en Mijnbouw, v. 73, p. 235–240.

Reamer, S.K., and Hinzen, K.-G., 2004, An earthquake catalog for the northern Rhine area, central Europe (1975–2002): Seismological Research Letters, v. 75, p. 713–725.

Regenauer-Lieb, K., 1999, Dilatant plasticity applied to Alpine collision: Ductile void growth in the intraplate area beneath the Eifel volcanic field: Journal of Geodynamics, v. 27, p. 1–21, doi: 10.1016/S0264-3707(97)00024-0.

Ritter, J.R.R., Jordan, M., Christensen, U.R., and Achauer, U., 2001, A mantle plume below the Eifel volcanic fields, Germany: Earth and Planetary Science Letters, v. 186, p. 7–14, doi: 10.1016/S0012-821X(01)00226-6.

Robel, F., 1958, Das Erdbeben im Neuwieder Becken vom 2 Oktober 1956: Sonderveröffentlungen des Geologischen Institutes der Universität zu Köln, 4: Rheinische Erdbeben, v. I, p. 3–10.

Robel, F., 1959, Die Erdbebenstation Bensberg bei Köln: Zeitschrift für Geophysik, v. 25, p. 16–32.

Robel, F., and Ahorner, L., 1958, Das Euskirchener Beben vom 5 August 1957: Sonderveröffentlungen des Geologischen Institutes der Universität zu Köln, 7: Rheinische Erdbeben, v. I, p. 11–15.

Rosenhauer, W., and Ahorner, L., 1994, Seismic hazard assessment for the Lower Rhine Embayment before and after the 1992 Roermond earthquake: Geologie en Mijnbouw, v. 73, p. 415–424.

Scherbaum, F., 1994, Modelling the Roermond earthquake of April 13, 1992, by stochastic simulation of its high frequency strong ground motion: Geophysical Journal International, v. 119, p. 31–43.

Scherbaum, F., Hinzen, K.-G., and Ohrnberger, M., 2003, Determination of shallow shear wave velocity profiles in the Cologne/Germany area using ambient vibrations: Geophysical Journal International, v. 152, p. 597–612, doi: 10.1046/j.1365-246X.2003.01856.x.

Schmedes, J., Hainzl, S., Reamer, S.K., Scherbaum, F., and Hinzen, K.-G., 2005, Moment release in the Lower Rhine Embayment, Germany: Seismological perspective of the deformation process: Geophysical Journal International, v. 160, p. 901–909.

Schmidt, J.F.J., 1879, Studien über Erdbeben: Leipzig, 360 p.

Schmincke, H.-U., 1970, "Base surge"—Ablagerungen des Laacher See-Vulkans: Der Aufschluss, v. 21, p. 350–364.

Schmincke, H.-U., Park, C., and Harms, E., 1999, Evolution and environmental impacts of the eruption of Laacher See Volcano (Germany) 12,900 a BP: Quaternary International, v. 61, p. 61–72, doi: 10.1016/S1040-6182(99)00017-8.

Schwarzbach, M., 1951, Erdbeben des Rheinlandes: Kölner Geologische Hefte, v. 1, p. 3–28.

Sieberg, A., 1940, Beiträge zum Erdbebenkatalog Deutschlands und angrenzender Gebiete für die Jahre 58 bis 1799: Mitteilungen des Deutschen Erdbebendienstes, Jena, v. 2, p. 112.

Sponheuer, W., 1952, Erdbebenkatalog Deutschlands und angrenzender Gebiete für die Jahre 1800 bis 1899: Mitteilungen des Deutschen Erdbebendienstes, Jena, v. 3, p. 195.

Stein, R.S., 2003, Earthquake conversations: Scientific American, v. 288, p. 72–79.

Stein, R.S., 2004, Tidal triggering caught in the act (Perspective): Science, v. 305, p. 1248–1249, doi: 10.1126/science.1100726.

Stein, S., and Newman, A., 2004, Characteristic and uncharacteristic earthquakes as possible artifacts: Applications to the New Madrid and Wabash seismic zones: Seismological Research Letters, v. 75, p. 173–187.

Techmer, A., Skupin, K., Schollmeyer, G., Pelzing, R., Dickhof, A., Salamon, M., Frechen, M., and Kloastermann, J., 2005, Optical luminescence dating for timing of paleoseismological activity at the Viersen fault, lower Rhine region (Germany), *in* Kümpel, H.-J., ed., Forschungsbericht 2004 und Aktualisiertes Forschungsprogramm 2005: Hannover, Geowissenschaftliche Gemeinschaftsaufgaben, 160 p.

Toda, S., Stein, R.S., Richards-Dinger, K., and Bozkurt, S., 2005, Forecasting the evolution of seismicity in southern California: Animations built on earthquake stress transfer: Journal of Geophysical Research, v. 110, B05S16, doi: 10.1029/2004JB003415.

Van Balen, R.T., Houtgast, R.F., Van der Wateren, F.M., Vandenberghe, J., and Bogaart, P.W., 2000, Sediment budget and tectonic evolution of the Maas catchment in the Ardennes and the Roer valley rift system: Global and Planetary Change, v. 27, p. 113–129, doi: 10.1016/S0921-8181(01)00062-5.

Van den Berg, M.W., 1994, Neotectonics of the Roer valley rift system: Style and rate of crustal deformation inferred from syn-tectonic sedimentation: Geologie en Mijnbouw, v. 73, p. 143–156.

Vanneste, K., and Verbeeck, K., 2001, Paleoseismological analysis of the Rurrand fault near Jülich, Roer valley graben, Germany: Coseismic or aseismic faulting history?: Netherlands Journal of Geosciences/Geologie en Mijnbouw, v. 80, p. 155–169.

Vanneste, K., Verbeeck, K., Camelbeeck, T., Paulissen, E., Meghraoui, M., Renardy, F., Jongmans, D., and Frechen, M., 2001, Surface-rupturing history of the Bree fault scarp, Roer valley graben: Evidence for six events since late Pleistocene: Journal of Seismology, v. 5, p. 329–359, doi: 10.1023/A:1011419408419.

Van Rummelen, F.H., 1945, Overzicht van de tusschen 600 en 1940 in Zuid—Limburg en omgeving waargenomen aardbevingen, en van aardbevingen welke mogelijk hier haren invloed kunnen hebben doen gelden. Overdruk uit Mededeel Jaarverslag Genealogisch Bureau 1942–1943, 130 p.

von Kamp, H., 1986, Geologische Karte von NRW: Blätter, Geological Survey Northrhien-Westphalia, v. C5102, C5106, scale 1:100,000.

Waldhauser, F., and Ellsworth, W.L., 2000, A double-difference earthquake location algorithm: Method and application to the northern Hayward fault: Bulletin of the Seismological Society of America, v. 90, p. 1353–1368, doi: 10.1785/0120000006.

Wörner, G., Staudigel, H., and Zindler, A., 1985, Isotopic constraints on open system evolution of the Laacher See magma chamber (Eifel, West Germany): Earth and Planetary Science Letters, v. 75, p. 37–49, doi: 10.1016/0012-821X(85)90048-2.

Yeats, R.S., Sieh, K., and Allen, C.R., 1997, The Geology of Earthquakes: New York, Oxford University Press, 568 p.

Youngs, R.R., and Coppersmith, K.J., 1985, Implication of fault slip rates and earthquake recurrence models to probabilistic seismic hazard assessment: Bulletin of the Seismological Society of America, v. 75, p. 939–964.

Ziegler, P.A., 1992, European Cenozoic rift system: Tectonophysics, v. 208, p. 91–111, doi: 10.1016/0040-1951(92)90338-7.

Ziegler, P.A., 1994, Cenozoic rift system of western and central Europe: An overview: Geologie en Mijnbouw, v. 73, p. 99–127.

Ziegler, P.A., 2001, Dynamic processes controlling development of rifted basins: Eucor-Urgent Publication, v. 1, p. 51.

MANUSCRIPT ACCEPTED BY THE SOCIETY 29 NOVEMBER 2006

The Geological Society of America
Special Paper 425
2007

Motion of Adria and ongoing inversion of the Pannonian Basin: Seismicity, GPS velocities, and stress transfer

Gábor Bada[†]

Department of Geophysics, Eötvös L. University, Pázmány P. st. 1/C, 1117 Budapest, Hungary, and *Research Centre for Integrated Solid Earth Science (ISES), Vrije Universiteit, De Boelelaan 1085, 1081 HV, Amsterdam, the Netherlands*

Gyula Grenerczy

FÖMI Satellite Geodetic Observatory, 2614 Penc, Hungary

László Tóth

Seismological Observatory, Hungarian Academy of Sciences, Meredek u. 18, 1112 Budapest, Hungary

Frank Horváth

Department of Geophysics, Eötvös L. University, Pázmány P. st. 1/C, 1117 Budapest, Hungary

Seth Stein

Department of Earth and Planetary Sciences, Northwestern University, 1850 Campus Drive, Evanston, Illinois 60208, USA

Sierd Cloetingh

Research Centre for Integrated Solid Earth Science (ISES), Vrije Universiteit, De Boelelaan 1085, 1081 HV Amsterdam, the Netherlands

Gábor Windhoffer[‡]

Department of Geophysics, Eötvös L. University, Pázmány P. st. 1/C, 1117 Budapest, Hungary

László Fodor

Geological Institute of Hungary, Stefánia út 14, 1143 Budapest, Hungary

Nicholas Pinter

Department of Geology, Southern Illinois University, Carbondale, Illinois 62901, USA

István Fejes

FÖMI Satellite Geodetic Observatory, 2614 Penc, Hungary

[†]E-mail: bada@ludens.elte.hu.
[‡]deceased.

Bada, G., Grenerczy, G., Tóth, L., Horváth, F., Stein, S., Cloetingh, S., Windhoffer, G., Fodor, L., Pinter, N., and Fejes, I., 2007, Motion of Adria and ongoing inversion of the Pannonian Basin: Seismicity, GPS velocities, and stress transfer, *in* Stein, S., and Mazzotti, S., ed., Continental Intraplate Earthquakes: Science, Hazard, and Policy Issues: Geological Society of America Special Paper 425, p. 243–262, doi: 10.1130/2007.2425(16). For permission to copy, contact editing@geosociety.org. ©2007 The Geological Society of America. All rights reserved.

ABSTRACT

We present data and models for the present-day stress and strain pattern in the Pannonian Basin and surrounding East Alpine–Dinaric orogens. Formation of the Pannonian Basin within the Alpine mountain belt started in the early Miocene, whereas its compressional reactivation has been taking place since late Pliocene–Quaternary time. Basin inversion is related to changes in the stress field from a state of tension during basin formation in the Miocene to a state of compression resulting from the convergence between the Adria microplate and the European plate. Seismicity indicates that deformation is mainly concentrated along Adria's boundaries where pure contraction (thrusting in Friuli and the southeastern Dinarides), often in combination with transform faulting (dextral transpression in the central Dinarides), is predominant. Tectonic stresses and deformation are transferred into the Pannonian Basin, resulting in a complex pattern of ongoing tectonic activity. From the margin of Adria toward the interior of the Pannonian Basin, the dominant style of deformation gradually changes from pure contraction, through transpression, to strike-slip faulting. Shortening in the basin system, documented by earthquake focal mechanisms, global positioning system (GPS) data, and the neotectonic habitat, has led to considerable seismotectonic activity and folding of the lithosphere. The state of recent stress and deformation in the Pannonian Basin is governed by the interaction of plate-boundary and intraplate forces, which include the counterclockwise rotation and N-NE–directed indentation of the Adria microplate ("Adria-push") as the dominant source of compression, in combination with buoyancy forces associated with differential topography and lithospheric heterogeneities.

Keywords: Pannonian Basin, seismicity, GPS, stress transfer.

INTRODUCTION

The kinematic and dynamic processes causing extensional basin formation and subsequent deformation of back-arc basins have long been a focus of tectonic studies. The Mediterranean system of back-arc basins represents a key example of these processes within the convergence zone between the African (Nubia) and Eurasian plates (e.g., Horváth and Berckhemer, 1982; Cloetingh et al., 1995; Jackson, 1994; Ziegler et al., 2001; Faccenna et al., 2004; Rosenbaum and Lister, 2004; Horváth et al., 2006). The complex Cenozoic structural history of the region is manifested in a mosaic of tectonic domains of different geometry, thermomechanical character, and deformation pattern (Fig. 1).

Structural styles of late-stage intraplate deformation in the Pannonian Basin illustrate models of basin inversion (e.g., Williams et al., 1989; Lowell, 1995). Inversion occurs when the regional tectonic stress field changes from tension, which controlled basin formation and subsidence, to compression, which results in contraction and flexure of the lithosphere associated with differential vertical movements (Ziegler et al., 1995). The Pannonian Basin has reached a more advanced stage of evolution with respect to other active Mediterranean back-arc basins, and its structural inversion has been taking place for the last few million years. Changes in the stress field have resulted in an extensive reactivation of preexisting fault zones as documented by

several structural investigations (e.g., Csontos et al., 2005; Fodor et al., 2005; Horváth et al., 2006; Bada et al., 2006), intraplate seismicity (Tóth et al., 2002), and active contraction shown by repeated global positioning system (GPS) measurements (Grenerczy et al., 2005). The main criteria for basin inversion (Turner and Williams, 2004), i.e., reversals in the sense of fault movements, a change in the polarity of structural relief, and exhumation of the synrift basin fill, are observed. However, since inversion of the Pannonian Basin is still in an early phase, these features are not yet fully developed. Hence, reconstruction of the last episodes of the stress and strain fields, including the recent tectonics of the area, is challenging and requires integration of data from various sources.

How tectonic stress and strain propagate from plate boundaries into continental areas is a key problem (Ziegler et al., 1995, 2002). It is generally accepted that the style and amount of intraplate deformation depend on the state of the confining stress and thermomechanical properties of the deforming medium. In zones of active continental collision like the central Mediterranean, changes in plate geometry and configuration, direction and rate of relative plate motions, and rheology can cause strong temporal and lateral variations of the strain field. Since rheology controls the response of the lithosphere to stresses, and thus the tectonic reactivation of sedimentary basins, characterization of rheological properties is also of key importance in constraining and

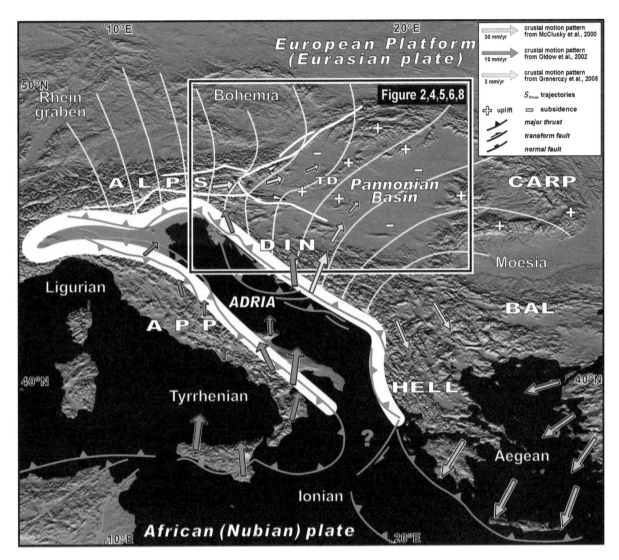

Figure 1. Topography and present-day tectonics in the central Mediterranean region. N-NE–directed motion and counterclockwise rotation of the Adriatic microplate, outlined by thick pale yellow line, result in collision in the southern Alps and dextral transpression in the Dinarides. Tectonic stresses and strain are transmitted into the Pannonian Basin, resulting in its ongoing structural inversion. Generalized pattern of crustal motions from global positioning system (GPS) studies is shown in a fixed Eurasian frame. Note the different scale for the velocities from different studies. Insert indicates area of detailed study. Structural elements for the Mediterranean, shown in red, are after Faccenna et al. (2004). Abbreviations: APP—Apennines; BAL—Balkanides; CARP—Carpathians; DIN—Dinarides; HELL—Hellenides; TD—Transdanubia.

quantifying tectonic models. This is also valid for the Pannonian region, which has been the subject of repeated tectonic activity and, hence, is an area of strain softening where tectonic units of different history and rheological properties are in close contact. Given the short-scale variations in the boundary conditions and the rigidity of the lithosphere, one can expect a complex pattern of present-day deformation manifested in changes of tectonic style and stress regime over short distances.

The focus of this paper is the active deformation of the Pannonian back-arc basin system and its vicinity. First, we outline a neotectonic framework through the main features of late-stage episodes of structural evolution. Next, we present data that

describe and quantify ongoing active deformation, including seismicity, GPS site velocities, and stress indicators. Finally, we discuss the kinematics and dynamics of basin inversion in a regional context focusing on the motion of the Adria microplate as the engine driving deformation in the Pannonian region.

NEOTECTONIC FRAMEWORK

The Pannonian Basin, located within the Alpine orogenic system (Fig. 1), is an excellent natural laboratory for active tectonic studies because of the availability of high-quality multidisciplinary data sets and its tectonic history. Alpine, Carpathian, and

Dinaric mountain belts surround this Neogene extensional basin. Its broader geological environment, the Mediterranean region, is a wide zone of convergence between the Eurasian and African plates. At present, Nubia—the portion of Africa west of the East African Rift, along which Africa began splitting 15–35 m.y. ago—is converging on Eurasia. Due to the complex kinematics of these two major plates, and the numerous microplates sandwiched between them, the region has gone through a polyphase deformation history since the opening of the Atlantic Ocean in Mesozoic times (Biju-Duval et al., 1977; Dercourt et al., 1986; Dewey et al., 1989; Şengör, 1993; Yilmaz et al., 1996). Extensional basins in this overall compressional setting are superimposed on former orogenic terranes and are associated with orogen-parallel displacement of internal blocks and oroclinal bending (Horváth and Berckhemer, 1982). Although their ages, structure, and tectonics show significant differences, there are several common features in their formation and evolution. Their similar positions behind a once- or still-active subduction zone suggest an essentially causal relationship (e.g., Wortel and Spakman, 2000; Faccenna et al., 2004). Other important geodynamic processes include the extensional collapse of gravitationally unstable orogenic wedges (Dewey, 1988), and the lateral extrusion of internal terranes (Şengör et al., 1985; Ratschbacher et al., 1991; Royden, 1993).

The behavior of the Adria microplate is fundamental to the structural history of the central Mediterranean because its motion has had a considerable effect on the evolution of the strain and stress pattern (Fig. 1). Late Cenozoic tectonic reconstructions indicate that a northward drift and simultaneous counterclockwise rotation of the Adria indenter resulted in frontal collision in the Alps and oblique convergence in the Dinarides, accommodating shortening of ~600 km since late Eocene times (ca. 40 Ma) (Dewey et al., 1989; Roeder and Bachmann, 1996; Schmid et al., 2004). On Adria's western side, the Apennines in the Pliocene-Quaternary underwent a switch from compression to extension due to the combined effect of the NE-directed motion of Adria and the consumption of subductable lithosphere beneath the main part of Italy (Stein and Sella, 2006). Space geodetic data (Ward, 1994; Oldow et al., 2002; Battaglia et al., 2004; D'Agostino et al., 2005; Grenerczy and Kenyeres, 2006; Grenerczy et al., 2005) and seismotectonic studies (Anderson, 1987; Anderson and Jackson, 1987; Console et al., 1993) indicate that Adria today moves independently from both the African (Nubian) and the Eurasian plates. However, its exact geometry and geographic extent, level of rigidity, degree of separation from Africa, and possible internal segmentation are still a matter of debate (Pinter and Grenerczy, 2006; Stein and Sella, 2006).

Intense shortening and crustal thickening at the north-northeastern front of Adria during continental collision in late Paleogene–early Miocene times resulted in gravitational instability of the Alpine-Dinaric orogen (Ratschbacher et al., 1989, 1991; Dewey, 1988; Frisch et al., 1998). Consequently, the assembly of continental blocks in the axial zone of Adria-Europe convergence disintegrated and experienced significant stretching, rigid body rotation, and translation mainly in the form of continental extru-

sion (Ratschbacher et al., 1991), which led to the formation of the Pannonian Basin during the early Miocene (e.g., Balla, 1984; Csontos et al., 1992; Horváth, 1993; Fodor et al., 1999). Plate tectonic reconstructions (Royden, 1988; Săndulescu, 1988), kinematic data (Fodor et al., 1999), and numerical modeling (Bada, 1999) highlight the predominant role of the Carpathian subduction in basin formation. Subduction led to significant back-arc lithospheric extension from the early to late Miocene due to trench-pull forces exerted on the overriding plate by the continuous rollback of the Carpathian arc. Due to the finite strength of the Pannonian lithosphere, tensional stresses were transmitted far west from the arc, so nearly the entire Pannonian Basin system extended significantly.

Unlike other Mediterranean back-arc basins, the Pannonian Basin rapidly reached a mature stage of evolution because extension ended due to the complete consumption of subductable lithosphere of the European foreland (Horváth, 1993). Thus, the basin became completely landlocked and constrained on all sides, so the continuous N-NE–directed indentation of Adria built up compressional stresses in the Pannonian lithosphere. As a result, the stress field has been changing from extension to compression, causing positive structural inversion since Pliocene times (Horváth and Cloetingh, 1996; Bada et al., 1998, 2001; Fodor et al., 1999; Horváth et al., 2006). This change in the regional stress field has resulted in fault reactivation, seismicity, and the development of surface topography (Horváth and Cloetingh, 1996; Bada et al., 1999; Gerner et al., 1999). The extended, hot, and, hence, weak Pannonian lithosphere has become prone to reactivation under relatively low compressional stresses. Due to its low rigidity and the intraplate compression concentrated in the thin elastic core of the crust, the area exhibits large-scale bending manifested in Quaternary subsidence and uplift anomalies that control the morphology within the basin system (Horváth and Cloetingh, 1996). This deformation has been interpreted as irregular lithospheric folding (Cloetingh et al., 1999), with wavelengths ranging from a few kilometers (local basin inversion) to hundreds of kilometers (whole lithospheric folding). Multiscale folding of the Pannonian lithosphere is manifested as short-wavelength vertical motions (i.e., uplifting and subsiding areas in close proximity) that define the landscape morphology and topographic features (Horváth and Cloetingh, 1996; Fodor et al., 2005; Ruszkiczay-Rüdiger et al., 2005; Pinter, 2005; Fig. 1). Paleostress data (Bada, 1999) indicate a characteristic short-scale variation of the stress directions and regimes in late Cenozoic times, reflected in changes of the dominant style of deformation (Horváth, 1995; Fodor et al., 1999). The earliest phase of compression at the southwestern margin of the basin system in the vicinity of the Alpine and Dinaric belts began in the late Miocene. In the Pliocene to Quaternary, the onset of inversion gradually advanced inward into the Pannonian Basin, farther from the Adria indenter. This shift in structural styles has not yet finished, as will be shown by contemporaneous stress and strain data. Although it is generally accepted that the behavior of Adria is key to understanding the active tectonic processes, the mode and rate of stress propagation and related deformation transfer are not fully understood.

SEISMOTECTONICS

Seismicity in the Pannonian region is moderate, with significant variations between different tectonic domains (Fig. 2; Zsíros, 2000; Tóth et al., 2002). The highest seismic activity is observed at the southern Alps–Dinarides transition in northern Italy and Slovenia, indicating significant present deformation near Adria's margin. In the Pannonian Basin, earthquake activity is more moderate and diffuse. To the northwest, at the eastern termination of the Alps, the linear trend of epicenters indicates active faulting along a SW-NE–trending shear zone, the Mur-Mürz-Žilina strike-slip fault.

Areas in the European foreland, including the Bohemian and Moesian Massifs (Fig. 1), show significantly lower levels of seismic activity, which points to strain localization in the active Alpine orogens and Mediterranean back-arc basins. The degree of strain concentration and the amount of deformation can be estimated from the spatial pattern of seismic energy release (Gerner et al., 1999; Tóth et al., 2002). The deformation in the Pannonian Basin is more intense than in the neighboring orogens, except

for the southern Alps and main part of the Dinarides. Interestingly, the Pannonian Basin appears to be far more active than the bulk of the Carpathians. These inferences may, however, be somewhat biased because the calculated seismic energy release is dominated by large but infrequent earthquakes. Thus, the values obtained by Gerner et al. (1999) for the Pannonian Basin, characterized by low to medium earthquake magnitudes, offer a conservative estimate for the level of seismicity.

Seismotectonic models have been published for the Alps and northern Dinarides, where a dense seismic network allows the precise determination of earthquake hypocenters (Anderson and Jackson, 1987; Slejko et al., 1989; Carulli et al., 1990; Favali et al., 1990; Del Ben et al., 1991; Console et al., 1993; Poljak et al., 2000). The seismotectonics of the Pannonian Basin were originally explained by simple models based upon delineation of seismically active zones at the boundaries of internal rigid blocks (Horváth, 1984; Gutdeutsch and Aric, 1988). These zones, however, have been always somewhat subjective because of the dispersed character of epicenter distribution. Due to inaccurate seismic and geological information, it is difficult to assign most

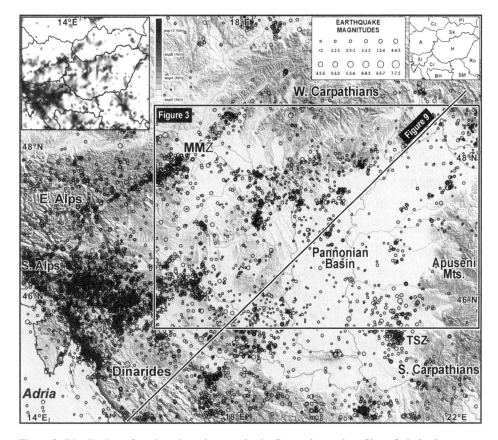

Figure 2. Distribution of earthquake epicenters in the Pannonian region. Size of circles is proportional to magnitude. Data are from the GeoRisk Earthquake Research Institute, Budapest, Hungary (Zsíros, 2000; Tóth et al., 2002). Upper-left inset shows cumulative seismic energy release calculated by Gerner et al. (1999). MMŽ—Mur-Mürz-Žilina fault zone; TSZ—Timişoara seismogenic zone. Upper-right inset indicates political boundaries. Country abbreviations: A—Austria; BH—Bosnia-Herzegovina; Cz—Czech Republic; H—Hungary; Cr—Croatia; Pl—Poland; Ro—Romania; Sk—Slovakia; Sl—Slovenia; SM—Serbia.

earthquakes to specific tectonic structures at an acceptable level of confidence. This is particularly the case for small events (M < 4). For large historical earthquakes, the difficulty probably comes mostly from inaccurate hypocenter information. Consequently, the surface expression, exact three-dimensional (3-D) geometry, mechanical properties, and reactivation potential of most seismoactive faults are poorly constrained. The diffuse pattern of epicenters is thought to be the result of the short time interval of earthquake instrumental records relative to long recurrence times. Thus, the observed seismicity is insufficient to fully describe the seismicity of the Pannonian Basin. Even so, seismotectonic models account for several features of seismicity. A main weakness of these models is, however, that the seismicity within discrete tectonic blocks is difficult to explain by block models that ignore internal deformation. Consequently, Gerner et al. (1999) explained the seismicity in the Pannonian Basin by the release of accumulated stress along existing weakness zones properly oriented with respect to the prevailing stress field. These zones correspond either to preexisting faults or other sites of compositional or thermal weaknesses. Deformation can occur anywhere if the resolved stress exceeds the local shear strength of the rock. Thus, neither earthquakes nor other forms of deformation are restricted to block boundaries, yielding a more distributed seismicity pattern.

Focal depths in the Pannonian region suggest that earthquakes are concentrated within the uppermost 20 km of the crust. Consistent with a rheological model of the area (Cloetingh et al., 2006), most events occur at depths of 6–15 km. The lack of earthquakes below 20 km argues for a weak Pannonian Basin with lithospheric strength considerably lower than in the surrounding areas. Earthquakes occur almost exclusively in the brittle upper crust, limited by the thermally controlled depth to the brittle-ductile transition. Given the high heat flow in the basin (Dövényi and Horváth, 1988), brittle behavior is restricted to shallow depths. A combination of deformation modes, i.e., brittle faulting and ductile flow, is likely at the brittle-ductile transition zone at a depth of ~10–15 km. On the other hand, the Pannonian lithosphere has extremely low rigidity, making it prone to repeated tectonic reactivation, which often occurs by aseismic deformation. This idea is supported by the Quaternary vertical motions commonly associated with large-scale buckling (Horváth and Cloetingh, 1996). The exact proportion of ductile versus brittle deformation is difficult to assess due to limitations of recurrence estimates for intraplate seismicity.

Various seismological databases give insight into earthquake recurrence rates on different time scales, including paleo-earthquakes ($<10^{-3}$ yr^{-1}), historical earthquakes (10^{-3}–10^{-2} yr^{-1}), instrumental data (10^{-2}–10^{-1} yr^{-1}), and site-specific instrumental data or local seismic monitoring ($>10^{-1}$ yr^{-1}). The earthquake catalogue of the Pannonian region goes back to pre-instrumental time (Zsíros, 2000). However, for historical events, the only source of information is from macroseismic observations of the felt intensity distribution. Hence, magnitudes have been estimated by assuming that they depend on epicentral intensity and focal depth according to the relation of $M_M = aI_0 + b(\log h) + c$, where M_M is the macroseismic magnitude, I_0 is the epicentral

intensity, h the focal depth in km, and a, b, and c are constants. The best-fitting values of these constants for the central Pannonian Basin, between 45.5°N and 49.0°N latitude and 16.0°E and 23.0°E longitude (rectangle in Fig. 2), are 0.68 ± 0.02, 0.96 ± 0.07, and -0.91 ± 0.10, respectively (Tóth et al., 2006). Consequently, the best-fitting relation using an average 12.6 km focal depth becomes $M_M = 0.68I_0 + 0.15$. The historic and instrumental data have been combined to estimate earthquake recurrence rates using the Gutenberg and Richter (1944) relation $\log N = a - b\mathrm{M}$, where N is the annual number of earthquakes with magnitude equal or greater than M. For the central Pannonian Basin, earthquake recurrence times approximately follow $\log N = 3.3 - 0.9\mathrm{M}$ for most magnitude ranges (Fig. 3). This corresponds to a return period of ~100 yr for M = 6 earthquakes, whereas M = 5 events occur every 20 yr on average. Based on the results of high-sensitivity monitoring in the last decade (Tóth et al., 2002, 2005), the average annual number of M = 3 and M = 2 earthquakes is 4 and 30, respectively. Completeness tests based on the "magnitude recurrence fit" (Tóth et al., 2006) show that the catalogue can be considered complete since 1600 for M ≥ 5.8, since 1700 for M ≥ 5.3, since 1800 for M ≥ 4.7, since 1850 for M ≥ 4.2, since 1880 for M ≥ 3.5, and since 1995 for M ≥ 2.0 events. Figure 3 also suggests possible deviations from the trend. Large earthquakes (M ~ 6) appear to occur less frequently than expected based on the rates of small ones (M ≤ 4), whereas the frequency of medium events (4.5 ≤ M ≤ 5.5) is slightly higher. This might be of tectonic importance, due to the thermomechanical character of the Pannonian lithosphere, or the finite dimensions of seismoactive faults, or an artifact owing to the limited time sampling of the earthquake catalogue (Stein and Newman, 2004).

CRUSTAL DEFORMATION FROM GPS DATA

Geodynamic processes, driving forces, and associated plate motions in the mobile Mediterranean system are reflected in the present-day surface deformation pattern. Direct measurement of crustal movements in the Pannonian Basin and its surroundings, using the GPS space geodetic technique, started in the early 1990s with the establishment of regional, national, and local programs (for an overview, see Fejes et al., 1993a, 1993b; Grenerczy, 2005). The spatial coverage and site density of any of these networks alone are insufficient to provide an intraplate velocity field at a resolution adequate for reliable (neo)tectonic interpretation. Consequently, velocity solutions from national-scale networks were combined, and the pattern of recent horizontal crustal motions relative to the European Platform was obtained for the central Mediterranean region (Fig. 1). Figure 4 shows GPS velocities for the Pannonian Basin and adjacent territories interpolated into a continuous velocity field using linear point kriging. The field illustrates the large-scale contemporaneous crustal motion pattern from the margins of Adria toward the rigid Eurasian plate interior.

The velocity field indicates a heterogeneous pattern, in both direction and rate of crustal motions (Grenerczy and Kenyeres, 2006; Grenerczy et al., 2005). The Adria microplate moves

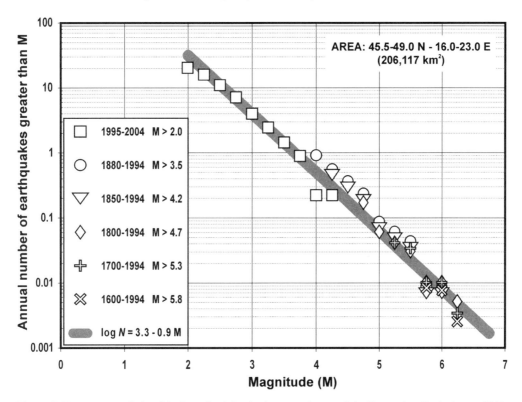

Figure 3. Recurrence relationship for seismicity in the central part of the Pannonian Basin (area: 45.5–49.0°N, 16.0–23.0°E, box in Fig. 2). The annual number of earthquakes, on a logarithmic scale, is plotted as a function of magnitude. The catalog time intervals differ as a function of magnitude.

northeast with respect to the stable European Platform and seems to be independent from both the Eurasian and African (Nubian) plates. Its northern parts move at ~2.5–3.5 mm/yr without significant differences between the eastern and western sectors. These values increase to 4–5 mm/yr further to the southeast along the Dinarides. Site velocities at the margins of Adria indicate counterclockwise rotation around an Euler pole in the western Alps (Grenerczy et al., 2005), in accord with seismotectonic studies (Anderson and Jackson, 1987), paleomagnetic analysis (Márton, 2005), and other space geodetic investigations (Ward, 1994; Calais et al., 2002; Battaglia et al., 2004; D'Agostino et al., 2005). However, the present data are too sparse to show whether Adria moves as a single rigid microplate or deforms internally (Calais et al., 2002; Oldow et al., 2002; Battaglia et al., 2004).

The northeastward motion of the Adria microplate is transferred far beyond its boundaries, causing a complex strain pattern through the Alps, Dinarides, and Pannonian Basin (Fig. 4). At the northern tip of Adria, calculated velocities suggest northward indentation into the Alpine chain at an average convergence rate of 2.5 mm/yr. It appears that this northward motion is completely absorbed in the Alps, resulting in 30 ppb/yr strain rate over the southern Alps, where the bulk of contraction is concentrated (see also D'Agostino et al., 2005), as supported by the high seismic activity (Fig. 2). In contrast, N-S shortening is rather limited in the eastern Alps. Instead, GPS data indicate

orogen-parallel extrusion of crustal blocks to the E-NE, toward the interior in the Pannonian Basin (Grenerczy et al., 2000). This pattern is a persistent feature of the tectonic history in the eastern Alps–Pannonian Basin, where lateral extrusion of crustal and lithospheric flakes due to the gravitational collapse of the Alpine orogen was closely associated with the formation of the Pannonian Basin (e.g., Ratschbacher et al., 1991; Horváth, 1993; Frisch et al., 1998; Bada, 1999; Fodor et al., 1999). Eastward motion of these units diminishes toward the east, suggesting considerable intraplate deformation within the Pannonian Basin (Grenerczy, 2002). At the northeastern edge of Adria, the central Dinarides take up 3–4 mm/yr convergence. Unlike the Alps, deformation is transmitted across the Dinarides and into the Pannonian Basin without a major change in the direction of motion. However, the velocity gradually decreases toward the northeast, with distance from Adria. About ~1–1.5 mm/yr crustal motion is absorbed in the Pannonian Basin, causing around 4 ppb/yr of contraction. However, deformation is not uniform, and significant strain partitioning prevails across the Pannonian Basin. Most deformation is concentrated in the western and central parts of the basin, whereas the eastern areas and most parts of the Carpathians show no detectable horizontal motion. Significant contraction is also shown by systematic shortening of GPS baseline lengths within the basin (Grenerczy and Bada, 2005). The shortening rate is around 1–2 mm/yr, consistent with the model of structural inversion in the

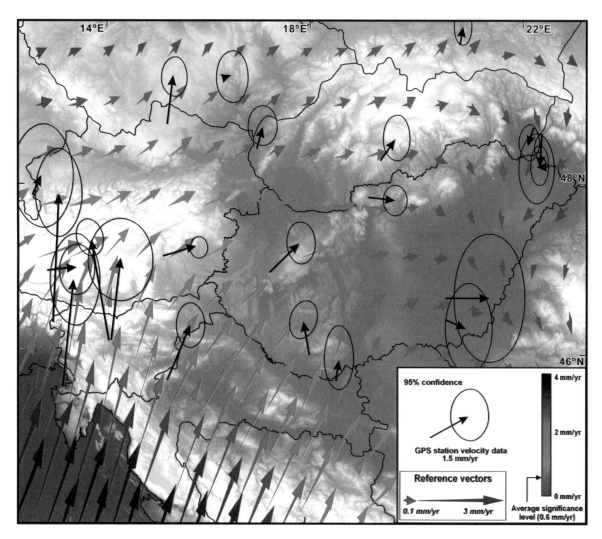

Figure 4. Global positioning system (GPS) site velocities and interpolated GPS velocity field in the Pannonian region with respect to stable Eurasia (after Grenerczy et al., 2005). The velocities show the motion of crustal units to the E-NE at decreasing rates away from the margins of Adria. Interpolation was done using a larger-scale data set.

Pannonian Basin. The current resolution of the GPS site velocities is, however, insufficient to identify and quantify localized deformation along large-scale shear zones or individual fault segments. Thus, the relative motion between basement blocks is not yet well constrained and requires further GPS site densification. Consequently, the style of local deformation and fault kinematics are presently inferred from the analysis of stress indicators, seismotectonic studies, or structural investigations.

STRESS DIRECTIONS AND REGIMES IN THE PANNONIAN REGION

The rheological weakness of the Pannonian Basin and its tectonic position near the Europe-Adria collision zone make it a sensitive recorder of changes in lithospheric stress induced by near- and far-field plate tectonic processes. To constrain the dynamics of basin inversion and the sources of active deformation in the Pannonian region, efforts have been made to infer the state of tectonic stress by collecting contemporary stress indicators (Gerner et al., 1999). The stress database, which has been compiled according to the conventions of the World Stress Map Project (Zoback, 1992) since the early 1990s and updated regularly, contains several hundred entries. These data include 198 earthquake focal mechanism solutions (FMS), 223 borehole breakouts, and 19 in situ stress measurements. Besides the main trends of stress directions (Fig. 5), the areal distribution of tectonic regimes, mainly based on focal mechanism solution data, was also analyzed (Fig. 6). The quantity of data is adequate for statistical analysis to determine mean horizontal stress directions and stress regimes. The density of data is, however, inadequate to resolve the regional pattern of vertical stress deviations, which is therefore not addressed in this paper.

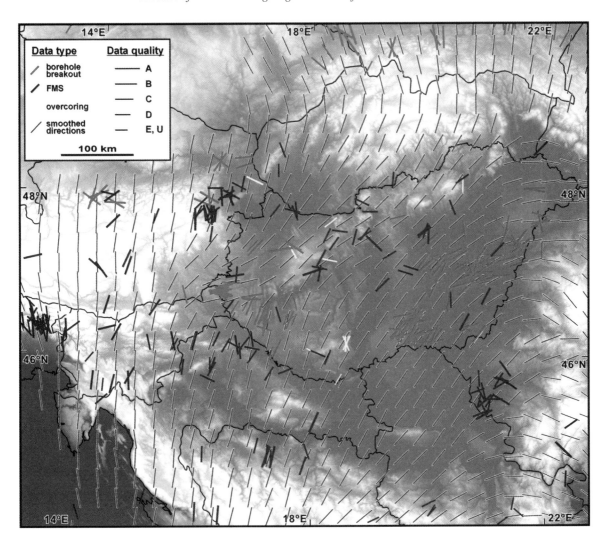

Figure 5. Contemporary stress directions in the Pannonian region. Data on the orientation of the maximum horizontal stress axis (S_{Hmax}) come from earthquake focal mechanism solutions (FMS), borehole breakout analyses, and in situ stress measurements. The smoothed stress pattern of the displayed data set, obtained by the algorithm of Hansen and Mount (1990), is also plotted.

Horizontal Stress Directions

The Pannonian Basin and surroundings show significant lateral variations in maximum horizontal stress orientations (S_{Hmax}) (Fig. 5). As recognized by Müller et al. (1992) and Rebaï et al. (1992), and confirmed by Bada et al. (1999) and Gerner et al. (1999), the present-day stress pattern in this region is considerably more heterogeneous than in the stable continental areas of the European Platform. This complexity is typical for mobile areas in orogenic belts, such as the whole Mediterranean region.

The most characteristic feature of the contemporaneous stress field is the gradual clockwise rotation of S_{Hmax} directions from west to east along the margins of Adria illustrated by the smoothed stress directions (Fig. 5). The alignment of compression in the southern Alps is NNW-SSE, becoming N-S further north in the central part of the eastern Alps, and to the east in the

Alps-Dinarides junction zone in Slovenia. At the Alps-Carpathians transition, the dominant direction of S_{Hmax} is NNE-SSW to NE-SW. Further south, the radial pattern of maximum horizontal stress can be traced along the Dinaric belt as S_{Hmax} orientations gradually change from NNE-SSW in Croatia, to NE-SW in Bosnia and Serbia, and then to E-W and WNW-ESE toward the southern Carpathians in Romania.

The regional fan-like alignment of compression is also detectable inside the Pannonian Basin. Stress indicators, mainly borehole breakout data (Fig. 5), suggest that S_{Hmax} orientation changes from NNE-SSW in the southwestern parts to NE-SW in the east. Whereas a rather uniform stress direction prevails in the eastern Pannonian Basin, the western areas show rapid lateral changes (Windhoffer et al., 2001). In western Hungary, maximum horizontal stress becomes oriented nearly E-W, close to areas with NNE-SSW stress directions more to the south.

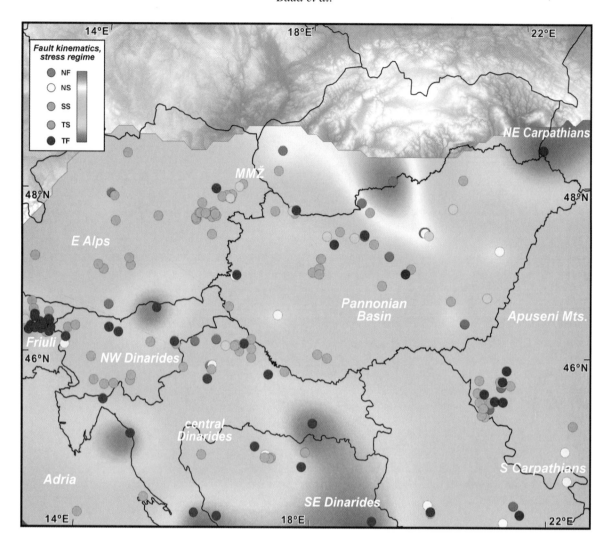

Figure 6. Stress regime and style of faulting in the Pannonian region obtained from earthquake focal mechanism solutions. Focal mechanisms are classified as normal faulting (NF), transtension (NS), strike-slip faulting (SS), transpression (TS), or thrust faulting (TF) and are plotted as colored circles. Main trends of the dominant tectonic style, calculated from the interpolation of focal mechanism solutions, are shown by corresponding coloring. MMŽ—Mur-Mürz-Žilina fault zone.

This short-wavelength stress deviation is likely to be related to the lateral motion of crustal blocks squeezed out from the eastern Alps (Bada et al., 2001), as also shown by GPS data. The alignment of maximum horizontal stress closely matches the direction of ongoing crustal motions in most parts of the study area (cf. Figs. 4 and 5). Differences appear only in the southern Alps, where the NNW-SSE alignment of S_{Hmax} deviates ~20°–30° from crustal motion directions derived from GPS measurements.

Stress Regimes from Earthquake Focal Mechanisms

The dominant style of deformation and stress regimes were determined from earthquake focal mechanisms derived using first-motion analysis for the Pannonian Basin and surrounding areas (Gerner et al., 1999; Marović et al., 2002; Tóth et al.,

2002, and references therein). Fault-plane solutions represent slip on the fault during an earthquake, and they yield the pressure (*P*) and tension (*T*) axes of maximum shortening and extension, respectively. Stress directions were inferred on the assumption that the most compressive principal stress, σ_1, falls near the *P* axis, whereas the least compressive principal stress, σ_3, closely coincides with the *T* axis. As argued by McKenzie (1969), *P* and *T* axes do not necessarily coincide with stress directions when earthquakes occur along preexisting faults, which is often the case in the Pannonian region. Regional compilations, however, show that the average orientation of the *P* axis provides a reasonably good indication of the orientation of σ_1 (Sbar and Sykes, 1973). In addition, earthquake *P* and *T* axes often correlate well with axes of geodetic strain (Molnar et al., 1973; Klosko et al., 2002; England, 2003), justifying comparison of GPS data with distributed earthquake focal mechanisms.

Following Anderson (1951), the orientation of principal stress axes with respect to the horizontal plane determines the style of faulting. Accordingly, three main stress regimes are defined: normal faulting (NF) when P (σ_1) is vertical and T (σ_3) is horizontal; strike-slip (SS) when both P (σ_1) and T (σ_3) are horizontal; and thrust faulting (TF) when P (σ_1) is horizontal and T (σ_3) is vertical. The World Stress Map Project also uses transtension (NS) and transpression (TS), which are combinations of strike-slip faulting with normal and thrust faulting, respectively (Zoback, 1992). We applied a similar categorization scheme by adopting cutoff values for plunges of P, T, and B axes (Zoback, 1992). Each fault-plane solution was assigned one of these five stress regime categories and plotted by color circles (Fig. 6). Since focal mechanism solution data are irregularly distributed in the Pannonian region, the lateral variation of the stress regimes was estimated by interpolation to areas with few or no data using the kriging method. For calculation, the quality of the stress data, according to the World Stress Map Project quality ranking scheme (Zoback and Zoback, 1989), was taken into account, and the regime assignments were converted to numerical values. Each focal mechanism solution data entry was assigned a number: NF-1, NS-2, SS-3, TS-4, TF-5 (after Müller et al., 1997). It should be noted that an interpolated strike-slip regime can result from either strike-slip fault-plane solutions or averaging of normal and reverse solutions. Despite these limitations, the regional distribution of stress regimes turned out to be useful and consistent with other tectonic and geodetic data.

Similar to the stress directions, tectonic regimes in the Pannonian region show significant lateral variations (Fig. 6). Small-scale changes in the stress regime, over distances of a few tens to hundreds of kilometers, indicate complex tectonic processes and/or a highly heterogeneous crust. The dominant style of deformation at the margins of Adria is thrusting, often in combination with strike-slip faulting, reflecting convergence between Adria and the Alps-Dinarides. Variations within the Dinarides may be due to the angular difference between the mean direction of tectonic compression resulting from Adria motion and the main structural trends. In other words, oblique convergence in the central Dinarides leads to dextral transpression along inherited NW-SE–trending fault zones associated with lateral motion of crustal blocks. On the other hand, compression in the Friuli–northwestern Dinarides and the southeastern Dinarides is perpendicular to the main structural trends, resulting mainly in reverse faulting with only minor strike-slip faulting (cf. Figs. 4, 5, and 6). In the eastern Alps and particularly along the Mur-Mürz-Žilina shear zone, pure strike-slip faulting prevails, which confirms that most shortening between Adria and the European Platform is absorbed in the narrow Friuli zone within the southern Alps (D'Agostino et al., 2005; Grenerczy et al., 2005).

The stress regime changes gradually from the edge of Adria toward the interior of the Pannonian Basin. Accordingly, the dominant style of deformation shifts from transpression in the south and southwest to strike-slip faulting in the central and southeastern part of the basin. Some of the transferred stresses are released en route, so this effect gets weaker farther from Adria. This is supported by GPS data, which show that the degree of shortening is decreasing toward the northeast, in the direction of the Pannonian Basin center (Grenerczy and Bada, 2005). We therefore suggest that inversion of the Pannonian Basin system may be in a more advanced stage in its western and southern parts. Fault-plane solutions in the northeastern Carpathians (some outside the area of Fig. 6) make this simple pattern more complicated, implying that inversion is somewhat hampered in the eastern Pannonian Basin, perhaps due to sublithospheric processes and thermal effects resulting from an asthenospheric dome (Huismans et al., 2001). If this is the case, compression may be transmitted across the Pannonian lithosphere, resulting in the propagation of shortening as far as the northeastern part of the Carpathian orogen, which is the northeast edge of the contact zone between the Mediterranean mobile belt and the southern edge of stable Europe.

Stress Inversion of Earthquake Fault-Plane and Slip Data

A more quantitative investigation of the stress field was carried out by inverting earthquake mechanisms. We determined the principal stress directions and the tectonic regime, the latter from the shape and orientation of the stress ellipsoid. Fault-plane solutions provide the orientation of the nodal (fault and auxiliary) planes, the slip vector along the fault, and the sense of slip. By assuming that fault slip during an earthquake occurs in the direction of maximum resolved shear stress (Bott, 1959), one can determine a best-fitting stress tensor responsible for the slip along that particular fault (Gephart and Forsyth, 1984). The method is similar to the inversion of fault-slip data in paleostress investigations (Angelier, 1979). Because first-motion focal mechanisms cannot distinguish between the fault and auxiliary planes, both planes were utilized. The inversion yields four parameters: the orientation of three principal stress axes ($\sigma_1 \geq \sigma_2 \geq \sigma_3$) and their relative magnitudes expressed by the stress ratio factor $R = (\sigma_2 - \sigma_3)/(\sigma_1 - \sigma_3)$. We employed the TENSOR program (Delvaux, 1993) to find these four parameters. This method, elaborated originally by Angelier and Mechler (1977) and adopted by Delvaux (1993), is especially suitable for the inversion of focal mechanisms because orthogonal planes corresponding to the fault and nodal planes are used to identify the compressional and extensional quadrants. The separation of focal mechanism solution data into subsets and a first estimation of the stress tensor were performed using this method. Subsequently, a more accurate calculation was carried out for each regional group of mechanisms by applying the rotational optimization method (Delvaux and Sperner, 2003), which minimizes the slip deviation α, which is the angular misfit between the theoretical and actual slip directions on the fault plane.

Inversion of fault-plane solutions yielded seven regional data subsets (Fig. 7) that correspond to stress provinces characterized by rather uniform stress orientations (directions of $\sigma_1 \geq \sigma_2 \geq \sigma_3$) and regimes ($R$ values):

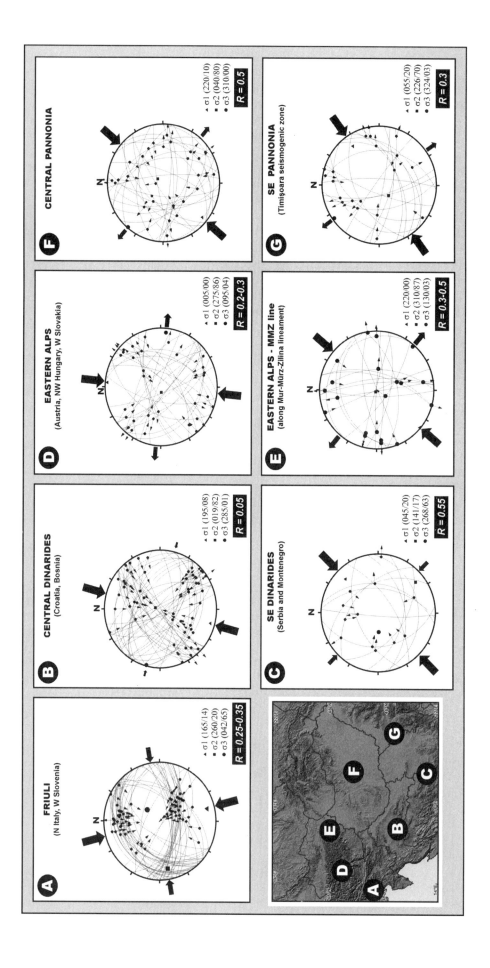

Figure 7. Results of stress inversion of focal mechanism solutions in the Pannonian region. Seven stress provinces are characterized by consistent stress directions and regimes. Stereograms show fault and slip orientation from focal mechanisms, and the direction of the principal stress axes $\sigma_1 \geq \sigma_2 \geq \sigma_3$. The shape of the stress ellipsoids is described by the ratio of the principal stress magnitudes, expressed by $R = (\sigma_2 - \sigma_3)/(\sigma - \sigma_3)$.

1. The Friuli zone in northern Italy (Fig. 7A) is characterized by predominantly ENE-WSW–oriented reverse faults that define a pure thrust-faulting stress regime (Fig. 6). The azimuth of σ_1 is 165° and $R = 0.25$–0.35.

2. The central Dinarides (Fig. 7B) is dominated by NW-SE– and NE-SW–directed dextral and sinistral faults, respectively, often in combination with thrusting that argues for NNE-SSW–directed compression. The low value of the R indicates that σ_2 and σ_3 are close in magnitude, which is typical for transpression (Fig. 6). The azimuth of σ_1 is 195° and $R = 0.05$.

3. Even though only a few focal mechanism solution data are available, so stress inversion is unreliable in statistical terms, the southeastern Dinarides are characterized by pure contraction (Fig. 6) with NW-SE–oriented reverse faults (Fig. 7C). The azimuth of σ_1 is 045° and $R = 0.55$.

4. In the eastern Alps, NW-SE– and NE-SW–directed strike-slip faults are activated in dextral and sinistral senses, respectively, defining a strike-slip faulting stress regime (Fig. 6). The maximum horizontal compression, coinciding with σ_1, is N-S directed (Fig. 7D). The azimuth of σ_1 is N005° and $R = 0.2$–0.3.

5. Stress directions near the seismoactive Mur-Mürz-Žilina shear zone exhibit a 35° clockwise deviation from the rest of the eastern Alps, while both areas show mainly strike-slip motions (Fig. 6). The orientation of σ_1 is close to NE-SW, and there is evidence of reactivation on a set of E-W–striking sinistral and N-S–trending dextral faults, which constitute the principal displacement zone (Fig. 7E). The azimuth of σ_1 is N220° and $R = 0.3$–0.5.

6. Most of the central Pannonian Basin is characterized by a strike-slip faulting stress regime (Fig. 6), and the maximum principal stress axis is oriented NE-SW (Fig. 7F). Although the R value indicates a stable stress field, the diverse focal mechanisms reflect a heterogeneous stress regime with a deformation style that varies systematically from SW to NE (Fig. 6). In addition, during stress inversion, a relatively large tolerance for α had to be introduced to obtain a single stress tensor. Interestingly, the orientation of the maximum principal stress (σ_1) appears to be rather constant across the basin interior. The azimuth of σ_1 is N220° and $R = 0.5$.

7. The Timişoara seismogenic zone in the southeast Pannonian Basin is dominated by ENE-WSW–oriented sinistral and roughly N-S–directed dextral faulting, suggesting the dominance of a strike-slip faulting stress regime (Fig. 7G). The azimuth of σ_1 is N055° and $R = 0.3$.

The reduced stress tensors for the Mur-Mürz-Žilina lineament and the central and southeastern parts of the Pannonian Basin are quite similar. One may therefore argue for a single stress province in these areas, at a distance from Adria's margins. Separation of fault-plane solutions, however, seems reasonable due to the geographical clustering of epicenters (Fig. 6).

DISCUSSION

Structural Inversion in the Pannonian Basin

Earthquake data, the GPS velocity field, and the present-day stress pattern suggest that active contraction is taking place in the Pannonian lithosphere. Earthquake data indicate that the most intense deformation is at the contact zone between Adria and the Alpine-Dinarides orogen. Our results suggest that compressional tectonic stresses are transferred far from the margins of Adria, resulting in a complex stress and strain pattern in the Pannonian region (Figs. 8 and 9). The direction of compression closely matches the direction of crustal motions in most parts of the study area (Figs. 4 and 5). In the Friuli zone of the southern Alps, however, an angular difference of ~20°–30° occurs between the calculated GPS motions and the direction of tectonic compression inferred from P axis orientations and the results of stress inversion. This mismatch, if real, may reflect a difference between horizontal surface motions shown by GPS measurements and subsurface shortening inferred from earthquake mechanisms.

The high level of compression concentrated in the elastic-brittle core of the Pannonian lithosphere is likely to be comparable to its integrated rheological strength (Figs. 8 and 9C). Basin modeling shows that an increase in the level of compressive stress during Pliocene-Quaternary times can explain accelerated subsidence in the center of the Pannonian Basin and uplift of its flanks. Therefore, both observations (Horváth et al., 2006) and modeling (Horváth and Cloetingh, 1996) indicate that stresses can cause differential vertical movements across the basin-orogen system. Several flat-lying, low-altitude areas (e.g., Great Hungarian Plain, Sava and Drava troughs) have been continuously subsiding since the onset of basin formation in the early Miocene and were completely filled with a 500–2500-m-thick alluvial-lacustrine sequence during the Pliocene-Quaternary. In contrast, the periphery of the basin system and the internal mountain ranges in Transdanubia have uplifted and significantly eroded since late Miocene–Pliocene times. Fission-track studies, exposure dating, and morphotectonic analysis suggest that this process started in the Carpathians and Apuseni Mountains during the Miocene (Sanders et al., 1999), followed by an early phase of uplift in southern Transdanubia in the Pliocene, and then the uplift of northern Transdanubia during the Pleistocene (Fodor et al., 2005; Ruszkiczay-Rüdiger et al., 2005). A similar pattern was recognized for the evolution of the stress fields during the latest Miocene through Quaternary. Paleostress studies (Bada, 1999; Fodor et al., 1999) indicate that inversion started in the southwestern regions of the Pannonian Basin at the end of the Miocene. The onset of structural inversion gradually migrated from southwest to northeast, toward more internal parts of the system, first to the main part of Transdanubia and then the western parts of the Great Hungarian Plain. The earlier start of basin inversion closer to Adria is manifested in a more pronounced structural expression, including significant vertical surface motions,

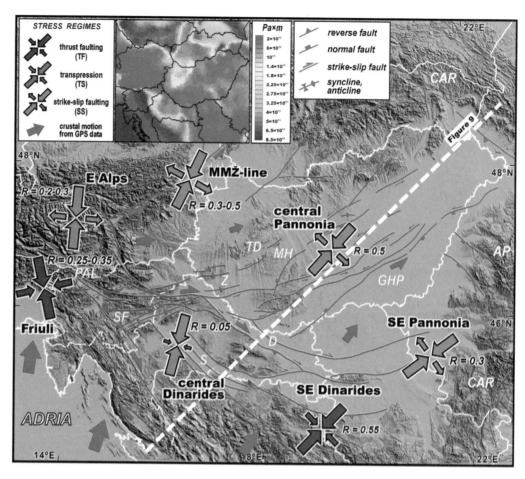

Figure 8. Contemporary stress field in the Pannonian region, pattern of horizontal crustal motions, and the location and kinematics of active structures. Horizontal stress directions are shown by arrows, whereas stress regimes are indicated by coloring, relative size of stress arrows, and the *R* value. Inset shows integrated lithospheric strength of the system, revealing high rheological contrast between the Pannonian Basin and neighboring orogenic terrains (after Cloetingh et al., 2005b). Structural elements were compiled partly after Horváth and Tari (1988), Pogácsás et al. (1989), Prelogović et al. (1998), Tomljenović and Csontos (2001), Bada et al. (2003), Wórum and Hámori (2004), Decker et al. (2005), Fodor et al. (2005), and Vrabec and Fodor (2005). Abbreviations: AP—Apuseni Mountains; CAR—Carpathians; D—Drava trough; GHP—Great Hungarian Plain; MH—Mid-Hungarian fault system; PAL—Periadriatic line; S—Sava trough; SF—Sava folds; TD—Transdanubia; Z—Zala basin; GPS—global positioning system.

whereas related structural features are far less developed in the eastern areas. For instance, structural cross sections suggest that fold amplitudes decrease from west to east from the Sava fold belt through the Zala basin to southern Transdanubia (Horváth, 1995; Placer, 1999; Márton et al., 2002; Bada et al., 2006), with earliest phase of folding occurring in the latest Miocene (Fig. 8). No such pronounced compressional structures are present in the Great Hungarian Plain. Instead, active deformation is taking place along preexisting faults and shear zones, and these structures are reactivated mainly in a strike-slip sense.

Stress Transfer and Ongoing Deformation

Present-day propagation of tectonic stress from the margins of Adria, the main engine of active deformation in the central Mediterranean, results in orthogonal convergence in the Friuli area of the southern Alps and in the southeastern Dinarides (Figs. 8 and 9A). In the central Dinarides, convergence is oblique to the main structural trends, causing a combination of dextral strike-slip and reverse faulting and associated lateral motion of crustal blocks (see also Ilić and Neubauer, 2005). Much of Adria's convergence is taken up by this Dinaridic transpressional corridor, which is sandwiched between Adria and the Pannonian Basin. Stress inversion predicts short-wavelength variation of tectonic regimes here due to the similar magnitudes of, and hence, the possible flip-flop between the two least principal stresses, σ_2 and σ_3.

The eastern Alps are almost exclusively characterized by strike-slip faulting with no major shortening north of the Periadriatic line, which separates the eastern and southern Alps. Instead,

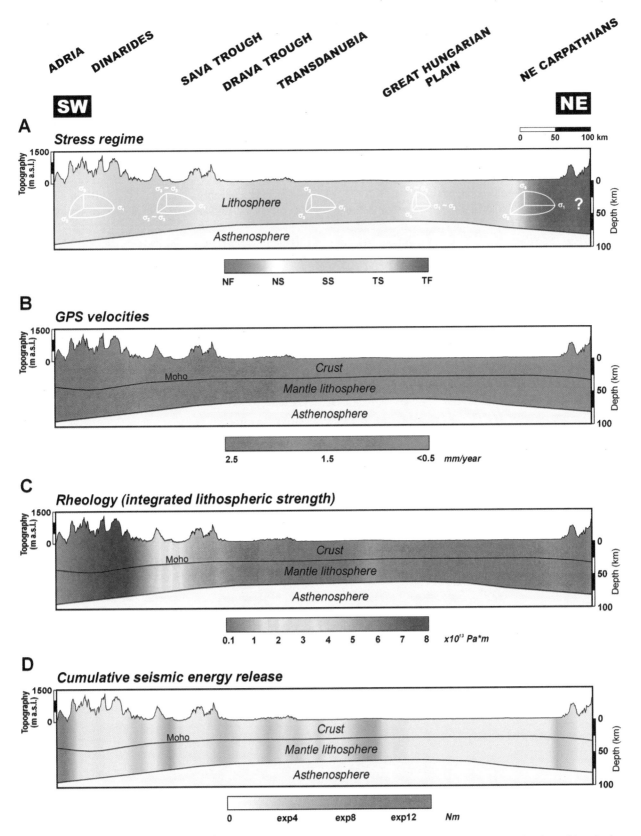

Figure 9. Transects across the center of the Pannonian Basin from the Adriatic microplate in the southwest to the Carpathians in the northeast (a.s.l.—above sea level). (A) Change of stress regimes determined from earthquake focal mechanisms. Focal mechanisms are classified as normal faulting (NF), transtension (NS), strike-slip faulting (SS), transpression (TS), or thrust faulting (TF). (B) Horizontal surface deformation rate inferred from global positioning system (GPS) measurements. (C) Rheology of the lithosphere based on strength modeling. (D) Cumulative seismic energy release. Vertical variation of these parameters is not considered. Depth of Moho and thickness of lithosphere are from Horváth et al. (2006). For location of transects, see Figures 2 and 8.

the GPS velocities (Grenerczy et al., 2000, 2005) and horizontal stress deviations in the western Pannonian Basin indicate that tectonic units are squeezed out from the axial zone of the Alpine orogen to the E-NE (cf. Figs. 4 and 5). Lateral extrusion of crustal flakes toward the basin interior occurs along the sinistral Mur-Mürz-Žilina fault zone in the north (Decker et al., 2005) and a more complex transpressional corridor adjoining the Mid-Hungarian fault system in the south (e.g., Tomljenović and Csontos, 2001; Bada et al., 2003; Wórum and Hámori, 2004; Fodor et al., 2005; Vrabec and Fodor, 2006). In contrast, compression directly propagates from Adria's margin through the Dinarides into the Pannonian Basin, as indicated by the stress regime distribution (Figs. 6 and 9A) and the present structural pattern (Fig. 8). Focal mechanisms (Gerner et al., 1999; Tóth et al., 2002) support the conclusion that active deformation in the Pannonian Basin is primarily controlled by the reactivation of preexisting (Miocene) extensional faults, mainly with strike-slip character. In the western part of the basin, reverse faulting results in folding of the Neogene to Quaternary strata. More to the east, the style of deformation becomes strike-slip faulting with transpressional (local shortening) and, in the northeastern part of the Great Hungarian Plain, transtensional (local extension) character. The influence of Adria motion is less and less pronounced toward the eastern part of the Pannonian Basin where asthenosphere-lithosphere interaction may take over, resulting in limited ongoing lithospheric extension (Becker, 1993; Huismans et al., 2001).

Possible sources of compression driving inversion in the Pannonian Basin have been investigated by numerical modeling (Bada et al., 1998, 2001). Results suggest that the recent stress in the Pannonian Basin is primarily controlled by the counterclockwise rotation and N-NE motion of the essentially rigid Adriatic microplate. This "Adria-push" combines with buoyancy forces associated with elevated topography and crustal thickness variations. The Alps are confined between the indenting Adria microplate and an extremely rigid buffer, the Bohemian Massif, to the north (Fig. 1). Considerable gravitational potential energy is stored in the elevated and thickened crust of the Alps, so the region is characterized by reduced compression, as is often the case in mountain belts (Bird, 1991). Consequently, eastward motion of the Alpine orogen, either by gravitational spreading (Bada et al., 2001) or lateral escape (Robl and Stüwe, 2005), takes place in the direction of minimum gravitational potential energy, toward the interior of the Pannonian Basin.

Lateral extrusion of the eastern Alps has been a popular model for the formation of the Pannonian Basin (e.g., Ratschbacher et al., 1989, 1991; Dewey, 1988; Frisch et al., 1998; Horváth et al., 2006). The GPS data, stress indicators, and the distribution of seismicity indicate that this scenario applies to the present-day tectonics of the western part of the Pannonian Basin. Due to orogen-parallel extrusion of crustal units, mass transfer to the E-NE causes considerable internal deformation in the Pannonian Basin. This is indicated by an eastward decrease of GPS velocities (Grenerczy, 2002), the high level of seismic energy release (Gerner et al., 1999; Tóth et al., 2002), and differential

vertical motions likely associated with buckling of the lithosphere (Horváth and Cloetingh, 1996; Fodor et al., 2005; Ruszkiczay-Rüdiger et al., 2005). The situation differs along Adria's eastern margin (Fig. 9), where compressional stresses propagate orthogonally through the Dinarides into the Pannonian Basin. Although most of Adria's motion is absorbed in the Dinarides, as suggested by GPS data (Grenerczy et al., 2005), shortening is also detectable inside the Pannonian Basin at an overall rate of ~1–2 mm/yr. This rate decreases toward the northeast (Fig. 9B), in good agreement with the areal distribution of stress regimes (Fig. 9A).

Lithospheric Strength and Rate of Deformation

The thermomechanical character of the Pannonian lithosphere is crucial in determining the style and rate of active deformation. The rheological structure of continental lithosphere can be regarded as a two-layered viscoelastic beam. Its response to compressional stresses depends on the thickness, strength, and spacing of the two competent layers, the stress magnitudes, the strain rates, and, most critically, the thermal regime. Because the structure of continental lithosphere is heterogeneous, its weakest part starts to yield first once the stress reaches its strength limit, causing strain localization. Pronounced lateral heterogeneities exist in the mechanical properties of the European lithosphere (Cloetingh et al., 2005). Modeling indicates that due to its high heat flow and attenuated lithosphere, the Pannonian Basin is one of the weakest parts of the European continent with important lateral strength variation (Cloetingh et al., 2006; Figs. 8 and 9C). Pronounced lithospheric weakness since Cretaceous times has led to a high degree of strain concentration, mainly in the form of repeated fault reactivation. Once formed, young and warm sedimentary basins like the Pannonian are very sensitive to the fluctuations of intraplate stresses. The more attenuated and warmer the lithosphere is, the more easily it deforms. Because the basin lithosphere is normally characterized by the highest temperatures at the end of or shortly after rifting, a stress increase at this time can easily result in structural inversion.

The high heat flow of the Pannonian Basin (Dövényi and Horváth, 1988) implies considerably lower lithospheric strength than in the surroundings. On the other hand, stresses in the lithosphere are amplified beneath basins (Kusznir and Bott, 1977) because tectonic stresses are supported by thinner lithosphere. This would lead to stress concentration in the Pannonian lithosphere, which has thinner crust than the Alps and Carpathians (Horváth et al., 2006). Actual stress magnitudes are unknown, making the quantification of stress-strain relation problematic. However, the seismicity in the Pannonian Basin implies that stress magnitudes are large enough to induce earthquakes on preexisting faults and, to a lesser extent, failure of intact rocks. Seismicity pattern in the Pannonian Basin, both in terms of epicenter distribution (Fig. 2) and the cumulative seismic energy release (Gerner et al., 1999; Tóth et al., 2002; Fig. 9D), can be fairly well explained by the release of accumulated stress at weakness zones properly oriented with respect to the stress field when the tectonic stress exceeds the shear strength of an inherited fault zone or the yield strength of intact rocks.

Distributed seismicity argues for the internal deformation of principal tectonic units constituting the basement of the Pannonian Basin (Gerner et al., 1999), rather than motion along the boundaries of large-scale fully rigid blocks (Gutdeutsch and Aric, 1988). Areas of concentrated seismic activity can be delineated by the 3-D structural mapping of preexisting fault zones and tectonic modeling of their reactivation potential in the prevailing stress field. The first results of such combined analysis (Windhoffer et al., 2005) argue for the repeated reactivation of basement structures in the Pannonian lithosphere under a relatively stable stress field. The map of active structures (Fig. 8) indicates that seismoactive faulting is mainly concentrated along more internal shear zones rather than at major tectonic boundaries such as the Mid-Hungarian fault system. However, resolving the diffuse versus localized mode of deformation in the Pannonian Basin is hampered by the scarcity of GPS networks, the limitations of seismic catalogues, and the inadequate knowledge of major seismotectonic zones.

The low to medium levels of seismicity (Fig. 3) and GPS velocities (Figs. 4 and 9B) in the Pannonian Basin offer additional possible insight. An easily deformable lithosphere (Fig. 9C) and the expected high level of compressional stresses concentrated in it would argue for intense deformation. However, the data do not support high strain rates. The lack of large (M7) earthquakes in the Pannonian Basin may refer to a short earthquake catalogue. However, if this is not simply an artifact of the short earthquake history (Stein and Newman, 2004), the dominance of smaller-magnitude earthquakes may be explained by the weak rheology of the Pannonian lithosphere. Due to limited lithospheric strength, stresses cannot build up but are released mainly in the form of small- to medium-magnitude earthquakes. Despite the scarcity of M7 events, considerable seismic energy is released in the basin (Gerner et al., 1999; Tóth et al., 2002; Fig. 9D), highlighting the intensity of ongoing tectonic processes relative to the surrounding orogens, except for the southern Alps–Dinarides belt where most deformation is concentrated.

CONCLUSIONS

An integrated study of the stress and strain fields in the Pannonian Basin and its surroundings suggests the following main conclusions:

1. The basin is characterized by a medium level of tectonic activity, intermediate between levels observed in the "active" Alpine orogen and the "passive" European intraplate areas.

2. The basin is subject to compressional tectonic stresses and thus has been inverting since late Pliocene–Quaternary times.

3. From the frontal zone of "Adria-push" in the Dinarides toward the interior of the Pannonian Basin, the dominant style of deformation gradually changes from pure contraction through transpression to strike-slip faulting.

4. GPS data, stress indicators, and the distribution of seismicity indicate that lateral extrusion of the eastern Alps has a significant role in the active tectonics of the Pannonian Basin.

5. GPS data indicate that shortening is taking place at a strain rate of ~4 ppb/yr averaged over the entire Pannonian Basin, consistent with the observed pattern and rates of structural inversion.

6. Surface expression, geometry, mechanical properties, and reactivation potential of seismotectonic zones in the Pannonian Basin are poorly constrained. Most active structures are controlled by the reactivation of preexisting basement faults and shear zones. In the west, reverse faulting results mainly in folding of the Neogene to Quaternary strata. Toward the east, the style of deformation becomes strike-slip faulting with either transpressional or transtensional character.

ACKNOWLEDGMENTS

We thank reviewers, Claudio Faccenna and Franz Neubauer, and volume editor, Stéphane Mazzotti, for their useful comments and suggestions. Financial support was provided by the Hungarian Scientific Research Fund (OTKA) projects no. F043715, T034928, F61211, T042900, and NK60445, and the Netherlands Research Centre for Integrated Solid Earth Science (ISES). The European Union is thanked for supporting global positioning system (GPS) programs through EU grant EVK2-CT-2002-00140. L. Fodor thanks the Hungarian Academy of Sciences for his Bolyai János scholarship, and the Geological Institute of Hungary for supporting neotectonic research. We thank D. Delvaux for the use of his TENSOR program.

REFERENCES CITED

Anderson, E.M., 1951, The Dynamics of Faulting: Edinburgh, UK, Oliver and Boyd Ltd., 206 p.

Anderson, H., 1987, Is the Adriatic an African promontory?: Geology, v. 15, p. 212–215, doi: 10.1130/0091-7613(1987)15<212:ITAAAP>2.0.CO;2.

Anderson, H., and Jackson, J., 1987, Active tectonics of the Adriatic region: Geophysical Journal of the Royal Astronomical Society, v. 91, p. 937–983.

Angelier, J., 1979, Determination of the mean principal directions of stresses for a given fault population: Tectonophysics, v. 56, p. 17–26, doi: 10.1016/0040-1951(79)90081-7.

Angelier, J., and Mechler, P., 1977, Sur une méthode graphique de recherche des contraintes principales également utilisable en tectonique et en seismologie: La méthode des dièdres droits: Bulletin de la Société Géologique de France, v. 7, p. 1309–1318.

Bada, G., 1999, Cenozoic stress field evolution in the Pannonian Basin and surrounding orogens: Inferences from kinematic indicators and finite element stress modelling [Ph.D. thesis]: Amsterdam, the Netherlands, Vrije Universiteit, 204 p.

Bada, G., Gerner, P., Cloetingh, S., and Horváth, F., 1998, Sources of recent tectonic stress in the Pannonian region: Inferences from finite element modelling: Geophysical Journal International, v. 134, p. 87–102, doi: 10.1046/j.1365-246x.1998.00545.x.

Bada, G., Horváth, F., Fejes, I., and Gerner, P., 1999, Review of the present-day geodynamics of the Pannonian Basin: Progress and problems: Journal of Geodynamics, v. 27, p. 501–527, doi: 10.1016/S0264-3707(98)00013-1.

Bada, G., Horváth, F., Cloetingh, S., Coblentz, D.D., and Tóth, T., 2001, The role of topography induced gravitational stresses in basin inversion: The case study of the Pannonian Basin: Tectonics, v. 20, p. 343–363, doi: 10.1029/2001TC900001.

Bada, G., Fodor, L., Windhoffer, G., Ruszkiczay-Rüdiger, Zs., Sacchi, M., Dunai, T., Tóth, L., Cloetingh, S., and Horváth, F., 2003, Lithosphere dynamics and present-day deformation pattern in the Pannonian Basin: Geophysical Research Abstracts, v. 5, no. 05772.

Bada, G., Horváth, F., Tóth, L., Fodor, L., Timár, G., and Cloetingh, S., 2006, Societal aspects of ongoing deformation in the Pannonian region, *in* Pinter, N., Grenerczy, Gy., Weber, J., Stein, S., and Medak, D., eds., The Adria Microplate: GPS Geodesy, Tectonics, and Hazards: Dordrecht, The Netherlands, Springer-Verlag, NATO Advanced Research Workshop Series, v. IV/61, p. 385–402.

Balla, Z., 1984, The Carpathian loop and the Pannonian Basin: A kinematic analysis: Geophysical Transactions, v. 30, p. 313–353.

Battaglia, M., Murray, M.H., Serpelloni, E., and Burgmann, R., 2004, The Adriatic region: An independent microplate within the Africa-Eurasia collision zone: Geophysical Research Letters, v. 31, doi: 10.1029/2004GL019723.

Becker, A., 1993, Contemporary state of stress and neotectonic deformation in the Carpathian-Pannonian region: Terra Nova, v. 5, p. 375–388.

Biju-Duval, B.J., Dercourt, J., and Le Pichon, X., 1977, From the Tethys ocean to the Mediterranean Sea: A plate tectonic evolution of the Western Alpine System, *in* Biju-Duval, B.J., and Montadert, L., eds., International Symposium on the Structural History of the Mediterranean Basins: Paris, France, Ed. Technip, p. 143–164.

Bird, P., 1991, Lateral extrusion of lower crust from under high topography, in the isostatic limit: Journal of Geophysical Research, v. 96, p. 10,275–10,286.

Bott, M.H.P., 1959, The mechanics of oblique slip faulting: Geological Magazine, v. 96, p. 109–117.

Calais, E., Nocquet, J.-M., Jouanne, F., and Tardy, M., 2002, Current strain regime in the western Alps from continuous global positioning system measurements, 1996–2001: Geology, v. 30, p. 651–654, doi: 10.1130/0091-7613(2002)030<0651:CSRITW>2.0.CO;2.

Carulli, G.B., Nicolich, R., Rebez, A., and Slejko, D., 1990, Seismotectonics of the Northwest External Dinarides: Tectonophysics, v. 179, p. 11–25, doi: 10.1016/0040-1951(90)90353-A.

Cloetingh, S., D'Argenio, B., Catalano, R., Horváth, F., and Sassi, W., eds., 1995, Interplay of extension and compression in basin formation: Tectonophysics, v. 252, p. 1–484, doi: 10.1016/0040-1951(95)00105-0.

Cloetingh, S., Burov, E., and Poliakov, A., 1999, Lithosphere folding: Primary response to compression? (from central Asia to Paris basin): Tectonics, v. 18, p. 1064–1083, doi: 10.1029/1999TC900040.

Cloetingh, S., Ziegler, P., Beekman, F., Andriessen, P., Maţenco, L., Bada, G., Garcia-Castellanos, D., Hardebol, N., Dezes, P., and Sokoutis, D., 2005, Lithospheric memory, state of stress and rheology: Neotectonic controls on Europe's intraplate continental topography: Quaternary Science Reviews, v. 24, p. 241–304, doi: 10.1016/j.quascirev.2004.06.015.

Cloetingh, S., Bada, G., Maţenco, L., Lankreijer, A., Horváth, F., and Dinu, C., 2006, Neotectonics of the Pannonian-Carpathian system: Inferences from thermomechanical modelling, *in* Gee, D., and Stephenson, R., eds., European Lithosphere Dynamics: Geological Society [London] Memoir 32, p. 207–221.

Console, R., Di Giovambattista, R., Favalli, P., Presgave, B.W., and Smriglio, G., 1993, Seismicity of the Adriatic microplate: Tectonophysics, v. 218, p. 343–354, doi: 10.1016/0040-1951(93)90323-C.

Csontos, L., Nagymarosy, A., Horváth, F., and Kováč, M., 1992, Tertiary evolution of the Intra-Carpathian area: A model: Tectonophysics, v. 208, p. 221–241, doi: 10.1016/0040-1951(92)90346-8.

Csontos, L., Magyari, Á., Van Vliet-Lanoë, B., and Musitz, B., 2005, Neotectonics of the Somogy Hills (part II): Evidence from seismic sections: Tectonophysics, v. 410, p. 63–80, doi: 10.1016/j.tecto.2005.05.049.

D'Agostino, N., Cheloni, D., Mantenuto, S., Selvaggi, G., Michelini, A., and Zuliani, D., 2005, Strain accumulation in the southern Alps (NE Italy) and deformation at the northeastern boundary of Adria observed by CGPS measurements: Geophysical Research Letters, v. 32, p. L19306, doi: 10.1029/2005GL024266.

Decker, K., Peresson, H., and Hinsch, R., 2005, Active tectonics and Quaternary basin formation along the Vienna Basin transform fault: Quaternary Science Reviews, v. 24, p. 305–320, doi: 10.1016/j.quascirev.2004.04.012.

Del Ben, A., Finetti, I., Rebez, A., and Slejko, D., 1991, Seismicity and seismotectonics at the Alps-Dinarides contact: Bollettino di Geofisica Teorica ed Applicata, v. 33, p. 155–176.

Delvaux, D., 1993, The TENSOR program for paleostress reconstruction: Examples from the East African and the Baikal rift zones: Terra Abstracts, v. 5, p. 216.

Delvaux, D., and Sperner, B., 2003, New aspects of tectonic stress inversion with reference to the TENSOR program, *in* Nieuwland, D., ed., New

Insights into Structural Interpretation and Modelling: Geological Society [London] Special Publication 212, p. 75–100.

Dercourt, J., Zonenshain, L.P., Ricou, L.E., Kazmin, V.G., Le Pichon, X., Knipper, A.L., Grandjacquet, C., Sbortshikov, I.M., Geyssant, J., Lepvrier, C., Pechersky, D.H., Boulin, J., Sibuet, J.C., Savostin, L.A., Sorokhtin, O., Westphal, M., Bazhenov, M.L., Lauer, J.P., and Biju-Duval, B., 1986, Geological evolution of the Tethys belt from Atlantic to the Pamir since Lias: Tectonophysics, v. 123, p. 241–315, doi: 10.1016/0040-1951(86)90199-X.

Dewey, J.F., 1988, Extensional collapse of orogens: Tectonics, v. 7, p. 1123–1139.

Dewey, J.F., Helman, M.L., Turco, E., Hutton, D.H.W., and Knott, S.D., 1989, Kinematics of the western Mediterranean, *in* Coward, M.P., Dietrich, D., and Park, R.G., eds., Alpine Tectonics: Geological Society [London] Special Publication 45, p. 265–283.

Dövényi, P., and Horváth, F., 1988, A review of temperature, thermal conductivity and heat flow data from the Pannonian Basin, *in* Royden, L.H., and Horváth, F., eds., The Pannonian Basin: Tulsa, Oklahoma, American Association of Petroleum Geologists Memoir 45, p. 195–233.

England, P., 2003, The alignment of earthquake T-axes with the principal axes of geodetic strain in the Aegean region: Turkish Journal of Earth Sciences, v. 12, p. 47–53.

Faccenna, C., Piromallo, C., Crespo-Blan, A., Jolivet, L., and Rossetti, F., 2004, Lateral slab deformation and the origin of the western Mediterranean arcs: Tectonics, v. 23, p. TC1012, doi: 10.1029/2002TC001488.

Favali, P., Mele, G., and Mattietti, G., 1990, Contribution to the study of the Apulian microplate geodynamics: Memorie della Società Geologica Italiana, v. 44, p. 71–80.

Fejes, I., Borza, T., Busics, I., and Kenyeres, A., 1993a, Realization of the Hungarian Geodynamic GPS Reference Network: Journal of Geodynamics, v. 18, p. 145–152, doi: 10.1016/0264-3707(93)90036-6.

Fejes, I., Barlik, M., Busics, I., Packelski, W., Rogowsky, J., Sledzinsky, J., and Zielinsky, J., 1993b, The Central Europe Regional Geodynamics Project, *in* Proceedings of the 2nd International Seminar on "GPS in Central Europe": Penc, Hungary, Hungarian Academy of Sciences, April 27–29, 1993, p. 106–115.

Fodor, L., Csontos, L., Bada, G., Benkovics, L., and Győrfi, I., 1999, Tertiary tectonic evolution of the Carpatho-Pannonian region: A new synthesis of paleostress data, *in* Durand, B., Jolivet, L., Horváth, F., and Séranne, M., eds., The Mediterranean Basins: Tertiary Extension within the Alpine Orogen: Geological Society [London] Special Publication 156, p. 295–334.

Fodor, L., Bada, G., Csillag, G., Horváth, E., Ruszkiczay-Rüdiger, Zs., Horváth, F., Cloetingh, S., Palotás, K., Síkhegyi, F., and Timár, G., 2005, An outline of neotectonic structures and morphotectonics of the western and central Pannonian Basin: Tectonophysics, v. 410, p. 15–41, doi: 10.1016/j.tecto.2005.06.008.

Frisch, W., Kuhleman, J., Dunkl, I., and Brügel, A., 1998, Palinspastic reconstruction and topographic evolution of the eastern Alps during the late Tertiary tectonic extrusion: Tectonophysics, v. 297, p. 1–15, doi: 10.1016/S0040-1951(98)00160-7.

Gephart, J., and Forsyth, D.W., 1984, An improved method for determining the regional stress tensor using earthquake focal mechanism data: Application to the San Fernando earthquake sequence: Journal of Geophysical Research, v. 89, p. 9305–9320.

Gerner, P., Bada, G., Dövényi, P., Müller, B., Oncescu, M.C., Cloetingh, S., and Horváth, F., 1999, Recent tectonic stress and crustal deformation in and around the Pannonian Basin: Data and models, *in* Durand, B., Jolivet, L., Horváth, F., and Séranne, M., eds., The Mediterranean Basins: Tertiary Extension within the Alpine Orogen: Geological Society [London] Special Publication 156, p. 269–294.

Grenerczy, G., 2002, Tectonic processes in the Eurasian-African plate boundary zone revealed by space geodesy, *in* Stein, S., and Freymueller, J.T., eds., Plate Boundary Zones: American Geophysical Union Geodynamics Monograph 30, p. 67–86.

Grenerczy, G., 2005, Crustal motions from space geodesy: A review from EPN, CEGRN, and HGRN data, *in* Fodor, L., and Brezsnyánszky, K., eds., Applications of GPS in Plate Tectonics in Research on Fossil Energy Resources and in Earthquake Hazard Assessment: Budapest, Hungary, Occasional Papers of the Geological Institute of Hungary, v. 204, p. 31–34.

Grenerczy, Gy., and Bada, G., 2005, GPS baseline length changes and their tectonic interpretation in the Pannonian Basin: Geophysical Research Abstracts, v. 7, no. 04808.

Grenerczy, Gy., and Kenyeres, A., 2006, Crustal deformation between Adria and the European Platform from space geodesy, *in* Pinter, N., Grenerczy, Gy., Weber,

J., Stein, S., and Medak, D., eds., The Adria Microplate: GPS Geodesy, Tectonics, and Hazards: Dordrecht, The Netherlands, Springer-Verlag, NATO Advanced Research Workshop Series, v. IV/61, p. 321–334.

Grenerczy, Gy., Kenyeres, A., and Fejes, I., 2000, Present crustal movement and strain distribution in Central Europe inferred from GPS measurements: Journal of Geophysical Research, v. 105, p. 21,835–21,846, doi: 10.1029/2000JB900127.

Grenerczy, Gy., Sella, G.F., Stein, S., and Kenyeres, A., 2005, Tectonic implications of the GPS velocity field in the northern Adriatic region: Geophysical Research Letters, v. 32, L16311, doi: 10.1029/2005GL022947.

Gutdeutsch, R., and Aric, K., 1988, Seismicity and neotectonics of the East Alpine–Carpathian and Pannonian area, *in* Royden, L.H., and Horváth, F., eds., The Pannonian Basin: Tulsa, Oklahoma, American Association of Petroleum Geologists Memoir 45, p. 183–194.

Gutenberg, B., and Richter, C.F., 1944, Frequency of earthquakes in California: Bulletin of the Seismological Society of America, v. 34, p. 185–188.

Hansen, K.N., and Mount, V.S., 1990, Smoothing and extrapolation of crustal stress orientation measurements: Journal of Geophysical Research, v. 95, p. 1155–1166.

Horváth, F., 1984, Neotectonics of the Pannonian Basin and the surrounding mountain belts: Alps, Carpathians and Dinarides: Annales Geophysicae, v. 2, p. 147–154.

Horváth, F., 1993, Towards a mechanical model for the formation of the Pannonian Basin: Tectonophysics, v. 226, p. 333–357, doi: 10.1016/0040-1951(93)90126-5.

Horváth, F., 1995, Phases of compression during the evolution of the Pannonian Basin and its bearing on hydrocarbon exploration: Marine and Petroleum Geology, v. 12, p. 837–844, doi: 10.1016/0264-8172(95)98851-U.

Horváth, F., and Berckhemer, H., 1982, Mediterranean backarc basins, *in* Berckhemer, H., and Hsü, K., eds., Alpine-Mediterranean Geodynamics: American Geophysical Union Geodynamics Monograph 7, p. 141–173.

Horváth, F., and Cloetingh, S., 1996, Stress-induced late-stage subsidence anomalies in the Pannonian Basin: Tectonophysics, v. 266, p. 287–300, doi: 10.1016/S0040-1951(96)00194-1.

Horváth, F., and Tari, G., 1988, A Neogén szerkezetfejlődés és a szerkezeti elemek szerepe a szénhidrogén migrációban és csapdázódásban: Budapest, Hungary, Kőolajkutató Vállalat, 155 p.

Horváth, F., Bada, G., Szafián, P., Tari, G., Ádám, A., and Cloetingh, S., 2006, Formation and deformation of the Pannonian Basin: Constraints from observational data, *in* Gee, D., and Stephenson, R., eds., European Lithosphere Dynamics: Geological Society [London] Memoir 32, p. 191–206.

Huismans, R., Podladchikov, Y., and Cloetingh, S., 2001, Dynamic modeling of the transition from passive to active rifting, application to the Pannonian Basin: Tectonics, v. 20, p. 1021–1039, doi: 10.1029/2001TC900010.

Ilić, A., and Neubauer, F., 2005, Tertiary to recent oblique convergence and wrenching of the Central Dinarides: Constraints from a palaeostress study: Tectonophysics, v. 410, p. 465–484, doi: 10.1016/j.tecto.2005.02.019.

Jackson, J., 1994, Active tectonics of the Aegean region: Annual Review of Earth and Planetary Sciences, v. 22, p. 239–271, doi: 10.1146/annurev.ea.22.050194.001323.

Klosko, E.R., Stein, S., Hindle, D., Dixon, T., and Norabuena, E., 2002, Comparison of geodetic, geologic, and seismological observations in the Nazca–South American plate boundary zone, *in* Stein, S., and Freymueller, J.T., eds., Plate Boundary Zones: American Geophysical Union Geodynamics Monograph 30, p. 123–133.

Kusznir, N.J., and Bott, M.H.P., 1977, Stress concentration in the upper lithosphere caused by underlying visco-elastic creep: Tectonophysics, v. 43, p. 247–256, doi: 10.1016/0040-1951(77)90119-6.

Lowell, J.D., 1995, Mechanics of basin inversion from worldwide examples, *in* Buchanan, J.G., and Buchanan, P.G., eds., Basin Inversion: Geological Society [London] Special Publication 88, p. 39–57.

Marović, M., Djoković, I., Pešić, L., Radovanović, S., Toljić, M., and Gerzina, N., 2002, Neotectonics and seismicity of the southern margin of the Pannonian Basin in Serbia, *in* Cloetingh, S., Horváth, F., Bada, G., and Lankreijer, A., eds., Neotectonics and Surface Processes: The Pannonian Basin and Alpine/Carpathian System: Katlenburg-Lindau, Germany, European Geosciences Union, St. Mueller Special Publication Series, v. 3, p. 277–295.

Márton, E., 2005, Paleomagnetic evidence for Tertiary counterclockwise rotation of Adria with respect to Africa, *in* Pinter, N., Grenerczy, Gy., Weber, J., Stein, S., and Medak, D., eds., The Adria Microplate: GPS Geodesy, Tectonics, and Hazards: Dordrecht, The Netherlands, Springer-Verlag, NATO Advanced Research Workshop Series, v. IV/61, p. 71–79.

Márton, E., Fodor, L., Jelen, B., Márton, P., Rifelj, H., and Kevrić, R., 2002, Miocene to Quaternary deformation in NE Slovenia: Complex paleomagnetic and structural study: Journal of Geodynamics, v. 34, p. 627–651, doi: 10.1016/S0264-3707(02)00036-4.

McClusky, S.C., Balassanian, S., Barka, A.A., Ergintav, S., Georgiev, I., Gürkan, O., Hamburger, M., Hurst, K., Kahle, H., Kastens, K., Kekelidse, G., King, R., Kotzev, V., Lenk, O., Mahmoud, S., Mishin, A., Nadaria, M., Ouzounis, A., Paradisissis, D., Peter, Y., Prilepin, M., Reilinger, R.E., Sanli, I., Seeger, H., Teableb, A., Toksöz, N., and Veis, G., 2000, Global positioning system constraints on plate kinematics and dynamics in the eastern Mediterranean and Caucasus: Journal of Geophysical Research, v. 105, p. 5695–5719, doi: 10.1029/1999JB900351.

McKenzie, D.P., 1969, The relation between fluid injection, plane solutions for earthquakes and the directions of the principal stresses: Bulletin of the Seismological Society of America, v. 56, p. 591–601.

Molnar, P., Fitch, T., and Wu, F., 1973, Fault plane solutions of shallow earthquakes in Asia: Earth and Planetary Science Letters, v. 19, p. 101–112, doi: 10.1016/0012-821X(73)90104-0.

Müller, B., Zoback, M.L., Fuchs, K., Mastin, L., Gregersen, S., Pavoni, N., Stephansson, O., and Ljunggren, C., 1992, Regional pattern of tectonic stress in Europe: Journal of Geophysical Research, v. 97, p. 11,783–11,803.

Müller, B., Wehrle, V., Zeyen, H., and Fuchs, K., 1997, Short-scale variations of tectonic regimes in the western European stress province north of the Alps and Pyrenees: Tectonophysics, v. 275, p. 199–219, doi: 10.1016/S0040-1951(97)00021-8.

Oldow, J.S., Ferranti, L., Lewis, D.S., Campbell, J.K., D'Argenio, B., Catalano, R., Pappone, G., Carmignani, L., Conti, P., and Aiken, C.L.V., 2002, Active fragmentation of Adria, the north African promontory, central Mediterranean orogen: Geology, v. 30, p. 779–782, doi: 10.1130/0091-7613(2002)030<0779:AFOATN>2.0.CO;2.

Pinter, N., 2005, Applications of tectonic geomorphology for deciphering active deformation in the Pannonian Basin, Hungary, *in* Fodor, L., and Brezsnyánszky, K., eds., Applications of GPS in Plate Tectonics in Research on Fossil Energy Resources and in Earthquake Hazard Assessment: Budapest, Hungary, Occasional Papers of the Geological Institute of Hungary, v. 204, p. 25–51.

Pinter, N., and Grenerczy, Gy., 2006, Recent advances in peri-Adriatic geodynamics and future research directions, *in* Pinter, N., Grenerczy, Gy., Weber, J., Stein, S., and Medak, D., eds., The Adria Microplate: GPS Geodesy, Tectonics, and Hazards: Dordrecht, The Netherlands, Springer-Verlag, NATO Advanced Research Workshop Series, v. IV/61, p. 1–20.

Placer, L., 1999, Structural meaning of the Sava folds: Geologija, v. 41, p. 223–255.

Pogácsás, Gy., Lakatos, L., Barvitz, A., Vakarcs, G., and Farkas, Cs., 1989, Pliocene-Quaternary strike-slip faults in the Great Hungarian Plain: Általános Földtani Szemle, v. 24, p. 149–169.

Poljak, M., Živčić, M., and Zupančič, P., 2000, The seismotectonic characteristics of Slovenia: Pure and Applied Geophysics, v. 157, p. 37–55, doi: 10.1007/PL00001099.

Prelogović, E., Saftić, B., Kuk, V., Velić, J., Dragaš, M., and Lučić, D., 1998, Tectonic activity in the Croatian part of the Pannonian Basin: Tectonophysics, v. 297, p. 283–293, doi: 10.1016/S0040-1951(98)00173-5.

Ratschbacher, L., Frisch, W., Neubauer, F., Schmid, S.M., and Neugebauer, J., 1989, Extension in compressional orogenic belts: The eastern Alps: Geology, v. 17, p. 404–407, doi: 10.1130/0091-7613(1989)017<0404: EICOBT>2.3.CO;2.

Ratschbacher, L., Frisch, W., Linzer, H.-G., and Marle, O., 1991, Lateral extrusion in the eastern Alps, part 2: Structural analysis: Tectonics, v. 10, p. 257–271.

Rebaï, S., Philip, H., and Taboada, A., 1992, Modern tectonic stress field in the Mediterranean region: Evidence for variation in stress directions at different scale: Geophysical Journal International, v. 110, p. 106–140.

Robl, J., and Stüwe, K., 2005, Continental collision with finite indenter strength: 2. European eastern Alps: Tectonics, v. 24, p. TC4014, doi: 10.1029/2004TC001741.

Roeder, D., and Bachmann, G., 1996, Evolution, structure and petroleum geology of the German Molasse Basin, *in* Ziegler, P.A., and Horváth, F., eds., Peri-Tethys Memoir 2: Structure and Prospects of Alpine Basins and Forelands: Mémoires du Muséum National d'Histoire Naturelle, v. 170, p. 263–284.

Rosenbaum, G., and Lister, G.S., 2004, Neogene and Quaternary rollback evolution of the Tyrrhenian Sea, the Apennines, and the Sicilian Maghrebides: Tectonics, v. 23, p. TC1013, doi: 10.1029/2003TC001518.

Royden, L.H., 1988, Late Cenozoic tectonics of the Pannonian Basin system, *in* Royden, L.H., and Horváth, F., eds., The Pannonian Basin: Tulsa, Oklahoma, American Association of Petroleum Geologists Memoir 45, p. 27–48.

Royden, L.H., 1993, Evolution of retreating subduction boundaries formed during continental collision: Tectonics, v. 12, p. 629–638.

Ruszkiczay-Rüdiger, Zs., Dunai, T., Bada, G., Fodor, L., and Horváth, E., 2005, Middle to late Pleistocene uplift rate of the Hungarian Mountain Range at the Danube Bend (Pannonian Basin), using in situ produced ^{3}He: Tectonophysics, v. 410, p. 173–187, doi: 10.1016/j.tecto.2005.02.017.

Sanders, C., Andriessen, P., and Cloetingh, S., 1999, Life cycle of the East Carpathian orogen: Erosion history of a doubly vergent critical wedge assessed by fission track thermochronology: Journal of Geophysical Research, v. 104, p. 29,095–29,112, doi: 10.1029/1998JB900046.

Săndulescu, M., 1988, Cenozoic tectonic history of the Carpathians, *in* Royden, L.H., and Horváth, F., eds., The Pannonian Basin: Tulsa, Oklahoma, American Association of Petroleum Geologists Memoir 45, p. 17–25.

Sbar, M.L., and Sykes, L.R., 1973, Contemporary compressive stress and seismicity in eastern North America: An example of intra-plate tectonics: Geological Society of America Bulletin, v. 84, p. 1861–1882, doi: 10.1130/0016-7606(1973)84<1861:CCSASI>2.0.CO;2.

Schmid, S.M., Fügenschuh, B., Kissling, E., and Schuster, R., 2004, Tectonic map and overall architecture of the Alpine orogen: Eclogae Geologicae Helvetiae, v. 97, p. 93–117, doi: 10.1007/s00015-004-1113-x.

Şengör, A.M.C., 1993, Some current problems on the tectonic evolution of the Mediterranean during the Cainozoic, *in* Boschi, E., et al., eds., Recent Evolution and Seismicity of the Mediterranean Region: Rotterdam, The Netherlands, Kluwer Academic Publishers, p. 1–51.

Şengör, A.M.C., Görür, N., and Saroglu, F., 1985, Strike-slip faulting and related basin formation in zones of tectonic escape: Turkey as a case study, *in* Biddle, K.T., and Christie-Blick, N., eds., Strike-slip deformation, Basin Formation, and Sedimentation: Society of Economic Paleontologists and Mineralogists Special Publication 37, p. 227–264.

Slejko, D., Carulli, G.B., Nicholic, R., Rebez, A., Zanferrari, A., Cavallin, A., Doglioni, C., Carraro, G., Castaldini, D., Iliceto, V., Semenza, E., and Zanolla, C., 1989, Seismotectonics of the eastern southern Alps: A review: Bollettino di Geofisica Teorica ed Applicata, v. 31, p. 109–136.

Stein, S., and Newman, A., 2004, Characteristic and uncharacteristic earthquakes as possible artifacts: Applications to the New Madrid and Wabash seismic zones: Seismological Research Letters, v. 75, p. 173–187.

Stein, S., and Sella, G.F., 2006, Pleistocene change from convergence to extension in the Apennines as a consequence of Adria microplate motion, *in* Pinter, N., Grenerczy, Gy., Weber, J., Stein, S., and Medak, D., eds., The Adria Microplate: GPS Geodesy, Tectonics, and Hazards: Dordrecht, The Netherlands, Springer Verlag, NATO Advanced Research Workshop Series, v. IV/61, p. 21–33.

Tomljenović, B., and Csontos, L., 2001, Neogene–Quaternary structures in the border zone between Alps, Dinarides and Pannonian Basin (Hrvatsko zagorje and Karlovac basin, Croatia): International Journal of Earth Sciences, v. 90, p. 560–578, doi: 10.1007/s005310000176.

Tóth, L., Mónus, P., Zsíros, T., and Kiszely, M., 2002, Seismicity in the Pannonian Region—Earthquake data, *in* Cloetingh, S., Horváth, F., Bada, G., and Lankreijer, A., eds., Neotectonics and Surface Processes: The Pannonian Basin and Alpine/Carpathian System: Katlenburg-Lindau, Germany, European Geosciences Union, St. Mueller Special Publication Series, v. 3, p. 9–28.

Tóth, L., Győri, E., Mónus, P., and Zsíros, T., 2006, Seismic hazard in the Pannonian region, *in* Pinter, N., Grenerczy, Gy., Weber, J., Stein, S., and Medak, D., eds., The Adria Microplate: GPS Geodesy, Tectonics, and Hazards: Dordrecht, The Netherlands, Springer-Verlag, NATO Advanced Research Workshop Series, v. IV/61, p. 369–384.

Turner, J.P., and Williams, G.A., 2004, Sedimentary basin inversion and intra-plate shortening: Earth-Science Reviews, v. 65, p. 277–304, doi: 10.1016/j.earscirev.2003.10.002.

Vrabec, M., and Fodor, L., 2006, Late Cenozoic tectonics of Slovenia: Structural styles at the northeastern corner of the Adriatic microplate, *in* Pinter, N., Grenerczy, Gy., Weber, J., Stein, S., and Medak, D., eds., The Adria Microplate: GPS Geodesy, Tectonics, and Hazards: Dordrecht, The Netherlands, Springer-Verlag, NATO Advanced Research Workshop Series, v. IV/61, p. 151–168.

Ward, S.N., 1994, Constraints on the seismotectonics of the central Mediterranean from very long baseline interferometry: Geophysical Journal International, v. 117, p. 441–452.

Williams, G.D., Powell, C.M., and Cooper, M.A., 1989, Geometry and kinematics of inversion tectonics, *in* Cooper, M.A., and Williams, G.D., eds., Inversion Tectonics: Geological Society [London] Special Publication 44, p. 3–15.

Windhoffer, G., Bada, G., Dövényi, P., and Horváth, F., 2001, New crustal stress determinations in Hungary from borehole breakout analysis: Földtani Közlöny, v. 131, p. 541–560.

Windhoffer, G., Bada, G., Nieuwland, D., Wórum, G., Horváth, F., and Cloetingh, S., 2005, On the mechanics of basin formation in the Pannonian Basin: Inferences from analogue and numerical modelling: Tectonophysics, v. 410, p. 389–415, doi: 10.1016/j.tecto.2004.10.019.

Wortel, M.J.R., and Spakman, W., 2000, Subduction and slab detachment in the Mediterranean-Carpathian region: Science, v. 290, p. 1910–1917, doi: 10.1126/science.290.5498.1910.

Wórum, G., and Hámori, Z., 2004, A BAF kutatás szempontjából releváns a MOL Rt. által készített archív szeizmikus szelvények újrafeldolgozása: Pécs, Hungary, Mecsekérc Rt., 39 p.

Yilmaz, P.O., Norton, I.O., Leary, D., and Chuchla, R.J., 1996, Tectonic evolution and paleogeography of Europe, *in* Ziegler, P.A., and Horváth, F., eds., Peri-Tethys Memoir 2: Structure and Prospects of Alpine Basins and Forelands: Mémoires du Muséum National d'Histoire Naturelle, v. 170, p. 47–60.

Ziegler, P.A., Cloetingh, S., and Van Wees, J.D., 1995, Dynamics of intra-plate compressional deformation: The Alpine foreland and other examples: Tectonophysics, v. 252, p. 7–59, doi: 10.1016/0040-1951(95)00102-6.

Ziegler, P.A., Cavazza, W., Robertson, A.H.F., and Crasquin-Soleau, S., eds., 2001, Peri-Tethys Memoir 6: Peri-Tethyan Rift/Wrench Basins and Passive Margins: Mémoires du Muséum National d'Histoire Naturelle, v. 186, p. 762.

Ziegler, P.A., Bertotti, G., and Cloetingh, S., 2002, Dynamic processes controlling foreland development—The role of mechanical (de)coupling of orogenic wedges and forelands, *in* Bertotti, G., Schulmann, K., and Cloetingh, S., eds., Continental Collision and the Tectono-Sedimentary Evolution of Forelands: Katlenburg-Lindau, Germany, European Geosciences Union, St. Mueller Special Publication Series, v. 1, p. 17–56.

Zoback, M.L., 1992, First and second order patterns of stress in the lithosphere: The World Stress Map Project: Journal of Geophysical Research, v. 97, p. 11,703–11,728.

Zoback, M.L., and Zoback, M., 1989, Tectonic stress field of the conterminous United States, *in* Pakiser, L., and Mooney, W.D., eds., Geophysical Framework of the Continental U.S.: Geological Society of America Memoir 172, p. 523–539.

Zsíros, T., 2000, Hungarian Earthquake Catalogue (456–1995): Budapest, Hungary, Geodetic and Geophysical Research Institute, Hungarian Academy of Sciences, 495 p.

MANUSCRIPT ACCEPTED BY THE SOCIETY 29 NOVEMBER 2006

The Geological Society of America
Special Paper 425
2007

Toward a better model of earthquake hazard in Australia

Mark Leonard
Geoscience Australia, Risk Research Group, P.O. Box 378, Canberra, ACT 2601, Australia
David Robinson
Geoscience Australia, Risk Research Group, P.O. Box 378, Canberra, ACT 2601, Australia, and
The Australian National University, Research School of Earth Sciences, Canberra ACT 0200, Australia
Trevor Allen
John Schneider
Dan Clark
Trevor Dhu
David Burbidge
Geoscience Australia, Risk Research Group, P.O. Box 378, Canberra, ACT 2601, Australia

ABSTRACT

The tectonic setting of Australia has much in common with North America east of the Rocky Mountains because stable continental crust makes up the whole continent. The seismicity is still sufficient to have caused several damaging earthquakes in the past 50 yr. However, uncertainties in the earthquake catalogue limit the reliability of hazard models. To complement traditional hazard estimation methods, alternative methods such as paleoseismic, geodynamic numerical models and high-resolution global positioning system (GPS) are being investigated. Smoothed seismicity analysis shows that seismic recurrence varies widely across Australia. Despite the limitations of the catalogue, comparisons of regional strain rates calculated from the seismicity are consistent with data derived from geodetic techniques. Recent paleoseismic studies, particularly those examining high-resolution digital elevation models, have identified many potential prehistoric fault scarps. Detailed investigation of a few of these scarps suggests that the locus of strain release is migratory on a time scale an order of magnitude greater than the instrumental seismic catalogue, consistent with Australia's low-relief landscape. Numerical models based on the properties of the Australian plate provide alternative constraints on long-term crustal deformation.

Two attenuation models for Australia have recently been developed. Because Australia is an old, deeply weathered continent that has experienced little Holocene glaciation, it has very little material comparable to North American "hard rock" site classification. The combination of relatively low attenuation crust under widespread thick weathered regolith makes the use of ground-motion and site response models derived from Australian data vital for Australian hazard assessment. Risk modeling has been used to assess sensitivities associated with variations in both source and ground-motion models. Systematic analyses allow the uncertainty in these models to be quantified. Uncertainty in most input models contributes a 30%–50% variation in the predicted loss. Where a city lies in a thick sedimentary basin, such as Perth, uncertainties in the behavior of the basin can result in a 500% variation in predicted loss.

Keywords: earthquake, hazard, risk, Australia.

Leonard, M., Robinson, D., Allen, T., Schneider, J., Clark, D., Dhu, T., and Burbidge, D., 2007, Toward a better model of earthquake hazard in Australia, *in* Stein, S., and Mazzotti, S., ed., Continental Intraplate Earthquakes: Science, Hazard, and Policy Issues: Geological Society of America Special Paper 425, p. 263–283, doi: 10.1130/2007.2425(17). For permission to copy, contact editing@geosociety.org. ©2007 The Geological Society of America. All rights reserved.

INTRODUCTION

This paper provides an overview of some of the current earthquake research being conducted in Australia, mainly that at Geoscience Australia. Assessment of earthquake hazard in any region depends on three key elements; (1) a model of earthquake occurrence (a source or seismicity model), (2) a model of how ground shaking varies (or attenuates) as it propagates through the crust, and (3) a model of how the near-surface regolith at a site modifies the observed ground motions (i.e., site response). The research discussed in this paper is divided into three sections that discuss these areas and a section that evaluates the effects of parameter uncertainty in these three areas on loss estimation (probable maximum loss curves) via case studies of earthquake risk in the Newcastle–Lake Macquarie and Perth areas.

Australia is a relatively old and stable continental region with relatively low seismic hazard by world standards because of its distance from any major tectonic plate boundaries. However, Earth's crust beneath Australia transmits seismic waves relatively efficiently, so even moderate-size earthquakes can do damage over a broad area. Although the present Australian building code generally accounts for modest earthquake risk, many structures, such as older masonry buildings, are vulnerable to damage or failure in moderate-size earthquakes. These factors together with large urban concentrations of population and infrastructure (e.g., Melbourne and Sydney) mean that the potential for catastrophic impact from a relatively moderate earthquake cannot be ignored.

Present models used to estimate earthquake risk in Australia are highly uncertain because of fundamental limitations in our understanding of several major attributes of Australian earthquakes, specifically, the relationship of seismicity (earthquake locations and magnitudes) to faults and present-day tectonics. Present-day estimates of future patterns are based on statistical analyses of, at most, the past 100 yr of data, which comprise a very short and sparse data set of heterogeneous quality. During this short period, several large events have occurred in areas that had no known history of earthquake activity, including the Tennant Creek earthquake sequence on 22 January 1988 (M_S 6.3, 6.4, and 6.7) (Fig. 1). The fact that the return period for large earthquakes on these faults can be on the order of a hundred-thousand years presents a significant problem in the assessment of hazard. By combining the prehistoric seismicity record and direct and modeled estimates of deformation rates with the contemporary seismic record, we can improve the reliability of the source zonation and subsequent hazard models.

Another important issue is the relationship between magnitude and ground-shaking potential, which is a function of the transmission mechanism of seismic waves through Earth. Models are just now being developed to adequately predict ground shaking in Australia, including near-surface site amplification in sediments underlying major urban centers.

EARTHQUAKE OCCURRENCE

Paleoseismicity

Combining Research into Quaternary Geology, Tectonics, and Geomorphology

A range of ongoing research projects will improve our understanding of the paleoseismicity of Australia. These include:

- searches for and characterization of prehistoric fault scarps using both traditional geological techniques and new high-resolution digital elevation models (DEMs);
- geological investigations, including trenching of fault scarps, to obtain information regarding the timing and size of prehistoric earthquake events; and
- fault mapping/imaging in three dimensions using geophysical data (aeromagnetics, seismic surveys, and ground-penetrating radar).

Large earthquakes can result in significant modification to the landscape, such as the development of a fault scarp, or the diversion of drainage lines (e.g., McCalpin, 1996; Clark and McCue, 2003). Recurrent seismicity can lead to the development of major relief, such as a mountain range. Both instances leave a characteristic mark in the basement geology, even if the tectonic regime is relatively young in geological terms. Therefore, geomorphological, geological, and geophysical records can provide important first-order constraints on the development of seismicity models.

In recent times, high-resolution digital elevation models (DEMs) have emerged as an important tool for finding and characterizing earthquake-related geomorphology, particularly fault scarps. Unlike air photo interpretation, shaded-relief images derived from DEMs may be used to enhance lineaments by simulating topographic illumination from a variety of azimuths and elevations. This approach has been used to investigate known faults in areas of high tectonic activity (e.g., Oguchi et al., 2003), but to our knowledge, this is the first time it has been used for broad-scale reconnaissance. To test these methods in a continental intraplate setting, a reconnaissance investigation of two new DEM data sets was undertaken, covering a large portion of southwest and central Western Australia (Fig. 2). The Shuttle Radar Topography Mission (SRTM) has a 90 m resolution (3 arc-second), and can discriminate relief of down to ~3 m, and the Western Australian Department of Land Administration model (DOLA) was generated using digital photogrammetry with a 10 m pixel size and an expected vertical accuracy of ±1.5 m. Preliminary results indicate the discovery of 33 likely fault scarps (ranging in length from ~15 km to over 45 km, and from ~1.5 m to 20 m in height) of probable Quaternary age (Clark and Collins, 2004; Clark, 2005). While only one of these newly discovered scarps has been verified by trenching (Estrada et al., 2006), more than 20 have been confirmed on the ground.

Of the 60 possible fault scarps now known from this area, only two have been the subject of detailed paleoseismic investigation to determine the timing of the most recent events and recurrence for large events (Hyden—Crone et al., 2003; Lort

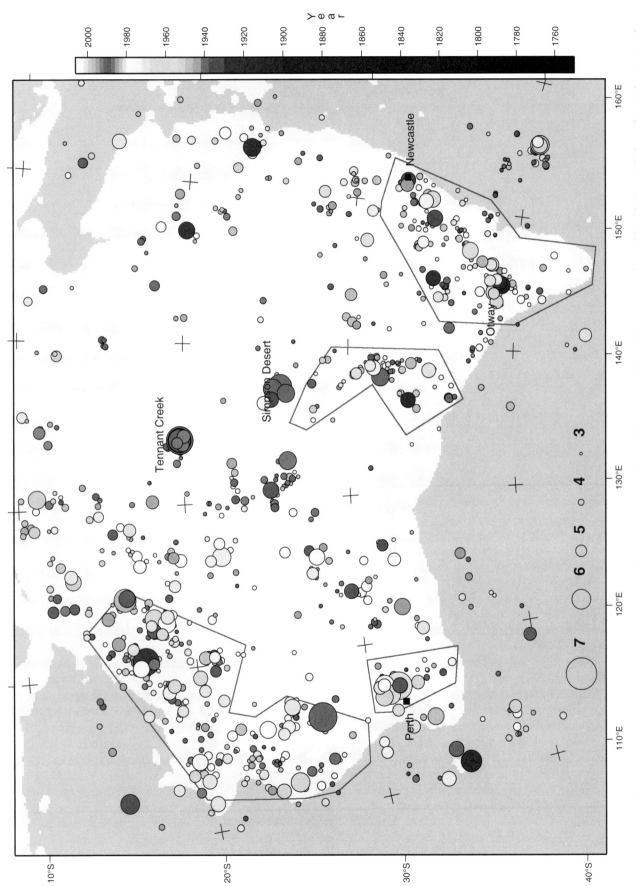

Figure 1. Seismicity of Australia showing all earthquakes of magnitude 3.5 or greater since 1974 and magnitude 5.5 or greater since 1909. 1975 and 1910 are the catalogue completeness dates for magnitude 3.5 and 5.5 earthquake, respectively. The ocean color change is at a depth of 2.2 km, which is approximately the edge of the continental shelf. The four regions referred to in the text, northwest Australia, southwest Australia, south Australia, and southeast Australia, are outlined in purple.

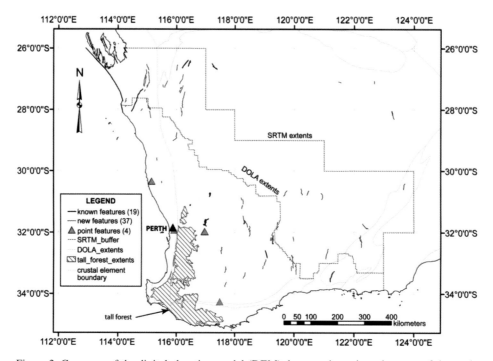

Figure 2. Coverage of the digital elevation model (DEM) data sets investigated as part of the study of Clark (2005). Known and new neotectonic features are superposed onto the map in black and red, respectively. Similarly, point features with black margins were known prior to this study and those with red margins are new. Crustal element boundaries are after Shaw et al. (1996). The central crustal element in the figure is Archean in age and is known as the Yilgarn craton. SRTM—Shuttle Radar Topography Mission; DOLA—Department of Land Administration.

River—T. Crone, 2003, personal commun.). The likely timing of the most recent event on two additional scarps has also been determined (Mt. Narryer east and west scarps, Williams, 1979), and three of the scarps reflect historic events (Meckering, Calingiri, and Cadoux; Gordon and Lewis, 1980; Lewis et al., 1981). Though the data are sparse, results suggest that tens of thousands of years typically separate surface-rupturing events on a given fault and that activity is highly episodic (e.g., Crone et al., 1997, 2003).

If proven, each of the scarps identified represents at least one Quaternary earthquake of magnitude 6 or greater. If these scarps are in fact earthquake faults, then they provide important clues to the crustal deformation mechanisms in the Yilgarn craton (Fig. 2). For example, their orientation and dominant reverse mechanism (a mechanism could often be implied from the DEMs) are consistent with the direction of the contemporary seismic stress field inferred from in situ (e.g., Hillis and Reynolds, 2003, 2003) and earthquake-derived (Clark and Leonard, 2003) stress data. The spatial distribution of scarps is consistent with uniformly distributed strain across the Yilgarn craton at a time scale of tens of thousands of years. Many, if not most, of the scarps are not associated with contemporary seismicity, suggesting that strain release at longer time scales (e.g., hundreds to thousands of years) is migratory. In the few instances where high-resolution aeromagnetic data coincide with the location of a scarp (e.g.,

Hyden, Meckering, Cadoux, Lort River), the ruptures are seen to exploit preexisting crustal weaknesses.

In general, the surface (or geomorphic) expression of the scarps revealed in the DEM study and in paleoseismic investigations is subtle, and evidence is preserved for only one or two large Quaternary earthquake events on any given fault. Geological and geophysical studies of the 1968 Meckering earthquake scarp (Gordon and Lewis, 1980; Dentith and Featherstone, 2003; Clark et al., 2003) and the prehistoric Hyden scarp (e.g., Crone et al., 2003; Clark, 2006) suggest that Quaternary earthquakes occur on a family of faults within a region, rather than on individual large faults (see also Crone et al., 1997), which confirms geomorphologic and trenching observations and implies that strain is distributed on a local as well as cratonwide scale. As a consequence, uplift at Earth's surface is distributed across a wide area, the rates of erosion compete with rates of uplift, and the landscape remains essentially flat. Furthermore, these observations are consistent with the idea that large earthquakes are episodic within any given area.

In contrast to southwest Australia, the rugged and youthful topography of the Mt. Lofty–Flinders Ranges of South Australia preserves evidence of a very different style of seismicity. Here, rates of uplift exceed rates of erosion, and large offsets on individual structures are clearly the result of many major earthquakes (e.g., Sandiford, 2003a; Quigley et al., 2005). Although it is difficult in most cases to know the size and timing of these prehistoric events, it is possible to estimate the long-term rates of slip or movement on these faults from the geologic record (e.g., strata, marine strandlines, riverine terraces, drainage diversion). For example, slip rates on faults bounding the Mt. Lofty Ranges in South Australia, which are responsible for the generation of significant relief (many tens of meters), have been estimated at 0.02–0.03 mm/yr averaged over the past 3–5 m.y. (Sandiford, 2003a). The same author reports slip rates of 0.01–0.02 mm/yr from faults in the central Murray Basin and rates of up to 0.1 mm/yr from the Otway Ranges in Victoria, averaged over a similar time period (Sandiford, 2003a, 2003b).

The Lake Edgar fault in southwest Tasmania (Fig. 3) is an example of dating of individual events on an intraplate fault (Cupper et al., 2004; Clark, 2006). Geomorphic mapping of the ~30-km-long and up to 8-m-high Lake Edgar fault scarp suggests that three large surface-rupturing events with vertical displacements of 2.4–3.1 m have occurred in late Quaternary times. Displacement of this magnitude implies earthquake magnitudes on the order of M_w 6.8–7.0. Optically stimulated luminescence (OSL) dates from a sequence of three fluvial terraces associated with scarp incision provide brackets on the age of these events: ca. 17–18 ka, ca. 25–28 ka, and ca. 48–61 ka. Estimates for the average slip rate calculated for the two complete seismic cycles range from 0.17 to 0.20 mm/yr (unweighted mean). Taken at face value, the slip rate on the Lake Edgar fault is large compared to those inferred for other intracratonic faults in Australia, even where significant fault-controlled relief is evident. However, it is not possible to tell if the slip rate calculated from the last two complete seismic cycles is representative of the long-term (i.e., hundreds of thousands of years) slip rate on the fault, since there is no evidence for previous events.

Not all evidence for prehistoric earthquakes is as dramatic as the Lake Edgar fault scarp. Other indicators include originally horizontal strata that are now tilted or folded (e.g., Nullarbor scarps, Western Australia; Lowry, 1972), the deformed shorelines of lakes, uplifted marine and fluvial terraces (e.g., Macquarie Harbour marine terraces, Tasmania; Sandiford, 2004), and liquefaction deposits (e.g., Robe South Australia, Meckering Western Australia; Cummins et al., 2003). Geoscience Australia is currently compiling a database of geologic indicators of recent earthquake or neotectonic activity (Clark, 2004). The database now contains over 160 independent instances of evidence for possible Quaternary deformation.

The completeness of the inventory of neotectonic features varies considerably across Australia. The southwest of Western Australia is the most complete. It is likely that scarps related to most events of magnitude greater than 6.5 that have occurred in the last several tens of thousands of years (perhaps a hundred thousand or more) have been captured. Similarly, a large proportion of the most active faults in the Flinders Ranges–Mt. Lofty Ranges of South Australia has been captured. The catalogue is grossly incomplete in most other regions of Australia due a combination of factors including poor potential for scarp preservation and lack of suitable data sets with which to find scarps (e.g., high-resolution digital elevation models). In the short-term, this deficiency is being partly addressed through ongoing study of the Shuttle Radar Tomography Mission 90 m digital elevation model. Preliminary results from appraisal of a subset of this data set are promising, and more than 20 new post–middle Miocene scarps have been identified on the Nullarbor Plain of eastern Western Australia and western South Australia. However, the resolution of the data is such that only multiple-event scarps (vertical relief greater than 3 m) are readily visible. This is a significant limitation in eastern Australia, where rates of erosion are comparable to rates of relief generation on active faults. As an example, if just the Shuttle Radar DEM had been available instead of the 10-m-resolution data, only half of the new scarps identified in the southwest of Western Australia would have been identified.

Figure 4 shows the distribution of neotectonic features superimposed on a map of historic seismicity. It appears from this comparison that there are broad bands or regions characterized by the occurrence of scarps and seismicity. If the date of the most recent event and the recurrence interval between events can be determined for these features, they will provide valuable constraints on the seismicity model that underlies the seismic hazard of Australia.

Seismicity of Australia

Geoscience Australia's catalogue of Australian earthquakes contains ~27,000 earthquakes, of which 17,000 are considered main shocks and 10,000 aftershocks (Leonard, 2007). The earliest documented earthquake in Australia, in June 1788, was felt in Sydney Cove five months after European settlement. The catalogue is reasonably complete above magnitude 5.0–5.5 since 1910 and complete for magnitude 3.5 or greater since 1975. Generally, southern Australia has earlier dates of completion than northern Australia, and much of southern Australia is complete above magnitude 5.0 since 1880. In contrast, for parts of northern Australia, this level of completion is not reached until as late as 1960. The 131 events of magnitude 5 or greater since 1910 and 833 events of magnitude 3.5 or greater since 1975 are shown in Figure 1. These earthquakes are distributed over the entire continent, and only a few areas have been essentially aseismic (e.g., no earthquakes above magnitude 3.5) over the past 30 yr.

Based on the spatial coherence of the seismicity, four regions of sustained enhanced seismicity have been identified and are shown in Figure 1. They are southern Australia (SA), southeast Australia (SEA), northwest Australia (NWA), and southwest Australia (SWA). There are other areas of concentrated seismic activity, but these have either been active for less than 50 yr or are of an area smaller than 200,000 km². On average, the SA region has a magnitude 5 or larger earthquake every 10 yr, SEA every 3 yr, NWA every 1.2 yr, and SWA every 4.5 yr; Australia has 1.3 earthquakes above magnitude 5 per year. When aftershocks are removed, Australia has ~1.1 earthquake above magnitude 5 per year (Fig. 5).

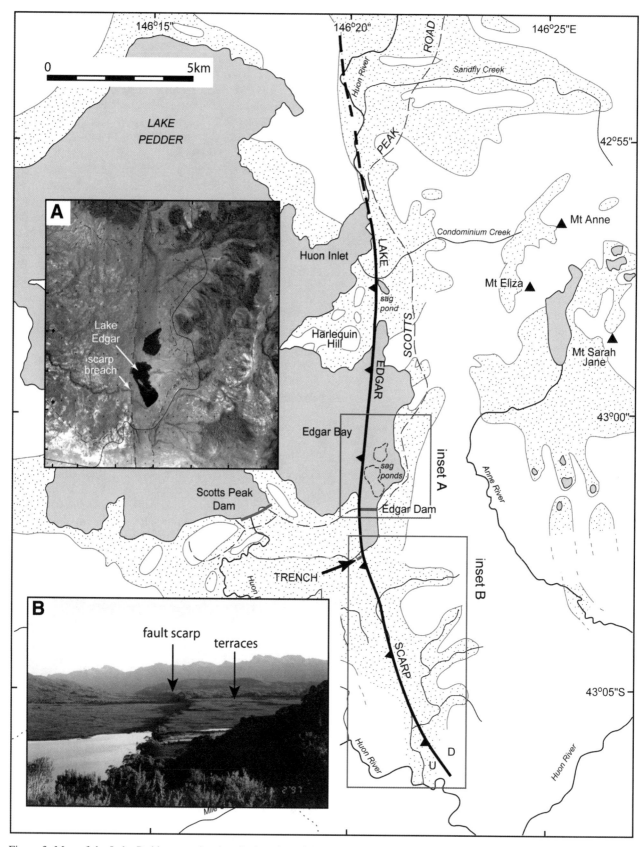

Figure 3. Map of the Lake Pedder area showing the location of the Lake Edgar fault scarp. Areas of significant Quaternary sediment accumulation (>1 m) have been added in stipple. (A) Fluvial fans south of Edgar Dam (looking south along the fault scarp). (B) Lake Edgar and its companion lake prior to flooding. The current impoundment level is indicated. Note that westerly flowing drainage breaches the scarp proximal to the western margin of Lake Edgar. U—upthrown side, D—downthrown side.

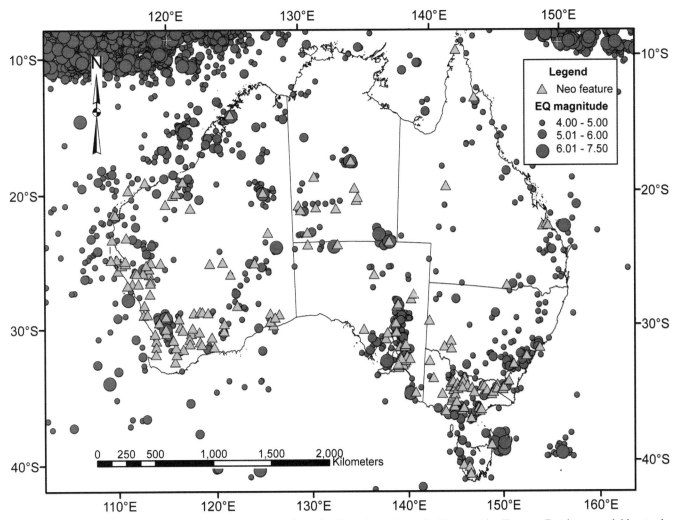

Figure 4. Location of instances of Quaternary deformation from the Geoscience Australia Neotectonics Features Database overlaid onto the catalogue of historic seismicity. EQ—earthquake.

In the late 1980s and early 1990s, several local magnitude scales were developed, including: South Australia (Greenhalgh and Parham, 1986; Greenhalgh and Singh, 1986), Western Australia (Gaull and Gregson, 1991), and southeast Australia (Michael-Leiba and Malafant, 1992; Wilkie et al., 1993). Prior to these dates, Australian earthquake magnitudes were calculated using the Richter (1935) relation. At 500 km, a typical source-receiver distance for Australian earthquake recordings, Richter's M_L magnitudes overestimate SA, WA, and SEA magnitudes by 0.42, 0.55, and 0.62 magnitude units, respectively. Since magnitudes of the pre-1990 catalogue have not been recalculated, magnitudes in the Australian earthquake catalogue are disparate for pre- and post-1990 values. For events around magnitude 5.5 and greater, m_b and M_S scales are commonly used. These scales are unlikely to correlate exactly with the extrapolation of M_L. Consequently, the variations in magnitude estimates increase the uncertainty in estimates of recurrence relations. The most likely effect of using the earlier earthquakes overestimates of

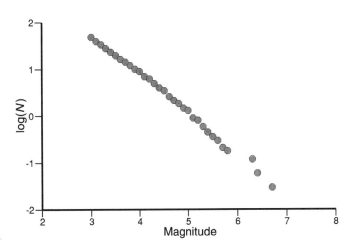

Figure 5. Recurrence relations for all of Australia based on the past 33 yr of data but normalized to 100 yr.

the earthquake activity (or the *a* value) is that their reported M_L values will on average be higher. Although the effect on the *b* value should be minor, overestimating *b* is possible due to the fact that a larger average epicentral distance was used to estimate the magnitude of the larger earthquakes. Geoscience Australia is currently collating as many older readings and digital waveforms as is feasible in order to build a comprehensive earthquake database. This will be used to revise magnitude scales and produce an updated catalogue using consistent magnitudes and locations. Where original data are not available, empirical relations will be developed to convert old M_L to new M_L, similar to the M_W scales that are being developed in conjunction with the ground-attenuation research discussed below.

Leonard (2007) undertook a recurrence analysis in which Australia was divided into 85 km square cells and *a* and *b* values were calculated using the weighted average of the cell and its eight neighbors. Due to the low level of seismicity for much of Australia, only ~40% of the cells had statistically significant results. The remaining cells were left with null values. About two-thirds of the events had magnitudes calculated using the original Richter (1935) magnitude scale, and one-third used Australian specific scales. The results of the recurrence analysis are shown in Figures 6 and 7. As expected, there is a clear relationship between the seismicity (Fig. 1) and areas of high *a* (Fig. 6). The Gutenberg-Richter relation was used to estimate the probability of a magnitude 5.0 or greater earthquake occurring during 100 yr in that cell, normalized to 10,000 km^2 (Fig. 7). The areas of high seismicity calculated using this smoothed seismicity approach compare well with the GSHAP (Global Seismic Hazard Assessment Program) hazard map (Giardini, 1999) for Australia. The areas where the magnitude 5.0 probability is greater than 0.4 are similar to the areas where GHAP predicts 10% or greater probability of exceeding a ground velocity of 0.8 m/s in 50 yr.

Outside the four regions discussed already (SWA, NWA, SA, and SEA), there are numerous smaller areas of enhanced seismicity, which tend to appear as hotspots on earthquake hazard maps. For most of these hotspots, the seismicity appears not to have been continuous over the past 50 yr, whereas the seismicity in the NWA, SA, and SEA regions has been continuous for at least a century. Outside these three regions, the pattern of seismicity across most of Australia appears to be spatially and temporally clustered. Many of the earthquakes of magnitude 5.5 and greater appear to be episodic, with most earthquakes above M6.0 having episodic characteristics. Examples of this episodic behavior include the Flinders Island earthquakes in the 1880s, Robe in the 1890s, Warooka in the 1900s, Simpson Desert in the 1930s and 1940s, SWA since the 1960s, Lake Mackay in the 1970s, and Tennant Creek in the 1980s (Leonard, 2007).

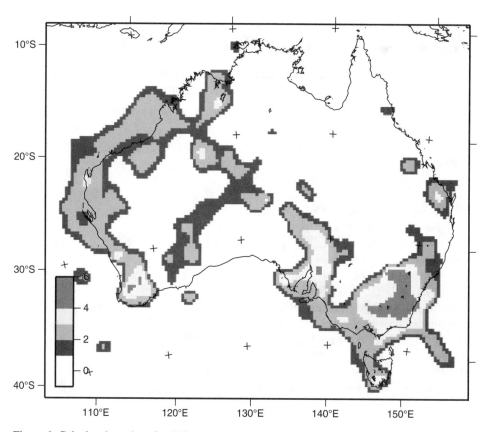

Figure 6. Calculated *a* values for 85 km square cells. Where there were insufficient earthquakes to calculate a reliable estimate, no value was calculated.

While each seismic episode is unique, a few generalizations can be made. These regions have all had periods of high activity associated with at least one magnitude >6.0 earthquake, which were preceded by a period of very low seismicity and/or followed by a period of very low seismicity. The seismicity during the periods of low seismicity is more than an order of magnitude lower than during the period of high activity. The period of high activity, with large earthquakes, is typically 1–10 yr, and the largest earthquakes often have aftershock sequences. A period of moderate activity typically lasts from a few years to a few decades following the large earthquakes. Some episodes (e.g., SWA and Tennant Creek) had a period of moderate foreshock activity a few years before the very large earthquakes. Based on the few data available for large earthquakes (e.g., Crone et al., 1997, 2003; Clark and McCue, 2003; Clark, 2006), recurrence intervals between main shocks of 10 k.y. to 100 k.y. are typical.

Each earthquake event represents a finite amount of deformation (i.e., slip on a fault). This deformation can be estimated from the amount of energy released during an earthquake, which is derived from its magnitude (Kostrov, 1974; Hanks and Kanamori, 1979; Johnston, 1994, 1996). Other measures of deformation can be obtained from GPS (Tregoning, 2003) and satellite laser ranging (Smith and Kolenkiewicz, 1990). Leonard (2007) compared the various estimates for the east-west com-

pression across southern Australia and concluded that the compression rate across southern Australia is 0.0–2.65 mm/yr, and the most likely range is 0.5–1.0 mm/yr.

The fixed seismic networks in Australia locate only 4% of current earthquakes to within 2 km (2σ error), 14% within 3 km, and 32% within 5 km. All the events located to within 3 km occur in the SWA, SA, and SEA regions. In recent years, several dense temporary networks have been deployed to acquire data in order to accurately locate earthquakes and ground-motion studies (see following). Data from both the Burakin deployment, within the SWA region, and the Flinders Ranges deployment, within the SA region, have not been fully processed yet. Once complete, the data should enable some basic questions to be answered, such as: "Are the earthquakes associated with mapped geological faults or tectonic boundaries?" This will have implications for seismic source zonation.

Strain-Rate Modeling

In order to verify the seismically derived estimates of regional strain, direct measurements of deformation are being undertaken in both southwest Western Australia and central South Australia. The approach is to deploy arrays of ~50 GPS receivers and record their positions over a period of one month. This process yields

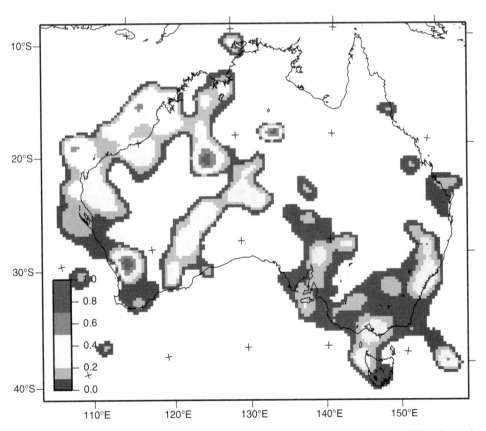

Figure 7. Estimated probability of a magnitude 5 or greater earthquake occurring in 100 yr for each 85 km^2 cell.

the locations of monuments to an internal precision of better than 0.1 mm (Featherstone et al., 2003). It is estimated that it will take at least 3–5 yr for the ground surface to deform sufficiently to warrant a return campaign. These measurements, along with other direct measures, such as satellite laser ranging, will be compared with seismically, geodynamically, and geologically derived deformation rates. Insight into the long-term behavior of seismicity is likely to follow from this comparison.

Evaluation of stress in Earth's crust from studies of the focal mechanisms of past earthquakes (Clark and Leonard, 2003) has demonstrated that the deeper seismogenic stress field is consistent with the contemporary seismic stress field inferred from shallow in situ data (e.g., Hillis and Reynolds 2003) and surface neotectonic activity (Clark and Bodorkos, 2004). The consistency of the stress field provides valuable constraints on stress and strain models.

One way to geodynamically model a plate is to use thin-shell finite-element models. Thin-plate models reduce the problem from three dimensions to two dimensions by vertically integrating the properties of the lithosphere. Different parts of the plate have different strengths depending on the local geotherm and material properties. The output of the model includes estimates of the principal horizontal stress, long-term strain rate, and depth to brittle-ductile transition (i.e., the maximum depth for earthquake nucleation).

An example of the neotectonic strain rate model for the whole of the Australian plate is shown in Figure 8. The model has a spatially varying strength based on the local geotherm and material properties. The parameters of the model (e.g., density, thermal conductivity, forces exerted on the plate margins, etc.) have been chosen to minimize the difference between the output of the model and a variety of geophysical observations (depth to the Mohorovicic discontinuity, principal horizontal stress azimuths, stress regime, GPS velocity data, and rate of spreading of the mid-ocean ridges along the southern margins of the plate). The model shown had the minimum misfit of all the models tried. In general, the model gives surface velocities within 1 mm/yr of the observed values and an azimuth of the principal stress directions within a few tens of degrees of the observations.

Once we have an estimate of the long-term tectonic strain rate, then the average rate of seismic moment release can be deduced from Kostrov's formula (Kostrov, 1974). Given some known (or estimated) distribution of earthquakes (e.g., a Gutenburg-Richter distribution), we can estimate an upper limit to the average number of earthquakes above a given size in a particular area and therefore the seismic hazard. This method bypasses the problem of an insufficiently long record of the seismicity to estimate the hazard and therefore allows us to calculate the hazard based solely on our knowledge of the physical properties of the plate and the forces acting on it.

These strain models give an estimate of the location of currently accumulating stress, which is an indicator of the locations earthquakes are likely to occur in the future. Where these correlate with areas that the prehistoric and contemporary seismic records also indicate had higher levels of seismicity, we can have confidence in the reliability of the source zonations.

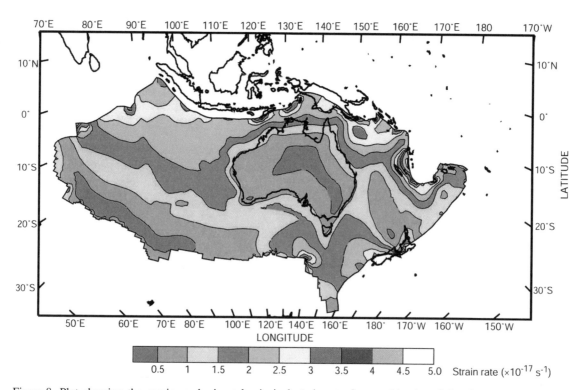

Figure 8. Plot showing the maximum horizontal principal strain rate from a thin-plate finite-element model (from Burbidge, 2004).

GROUND MOTION

Attenuation

Very little is known concerning the appropriate ground-motion model for Australia. The first attenuation model developed using only Australian data (Gaull et al., 1990) was based on seismic intensity data. These data are obtained from personal perceptions of shaking and damage. A conversion factor was developed for estimating peak ground acceleration (PGA), but this conversion is fraught with uncertainty. For this reason, earthquake risk assessments in Australia typically adopt spectral ground-motion models from other stable continental regions such as eastern North America (e.g., Dhu and Jones, 2002; Jones et al., 2005). However, there has been very little analysis undertaken to show whether eastern North America models (e.g., Atkinson and Boore, 1997; Toro et al., 1997) are applicable to Australian conditions. Moreover, there has been some conjecture that ground motion in some parts of Australia resembles that of active tectonic regions and may be better described by models such as that of Sadigh et al. (1997), which was developed from shallow Californian earthquake data.

Using attenuation models derived from limited seismic intensity and weak-motion data from central Australia (i.e., Bowman and Kennett, 1991), Bakun and McGarr (2002) observed that crustal attenuation falls somewhere between eastern North America and western North America. It is recognized that attenuation across Australia cannot be captured with a single ground-motion model. Rather, the attenuation of seismic wave energy varies transversely across the continent, with relatively low attenuation in the Archean and Proterozoic terranes of western and central Australia and higher attenuation in the younger Paleozoic terranes of eastern Australia (e.g., Gaull et al., 1990).

Australian Ground-Motion Data

Ground-motion prediction models for the Australian crust have been difficult to quantify in the past due to a lack of ground-motion data from moderate-to-large local earthquakes. Two key data sets have recently been compiled to derive empirical ground-motion attenuation models: one from data recorded in the Paleozoic crust of southeastern Australia and the other from the Archean shield regions of southwestern Western Australia (Fig. 1).

Due to the development of much of the nation's infrastructure and higher-than-average earthquake activity, the seismograph network in southeastern Australia is well developed. Earthquake data were primarily recorded and located by the Environmental Systems and Services, Seismology Research Centre, Melbourne, which has monitored earthquake activity in the region since the mid 1970s. Most of the earthquakes selected were well-located, with well-constrained focal depths with uncertainties of approximately ±4 km or less. The events occurred from 1993 to 2004 and had moment magnitudes of $2.0 \leq M_W \leq 4.7$. Data in this region have a good spatial distribution (to a distance of approximately

700 km from the source) and comprise approximately 1220 ground-motion records, with typical sampling rates of 100 Hz.

Data used for the empirical southwestern Western Australia ground-motion model were obtained from the Burakin earthquake sequence of 2001–2002. These data provided the largest resource of high sample-rate data in southwestern Western Australia that are useful for spectral studies. Following the onset of activity at Burakin in September 2001, Geoscience Australia deployed a temporary seismic network in the region. In the six-month period following commencement of seismicity, a total of six earthquakes of magnitude M_W 4.0 or greater, coupled with some 18,000 smaller events, occurred in the region (Leonard, 2003a). Of the many hundreds of located events, 69 earthquakes of magnitude $2.2 \leq M_W \leq 4.6$ were analyzed. The data set is composed of almost 400 ground-motion records, including strong-motion data for the earthquakes greater than M_W 4.0 at hypocentral distances less than 10 km. All earthquakes in this data set were located with focal depths shallower than 2.7 km (Allen et al., 2006).

The clustered nature of the earthquake sources means that the recorded data are also spatially clustered, which may introduce azimuthal- and site-dependent biases to any empirical model. Spectral analysis of the Burakin southwestern Western Australia data set indicates that average corner frequencies for these events are relatively low, particularly for the smaller events (Allen et al., 2006). These imply low stress drops for lower magnitudes (M_W < 4.0) and increasing stress drops for larger magnitudes. It may not be surprising that the stress drops are low for the low-magnitude events. The larger events may represent faulting of previously unfractured rock or healed fault asperities, whereas the smaller events are adjustment events or aftershocks that occur on recently faulted surfaces under relatively low deviatoric stresses (Allen et al., 2006). Therefore, the shallow, low-magnitude events do not generate the high-frequency ground motions that would typically be expected for intraplate earthquakes (e.g., Allen et al., 2004). It should be noted that the moment magnitudes were estimated using attenuation parameters derived from recent Australian ground-motion studies (e.g., Allen et al., 2006).

Earthquake data used in this study have been compiled into the Australian Ground-Motion Database, which is freely available from Geoscience Australia. It is hoped that this information will provide a valuable resource for ground-motion attenuation studies in intraplate regions.

Empirical Ground-Motion Prediction Models for Australia

The development of empirical ground-motion attenuation models for Australia largely follows the methods adopted by Atkinson (2004a) for eastern North America. Detailed analysis of the southeastern Australia data set indicates that the decay of low-frequency Fourier spectral amplitudes is described by a tri-linear geometrical spreading operator. The subsequent decay of low-frequency (less than 2 Hz) spectral amplitudes can be approximated by the coefficient of $R^{-1.3}$ (where R is hypocentral distance) within 90 km of the seismic source (Allen et al., 2007).

From 90 to 160 km, the southeastern Australia data set indicates a zone in which the seismic coda appears to be affected by crustal reflections and refractions. In this distance range, geometrical attenuation is approximated as $R^{+0.1}$. Beyond 160 km, low-frequency seismic energy attenuates rapidly as $R^{-1.6}$. The apparent high geometrical attenuation at larger distances is attributed to the well-established velocity gradient in southeastern Australia (Collins et al., 2003), which allows dispersion of *Lg*-wave energy into the mantle (e.g., Bowman and Kennett, 1991; Atkinson and Mereu, 1992). Since the geometrical spreading parameters are chosen to model the attenuation of low-frequency seismic energy, these parameters implicitly incorporate some anelastic attenuation effects. The coefficient $R^{-1.6}$ for $R > 160$ km is the attenuation parameter that minimized the uncertainties in the calculation of the corresponding frequency-dependent anelastic attenuation parameter ($Q[f]$) (Allen et al., 2007).

The short hypocentral distance range ($10 < R < 170$ km), coupled with the limited number and spatially clustered nature of the southwestern Western Australia data set, caused some ambiguity in interpreting the shape of the geometrical attenuation curve, particularly for $R > 100$ km. Consequently, this made it difficult to quantify a complex attenuation model similar to that for southeastern Australia. For $R < 80$ km, the low-frequency geometrical attenuation for the southwestern Western Australia data set is approximated by $R^{-1.0}$ (Allen et al., 2006). Beyond the postcritical distance of 80 km, we assume theoretical cylindrical attenuation of $R^{-0.5}$ (e.g., Herrmann and Kijko, 1983). The postcritical distance is an approximation based on the estimates of twice the crustal thickness for the region (Dentith et al., 2000).

Figure 9 compares Fourier ground-motion prediction models for southeastern Australia (Allen et al., 2007), southwestern Western Australia (Allen et al., 2006), and eastern North America (Atkinson, 2004a, 2004b) at hypocentral distances of 1, 10, 50, and 100 km from an earthquake source of M_w 4.5. The southeastern Australia and eastern North America models are relatively similar because the attenuation is essentially the same over this distance range (attenuation for eastern North America is described as $R^{-1.3}$ for $R < 70$ km; Atkinson, 2004a). Due to the higher atten-

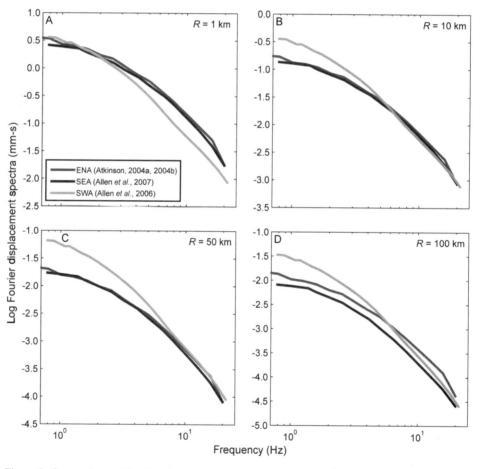

Figure 9. Comparison of Fourier displacement spectra at hypocentral distances of (A) 1 km, (B) 10 km, (C) 50 km, and (D) 100 km for an earthquake of M_w 4.5. Note that for all three models, source spectra (i.e., $R = 1$ km) converge at low frequency, indicating that estimates of M_w are consistent in A. In figures B–D, we observe the effect of lower rates of attenuation in southwestern Western Australia. ENA—eastern North America; SEA—southeastern Australia; SWA—southwestern Western Australia.

uation observed in southeastern Australia for $R > 70$ km, the models differ considerably at larger distances (Fig. 10). Low-frequency spectral amplitudes calculated for the southwestern Western Australia model are significantly higher with increasing source-receiver distance than for both southeastern Australia and eastern North America.

Residuals of observed and modeled ground motions for southeastern Australia and southwestern Western Australia at a frequency of 2 Hz are plotted against hypocentral distance in Figures 11 and 12, respectively. The eastern North America model (Atkinson, 2004a, 2004b) is also applied to data from each region and the residuals are compared. At shorter hypocentral distances ($R < 70$ km), the southeastern Australia and eastern North America models are roughly equivalent for all frequencies as demonstrated by the similarity of Fourier displacement spectra (Figs. 9A–9C). At larger hypocentral distances ($R > 70$ km), the Atkinson (2004a, 2004b) model for eastern North America overestimates ground motions significantly. For the southwestern Western Australia model, Fourier spectral amplitude residuals at 2 Hz are generally quite low over the observed hypocentral distance range (Fig. 12). The Atkinson (2004a, 2004b) model appears to slightly underestimate recorded ground motions at short hypocentral distances ($R < 70$ km) when applied to southwestern Western Australia data, particularly for lower frequencies (less than 2 Hz).

Comparison of Australian and Eastern North America Ground-Motion Models

Results indicate that the frequency-dependent characteristics of earthquake-generated ground motion differ significantly for events in southeastern Australia compared to southwestern Western Australia (Fig. 9). Earthquake source spectra (i.e., $R = 1$ km) in southeastern Australia contain higher levels of short-period energy than southwestern Western Australia and are more consistent with eastern North America spectral models. In

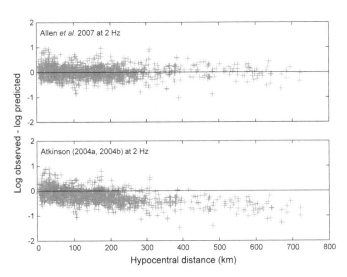

Figure 11. Log residuals of modeled southeastern Australia (SEA) ground motions plotted against hypocentral distance at 2 Hz. Data are compared with the attenuation model of Atkinson (2004a, 2004b). At distances greater than ~70 km, Atkinson's (2004a, 2004b) model overestimates ground motions when applied to southeastern Australia data.

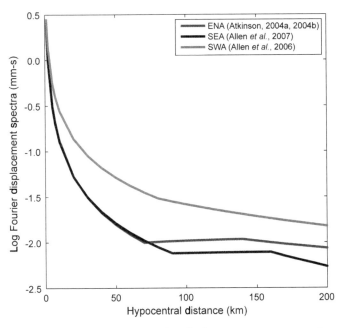

Figure 10. Comparison of Fourier displacement spectra at a frequency of 1.0 Hz for hypocentral distances to 200 km for an earthquake of M_w 4.5. It can be observed that attenuation in southeastern Australia (SEA) and eastern North America (ENA) are very similar for $R < 70$ km. Minor variations in this range are due to differences in anelastic attenuation (i.e., the seismic quality factor, Q). Beyond this distance, we observe higher rates of ground-motion attenuation in southeastern Australia. SWA—southwestern Australia.

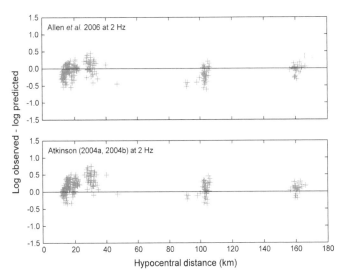

Figure 12. Residuals of modeled southwestern Western Australia (SWA) ground motions plotted against hypocentral distance at 2 Hz. At low frequencies, the Atkinson (2004a, 2004b) model appears to slightly underestimate ground motion at short hypocentral distances when applied to southwestern Western Australia data.

contrast, longer-period ground motion appears to attenuate less with distance in southwestern Western Australia than in southeastern Australia. This is due to the lower geometrical attenuation observed for southwestern Western Australia. The deficiency of short-period energy in southwestern Western Australia source spectra is due to the low stress drop nature of the events (Allen et al., 2006). This is particularly apparent for smaller magnitudes. Southwestern Western Australia source spectra begin to converge with southeastern Australia and eastern North America models at increasing magnitudes (Fig. 9A). We are therefore concerned that an attenuation model based on the southwestern Western Australia data set may have limited application for predicting ground motions of isolated crustal earthquakes with higher stress drop, particularly for frequencies greater than 2 Hz.

Given the similarity of the geometrical attenuation between the southeastern Australia and eastern North America models, we observe that there is no major difference in spectral amplitudes at short source-receiver distances. Beyond 70 km, however, the southeastern Australia ground-motion model indicates higher attenuation than in eastern North America. Although this difference is important in terms of magnitude estimation, it may not be overly significant in terms of risk because we do not expect to see a major contribution to damage from earthquakes beyond this hypocentral distance range. This suggests that using modern eastern North American ground-motion prediction equations (e.g., Atkinson and Boore, 2006) for earthquake hazard and risk studies in southeastern Australia could serve as sufficient proxies in the absence of reliable Australian ground-motion prediction models, particularly for short return periods. This is because at shorter return periods, disaggregated hazard is governed by moderate-sized events at close range (e.g., Jones et al., 2005). However, for longer return periods, where there is a greater likelihood of a maximum credible earthquake (typically assumed to be approximately M_W 7.2 for Australia), eastern North America attenuation models will overestimate probabilistic hazard in southeastern Australia since there will be a significant contribution to hazard at larger source-receiver distances.

This work provides a framework for regional ground-motion prediction equations and indicates that we cannot simply rely on ground-motion prediction models from other regions for Australian earthquake hazard and risk assessments. Source and attenuation parameters derived in these empirical studies are currently being used as key inputs to stochastic methods to predict ground motions for larger-magnitude events.

Site Response

Irrespective of the choice of attenuation model, there is a need to incorporate local geology into any earthquake risk assessment. For example, the presence of regolith (i.e., soils, geological sediments, and weathered rock material) can dramatically increase the ground shaking during an earthquake. A common approach to incorporating the effect of local geology is to adopt a model based on a veneer of regolith overlying hard

rock (e.g., Dhu and Jones, 2002). However, generic hard rock attenuation models may also need to be modified for variations in crustal geology. This is particularly relevant where the local crustal geology is markedly different from that assumed in the generic attenuation model.

Modeling of site response in regions with softer crustal geology is often hindered by a lack of detailed shear-wave velocity data at depth. For example, Geoscience Australia has recently completed a major earthquake risk assessment in Perth (Jones et al., 2005). Detailed geotechnical data obtained for the region's regolith formed the basis of a numerical model of the site response. This model was based on a similar approach to the work conducted for previous Australian site response models (e.g., Dhu and Jones, 2002), and it incorporated a great deal of detailed information and variability in the region's regolith. However, the Perth study region is located in the deep (~15 km) sedimentary Perth Basin (Jones et al., 2005), for which there is very little shear-wave velocity information available. Consequently, the numerical model of site response incorporated "best estimates" of shear-wave velocity and material damping.

An alternative approach to developing site response models uses observations. Ground motions recorded during the 2001–2002 Burakin earthquake sequence (Leonard, 2003b) have been used to create an empirical model for Perth (Jones et al., 2005). Specifically, two ground motions were recorded from a magnitude M_W 4.4 earthquake located ~160 km and 190 km northwest of the Perth, respectively. The recording at station PIG4, within the Yilgarn craton, is assumed to be unaffected by the crust and regolith that modify the ground shaking in the Perth Basin. In contrast, the recording at station EPS is located in the center of the Perth Basin and is assumed to incorporate the effects of the region's regolith and crustal geology. These two recordings were distance corrected so that the ratio of the ground motion provides an empirical model for the effects of local geology.

A comparison of the numerical and empirical site response models demonstrates that the numerical model tends to predict significantly more amplification of ground shaking than the empirical model for periods greater than 0.1 s (Fig. 13). There are numerous explanations for this discrepancy, ranging from possible inaccuracies in the crustal velocity structure used in the numerical models through to the fact that the empirical model was derived from small earthquakes at large distances. As discussed later in this paper, the differences in these models have a dramatic impact on estimates of earthquake risk in the Perth region.

Geoscience Australia is currently focused on developing national-scale earthquake risk assessments. The fact that Australia is an old continent means that virtually all of Australia is weathered to some degree. Moreover, since very little of Australia has experienced glaciation in the Holocene, as opposed to North America and Europe, there is essentially no "hard rock" equivalent site category for Australian site conditions. Hence, site response will form an important component of any earthquake risk assessment. A site response model based on a veneer

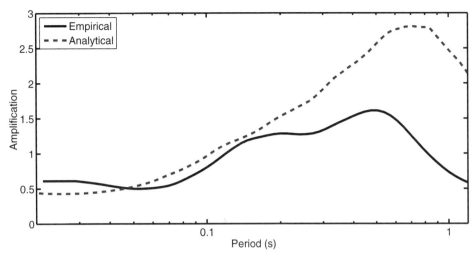

Figure 13. Empirical and numerical site response models for Perth, Western Australia.

of regolith over hard rock is realistic in the majority of Australia's inhabited regions. However, more care must be used in regions such as Perth to accurately incorporate complex crustal geology in ground-motion models.

Risk Modeling

Risk modeling provides a tool to test the sensitivity of earthquake risk assessments to key hazard inputs such as magnitude, source zones, attenuation, and site amplification. Risk assessments for the Newcastle–Lake Macquarie region and Perth city demonstrate the need for research to improve earthquake hazard models.

Figure 14 illustrates the computational flow chart for the probabilistic seismic risk assessment (PSRA) adopted in this study. "Hazard Data and Parameters" includes a set of seismic source zones that describe the seismicity of the region, the attenuation relationships for modeling the propagation of motion, and

the amplification models for estimating local site effects. "Elements at Risk & Vulnerability" includes an exposure database that describes a portfolio of buildings, engineering models describing building response (or capacity) to earthquake-induced motion, and a set of fragility curves that describe the vulnerability for different classes of buildings. The PSRA incorporates natural variability (or aleatory uncertainty) by treating key parameters such as earthquake magnitude or ground motion as random variables. "Synthetic Event Generation" is undertaken through stratified Monte Carlo sampling using the source model (e.g., Robinson et al., 2005; Patchett et al., 2005). "Propagation & Amplification" of earthquake-induced ground motion is achieved by using attenuation (e.g., Toro et al., 1997; Atkinson and Boore, 1997) and amplification (e.g., Dhu and Jones, 2002) models. The "Physical Impact" of the earthquake is estimated through the capacity spectrum method (FEMA, 1999; Kircher et al., 1997a), and the "Direct Financial Loss" is modeled using reconstruction costs derived from insurance loss data as well as industry construction

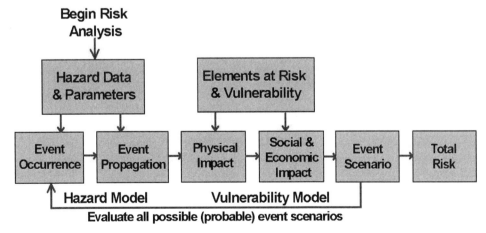

Figure 14. Flow chart describing the probabilistic seismic risk analysis.

data (Kircher et al., 1997b). This is an extension of probabilistic seismic hazard analysis (PSHA), in which only the "Synthetic Event Generation" and the "Propagation and Amplification" stages are considered. As with the PSRA, a loop is conducted over all "probable" events, and the results are aggregated to produce hazard estimates. Robinson et al. (2006a) provided a brief overview of the process for PSRA, and Robinson et al. (2005) provided a detailed description. For this paper, we consider only the direct cost of repair or replacement for buildings and not the indirect costs associated with factors such as loss of contents, business interruption or accommodation, loss of lifelines, death or injury, or overall social or economic impact.

Newcastle–Lake Macquarie is located on the eastern coast of Australia, ~170 km north of Sydney. The 1989 Newcastle earthquake was the most damaging and costly earthquake in Australia since European settlement. Beyond the insured losses of $862 million (IDRO, 2002) or roughly 5% of building value (Edwards et al., 2004), estimates of economic losses range from $1 billion (Melchers, 1990) to more than $4 billion (Bureau of Transport Economics, 2001). In addition, there were 12 deaths and 100–120 serious injuries reported (Melchers, 1990). The Newcastle–Lake Macquarie case study considers earthquake-induced damage to 6305 sample buildings with survey factors used to represent the total building stock of roughly 102,000. The "Hazard Data and Parameters" were defined by three earthquake source zones (Dhu and Jones, 2002), two attenuation models, T97 (Toro et al., 1997) and AB97 (Atkinson and Boore, 1997), and an amplification model (Dhu and Jones, 2002). This case study was used to evaluate the sensitivity of risk assessments to earthquake magnitude, earthquake frequency of occurrence, and ground-motion attenuation.

The city of Perth is located in the southwest of Australia, ~150 km from the southwest seismic zone in Figure 1. As discussed already, the southwest seismic zone is one of the most seismically active regions of Australia, and it has experienced at least three earthquakes of $M_L \geq 5.9$ in the past 40 yr. The case study considers damage to 4534 sample buildings with the results scaled to represent the total building stock of roughly 354,112 buildings in the region. The "Hazard Data and Parameters" were defined by five earthquake source zones (Jones et al., 2005), the T97 attenuation model, and the two site amplification models discussed already.

Risk estimates for the sensitivity study were illustrated using risk exceedance curves, average annualized loss values, and scenario loss for the 1989 Newcastle earthquake. Risk exceedance (RE) curves present loss values for the building stock that are expected to be exceeded as a function of varying levels of annualized probability. They are analogous to the use of hazard curves for interpreting earthquake hazard (e.g., Frankel et al., 2000). The probability levels are related in a one-to-one fashion to return periods via the Poisson relationship (e.g., Robinson et al., 2005). Average annualized loss (AAL) represents the average loss that is expected to occur per year, computed by summing all loss represented by the RE curve and normalizing it by the number of years over which it is expected to occur (e.g., Robinson et al., 2006a). Scenario loss values were computed by considering the Newcastle 1989 event and modeling the loss to building stock that occurred as a result.

Results

Figure 15 presents the RE curve and AAL for the Newcastle–Lake Macquarie earthquake study when the T97 attenuation model is used. This curve is labeled here and in subsequent figures as "Original." The square corresponds to the modeled Newcastle 1989 event scenario and illustrates a loss of 6.5% of the building stock or roughly $1.1 billion in 1989 dollars. The probability of exceedance of this loss is 0.0007 per year, for a return period of 1435 yr. Note that this event is modeled as a "moment magnitude" 5.35 earthquake, which corresponds to a Richter magnitude 5.6. When the magnitude of the 1989 event is increased by one-quarter unit from moment magnitude 5.35 to 5.6, the loss increases by 70% (Fig. 15). The return period of the 1989 $M_W = 5.6$ event loss also increases to 3008 yr. Figure 15 also shows the effect of increasing the maximum or upper-bound magnitude in each of the source zones by a magnitude unit, from M_W 6.5 to M_W 7.5. In this case, there is an increase in the loss of 37% at the 0.002-per-year probability (500 yr) RE curve and an increase in AAL of 44%. As annual probabilities decrease, the divergence in curves increases, with losses doubling the "Original" curve at roughly 1/10,000 yr. These results clearly depict the importance of magnitude in earthquake risk assessment.

The "Original" RE curve was developed using three earthquake source zones, including two zones of increased activity in the immediate vicinity of Newcastle. When the two high-activity zones are merged into the larger background area, the average annual loss decreases by 53%. The 500 yr loss decreases by 43%, and the return period of the 1989 event loss increases by 68% to 2748 yr (Fig. 16). It is not always easy to determine the best combination of source zones for a risk assessment. This case demonstrates how a modification of the source zones can dramatically influence a risk assessment.

Figure 17 presents the results for three different combinations of viable attenuation models corresponding to T97, AB97, and T97 + AB97. In the case of T97 + AB97, the building damage is evaluated for both the T97 and AB97 attenuation models independently, and the average damage considered. The RE curve corresponding to AB97 illustrates a decrease in loss of 67% at the 500 yr level and a decrease in annualized loss of 68%. The Newcastle 1989 loss decreases by 78% with a return period of 827 yr. When the two attenuation models are used together (T97 + AB97), the results fall between the individual T97 and AB97 results. The 500 yr loss decreases from the "Original" by 34%, the AAL by 35%, and the Newcastle 1989 loss by 40%. Figure 17 demonstrates the sensitivity of risk assessments to the selection of attenuation model, which is consistent with the level of understanding (epistemic uncertainty) of ground-motion attenuation in this region of Australia.

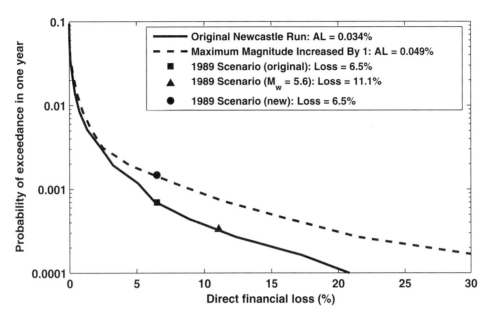

Figure 15. Modeled risk exceedance (RE) curve for Newcastle and Lake Macquarie (case 1) showing estimated maximum losses as a function of probability exceedance levels when the Toro et al. (1997) (T97) attenuation model is used. The annualized loss (AL; shown in the legend) represents the total loss expected per year when all events in the RE curve are considered. Modeled loss associated with the Newcastle 1989 event is also shown by the square and triangle when the event is considered to have magnitude Mw = 5.35 (case 1) and Mw = 5.6 (case 2), respectively. The RE curve and its associated square, labeled as "original," were used as reference estimates for the Newcastle sensitivity study. The dashed line corresponds to case 3. It represents the RE curve when the maximum magnitude is increased by 1. The annualized loss and the probability of the Newcastle 1989 event (circle) both increase for case 3.

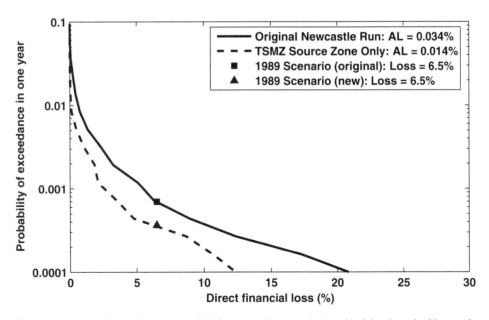

Figure 16. Comparison of cases 1 and 4 demonstrating a reduction in risk when the Newcastle triangle source zone is no longer considered as a separate zone and its seismicity merged with the Tasman Sea margin zone (TSMZ).

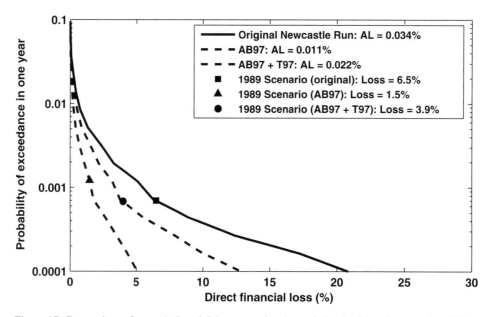

Figure 17. Comparison of cases 1, 5, and 6 demonstrating the variation in risk estimates when different attenuation models are used. The Atkinson and Boore (1997) (AB97) attenuation model leads to (1) lower levels of loss for the Newcastle 1989 event and (2) lower risk for the region than the Toro et al. (1997) (T97) attenuation model. Estimates using both attenuation models (T97 + AB97) fall between the two individual estimates.

The Perth case study demonstrates the sensitivity of a basin containing thick sediments to amplification factors by illustrating the different risk results for empirically and numerically derived amplification factors (Fig. 18). Using the numerically derived amplification factors, the 500 yr loss increased by 249% and the AAL increased by 506%.

The results of the sensitivity analysis are summarized in Table 1. The return period for modeled Newcastle 1989 event and the change in event loss are presented for all of the Newcastle cases. The changes with respect to best estimates for 500 yr loss and AAL are presented for each case.

Discussion

A comparison between cases 1 and 2 indicates that a scenario simulation is sensitive to earthquake magnitude. The change in magnitude of one-quarter unit is typically less than the width of a 95% confidence interval for the moment magnitude of an earthquake. As discussed already, the magnitudes of most earthquakes in the Australian catalogue have uncertainties of at least one-quarter of a magnitude unit. This comparison highlights the need for ongoing research into constraining magnitude estimates from limited network data.

Cases 1 and 3 reveal the sensitivity of PSRA to source-zone upper-bound magnitudes. Ideally, the upper-bound magnitude for a source zone should be supported by data (historic or prehistoric) or by some other means such as stress-field modeling. Improvements to neotectonic data and earthquake modeling will allow more-constrained estimates of upper-bound magnitude.

Classifications of earthquake source zones rely on a combination of instrumental and historical earthquake catalogues with geological and neotectonics information. The earthquake source zones for the Newcastle–Lake Macquarie study were based on a consensus of expert opinion; these zones were defined on the bases of elevated levels of historic seismicity and evidence for an active fault offshore of Newcastle. A background zone covering the seismicity of most of southeastern Australia was also included. This zone has substantially lower seismicity, where earthquakes in the magnitude 5.0–5.5 level occur at one-third the rate of the two Newcastle-area zones. A comparison between cases 1 and 4 illustrates that the manner in which seismicity is interpreted and spatially averaged within zones has significant implications for loss and recurrence interval estimates. Better estimates of the dimension and recurrence rates of seismic zones, based on sound seismological, geological, and neotectonic models will improve the reliability of the seismic source models.

Cases 1, 5, and 6 indicate that the choice of attenuation model can have a significant impact on risk estimates. The attenuation models used for these cases were derived for eastern North America. It is anticipated that Australian risk assessments will improve as Australian attenuation models become available. However, it should be noted that T97 and AB97, derived for the same region, lead to quite different risk estimates when applied to the Newcastle–Lake Macquarie region. While, as discussed already, Geoscience Australia is developing Australia-specific attenuation models, the differences

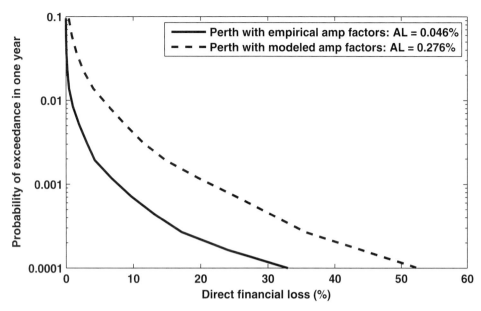

Figure 18. Comparison of cases 7 and 8 demonstrating different levels of risk when empirical and modeled amplification factors are considered.

TABLE 1. SUMMARY OF SENSITIVITY ANALYSIS

Case	1989 return period	Relative change (%)		
$M_w = 5.35$	(yr)	1989 loss	500 yr loss	AAL
1. Newcastle "Original"	1435	0*	0*	0*
2. 1989 as $M_w = 5.6$	3008	70	0	0
3. Mmax from 6.5 to 7.5	679	0	37	44
4. Decrease in event frequency	2748	0	−43	−59
5. AB97	827	−78	−67	−68
6. AB97 + T97	1462	−39	−34	−35
7. Perth empirical amp factors			0*	0*
8. Perth numerical amp factors			249	506
Note: The reference values are identified by an asterisk. AAL—average annualized loss.				

among cases 1, 5, and 6 illustrate that attenuation estimates will need to be revisited periodically as more data and new theories become available.

Site response forms an important component of Australian earthquake risk assessments due to the high level of weathering and presence of regolith. Robinson et al. (2006b) described a technique for computing, and estimating uncertainty in, amplification factors that is useful in situations where a veneer of regolith overlies hard rock. Cases 7 and 8 illustrate that a site response model based on a veneer of regolith over hard rock is not realistic in the city of Perth. The high level of variation results from a failure of the numerical site-class and ground-motion combination to adequately account for a deep basin of weathered sediment. In other cities, such as Newcastle, it is appropriate to assume a veneer of regolith over hard rock, and the variations in risk assessments will be considerably less that those shown for Perth. For example, a similar comparison of empirical and numerical models in Newcastle would result in variations an order of magnitude smaller than those presented for Perth. Careful development of

site response models that are consistent with associated ground-motion (attenuation) models is essential in order to accurately model earthquake risk in Australia.

We do not suggest that the alternative models used in this sensitivity analysis are all necessarily realistic or viable. They do, however, demonstrate the effects of some of the key controlling parameters on loss estimates. The results also illustrate some of the dangers in comparing loss results from "black boxes" where model parameters and computational algorithms are not readily understood. Similar results were presented in Newman et al. (2001) for the New Madrid area that demonstrated the sensitivity of earthquake hazard maps to attenuation, recurrence rate, and maximum magnitude.

CONCLUSION

We have demonstrated in this paper that improvements in the estimates of earthquake hazard in Australia require the development of improved seismicity, attenuation, and site response models. In stable continental crust, particularly where only short instrumental and historical records are available, the existing earthquake catalogue is not sufficient to constrain the key seismic source zone parameters (zone boundaries, recurrence rates, maximum magnitude, and depth). Improvements to these will come from a combination of seismological research (e.g., magnitude, location, depth, fault mechanisms) and geological (e.g., paleoseismicity, geomorphology) and geophysical/geodynamic (e.g., potential field analysis, numerical modeling) research. Incorporating regionally specific attenuation is very important, and applying one stable continental regions attenuation relationship to another area results in poorly constrained hazard estimates. For Australia, where the widespread deep weathering challenges the concept of hard rock attenuation models, this is particularly true. Site response can have both amplification and deamplification effects. In the Newcastle-Macquarie region, the site effects are amplifying, up to four times, whereas in Perth, the thick sedimentary basin has a deamplifying effect. Again numerical and, where seismic data is available, empirical site response models are essential for estimating earthquake hazard and risk, particularly in urban areas. Ongoing research is being undertaken in Australia to address all these short-comings in the current generation of earthquake hazard and risk models.

REFERENCES CITED

Allen, T.I., Gibson, G., Brown, A., and Cull, J.P., 2004, Depth variation of seismic source scaling relations: Implications for earthquake hazard in southeastern Australia: Tectonophysics, v. 390, p. 5–24, doi: 10.1016/j.tecto.2004.03.018.

Allen, T.I., Dhu, T., Cummins, P.R., and Schneider, J.F., 2006, Empirical attenuation of ground-motion spectral amplitudes in southwestern Western Australia: Bulletin of the Seismological Society of America, v. 96, p. 572–585, doi: 10.1785/0120040238.

Allen, T.I., Cummins, P.R., Dhu, T., and Schneider J.F., 2007, Attenuation of ground-motion spectral amplitudes in southeastern Australia: Bulletin of the Seismological Society of America, v. 97, p. 1279–1292.

Atkinson, G.M., 2004a, Empirical attenuation of ground-motion spectral amplitudes in southeastern Canada and the northeastern United States: Bulletin of the Seismological Society of America, v. 94, p. 1079–1095, doi: 10.1785/0120030175.

Atkinson, G.M., 2004b, Erratum to "Empirical attenuation of ground-motion spectral amplitudes in southeastern Canada and the northeastern United States": Bulletin of the Seismological Society of America, v. 94, p. 2419–2423, doi: 10.1785/0120040161.

Atkinson, G.M., and Boore, D.M., 1997, Some comparisons between recent ground-motion relations: Seismological Research Letters, v. 68, p. 24–40.

Atkinson, G.M., and Boore, D.M., 2006, Earthquake ground-motion predictions for eastern North America: Bulletin of the Seismological Society of America, v. 96, p. 2181–2205.

Atkinson, G.M., and Mereu, R.F., 1992, The shape of ground motion attenuation curves in southeastern Canada: Bulletin of the Seismological Society of America, v. 82, p. 2014–2031.

Bakun, W.H., and McGarr, A., 2002, Differences in attenuation among the stable continental regions: Geophysical Research Letters, v. 29, no. 23, p. 2121.

Bowman, J.R., and Kennett, B.L.N., 1991, Propagation of *Lg* waves in the North Australian craton: Influence of crustal velocity gradients: Bulletin of the Seismological Society of America, v. 81, p. 592–610.

Burbidge, D.R., 2004, Thin-plate neotectonic models of the Australian plate: Journal of Geophysical Research, v. 109.

Bureau of Transport Economics, 2001, Economic Costs of Natural Disasters in Australia: Canberra, Bureau of Transport Economics, 103 p.

Clark, D.J., 2004, Compilation of Evidence for Neotectonic Deformation within Australia: Implications for Seismic Hazard Assessment: Canberra, Geoscience Australia, 67 p.

Clark, D.J., 2005, Identification of Quaternary faults in southwest and central western Western Australia using DEM-based hill shading: AUSGEO News, v. 78, p. 8–10.

Clark, D.J., 2006, A seismic source zone model based on neotectonic data: Canberra, Australia, Proceedings of the Australian Earthquake Engineering Society 2006 Meeting, p. 69–76.

Clark, D.J., and Bodorkos, S., 2004, Fracture systems in granite pavements of the eastern Pilbara craton, Western Australia: Indicators of neotectonic activity?: Australian Journal of Earth Sciences, v. 51, p. 831–846, doi: 10.1111/j.1400–0952.2004.01088.x.

Clark, D.J., and Collins, C., 2004, Identification of Quaternary faults in southwest Western Australia using DEM-based hill shading: Eos (Transactions, American Geophysical Union), v. 85, no. 47, p. 339.

Clark, D.J., and Leonard, M., 2003, Principal stress orientations from multiple focal plane solutions: New insight in to the Australian intraplate stress field: Geological Society of Australia Special Publication 22 and Geological Society of America Special Paper 372, p. 91–105.

Clark, D.J., and McCue, K., 2003, Australian palaeoseismology: Towards a better basis for seismic hazard estimation: Annales de Geophysique, v. 46, p. 1087–1105.

Clark, D.J., Dentith, M., and Leonard, M., 2003, Linking earthquakes to geology: Contemporary deformation controlled by ancient structure, *in* Reddy, S.M., Fitzsimmons, I.C.W., and Collins, A.S., eds., Specialist group on tectonics and Structural Geology Field Meeting, Kalbarri, 22–26 September 2003: Geological Society of Australia, Abstracts, v. 72, p. 90.

Collins, C.D.N., Drummond, B.J., and Nicoll, M.G., 2003, Crustal thickness patterns in the Australian continent, *in* Hillis, R.R., and Muller, D., eds., Evolution and Dynamics of the Australian Plate: Geological Society of Australia Special Publication 22 and Geological Society of America Special Paper 372, p. 121–128.

Crone, A.J., Machette, M.N., and Bowman, J.R., 1997, Episodic nature of earthquake activity in stable continental regions revealed by palaeoseismicity studies of Australian and North American Quaternary faults: Australian Journal of Earth Sciences, v. 44, p. 203–214.

Crone, A.J., de Martini, P.M., Machette, M.N., Okumura, K., and Prescott, J.R., 2003, Paleoseismicity of aseismic Quaternary faults in Australia: Implications for fault behaviour in stable continental regions: Bulletin of the Seismological Society of America, v. 93, p. 1913–1934, doi: 10.1785/0120000094.

Cummins, P., Clark, D., Collins, C., Tuttle, M., and Van Arsdale, R., 2003, The potential for paleoliquefaction studies to contribute to Australia's earthquake hazard map: Reno, Nevada, XVI International Quaternary Association Congress Program with Abstracts, paper 19-9.

Cupper, M., Clark, D.J., Sandiford, M., and Kiernan, K., 2004, Geochronology of periglacial deposits in southwest Tasmania, *in* Haberle, S., ed., Proceedings of the 2004 Australian Quaternary Association Meeting: Cradle Mountain, Tasmania, Australian Quaternary Association, p. 15.

Dentith, M.C., and Featherstone, W.E., 2003, Controls on intra-plate seismicity in southwestern Australia: Tectonophysics, v. 376, p. 167–184.

Dentith, M.C., Dent, V.C., and Drummond, B.J., 2000, Deep crustal structure in the southwestern Yilgarn craton, Western Australia: Tectonophysics, v. 325, p. 227–255, doi: 10.1016/S0040–1951(00)00119–0.

Dhu, T., and Jones, T., eds., 2002, Earthquake Risk in Newcastle and Lake Macquarie: Canberra, Geoscience Australia, Record 2002/15, 271 p.

Edwards, M.R., Robinson, D.J., McAneney, K.J., and Schneider, J., 2004, Vulnerability of residential structures in Australia, *in* 13th World Conference on Earthquake Engineering: Vancouver, paper no. 2985.

Estrada, B., Clark, D., Wyrwoll, K.-H., and Dentith, M., 2006, Paleoseismic investigation of a recently identified Quaternary fault in western Australia: The Dumbleyung Fault: Canberra, Australia, Proceedings of the Australian Earthquake Engineering Society 2006 Meeting, p. 189–194.

Featherstone, W.E., Penna, N.T., Leonard, M., Clark, D.J., Dawson, J., Dentith, M.C., Darby, D., and McCarthy, R., 2003, GPS-geodetic deformation monitoring of the southwest seismic zone of Western Australia: Epoch one: Journal of the Royal Society of Western Australia, v. 87.

Federal Emergency Management Agency (FEMA), 1999, HAZUS99: Technical Manual: Washington D.C., Federal Emergency Management Agency, vi, 478 p.

Frankel, A.D., Mueller, C.S., Barnhard, T.P., Leyendecker, E.V., Wesson, R.L., Harmsen, S.C., Klein, F.W., Perkins, D.M., Dickman, N.C., Hanson, S.C., and Hopper, M.G., 2000, U.S. Geological Survey National seismic hazard maps: Earthquake Spectra, v. 16, no. 1, p. 1–19.

Gaull, B.A., and Gregson, P.J., 1991, A new local magnitude scale for Western Australia: Australian Journal of Earth Sciences, v. 38, p. 251–260.

Gaull, B.A., Michael, L.M.O., and Rynn, J.M.W., 1990, Probabilistic earthquake risk maps of Australia: Australian Journal of Earth Sciences, v. 37, p. 169–187.

Giardini, D., 1999, The Global Seismic Hazard Assessment Program (GSHAP) 1992–1999: Anali di Geofisica Special Edition, v. 42, p. 957–974.

Gordon, F.R., and Lewis, J.D., 1980, The Meckering and Calingiri earthquakes October 1968 and March 1970, Western Australia: Geological Survey Bulletin, v. 126, 229 p.

Greenhalgh, S.A., and Parham, R.T., 1986, The Richter earthquake magnitude scale in South Australia: Australian Journal of Earth Sciences, v. 33, p. 519–528.

Greenhalgh, S.A., and Singh, R., 1986, A revised magnitude scale for South Australian earthquakes: Bulletin of the Seismological Society of America, v. 76, no. 3, p. 757–769.

Hanks, T.C., and Kanamori, H., 1979, A moment magnitude scale: Journal of Geophysical Research, v. 84, p. 2348–2350.

Herrmann, R.B., and Kijko, A., 1983, Modelling some empirical vertical component *Lg* relations: Bulletin of the Seismological Society of America, v. 73, p. 157–171.

Hillis, R.R., and Reynolds, S.D., 2003, In situ stress field of Australia: Geological Society of Australia Special Publication 22 and Geological Society of America Special Paper 372, p. 49–58.

Insurance Disaster Response Organisation (IDRO), 2002, Insurance Disaster Response Organisation: http://www.idro.com.au.

Johnston, A.C., 1994, Seismotectonic interpretations and conclusions from the stable continental region seismicity database: Electric Power Research Instrument Report TR-102261–V1, 103 p.

Johnston, A.C., 1996, Seismic moment assessment of earthquakes in stable continental regions: Geophysical Journal International, v. 124, p. 381–414 (Part I); v. 125, p. 639–678 (Part II); v. 126, p. 314–344 (Part III).

Jones, T., Middelmann, M., and Corby, N., eds., 2005, Natural Hazard Risk in Perth, Western Australia: Canberra, Geoscience Australia–Australian Government, 352 p.

Kircher, C.A., Nassar, A.A., Kustu, O., and Holmes, W.T., 1997a, Development of building damage functions for earthquake loss estimation: Earthquake Spectra, v. 13, p. 663–682, doi: 10.1193/1.1585974.

Kircher, C.A., Reithermann, R.K., Whitman, R.V., and Arnold, C., 1997b, Estimation of earthquake losses to buildings: Earthquake Spectra, v. 13, p. 703–720, doi: 10.1193/1.1585976.

Kostrov, B., 1974, Seismic moment and energy of earthquakes, and seismic flow of rock: Izvestiya Academy of Science, USSR: Physics of Solid Earth, v. 1, p. 23–40.

Leonard, M., 2003a, Respite leaves Burakin quaking in anticipation: AUSGEO News, v. 70, p. 5–7.

Leonard, M., 2003b, Small moves towards a big event in South Australia: AUSGEO News, v. 70, p. 4.

Leonard, M., 2007, One hundred years of earthquake recording in Australia: Bulletin of the Seismological Society of America (in press).

Lewis, J.D., Daetwyler, N.A., Bunting, J.A., and Moncrieff, J.S., 1981, The Cadoux Earthquake: Western Australia Geological Survey Report 1981/11, 133 p.

Lowry, D., 1972, Madura-Burnabbie Australia: Geological series—Explanatory notes: Geological Survey of Western Australia, v. SH 52–13, SI 52–1, scale 1:250,000, 14 p.

McCalpin, J., 1996, Paleoseismology: San Diego, Academic Press, 588 p.

Melchers, R.E., 1990, Newcastle Earthquake Study: Canberra, ACT, Australia, The Institute of Engineers, 155 p.

Michael-Leiba, M., and Malafant, K., 1992, A new local magnitude scale for southeastern Australia: Bureau of Mineral Resources Journal of Australian Geology and Geophysics, v. 13, p. 201–205.

Newman, A., Schneider, J., Stein, S., and Mendez, A., 2001, Uncertainties in seismic hazard maps for the New Madrid seismic zone and implications for seismic hazard communication: Seismological Research Letters, v. 72, p. 647–663.

Oguchi, T., Aoki, T., and Matsuta, N., 2003, Identification of an active fault in the Japanese Alps from DEM-based hill shading: Computers & Geosciences, v. 29, p. 885–891, doi: 10.1016/S0098–3004(03)00083–9.

Patchett, A., Robinson, D., Dhu, T., and Sanabria, A., 2005, Investigating earthquake risk models and uncertainty in probabilistic seismic risk analyses: Geoscience Australia Record, v. 2005, no. 02, p. 77.

Quigley, M., Cupper, M., and Sandiford, M., 2005, Quaternary faults of southern Australia: Palaeoseismicity, slip rates and origin: Australian Journal of Earth Sciences, v. 53, p. 285–301.

Richter, C.F., 1935, An instrumental earthquake magnitude scale: Bulletin of the Seismological Society of America, v. 25, p. 1–32.

Robinson, D., Fulford, G., and Dhu, T., 2005, EQRM: Geoscience Australia's Earthquake Risk Model: Technical Manual: Version 3.0: Canberra, Geoscience Australia, Record 2005/01, 142 p.

Robinson, D., Dhu, T., and Schneider, J., 2006a, Practical probabilistic seismic risk analysis: A demonstration of capability: Seismological Research Letters, v. 77, p. 453–459.

Robinson, D., Dhu, T., and Schneider, J., 2006b, SUA: A computer program to compute regolith site-response and estimate uncertainty for probabilistic seismic hazard analyses: Computers and Geosciences, v. 32, p. 109–123, doi: 10.1016/j.cageo.2005.02.017.

Sadigh, K., Chang, C.Y., Egan, J.A., Makdisi, F., and Youngs, R.R., 1997, Attenuation relationships for shallow crustal earthquakes based on California strong motion data: Seismological Research Letters, v. 68, p. 180–189.

Sandiford, M., 2003a, Neotectonics of southeastern Australia: Linking the Quaternary faulting record with seismicity and in situ stress, *in* Hillis, R.R., and Muller, D., eds., Evolution and Dynamics of the Australian Plate: Geological Society of Australia Special Publication 22 and Geological Society of America Special Paper 372, p. 101–113.

Sandiford, M., 2003b, Geomorphic constraints on the late Neogene tectonics of the Otway Ranges: Australian Journal of Earth Sciences, v. 50, p. 69–80, doi: 10.1046/j.1440–0952.2003.00973.x.

Sandiford, M., 2004, Macquarie Harbour marine terraces image gallery: http://jaeger.earthsci.unimelb.edu.au/Images/Geological/geological.html, last accessed 11 May 2007.

Shaw, R.D., Wellman, P., Gunn, P.J., Whitaker, A.J., Tarlowski, C., and Morse, M., 1996, Guide to using the Australian crustal elements map: Australian Geological Survey Organisation, Record 1996, no. 30, 44 p.

Smith, D.E., and Kolenkiewicz, R., 1990, Tectonic motion and deformation from satellite laser ranging to LAGEOS: Journal of Geophysical Research, v. 95, p. 22,013–22,041.

Toro, G.R., Abrahamson, N.A., and Schneider, J.F., 1997, Model of strong ground motions from earthquakes in central and eastern North America; best estimates and uncertainties: Seismological Research Letters, v. 68, p. 41–57.

Tregoning, P., 2003, Is the Australian plate deforming? A space geodetic perspective, *in* Hillis, R.R., and Muller, D., eds., Evolution and Dynamics of the Australian Plate: Geological Society of Australia Special Publication 22 and Geological Society of America Special Paper 372, p. 41–48.

Wilkie, J., Gibson, G., and Wesson, V., 1993, Application and extension of the ML earthquake magnitude scale in the Victorian region: AGSO Journal of Australian Geology & Geophysics, v. 14, p. 35–46.

Williams, I.R., 1979, Recent fault scarps in the Mount Narryer area, Byro: Western Australia Geological Survey Annual Report 1978, v. 51–55, scale 1:250,000, 1 sheet.

MANUSCRIPT ACCEPTED BY THE SOCIETY 29 NOVEMBER 2006

The Geological Society of America
Special Paper 425
2007

The seismicity of the Antarctic plate

Anya M. Reading[†]

School of Earth Sciences, University of Tasmania, Private Bag 79, Hobart, TAS, 7001, Australia

ABSTRACT

Earthquakes occur in Antarctica. The previously held notion that Antarctica is essentially aseismic has been disproved by using records from established Global Seismic Network stations and recently deployed temporary stations on the Antarctic continent. However, the seismicity observed in Antarctica is very low in comparison with other continental intraplate regions. This contribution critically reviews magnitude threshold levels for recorded earthquakes and the available earthquake hypocenter data for Antarctica and the surrounding oceans. Patterns are identified in the distribution of Antarctic earthquakes and the deformation of the Antarctic plate, and the interplay between tectonic and ice-related forces controlling this distribution is discussed.

In the continental intraplate region of Antarctica, earthquakes occur in three settings. Two are likely to have distributions with a tectonic control (although the level may be suppressed by ice cover)—those in the Transantarctic Mountains and scattered events in the interior. Finally, seismicity in the coastal zone and continental margin is likely to be most strongly controlled by the interaction between glacial isostatic adjustment and lithospheric thickness, with a regional tectonic component in some locations.

Keywords: Antarctic, seismicity, passive margin, glacial isostatic adjustment, magnitude threshold.

INTRODUCTION

The seismicity of Antarctica may have several different controlling factors: the tectonic forces that operate across Earth's plates, the underlying tectonic structure, including contrasting tectonic province boundaries and major faults, and the forces due to the loading and partial unloading of the ice cover that dominates the continent. These possible influences are summarized in the following introductory sections.

Analysis of the spatial distribution of Antarctic earthquakes has been decades behind that of other continents owing to the paucity of Global Seismic Network stations in the Southern Hemisphere and the difficulties of operating temporary seismic networks in the inhospitable Antarctic interior. Analysis of source mechanisms is even further behind owing to the low magnitudes of Antarctic earthquakes and the time-consuming process of obtaining data from several different national agencies before a single event can be studied. Efforts such as the webbased AnSWeR,

[†]E-mail: anya.reading@cutas.edu.au.

Reading, A.M., 2007, The seismicity of the Antarctic plate, *in* Stein, S., and Mazzotti, S., ed., Continental Intraplate Earthquakes: Science, Hazard, and Policy Issues: Geological Society of America Special Paper 425, p. 285–298, doi: 10.1130/2007.2425(18). For permission to copy, contact editing@geosociety.org.

the Antarctic Seismic Web resource (current URL available from the author by e-mail), have at least provided researchers with a means of locating current seismic stations in Antarctica and the agencies that hold waveform data. However, the need for all Antarctic nations to contribute data from permanent and temporary stations to a central data management center on a routine basis is very apparent. Improved coordination may arise from international efforts associated with the International Polar Year of 2007–2008.

This contribution presents available data, insights into missing data, and provides a systematic overview of Antarctic seismicity and the different controlling factors underlying the main features of the observed distribution.

Tectonic Setting of the Antarctic Plate and Associated Stress

Antarctica has a unique plate tectonic setting—it is almost totally surrounded by passive continental margins (Fig. 1A). Framing the continent, vast areas of oceanic crust and submarine plateaus extend to a plate margin that consists mostly of spreading ridges (Fig. 1B; Smith and Sandwell, 1997; Hayes, 1991).

This study includes events occurring in the oceanic part of the plate in order to make sense of the continental intraplate seismicity. Earthquakes in the deep ocean (Wiens and Stein, 1983) and on passive continental margins (Stein et al., 1989) are potential indicators of rheological properties, stress accumulation, and release. It is assumed that tectonic stress transfer from one plate to another across plate margins is of lesser importance than the intraplate stress transfer and that stress transfer from ocean to continent or vice versa across passive margins may be significant under some circumstances.

The Continent of Antarctica

East Antarctica is a stable cratonic region (Tingey, 1991) where the details of its internal structure are substantially obscured by ice. The earlier model of a mostly Archean crust has been amended in the light of new geochronological results from several of the continents that made up East Gondwana. Paleoproterozoic and Proterozoic mobile belts run roughly perpendicular to the East Antarctic coastline, but their location in the interior of Antarctica is not constrained (Fitzsimons, 2003, 2000). On the Ross Sea margin of East Antarctica, the major physiographic feature

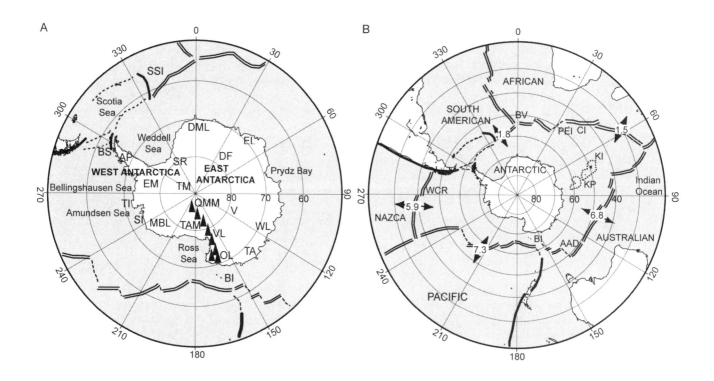

Figure 1. (A) Geographical locations and major physiographic features of Antarctica. SSI—South Sandwich Islands, BS—Bransfield Strait, AP—Antarctic Peninsula, TI—Thurston Island, SI—Siple Island, MBL—Marie Byrd Land, TAM—Transantarctic Mountains, VL—Victoria Land, OL—Oates Land, V—Vostok, Lake Vostok, TA—Terre Adelie, WL—Wilkes Land, EL—Enderby Land, DF—Dome Fuji, DML—Dronning Maud Land, SR—Shackleton Range, TM—Thiel Mountains, EM—Ellsworth Mountains, QMM—Queen Maud Mountains. (B) The Antarctic plate and surrounding plates. BV—Bouvet triple junction, WCR—West Chile Rise, BI—Balleny Islands, AAD—Australian-Antarctic discordance, KP—Kerguelen Plateau, KI—Kerguelen Islands, CI—Crozet Islands, PEI—Prince Edward Islands. Figures give divergence rates in cm/yr (Rundquist and Sobolev, 2002). Plate margins are indicated as follows: divergent—double line, convergent—single thick line, other—dotted line.

of the continent, the Transantarctic Mountains, are exposed. They extend from the Ross Sea side of the South Pole over 2000 km northward, along approximately 160°E longitude, through Victoria Land to Oates Land. Here, the metamorphosed sediments of the Robertson Bay, Bowers, and Wilson terranes are exposed (Tingey, 1991, and references therein). They are bounded by major tectonic structures, such as the Lanterman fault, that were active prior to and through the Cenozoic (Capponi et al., 2002). Cenozoic seafloor spreading in the western Ross Sea is thought to have propagated southward into continental Antarctica, possibly acting as a trigger for the flexural uplift that formed the Transantarctic Mountains (Fitzgerald, 2002).

West Antarctica is generally younger and consists of crustal blocks that came together with the assembly of Gondwana, and it was subject to later, Mesozoic, volcanism and deformation (Vaughan and Storey, 1997). Much of West Antarctica lies below sea level, while the sub-ice elevation of East Antarctica is mostly around 1–2 km (Lythe et al., 2001). The contrast between West and East Antarctica is also evident at depth. Images derived from seismic data using tomographic techniques show thinner, warmer crust beneath West Antarctica and much thicker, colder crust beneath East Antarctica (Ritzwoller et al., 2001; Morelli and Danesi, 2004; Bannister et al., 2003). The West Antarctic rift system, active following the breakup of Gondwana (Fitzgerald, 2002), runs along the Ross Sea margin of East Antarctica, adjacent to the Transantarctic Mountains, but its location through West Antarctica is unknown. The low density of available earthquake-to-station raypaths across Antarctica limits the resolution in current models. However, more precise lithospheric thicknesses in East and West Antarctica are likely to be determined in the near future using data from current deployments of temporary seismic instruments in the interior (by the author and others). Accurate lithospheric structure is necessary to constrain rheological modeling, e.g., of large-scale glacial isostatic adjustment. Airborne geophysical surveys are also making significant progress toward delineating major structures such as the West Antarctic rift system (Behrendt et al., 1996).

There is a paucity of stress measurements on the Antarctic continent itself and hence a near-total lack of direct information on the stress state of the Antarctic plate. However, field investigations have begun to determine stress directions within the Antarctic plate by direct and indirect means (e.g., Paulsen and Wilson, 2002).

The Ocean Surrounding Antarctica

The oceanic regions of the Antarctic plate are shown in Figure 1B, together with the spreading rates of surrounding ridges (two significant figures were taken from Rundquist and Sobolev [2002] and references therein). Spreading is slow at the South American and African plate boundaries (1.8 and 1.5 cm/yr, respectively) and faster at the margins with the Australian, Pacific, and Nazca plates (6.8, 7.3, and 5.9 cm/yr). In two places, the plate margin is close to the Antarctic continent. The Balleny

Islands region is one such location, where the circum-Antarctic spreading ridge and the region south of Macquarie Island (Cande and Stock, 2004) are characterized by high seismicity. A great earthquake, $M_w = 8.1$, took place in this region on 25 March 1998 within the oceanic lithosphere of the Antarctic plate (Toda and Stein, 2000; Antonioli et al., 2002) and not at the plate boundary. However, the influence of the plate boundary is clearly very strong in this region, which could be regarded as a diffuse plate-boundary zone. As such, analyses of focal mechanisms from this region are not included in this review.

North of the Antarctic Peninsula, a short section of active margin (Galindo-Zaldivar et al., 2004) occurs close to the continent. Subduction along the peninsula ceased progressively, from south to north, during the Tertiary, and only the Bransfield Strait remains active at the present day (Larter and Barker, 1991). Analysis of focal mechanisms from this region is also outside the scope of this review. The active tectonics of the nearby Scotia Sea include subduction of the Atlantic plate beneath the Scotia plate at the South Sandwich Island Arc (Larter et al., 1998). The Antarctic plate itself is not significantly affected by this subduction.

Around most of Antarctica, there is a large distance between the ocean ridge and the Antarctic continental rise. In the Indian Ocean sector, the Antarctic plate is dominated by the Kerguelen Plateau (Frey et al., 2000). In the Pacific Ocean, north of the Bellingshausen Sea, the spreading margin of the West Chile Ridge is similarly a great distance from the Antarctic continental rise, although this region is characterized by deep waters and contains no such large submarine plateaus. The Australian-Antarctic discordance (AAD) is a notable topographic feature of the ocean floor in the region south of Australia, where the elevation of the ridge crest is suppressed (Marks et al., 1999).

Observed Deformation of the Antarctic Plate

The continental region of the Antarctic plate is deforming very slowly; surveyed locations across East Antarctica (Fig. 1A) have negligible relative horizontal motion rates of less than 2 mm/yr (Bouin and Vigny, 2000; Negusini et al., 2005). This is consistent with the low rates of observed seismicity in the continental interior. In West Antarctica, relative motions are greater, on the order of 5 mm/yr, especially in the northern part of the Antarctic Peninsula (Dietrich et al., 2004), which is subject to active local tectonics. The Antarctic Peninsula may also be more affected by glacial rebound than East Antarctica (James and Ivins, 1998). Relative horizontal motion between East and West Antarctica is also very small, 1–2 mm/yr (Dietrich et al., 2004; Donnellan and Luyendyk, 2004). Observations of vertical motion using global positioning system (GPS) measurements require a longer stream of higher-quality data. Uplift rates have recently been determined to be 4.5 ± 2.3 mm/yr in the northern Transantarctic Mountains (Raymond et al., 2004) and 12 ± 4 mm/yr in Marie Byrd Land (Donnellan and Luyendyk, 2004). In East Antarctica, uplift rates at stations east and west of the Lambert Glacier are less than 2 ± 0.5 mm/yr (Tregoning et al., 2004).

Stress Due to Ice Loading and Unloading

Johnston (1994) provided a wealth of detail on the behavior of stable continental regions of Earth's surface. Notwithstanding deficiencies in Antarctic data (discussed in the next section), all other continents exhibit a higher level of observed seismicity than Antarctica. This is illustrated by the recorded seismicity for southern Australia and southern Africa displayed in Figure 2. Low Antarctic seismicity is consistent with the low values of deformation observed in many of the GPS surveys. Several authors (e.g., Johnston, 1987) have noted that ice-covered intraplate regions show suppressed seismicity. It is proposed that the likelihood of brittle failure in the crust may be changed by the vertical stress due to the overburden of the ice. In a compressive regime, an increase in vertical stress would inhibit failure. Additionally,

ice or permafrost may prevent water from percolating into the crust: pore pressure would remain low, again inhibiting failure. On removal of the ice cover, seismicity may be influenced by the strain component associated with glacial isostatic adjustment or rebound of the lithosphere. James and Ivins (1998) discussed the likely rebound associated with the current retreat of the Antarctic ice sheets. The failure of Earth's crust due to such motion has been discussed (for example) in the context of present-day Fennoscandia by Muir-Wood (1989) and in more general terms by Schultz and Zuber (1994). A more recent paper by Ivins et al. (2003) extended the stress prediction (and hence likelihood of crustal failure) analysis. It is possible that glacial isostatic adjustment could have been a significant influence behind the great 25 March 1998 earthquake within the Antarctic oceanic crust (Ivins et al., 2003; Kreemer and Holt, 2000; Tsuboi et al., 2000).

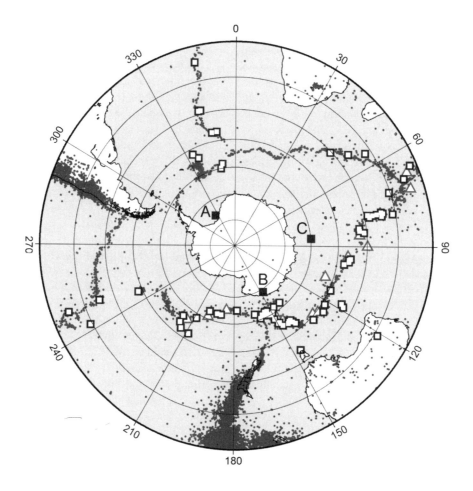

Figure 2. A snapshot of missing seismicity of the Southern Hemisphere > 20°S. Pale open triangles show events overlooked by the International Seismological Centre (ISC) for 1995–1998 and identified by Rouland (2005, personal commun.); dark open and filled squares show events overlooked for 1999 (all with M_s > 3.7) and identified using an improved method by Rouland et al. (2003). Filled squares indicate intraplate events and are included in Figures 3 and 4. Information for the intraplate events is included in Table 1. Small filled circles indicate plate boundaries delineated by background seismicity (1995–1999) as recorded in the *ISC Bulletin* with m_b > 3.5.

DATA

The *ISC Bulletin* (International Seismological Centre) underestimates the true seismicity of the high-latitude Southern Hemisphere. This effect is shown in Figure 2, which includes both events for 1995–1999 not included by the ISC and those listed by the ISC south of 20°S. Note that throughout this discussion, the quoted magnitude scale (for explanations, see Stein and Wysession, 2003) is that used by the given source. The *ISC Bulletin* (and other lists for the most recent events), together with the various additional sources of data in this study, contains too many inconsistencies in event data from Antarctica for detailed conclusions to be drawn about magnitudes. Hence, those quoted are only approximations for most events.

The occurrence of earthquakes in continental Antarctica was confirmed as late as the 1980s (Adams et al., 1985; Adams and Akoto, 1986). The small number of seismic observatories on the Antarctic continent and the vast distance across the Southern Ocean to other low-noise recording sites contributed to the misconception that Antarctica was essentially aseismic. During the 1990s, several new Antarctic seismic observatories were installed, and a number of preexisting observatories were linked to international data distribution centers. Kaminuma (1994) reported on the increasing number of Antarctic earthquakes that were being recorded, and Rouland et al. (1992) began using Geoscope data to investigate previously undetected earthquakes. Reading (2002) included reported earthquakes from temporary deployments or upgraded observatory facilities in an attempt to produce a seismicity map that included all reported Antarctic earthquakes to date. Inclusion of small-magnitude, locally recorded events (see also Robertson et al., 2002; Kaminuma et al., 1998) was valuable in confirming that many parts of Antarctica are subject to limited seismic deformation, but it also raised new problems because the identification of areas showing microseismicity was strongly biased by the location of low-noise permanent and temporary seismic stations. A later paper (Reading, 2005) excluded the microseismicity ($m_b < 3.5$) and introduced the need to consider the very large earthquakes (and hence, large strains) that were supported by the oceanic part of the Antarctic plate. This seems to be a feature of Antarctic plate oceanic seismicity, although the repeat times for these great earthquakes are likely to be hundreds of years and so cannot be inferred from present records. Even such large events may not be present in the historical record if they occurred offshore and/or at high latitude. Locally recorded earthquakes have now been identified in Marie Byrd Land (all with $M_L < 3.5$; Winberry and Anandakrishnan, 2003). Previously, this was the only extended region with no recorded seismicity (Reading, 2002). The local recordings suggest that intraplate seismicity does, in fact, exist across Antarctica but at a much lower level than in comparable stable continental regions. The magnitude cutoff for Figure 3, $m_b = 3.5$, means that events from Marie Byrd Land have not been shown in order to avoid skewing the summary map with locally recorded microseismicity. Rouland et al. (2003) reported that earthquakes with surprisingly

large magnitudes (up to $M_S = 5.2$) that occurred in the Southern Hemisphere high latitudes were overlooked by standard procedures. Many tens of events were missed in a single year (1999), confirming that threshold levels for a complete record of seismicity are as high as $m_b = 5.3$.

While the number of earthquakes in the *ISC Bulletin* that are located in Antarctica has increased in recent decades, the assumed recording threshold has also risen as it becomes clear that many earthquakes are likely to have been overlooked. Okal (1981) reported the improvement in the Antarctic plate earthquake detection threshold from $m_b = 6$ to an optimistic $m_b = 4.9$ after the network improvements of the early 1960s. The current global average (about $m_b = 4.4$) reported by the National Earthquake Information Center (NEIC) (Ringdal, 1986) compares with the more likely value of $m_b = 5.3$ in Antarctica (Sipkin et al., 2000; Rouland et al., 2003). Given the power-law relationship between seismicity and earthquake magnitude, an upper limit of earthquake occurrence in Antarctica is up to an order of magnitude greater than the value reflected by recorded data to date.

OBSERVED SEISMICITY DISTRIBUTION

Figure 3 highlights the distribution of recorded ($m_b > 3.5$) seismicity across the Antarctic plate for the past 25 yr against a background of seismicity at ocean ridges and across surrounding continents. The few available focal mechanisms are shown in Figure 4. The highest magnitudes determined for Antarctic continental intraplate earthquakes are approximately $m_b = 4.5$, although some earthquakes in the coastal/continental rise region of Wilkes Land reach $m_b > 5.0$ (Reading, 2002; Table 1). Since few or no earthquakes have been recorded in the continental intraplate region with $m_b > 4.5$, it is likely that the known seismicity is under-represented by less than the order of magnitude indicated as an upper limit in the previous section, but it would be wise to remain cautious about underestimating Antarctic seismicity. In the oceanic regions, there is no correlation between seismicity and spreading rates of adjacent ridges. Because the generalizations here are based on very small numbers of earthquakes, the distribution patterns may need to be revised slightly in the light of a longer recording time and/or lower threshold for observed activity. Nevertheless, across the Antarctic plate, localized or extended seismic activity occurs in the following areas:

Transantarctic Mountains

This is the most seismically active region in continental Antarctica. The belt of larger earthquakes begins at the Queen Maud Mountains (Fig. 3; Table 1, group 01), the southernmost region of significant topography associated with the main Transantarctic Mountain range. Events occurring along the Ross Sea margin through Victoria Land (Fig. 3; Table 1, group 02) are recorded relatively frequently in the *ISC Bulletin*, and microseismicity clusters are known to be associated with some major glaciers (S. Bannister, 2005, personal commun.). Although some

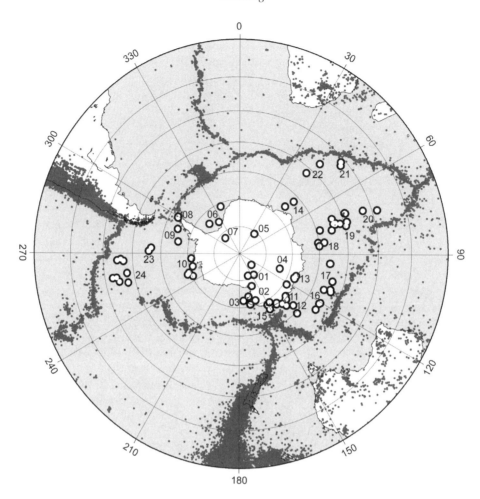

Figure 3. The seismicity of the Antarctic plate and surrounding regions (1980–February 2005) for $m_b > 3.5$. Open circles highlight earthquakes occurring within the Antarctic plate that are sufficiently far from plate boundaries or are sufficiently well located to be termed "intraplate." Intraplate events are grouped to assist identification in discussion. Group numbers correspond to those given in Table 1. Noncatalogued events for 1999 are included as identified by Rouland et al. (2003). Small filled circles indicate plate boundaries delineated by background seismicity as recorded in the *ISC Bulletin* with $m_b > 3.5$.

of these events are probably icequakes rather than earthquakes, there is a strong possibility that basal melt and/or altered pore-pressure conditions may cause small earthquakes in the same regions. There is no recorded seismicity associated with the isolated topography of locations such as the Thiel Mountains or the Ellsworth Mountains. Recorded seismicity extends northward to Oates Land, where the Transantarctic Mountains meet the Southern Ocean. Earthquakes are also recorded at locations in the adjacent western Ross Sea (Fig. 3; Table 1, group 03), where there are numerous Cenozoic and other fault systems. A focal mechanism determined for the largest Ross Sea event (Fig. 4C) indicates near-vertical dip-slip deformation. This is consistent with the continuing uplift of the Transantarctic Mountains, although more focal mechanism determinations are required before robust conclusions may be drawn. Listed events that occurred prior to 1980 and that were located in the Victoria Land–Terre Adelie inte-

rior (noted as unreliable by Reading, 2002) are now recognized as mislocated events (owing to later phases being picked as first arrivals) from the South Sandwich Islands. No recorded events are spatially associated with the trend of the West Antarctic rift system.

Lake Vostok

Although earthquakes within the interior of Antarctica are very rare, small numbers of hypocenters are located along an extrapolation of the axis of what is now known to be a 200-km-long subglacial lake. A single hypocenter exists from an event with $m_b > 3.5$ (Fig. 3; Table 1, 04). Geophysical surveys (Studinger et al., 2003a) suggest that the lake lies at a major structural boundary and is tectonically controlled. Confirmation (albeit circumstantial) of the previously recorded seismicity associated with Lake Vostok came with the local recording of a

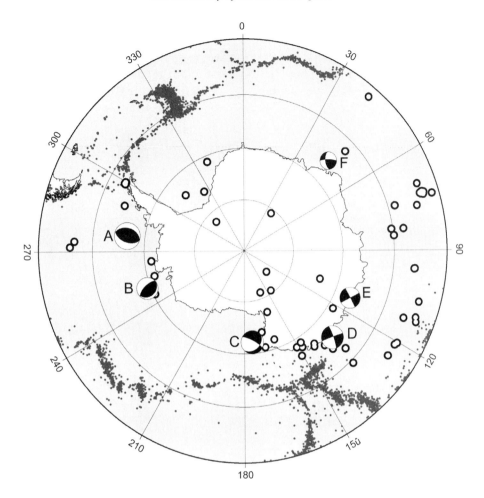

Figure 4. Double-couple components of focal mechanisms for Antarctic intraplate earthquakes (A–E) listed in the Harvard Centroid Moment Tensor (CMT) catalogue 1976–2005. Focal mechanism F is taken from Negishi et al. (1998). Event details are listed in Table 1, and other symbols are as Figure 3.

$M_W = 3.2$ earthquake that occurred on 05 January 2001 followed by a smaller aftershock ~4 min later (Studinger et al., 2003b).

Dome Fuji

An isolated event has also occurred at high latitudes in the East Antarctic interior south of Dome Fuji (Fig. 3; Table 1, 05). This is not associated with any known tectonic or topographic feature (see the BEDMAP compilation, Lythe et al., 2001), and the details of the sub-ice structure remain substantially unknown. The Japanese Antarctic Research Expedition has plans for a multidisciplinary field campaign in this region in the future.

Shackleton Range

A further isolated event from the Antarctic interior occurred to the west of the Shackleton Range (Fig. 3; Table 1, 07), which was probably associated with the tectonics of the Weddell Sea.

Weddell Sea

Seismicity in the Weddell Sea has been confirmed by records from an array located near Neumayer station on the Dronning Maud Land coast. There are two types of events, some are located offshore, often near the base of the continental rise (Fig. 3; Table 1, group 06), and others are located onshore (m_b < 3.5), in clusters associated with glacial features (A. Eckstaller, 2005, personal commun.). This distribution, also seen in Oates Land, is typical of Antarctic events, and the characteristics and possible controls underlying each event type are discussed in a later section.

Antarctic Peninsula

Recorded earthquakes in the northern part of the peninsula are associated with subduction in the Bransfield Strait (Fig. 3; Table 1, group 08; Robertson et al., 2002). A few are

F-E	Cat	Date	Time	Latitude (°S)	Longitude (°E/W)	Dep	Mag	Note
01 Transantarctic Mts (Interior)								
729	NEIC	2003/10/21	07:16:00.48	83.902	132.285	28.8	m_b 4.5	
729	ISC	2001/05/03	09:15:51.86	80.5517	145.2974	0	m_b 3.5	
729	ISC	1998/03/25	03:46:28.10	71.7408	160.6958	0	m_b 4.2	
729	ISC	1996/01/16	15:58:18.09	81.2163	158.4573	10	m_b 4.3	
02 (Victoria Land)								
727	ISC	1999/01/31	13:05:57.50	77.1769	158.9384	0	m_b 3.5	
727	ISC	1997/05/20	16:46:56.03	73.7422	167.6956	10	m_b 4.0	
727	ISC	1995/03/31	03:32:44.59	70.7982	167.4378	10	m_b 4.2	
03 (Ross Sea)								
728	ISC	1993/05/31	08:34:22.75	72.4602	174.8041	10	m_b 5.3	F4C
04 Interior (Vostok)								
729	ISC	1996/01/28	04:55:32.17	73.729	109.6577	0	m_b 3.8	
05 Interior (Fuji)								
729	ISC	1982/11/04	00:14:19.29	80.7894	36.8619	0	m_b 4.5	
06 Weddell Sea								
157	ISC	2002/03/20	22:40:03.84	74.0149	−46.8079	10	m_b 3.9	
157	ISC	1997/12/25	22:13:07.12	71.1596	−22.8183	33	m_b 3.9	
157	RD03	1999/07/16	09:36:28.00	75.9000	−34.4000	32	M_s 4.2	F2A
07 (Interior)								
729	ISC	1995/01/12	04:26:03.84	82.0441	−44.1	10	m_b 4.7	
08 Antarctic Peninsula								
155	ISC	1982/12/13	02:50:47.99	63.1493	−60.9981	3	m_b 5.8	
155	ISC	1982/12/12	20:29:52.65	63.0096	−60.8606	35.0	m_b 5.0	
155	ISC	1982/12/12	19:57:33.69	63.0109	−60.8841	33.8	m_b 5.2	
155	ISC	1982/12/12	17:06:59.02	63.0469	−60.1902	45.7	m_b 5.8	
09 (S. Pacific)								
692	ISC	1996/06/07	08:31:19.75	64.7062	−69.4853	20.3	m_b 4.4	
692	ISC	1989/07/22	21:35:53.79	66.0393	−79.7449	10	m_b 5.1	
692	ISC	1977/02/05	03:29:19.37	66.4891	−82.4527	30.8	m_b 6.2	F4A
10 Bellingshausen/Amundsen Sea								
692	ISC	1997/01/18	05:01:32.63	71.3234	−95.8196	29.7	m_b 4.5	
692	ISC	1996/03/06	08:33:53.23	69.4225	−110.2667	10	m_b 5.6	F4B
692	ISC	1989/10/10	06:04:45.29	70.3504	−115.0141	10	m_b 5.3	
692	ISC	1984/06/15	04:18:57.97	68.6615	−111.4648	10	m_b 5.1	
692	ISC	1980/05/17	08:42:37.80	71.4458	−104.9073	10	m_b 4.8	
11 Terre Adelie Coast								
729	ISC	2001/03/11	23:54:55.80	68.4828	150.6979	0	m_b 3.7	
729	ISC	2000/12/07	22:45:51.93	67.1854	142.1101	0	m_b 4.1	
729	ISC	2000/02/13	07:17:15.95	66.6593	131.4856	0	M_L 4.2	
729	ISC	1998/03/25	10:50:56.80	66.7597	142.9031	0	M_L 4.1	
729	RD03	1999/06/05	04:10:19.00	68.9000	147.6000	0	M_s 3.7	F2B
12 (S. of Australia)								
437	NEIC	2005/02/24	20:33:46.05	65.752	133.463	10	m_b 4.9	
437	NEIC	2005/02/23	07:36:51.40	65.832	133.677	10	m_b 4.7	
437	NEIC	2005/02/23	04:44:33.02	65.776	133.521	10	m_b 4.8	
437	NEIC	2005/02/23	02:18:45.03	65.719	133.496	10	m_b 4.7	
437	NEIC	2005/02/22	23:14:17.82	65.71	133.289	10	m_b 5.7	F4D
437	ISC	2002/02/27	02:49:58.58	59.4349	134.886	10	m_b 3.8	
437	ISC	2001/02/05	02:30:14.35	65.6026	138.6153	0	m_b 3.9	
437	ISC	1998/03/25	15:55:28.70	64.0287	136.8187	0	m_b 3.6	
437	ISC	1996/03/28	07:23:19.51	62.3513	132.905	33	m_b 4.2	
437	ISC	1981/12/15	16:08:19.02	65.7418	133.6137	10	m_b 4.9	
13 Wilkes Land Coast								
729	ISC	2000/08/05	02:27:24.27	67.113	110.6795	0	m_b 3.9	
729	ISC	1998/10/03	07:21:41.00	67.2645	113.7197	0	m_b 3.9	
729	ISC	1984/05/19	10:21:32.14	67.4571	112.8474	0	m_b 4.5	
729	ISC	1984/05/19	04:01:15.79	67.4908	112.9766	33	m_b 5.1	F4E
729	ISC	1983/09/20	03:27:15.46	68.8726	122.1174	0	m_b 4.4	
14 Enderby Land Coast								
425	ISCJ	1996/09/25	14:59:50.90	65.5647	44.7025	10	m_b 4.6	F4F
425	ISC	1985/07/27	07:18:50.40	62.0759	46.819	10	m_b 4.8	

(continued)

TABLE 1. EVENT INFORMATION FOR ANTARCTIC INTRAPLATE EARTHQUAKES
(1980–FEBRUARY 2005) (*Continued*)

F-E	Cat	Date	Time	Latitude (°S)	Longitude (°E/W)	Dep	Mag	Note
15 Balleny Islands								
702	ISC	1998/03/25	10:34:38.00	67.9897	148.1364	0	m_b 3.7	
702	ISC	1998/03/25	05:30:27.30	66.6158	150.4941	0	M_L 4.5	
16 S. of Australia								
437	ISC	2003/03/15	18:49:13.46	55.0067	120.3532	10	m_b 4.8	
437	ISC	1997/11/05	19:02:20.60	55.2354	121.1092	0	m_b 3.8	
437	ISC	1996/02/03	14:47:23.23	55.1997	125.1794	10	m_b 4.4	
17 S. Indian Ocean								
435	ISC	2003/05/24	14:23:32.98	53.4173	111.9902	0	m_b 3.9	
425	ISC	1999/01/19	12:42:35.00	55.877	95.869	10	m_b 4.1	
425	ISC	1999/01/19	03:16:45.14	56.001	95.96	10	m_b 4.5	
425	ISC	1997/09/29	02:26:13.21	55.9121	112.2894	10	m_b 4.0	
425	ISC	1996/12/16	20:56:59.70	54.02	105.91	0	m_b 3.8	
435	ISC	1996/05/26	04:35:07.60	53.74	110.77	0	m_b 3.7	
18 S. Kerguelen Plateau								
436	ISC	2002/10/19	15:42:42.42	58.712	74.1245	10	m_b 3.7	
436	ISC	1997/07/22	18:49:56.46	60.2497	82.1723	10	m_b 4.6	
436	ISC	1981/04/06	21:53:21.57	57.9924	82.4966	0	m_b 4.7	
436	RD03	1999/06/17	03:34:51.00	59.9000	84.6000	0	M_s 4.2	F2C
19 Kerguelen Islands (m_b > 5.2)								
433	ISC	1994/05/15	03:44:57.87	48.9323	73.7541	10	m_b 5.8	
433	ISC	1993/03/10	21:24:04.60	53.2079	72.6197	10	m_b 5.2	
433	ISC	1993/03/02	17:26:59.56	53.5318	72.6834	10	m_b 5.3	
433	ISC	1993/02/20	20:32:17.95	53.1282	73.1157	10	m_b 5.3	
433	ISC	1993/02/20	01:57:09.26	52.8364	73.2109	10	m_b 5.3	
433	ISC	1993/02/19	04:01:02.31	52.8629	72.6369	10	m_b 5.4	
433	ISC	1993/02/18	19:35:36.46	53.098	72.6156	10	m_b 5.6	
433	ISC	1993/02/14	16:48:22.39	52.9147	73.3199	10	m_b 5.2	
433	ISC	1993/01/25	05:59:30.91	53.0243	72.9816	10	m_b 5.7	
433	ISC	1993/01/20	23:09:50.16	52.8013	73.0386	10	m_b 5.3	
433	ISC	1992/12/18	12:02:19.85	52.9501	72.8527	10	m_b 5.5	
20 Indian Ocean (N. of Kerguelen)								
429	ISC	1999/01/24	05:35:25.96	42.779	71.035	10	m_b 4.2	
425	ISC	1997/10/14	04:24:23.10	38.3708	72.6638	0	m_b 3.8	
21 Crozet Islands								
432	ISC	1999/01/27	00:06:28.60	42.0142	49.6773	0	m_b 4.1	
432	ISC	1998/03/28	16:13:59.38	41.3219	48.6626	10	m_b 3.7	
22 Prince Edward Islands								
431	ISC	1998/05/11	01:13:36.94	46.5915	42.6519	10	m_b 4.0	
431	ISC	1996/02/28	10:03:07.47	51.8735	40.3801	10	m_b 5.4	
23 S. Pacific Ocean (E. of Drake Passage)								
692	ISC	1998/11/08	18:08:33.70	55.7501	−88.6054	0	m_b 4.1	
692	ISC	1998/11/08	05:37:20.54	56.518	−86.651	10	m_b 4.4	
24 S. Pacific Ocean (E. of Tierra del Fuego) (m_b > 5.2)								
692	ISC	2002/03/07	22:48:28.00	44.4999	−102.8999	10	m_b 5.5	
692	ISC	1995/03/29	14:39:49.10	42.4581	−100.792	10	m_b 5.2	
692	ISC	1994/12/27	00:43:24.03	45.8186	−92.8045	10	m_b 5.1	
692	ISC	1993/10/28	02:53:17.08	46.7488	−93.763	10	m_b 5.5	
692	ISC	1993/05/22	03:11:37.10	47.41	−104.235	10	m_b 5.0	

Note: Data are taken from the International Seismological Centre (ISC) *Bulletin*; information for more recent events is from National Earthquake Information Center (NEIC). Magnitudes listed are m_b, or, if this is unavailable, M_L. Three events overlooked by the ISC but identified by Rouland et al. (2003) are also included, for which M_s is listed (catalogue abbreviation, RD03). An event for which a focal mechanism was calculated using data from the Japanese station, Syowa (Negishi et al., 1998), is denoted by ISCJ. A single event occurring in 1977, for which a focal mechanism has been calculated, is also listed.

Abbreviations: GP—group number (as shown in Fig. 3); F-E—Flynn-Engdahl region or part of region (Young et al., 1996); location—geographic location shown in Figure 1; cat—catalogue source, as above; event date; time; latitude; longitude; depth; magnitude (as given by source); note—figure number and event label for overlooked events (Fig. 2) and those with focal mechanisms (Fig. 4).

approximately located on the continental rise in the southern Pacific–easternmost Bellingshausen Sea (Fig. 3; Table 1, group 09). A focal mechanism for the largest event in this region indicates thrust faulting (Fig. 4A) and is consistent with compression between the southern Pacific and Antarctic Peninsula. Locations outside the Weddell Sea are of limited accuracy: an estimate of the uncertainty is ±40 to ±120 km. For this reason, a detailed survey of the physiography of the source locations of these events is not yet possible.

Bellingshausen Sea and Amundsen Sea

Earthquakes occur in the offshore region of the west Bellingshausen Sea, in the vicinity of Thurston Island, approximately at the base of the continental rise. A notable cluster of seismicity is seen several hundred kilometers offshore, north of Siple Island, in the Amundsen Sea (Fig. 3; Table 1, group 10). Earthquakes also occur much further north, in the oceanic Antarctic plate, south of the West Chile Rise, but well away from the plate boundary (Fig. 3; Table 1, groups 23 and 24). There is evidence of compressive deformation in oceanic crust of the Bellingshausen Sea (Gohl et al., 1997). This deformation continues today (Okal, 1980), is consistent with the thrust component of the focal mechanism shown (Fig. 4B), and may be the mechanism for most of the seismicity away from the Antarctic continent.

Terre Adelie

In East Antarctica, in the region of Terre Adelie, seismic events are clustered along the coast (Fig. 3; Table 1, group 11). Although the locations are not well constrained, the active region is likely to be offshore, close to the continental rise. A focal mechanism (Fig. 4D) from one the largest and most recent earthquakes (22 February 2005) in this review shows strike-slip deformation. This is easier to reconcile with plate-boundary influences than with crustal motion due to glacial isostatic adjustment. However, Ivins et al. (2003) discussed the possible triggering by glacial isostatic adjustment of events with mechanisms related to regional tectonic stress. A few event hypocenters occur further offshore, as far as 60°S (Fig. 3; Table 1, group 12) and may be controlled by plate-margin processes, although they are a considerable distance from the diffuse zone of high seismicity associated with the Balleny Island region.

Wilkes Land

Events are also observed further along the coast of Wilkes Land (Fig. 3; Table 1, group 13). A focal mechanism from 1984 (Fig. 4E) again shows strike-slip deformation as discussed above. These moderate-sized events abruptly end west of Wilkes Land for reasons that are unclear. This change is unlikely to be an artifact of the distribution of recording stations since Mawson station

is located in the "quiet" region, which extends along the coast between Wilkes Land and Enderby Land.

Enderby Land

The coastal quiet zone is broken by a very isolated group of events that occurred offshore of Enderby Land (Fig. 3; Table 1, group 14). This is close to the Japanese Antarctic station of Syowa. Records from this station and others have been used to calculate the focal mechanism shown for this relatively small event (Negishi et al., 1998; Fig. 4F). This event, dominantly strike-slip with a fault plane dipping at 60°, could again reflect plate-boundary stress directions with a possible glacial isostatic adjustment trigger. However, more focal mechanisms are clearly needed before a robust determination of stress directions can be made.

Balleny Islands

The intraplate region northwest of the Balleny Islands has extremely high seismicity, as one would expect given its location close to the complex plate boundaries that form the triple junction between the Pacific, Australian, and Antarctic plates. Plate-boundary earthquakes have not been included in this review; however, a few events (Fig. 3; Table 1, group 15) in the Balleny Islands Flynn-Engdahl region may be more closely related to the coastal events of Terre Adelie. The oceanic crust is capable of sustaining very high strains (Choy and McGarr, 2002), as evidenced by the occurrence of the great earthquake of 25 March 1998 (Antonioli et al., 2002), and the relation between plate-boundary and intraplate processes in this region is clearly complex.

South of the Australian-Antarctic Ridge

Intraplate seismicity is observed between the Australian-Antarctic Ridge and the Antarctic continent (Fig. 3; Table 1, groups 16 and 17) in contrast to some of the other oceanic regions of the plate, offshore from Dronning Maud Land and Marie Byrd Land. The controls on this distribution are not clear.

Kerguelen Plateau

Scattered groups of intraplate earthquakes occur on the Kerguelen Plateau, in the southern Indian Ocean (Fig. 3; Table 1, groups 18 and 19). The anomalous material that makes up the plateau is older than the surrounding ocean floor and is likely associated with Indian Ocean tectonics at the breakup of Gondwana (Frey et al., 2000). Active tectonic processes are evident at volcanic Heard Island (south of Kerguelen Island). A focal mechanism from an earthquake that occurred north of Kerguelen Island in 1973 (Okal, 1981) shows an extensional mechanism of deformation that suggests that the northern seismicity may be related to the volcanic activity beneath the Kerguelen Plateau.

Seismicity continues to be observed between the plateau and the ridge at the plate margin (Fig. 3; Table 1, group 20).

Crozet Islands–Prince Edward Islands

Isolated events are located in the intraplate oceanic crust, in the vicinity of the Crozet Islands (Fig. 3; Table 1, group 21) and Prince Edward Islands (Fig. 3; Table 1, group 22). These are also regions of past and present oceanic volcanic activity. The region between Kerguelen and Crozet–Prince Edward Islands is very quiet.

DISCUSSION

The findings of Rouland et al. (2003), who reported overlooked events within the Antarctic intraplate region in a single year (1999), led us to ask: is the seismicity as low as it appears? Three intraplate events that occurred away from plate boundaries, with m_b = 3.5–3.9, were reported in the *ISC Bulletin* for 1999. The same number of intraplate events, three, was reported as overlooked. Although these numbers are small, they suggest that the actual seismicity may be at least twice that reported by the ISC. The magnitudes of the overlooked events were M_s = 3.7, 4.2, and 4.2 for Victoria Land, Weddell Sea, and Southern Kerguelen Plateau, respectively. Because these are significantly larger than the magnitudes of the Antarctic intraplate events reported that year by the ISC, the problem is not simply one of threshold magnitude. Rouland et al. (2003) suggested that automated procedures (not necessarily tuned for Antarctic events), simple clerical errors, and/or the physical properties (e.g., emergent onset) of some earthquake sources may cause some earthquakes to be overlooked. Lack of data redundancy in the Southern Hemisphere may also be a contributing problem. Coastal Antarctic and sub-Antarctic recording sites are prone to spells of very noisy conditions, so an event within recording range of a very small number of such stations may not eventually reach the *ISC Bulletin*.

Even if the incidence of earthquakes in continental Antarctica is twice as high as appears from the *ISC Bulletin*, Antarctic seismicity remains low. This is likely to be due in part to the low tectonic stresses acting on the Antarctic plate: the lack of extensive subducting margins, the distance between the surrounding ridge crests and the continent, and the ability of the oceanic crust to sustain very high levels of strain. It is also likely to be due to the effect of an ice load on the crust, which shifts the principal stresses away from a condition of brittle failure. The observed seismicity distribution is likely to be a highly suppressed version of the seismicity that would be apparent if there were no ice, with an overprint of the coastal seismicity due to glacial unloading.

Over the past few years, Antarctic seismology has matured from being a curiosity into a science with controlling mechanisms under discussion (Kaminuma, 2000). From the present, improved, observed distribution in the intraplate region, earthquakes appear to occur in three settings. Two are likely to have distributions with a tectonic, or relict tectonic structural, control

(although the level may be suppressed by ice cover): in the Transantarctic Mountains and the scattered events in the interior. In addition, earthquakes in the coastal zone and continental margin are likely to be most strongly controlled by the interaction between glaciogenic and tectonic forces. In addition to the possible controlling mechanisms already mentioned, i.e., slab pull, ridge-push, ice loading and ice unloading, topography and sub-ice topography may be additional controls on the distribution of seismicity. Stresses induced by topography and density variation can be of similar magnitude to those tectonically induced (Mareschal and Kuang, 1986).

The seismicity of the Transantarctic Mountains is due to a combination of the uplift of the range, which is accommodated in fault-bounded blocks (Fitzgerald, 2002; Raymond et al., 2004), and the effect of glacial unloading at the margin of the continent (Ivins et al., 2003). At the Ross Sea margin, there are good constraints on the glacial history (James and Ivins, 1998) and lithospheric structure (Bannister et al., 2003), which will enable the glaciogenic stresses to be modeled more accurately in the near future. Some microseismicity is due to the interaction between glaciers and the crust. Crustal strain may be released at points where the basal conditions of glaciers lead to changes in the magnitude and/or direction of the principal stress axes or the microseismicity may simply be a local phenomenon, dominated by icequakes.

The seismicity of the interior of the continent is due to tectonic forces but is nearly completely suppressed by ice cover. The few events that do occur appear to be located on tectonic boundaries that may provide a weak point, or contrast in stress regime, for seismic strain release. Sub-ice topography and density heterogeneities may also play a part, but there are too few recorded earthquakes and the structure of the interior is too poorly known to draw any but the most general conclusions at this time.

The seismicity of the continental margin is most likely to be dominated by the glacial rebound processes and the controls that lithospheric thickness and oceanic crust rheology and structure place on the spatial distribution of stress release (Ivins et al., 2003). The distribution of seismicity around the Antarctic coast lends weight to this suggestion; however, there are some additional features of the observed distribution. The seismicity adjacent to Oates Land may be influenced by the tectonics of the Balleny Island plate margin or it may be substantially due to uplift of the Transantarctic Mountains. The Oates Land–Ross Sea margin exposes many previously active faults that could facilitate strain partitioning and stress release between the oceanic plate and the northern Transantarctic Mountains. The focal mechanisms from the Terre Adelie and Wilkes Land coast imply that regional tectonic strain from the Balleny Island plate margin is being released, although the event(s) may be triggered by the uplift of the crust in glacial isostatic adjustment. From Wilkes Land to Enderby Land, the sudden lack of coastal seismicity is notable. Possible influences include increased distance from the plate boundary, mantle heterogeneity associated with the Australian-Antarctic discordance, and/or crustal and upper lithospheric boundaries through Wilkes Land (Fitzsimons, 2003). There could be a change in the nature of faults dating from the time of continental rifting so that they are not reactivated by a current stress regime (Stein et al., 1989). The seismicity of the Weddell Sea has relatively well-constrained hypocenters, which might make it a suitable test of the mechanisms proposed by Ivins et al. (2003) as soon as the glacial history of western Dronning Maud Land is sufficiently well known.

Scattered seismicity in the Kerguelen Plateau and Bellingshausen Sea implies deformation within the oceanic part of the plate. It is possible that this releases some of the strain due to ridge-push forces and provides a buffer for the Antarctic continent in the center of the plate. There are no currently recorded earthquakes with $m_b > 3.5$ in the interior of the Antarctic continent inland from the Kerguelen Plateau or Bellingshausen Sea. Alternatively, the relationship may be coincidental given that the interior seismicity is so low.

In order to further understand the interplay among the many factors underlying the distribution of so few earthquakes, we must improve the recording of earthquakes throughout the high latitudes of the Southern Hemisphere. The accuracy of hypocenter locations is important in understanding the setting of the earthquake with respect to features such as the continental rise. More focal mechanisms would also provide clearer insights into the style of failure of the crust and, where sufficient events occur, underlying stresses in the lithosphere.

Improved coverage of seismic data is also required in order to apply the more advanced seismological techniques that we now take for granted in other continents. These include higher-resolution mapping of the seismic structure and, hence, the depth and temperature of the lithosphere and underlying mantle. When the records of earthquakes and our knowledge of Earth structure are improved in this way, we can hope to effectively constrain tectonic and glacial isostatic adjustment models (e.g., Ivins et al., 2003).

SUMMARY

The seismicity of continental Antarctica may be at least twice as great as it appears from the *ISC Bulletin*. The seismicity distribution in the Transantarctic Mountains is dominated by tectonic and relict tectonic controls with some influence of ice unloading at the Ross Sea margin. The continental interior shows suppression of crustal failure due to tectonic stress by ice loading, and the low seismicity is controlled by the relict tectonic and physiographic structure beneath the ice. The continental margin shows seismicity that is likely to be dominated by glacial isostatic adjustment and lithospheric structure with a regional tectonic influence in some locations.

ACKNOWLEDGMENTS

Suggestions from two anonymous reviewers and the editor improved the manuscript. Discussions at ANTEC (Scientific Committee for Antarctic Research group of experts on Antarc-

tic Neotectonics) workshops and sessions have been invaluable in both finding earthquake data and developing the ideas in this contribution. Individuals who assisted in the recent compilation include Daniel Rouland, Sridhar Anandakrishnan, Stephen Bannister, Alfons Eckstaller, and Kevin McCue (Australian Seismological Centre). The International Seismology Centre (ISC) is acknowledged for its *Bulletin* and the fact that a missed event is something of note. Maps were produced using GMT (Generic Mapping Tools) (Wessel and Smith, 1991). Antarctic seismology owes a great debt to Robin Adams and Katsutada Kaminuma.

REFERENCES CITED

Adams, R.D., and Akoto, A.M., 1986, Earthquakes in continental Antarctica: Journal of Geodynamics, v. 6, p. 263–270, doi: 10.1016/0264-3707-(86)90043-8.
Adams, R.D., Hughes, A.A., and Zhang, B.M., 1985, A confirmed earthquake in continental Antarctica: Geophysical Journal of the Royal Astronomical Society, v. 81, p. 489–492.
Antonioli, A., Cocco, M., Das, S., and Henry, C., 2002, Dynamic stress triggering during the great 25 March 1998 Antarctic plate earthquake: Bulletin of the Seismological Society of America, v. 92, p. 896–903, doi: 10.1785/0120010164.
Bannister, S., Yu, Y., Leitner, B., and Kennett, B.L.N., 2003, Variations in crustal structure across the transition from West to East Antarctica, southern Victoria Land: Geophysical Journal International, v. 155, p. 870–884, doi: 10.1111/j.1365-246X.2003.02094.x.
Behrendt, J.C., Saltus, R., Damaske, D., McCafferty, A., Finn, C.A., Blankenship, D., and Bell, R.E., 1996, Patterns of late Cenozoic volcanic and tectonic activity in the West Antarctic rift system revealed by aeromagnetic surveys: Tectonics, v. 15, p. 660–676, doi: 10.1029/95TC03500.
Bouin, M., and Vigny, C., 2000, New constraints on Antarctic plate motion and deformation from GPS data: Journal of Geophysical Research, v. 105, p. 28,279–28,293, doi: 10.1029/2000JB900285.
Cande, S.C., and Stock, J.M., 2004, Pacific-Antarctic-Australia motion and the formation of the Macquarie plate: Geophysical Journal International, v. 157, p. 399–414, doi: 10.1111/j.1365-246X.2004.02224.x.
Capponi, G., Crispini, L., and Meccheri, M., 2002, Tectonic evolution at the boundary between the Wilson and Bowers terranes (northern Victoria Land, Antarctica): Structural evidence from the Mountaineer and Lanterman Ranges, in Gamble, J.A., Skinner, D.N.B., and Henrys, S., eds., Antarctica at the Close of a Millennium: Royal Society of New Zealand Bulletin, v. 35, p. 105–112.
Choy, G.L., and McGarr, A., 2002, Strike-slip earthquakes in the oceanic lithosphere: Observations of exceptionally high apparent stress: Geophysical Journal International, v. 150, p. 506–523, doi: 10.1046/j.1365-246X.2002.01720.x.
Dietrich, R., Rulke, A., Ihde, J., Lindner, K., Miller, H., Niemeier, W., Schenke, H.-W., and Seeber, G., 2004, Plate kinematics and deformation status of the Antarctic Peninsula based on GPS: Global and Planetary Change, v. 42, p. 313–321, doi: 10.1016/j.gloplacha.2003.12.003.
Donnellan, A., and Luyendyk, B.P., 2004, GPS evidence for a coherent Antarctic plate and for postglacial rebound in Marie Byrd Land: Global and Planetary Change, v. 42, p. 305–311, doi: 10.1016/j.gloplacha.2004.02.006.
Fitzgerald, P., 2002, Tectonics and landscape evolution of the Antarctic plate since the breakup of Gondwana, with an emphasis on the West Antarctic rift system and the TransAntarctic Mountains, in Gamble, J.A., Skinner, D.N.B., and Henrys, S., eds., Antarctica at the Close of a Millennium: Royal Society of New Zealand Bulletin, v. 35, p. 453–469.
Fitzsimons, I.C.W., 2000, A review of tectonic events in the East Antarctic Shield and their implications for Gondwana and earlier supercontinents: Journal of African Earth Sciences, v. 31, p. 3–23, doi: 10.1016/S0899-5362(00)00069-5.
Fitzsimons, I.C.W., 2003, Proterozoic basement provinces of southern and south-western Australia and their correlation with Antarctica, in Yoshida, M., Windely, B.F., and Dasgupta, S., eds., Proterozoic East Gondwana: Supercontinent Assembly and Breakup: Geological Society of London Special Publication 206, p. 93–130.

Frey, F.A., Coffin, M.F., Wallace, P.J., Weis, D., Zhao, X., Wise, S.W., Jr., Wahnert, V., Teagle, D.A.H., Saccocia, P.J., Reusch, D.N., Pringle, M.S., Nicolaysen, K.E., Neal, C.R., Muller, R.D., Moore, C.L., Mahoney, J.J., Keszthelyi, L., Inokuchi, H., Duncan, R.A., Delius, H., Damuth, J.E., Damasceno, D., Coxall, H.K., Borre, M.K., Boehm, F., Barling, J., Arndt, N., and Antretter, M., 2000, Origin and evolution of a submarine large igneous province: The Kerguelen Plateau and Broken Ridge, southern Indian Ocean: Earth and Planetary Science Letters, v. 176, p. 73–89, doi: 10.1016/S0012-821X(99)00315-5.

Galindo-Zaldivar, J., Gamboa, L., Maldonado, A., Nakao, S., and Bochu, Y., 2004, Tectonic development of the Bransfield Basin and its prolongation to the South Scotia Ridge, northern Antarctic Peninsula: Marine Geology, v. 206, p. 267–282, doi: 10.1016/j.margeo.2004.02.007.

Gohl, K., Nitsche, F., and Miller, H., 1997, Seismic and gravity data reveal Tertiary interplate subduction in the Bellingshausen Sea, southeast Pacific: Geology, v. 25, p. 371–374, doi: 10.1130/0091-7613-(1997)025<0371:SAGDRT>2.3.CO;2.

Hayes, D.E., 1991, Tectonics and age of the ocean crust: Circum-Antarctic to 30°S (map), in Hayes, D.E., ed., Marine Geological and Geophysical Atlas of the Circum-Antarctic to 30°S At 50°S, 1:11,674,000: American Geophysical Union, Antarctic Research Series, v. 54.

Ivins, E.R., James, T.S., and Klemann, V., 2003, Glacial isostatic stress shadowing by the Antarctic ice sheet: Journal of Geophysical Research, v. 108, no. B12, p. 2560, doi: 10.1029/2002JB002182.

James, T.S., and Ivins, E.R., 1998, Predictions of Antarctic crustal motions driven by present-day ice sheet evolution and by isostatic memory of the Last Glacial Maximum: Journal of Geophysical Research, v. 103, p. 4993–5017, doi: 10.1029/97JB03539.

Johnston, A.C., 1987, Suppression of earthquakes by large continental ice sheets: Nature, v. 330, p. 467–469, doi: 10.1038/330467a0.

Johnston, A.C., 1994, Seismotectonic interpretations and conclusions from the stable continental region earthquake database, in Schneider, J.F., ed., The Earthquakes of Stable Continental Regions: Palo Alto, California, Electric Power Research Institute, Report TR-102261, p. 4-1–4-102.

Kaminuma, K., 1994, Seismic activity in and around the Antarctic continent: Terra Antarctica, v. 1, Special Issue, p. 423–426.

Kaminuma, K., 2000, A revaluation of the seismicity in the Antarctic: Polar Geoscience, v. 13, p. 145–157.

Kaminuma, K., Kanao, M., and Kubo, A., 1998, Local earthquake activity around Syowa Station: Antarctica: Polar Geoscience, v. 11, p. 23–31.

Kreemer, C., and Holt, W.E., 2000, What caused the March 25, 1998, Antarctic plate earthquake?: Inferences from regional stress and strain rate fields: Geophysical Research Letters, v. 27, p. 2297–2300, doi: 10.1029/1999GL011188.

Larter, R.D., and Barker, P.F., 1991, Effects of ridge crest–trench collision on Antarctic-Phoenix spreading: Forces on a young subducting plate: Journal of Geophysical Research, v. 96, p. 19,583–19,609.

Larter, R.D., King, E.C., Leat, P.T., and Reading, A.M., 1998, South Sandwich slices reveal much about arc structure, geodynamics and composition: Eos (Transactions, American Geophysical Union), v. 79, no. 24, p. 281–285, doi: 10.1029/98EO00207.

Lythe, M.B., Vaughan, D.G., and the BEDMAP Consortium, 2001, BEDMAP: A new ice thickness and subglacial topographic model of Antarctica: Journal of Geophysical Research, v. 106, p. 11,335–11,351, doi: 10.1029/2000JB900449.

Mareschal, J.-C., and Kuang, J., 1986, Intraplate stresses and seismicity: The role of topography and density heterogeneities: Tectonophysics, v. 132, p. 153–162, doi: 10.1016/0040-1951(86)90030-2.

Marks, K.M., Stock, J.M., and Quinn, K.J., 1999, Evolution of the Australian-Antarctic discordance since Miocene time: Journal of Geophysical Research, v. 104, no. B3, p. 4967–4981, doi: 10.1029/1998JB900075.

Morelli, A., and Danesi, S., 2004, Seismological imaging of the Antarctic continental lithosphere: A review: Global and Planetary Change, v. 42, p. 155–165, doi: 10.1016/j.gloplacha.2003.12.005.

Muir-Wood, R., 1989, Extraordinary deglaciation reverse faulting in northern Fennoscandia, in Gregersen, S., and Basham, P.W., eds., Earthquakes at North American Passive Margins: Neotectonics and Postglacial Rebound: NATO Advanced Studies Institute Series, ser. C, v. 266, p. 141–174.

Negishi, H., Nogi, Y., and Kaminuma, K., 1998, An intraplate earthquake that occurred near Syowa Station, East Antarctica: Polar Geoscience, v. 11, p. 32–41.

Negusini, M., Mancini, F., Gandolfi, S., and Capra, A., 2005, Terra Nova Bay GPS permanent station (Antarctica): Data quality and first attempt in the evaluation of regional displacement: Journal of Geodynamics, v. 39, p. 81–90, doi: 10.1016/j.jog.2004.10.002.

Okal, E.A., 1980, The Bellingshausen Sea earthquake of February 5, 1977: Evidence for ridge-generated compression in the Antarctic plate: Earth and Planetary Science Letters, v. 46, p. 306–310, doi: 10.1016/0012-821-X(80)90016-3.

Okal, E.A., 1981, Intraplate seismicity of Antarctica and tectonic implications: Earth and Planetary Science Letters, v. 52, p. 397–409, doi: 10.1016/0012-821X(81)90192-8.

Paulsen, T., and Wilson, T., 2002, Volcanic cone alignments and the intraplate stress field in the Mount Morning region, South Victoria Land, Antarctica: Denver, Geological Society of America Annual Meeting Abstracts with Programs, October 2002, v. 34, no. 6, p. 437.

Raymond, C.A., Ivins, E.R., Heflin, M.B., and James, T.S., 2004, Quasi-continuous global positioning system measurements of glacial isostatic deformation in the northern Transantarctic Mountains: Global and Planetary Change, v. 42, p. 295–303, doi: 10.1016/j.gloplacha.2003.11.013.

Reading, A.M., 2002, Antarctic seismicity and neotectonics, in Gamble, J.A., Skinner, D.N.B., and Henrys, S., eds., Antarctica at the Close of a Millennium: Royal Society of New Zealand Bulletin, v. 35, p. 479–484.

Reading, A.M., 2006, On seismic strain-release within the Antarctic Plate, in Futterer, D.K., Damaske, D., Kleinschmidt, G., Miler, H., Tessensohn, F., eds., Antarctica—Contributions to Global Earth Sciences: Berlin, Heidelberg, New York, Springer-Verlag, p. 351–356.

Ringdal, F., 1986, Study of magnitudes, seismicity and earthquake detectability using a global network: Bulletin of the Seismological Society of America, v. 76, p. 1641–1659.

Ritzwoller, M.H., Shapiro, N.M., Levshin, A.L., and Leahy, G.M., 2001, Crustal and upper mantle structure beneath Antarctica and surrounding oceans: Journal of Geophysical Research, v. 106, p. 30,645–30,670, doi: 10.1029/2001JB000179.

Robertson, S.D., Wiens, D.A., Shore, P.J., Smith, G.P., and Vera, E., 2002, Seismicity and tectonics of the South Shetland Islands and Bransfield Strait from the SEPA broadband seismograph deployment, in Gamble, J.A., Skinner, D.N.B., and Henrys, S., eds., Antarctica at the Close of a Millennium: Royal Society of New Zealand Bulletin, v. 35, p. 549–553.

Rouland, D., Condis, C., Parmentier, C., and Souriau, A., 1992, Previously undetected earthquakes in the Southern Hemisphere located using long-period Geoscope data: Bulletin of the Seismological Society of America, v. 82, p. 2448–2463.

Rouland, D., Condis, C., and Roult, G., 2003, Overlooked earthquakes on and around the Antarctic plate: Identification and location of 1999 shallow depth events: Tectonophysics, v. 376, p. 1–17, doi: 10.1016/j.tecto.2003.08.006.

Rundquist, D.V., and Sobolev, P.O., 2002, Seismicity of mid-oceanic ridges and its geodynamic implications: A review: Earth-Science Reviews, v. 58, p. 143–161, doi: 10.1016/S0012-8252(01)00086-1.

Schultz, R.A., and Zuber, M.T., 1994, Observations, models, and mechanisms of failure of surface rocks surrounding planetary surface loads: Journal of Geophysical Research, v. 99, no. E7, p. 14,691–14,702, doi: 10.1029/94JE01140.

Sipkin, S.A., Person, W.J., and Presgrave, B.W., 2000, Earthquake bulletins and catalogues at the U.S. Geological Survey: National Earthquake Information Center Incorporated Research Institutions for Seismology Newsletter, v. 1.

Smith, W., and Sandwell, D., 1997, Measured and estimated seafloor topography (version 4.2): World Data Center A for Marine Geology and Geophysics, research publication RP-1, poster.

Stein, S., and Wysession, M., 2003, An Introduction to Seismology, Earthquakes, and Earth Structure: Malden, Massachusetts, Blackwell Publishing Ltd, 498 p.

Stein, S., Cloetingh, S., Sleep, N.H., and Wortel, R., 1989, Passive margin earthquakes, stresses and rheology, in Gregersen, S., and Basham, P.W., eds., Earthquakes at North American Passive Margins: Neotectonics and Postglacial Rebound: NATO Advanced Studies Institute Series, ser. C, v. 266, p. 231–259.

Studinger, M., Bell, R.E., Karner, G.D., Tikku, A.A., Holt, J.W., Morse, D.L., Richter, T.G., Kempf, S.D., Peters, M.E., Blankenship, D.D., Sweeney, R.E., and Rystrom, V.L., 2003a, Ice cover, landscape setting and geological framework of Lake Vostok, East Antarctica: Earth and Planetary Science Letters, v. 205, p. 195–210, doi: 10.1016/S0012-821X(02)01041-5.

Studinger, M., Karner, G.D., Bell, R.E., Levin, V., Raymond, C.A., and Tikku, A.A., 2003b, Geophysical models for the tectonic framework of the Lake

Vostok region, East Antarctica: Earth and Planetary Science Letters, v. 216, p. 663–677, doi: 10.1016/S0012-821X(03)00548-X.

Tingey, R.J., 1991, The Geology of Antarctica: Oxford Monographs on Geology and Geophysics: Oxford, Oxford University Press, Monographs on Geology and Geophysics, no. 17, 704 p.

Toda, S., and Stein, R.S., 2000, Did stress-triggering cause the large off-fault aftershocks of the 25 March 1998 M_w = 8.1 Antarctic plate earthquake?: Geophysical Research Letters, v. 27, no. 15, p. 2301–2304, doi: 10.1029/1999GL011129.

Tregoning, P., Morgan, P.J., and Coleman, R., 2004, The effect of receiver firmware upgrades on GPS vertical timeseries: Cahiers du Centre Europeen de Geodynamique et de Seismologie, v. 23, p. 37–46.

Tsuboi, S., Kikuchi, M., Yamanaka, Y., and Kanao, M., 2000, The March 25, 1998, Antarctic earthquake: Great earthquake caused by post-glacial rebound: Earth, Planets and Space, v. 52, p. 133–136.

Vaughan, A.P.M., and Storey, B.C., 1997, Mesozoic geodynamic evolution of the Antarctic Peninsula, *in* Ricci, C.A., ed., The Antarctic Region, Geological Evolution and Processes: Siena, Italy, Terra Antartica special edition, p. 373–382.

Wessel, P., and Smith, W.H.F., 1991, Free software helps map and display data: Eos (Transactions, American Geophysical Union), v. 72, p. 441, doi: 10.1029/90EO00319.

Wiens, D.A., and Stein, S., 1983, Age dependence of oceanic intraplate seismicity and implications for lithospheric evolution: Journal of Geophysical Research, v. 88, p. 6455–6468.

Winberry, J.P., and Anandakrishnan, S., 2003, Seismicity and neotectonics of West Antarctica: Geophysical Research Letters, v. 30, doi: 10.1029/2003GL018001.

Young, J.B., Presgrave, B.W., Aichele, H., Wiens, D.A., and Flinn, E.A., 1996, The Flinn-Engdahl regionalisation scheme: The 1995 revision: Physics of the Earth and Planetary Interiors, v. 96, p. 223–297, doi: 10.1016/0031-9201(96)03141-X.

MANUSCRIPT ACCEPTED BY THE SOCIETY 29 NOVEMBER 2006

The Geological Society of America
Special Paper 425
2007

Active tectonics and intracontinental earthquakes in China: The kinematics and geodynamics

Mian Liu
Youqing Yang
Department of Geological Sciences, University of Missouri, Columbia, Missouri 65211, USA

Zhengkang Shen
State Key Laboratory of Earthquake Dynamics, Institute of Geology, China Earthquake Administration, Beijing 100029, China

Shimin Wang
Department of Geological Sciences, University of Missouri, Columbia, Missouri 65211, USA

Min Wang
Institute of Earthquake Science, China Earthquake Administration, Beijing 100036, China

Yongge Wan
School of Disaster Prevention Techniques, Yanjiao, Beijing 101601, China

ABSTRACT

China is a country of intense intracontinental seismicity. Most earthquakes in western China occur within the diffuse Indo-Eurasian plate-boundary zone, which extends thousands of kilometers into Asia. Earthquakes in eastern China mainly occur within the North China block, which is part of the Archean Sino-Korean craton that has been thermally rejuvenated since late Mesozoic. Here, we summarize neotectonic and geodetic results of crustal kinematics and explore their implications for geodynamics and seismicity using numerical modeling. Quaternary fault movements and global positioning system (GPS) measurements indicate a strong influence of the Indo-Asian collision on crustal motion in continental China. Using a spherical three-dimensional (3-D) finite-element model, we show that the effects of the collisional plate-boundary force are largely limited to western China, whereas gravitational spreading of the Tibetan Plateau has a broad impact on crustal deformation in much of Asia. The intense seismicity in the North China block, and the lack of seismicity in the South China block, may be explained primarily by the tectonic boundary conditions that produce high deviatoric stresses within the North China block but allow the South China block to move coherently as a rigid block. Within the North China block, seismicity is concentrated in the circum-Ordos rifts, reflecting the control of lithospheric heterogeneity. Finally, we calculated the change of Coulomb stresses associated with 49 major (M ≥ 6.5) earthquakes in the North China block since 1303. The results show that ~80% of these events occurred in regions of increasing Coulomb stresses caused by previous events.

Keywords: earthquakes, China, GPS, geodynamics, modeling.

Liu, M., Yang, Y., Shen, Z., Wang, S., Wang, M., and Wan, Y., 2007, Active tectonics and intracontinental earthquakes in China: The kinematics and geodynamics, *in* Stein, S., and Mazzotti, S., ed., Continental Intraplate Earthquakes: Science, Hazard, and Policy Issues: Geological Society of America Special Paper 425, p. 299–318, doi: 10.1130/2007.2425(19). For permission to copy, contact editing@geosociety.org. ©2007 The Geological Society of America. All rights reserved.

INTRODUCTION

In King Jie's 10th year of the Xia Dynasty (1767 B.C.), an earth-quake caused interruption of the Yi and Lo Rivers. In the capital of Zhengxuen, buildings cracked and collapsed.

—State Records: Zhou Dynasty

This is one of the earliest written records of earthquakes in China. The Chinese catalog of historic earthquakes shows more than 1000 M ≥ 6 events since A.D. 23 (Ming et al., 1995). At least thirteen of these events were catastrophic (M ≥ 8). The 1556 Huaxian earthquake reportedly killed 830,000 people, making it the deadliest earthquake in human history (Ming et al., 1995). Modern earthquakes in China are intense and widespread. The best-known event is perhaps the 1976 Tangshan earthquake (M = 7.8), which killed ~250,000 people and injured millions (Chen et al., 1988).

The intense seismicity in China cannot be readily explained by plate tectonics theory, which predicts that earthquakes are concentrated within narrowly defined plate-boundary zones. As shown in Figure 1, most earthquakes in China occur within the interior of the Eurasian plate. In western China (approximately west of 105°E), seismicity is closely associated with the roughly E-W–trending fault systems resulting from the Indo-Asian collision. East of ~105°E, the influence of Indo-Asian collision is less clear. Major active fault zones there trend NE and NEE due to subduction of the Pacific plate under the Eurasian plate (Deng et al., 2002; Zhang et al., 2003). Active crustal motion on these faults, however, may be influenced by the Indo-Asian collision (Tapponnier and Molnar, 1977; Zhang et al., 2003). Most earthquakes in eastern China occur within the North China block, a geological province including the Ordos Plateau and surrounding rifts, the North China Plain, and the coastal regions. These events are commonly regarded as intraplate earthquakes because the North China block is in the interior of the Eurasian plate, within the Archean Sino-Korean craton, and thousands of kilometers away from plate boundaries. Because the North China block is one of the most densely populated areas in China, with vital economic and cultural centers, understanding earthquake hazards there is a pressing societal need.

Neotectonic studies in China, especially in the North China block, have been intensive in the past decades. Extensive global positioning systems (GPS) measurements in China have greatly

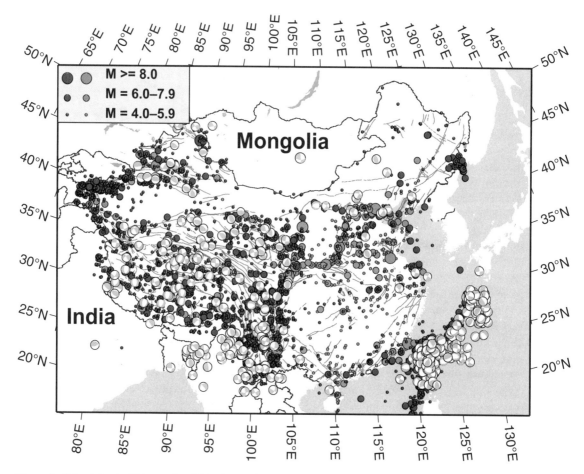

Figure 1. Seismicity in China and neighboring regions. Blue dots are the epicenters of historical earthquakes before 1900 A.D., and red dots are those from 1900 to 1990. The green fault-plane solutions are from the Harvard catalog (1976–2004) without scaling. Solid lines are active faults.

refined our knowledge of crustal kinematics. In this paper, we first summarize and analyze the GPS and neotectonic data in China to outline the crustal kinematics. We then explore the driving mechanisms and their interplay with lithospheric structures in controlling seismicity in China.

TECTONICS AND CRUSTAL KINEMATICS

Neotectonics

Neotectonic studies have shown a strong influence of plate-boundary processes on the diffuse crustal deformation in China and surrounding regions (Tapponnier and Molnar, 1977, 1979; Wesnousky et al., 1984; Ye et al., 1985; Burchfiel et al., 1991; Avouac and Tapponnier, 1993; Xu et al., 1993; Allen et al., 1998; Zhang et al., 1998, 2003). Figure 2 shows a simplified map of the major tectonic units and their Quaternary crustal motions based on fault slips and other neotectonic data (Ma, 1989; Deng et al., 2002). West of ~105°E, Quaternary tectonics is clearly controlled by the Indo-Asian collision, which has caused roughly N-S crustal contraction over a broad

region extending from the Himalayan front all the way to the Altai Mountains. Deformations are largely localized within the roughly E-W–oriented fault zones that separate the region into a hierarchy of tectonic units (Geological Institute, 1974). Within each unit, deformation is relatively coherent. The first-order tectonic units include the Himalayan-Tibetan Plateau, the Tarim block, and the Tianshan mountain belt.

The Himalayan-Tibetan Plateau is bounded on the southern side by the Indo-Eurasian plate-boundary fault zone, where Holocene slip rates are as high as 15–18 mm/yr (Lavé and Avouac, 2000) and seismicity is intense. The northern side of the plateau is marked by the sinistral Altyn Tagh–Qilian–Haiyuan fault system. Estimates of Holocene slip rates on these faults vary significantly among previous studies: ~4–30 mm/yr on the Altyn Tagh fault and ~3–19 mm/yr on the Haiyuan fault (Peltzer et al., 1989; Peltzer and Saucier, 1996; Deng et al., 2002; Lasserre et al., 2002). The Tarim Basin is a rigid block with little internal deformation or seismicity (Avouac et al., 1993; Lu et al., 1994; Allen et al., 1999; Molnar and Ghose, 2000; Kao et al., 2001; Yang and Liu, 2002). The Tianshan mountain belt has been rejuvenated by the Indo-Asian collision since the Tertiary.

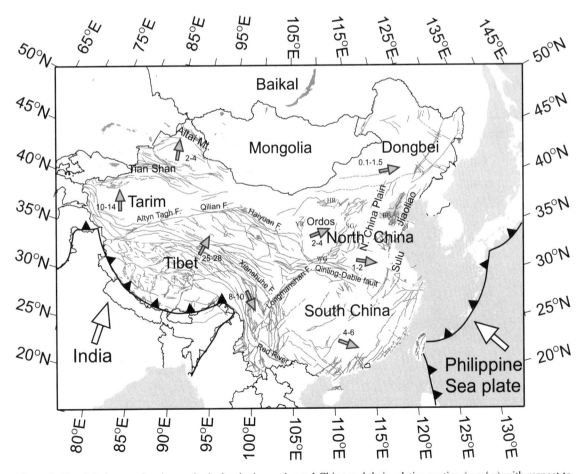

Figure 2. Simplified map of major geological units in continental China and their relative motion (mm/yr) with respect to stable Siberia, based on Quaternary fault-slip rates and other neotectonic data (after Ma, 1989; Deng et al., 2002). Thin lines are active faults. WG—Weihe graben; SG—Shanxi graben; YR—Yinchuan rift; HR—Hetao rift; BB—Bohai Basin.

Across the Tianshan mountain belt, active crustal shortening is 15–7 mm/yr estimated from balanced crustal sections. The amount of shortening decreases from west to east along the mountain belt (Deng et al., 2002). The Tianshan mountain belt is bounded by thrust and strike-slip faults with intense seismicity, manifesting the far-field impact of the Indo-Asian collision.

East of ~105°E, Cenozoic fault systems are predominantly oriented NNE and NWW, reflecting the influence from both the Indo-Eurasian collision and subduction of the Pacific and the Philippine Sea plates (Deng et al., 2002; Zhang et al., 2003). The rates of Quaternary crustal deformation are much lower than in western China. Major deformation and seismicity occur within the North China block. The western part of the North China block includes the stable Ordos Plateau and the surrounding rift systems: the Yinchuan rift basins to the west, the Hetao rift zone to the north, the Shanxi graben to the east, and the Weihe graben to the south. These rift zones initiated perhaps as early as the Miocene but developed mainly during Pliocene time (Zhang et al., 1998). Neotectonic evidence shows 2–6 mm/yr lateral motions on these fault zones (Deng et al., 2002) and about ~1.2 mm/yr extension across the Shanxi graben (Zhang et al., 1998). Historic records show three M = 8 and more than 30 M ≥ 6 earthquakes in these rift zones. East of the Ordos system, there is the North China Plain, which is a region of Mesozoic-Cenozoic rift basins and uplift structures crosscut by a system of NNE- and NWW-oriented fault zones, on which many large modern earthquakes have occurred, including the 1976 Tangshan earthquake. The North China Plain is separated from the Bohai Basin and other coastal regions (collectively called the Jiaoliao block for the northern part and the Sulu block for the southern part) by the Tanlu fault zone, a major structure in east Asia and the locus of numerous large earthquakes, including the 1668 M = 8.5 Tancheng event.

North of the North China block, there is the relatively stable Siberian shield, where the major Quaternary crustal deformation is extension across the Baikal rift zone. Within China, this region is called the Dongbei or Mongolian-Alashan block; here, the Quaternary crustal deformation and seismicity are weak. The South China Block is south of the North China block and is separated from it by the Qingling-Dabie fault zone and from the Tibetan Plateau by the Longmanshan–Xianshuhe–Red River fault system. Within the South China block, Quaternary deformation is minor, and seismicity is quiescent relative to the North China block.

GPS Measurements

Extensive GPS measurements in the past two decades have provided many details of crustal motion in China and surrounding regions. These studies include: Bilham et al. (1997) and Paul et al. (2001) for the central Himalayas, Abdrakhmatov et al. (1996) and Reigber et al. (1999) for the central and western Tianshan, King et al. (1997) and Chen et al. (2000) for the east borderland of the Tibetan Plateau, Shen et al. (2000) for North

China, Calais et al. (1998) for the Lake Baikal area, Bendick et al. (2000) and Shen et al. (2001) for the central Altyn Tagh fault, Shen et al. (2001) and Reigber et al. (2001) for the Tarim Basin and Qaidam Basin, Wang et al. (2001) and Zhang et al. (2004) for the interior of Tibetan Plateau, Banerjee and Bürgmann (2002) for western Himalayas and the Karakoram fault, Michel et al. (2001) for the Sandaland block, Vigny et al. (2003) for the Sagaing fault, Calais et al. (2003) for Mongolia, and Chen et al. (2004) for southern Tibet. A consistent regional GPS velocity field over continental China has emerged from the Crustal Motion Observation Network of China (CMONOC), established in 1998 by the State Seismological Bureau of China (now the Chinese Earthquake Administration). The CMONOC is composed of 25 continuous stations and ~1000 survey mode stations. The survey mode stations were observed in 1999, 2001, and 2004, with 3–5 24 h sessions per site and at least a couple of dozens of stations surveyed simultaneously. Station velocities obtained on the 1999 and 2001 data are shown in Figure 3. Additional GPS data sets from Bilham et al. (1997) and Paul et al. (2001) across the central Himalaya, and from Wang et al. (2001) along a north-south profile across eastern Tibet, were used in Figure 3 to fill in regions where the CMONOC network has no coverage.

The composite data set provides a detailed picture of crustal deformation for most parts of continental China. The general pattern of crustal motion and deformation is remarkably consistent with that derived from neotectonic studies (Fig. 2), with some noticeable differences. Present convergence between the India and Eurasia plates is ~36 mm/yr at the west Himalaya syntaxis and 40 mm/yr at the east Himalaya syntaxis, respectively (Paul et al., 2001); these rates are significantly lower than the ~50 mm/yr relative motion predicted by the NUVEL-1 model (DeMets et al., 1990, 1994), which was based on marine magnetic data for the past 3 m.y. Nearly half of the convergence is absorbed across the Himalayas (Bilham et al., 1997; Wang et al., 2001), and the rest is partitioned between rather uniform shortening across the Tibetan Plateau (Wang et al., 2001) and Tianshan (Abdrakhmatov et al., 1996). The Tarim Basin rotates clockwise as a rigid block relative to Siberia at a rate of ~9 nanoradian/yr (Shen et al., 2001).

Along the major strike-slip fault zones within and around the Tibetan Plateau, the GPS measured slip rates are considerably lower than those derived from neotectonic studies, although some of the neotectonic estimates are disputed (Deng et al., 2002). For the Altyn Tagh fault, the GPS data indicate ~9 mm/yr on the central segment (Bendick et al., 2000; Shen et al., 2001) and ~7 mm/yr on the western (Karokash) segment (Shen et al., 2001), in comparison with 20–30 mm/yr from geological estimates (Tapponnier et al., 2001; Peltzer et al., 1989). Northeast of the plateau, ~7 mm/yr transpressional slip was determined across the Haiyuan fault (Shen et al., 2001), much slower than the 19 mm/yr slip determined geologically (Lasserre et al., 2002). Southwest of the plateau, GPS and InSAR (Interferometric Synthetic Aperture Radar) studies measured ~5 mm/yr and ~1 mm/yr left-slip across the Karakoram

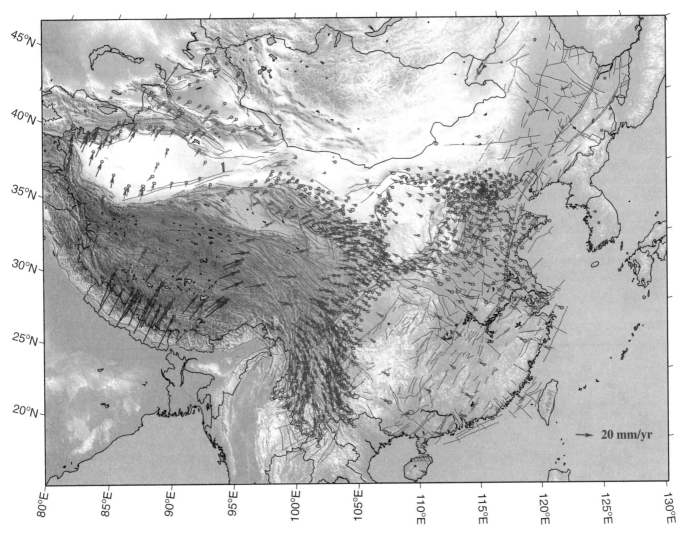

Figure 3. Horizontal velocity field in continental China, derived from global positioning system (GPS) data, with respect to stable Eurasia plate. Blue and black arrows are data from the Crustal Motion Observation Network of China (CMONOC) and non-CMONOC networks, respectively.

fault, respectively (Banerjee and Bürgmann, 2002), compared to ~11 mm/yr slip estimated from geomorphic studies (Chevalier et al., 2005). Such discrepancies may result from errors in the derived geological rates. Estimates by Chinese workers on some of these faults are much lower, closer to the GPS values (Zhang et al., 2003). On the other hand, some of the discrepancy may reflect the time scale–dependent crustal rheology and deformation (Liu et al., 2000; He et al., 2003).

Consistent with neotectonic data, GPS data indicate weak crustal deformation in east Asia. Extensive GPS measurements have been recorded for the North China block because of the intense seismicity and dense population there. By analyzing GPS data collected from a network during 1992–1996, Shen et al. (2000) found that regional deformation in North China is dominated by left-lateral slip (~2 mm/yr) across the E-SE–trending Zhangjiakou-Penglai seismic zone and extension (~4 mm/yr) across the N-NE–trending Shanxi rift. However, extension across

the Shanxi graben is not clear in the more complete data sets (He et al., 2003). GPS sites within the South China block are sparse because of the weak neotectonic activity and low seismicity. In general, the GPS data show low velocity gradients within the South China block, attesting to its stability. In the Lake Baikal region, GPS measurements since 1994 reveal ~4 mm/yr crustal extension across the Baikal rift in the NW-SE direction, normal to the elongated direction of the lake (Calais et al., 1998, 2003). In northeast China, GPS measurements are sparse, and the crustal motion seems insignificant.

Figure 4 shows the calculated horizontal strain rates based on an interpolated GPS velocity field (Shen et al., 2003). The highest strain rates, besides those along the Himalayas, are along the Xianshuihe fault, consistent with the high (~10 mm/yr) slip rates from GPS data (Wang et al., 2003a). Other regions of high strain rates include the Tibetan Plateau and the Tianshan-Altai mountain belts. East of 105°E, high rates are found around the Ordos block and in the North China Plain. The strain rates in South China are not well constrained because of the scarcity of GPS sites.

GEODYNAMICS

The GPS and neotectonic studies of continental China provide useful kinematic constraints for understanding the geodynamics of crustal deformation and seismicity. In this section, we present a suite of geodynamic models of different spatial scales, constrained by the GPS and neotectonic data, to explore (1) the major driving forces and critical boundary conditions for active crustal deformation in China and neighboring regions, (2) the cause of the contrasting seismicity between the North China and South China blocks, (3) intraplate seismicity in the circum-Ordos rift zones, and (4) Coulomb stress evolution and the triggering effects associated with major earthquakes in North China.

Driving Forces for Diffuse Asian Continental Deformation

As shown in Figure 2, China and the surrounding Asian continent are under the influence of tectonic processes from two plate boundaries: the indentation of the Indian plate into the Eurasian plate, and subduction of the Pacific and the Philippine Sea plates along the eastern margins of the Eurasian plate. Some workers have suggested a dominant role of the Indo-Asian collision in Cenozoic continental deformation in Asia (Molnar and Tapponnier, 1975; Tapponnier and Molnar, 1979; Tapponnier et al., 1982), but the temporal-spatial extent of the collisional effects is controversial. Others have argued that subduction along the eastern margins of the Asian continent has played a critical role in early Tertiary rifting and volcanism in eastern China (Northrup et al., 1995; Ren et al., 2002; Zhang et al., 2003). Numerous studies have attempted to reproduce the present crustal motions in various viscous thin-shell models

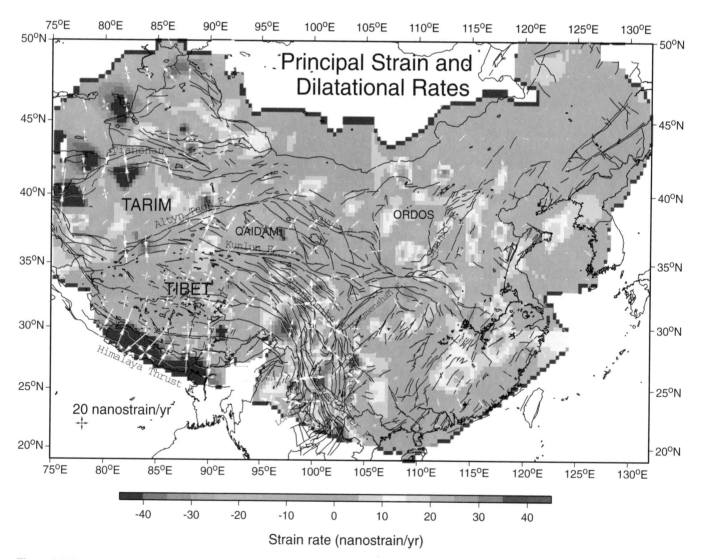

Figure 4. Principal strain (background) and dilatational strain (white arrows) rates derived from global positioning system (GPS) velocity data.

(Kong and Bird, 1996; Flesch et al., 2001). However, the relative roles of major driving forces and their temporal-spatial impacts through the Cenozoic remain uncertain.

To explore the roles of driving forces, tectonic boundary conditions, and lithospheric structure in active tectonics in China and surrounding regions, we have developed a preliminary three dimensional (3-D) finite-element model (Fig. 5). Main driving forces in the model include (1) the plate-boundary force from the Indo-Asian collision, (2) the plate-boundary force related to subduction around the east margins of the Asian continent, and (3) the gravitational buoyancy forces resulting from lateral mass variations within the Asian lithosphere and mantle, primarily reflected by topographic loading in the Tibetan Plateau. Although the effects of gravitational spreading of the Tibetan Plateau are sometimes lumped together with plate-boundary forces as an integral part of the Indo-Asian collision, there are major differences between these two. The plate-boundary force is related to the rate of plate convergence, which has been roughly steady state for the past ~50 m.y. (Molnar and Tapponnier, 1975; Patriat and Achache, 1984). The gravitational buoyancy force arises from the isostatically compensated topography, which has increased with time. Thus, we treat these two forces separately in the model.

The model assumes a power-law viscous fluid rheology, with the strain rate proportional to the cubic power of stress (Brace and Kohlstedt, 1980; Kirby and Kronenberg, 1987). Because of the large region modeled here, we constructed the 3-D finite-element model in spherical geometry to include the effects of the curvature of Earth's surface. The topographic loading is calculated using digital topography data, assuming local isostasy. The model crust sits on a viscous foundation. The vertical resistant force on the base of the crust is proportional to the vertical displacement of the crust and is a function of the effective viscosity of the underlying mantle. The indentation rate of the Indian plate into Asia is from the NUVEL-1A model (DeMets et al., 1990,

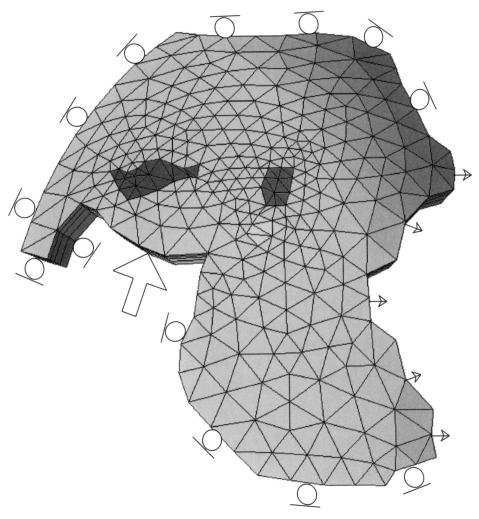

Figure 5. Finite-element mesh and boundary conditions of the continental-scale model. Areas of dark blue, pink, and light blue are the relatively stiff Tarim block, the Ordos Plateau, and the Sichuan Basin, respectively.

1994). A free-slip boundary condition (no shear stress and normal velocity) is assigned to the western (60°E), northern (70°N), and southeastern boundaries of the model domain (Fig. 5). On the eastern boundary, we use either the velocity boundary conditions based on GPS data (Wang et al., 2001) or geologically inferred plate-boundary kinematics (Northrup et al., 1995).

Figure 6A shows the predicted velocity field driven by the plate-boundary force from the collision alone. The resulting vertical and horizontal motions are largely limited to the Tibetan Plateau and surrounding regions, with little impact on regions east of 105°E. On the other hand, gravitational spreading of the Tibetan Plateau and other highlands in central Asia has a broad impact on active tectonics in Asia (Fig. 6B).

The predicted strain rates, particularly those from gravitational spreading, are highly sensitive to the rheology of the crust. A crustal effective viscosity in the range of 10^{22}–10^{23} Pa s is necessary to fit the GPS and neotectonic strain rates. These values are consistent with previous estimates of long-term continental deformation in Asia (England and Houseman, 1986; Flesch et al., 2001). A fixed boundary along the eastern margins of the Asian continent is used in both cases in Figure 6 to isolate the effects of the Indo-Eurasian plate-boundary force and the gravitational buoyancy force. Replacing this with a velocity boundary condition based on the GPS data has little impact on the predicted deformation field, suggesting that active tectonics in Asia are largely controlled by the plate-boundary force from the Indo-Asian collision and the gravitational buoyancy forces arising from the high topography of the Tibetan Plateau and surrounding regions. However, the situation may have been quite different in the early Cenozoic, when much of the Tibetan Plateau was not uplifted and the eastern margin of the Asian continent was dominated by back-arc spreading, possibly related to deceleration of the Pacific-Eurasian plate convergence and trench rollback (Northrup et al., 1995).

Figure 7 shows the results for a preferred model of present active tectonics driven jointly by the combined forces of Indo-Asian continental collision and gravitational spreading of the Tibetan Plateau, using present kinematic boundary conditions along the eastern margins of the Asian continent. The predicted uplift occurs mainly in and around the Tibet Plateau, and the horizontal velocities are generally comparable with the GPS data (Fig. 3). The predicted stress field, with NE-directed principal horizontal compressive stresses in much of Asia, generally agrees with the observed stress field (Ma, 1989).

In summary, active crustal deformation in China and surrounding regions is controlled by the indentation of the Indian plate and gravitational spreading of the Tibetan Plateau. The direct impact of the Indian indentation is limited to western China, whereas gravitational spreading of the Tibetan Plateau has broad impact on most of east Asia. The influence of plate-boundary forces along the subduction zones around the eastern margins of the Asian continent is minimal at present but may have played an important role in early Cenozoic. We did not explore the role of basal shear from asthenospheric flow for lack of constraints.

Contrasting Seismicity between the North and South China Blocks

Within east Asia, seismicity shows strong spatial variations (Fig. 1). The North China block is one of the most active intraplate seismic regions in the world, with intense historic and modern events. The South China block, on the other hand, is seismically quiescent (Fig. 8A).

The cause of the contrasting seismicity between the North China block and the South China block remains speculative. The North China block was part of a stable craton until the late Mesozoic, when the lithosphere was thermally thinned, resulting in widespread volcanism and rifting (Ye et al., 1987; Menzies and Xu, 1988; Liu et al., 2001; Ren et al., 2002; Xu et al., 2003). The volcanism waned in the late Cenozoic, but the lithosphere in the North China block remains abnormally thin (~80 km in places) (Ma, 1989). Seismic tomography shows broad and prominent low-velocity structures in the upper mantle under the North China block (Wang et al., 2003b; Liu et al., 2004; Huang and Zhao, 2006). Some workers have ascribed the high seismicity in the North China block to its lithospheric structures (Ma, 1989). Others have linked seismicity in east Asia to the far-field effects of the Indo-Asian collision (Tapponnier and Molnar, 1977), but the contrasting seismicity between the North China block and South China block remains unexplained. In this section, we summarize our recent study (Liu and Yang, 2005) of the contrasting seismicity between the North China and South China blocks.

The strain-rate field derived from GPS data (Fig. 4) shows relatively high strain rates around the Ordos plateau and the North China Plain, but also high-rate regions in the South China block. The results for the South China block may be biased by the localized regression algorithm used in deriving Figure 4, because the data are sparse there (Fig. 3). Here, we use a different approach to calculate the average strain rate, defined as, (ϕ and λ are longitude and latitude, respectively), within a sliding spatial window. The window has to contain at least three velocity sites for the strain rate to be determined. We require at least six sites in a grid window to reduce the impact of erratic sites. This requires an 8° × 8° window to cover all regions shown in Figure 8B. Within each window, linear regression is used to calculate the average velocity gradient. We then iterate with finer windows of 4° × 4°, 2° × 2°, and 1° × 1° to refine the strain rates in areas where the sites are sufficiently dense. The result shows a better correlation with seismicity. The Ordos Plateau is shown as a stable block with low strain rates. High strain rates are found in the North China Plain and around the Ordos Plateau. The South China block has low strain rates.

To explore the impact of the observed crustal kinematics and lithospheric structure on seismicity in the North China block and South China block, we developed a 3-D finite-element model to calculate the long-term stress states and strain energy in these regions. The North China block is subdivided into the Ordos Plateau, the North China Plain, and the Sulu block because of their distinctive tectonic histories. Displace-

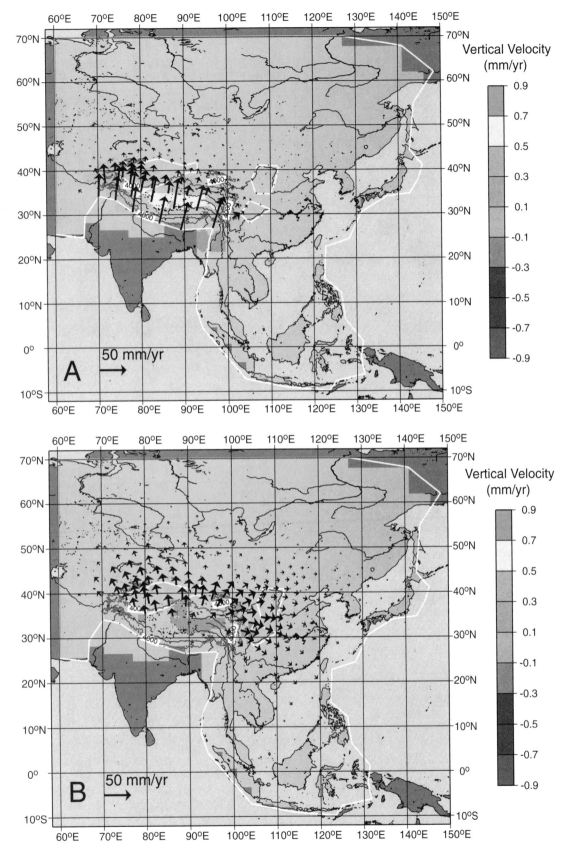

Figure 6. (A) Predicted surface horizontal (arrows) and vertical (background color) velocities caused solely by the compressive force on the Indo-Eurasian plate boundary. (B) Predicted surface horizontal and vertical velocities caused solely by gravitational spreading of the Tibetan Plateau and other regions of high topography. In both cases, a cubic power-law rheology was used: $\sigma = B\varepsilon^{1/3}$, $B = 10^{13}$ Pa s$^{1/3}$.

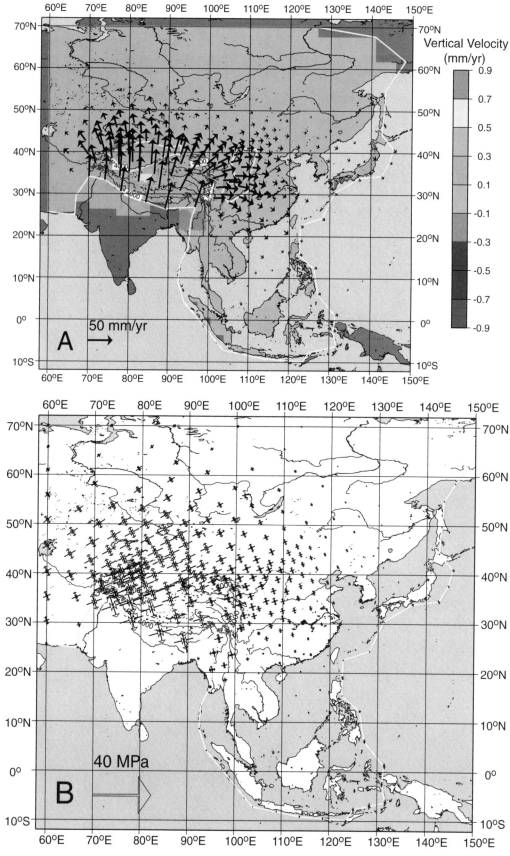

Figure 7. (A) Predicted surface vertical (background) and horizontal (arrows) velocities for a preferred case of combined driving forces (rheological parameters same as those in Fig. 6). (B) Surface horizontal maximum and minimum compressive stresses for the case in A.

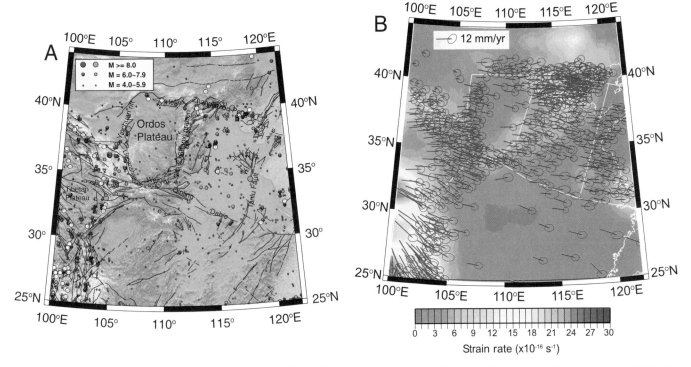

Figure 8. (A) Topographic relief and seismicity in eastern China. Legend is the same as in Figure 1. (B) Global positioning system (GPS) site velocities with respect to stable Eurasia. The error ellipses show the 95% confidence level. The data are from the Crustal Motion Observation Network of China (CMONOC) networks (He et al., 2003; Zhang et al., 2004). The background color shows the calculated average strain rates (see text). Thin lines show the coastlines, and the thick white lines show the simplified fault zones separating the tectonic units simulated in the numerical model (see Fig. 9). SCB—South China block; TP—eastern Tibetan Plateau; AM—Alashan-Mongolia shield. The North China block is subdivided into the Ordos Plateau (OP), North China Plain (NCP), and the Sulu block (SL).

ment boundary conditions, simplified from the GPS data, are applied to the edges of the model domain. The model crust has two layers with viscoelastic rheology.

Figure 9A shows the predicted stresses and the long-term strain energy, which is given by the product of the stress and strain tensors. Here, we assume a uniform crust for the North China block and South China block and no internal faults; thus, the higher shear stress and strain energy in the North China block than in the South China block are caused solely by the imposed kinematic boundary conditions. The North China block is compressed in the NE-NEE direction between the expanding Tibetan Plateau and the stable Alashan-Mongolian shield, while it moves relatively freely in the SE direction—hence the relatively large differential stresses and strain energy. Conversely, the South China block moves southeastward rather uniformly as a coherent block, resulting in little differential stresses and strain energy except near its margins. The contrast is more evident in Figure 9B, where we considered a stiff Ordos Plateau and a weak North China Plain and Sulu block as indicated by geophysical data (Ma, 1989; Liu et al., 2004). The internal fault zones that bound those tectonic units are simulated as rheological weak zones (Liu et al., 2002; Liu and Yang, 2003). The general patterns of stresses and strain energy are similar to those in Figure 9A, but some details of the model results, including

the low strain rates within the Ordos Plateau, and the NW-SE extension near the Shanxi graben, fit the observations better (Xu and Ma, 1992; Zhang et al., 2003).

Figure 9C shows the estimated seismic energy release since 23 A.D. derived from Chinese catalogs of historic earthquakes and modern events. We used the Gutenberg-Richter energy formula (Lay and Wallace, 1995) and approximated all magnitudes as M_s. The strain energy was averaged over a 20-km-thick seismogenic crust. The seismically released energy is two orders of magnitude lower than the predicted long-term strain energy (Fig. 9B), presumably because not all energy is released by earthquakes. The spatial pattern of seismic energy release is quite comparable with that in Figure 9B, although the records of historic events may be incomplete. This suggests that the intense seismicity in the North China block reflects a long-term pattern of stress accumulation and release that will continue into the future.

The Circum-Ordos Seismic Zones

Within the North China block, seismicity is clearly controlled by the heterogeneous lithospheric structure, best shown in the Ordos Plateau and the surrounding rift systems (Fig. 8A). The interior of the plateau has been stable through the Cenozoic. Deformation and seismicity are concentrated within the

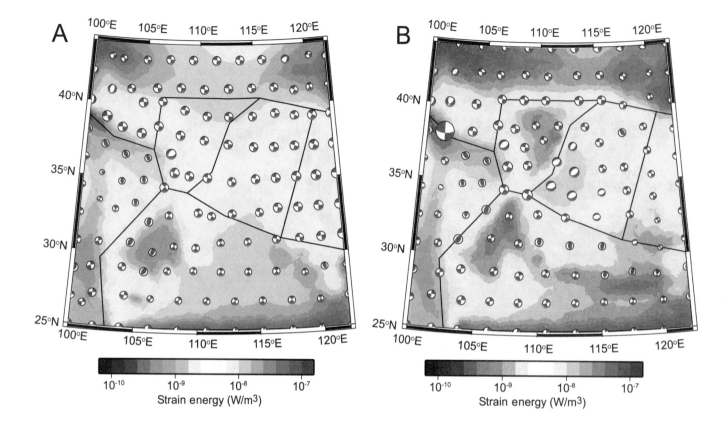

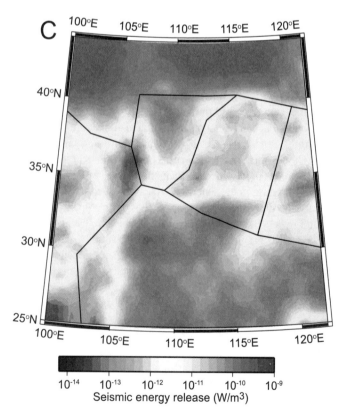

Figure 9. (A) Predicted steady-state strain energy (background color) and the deviatoric stresses ("beach balls") for a case of homogeneous crust (Young's modulus: 70 GPa; Poisson ratio: 0.25; viscosity: 5×10^{23} Pa s). The Tibet Plateau was assumed to be weaker (viscosity: 3×10^{22} Pa s). The three-dimensional stress states are represented by the lower-hemisphere stereographic projection; the maximum (σ_1) and minimum (σ_3) principal stresses bisect the white and shaded quadrants, respectively. (B) Results of a model of heterogeneous crust: the upper crustal viscosity is 1×10^{23} Pa s for the North China Plain (NCP) and the Sulu blocks and 1×10^{25} Pa s for the rest of the region. The lower crustal viscosity is 5×10^{22} Pa s for most regions but is 1×10^{22} Pa s for the Sulu block and the Tibetan Plateau. All fault zones are simulated as weak zones (viscosity: 1×10^{22} Pa s). (C) Calculated seismic energy release since 23 A.D. based on the Chinese earthquake catalog.

surrounding rift zones. The Shanxi graben, on the eastern side, is an over 700-km-long echelon of extensional basins that mainly developed in the Pliocene and Quaternary (Xu and Ma, 1992; Zhang et al., 1998). Nineteen M ≥ 6 historic earthquakes occurred here, including two M = 8 events, the 1303 Hongdong and the 1695 Linfen earthquakes. The Weihe graben, on the southern side of the plateau, is structurally connected to the Shanxi graben. The Weihe graben had a number of destructive earthquakes, including the deadly 1556 Huaxian earthquake (M = 8). Abundant ground fissures and Quaternary faulting indicate that this region is tectonically active today (Li et al., 2003). On the western and northern side of the plateau, there is the Yingchuan-Hetao rift, which also has experienced intense seismicity. The Liupanshan thrust belt on the southwestern side of the Ordos Plateau is the transition zone between the Tibetan Plateau and the North China block. Formation of the thrust belt started during the Pliocene, contemporaneous with left-slip motion on the Haiyuan fault (Zhang et al., 1991), best known for the 1920 Haiyuan earthquake (M = 8.7).

Although both neotectonics (Xu et al., 1993; Zhang et al., 1998) and historic seismicity indicate that the circum-Ordos rift zones have been active, only a few moderate-sized events have occurred in the past two hundred years, mainly near the northeastern side of the plateau (Fig. 8A). The GPS data, while indicating relative high strain rates around the Ordos Plateau (Figs. 4 and 8B), permit alternative interpretations. For example, based on the CMONOC GPS data collected between 1992 and 1996, Shen et al. (2000) found a 4 mm/yr extension rate across the Shanxi graben. However, using the same data set but with updated measurements extending to 2001, He et al. (2003) found no clear velocity jump across the Shanxi graben. This discrepancy with geological evidence of active extension across the graben may reflect data errors, or more likely, the time scale–dependent crustal rheology and deformation (Liu et al., 2000; He et al., 2003).

To explore the effects of time scale–dependent crustal kinematics and heterogeneous lithospheric structure, we built a local-scale geodynamic model similar to the regional-scale model for the North China and South China blocks (Fig. 10). In this local-scale model, the circum-Ordos rifts are simulated as rheological weak zones with finite widths. Figure 10 shows the predicted stresses and strain rates within the upper crust under the boundary load derived from interpolated GPS velocities. The lower maximum shear stress in the circum-Ordos rifts and other fault zones results from the lower viscosity assumed for these fault zones. The predicted stress states, such as the widespread extensional stresses in much of the North China Plain and the Mongolian shield, and the lack of thrust faulting in the Liupanshan region and northeastern Tibetan Plateau are inconsistent with Cenozoic structures and neotectonic data. To better fit the geological structures, we found it necessary to

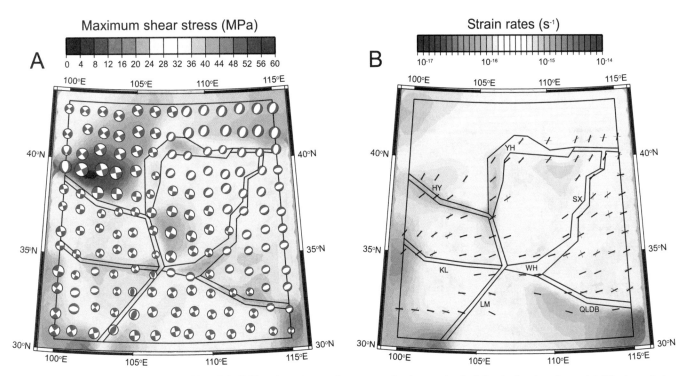

Figure 10. (A) Predicted stress states ("beach balls") and maximum shear stress (background color) for the local-scale model. The boundaries of the model domain are loaded by imposed velocities interpolated from the global positioning system (GPS) data. (B) Predicted strain rates (background color) and the direction of maximum horizontal compressive stress (black bars), compared with the observed stress orientations interpolated from the World Stress Map (http://world-stress-map.org). HY—Haiyuan fault; YH—Yinchuan-Hetao rifts; KL—Kunlun fault; WH—Weihe graben; SX—Shanxi graben; LM—Longmanshan fault; QLDB—Qinling-Dabie fault zone.

increase the velocity on the western side of the model domain, especially along the northeastern front of the Tibetan Plateau (Fig. 11). These requirements seem consistent with the fact that GPS velocities on the major strike-slip faults in this region are systematically lower than the geologically estimated slip rates. Thus, GPS site velocities, which are measured during a period of a few years and reflect the instantaneous velocity field, may not be representative of the long-term geological rates of crustal motion. Similar observations have been made in other regions (Liu et al., 2000; Friedrich et al., 2003). Note that in both cases the predicted direction of maximum horizontal compressive stresses is generally comparable to those observed, indicating that the directions of long-term crustal motion are close to those indicated by the GPS data. The fits are generally better in the western part of the model domain, perhaps reflecting the dominance of compressive stresses from the expanding Tibetan Plateau. The relatively poor fit in the North China Plain may reflect the effects of other processes, such as subduction and basal drag associated with mantle flow (Liu et al., 2004), which may be important there but are not included in this model.

Stress Evolution and Triggering of Large Earthquakes in North China

Intracontinental seismicity in China is spatially correlates to regions of high strain rates (Figs. 4 and 8B). The temporal patterns of these earthquakes, however, remain poorly understood. The Weihe and Shanxi grabens have experienced more

than 30 M ≥ 6 events in the past 2000 yr, but instruments have recorded no major events there in the past century, and modern seismicity in the North China block has been concentrated in the North China Plain (Fig. 1). Some explanations for the spatial-temporal evolution of seismicity may be provided by the changes of the Coulomb stresses associated with earthquakes, which have been shown to have significant influences on earthquake sequences in both interplate (King et al., 1994; Stein, 1999) and intraplate settings (Li et al., this volume). Shen et al. (2004) simulated the Coulomb failure stress change (ΔCFS) in North China since 1303 using 49 destructive events (M ≥ 6.5) from the Chinese catalog of historic earthquakes.

The ΔCFS on a fault plane is calculated as $\Delta\sigma_f = \Delta\tau + \mu\Delta\sigma_n$, where $\Delta\tau$ is the change of shear stress on the fault plane, $\Delta\sigma_n$ is the change of normal stress ($\Delta\sigma_n > 0$ indicates an increase in tension), and μ is the frictional coefficient ($\mu = 0.4$ for this study). Positive $\Delta\sigma_f$ moves a fault toward failure, and vice versa. The stress evolution was calculated using the viscoelastic codes by Zeng (2001). The earthquakes used in the simulation are shown in Figure 12. The locations, magnitudes, and intensity data are from the Chinese earthquake catalog (Ming et al., 1995). Because most of the earthquakes occurred on faults buried under sediments, we first derived an empirical relationship between the earthquake ground-shaking intensity distribution (and magnitude) and earthquake rupture parameters using modern instrumentally recorded strong earthquakes in North China, and then we used this empirical formula to infer fault rupture parameters of historical earthquakes, including the rupture length and the amount of slip. The

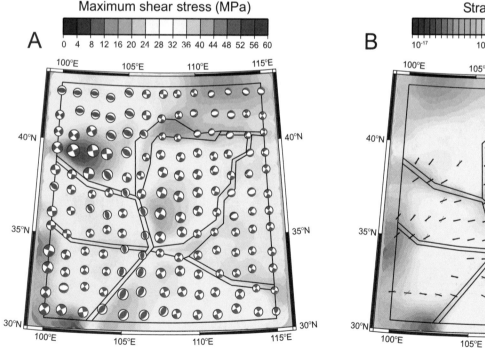
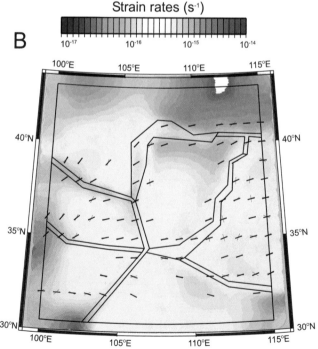

Figure 11. Results of the local-scale model similar to Figure 10, except faster eastward velocities (3 mm/yr higher than the global positioning system [GPS] velocities) are assumed along the eastern side of the Tibetan Plateau.

earthquake rake angles were derived from geologically determined fault parameters and seismically estimated orientations of regional tectonic stresses. Earthquake ruptures were assumed to span 2–20 km based on focal-depth distribution of small- to medium-sized earthquakes in the region. The incremental secular loading stresses were assumed to be depth invariant, and were calculated from the GPS velocity field assuming linear elasticity of the media (Shen et al., 2003).

The initial values of ΔCFS were assumed to be zero everywhere, and the simulations started with the 1303 Hongdong earthquake (M = 8.0). The stress evolution was then calculated with secular loading and for each of the sequential earthquakes using the geologically derived rapture parameters. Figure 13A shows one snapshot of the ΔCFS field after the 1888 Bohai earthquake and before the 1910 Huanghai earthquake. The ΔCFS was evaluated on vertical fault planes trending N40°E, which is the dominant fault geometry of the region. The triggering effects are suggested by the spatial correlation between regions of positive ΔCFS and the loci of major earthquakes since 1880. The 1976 Tangshan earthquake occurred in an area of increased ΔCFS that was the result of the 1679 Sanhe-Pinggu earthquake (M = 8) and secular tectonic loading, with an accumulated ΔCFS of ~1.8 bar. Our analysis shows that for all 49 earthquakes with M ≥ 6.5 that have occurred since 1303 in North China, 39 out of the 48 subsequent events occurred in regions of positive ΔCFS; the triggering rate is 81.3%. Figure 13B shows

present ΔCFS, modified from that in Figure 13A mainly by the 1910 Huanghai earthquake and 1976 Tangshan earthquake. The high risk areas include the Bohai Basin, the west segment of the northern Qinling fault, western end of the Zhangjiakou-Penglai seismic zone, and the Taiyuan Basin in the Shanxi graben.

DISCUSSION

Diffuse continental deformation has challenged one of the basic tenets of plate tectonics theory that predicts deformation and seismicity concentrations within narrowly defined plate-boundary zones, and this has been propelling a fundamental transition from kinematic descriptions of tectonic plates toward a dynamic understanding of lithospheric deformation (Molnar, 1988). Some of the intracontinental deformation reflects diffuse plate-boundary zones. Space-geodetic measurements show that many plate-boundary zones are characterized by diffuse deformation, and the diffuse plate-boundary zones are not limited to continents (Gordon and Stein, 1992). Other intracontinental deformation occurs within plate interiors, often associated with certain lithospheric heterogeneities, and shows no clear link to plate-boundary processes.

The active tectonics and seismicity in China provide some of the best examples of both diffuse plate-boundary zone deformation and intraplate tectonics. The Tibetan Plateau and most parts of western China are extreme examples of diffuse plate-boundary

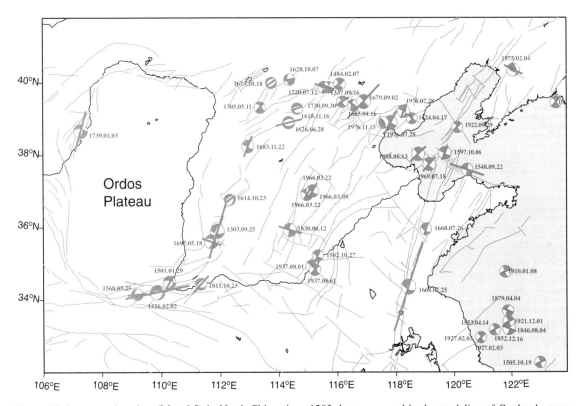

Figure 12. Large earthquakes (M ≥ 6.5) in North China since 1303 that were used in the modeling of Coulomb stress changes and their triggering effects.

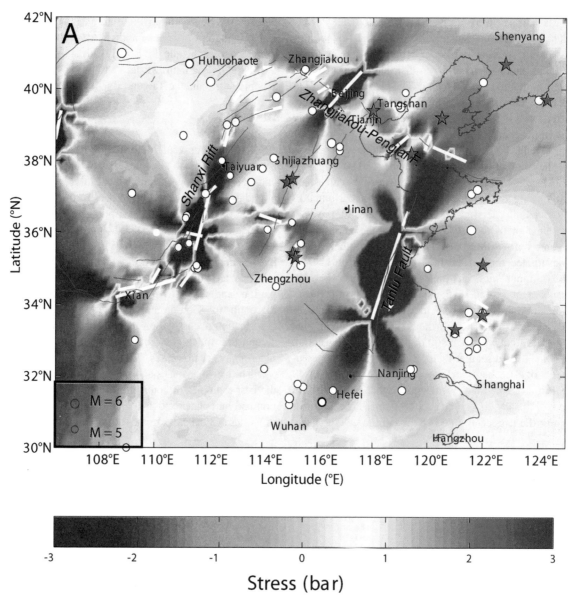

Figure 13 (*on this and following page*). (A) Predicted Coulomb stress changes in 1910. Stars and circles denote the epicenters of the subsequent M ≥ 6.5 and 6.5 ≥ M ≥ 5 earthquakes for 1910–1976, respectively.

zones, where continuous crustal deformation extends thousands of kilometers into the Asian continent from the Indo-Asian plate boundary. The deformation styles and orientations of geological structures are coherently related to the collisional plate-boundary processes. On the other hand, although the Indo-Asian collision may have influenced tectonics in eastern China and southeast Asia, those areas are not part of the diffuse Indo-Eurasian plate-boundary zone, as shown by their different type and style of deformation. While western China was under tectonic compression through much of the Cenozoic, eastern China experienced widespread rifting and basaltic volcanism (Ye et al., 1985; Ren et al., 2002; Liu et al., 2004). Thus, earthquakes in eastern China are commonly regarded as intraplate events. Crustal deformation in the North China

block, however, differs significantly from typical seismic zones within stable continents, such as the New Madrid seismic zone in central United States (see Li et al., this volume, and references therein), because the North China block has been thermally rejuvenated since the late Mesozoic and is no longer a stable continental block and the strain rates are one order of magnitude higher than those in the central and eastern United States (Newman et al., 1999; Gan and Prescott, 2001). Moreover, unlike the central and eastern United States, where seismicity shows no clear link to plate-boundary processes (Li et al., this volume), active crustal deformation and earthquakes in the North China block are strongly influenced by the compressive stress from the Indo-Asian collision and gravitational spreading of the Tibetan Plateau.

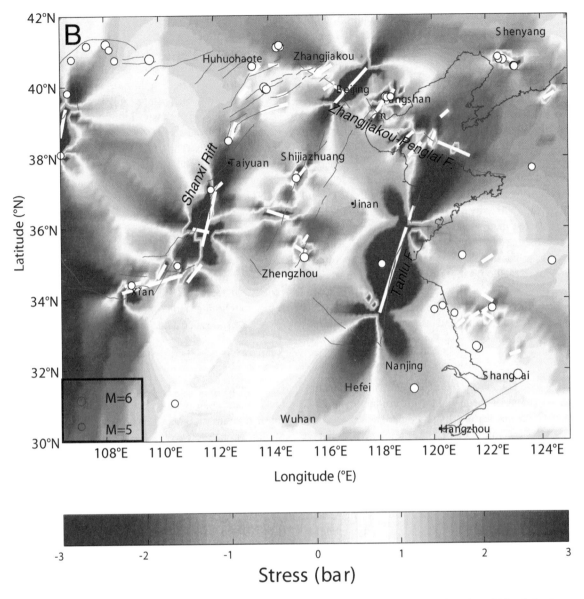

Figure 13 (*continued*). (B) Predicted Coulomb stress changes in 2004. White circles are earthquakes of M > 5 that have occurred between 1976 and present.

We have shown that the GPS site velocities provide a satisfactory first-order delineation of the regions of high strain rates and seismicity that is consistent with neotectonic data. However, along major strike-slip faults in western China, the GPS observed slip rates are consistently lower than those inferred from neotectonic studies. Although errors from processing GPS data collected from campaign-style measurements may not be excluded, the systematic discrepancy more likely results from different time scales represented by the GPS and neotectonic data (Liu et al., 2000). The GPS data, typically collected over a period of a few years, measure an instantaneous velocity field. During the period of GPS measurements, the faults are often partially or entirely locked. The neotectonic data, on the other hand, represent long-term, average slip rates that include both coseismic and interseismic displacements. Over dip-slip faults, the GPS velocity gradient is expected to be gentle if the fault is locked during the period of GPS measurements. The lack of a clear GPS velocity gradient across the Shanxi graben is consistent with the fact that no major earthquakes have occurred within the graben in the past 200 yr.

The analyses of crustal kinematics and dynamics presented in this study provide a geodynamic framework for understanding the variable distributions of seismicity in continental China. Both kinematics and lithospheric structures exert strong influences on seismicity. The Ordos Plateau and the Tarim block, for example, are located within regions of high rates of relative crustal motions; their lack of seismicity results mainly from their high lithospheric strength. The seismic quiescence within the South China block, on the other hand, may be largely attributed to the kinematic boundary conditions that allow the South China block to move coherently as a rigid block.

The seismic zones in China are well defined by seismicity, neotectonics, and now, to some extent, space-geodesy. The temporal patterns of the seismicity, however, remain poorly understood. For example, the Weihe and Shanxi grabens have experienced abundant large earthquakes in the past 2000 yr, but they have not seen major earthquakes in the past 200 yr. Modern seismicity within the North China block seems to have shifted to the North China Plain. We have attempted to explore the triggering effects of large earthquakes by modeling the associated Coulomb stress changes. Given the complexity of earthquake physics and incomplete knowledge of crustal structures in North China, we found it interesting that over 80% of M ≥ 6.5 events occurred in regions of increased Coulomb stresses caused by previous earthquakes. These results, albeit intrinsically limited, provide helpful information for future seismic hazard assessments in this part of China where population is dense and economy is booming.

CONCLUSIONS

1. Neotectonic data, mainly based on Quaternary fault slips, and GPS measurements indicate that the diffuse intracontinental deformation in China and surrounding regions and the associated seismicity are strongly influenced by the Indo-Asian collision. The effects of indentation of the Indian plate are largely limited to western China (west of 105°E). In contrast, the compressive stress arising from gravitational spreading of the Tibetan Plateau has a broader impact on active tectonics in much of central and east Asia. The subduction zones around the eastern margins of the Asian continent contribute little to the intracontinental deformation in the Asian continent at present but may have played a major role in the early Cenozoic, when most of the Tibetan Plateau was not fully uplifted and back-arc spreading was intense along the eastern margins of the Asian continent.

2. GPS data indicate that more than half of the present-day convergence (36–40 mm/yr) between the Indian and Eurasian plates is spread evenly across the Tibetan Plateau and the Tianshan, making western China and surrounding regions one of the widest diffuse plate-boundary zones. Seismicity in this part of the Asian continent is intense and is generally associated with crustal motion on a system of roughly E-W–oriented strike-slip and thrust faults driven by N-S–directed crustal shortening. Seismicity in eastern China is characterized by intense seismicity in the North China block and relative quiescence in the South China block. Geodynamic modeling suggests that this contrasting seismicity is primarily caused by the crustal kinematics and tectonic boundary conditions. The North China block experiences high deviatoric stresses because it is squeezed between the stable Siberian shield and the expanding Tibetan Plateau, while the crust is moving relatively freely southeastward. The South China block, facilitated by the surrounding strike-slip fault systems, moves more coherently as a rigid block, resulting in low internal deformation and seismicity.

3. Within the North China block, deformation and seismicity are largely controlled by the lateral heterogeneities of the lithospheric structure. High strain rates and intense seismicity within the circum-Ordos rifts result mainly from the weakness in the crust and perhaps mantle lithosphere in these rifts. In the North China Plain, seismicity is associated with a system of conjugate NNE and NWW faults. The predicted long-term strain energy pattern agrees with the spatial pattern of seismic strain energy released during the past 2000 yr, suggesting that the intense seismicity in North China is a long-term process that will continue in the future. However, the temporal pattern of large earthquakes within the North China block remains unclear. Preliminary calculations of earthquake-triggered Coulomb stress changes show that ~80% of large (M ≥ 6.5) earthquakes have occurred within regions of increased Coulomb stresses. At present, the high-risk regions include the Bohai Basin, the western segment of the northern Qinling fault, the western end of the Zhangjiakou-Penglai seismic zone, and the Taiyuan Basin in the Shanxi graben.

ACKNOWLEDGMENTS

This work is based on numerous projects with our collaborators. In particular, we acknowledge Li Yanxing and He Jiankun for their contribution to these projects. We thank An Yin and Seth Stein for constructive reviews, and Seth Stein for careful editing. Support for this work was provided by National Science Foundation (NSF) grant EAR-0207200, National Science Foundation of China grant 40228005, and the Research Council of the University of Missouri–Columbia.

REFERENCES CITED

Abdrakhmatov, K.Y., Aldazhanov, S.A., Hager, B.H., Hambuerger, M.W., Herring, T.A., Kalabaev, K.B., Makarov, V.I., Molnar, P., Panasyuk, S.V., Prilepin, M.T., Reilinger, R.E., Sadybakasov, I.S., Souter, B.J., Trapeznikov, A., Tsurkov, V.Y., and Zubovich, A.V., 1996, Relatively recent construction of the Tian Shan inferred from GPS measurements of present-day crustal deformation rates: Nature, v. 384, p. 450–453, doi: 10.1038/384450a0.

Allen, M.B., Macdonald, D.I.M., Xun, Z., Vincent, S.J., and Brouet, M.C., 1998, Transtensional deformation in the evolution of the Bohai Basin, northern China, *in* Holdsworth, R.E., Strachan, R.A., and Dewey, J.F., eds., Continental Transpressional and Transtensional Tectonics: Geological Society of London Special Publication 135, p. 215–229.

Allen, M.B., Vincent, S.J., and Wheeler, P.J., 1999, Late Cenozoic tectonics of the Kepingtage thrust zone: Interactions of the Tien Shan and the Tarim Basin, northwest China: Tectonics, v. 18, p. 639–654, doi: 10.1029/1999TC900019.

Avouac, J.-P., and Tapponnier, P., 1993, Kinematic model of active deformation in Central Asia: Geophysical Research Letters, v. 20, p. 895–898.

Avouac, J.-P., Tapponnier, P., Bai, M., You, H., and Wang, G., 1993, Active thrusting and folding along the northern Tien Shan and late Cenozoic rotation of the Tarim relative to Dzungaria and Kazakhstan: Journal of Geophysical Research, v. 98, p. 6755–6804.

Banerjee, P., and Bürgmann, R., 2002, Convergence across the northwest Himalaya from GPS measurement: Geophysical Research Letters, v. 29, doi: 10.1029/2002GL015184.

Bendick, R., Bilham, R., Freymueller, J.T., Larson, K., and Yin, G., 2000, Geodetic evidence for a low slip rate in the Altyn Tagh fault system: Nature, v. 404, p. 69–72, doi: 10.1038/35003555.

Bilham, R., Larson, K.M., Freymueller, J.T., Jouanne, F., Le Fort, P., Leturmy, P., Mugnier, J.L., Gamond, J.F., Glot, J.P., Martinod, J., Chaudury, N.L., Chitrakar, G.R., Gautam, U.P., Koirala, B.P., Pandey, M.R., Ranabhat, R., Sapkota, S.N., Shrestha, P.L., Thakuri, M.C., Timilsina, U.R., Tiwari, D.R., Vidal, G., Vigny, C., Galy, A., and de Voogd, B., 1997, GPS measurements of present-day convergence across the Nepal Himalaya: Nature (London), v. 386, p. 61–64.

Brace, W.F., and Kohlstedt, D.L., 1980, Limits on lithospheric stress imposed by laboratory experiments: Journal of Geophysical Research, v. 85, p. 6248–6252.

Burchfiel, B.C., Peizhen, Z., Yipeng, W., Weiqi, Z., Fangmin, S., Qidong, D., Molnar, P., and Royden, L., 1991, Geology of the Haiyuan fault zone, Ningxia-Hui autonomous region, China, and its relation to the evolution of the northeastern margin of the Tibetan Plateau: Tectonics, v. 10, p. 1091–1110.

Calais, E., Lesne, O., Deverchere, J., San'kov, V., Lukhnev, A., Miroshnitchenko, A., Buddo, V., Levi, K., Zalutzky, V., and Bashkuev, Y., 1998, Crustal deformation in the Baikal rift from GPS measurements: Geophysical Research Letters, v. 25, p. 4003–4006.

Calais, E., Vergnolle, M., and San'kov, V., 2003, GPS measurements of crustal deformation in the Baikal-Mongolia area (1994–2002): Implications for current kinematics of Asia: Journal of Geophysical Research, v. 108, no. B10, 2501, doi: 10.1029/2002JB002373.

Chen, Q., Freymueller, J., and Yang, Z., 2004, Spatially variable extension in southern Tibet based on GPS measurements: Journal of Geophysical Research, v. 109, no. B9, B09401, doi: 10.1029/2002JB002151.

Chen, Q., Freymueller, J.T., Xu, C., Jiang, W., Yang, Z., Wang, Q., and Liu, J., 2000, Active deformation in southern Tibet measured by GPS: Eos (Transactions, American Geophysical Union), v. 81, p. F1229–F1230.

Chen, Y., Tsoi, K.L., Chen, F.B., Gao, Z.H., Zou, Q.J., and Chen, Z.L., 1988, The Great Tangshan Earthquake of 1976: An Anatomy of Disaster: Oxford, Pergamon Press, 153 p.

Chevalier, M., Ryerson, F., and Tapponnier, P., 2005, Slip-rate measurements on the Karakorum Fault may imply secular variations in fault motion: Science, v. 307(5708), p. 411–414.

DeMets, C., Gordon, R.G., Argus, D.F., and Stein, S., 1990, Current plate motions: International Journal of Geophysics, v. 101, p. 425–478.

DeMets, C., Gordon, R.G., Argus, D.F., and Stein, S., 1994, Effect of recent revisions to the geomagnetic reversal time scale on estimates of current plate motion: Geophysical Research Letters, v. 21, p. 2191–2194, doi: 10.1029/94GL02118.

Deng, Q.D., Zhang, P.Z., Ran, Y.K., Yang, X.P., Min, W., and Chu, Q.Z., 2002, Basics characteristics of active tectonics of China (in Chinese): Science in China, v. 32, p. 1020–1030.

England, P.C., and Houseman, G.A., 1986, Finite strain calculations of continental deformation: 2. Comparison with the India-Asia collision: Journal of Geophysical Research, v. 91, p. 3664–3676.

Flesch, L.M., Haines, A.J., and Holt, W.E., 2001, Dynamics of the India-Eurasia collision zone: Journal of Geophysical Research, v. 106, p. 16,435–16,460, doi: 10.1029/2001JB000208.

Friedrich, A.M., Wernicke, B.P., Niemi, N.A., Bennett, R.A., and Davis, J.L., 2003, Comparison of geodetic and geological data from the Wasatch region, Utah, and implications for the spectral characteristics of Earth deformation at periods of 10 to 10 million years: Journal of Geophysical Research, v. 108, doi: 10.1029/2001JB000682.

Gan, W., and Prescott, W.H., 2001, Crustal deformation rates in central and eastern U.S. inferred from GPS: Geophysical Research Letters, v. 28, p. 3733–3736, doi: 10.1029/2001GL013266.

Geological Institute, Chinese Academy of Science, 1974, Preliminary study of the main features of tectonics in China and their development: Geological Sciences (in Chinese), v. 1, 105 p.

Gordon, R.G., and Stein, S., 1992, Global tectonics and space geodesy: Science, v. 256, p. 333–342, doi: 10.1126/science.256.5055.333.

He, J., Liu, M., and Li, Y., 2003, Is the Shanxi rift of northern China extending?: Geophysical Research Letters, v. 30, doi: 10.1029/2003GL018764.

Huang, J.L., and Zhao, D.P., 2006, High-resolution mantle tomography of China and surrounding regions: Journal of Geophysical Research-Solid Earth, v. 111, doi: 10.1029/2005JB004066.

Kao, H., Gao, R., Rau, R.J., Shi, D., Chen, R.Y., Guan, Y., and Wu, F.T., 2001, Seismic image of the Tarim Basin and its collision with Tibet: Geology, v. 29, p. 575–578, doi: 10.1130/0091-7613(2001)029<0575:SIOTTB>2.0.CO;2.

King, G.C.P., Stein, R.S., and Lin, J., 1994, Static stress changes and the triggering of earthquakes: Bulletin of the Seismological Society of America, v. 84, p. 935–953.

King, R.W., Shen, F., Burchfiel, B.C., Royden, L.H., Wang, E., Chen, Z., Liu, Y., Zhang, X., Zhao, J., and Li, Y., 1997, Geodetic measurement of crustal motion in southwest China: Geology, v. 25, p. 179–182, doi: 10.1130/0091-7613(1997)025<0179:GMOCMI>2.3.CO;2.

Kirby, S.H., and Kronenberg, A.K., 1987, Rheology of the lithosphere: Selected topics: Reviews of Geophysics, v. 25, p. 1219–1244.

Kong, X., and Bird, P., 1996, Neotectonics of Asia: Thin-shell finite-element models with faults, in Yin, A., and Harrison, T.M., eds., The Tectonic Evolution of Asia: World and Regional Geology Series: Cambridge, Cambridge University Press, p. 18–36.

Lasserre, C., Gaudemer, Y., and Tapponnier, P., 2002, Fast late Pleistocene slip rate on the Leng Long Ling segment of the Haiyuan fault, Qinghai, China: Journal of Geophysical Research, v. 107, no. B11, 2276, doi: 10.1029/2000JB000060.

Lavé, J., and Avouac, J.-P., 2000, Active folding of fluvial terraces across the Siwaliks Hills, Himalayas of central Nepal: Journal of Geophysical Research, v. 105(B3), p. 5735–5770, doi: 10.1029/1999JB900292.

Lay, T., and Wallace, T.C., 1995, Modern Global Seismology: San Diego, Academic Press, 521 p.

Li, X., Yan, W., Li, T., and Sun, G., 2001, Analysis on the movement tendency of Xi'an ground fissures: Journal of Engineering Geology (in Chinese), v. 9, p. 39–43.

Liu, J., Han, J., and Fyfe, W., 2001, Cenozoic episodic volcanism and continental rifting in northeast China and possible link to Japan Sea development as revealed from K-Ar geochronology: Tectonophysics, v. 339, p. 385–401, doi: 10.1016/S0040-1951(01)00132-9.

Liu, M., and Yang, Y., 2003, The collapse of Tibetan Plateau: Insights from 3-D finite element modeling: Journal of Geophysical Research, v. 108, 2361, doi: 10.1029/2002JB002248.

Liu, M., and Yang, Y., 2005, Contrasting seismicity between the North China and South China blocks: Kinematics and geodynamic: Geophysical Research Letters, v. 32, doi: 10.1029/2005GL023048.

Liu, M., Yang, Y., Stein, S., Zhu, Y., and Engeln, J., 2000, Crustal shortening in the Andes: Why do GPS rates differ from geological rates?: Geophysical Research Letters, v. 27, p. 3005–3008, doi: 10.1029/2000GL008532.

Liu, M., Yang, Y., Stein, S., and Klosko, E., 2002, Crustal shortening and extension in the central Andes: Insights from a viscoelastic model, in Stein, S., and Freymueller, J., eds., Plate Boundary Zone: Washington, D.C., American Geophysical Union, Geodynamics Series, v. 30, doi: 10.1029/030GD19.

Liu, M., Cui, X., and Liu, F., 2004, Cenozoic rifting and volcanism in eastern China: A mantle dynamic link to the Indo-Asian collision?: Tectonophysics, v. 393, p. 29–42, doi: 10.1016/j.tecto.2004.07.029.

Lu, H., Howell, D.G., Jia, D., Cai, D., Wu, S., Chen, C., Shi, Y., Valin, Z.C., and Guo, L., 1994, Kalpin transpression tectonics, northwestern Tarim Basin, western China: International Geology Review, v. 36, p. 975–981.

Ma, X., 1989, Atlas of Lithospheric Dynamics of China: Beijing, China Cartographic, 787 × 1092, 37 1/2 printed sheets.

Menzies, M.A., and Xu, Y., 1988, Geodynamics of the North China craton, in Flower, M.F.J., Chung, S.L., Lo, C.H., and Lee, T.Y., eds., Mantle Dynamics and Plate Interactions in East Asia: Washington, D.C., American Geophysical Union, Geodynamics Series, v. 27 p. 155–165.

Michel, G., Yu, Y., Zhu, S., Reigber, C., Becker, M., Reinhart, E., Simons, W., Ambrosius, B., Vigny, C., Chamot-Rooke, N., LePichon, X., Morgan, P., and Matheussen, S., 2001, Crustal motion and block behavior in SE-Asia from GPS measurements: Earth and Planetary Science Letters, v. 187, p. 239–244, doi: 10.1016/S0012-821X(01)00298-9.

Ming, Z.Q., Hu, G., Jiang, X., Liu, S.C., and Yang, Y.L., 1995, Catalog of Chinese Historic Strong Earthquakes from 23 A.D. to 1911 (in Chinese): Beijing, Seismological Publishing House, 514 p.

Molnar, P., 1988, Continental tectonics in the aftermath of plate tectonics: Nature, v. 335, p. 131–137, doi: 10.1038/335131a0.

Molnar, P., and Ghose, S., 2000, Seismic movements of major earthquakes and the rate of shortening across the Tien Shan: Geophysical Research Letters, v. 27, p. 2377–2380, doi: 10.1029/2000GL011637.

Molnar, P., and Tapponnier, P., 1975, Cenozoic tectonics of Asia: Effects of a continental collision: Science, v. 189, p. 419–426, doi: 10.1126/science.189.4201.419.

Newman, A., Stein, S., Weber, J., Engeln, J., Mao, A., and Dixon, T., 1999, Slow deformation and lower seismic hazard at the New Madrid seismic zone: Science, v. 284, p. 619–621, doi: 10.1126/science.284.5414.619.

Northrup, C.J., Royden, L.H., and Burchfiel, B.C., 1995, Motion of the Pacific plate relative to Eurasia and its potential relation to Cenozoic extension along the eastern margin of Eurasia: Geology, v. 23, p. 719–722, doi: 10.1130/0091-7613(1995)023<0719:MOTPPR>2.3.CO;2.

Patriat, P., and Achache, J., 1984, India-Eurasia collision chronology has implications for crustal shortening and driving mechanism of plates: Nature, v. 311, p. 615–621, doi: 10.1038/311615a0.

Paul, J., Burhmann, R., Gaur, V.K., Bilham, R., Larson, K., Ananda, M.B., Jade, S., Mukal, M., Anupama, T.S., Satyal, G., and Kumar, D., 2001, The motion and active deformation of India: Geophysical Research Letters, v. 28, p. 647–650, doi: 10.1029/2000GL011832.

Peltzer, G., and Saucier, F., 1996, Present-day kinematics of Asia derived from geologic fault rates: Journal of Geophysical Research, ser. B, Solid Earth and Planets, v. 101, p. 27,943–27,956, doi: 10.1029/96JB02698.

Peltzer, G., Tapponnier, P., and Amijio, R., 1989, Magnitude of late Quaternary left-lateral displacement along north edge of Tibet: Science, v. 246, p. 1285–1289, doi: 10.1126/science.246.4935.1285.

Reigber, C., Michel, G.W., Galas, R., Angermann, D., Klotz, J., Chen, J.Y., Papschev, A., Arslanov, R., Tzurkov, V.E., and Ishanov, M.C., 2001, New space geodetic constraints on the distribution of deformation in Central Asia: Earth and Planetary Science Letters, v. 191, p. 157–165, doi: 10.1016/S0012-821X(01)00414-9.

Ren, J., Tamaki, K., Li, S., and Zhang, J., 2002, Late Mesozoic and Cenozoic rifting and its dynamic setting in eastern China and adjacent areas: Tectonophysics, v. 344, p. 175–205, doi: 10.1016/S0040-1951(01)00271-2.

Shen, Z.-K., Zhao, C., Yin, A., and Jackson, D., 2000, Contemporary crustal deformation in east Asia constrained by Global Positioning System measurements: Journal of Geophysical Research, v. 105, p. 5721–5734, doi: 10.1029/1999JB900391.

Shen, Z.-K., Wang, M., Li, Y., Jackson, D.D., Yin, A., Dong, D., and Fang, P., 2001, Crustal deformation along the Altyn Tagh fault system, western China, from GPS: Journal of Geophysical Research, v. 106, p. 30,607–30,621, doi: 10.1029/2001JB000349.

Shen, Z.-K., Wang, M., Gan, W.-J., and Zhang, Z.-S., 2003, Contemporary tectonic strain rate field of Chinese continent and its geodynamic implications: Earth Science Frontiers, v. 10, suppl., p. 93–100.

Shen, Z., Wan, Y., Gan, W., Li, T., and Zeng, Y., 2004, Crustal stress evolution of the last 700 years in North China and earthquake sequence: Earthquake research in China (in Chinese), v. 20, p. 211–228.

Stein, R.S., 1999, The role of stress transfer in earthquake occurrence: Nature, v. 402, p. 605–609, doi: 10.1038/45144.

Tapponnier, P., and Molnar, P., 1977, Active faulting and tectonics in China: Journal of Geophysical Research, v. 82, p. 2905–2930.

Tapponnier, P., and Molnar, P., 1979, Active faulting and Cenozoic tectonics of Tien Shan, Mongolia and Baikal regions: Journal of Geophysical Research, v. 84, p. 3425–3459.

Tapponnier, P., Pelzer, G., LeDain, A.Y., Armijo, R., and Cobbold, P., 1982, Propagating extrusion tectonics in Asia: New insights from simple plasticine experiments: Geology, v. 10, p. 611–616, doi: 10.1130/0091-7613(1982)10<611:PETIAN>2.0.CO;2.

Tapponnier, P., Xu, Z., Roger, F., Meyer, B., Arnaud, N., Wittlinger, G., and Yang, J., 2001, Oblique stepwise rise and growth of the Tibet Plateau: Science, v. 294, p. 1671–1677.

Vigny, C., Socquet, A., Rangin, C., Chamot-Rooke, N., Pubellier, M., Bouin, M.-N., Bertrand, G., and Becker, M., 2003, Present-day crustal deformation around Sagaing fault: Journal of Geophysical Research, v. 108, doi: 10.1029/2002JB001999.

Wang, M., Shen, Z.K., Niu, Z., Zhang, Z., Sun, H., Gan, W., Wang, Q., and Ren, Q., 2003a, Contemporary crustal deformation and active blocks model of China mainland (in Chinese): Science in China, v. 33, suppl., p. 21–32.

Wang, Q., Zhang, P., Freymueller, J.T., Bilham, R., Larson, K.M., Lai, X., You, X., Niu, Z., Wu, J., Li, Y., Liu, J., Yang, Z., and Chen, Q., 2001, Present-day crustal deformation in China constrained by Global Positioning System measurements: Science, v. 294, p. 574–577, doi: 10.1126/science.1063647.

Wang, S., Hearn, T., Xu, Z., Ni, J., Yu, Y., and Zhang, X., 2003, Velocity structure of uppermost mantle beneath China continent from Pn tomography: Science in China Series-D, p. 143–150

Wesnousky, S.G., Jones, L.M., Scholz, C.H., and Deng, Q., 1984, Historical seismicity and rates of crustal deformation along the margins of the Ordos block: North China: Bulletin of the Seismological Society of America, v. 74, p. 1767–1783.

Xu, X., and Ma, X., 1992, Geodynamics of the Shanxi rift system, China: Tectonophysics, v. 208, p. 325–340, doi: 10.1016/0040-1951(92)90353-0.

Xu, X., Ma, X., and Deng, Q., 1993, Neotectonic activity along the Shanxi rift system, China: Tectonophysics, v. 219, p. 305–325, doi: 10.1016/0040-1951(93)90180-R.

Xu, Y.G., Menzies, M.A., Thirlwall, M.F., Huang, X., Liu, Y., and Chen, X., 2003, "Reactive" harzburgites from Huinan, NE China: Products of the lithosphere-asthenosphere interaction during lithospheric thinning?: Geochimica et Cosmochimica Acta, v. 67, p. 487–505, doi: 10.1016/S0016-7037(02)01089-X.

Yang, Y., and Liu, M., 2002, Cenozoic deformation of the Tarim Basin and its implications for collisional mountain building in the Tibetan Plateau and the Tian Shan: Tectonics, v. 21, p. 1059, doi: 10.1029/2001TC001300.

Ye, H., Shedlock, K.M., Hellinger, S.J., and Scluter, J.G., 1985, The North China Basin: An example of a Cenozoic rifted intraplate basin: Tectonics, v. 4, p. 153–170.

Ye, H., Zhang, B.T., and Ma, F.Y., 1987, The Cenozoic tectonic evolution of the great North China: Two types of rifting and crustal necking in the great North China and their tectonic implications: Tectonophysics, v. 133, p. 217–227, doi: 10.1016/0040-1951(87)90265-4.

Zeng, Y., 2001, Viscoelastic stress-triggering of the 1999 Hector Mine earthquake by the 1992 Landers earthquake: Geophysical Research Letters, v. 28, p. 3007–3010, doi: 10.1029/2000GL012806.

Zhang, P.Z., Burchfiel, B.C., Molnar, P., Zhang, W., Jiao, D., Deng, Q., Wang, Y.P., Royden, L., and Song, F.M., 1991, Amount and style of late Cenozoic deformation in the Liupan Shan area, Ningxia autonomous region, China: Tectonics, v. 10, p. 1111–1129.

Zhang, P.Z., Deng, Q., Zhang, G.M., Ma, J., Gan, W., Min, W., Mao, F., and Wang, Q., Strong earthquakes and crustal block motion in continental China: Science in China, v. 33, p. 12–20.

Zhang, P.Z., Shen, Z.K., Wang, M., Gan, W., Burgmann, R., Molnar, P., Wang, Q., Niu, Z., Sun, J., Wu, J., Hanrong, S., and Xinzhao, Y., 2004, Continuous deformation of the Tibetan Plateau from global positioning system data: Geology, v. 32, p. 809–812, doi: 10.1130/G20554.1.

Zhang, Y., Mercier, J.L., and Vergely, P., 1998, Extension in the graben systems around the Ordos (China), and its contribution to the extrusion tectonics of south China with respect to Gobi-Mongolia: Tectonophysics, v. 285, p. 41–75, doi: 10.1016/S0040-1951(97)00170-4.

Zhang, Y., Ma, Y., Yang, N., Shi, W., and Dong, S., 2003, Cenozoic extensional stress evolution in North China: Geodynamics, v. 36, p. 591–613, doi: 10.1016/j.jog.2003.08.001.

MANUSCRIPT ACCEPTED BY THE SOCIETY 29 NOVEMBER 2006

The Geological Society of America
Special Paper 425
2007

Seismic-reflection images of the crust beneath the 2001 M = 7.7 Kutch (Bhuj) epicentral region, western India

D. Sarkar[†]
K. Sain
P.R. Reddy
National Geophysical Research Institute, Hyderabad 500 007, India

R.D. Catchings
W.D. Mooney
U.S. Geological Survey, Mail Stop 977, Menlo Park, California 94025, USA

ABSTRACT

Three short (~35 km) seismic-reflection profiles are presented from the region of the 2001 Mw = 7.7 Bhuj (western India) earthquake. These profiles image a 35–45-km-thick crust with strong, near-horizontal reflections at all depths. The thickness of the crust increases by 10 km over a distance of ~50 km from the northern margin of the Gulf of Kutch to the earthquake epicenter. Aftershocks of the Bhuj earthquake extend to a depth of 37 km, indicating a cold, brittle crust to that depth. Our results show that all of these aftershocks are contained within the crust. Furthermore, there is no evidence for offsets in the crust-mantle boundary associated with deep (mantle) faulting. The existence of a thick (~45 km) and highly reflective crust at the epicentral zone may be indicative of crustal thickening due to the compressive regime of the past 55 m.y. Alternatively, this crustal thickening could be attributable to magmatic intrusions that date back to Mesozoic rifting associated with the breakup of Gondwanaland.

Keywords: India, seismic reflection, crustal thickness, Bhuj earthquake.

INTRODUCTION

The 26 January 2001, Mw 7.7 Bhuj, western India, earthquake had a focal depth of 21 ± 4 km (Kayal et al., 2002; Antolik and Dreger, 2003; Mandal et al., 2004). Aftershocks outlined an ENE-trending S-dipping reverse fault (45°–50° dip) extending to a depth of 37 km (Raphael and Bodin, 2002; Kayal et al.,

2002; Mishra and Zhao, 2003) with no obvious surface expression (Chandrasekhar et al., 2004; Rastogi, 2004). The largest amount of fault slip (10 m) was located close to the hypocenter and had an area of ~10 km × 20 km (Negishi et al., 2002). The area of the rupture zone (~40 km × 40 km) appears to be unusually small for a Mw 7.7 event, and suggests a high (13–25 MPa) static stress drop event (Negishi et al., 2002), similar to other

[†]E-mail: dipankars@rediffmail.com.

Sarkar, D., Sain, K., Reddy, P.R., Catchings, R.D., and Mooney, W.D., 2007, Seismic-reflection images of the crust beneath the 2001 M = 7.7 Kutch (Bhuj) epicentral region, western India, *in* Stein, S., and Mazzotti, S., ed., Continental Intraplate Earthquakes: Science, Hazard, and Policy Issues: Geological Society of America Special Paper 425, p. 319–327, doi: 10.1130/2007.2425(20). For permission to copy, contact editing@geosociety.org. ©2007 The Geological Society of America. All rights reserved.

intraplate earthquakes (Schulte and Mooney, 2005). A tomographic study of the aftershocks showed a distinctly high P-wave velocity and low S-wave velocity (and thus a high Poisson's ratio) at the lower-crustal hypocentral zone (Kayal et al., 2002). Mishra and Zhao (2003) suggested that the 2001 Bhuj earthquake was characterized by high crack density, saturation rate, and porosity in the depth range of 23–28 km. According to Mishra and Zhao (2003), these mid-crustal anomalies indicate the presence of a fluid-filled, fractured rock matrix, and would thus constitute a local zone of weakness.

The 2001 Bhuj earthquake occurred ~130 km southeast of the epicenter of the 1819 Allah Band earthquake (Mw 7.8), which also ruptured on a reverse fault (Fig. 1; Rajendran and Rajendran, 2001). Based on paleoliquefaction studies, Tuttle et al. (2001) reported that another similar-sized event occurred in the Allah Band region 800–1000 yr ago. Kaila et al. (1972) computed a recurrence interval of 200 yr for Kutch earthquakes, which is comparable to the time interval between the 1819 and

2001 events. In addition, a smaller (Mw 6.0) earthquake occurred in 1956 near Anjar (Fig. 1). This event also had a reverse mechanism (Chung and Gao, 1995), which indicates a prevailing compression-dominated regime.

Kutch seismicity, like that of the New Madrid seismic zone of the eastern United States, is typified by a relatively short (200–500 yr) recurrence interval of major earthquakes, which does not readily conform to the statistics inferred from the rate of microearthquake activity (Mandal and Rastogi, 2005). However, Stein and Newman (2004) argued that the calculated recurrence interval of the larger New Madrid seismic zone events is consistent with microseismic activity if the magnitudes of the larger events are not overestimated (Stein and Newman, 2004). The seismicity of the New Madrid seismic zone is considered to be a classic example of intraplate seismicity related to a reactivated rift (Mooney et al., 1983; Chiu et al., 1992; Liu and Zoback, 1997; Newman et al., 1999; Schulte and Mooney, 2005; Li et al., 2007; McKenna et al., 2006). It has been hypothesized that the North

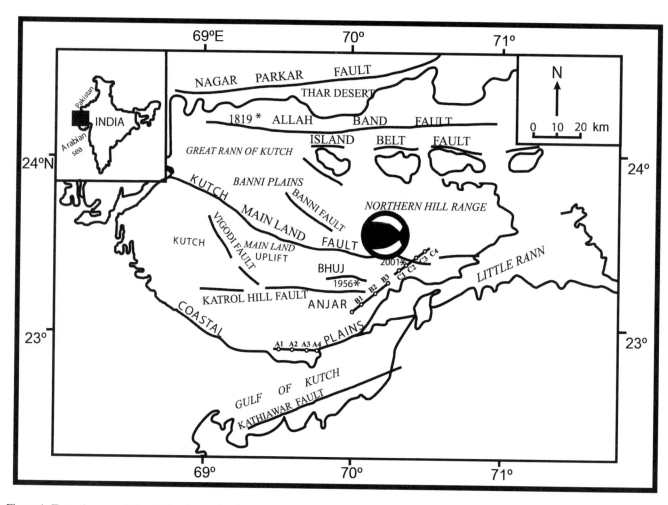

Figure 1. Tectonic map of Kutch showing major fault systems (adapted from Malik et al., 2000). The epicenters of the 2001 Bhuj earthquake (23.36°N, 70.34°E) as well as the two earlier events of 1819 and 1956 are indicated by asterisks. The focal mechanism of the 2001 earthquake is also shown. Three seismic lines (A, B, C) are shown as thick solid lines with corresponding shot points (open circles). Inset shows the study area.

American ambient stress field reactivates faults associated with the Reelfoot rift (Grana and Richardson, 1996).

Although the Kutch region has been rather well-explored by the oil industry, most studies have been limited to the shallow sedimentary formations and the Deccan volcanics overlying the granitic basement, and we still lack a well-determined crustal velocity model for this region. We have some knowledge of the geologically mapped faults but insufficient information regarding the disposition of blind faults in the crust. Such data are necessary in order to differentiate between the crustal structure of the seismogenic and nonseismogenic areas of the Kutch region.

In this study, we seek to determine the deep seismic structure of the crust at and adjacent to the epicenter of the Bhuj earthquake. Data from three seismic profiles within the region have been processed to determine crustal reflectivity patterns and to constrain the lateral variations in the crust that may be significant for understanding Kutch seismicity. With these objectives, we have depth-migrated seismic-refraction data collected in 1997 (NGRI, 2000) to produce a depth section similar to multichannel seismic data.

TECTONIC SETTING OF THE KUTCH REGION

The Kutch region forms a crucial tectonic segment of the western margin of the Indian subcontinent, and it falls within a zone of high seismic potential. The Kutch rift basin is flanked by the Nagar Parkar fault in the north and the Kathiawar fault in the south (Fig. 1). The Kutch basin contains several major E-W–trending faults, such as the Kutch Mainland fault on the northern fringe and the Katrol Hill fault in the central part of the Kutch Mainland uplift (Biswas, 1987; Malik et al., 2000).

The Kutch rift basin owes its origin to the Mesozoic breakup of Gondwanaland. Normal faulting within this extensional regime produced a number of prominent horsts and grabens. A change from rift-related extension to N-S compression probably occurred around 55 ± 1 Ma, subsequent to the collision of India with Eurasia (Ni and Barazangi, 1984; Gaetani and Garzanti, 1991; Klootwijk et al., 1992; Garzanti et al., 1996; Rowley, 1996; Leech et al., 2005). Most focal mechanism studies conducted in this area indicate E-W–striking reverse faulting and generally reflect N-S compressive stress (Rajendran and Rajendran, 2001). This region is subject to a high and regional stress field because of its proximity to the Indian-Arabian-Asian triple junction, located ~500 km to the west (Gupta et al., 2001), and is also affected by the plate-boundary zone (Li et al., 2002; Stein et al., 2002). The huge sediment load in the neighborhood of the Indus Delta is an additional source of stress on the Indian plate (Seeber et al., 2001). Since this is an active process of sediment loading, this stress has the opposite sign of the deglaciation hypothesis for the New Madrid seismic zone (Grollimund and Zoback, 2001).

REPROCESSING OF 1997 SEISMIC-PROFILING DATA

Seismic-refraction data were collected using two 60 channel Texas Instruments model DFS-V recording systems (NGRI,

2000). The geophone spacing was 100 m, and the shot interval was 7–8 km. Holes drilled to a depth of 20–30 m were used to detonate 50–500 kg explosives, depending upon the shot-receiver distances. The objective of the seismic survey was to delineate the basement configuration and to identify buried sediments that are often masked under higher-velocity Deccan volcanics. The refraction data, however, indicated some basement faults, either directly observable in the form of fault-plane reflections or in terms of an abrupt change in the basement depth. Here, we reprocessed the data to image the middle and lower crust.

The data presented here consist of three line segments: A, B, and C with 4, 3, and 4 shots, respectively (Fig. 1). We reprocessed these profiles using standard two-dimensional (2-D) Kirchhoff prestack depth migration from commercial software package ProMAX installed on a Sun Workstation. Prestack depth migration has become an important tool for obtaining quality images from seismic data acquired in geologically complex regions (Yilmaz and Doherty, 1987; Milkereit et al., 1990; Lafond and Levander, 1995; Audebert et al., 1997; Zelt et al., 1998; Pilipenko et al., 1999). The depth migration consists of an estimation of the velocity-depth model and the appropriate Green's function, which are used to relate times and amplitudes from each surface location to a region of subsurface points. Seismic traveltimes of selected phases are generally used to derive a large-wavelength velocity model, which, in turn, is used in prestack depth migration to derive the fine-scale structural image of a region.

For depth migration, we require an appropriate velocity model. Refraction velocity models derived from the same data set are only available for the basement (NGRI, 2000). In the absence of deeper refraction data, we calculated the subbasement models from the corresponding shot gathers. Application of automatic gain control and suitable band-pass filters enhanced the signals, which allowed us to identify intracrustal and Moho reflections.

Not all shot gathers allowed visual identification of deep reflection phases. The granitic basement, the Moho, and a third intracrustal phase were occasionally observed. We used the data from shot C3 to calculate the velocity-depth section for line C. Even though some intrabasin reflections were observed, they were used for this modeling. A mean sedimentary velocity, as obtained from earlier refraction analysis, was applied as far down as the basement level. The traveltimes of these phases were read and inverted to build a velocity model using a damped least-square technique (Sain and Kaila, 1994). The derived velocity-depth functions (Fig. 2) were sampled on a 200 m (horizontal) by 200 m (vertical) grid and used in the prestack depth migration for imaging crustal features. Since the migration procedure is sensitive to the crustal velocities used, particularly overestimations, a discontinuous velocity increase across the Moho was purposefully ignored. Instead, a smooth increase of crustal velocities was applied to avoid possible major distortions in the migrated section. On shot gathers from lines A and B, it was difficult to observe any clear and consistent phases. Therefore, we used the lower-crustal velocities for line C for the migration of data from lines A and B. Care was taken to process only subbasement

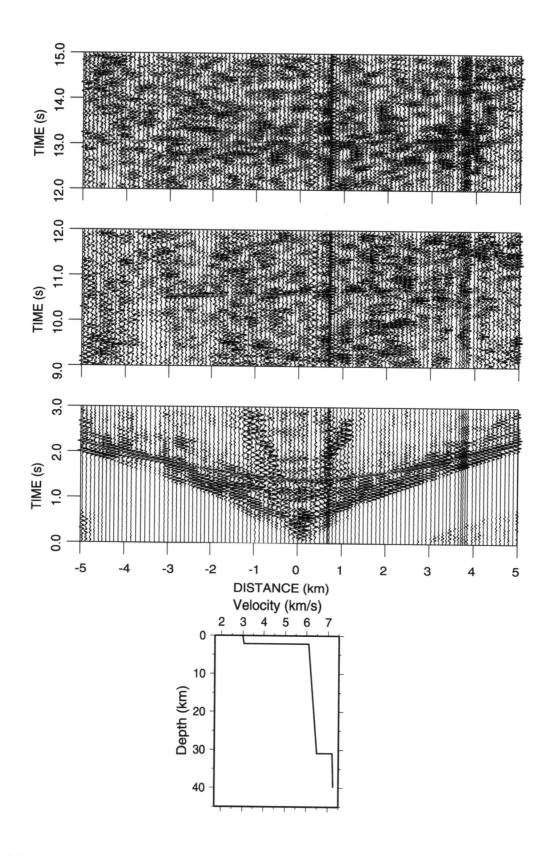

Figure 2. Record section (in three time windows) showing intracrustal reflections for shot point C3, along with corresponding velocity model that was used in prestack depth migration.

reflection data because the upper sectional details are proprietary to natural resource exploration.

RESULTS AND DISCUSSION

The nearly 35-km-long southern coastal segment (line A) depicts strong reflectivity throughout the entire crust, starting from the reflections at a depth of 5–6 km (Fig. 3). A clear W-dipping (20–25 km depth) horizon with a strong band of reflections marks the top of the prominent lower crust. This is noted in Figure 3 at 25 km depth from ~2 km east of shot point (SP) A1 and continues to between SPs A2 and A3. The Moho, the bottom of the lower-crustal reflective zone, shows a high-reflectivity signature and is nearly horizontal along the entire profile at an average depth of 35 km. The crust is evidently thinner here than most other crustal depth estimates (40–42 km) for the western coast of India (Sarkar et al., 2001, 2003). There is noticeable but relatively incoherent reflectivity in the mantle at a depth of 10–15 km below the Moho.

The seismic section along line B, which crosses a part of the E-W–trending Kutch Mainland Uplift, depicts reflectivity at depths of ~5, 10, 18, and 27–29 km, the latter of which represents the top of the lower crust (Fig. 4). N-dipping subhorizontal Moho reflection bands are weakly observed at 37–40 km depth. There is also a strong subcrustal reflection horizon at 50 km. The ~40-km-thick crust of the northern portion of the profile is consistent with the tectonic model of Stein et al. (2002), and it has been host to a number of historical earthquakes. A blind basement fault, nearly coinciding with the geologically known Kutch Hills fault, was inferred from earlier refraction analysis (NGRI,

2000) at ~10 km north of the town of Anjar. This fault could be associated with the 1956 Mw = 6.0 Anjar earthquake, a rupture that showed 1 m of offset along the fault and was the most severe earthquake prior to the 2001 Bhuj event.

Seismic line C is located ~15 km east of the 2001 Bhuj earthquake epicenter. This profile shows strong reflectivity in the entire crust, with reflection horizons at average depths of 5, 10, and 19 km (Fig. 5). The reflection Moho is clearly imaged as the base of the reflective crust, and it has a significant northward dip, from 38 to 45 km, within a distance of 35 km. A seismic reflection within the upper mantle is visible at a depth of 56–58 km near the southern end of the profile. This reflection gets partially obliterated due to the diffused and disturbed reflectivity farther north. The central portion of the profile, between distances of 10 and 20 km, shows a significant increase in crustal reflectivity, while the Moho appears as a thick reflection band.

The 2001 Bhuj earthquake had a S-dipping fault plane and a focal depth of 21 ± 4 km, and it occurred very close to the southern edge of increased reflectivity. The proximity of the fault plane with this reflective portion of the crust may not be coincidental. The laterally varying crustal reflectivity pattern exhibits enhanced reflectivity of the middle and lower crust beneath the epicenter of the 2001 earthquake. We hypothesize that the reflective zone consists of anastomosing shear zones created within a highly compressive stress regime (Lueschen et al., 1987; Hamilton, 1989; Smithson and Johnson, 1989; Mooney and Meissner, 1992). This observation of high reflectivity, along with the observance of a high Poisson's ratio (Kayal et al., 2002; Mishra and Zhao, 2003) in the hypocentral region, may also be indicative of enhanced fluid concentrations in a highly sheared zone (Mishra

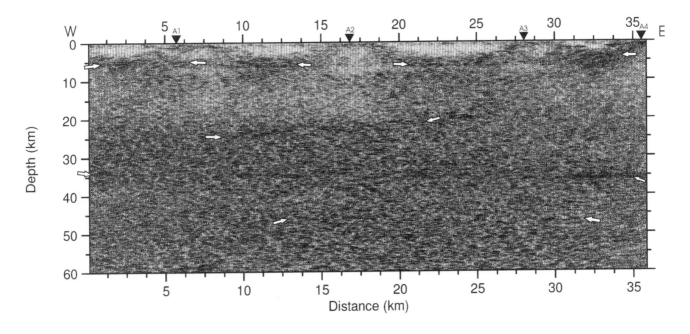

Figure 3. Prestack depth-migrated section for line A (Fig. 1). Prominent reflection boundaries are indicated by white arrows. Shot locations are indicated by inverted triangles along the axis. The top of the reflective lower crust is at a depth of ~22 km, and the Moho is at a depth of 35 km.

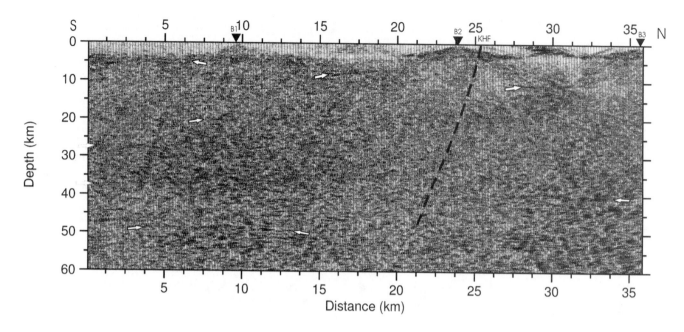

Figure 4. Prestack depth-migrated section for line B (Fig. 1). Prominent reflection boundaries are indicated by white arrows. Shot locations are indicated by inverted triangles along the axis. The Katrol Hill fault (KHF) is indicated by a thick dashed line but is not clearly imaged by the data. The Moho is at a depth of 38–40 km.

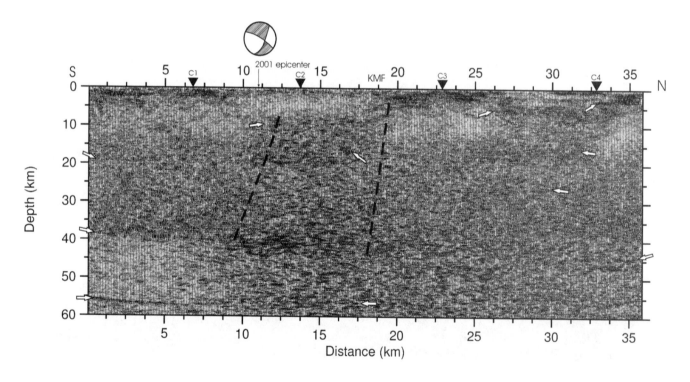

Figure 5. Prestack depth-migrated section for line C (Fig. 1). Prominent reflection boundaries are indicated by white arrows. Shot locations are indicated by inverted triangles along the axis. The hypocenter of the 2001 Bhuj earthquake along with the focal mechanism, as viewed on a vertical cross section, is superimposed on the depth section. Though not clearly imaged by the data, the Kutch Mainland fault (KMF) and the fault associated with the 2001 Bhuj earthquake are indicated by thick dashed lines. The Moho is at a depth of 38–45 km. A strong mantle reflector is imaged at a depth of 55 km.

and Zhao, 2003; Mandal and Rastogi, 2005). High fluid pressures would have the effect of reducing the effective normal stress on the Bhuj fault plane and would thus reduce the shear resistance of the fault (Sibson, 1994, 2002).

Figure 6 shows all three seismic-reflection profiles in three dimensions, along with the most prominent reflectors. The thinner (35 km) crust in the aseismic southern crustal profile (A) and the thick (45 km) crust in the northern segment (C) suggest that the Kutch Mainland uplift is associated with a crustal root. One possible tectonic explanation for such a thickened crust under the epicentral region (the Kutch Mainland uplift) is the prevailing N-S compression generated by the India-Eurasia collision. Alternatively, the crust may have been thickened by igneous intrusions during rift formation, signatures of which are visible in the form of strong sub-Moho reflectors. The aftershocks from the Bhuj earthquake and the location of the Kutch Mainland fault are also shown.

CONCLUSIONS

Three 35-km-long seismic-reflection profiles from the region of the 2001 M = 7.7 Bhuj earthquake provide images of the regional crustal structure and depth to the Moho. A zone of high reflectivity is observed in the lower crust starting at a depth of ~22 km. The crust-mantle boundary, defined as the base of the zone of strong reflectivity, deepens from 35 km at the coast to 45 km in the immediate epicentral region. Thus, the crustal thickness in the Bhuj epicentral region is approximately equal to the average value for continental India (40–44 km; Kaila and Sain, 1997). This observation contradicts the suggestion that seismic activity in the Kutch region is due to thin, rifted crust, such as that found along the East African Rift (Mechie et al., 1994; Mooney and Christensen, 1994). The 45 km crustal thickness compares favorably with the 42–46 km crustal thickness of the New Madrid seismic zone. Furthermore, the 22 km depth to the top of the reflective lower crust in the Bhuj region agrees well with the 26 km depth of the 7.3 km/s lower crust beneath New Madrid (Mooney et al., 1983). Thus, the Bhuj region appears to have a rifted crustal structure that is comparable to the New Madrid rift, but not the East African Rift.

High-angle reverse faults and near-vertical strike-slip faults are rarely imaged on seismic-reflection profiles. It is therefore not surprising that these data do not image the fault associated with the 2001 Bhuj earthquake (Biswas, 2005). The geometry of this

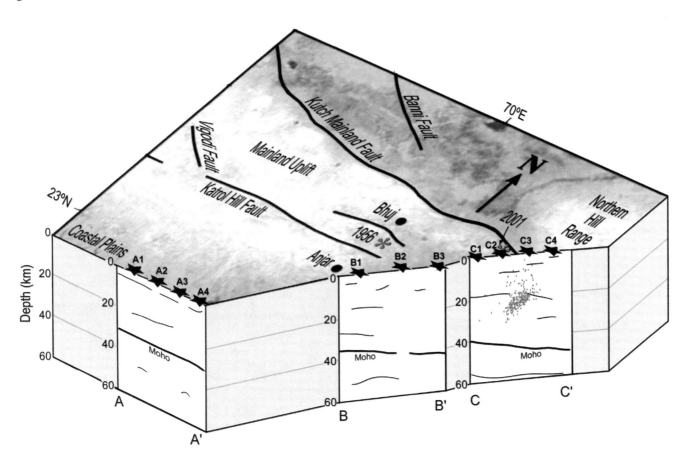

Figure 6. Pie-slice diagram of the Bhuj epicentral region. The three seismic-reflection profiles (A, B, C) are shown along with prominent reflectors. The aftershocks from the Bhuj earthquake have also been plotted as red dots, showing the location of the Kutch Mainland fault.

fault is best defined by the aftershocks, as reported by Kayal et al. (2002), Negishi et al. (2002), and Mishra and Zhao (2003). These aftershocks reach depths as great as 37 km, and our seismic profiles demonstrate that all of this seismicity is contained within the crust. However, there is no evidence on our seismic profiles for offsets in the crust-mantle boundary. On the contrary, the Moho appears to be flat, with a smooth dip from the coast to the interior of the continent.

The presence of seismic reflections some 10–15 km below the Moho is an unexpected and relatively rare observation on such profiles. We interpret these sub-Moho reflections as either mantle shear zones or mafic igneous intrusions that are associated with the rifting that occurred as India separated from Africa in the late Mesozoic.

The crustal properties associated with the epicentral region of the 2001 Kutch earthquake are significant in light of the regional seismicity. Our seismic image allows us to distinguish between the seismic structure at and adjacent to the 2001 Bhuj earthquake epicenter. Furthermore, a high Poisson's ratio, along with a laterally varying crustal velocity structure, has previously been linked to fluid-filled and highly fractured fault segments (Kayal et al., 2002; Mishra and Zhao, 2003), which are a plausible explanation for the high seismic moment release in the Kutch region.

ACKNOWLEDGMENTS

We thank the directors of the National Geophysical Research Institute and U.S. Geological Survey for their support of collaborative work and approval of this publication. We furthermore acknowledge I.M. Artemieva, G.S. Chulick, S. Detweiler, M. Coble, and S.T. McDonald for critical reviews of the manuscript. D. Eaton, H. Sato, S. Seth, S. Mazzotti, and an anonymous reviewer have provided constructive criticisms to improve this paper. Oil Industry Development Board (India) sponsored the Crustal Structure Studies Project (P.I.P.R. Reddy) of NGRI for undertaking the shallow seismic-refraction survey in Kutch to delineate its basinal structure and basement configuration. We thank the field party chief B. Rajendra Prasad and other members of the C.S.S. field party for data collection. Imaging of subbasement deep structure of two crustal segments was carried out under an India-U.S. bilateral collaboration between NGRI and the U.S. Geological Survey.

REFERENCES CITED

Antolik, M., and Dreger, D., 2003, Rupture process of the 26 January 2001 Mw 7.6 Bhuj, India, earthquake from teleseismic broadband data: Bulletin of the Seismological Society of America, v. 93, p. 1235–1248, doi: 10.1785/0120020142.

Audebert, F., Nichols, D., Rekdal, T., Biondi, B., Lumley, D., and Urdaneta, H., 1997, Imaging complex geologic structure with single-arrival Kirchhoff prestack depth migration: Geophysics, v. 62, p. 1533–1543, doi: 10.1190/1.1444256.

Biswas, S.K., 1987, Regional tectonic framework, structure, and evolution of the western marginal basins of India: Tectonophysics, v. 135, p. 307–327, doi: 10.1016/0040-1951(87)90115-6.

Biswas, S.K., 2005, A review of structure and tectonics of Kutch Basin, western India, with special reference to earthquakes: Current Science, v. 88, p. 1592–1600.

Chandrasekhar, D.V., Mishra, D.C., Singh, B., Vijayakumar, V., and Bürgmann, R., 2004, Source parameters of the Bhuj earthquake, India of January 26, 2001, from height and gravity changes: Geophysical Research Letters, v. 31, p. L19608, doi: 10.1029/2004GL020768.

Chiu, J.M., Johnston, A.C., and Yang, Y.T., 1992, Imaging of the active faults of the central New Madrid seismic zone using PANDA array data: Seismological Research Letters, v. 63, p. 375–393.

Chung, W.P., and Gao, H., 1995, Source parameters of the Anjar earthquake of July 21, 1956, India, and its seismotectonic implications for the Kutch rift basin: Tectonophysics, v. 242, p. 281–292, doi: 10.1016/0040-1951-(94)00203-L.

Gaetani, M., and Garzanti, E., 1991, Multicyclic history of the northern India continental margin (northwestern Himalaya): American Association of Petroleum Geologists Bulletin, v. 75, p. 1427–1446.

Garzanti, E., Critelli, S., and Ingersoll, R.V., 1996, Paleogeographic and paleotectonic evolution of the Himalayan range as reflected by detrital modes of Tertiary sandstones and modern sands (Indus transect, India and Pakistan): Geological Society of America Bulletin, v. 108, p. 631–642, doi: 10.1130/0016-7606(1996)108<0631:PAPEOT>2.3.CO;2.

Grana, J.P., and Richardson, R.M., 1996, Tectonic stress within the New Madrid seismic zone: Journal of Geophysical Research, v. 101, p. 5445–5458, doi: 10.1029/95JB03255.

Grollimund, B., and Zoback, M.D., 2001, Did glaciation trigger intraplate seismicity in the New Madrid seismic zone?: Geology, v. 29, p. 175–178, doi: 10.1130/0091-7613(2001)029<0175:DDTISI>2.0.CO;2.

Gupta, H.K., Purnachandra Rao, N., Rastogi, B.K., and Sarkar, D., 2001, The deadliest intraplate earthquake: Science, v. 291, p. 2101–2102, doi: 10.1126/science.1060197.

Hamilton, W.B., 1989, Crustal processes of the United States, in Pakiser, L.C., and Mooney, W.D., eds., Geophysical Framework of the Continental United States: Boulder, Colorado, Geological Society of America Memoir 172, p. 760–780.

Kaila, K.L., and Sain, K., 1997, Variation of crustal velocity structure in India as determined from DSS studies and their implications on regional tectonics: Journal of the Geological Society of India, v. 49, p. 395–407.

Kaila, K.L., Gaur, V.K., and Narain, H., 1972, Quantitative seismicity maps of India: Bulletin of the Seismological Society of America, v. 62, p. 1119–1132.

Kayal, J.R., Zhao, D., Mishra, O.P., De, R., and Singh, O.P., 2002, The 2001 Bhuj earthquake: Tomographic evidence for fluids at the hypocenter and its implications for rupture nucleation: Geophysical Research Letters, v. 29, p. 2152–2155, doi: 10.1029/2002GL015177.

Klootwijk, C.T., Gee, J.S., Peirce, J.W., Smith, G.M., and McFadden, P.L., 1992, An early India-Asia contact: Paleomagnetic constraints from Ninetyeast Ridge: Geology, v. 20, p. 395–398, doi: 10.1130/0091-7613(1992)020<0395:AEIACP>2.3.CO;2.

Lafond, C.F., and Levander, A., 1995, Migration of wide-aperture onshore-offshore seismic data, central California: Seismic images of late stage subduction: Journal of Geophysical Research, v. 100, p. 22,231–22,243, doi: 10.1029/95JB01968.

Leech, M., Singh, S., Jain, A.K., Klemperer, S.L., and Manickavasagam, R.M., 2005, The onset of India-Asia continental collision: Early, steep subduction required by the timing of UHP metamorphism in the western Himalaya: Earth and Planetary Science Letters, v. 234, no. 1–2, p. 83–97, doi: 10.1016/j.epsl.2005.02.038.

Li, Q., Liu, M., and Yang, Y., 2002, The 01/26/2001 Bhuj earthquake: Intraplate or interplate?, in Stein, S., and Freymueller, J., eds., Plate Boundary Zones: Washington, D.C., American Geophysical Union Geodynamic Monograph 30, p. 255–264.

Li, Q., Liu, M., Zhang, Q., and Sandoval, E., 2007, Stress evolution and seismicity in the central-eastern USA: Insights from geodynamic modeling, in Stein, S., and Mazzotti, S., eds., Continental Intraplate Earthquakes: Geological Society of America Special Paper 425, doi: 10.1130/2007.2425(11).

Liu, L., and Zoback, M.D., 1997, Lithospheric strength and intraplate seismicity in the New Madrid seismic zone: Tectonics, v. 16, p. 585–595, doi: 10.1029/97TC01467.

Lueschen, E., Wenzel, F., Sandmeier, K.J., Menges, D., Ruehl, T., Stiller, M., Janoth, W., Keller, F., Soellner, W., Thomas, R., Krohe, A., Stenger, R., Fuchs, K., Wilhelm, H., and Eisbacher, G., 1987, Near-vertical and wide-angle seismic surveys in the Black Forest, SW Germany: Journal of Geophysics, v. 62, p. 1–30.

Malik, J.N., Sohoni, P.S., Merh, S.S., and Karanth, R.V., 2000, Palaeoseismology and neotectonism of Kachchh, western India, *in* Okumura, K., Goto, H., and Takada, K., eds., Active Fault Research for the New Millennium: Proceedings of the Hokudan International Symposium and School on Active Faulting: Hyogo, Japan, Hokudan Co. Ltd., p. 251–259.

Mandal, P., and Rastogi, B.K., 2005, Self-organized fractal seismicity and b value of aftershocks of the 2001 Bhuj earthquake in Kutch (India): Pure and Applied Geophysics, v. 162, no. 1, p. 53–72, doi: 10.1007/s00024-004-2579-1.

Mandal, P., Rastogi, B.K., Satyanarayana, V.S., and Kousalya, M., 2004, Results from local earthquake velocity tomography: Implications toward the source process involved in generating the 2001 Bhuj earthquake in the lower crust beneath Kachchh (India): Bulletin of the Seismological Society of America, v. 94, no. 2, p. 633–649, doi: 10.1785/0120030056.

McKenna, J., Stein, S., and Stein, C., 2006, Is the New Madrid seismic zone hotter and weaker than its surroundings?, *in* Stein, S., and Mazzotti, S., eds., Continental Intraplate Earthquakes: Geological Society of America Special Paper 425, doi: 10.1130/2007.2425(12).

Mechie, J., Keller, G.R., Prodehl, C., Gaciri, S., Braile, L.W., Mooney, W.D., Gajewski, D., and Sandmeier, K.J., 1994, Crustal structure beneath the Kenya Rift from axial profile data, *in* Prodehl, C., Keller, G.R., and Khan, M.A., eds., Crustal and Upper Mantle Structure of the Kenya Rift: Tectonophysics, v. 236, p. 179–199.

Milkereit, B., Epili, D., Green, A.G., Mereu, R.F., Morel-a-l'Huissier, P., 1990, Migration of wide-angle seismic reflection data from the Grenville Front in Lake Huron: Journal of Geophysical Research, v. 95, p. 10,987–10,998.

Mishra, O.P., and Zhao, D., 2003, Crack density, saturation rate and porosity at the 2001 Bhuj, India, earthquake hypocenter: A fluid-driven earthquake?: Earth and Planetary Science Letters, v. 212, no. 3–4, p. 393–405, doi: 10.1016/S0012-821X(03)00285-1.

Mooney, W.D., and Christensen, N.I., 1994, Composition of the crust beneath the Kenya Rift: Tectonophysics, v. 236, p. 391–408, doi: 10.1016/0040-1951(94)90186-4.

Mooney, W.D., and Meissner, R., 1992, Multi-genetic origin of crustal reflectivity: A review of seismic reflection profiling of the continental lower crust and Moho, *in* Fountain, D.M., Arculus, R., and Kay, R.W., eds., The Lower Continental Crust: Amsterdam, Elsevier, p. 45–79.

Mooney, W.D., Andrews, M.C., Ginzburg, A., Peters, D.A., and Hamilton, R.M., 1983, Crustal structure of the northern Mississippi Embayment and a comparison with other continental rift zones: Tectonophysics, v. 94, p. 327–348, doi: 10.1016/0040-1951(83)90023-9.

National Geophysical Research Institute (NGRI), 2000, Integrated geophysical studies for hydrocarbon exploration, Kutch, India: Hyderabad, India, Technical Report NGRI-2000-EXP-296, 195 p.

Negishi, H., Mori, J., Sato, T., Singh, R., Kumar, S., and Hirata, N., 2002, Size and orientation of the fault plane for the 2001 Gujarat, India, earthquake (Mw7.7) from aftershock observations: A high stress drop event: Geophysical Research Letters, v. 29, no. 20, p. 1949–1952, doi: 10.1029/2002GL015280.

Newman, A., Stein, S., Weber, J., Engeln, J., Mao, A., and Dixon, T., 1999, Slow deformation and low seismic hazard at the New Madrid seismic zone: Science, v. 284, p. 619–621, doi: 10.1126/science.284.5414.619.

Ni, J., and Barazangi, M., 1984, Seismotectonics of the Himalayan collision zone: Geometry of the underthrusting Indian plate beneath the Himalaya: Journal of Geophysical Research, v. 89, no. B2, p. 1147–1163.

Pilipenko, V.N., Pavlenkova, N.I., and Luosto, U., 1999, Wide-angle reflection migration technique with an example from the POLAR profile (northern Scandinavia): Tectonophysics, v. 308, p. 445–457, doi: 10.1016/S0040-1951(99)00144-4.

Rajendran, C.P., and Rajendran, K., 2001, Characteristics of deformation and past seismicity associated with the 1819 Kutch earthquake, northwestern India: Bulletin of the Seismological Society of America, v. 91, p. 407–426, doi: 10.1785/0119990162.

Raphael, A., and Bodin, P., 2002, Relocating aftershocks of the 26 January 2001, Bhuj earthquake in western India: Seismological Research Letters, v. 73, no. 3, p. 417–418.

Rastogi, B.K., 2004, Damage due to the Mw 7.7 Kutch, India, earthquake of 2001: Tectonophysics, v. 390, no. 1–4, p. 85–103, doi: 10.1016/j.tecto.2004.03.030.

Rowley, D.B., 1996, Age of initiation of collision between India and Asia: A review of stratigraphic data: Earth and Planetary Science Letters, v. 145, p. 1–13, doi: 10.1016/S0012-821X(96)00201-4.

Sain, K., and Kaila, K.L., 1994, Inversion of wide angle seismic reflection travel times with a damped least squares technique: Geophysics, v. 59, p. 1735–1744, doi: 10.1190/1.1443560.

Sarkar, D., Chandrakala, K., Padmavathi Devi, P., Sridhar, A.R., Sain, K., and Reddy, P.R., 2001, Crustal velocity structure of western Dharwar craton, south India: Journal of Geodynamics, v. 31, p. 227–241, doi: 10.1016/S0264-3707(00)00021-1.

Sarkar, D., Ravikumar, M., Saul, J., Kind, R., Raju, P.S., Chadha, R.K., and Shukla, A.K., 2003, A receiver function perspective of the Dharwar craton (India) crustal structure: Geophysical Journal of the Interior, v. 154, p. 205–211.

Schulte, S.M., and Mooney, W.D., 2005, An updated global earthquake catalogue for stable continental regions: Reassessing the correlation with ancient rifts: Geophysical Journal of the Interior, v. 161, p. 707–721, doi: 10.1111/j.1365-246X.2005.02554.x.

Sibson, R.H., 1994, Crustal stress, faulting, and fluid flow, *in* Parnell, J., ed., Geofluids: Origins, Migration and Evolution of Fluids in Sedimentary Basins: Geological Society of London Special Publication 78, p. 69–84.

Sibson, R.H., 2002, Geology of the crustal earthquake source, *in* Lee, W.H.K., Kanamori, H., Jennings, P.C., and Kisslinger, C., eds., International Handbook of Earthquake and Engineering Seismology, Part A: Amsterdam, Academic Press, p. 455–473.

Smithson, S.B., and Johnson, R.A., 1989, Crustal structure of the western US based on reflection seismology, *in* Pakiser, L.C., and Mooney, W.D., eds., The Geophysical Framework of the Continental United States: Geological Society of America Memoir 172, p. 345–368.

Stein, S., and Newman, A., 2004, Characteristic and uncharacteristic earthquakes as possible artifacts: Applications to the New Madrid and Wabash seismic zones: Seismological Research Letters, v. 75, p. 173–187.

Stein, S., Sella, G., and Okal, E., 2002, The January 26, 2001, Bhuj earthquake and the diffuse western boundary of the Indian plate, *in* Stein, S., and Freymueller, J., eds., Plate Boundary Zones: Washington, D.C., American Geophysical Union Geodynamics Monograph 30, p. 243–254.

Tuttle, M., Johnston, A., Patterson, G., Tucker, K., Rajendra, C.P., Rajendran, K., Thakkar, M., and Schweig, E., 2001, Liquefaction induced by the 2001 Republic Day earthquake, India: Seismological Research Letters, v. 72, no. 3, p. 397.

Yilmaz, O., and Doherty, S.M., 1987, Seismic Data Processing: Tulsa, Oklahoma, Society of Exploration Geophysicists, 440 p.

Zelt, B.C., Talwani, M., and Zelt, C.A., 1998, Prestack depth migration of dense wide-angle seismic data: Tectonophysics, v. 286, p. 193–208, doi: 10.1016/S0040-1951(97)00265-5.

Manuscript Accepted by the Society 29 November 2006

The Geological Society of America
Special Paper 425
2007

Challenges in seismic hazard analysis for continental interiors

Gail M. Atkinson

Department of Earth Sciences, University of Western Ontario, London, Ontario, N6A 5B6, Canada

ABSTRACT

Seismic hazard zoning maps are a relatively simple and transparent consequence of the patterns of historical seismicity. Over time, we have refined our understanding of where and why earthquakes occur in continental interiors, and we have improved our characterization of the resulting ground motions and their probabilities. We have begun to understand the important role of uncertainty in seismic hazard analysis. However, there are still significant shortcomings in our treatment of uncertainty that are particularly pronounced for midplate regions. The same lack of knowledge that causes our uncertainty of the hazard also prevents us from accurately quantifying that uncertainty. Are the resulting seismic hazard maps reasonable for use in building codes, in light of this uncertainty?

In order to address this, a simple probability-based areal test of seismic hazard maps can be conducted that employs correlations between modified Mercalli intensity and ground-motion amplitude, in combination with the historic record of seismicity over the past 200 yr. This test shows that hazard maps are reflecting the potential for repeats of the largest historical events in areas that currently experience moderate seismicity. In any 50 yr time period, these areas will experience or exceed the expected area for damaging ground motions if there has been at least one event of M > 6.7 anywhere in eastern North America (or two events of M 6.5). If there have not been any large events, the damage area will be much less than predicted.

Keywords: seismic hazard, continental interiors, seismicity, ground motions.

HISTORICAL OVERVIEW OF HAZARD ANALYSIS

Seismic hazard analysis has been an element of good engineering design practice in modern countries for many decades; it is an integral component of building codes and standards for design of critical structures such as dams, offshore structures, and nuclear power plants. This discussion paper begins with a historical review of hazard analysis and its incorporation into building codes and then discusses some current issues that affect our ability to characterize seismic hazard in continental interiors.

Causes and Distribution of Seismic Hazard

Over 90% of the world's seismicity occurs within relatively narrow bands where two or more of the tectonic plates that make up Earth's lithosphere slide past or collide with each other. In plate-margin regions, seismotectonic processes are relatively well understood. Strain energy is accumulated by the relative motion of the plates, and it is released by seismic slip along plate-boundary faults. For crustal earthquakes (e.g., those along the San Andreas fault system), the faulting often ruptures

Atkinson, G.M., 2007, Challenges in seismic hazard analysis for continental interiors, *in* Stein, S., and Mazzotti, S., ed., Continental Intraplate Earthquakes: Science, Hazard, and Policy Issues: Geological Society of America Special Paper 425, p. 329–344, doi: 10.1130/2007.2425(21). For permission to copy, contact editing@geosociety.org. ©2007 The Geological Society of America. All rights reserved.

the ground surface during large earthquakes. The magnitudes of observed earthquakes, their rupture dimensions, and frequency of occurrence can be directly related to rates of slip and strain energy accumulation. This provides a valuable physical basis for interpreting seismicity.

In regions far removed from plate boundaries, including most continental regions, seismicity tends to be more diffuse and infrequent. Nevertheless, large and damaging earthquakes do occur in midplate regions, for example, the 1811–1812 New Madrid earthquakes of M > 7.5, or the devastating 2001 M 7.6 Bhuj, India, earthquake (where M is moment magnitude). The causative mechanisms of midplate earthquakes are often ambiguous. In general, earthquakes within stable continental interiors relieve long-term internal plate stresses that are driven by distant plate interactions. The locations where stresses are relieved are usually zones of weakness of large crustal extent, typically pre-existing faults left behind by older episodes of tectonism. Previously rifted or extended crust is believed to be of particular importance in localizing seismicity (Johnston et al., 1994). Because the earthquake-generation process is indirect, and potential zones of weakness are widespread, seismicity is often diffuse, occurring in broad regional zones rather than along narrow well-defined faults. The events may take place on a series of buried crustal faults in locations that cannot be readily foreseen. Furthermore, midplate earthquakes do not often cause surface rupture. A global overview of large events in stable continental interiors (Johnston et al., 1994) revealed that of 452 earthquakes with M > 5, including 17 events of M > 7, there were only 7 cases of surface rupture. Even the M 7.6 Bhuj, India, earthquake did not cause surface rupture. In eastern North America, there is only one known case of surface rupture during an historic earthquake, that of the M = 6 1989 Ungava, Quebec, earthquake (Adams et al., 1991). The lack of surface rupture makes geological investigations of midplate earthquake hazards very challenging. A further complicating factor is that faults in midplate regions may exhibit time-varying behavior, with periods of activity that last thousands or tens of thousands of years interspersed with periods of inactivity that are an order of magnitude longer in duration (Crone et al., 2003).

Because it is not generally possible to delineate and characterize the causative structures of seismicity in a deterministic fashion, probabilistic analyses form the basis for seismic hazard analysis in most continental interiors. One shortcoming of hazard analyses in midplate regions, either probabilistic or deterministic, is that potentially hazardous faults that have been quiescent in the last few thousand years may be difficult or even impossible to identify and characterize.

The Basics of Seismic Hazard Analysis

How has seismic hazard been accommodated in engineering design? The original impetus for inclusion of seismic provisions in building codes in North America was the engineering experience gained in the aftermath of the 1933 Long Beach, California, earthquake. This earthquake served as a wake-up call to engi-

neers: many schools, in particular, were damaged during the Long Beach event, and casualties would have been much heavier had the earthquake occurred while school was in session. A tradition was established in which experience gained in significant earthquakes is incorporated into subsequent updates to the building codes.

Through the 1940s and 1950s, seismic design provisions in building codes tended to be based on qualitative evaluations of hazard. Later, quantitative seismic hazard maps based on probabilistic analyses were introduced. In Canada, for example, a defining moment for seismic design philosophy came in 1970 with the inclusion of the first national probabilistic seismic hazard map. This map was based on the work of Milne and Davenport (1969), who used extreme value statistics to calculate a gridded map of peak ground acceleration (PGA) having an annual exceedance probability of 0.01 (100 yr return period). Under the assumption that earthquake arrivals are Poisson distributed, they wrote an expression for the largest shock amplitude experienced at a site per year, which had the form of a type II extreme value distribution (Gumbel, 1954). They also developed a related amplitude recurrence method, based on counting the annual number of exceedances of a specified acceleration at a site. The inherent assumption was that, broadly, the past level of earthquake activity at a point is statistically representative of the future, and hence the recurrence times may be treated probabilistically.

Since the 1970s, seismic hazard maps have been developed for building code applications based on a probabilistic approach. Around the same time that Milne and Davenport (1969) were developing their seismic hazard maps of Canada, Cornell (1968, 1971) was developing a somewhat different methodology, which was coded into a FORTRAN algorithm by McGuire (1976). In the Cornell-McGuire method, the spatial distribution of earthquakes is described by seismic source zones, which may be either areas or faults. The source zones are defined based on seismotectonic information. An active fault is defined as a line source; geologic information may be used, in addition to historical seismicity, to constrain the sizes of events and their rates of occurrence on the fault. Areas of diffuse seismicity, where earthquakes are occurring on a poorly understood network of buried faults, are represented as areal source zones (e.g., polygons in map view); historical seismicity is used to establish the rates of earthquake occurrence for earthquakes of different magnitudes. The exponential relation of Gutenberg and Richter (Richter, 1958), asymptotic to an upper-bound magnitude (Mx), is used to describe the magnitude recurrence statistics in most cases, although for specific faults, a characteristic earthquake model (Schwartz and Coppersmith, 1984) may be used. The upper magnitude bound for the recurrence relations, Mx, is a limit for integration in the hazard analysis, and it represents the magnitude above which the probability of occurrence is 0. Mx values may be defined from geological information in the case of well-understood active faults. For areal source zones, Mx is usually based on the largest observed magnitudes in similar tectonic regions worldwide. The rationale for this approach is that the historical time period is too short to establish Mx empirically for any particular source zone; by using a global seismicity database for similar regions, we essentially substitute space for time in extending the seismicity database. Thus, an Mx value for unrifted midplate regions would be about M 7, while Mx for rifted midplate regions such as the St. Lawrence Valley in eastern Canada or the New Madrid seismic zone in the central United States would be about M 7.5 to M 7.8 (Johnston et al., 1994; Bakun and Hopper, 2004b). For example, the Bhuj, India, event might suggest an upper limit of at least M 7.6 for rifted midplate tectonic regions (Bodin and Horton, 2004). The spatial distribution of earthquakes within each source area is usually assumed to be random (i.e., uniformly distributed), although other treatments are possible.

Ground-motion relations provide the link between earthquake occurrence within a zone and ground shaking at a site. Ground-motion relations are equations that specify the median amplitude of a ground-motion parameter, such as peak ground acceleration or response spectra, as a function of earthquake magnitude and distance; these relations also specify the distribution of ground-motion amplitudes about the median value (i.e., variability). (Note: the original formulations of Milne and Davenport [1969] and Cornell [1968] did not include ground-motion variability; this was added in refinements to the methods by Davenport [1972] and Cornell [1971], respectively.) To compute the probability of exceeding a specified ground-motion amplitude at a site, hazard contributions are integrated over all magnitudes and distances, for all source zones, according to the total probability theorem (in practice, sensible limits are placed on the integration range for computational efficiency). Thus, the mean annual rate (λ) of exceeding a specific shaking level, x, at a site is:

$$\lambda(X \geq x) = \sum_{i=1}^{I} \upsilon_i \iint f_i(m) f_i(R|m) P(X \geq x|m, R) dR dm,$$

where υ_i is the mean annual rate of the ith source, m is earthquake magnitude, R is the distance to the site, $f_i()$ represents a probability density function, and $P()$ stands for the probability of the argument (Reiter, 1990). Calculations are performed for a number of ground-motion amplitudes, and interpolation is used to find the ground motions associated with the chosen probability levels. The basic procedures have been described by EERI (Earthquake Engineering Research Institute) Committee on Seismic Risk (1989), the U.S. National Research Council Panel on Seismic Hazard Analysis (1988), Reiter (1990), and the Senior Seismic Hazard Analysis Committee (SSHAC) (1997). Because of its ability to incorporate both seismicity and geologic information, the Cornell-McGuire method quickly became widely used and popular in applications throughout the world. Its application to seismic zoning in Canada has been described by Basham et al. (1982, 1985), Adams et al. (1999), Adams and Halchuk (2003, 2004), and Adams and Atkinson (2003).

The ability to incorporate geologic information through the definition of seismic source zones appears to be a significant advance offered by the Cornell-McGuire method, as compared to the more statistically based methods pioneered by Milne and Davenport (1969). However, the amplitude recurrence distribution of Milne and Davenport (1969) and the Cornell (1968) method are actually rather similar. The division of a region into uniform zones of occurrence (as in the Cornell approach) is really a type of spatial smoothing that is applied before the numerical analysis is performed (Atkinson et al., 1982). If the data for the amplitude recurrence distribution analysis are smoothed over an identical area, then the results of the two analyses should agree. The sequence and manner in which the data are smoothed appear to constitute the real difference between the two approaches.

There is an advantage to using the Cornell approach when zones of earthquake occurrence can be delineated on the basis of independent geological evidence. In this case, the method includes important additional information that influences the seismic hazard. In most cases, however, the definition of the source zones is strongly influenced by the historical seismicity patterns, and the definition of source zones is simply a smoothing over concentrations of seismicity.

The definition of source zones also suffers from high subjectivity. In the 1980s and 1990s, the U.S. nuclear industry was struggling with the consequences of this problem as it aimed to reassess the seismic safety of existing nuclear power plants throughout the eastern United States. Teams of seismological consultants were tasked to develop a range of seismic source models to express the wide range of competing views. Through this process, the role of uncertainty in interpretation of geological and tectonic data was illuminated, and its effects on seismic hazard results were defined (EPRI, 1986). The essential questions are: over what area(s) should seismicity be smoothed? What geologic information should be used to determine the extent of such smoothing? How do we evaluate whether a set of defined source zones is "right" or even reasonable?

In view of these discussions, and the lack of a satisfactory resolution, the U.S. Geological Survey (Frankel et al., 1996) decided to develop a methodology for their national seismic hazard mapping program that would eliminate the need to define seismic source zones. Frankel's method is similar in concept to the smoothed amplitude recurrence method, although it is also different in some respects. Because of the difficulty of objectively defining seismic source zones, Frankel et al. (1996, 1999) chose to base the probabilistic amplitude calculations for regions far from identified active faults on smoothed historical seismicity, in which various scale lengths for the smoothing are considered. These scale lengths for the smoothing essentially take the place of seismogenic source zones.

At present, the Cornell-McGuire method is the most widely used method for site-specific analysis worldwide, and it is used in the Canadian national seismic hazard maps (Basham et al., 1982; Adams et al., 1999). The problems involved in the subjective definition of source zones were addressed in the latest maps (Adams et al., 1999; Adams and Halchuk, 2003) by using a range of possible models to define the associated uncertainty. The smoothed seismicity method, in combination with the separate treatment of known active fault sources, was used in the U.S. national seismic hazard maps (Frankel et al., 1996, 1999). These differences in approach are partly responsible for some of the discrepancies observed in seismic hazard maps at the Canada–U.S. border (Halchuk and Adams, 1999). For site-specific analysis of critical facilities, the Cornell-McGuire approach, with a thorough treatment of uncertainties in all input parameters, is the most widely used and accepted approach (e.g., McGuire et al., 2002).

An Informal Evaluation of How Far We Have Come

With significant advances in knowledge and methodology over the past few decades, as discussed in more detail in the next section, one might expect current seismic hazard maps to look very different from the first probabilistic seismic hazard maps produced decades ago. It is interesting to examine the extent to which our advances have influenced seismic zoning. As an illustration, I compare the most recent seismic zoning map for the moderately active midplate region of eastern Canada (Adams

and Halchuk, 2003) to the equivalent map drawn by Milne and Davenport 30 yr earlier. The current map was developed by the Geological Survey of Canada over the past 10 yr or so (Adams et al., 1999; Halchuk and Adams, 1999). It is based on the Cornell-McGuire method, and it maps 5% damped pseudoacceleration at selected periods for a probability level of 2% in 50 yr. It includes a relatively heavy weighting of geological factors believed to influence the likely locations of future large events. Figure 1 superimposes the latest seismic hazard results from these maps for eastern Canada, for a natural period of 0.2 s (from Adams et al., 1999), on the Milne and Davenport (1969) contours. I number the contours one through four to reflect the relative acceleration amplitudes associated with each contour, where each increase by one represents roughly a factor of two increase in amplitude (on both the Adams et al. and Milne and Davenport maps). The reason that relative amplitudes are plotted is that this is the best way to see the overall impact of the maps on seismic design levels. There are many differences in the plotted ground-motion parameters and how they are implemented in the design process. A longer return period

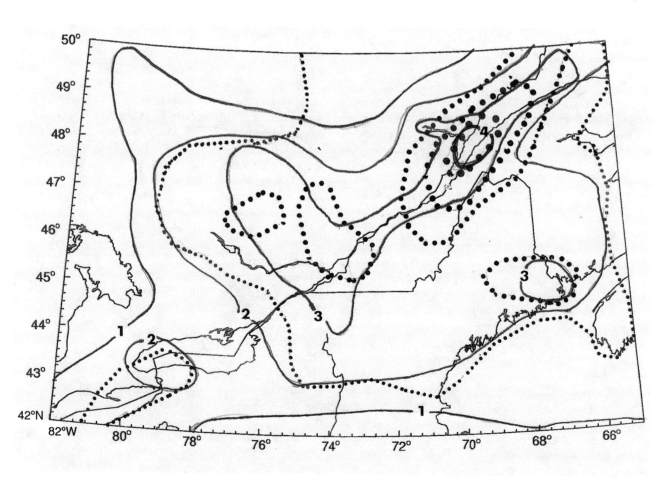

Figure 1. Comparison of seismic amplitude contours defined by Milne and Davenport (1969; dotted lines, based on 1/100 peak ground acceleration, where 1, 2, 3, and 4 are 3%g, 6%g, 10%g, and 20%g, respectively) to those defined by Adams and Halchuk (2003; solid lines, corresponding to 0.2 s spectral acceleration of 16%g, 32%g, 60%g, and 120%g, respectively, for a probability of 1/2500). Charlevoix region is the only area enclosed by contour 4 (green solid line and heavy dotted line). (Geographical note: St. Lawrence River crosses the figure from northeast to southwest, with Ottawa Valley following the Ottawa River as it leads off from the St. Lawrence near the center of the figure.)

for the input parameters implies larger input ground motions, but these are balanced against other factors that are used to calculate the seismic loads (such as ductility factors by which to divide the loads, and so on). With each new seismic map development, there has been a tendency to "calibrate" the code provisions back to a previous version. The calibrations have been based on the principle that the seismic forces should be equivalent, in an average way across the country, to those used in the previous version of the code (e.g., Heidebrecht et al., 1983). (Note: specific changes are made as required to accommodate deficiencies in practice identified from engineering experience in earthquakes.) This ensures that the overall level of seismic protection, which is believed to be adequate on balance (though subject to refinement to correct identified deficiencies), is maintained. It also acknowledges that significant changes in the overall concepts of seismic design, which would cause a real change in level of protection, evolve over a longer time frame than do changes in the evaluated levels of ground motion for a stated probability. Thus, the real importance of the seismic zoning maps in the design process,

at least for building code applications, is in establishing *relative* levels of seismic ground motion.

As seen in Figure 1, the similarities between the 1969 contours (dotted lines) and the 1999 contours (solid lines) are more striking than the differences. In both cases, the region of highest hazard (4) is confined to the Charlevoix seismic zone, where the most recent maps indicate a more tightly defined area of highest hazard. The newer maps feature smoother contours along the St. Lawrence, which result from smoothing the seismicity over broader geologic regions (in this case, an ancient rifted margin). Moderate hazard (2–3) is indicated throughout the St. Lawrence and Ottawa valleys, with a consistent pocket of elevated hazard near the border of New Brunswick with Maine.

These maps, which were prepared 30 yr apart and used different methods and different databases, reveal striking and persistent similarities, and differences that are not very marked. The reason for this can be appreciated by referring to the historical seismicity in the region, as plotted in Figure 2. Seismicity is concentrated in diffuse but reasonably well-defined clusters: along the Ottawa

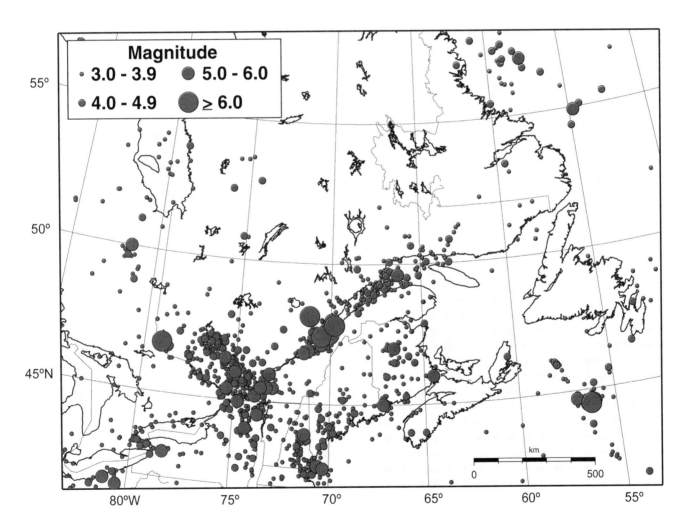

Figure 2. Historical seismicity of eastern Canada, from the Geological Survey of Canada (S. Halchuk). Note the correspondence between clusters of seismicity and areas of highest hazard shown on Figure 1.

and St. Lawrence valleys, near the New Brunswick–Maine border, and to a lesser extent near the western end of Lake Ontario. The largest historical events have been in the Charlevoix region. All of the seismic hazard maps of Canada, from 1970 to the present, strongly reflect these distributions. The more recent earthquake data indicate that some of these clusters are more tightly defined than was apparent from the older, less precise data; hence, the more recent maps feature tighter hazard contours in some areas. The underlying reality is that while methods and data have been refined, our overall understanding of seismic hazards in eastern Canada as applied to the National Building Code has not changed that much since the original work of Milne and Davenport (1969). While this example is specific to eastern Canada, the same general principle applies over many regions. Our seismic hazard zoning maps are a relatively simple and transparent consequence of the patterns of historical seismicity. What has improved over time is that we have refined our basic understanding of where and why earthquakes occur, and improved our characterization of the resulting ground motions and their probabilities. In the next section, I look at these advances for continental interiors in more detail, and then I examine current shortcomings and issues.

ADVANCES IN SEISMIC HAZARD ANALYSIS

Several significant advances to seismic hazard analysis for continental interiors over the past 10 yr or so are worth noting. The most helpful advances from a practical point of view have been the following:

Resolution of Some Common Misconceptions

There are two common misconceptions that have plagued probabilistic seismic hazard applications for decades. The first misconception hindered a well-reasoned trend to base seismic zoning maps and standards for critical structures on ground-motion values with low probabilities. In typical North American building codes, earthquake design provisions in the 1970s were based on ground motions with a 100 yr return period (0.01 per annum)—a relatively frequent occurrence. In the 1980s, there was a shift to a 500 yr return period (10% in 50 yr). The latest codes, such as the National Building Code of Canada (as of 2005) and the NEHRP (National Earthquake Hazards Reduction Program) guidelines for building codes in the United States (from 1997 onward), are based on a 2500 yr return period (2% in 50 yr). Ground-motion probabilities for design of critical structures have likewise drifted downward, from ~5% in 50 yr to values as low as 0.1% in 100 yr. An argument often advanced against this trend is that low probability hazard estimates are an extrapolation of a short historical record: "100 years of data are extrapolated to return periods of thousands of years." In fact, the low probability of the calculated ground motions results from breaking the problem into component parts, where the result is the product of the components (U.S. National Research Council Panel on Seismic Hazard Analysis, 1988). It is the ground motion

at a site that has a very low probability, not the event itself. For example, suppose we have a region that has experienced 10 potentially damaging (M > 5) earthquakes in the past 100 yr. The probability (per annum) of occurrence of an event of M > 5 is 0.1. If a M > 5 event occurs, we know from both regional and global recurrence models that the conditional probability of its magnitude being 6 or larger is ~0.1. Based on the total area of the subject region, the probability of the event being within 50 km of the site of interest is, say, 0.02. Finally, the probability of ground motions exceeding a certain target, given all of the above, is 0.5. The total probability of exceeding the ground-motion target is thus the product $(0.1)(0.1)(0.02)(0.5) = 10^{-4}$, or a "return period" of 10,000 yr. The dominant factor that lowers the probability of damaging ground motions is the sparse spatial distribution of events; in this sense, the low probability is more nearly an interpolation in space than an extrapolation in time. A growing recognition of this basic nature of low probability hazard analysis has improved our ability to utilize hazard analyses effectively to solve the engineering problems of interest, which necessarily involve design to rare earthquake ground motions rather than common occurrences.

Another misconception has revolved around the role of uncertainty. The results of seismic hazard analyses are subject to large uncertainty due to our limited knowledge of the component processes and large uncertainties in their interpretation; these uncertainties may become particularly pronounced at low probabilities. Because probabilistic hazard analyses are known to be subject to large uncertainty, there has been a significant tendency to rely on a "deterministic" approach to hazard, in which a design earthquake of a specified magnitude and location is used to determine the resulting ground motions. It is now widely understood, however, that uncertainty is inherent to the physical processes involved and is not specific to probabilistic analysis. Even in situations where the magnitude and locations of future earthquakes are indeed relatively deterministic (such as near a well-documented active fault), the ground motions remain uncertain. Thus, deterministic analyses are subject to the same fundamental uncertainties, be they aleatory or epistemic, as are probabilistic analyses. They also suffer from the additional problem that nobody knows the likelihood of the determined outcome.

I do not mean to suggest that uncertainty is not an important limitation to seismic hazard analysis. Later in the paper, I will discuss some significant shortcomings in our understanding of uncertainty, particularly as they apply to continental interiors. However, uncertainty cannot be understood or mitigated by treating processes that are largely stochastic in nature as deterministic events. Uncertainty is a critical area of hazard analysis where major advances have been made but which still requires much improvement in our understanding.

Treatment of Uncertainty

The proper treatment of uncertainty in hazard analysis is an area where significant advances have been made over the last decade. It has been recognized that it is important to distinguish between randomness in process (aleatory uncertainty) and uncertainty in knowledge (epistemic uncertainty) (McGuire and Toro, 1986). Randomness is physical variability that is inherent to the unpredictable nature of future events, an example of which is the scatter of ground-motion values about a median regression line. Randomness cannot be reduced by collecting additional information. Epistemic uncertainty arises from our incomplete knowledge of the physical mechanisms that control the random phenomena; it can be reduced by collecting additional information. This separation is clear-cut in principle, but in practice, the distinction is somewhat arbitrary. Anderson et al. (2000) provided arguments that the epistemic uncertainty is in fact the dominant component, but it is often cast instead as aleatory uncertainty, thus skewing hazard results. They pointed out that mean hazard maps at low probabilities would be significantly altered if the uncertainties were redistributed between aleatory and epistemic categories.

The seismic hazard maps developed for previous building codes (e.g., Basham et al., 1985) incorporated aleatory uncertainty (e.g., the variability in the ground-motion relations) but were known to be sensitive to epistemic uncertainty. In recent years, a formal method of handling this uncertainty has been developed (McGuire and Toro, 1986; Toro and McGuire, 1987a; McGuire et al., 2001), using a logic-tree approach. Each input variable to the analysis is represented by a discrete distribution of values, and subjective probabilities are used to describe the credibility of each possible assumption. Each possible combination of inputs produces a different output, so that a typical application of the process would produce thousands of possible results. The uncertainty in results can then be expressed by displaying a mean or median curve and fractiles that show the confidence with which the estimates can be made (e.g., EPRI, 1986; Toro and McGuire, 1987a; Bernreuter et al., 1985; McGuire, 1995). The use of a logic-tree approach to investigate and quantify uncertainty in seismic hazard estimates is a significant advance in methodology that has been explored for some building code hazard maps (e.g., Adams et al., 1999; Adams and Halchuk, 2003; Frankel et al., 1999) and is now widely used in site-specific analyses for critical structures throughout North America (e.g., McGuire et al., 2001).

Uniform Hazard Spectra

Another major change in the methodology of specifying ground motions for use in engineering design involves the use of the "uniform hazard spectrum." In 1970s and into the 1980s, seismic hazard maps presented expected levels of peak ground acceleration (PGA), and sometimes peak ground velocity (PGV). Similarly, most standards for critical facilities were based on evaluation of PGA and/or PGV at that time. For engineering design, a much more useful description of ground motion is a response spectrum (typically PSA, the pseudoacceleration spectrum), which defines the response of a damped single-degree-of-freedom oscillator to an earthquake accelerogram as a function

of the oscillator's natural period. The response spectrum contains information about both the amplitude and frequency content of the ground motion, as well as indirect information regarding its duration. In the past, the response spectrum used for engineering design was constructed by scaling a standard spectral shape to the site-specific PGA and/or PGV (e.g., Newmark and Hall, 1982). In the last 10–15 yr, it has become standard seismological practice to instead develop a uniform hazard spectrum. The underlying probabilistic seismic hazard calculation is the same. However, in the uniform hazard spectra methodology, the hazard analysis computes expected response spectral ordinates for a number of oscillator periods (McGuire, 1977, 1995). This eliminates the need to use standard spectral shapes scaled to an index parameter, such as PGA, and thus provides a more site-specific description of the earthquake spectrum; it also ensures a uniform hazard level for all spectral periods. This has been a natural evolution of seismic hazard methodology, made possible by improved ground-motion relations for spectral parameters.

Uniform hazard spectra computations, coupled with abundant new ground-motion data, have revealed that the scaled-spectrum approach used in past codes overestimated response spectra for intermediate periods for some types of earthquakes by a very significant margin (Atkinson, 1991). This is because the standard spectral shape was a description of ground motions for earthquakes in California within a limited magnitude and distance range. It is now well known that the shape of earthquake spectra is actually a function of magnitude and distance, and it varies regionally (e.g., Atkinson and Boore, 1997). In recent seismic hazard maps and other applications, a uniform hazard spectra approach is routinely used to overcome previous shortcomings of the scaled-spectrum approach and more accurately describe the site-specific frequency content of the expected ground motions. This concept has been taken even further in applications that develop uniform reliability spectra, which also consider the slope of the hazard curves at the probability levels of interest and the structural reliability goals (McGuire et al., 2002).

Figure 3 provides a typical uniform hazard spectrum, for Toronto, Ontario, for a probability level of 0.0001 per annum, in which uncertainty is included by plotting various fractiles (this curve was adopted from studies of facilities nearby). Considering all modeled combinations of input parameters, 50% of the results lie below the median curve, while 84% of results lie below the 84th percentile curve. Thus, for example, we can be 84% confident that the true hazard curve lies below the 84th percentile. Note that the mean uniform hazard spectrum, obtained by weighting each result by its probability of being the "correct" model, is significantly higher than the median.

Lower Probability Level for Computations

Another major trend in seismic hazard analysis is the lowering of the probability level for which the ground motion is being evaluated. Most modern codes are based on motions with a probability of 0.000404 per annum (2% in 50 yr). This change

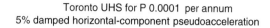

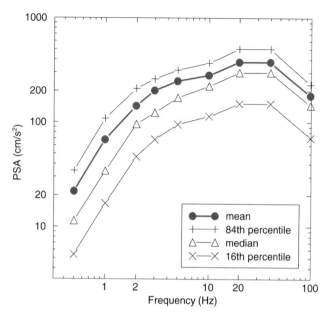

Figure 3. Example of uniform hazard spectrum (UHS) for Toronto for a probability of 0.0001 per annum, showing various fractiles of the results that represent epistemic uncertainty in the input parameters to the seismic hazard analysis. PSA—pseudoacceleration.

was motivated by studies over the last 10–20 yr that have shown that the best way to achieve uniform reliability across regions is by basing the seismic design on amplitudes that have a probability that is close to the target reliability level (e.g., Whitman, 1990). The reason is that the slope of the hazard curve—the rate at which ground-motion amplitudes increase as probability decreases—varies regionally. In active regions like California, ground-motion amplitudes may grow relatively slowly as probability is lowered from 1/100 to 1/1000; this is because the 1/100 motion may already represent nearby earthquakes close to the maximum magnitude. In inactive regions, 1/100 motions are small but grow steadily as the probability level is lowered. Thus, there is no single "factor of safety" that could be applied to motions calculated at, say, 1/100 per annum, that would provide design motions for a desired reliability of, say, 1/1000 per annum in both regions. The concept is illustrated in Figure 4. For uniform reliability across regions with differing seismic environments, the seismic hazard parameters on which the design is based should be calculated somewhere near the target reliability level. As discussed by Heidebrecht (2003), it is believed that this target level for seismic design of common structures corresponds to ground motions with a probability of ~2% in 50 yr. For critical structures such as dams or nuclear power plants, the target probability level is even lower. In the latest seismic hazard maps of Canada (Adams and Halchuk, 2003) and the United States (Frankel et al., 1996,

1999, 2002; to present), ground motions are calculated for an exceedance probability of 2% in 50 yr.

The rationalization of probability level for hazard computations, driven by an understanding of the regional variability in the slope of ground motion versus probability and its implication for seismic design, has been a significant advance that should lead to safer structures in regions where large earthquakes happen only rarely. However, this change has been controversial, and not all agree that higher levels of seismic design are warranted in regions where earthquakes are relatively infrequent. For example, Stein et al. (2003) argued that the use of 2% in 50 yr probability motions in the U.S. national hazard maps have resulted in a situation whereby buildings in Memphis, Tennessee, must effectively meet California seismic safety standards (at least for high-frequency structures), despite the obvious differences in seismicity levels in the two regions. The arguments of Stein et al. (2003), and rebuttal by Frankel (2003), are a good example of this debate.

Better Understanding and Definition of the Concept of "Design Earthquake"

A useful exercise to understand the results of a probabilistic seismic hazard analysis is to "disaggregate" the hazard. What the hazard analysis provides is an estimate of ground motion for a certain probability level. This ground motion represents a composite of contributions to hazard from earthquakes of all magnitudes at all distances (rather than a single design earthquake). By mathematically disaggregating the hazard, we evaluate the relative contributions of earthquakes of various magnitudes and distances to the calculated hazard. This allows us to define one or more "design earthquakes" that contribute strongly to hazard and that will reproduce the calculated ground motions (McGuire,

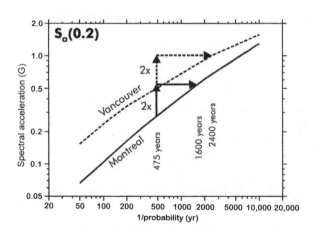

Figure 4. The effect of different slopes of the ground-motion probability curve, from Adams and Halchuk (2003). Note that if a 1/475 (10% in 50 yr) spectral acceleration (at a period of 0.2 s, for 5% damping) for Vancouver, British Columbia (an active area), were multiplied by a safety factor of 2, the resulting motion would have a probability of 1/2400. In Montreal (a less active area), the same exercise would result in a higher probability of 1/1600.

1995). Such design earthquakes are useful in engineering applications. Figure 5 shows the results of a typical disaggregation, in this case, for spectral acceleration (PSA, 5% damped horizontal component) with a natural period of 0.2 s, at Montreal, at the 2% in 50 yr probability level. The PSA at a natural period of 0.2 s is a good engineering measure of the ground motion that a typical low-rise building would "feel" during an earthquake. Figure 5 shows that the hazard at Montreal for this probability is dominated by earthquakes that are about M 6.5 and occur within 50 km of the city (see Halchuk and Adams, 2004).

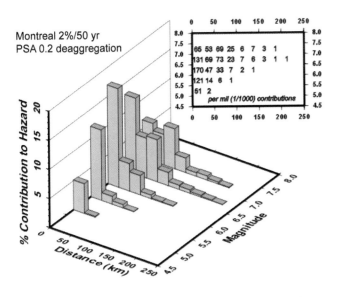

Figure 5. Hazard disaggregation for 2%/50 yr median PSA (0.2 s) at Montreal, from Adams and Halchuk (2003). PSA—pseudoacceleration.

Progress in understanding and evaluating appropriate design earthquakes has been an important and practical advance. It has improved our ability to utilize the results of a hazard analysis to help engineers design and analyze structures with appropriate input ground motions—motions that are characteristic of the type of earthquakes that the structure might actually experience. Structures are no longer designed to a scaled version of the famous 1933 El Centro record. Instead, appropriate real records in the magnitude-distance range that contributes most to hazard may be selected from a large catalogue, if such records exist. If the records do not exist for the region of interest, as is the case for most continental interiors, they may be generated by a simulation methodology (e.g., Atkinson and Beresnev, 1998, 2002; Saikia and Somerville, 1997; Hartzell et al., 1999). Alternatively, hybrid techniques may be used that combine the advantages of real recordings with the flexibility of simulation methodologies. For example, the amplitude spectrum for a desired magnitude, distance, and site condition, as based on a seismological model,

may be used in combination with the phase spectrum from a real recording (with some limits on the magnitude and distance to ensure a reasonable phase spectrum), in order to simulate a realistic time history (McGuire et al., 2001).

Advances in the understanding of the design earthquake concept have been linked with advances in the treatment of uncertainty and the implementation of the uniform hazard spectra concept to provide a powerful means of specifying design earthquake motions for important projects. The state-of-the-art in this regard is given in detail by McGuire et al. (2001, 2002). The development of seismic ground motions begins with a probabilistic seismic hazard analysis that draws from the current knowledge of earthquake sources, recurrence statistics, and ground-motion relations, according to the SSHAC (1997) methodology. This includes a thorough evaluation of all sources of epistemic uncertainty. The hazard analysis will result in response spectra over a broad range of frequencies (e.g., 0.2–100 Hz) and a wide probability range (e.g., 10^{-2} to 10^{-5} per annum). Disaggregation is then performed on the mean hazard for the target probability level, by both magnitude and distance, to determine the relative hazard contributions at 1 and 10 Hz. If multiple ground-motion relations have been used to characterize epistemic uncertainty, then the disaggregation may be done with each of these, weighted by the subjective probabilities that are used in the hazard analysis. McGuire et al. (2001) developed a catalogue of time histories, categorized by magnitude, distance, and site condition, to use in developing appropriate time histories to match the target spectra for the design events identified by the disaggregation. They also provided detailed procedures that can be used to ensure that the developed time histories match the target spectra in a rigorous way. An example application of these procedures for both typical eastern and western North American sites is given in McGuire et al. (2002).

Better Understanding of Earthquake Ground-Motion Processes

Over the last 5–10 yr, there has been a remarkable increase in recorded ground-motion data from earthquakes in all parts of North America, and from earthquakes in Japan and Taiwan. The ground-motion database has improved due to a combination of developments in seismometry and increased deployments of instruments. In well-instrumented regions, such as Japan, Taiwan, and California, there are now thousands of available strong-motion recordings. From these, we can develop a much better empirical characterization of ground-motion generation and propagation. We can determine the distribution of slip on faults and characterize factors that profoundly influence ground motion, such as directivity, near-fault displacements, and basin effects (e.g., Graves et al., 1998; Somerville et al., 1997). We can develop more robust empirical relations that are based on thousands of recordings (e.g., Boore et al., 1997; Atkinson and Boore, 2003). Even in regions where strong-motion and seismographic networks are relatively sparse, there are now thousands of useful recordings, although most of these are for small to moderate events at fairly large distances. These records, coupled with advancements in ground-motion modeling techniques (e.g., Pitarka et al., 2000; Beresnev and Atkinson, 2002), are improving our ability to understand and model ground-motion processes and will ultimately lead to refinements in future hazard evaluations. On the other hand, many puzzles still remain. For example, the well-recorded 1999 Chi-Chi Taiwan earthquake resulted in ground motions that were much smaller than would be predicted based on previous observations (Boore, 2001), while the 1988 Saguenay, Quebec, earthquake resulted in ground motions much larger than would be predicted based on previous observations (Atkinson and Boore, 1998). As yet, we do not really understand whether "anomalous" earthquakes are truly anomalous or are simply misunderstood in the context of our present models and approaches. Furthermore, we have not yet come to grips with complicating factors like the potential for highly asymmetric ground motions, such as on the hanging wall versus the footwall of thrust faults (Anderson et al., 2000).

Finally, with more seismic instrumentation and analyses, our understanding of seismicity patterns and magnitude recurrence statistics has gradually evolved over time, improving our understanding of seismic hazard.

DISCUSSION

Shortcomings in Application of Seismic Hazard Analysis

One area in which much progress remains to be made is in how we fully characterize and utilize uncertainty, given that we don't know all of the relevant parameters. This problem is particularly pronounced in continental interiors. As pointed out by Field et al. (2003), to ensure a stable hazard analysis process, we would ideally like to start by identifying all viable hypotheses and including them in an initial hazard model. Over time, as models are rejected, they could be removed from the full range of options. Under this approach, hazard analyses might not change dramatically over time, in contrast to the case in which we start with only a few well-known models, than add new models as new information emerges. However, such an ideal approach presupposes that we can identify all viable hypotheses at this time, which seems unlikely. As a cautionary tale, Field et al. (2003) pointed to an apparent paradox: in southern California, one of the most data-rich regions in the world for seismic hazard analysis, the multiple forecasts generated by the activities of a large working group in that area (Field et al., 2000, and papers therein) appear to show that, after extensive study, hazard estimates are more uncertain than previously thought. In other words, a larger amount of study of uncertainty led to a larger apparent uncertainty. To cite a specific example, even if we obtain further information on site characteristics by determining the shear-wave velocity profiles of all sites, we apparently do not significantly reduce uncertainty in site response (Field et al., 2000). There are significant other pieces of evidence, notably the precarious rock studies of Brune (1996, 2001), that also suggest that we may not understand the actual influence of uncertainty on seismic hazard.

The underlying problem is that a lack of knowledge is, by its very nature, not amenable to accurate quantification. Current probabilistic seismic hazard analysis (PSHA) does its best by trying to capture some of the obvious culprits, but we may be fooling ourselves into thinking we know more than we really do.

An example of this problem is the manner in which uncertainty in seismic source models in eastern Canada is handled at present in the national seismic hazard maps, presented by Adams and Halchuk (2003). The national hazard maps aim to estimate uncertainty in source models by considering two end members of a family of models. One end member assumes that future earthquakes will be concentrated in areas of past historical seismicity (the H model). The other end member asserts that the risk of large earthquakes should be considered uniform over a broad series of faults along the St. Lawrence, Saguenay, and Ottawa valleys, which formed several hundred million years ago during rifting and opening of the Iapetus Ocean. The argument is that these deep-seated rift faults are believed to be potential sources of weakness that could be reactivated by the current high horizontal compressive stress field. Several investigators have shown that large earthquakes in eastern North America occur preferentially within such zones (Kumarapeli and Saull, 1966; Adams and Basham, 1989; Johnston et al., 1994; Adams et al., 1999). Global studies indicate that, within stable continental interiors, 70% of earthquakes of M > 5 and all events of M > 7 occur within such rift zones (Johnston et al., 1994). Current seismic hazard evaluations for eastern Canada draw heavily on this concept. To capture this possibility, an alternative model (the R model) was drawn to encompass such rift fault features. Figure 6 shows the two models.

The idea is fine in concept, but the definition of the geologic zones is inevitably biased by past seismicity patterns, and thus it is less complete than it first appears. Figure 7 shows the original definition of rifted areas based on the geologic work of Kumarapeli and Saull (1966). Note that the rifted areas extend up the Saguenay graben, down Lake Champlain, and also west from the Ottawa valley, past Lake Nipissing. The original definition of the R model of seismicity, by Adams and Basham (1989), considered this information but was also influenced by contemporary seismicity. Thus, their original concept, shown in Figure 8, did not include the failed arm that extends up the Saguenay graben, because as of 1988 when the model was drawn, this arm was dormant and did not appear to be all that significant. This arm was added to the model after the occurrence of the 1988 Saguenay earthquake, despite previous knowledge of the rift features. Ironically, the original version was proposed in 1988 and was published just after the Saguenay earthquake occurred (Adams and Basham, 1989). A speculative arm west through Lake Ontario was proposed in the 1989 version based on a weak seismicity trend. It did not have a geologic basis in the Kumarapeli and Saull model but was proposed based on observations of some post-Ordovician faulting, which appeared similar in age and style to the Ottawa–St. Lawrence–Champlain faulting, though different in scale, and other work by Woolard concerning an arc of seismicity extending from the St. Lawrence through New Madrid

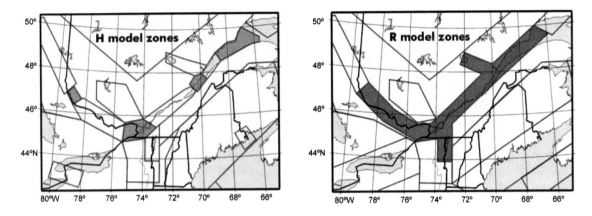

Figure 6. H and R models for eastern Canada (Adams and Halchuk, 2003). In the H model, seismicity is concentrated in the red zones, while in the R model, it is smoothed over the larger purple zone.

(J. Adams, 2004, personal commun.). Similarly, the extent of the current zone shown in Figure 6 is open to question. A westward extension of the rift past Lake Nipissing was not drawn in the current model due to the lack of seismicity extending from North Bay to Sudbury and a view that the geologic basis was weak (J. Adams, 2004, personal commun.). Such an inclusion would increase the ground-motion levels at locations such as Sudbury, Ontario, by about a factor of three (based on a rough calculation).

Other factors were also considered. For example, it was felt undesirable to spread the rift model out over too large an area (such as across southern Labrador), because this would dilute the hazard elsewhere (J. Adams, 2004, personal commun.). Presumably, if the next big eastern earthquake is in one of these excluded regions, future versions of this model will be revised. In the meantime, however, is it fair to say we are capturing uncertainty in seismotectonics through our "end member" models? And what

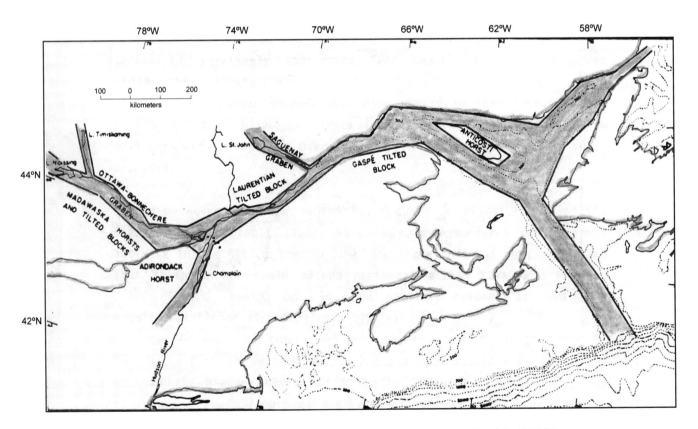

Figure 7. Rifted areas of eastern North America, according to Kumarapeli and Saull (1966).

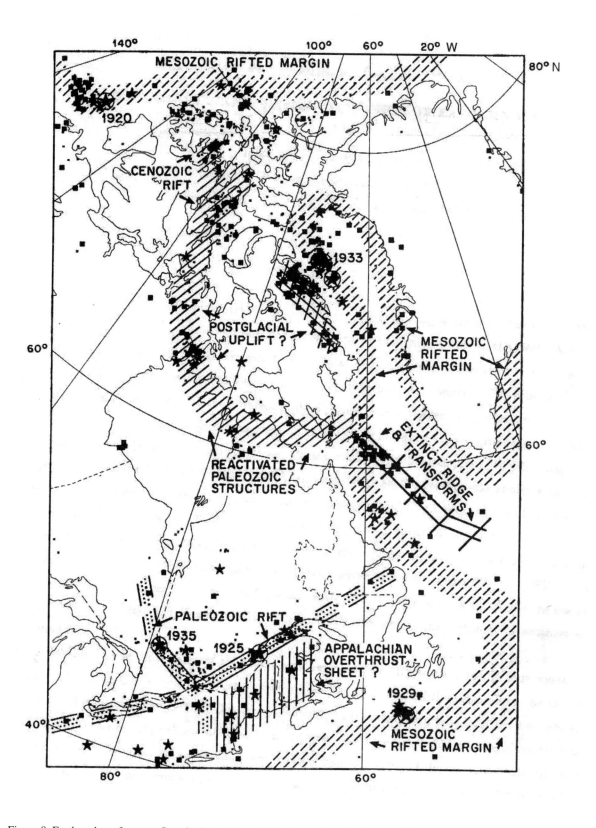

Figure 8. Earthquakes of eastern Canada (M > 3 since 1970; M > 4 since 1960; M > 5 since 1940; M > 6 since 1900) together with an interpretative framework for the cause of the seismicity proposed by Adams and Basham (1989), then later revised as per Figure 6.

about other possible explanations for features controlling seismicity, about which we know too little to warrant delineating models?

A more general problem in evaluating uncertainty is related to the fact that our current information is necessarily incomplete. Obviously, the uncertainty that is implied by incomplete information limits our ability to interpret this uncertainty. Thus, new information will continuously lead to revisions in our estimates of mean or median hazard and also our evaluations of uncertainty. Quite often, new information lies outside of the previously calculated uncertainty bounds for the parameters in question because the results of various studies or events could not be anticipated. A classic example is the ground-motion amplitudes from the 1988 Saguenay, Quebec, earthquake (M 5.8). This was the largest eastern North American earthquake to occur in 50 yr. Recorded ground motions exceeded the predictions of ground-motion relations that had just been developed for the region (Boore and Atkinson, 1987; Toro and McGuire, 1987b) by about a factor of four at high frequencies. Since such a large factor implied motions two standard deviations above the predicted median, these observations immediately cast doubt on the accuracy of the relations. Thus, new information has the potential to greatly change our evaluation of seismic hazard, and the revised parameters may lie far outside what we had calculated as their uncertainty bounds.

These examples are not presented as a critique of the work that has been done for current seismic hazard maps, but rather to illustrate our inability to accurately capture uncertainty. The same factors that create epistemic uncertainty also limit our ability to characterize it. This is a fundamental limitation with the evaluation of uncertainty that cannot be easily redressed. It should be recognized, then, that analyses of uncertainty, though useful, are inherently limited in their scope. I suspect that uncertainty is most often understated because we can't model tomorrow's surprises. However, Anderson et al. (2000) provided good reasons why uncertainty may also be overestimated, by the mixing of aleatory and epistemic uncertainty.

Since the mean hazard curves are partly a function of the amount of uncertainty, it would appear to follow that mean hazard is underestimated if there is a tendency to underestimate uncertainty. On the other hand, the proposition that the mean hazard is indeed raised by the effects of uncertainty may not be entirely justified. For example, Brune (1996) found that the distribution of precarious rocks in southern California was not consistent with the large mean values of ground motions predicted by PSHA studies. However the distribution of these rocks appeared to be consistent with hazard maps of Wesnousky (1986), which use only the median value for attenuation of peak ground motion and thus ignore even the aleatory uncertainty (Anderson et al., 2000). The question is raised as to what is the appropriate role of uncertainty in forming seismic hazard estimates (Anderson et al., 2000). In my view, this question has not been adequately answered.

For reasons such as those discussed here, the median is sometimes suggested as an alternative to the mean; the median is inherently more stable with respect to the influence of uncertainty and will change less over time as our uncertainty changes.

This was a key factor in the decision to base the seismic hazard maps of Canada on the median, for example (J. Adams, 2004, personal commun.). However, this is not an ideal solution either because the median is not the expected value. Moreover, since the median is significantly less than the mean (sometimes by a factor of two), it potentially underestimates the true hazard if we believe that our uncertainties are indeed real. Thus, there are significant outstanding issues in seismic hazard analysis with respect to the complete characterization and utilization of estimates of uncertainty. At the present time, I believe that resources are more appropriately expended on fundamental studies and data analyses that will actually reduce uncertainty, as opposed to exercises that aim to quantify and utilize it. For every dollar that is spent trying to quantify uncertainty, we should spend 10 dollars collecting and analyzing data that would reduce uncertainty.

Testing Seismic Hazard Analysis Results

In view of the uncertainties discussed here, do national seismic hazard maps provide a valid representation of expected ground-motion levels? Is it reasonable to require that buildings be designed to accommodate these ground motions? Although this is a complex issue, a simple probability-based test of the mapped ground motions can be made. Current maps are based on a 2% probability of exceedance in 50 yr. In a 50 yr time period, then, we should expect ~2% of the area on the map to reach or exceed the predicted level of ground motion; this expectation provides the basis for an area-based test of the hazard estimates (e.g., Ward, 1995). In eastern North America, the most direct and feasible test along these lines is based on modified Mercalli intensity (MMI). Based on correlations between intensity and instrumental ground motion, we can estimate the areas on the hazard maps that would be expected to experience a specified intensity level in a 50 yr period. I chose an intensity level of MMI = VII for this test because it is a damaging level of ground motion for which we have a reasonably complete record over the last 100–200 yr. The instrumental ground-motion parameter chosen to relate to intensity was the 5 Hz PSA (pseudoacceleration) value because this was the ground-motion parameter of most relevance to typical low-rise engineered structures that dominate historical eastern North America damage reports. Correlations based on California strong-motion data (Atkinson and Sonley, 2000) have shown that the average 5 Hz PSA value of 38%g is associated with MMI VII. The value might be somewhat lower in eastern North America (Kaka and Atkinson, 2004), but there are too few eastern North America data at this intensity level to properly assess this factor; thus, the California-based value was adopted.

On Figure 9, I shade the area of the 1996 national seismic hazard map of the U.S. Geological Survey (Frankel et al., 1996) that corresponds to an expected MMI VII (for 2% probability of exceedance in 50 yr). This area is the mapped area that has an expected PSA (5 Hz) > 30%g for NEHRP B/C boundary site conditions (defined as shear-wave velocity of 760 m/s). According to Joyner and Boore (2000), the amplification factor to convert this reference ground condition to a typical soil site condition (NEHRP D), such as would be found at most sites, is ~1.3; thus the 30%*g* level for B/C conditions corresponds to ~38%*g* for soil sites, or MMI = VII. The shaded area on Figure 9 sums to a total area of 1.05×10^6 km^2.

In a 50 yr time period, we should expect 2% of the shaded area of Figure 9 to experience MMI VII, or an area of 21,000 km^2. Let's compare this to our experience over the past 200 yr. The symbols on Figure 9 show the locations of events of moment magnitude M > 5 in the past 200 yr. The data were compiled from the U.S. Geological Survey catalog that was used to make the maps (Frankel et al., 1996), updated to 2005 using the IRIS (Incorporated Research Institutes in Seismology) database (www.iris.edu/quakes/catalogs.htm; accessed 4 July 2005). All magnitudes were initially converted to moment magnitude from *Lg* magnitude (mb*Lg* or mN*Lg*) using the empirical relation of Atkinson (1993) (M = 0.98 mb*Lg* − 0.39). Those events for which a better estimate of moment was available based on special studies were then corrected to the appropriate M. These M values were based on the studies of Bakun and Hopper (2004a, 2004b) for the central United States, supplemented by values from Atkinson and Hanks (1995) for recent events. The estimated area of MMI

VII for each event was based on its moment magnitude, using the empirical data of Johnston (1996) for stable continental interiors; the area versus moment data of Johnston follow the relation: log area(VII) = −23.5 + 6.849M − 0.4035M^2. The size of symbols used on Figure 9 for the historical events approximately equals these average areas.

In a 50 yr time period, we can simply sum the areas that have experienced MMI VII, since there is very little overlap between events (the 1811–1812 events are counted only once, as an event of M 7.8). For the last four 50 yr time periods, we obtain the following total areas for MMI VII:

1805–1854 = 237,000 km^2,
1855–1904 = 38,000 km^2,
1905–1954 = 7500 km^2,
1955–2004 = 1200 km^2.

Note the large differences in area in these time windows, and the apparent decreasing trend with time, despite the greater reporting of events in the past 50 yr. This is due to the overwhelming influence of the largest events, the 1812 New Madrid earthquake (M 7.8) in the first window, and the 1886 Charleston earthquake (M 6.9) in the second window. Consequently, in the first two time windows, the expected area of 21,000 km^2 for

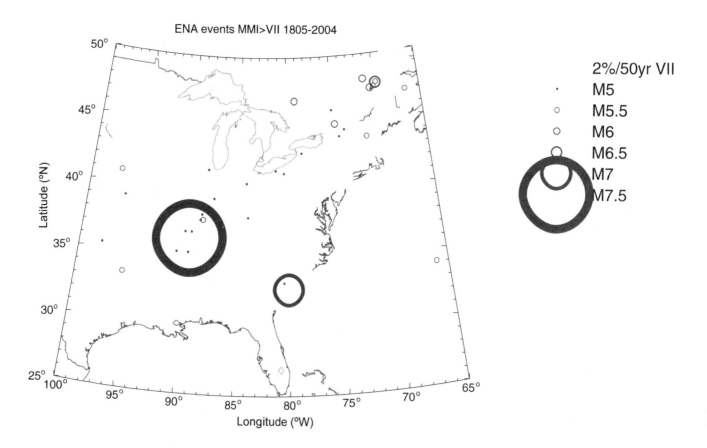

Figure 9. Historical seismicity of eastern North America (ENA) in the past 200 yr. Only events of M > 5 are shown, and symbol sizes are approximately equal to area that experienced MMI VII or greater. Shaded area is predicted to have a 2% chance of experiencing MMI VII or greater in a 50 yr time period, based on interpretation of the 1996 U.S. Geological Survey national hazard maps (Frankel et al., 1996).

MMI VII based on the national hazard maps is greatly exceeded, while in the second two time periods, the damage area is much less than predicted. The hazard maps are reflecting the potential for repeats of the largest historical events in areas that currently experience moderate seismicity. In any 50 yr time period, eastern North America will experience or exceed the expected MMI VII area if there is at least one event of M > 6.7 anywhere in eastern North America (or two events of M 6.5). If there are not any large events, the area of MMI VII will be much less than predicted.

The natural disasters of the last decade, such as the tsunami caused by the M 9 2004 Sumatran earthquake, and the massive loss of life in the M 7.6 2001 Bhuj, India, earthquake, have emphasized the importance of providing at least some protection against natural events that we know happen but that have long repeat times. In this context, it is prudent to require that in regions such as New Madrid, Charleston, and the St. Lawrence valley, structures should be designed to withstand the motions that would result from repeats of the larger historical earthquakes, even though such events occur infrequently. The difficulty in gaining widespread support for this position arises from its immediate economic consequences, but we have not experienced these events in our lifetimes. What will the experience be in the next 50 yr?

ACKNOWLEDGMENTS

I thank John Adams for comments on an early draft of this manuscript, and Stéphane Mazzotti and two anonymous reviewers for their suggestions. This paper is dedicated to Alan G. Davenport.

REFERENCES CITED

Adams, J., and Atkinson, G., 2003, Development of seismic hazard maps for the 2003 National Building Code of Canada: Canadian Journal of Civil Engineering, v. 30, p. 255–271, doi: 10.1139/l02-070.

Adams, J., and Basham, P., 1989, Seismicity and seismotectonics of eastern Canada: Geoscience Canada, v. 16, p. 3–16.

Adams, J., and Halchuk, S., 2003, Fourth Generation Seismic Hazard Maps of Canada: Values for Over 650 Canadian Localities Intended for the 2005 National Building Code of Canada: Geological Survey of Canada Open-File 4459, 150 p.

Adams, J., and Halchuk, S., 2004, Fourth generation seismic hazard maps for the 2005 National Building Code of Canada, *in* Proceedings of the 13th World Conference on Earthquake Engineering: Vancouver, Canada, paper 2502 on CD-ROM.

Adams, J., Wetmiller, R., Hasegawa, H., and Drysdale, J., 1991, The first surface faulting from a historical intraplate earthquake in North America: Nature, v. 352, p. 617–619, doi: 10.1038/352617a0.

Adams, J., Weichert, D., and Halchuk, S., 1999, Trial Seismic Hazard Maps of Canada—1999: 2%/50 Year Values for Selected Canadian Cities: Geological Survey of Canada Open-File 3724, 100 p. (See also http://www.seismo.nrcan.gc.ca/EarthquakesCanada.html.)

Anderson, J., Brune, J., Anooshehpoor, R., and Ni, S., 2000, New ground motion data and concepts in seismic hazard analysis: Current Science, v. 79, p. 1278–1290.

Atkinson, G., 1991, Use of the uniform hazard spectrum in characterizing expected levels of seismic ground shaking, *in* Proceedings of the 6th Canadian Conference on Earthquake Engineering: Toronto, p. 469–476.

Atkinson, G., 1993, Source spectra for earthquakes in eastern North America: Bulletin of the Seismological Society of America, v. 83, p. 1778–1798.

Atkinson, G., and Beresnev, I., 1998, Compatible ground-motion time histories for new national seismic hazard maps: Canadian Journal of Civil Engineering, v. 25, p. 305–318, doi: 10.1139/cjce-25-2-305.

Atkinson, G., and Beresnev, I., 2002, Ground motions at Memphis and St. Louis from M7.5 to 8 earthquakes in the New Madrid seismic zone: Bulletin of the Seismological Society of America, v. 92, p. 1015–1024, doi: 10.1785/0120010203.

Atkinson, G., and Boore, D., 1997, Some comparisons of recent ground motion relations: Seismological Research Letters, v. 68, p. 24–40.

Atkinson, G., and Boore, D., 1998, Evaluation of models for earthquake source spectra in eastern North America: Bulletin of the Seismological Society of America, v. 88, p. 917–934.

Atkinson, G., and Boore, D., 2003, Empirical ground-motion relations for subduction zone earthquakes and their application to Cascadia and other regions: Bulletin of the Seismological Society of America, v. 93, p. 1703–1729, doi: 10.1785/0120020156.

Atkinson, G., and Hanks, T., 1995, A high-frequency magnitude scale: Bulletin of the Seismological Society of America, v. 85, p. 825–833.

Atkinson, G., and Sonley, E., 2000, Empirical relationships between modified Mercalli intensity and response spectra: Bulletin of the Seismological Society of America, v. 90, p. 537–544, doi: 10.1785/0119990118.

Atkinson, G., Davenport, A., and Novak, M., 1982, Seismic risk to pipelines with application to northern Canada: Canadian Journal of Civil Engineering, v. 9, p. 248–264.

Bakun, W., and Hopper, M., 2004a, Historical seismic activity in the central United States: Seismological Research Letters, v. 75, p. 564–574.

Bakun, W., and Hopper, M., 2004b, Magnitudes and locations of the 1811–1812 New Madrid, Missouri, and the 1886 Charleston, South Carolina, earthquakes: Bulletin of the Seismological Society of America, v. 94, p. 64–75, doi: 10.1785/0120020122.

Basham, P., Weichert, D., Anglin, F., and Berry, M., 1982, New Probabilistic Strong Seismic Ground Motion Maps of Canada—A Compilation of Earthquake Source Zones, Methods and Results: Ottawa, Earth Physics Branch Open-File Report 82-33, 95 p.

Basham, P., Weichert, D., Anglin, F., and Berry, M., 1985, New probabilistic strong seismic ground motion maps of Canada: Bulletin of the Seismological Society of America, v. 75, p. 563–595.

Beresnev, I., and Atkinson, G., 2002, Source parameters of earthquakes in eastern and western North America based on finite-fault modeling: Bulletin of the Seismological Society of America, v. 92, p. 695–710, doi: 10.1785/0120010101.

Bernreuter, D., Savy, J., Mensing, R., Chen, C., and Davis, B., 1985, Seismic Hazard Characterization of the Eastern United States: Berkeley, Lawrence Livermore National Lab, University of California, 220 p.

Bodin, P., and Horton, S., 2004, Source parameters and tectonic implications of aftershocks of the M_w 7.6 Bhuj earthquake of January 26, 2001: Bulletin of the Seismological Society of America, v. 94, p. 818–827, doi: 10.1785/0120030176.

Boore, D., 2001, Comparisons of ground motions from the 1999 Chi-Chi earthquake with empirical predictions largely based on data from California: Bulletin of the Seismological Society of America, v. 91, p. 1212–1217, doi: 10.1785/0120000733.

Boore, D., and Atkinson, G., 1987, Stochastic prediction of ground motion and spectral response parameters at hard-rock sites in eastern North America: Bulletin of the Seismological Society of America, v. 77, p. 440–467.

Boore, D., Joyner, W., and Fumal, T., 1997, Equations for estimating horizontal response spectra and peak acceleration from western North American earthquakes: A summary of recent work: Seismological Research Letters, v. 68, p. 128–153.

Brune, J., 1996, Precariously balanced rocks and ground-motion maps for southern California: Bulletin of the Seismological Society of America, v. 86, p. 43–54.

Brune, J., 2001, Shattered rock and precarious rock evidence for strong asymmetry in ground motions during thrust faulting: Bulletin of the Seismological Society of America, v. 91, p. 441–447, doi: 10.1785/0120000118.

Cornell, C., 1968, Engineering seismic risk analysis: Bulletin of the Seismological Society of America, v. 58, p. 1583–1606.

Cornell, C., 1971, Probabilistic analysis of damage to structures under seismic loads, *in* Howells, D., Haigh, I., and Taylor, C., eds., Dynamic Waves in Civil Engineering: London, Wiley-Interscience, p. 473–493.

Crone, A., De Martini, P., Machette, M., Okumura, K., and Prescott, J., 2003, Paleoseismicity of two historically quiescent faults in Australia: Implications for fault behavior in stable continental regions: Bulletin of the Seismological Society of America, v. 93, p. 1913–1934, doi: 10.1785/0120000094.

Atkinson

Davenport, A., 1972, A Statistical Relationship between Shock Amplitude, Magnitude and Epicentral Distance and its Application to Seismic Zoning: London, Ontario, University of Western Ontario Engineering Science Research Report BLWT-4-72, 35 p.

EERI (Earthquake Engineering Research Institute) Committee on Seismic Risk, 1989, The basics of seismic risk analysis: Earthquake Spectra, v. 5, p. 675–702, doi: 10.1193/1.1585549.

EPRI (Electric Power Research Institute), 1986, Seismic Hazard Methodology for the Central and Eastern United States: Palo Alto, California, Electric Power Research Institute, Report NP-4726, 320 p.

Field, N., and the SCEC (Southern California Earthquake Center) Phase III Working Group, 2000, Accounting for site effects in probabilistic seismic hazard analyses of southern California: Overview of the SCEC Phase III report: Bulletin of the Seismological Society of America, v. 90, p. S1–S31, doi: 10.1785/0120000512.

Field, N., Jordan, T., and Cornell, C., 2003, OpenSHA: A developing community-modeling environment for seismic hazard analysis: Seismological Research Letters, v. 74, p. 406–419.

Frankel, A., 2003, Reply to "Should Memphis build for California's earthquakes?": Eos (Transactions, American Geophysical Union), v. 84, p. 19, 186.

Frankel, A., Mueller, C., Barnhard, T., Perkins, D., Leyendecker, E., Dickman, N., Hanson, S., and Hopper, M., 1996, National Seismic Hazard Maps, June 1996: U.S. Geological Survey Open-File Report 96-532, 100 p.

Frankel, A., Mueller, C., Barnhard, T., Perkins, D., Leyendecker, E., Dickman, N., Hanson, S., and Hopper, M., 1999, National seismic hazard mapping project: http://geohazards.cr.usgs.gov (accessed 10/2004).

Graves, R., Pitarka, A., and Somerville, P., 1998, Ground motion amplification in the Santa Monica area: Effects of shallow basin-edge structure: Bulletin of the Seismological Society of America, v. 88, p. 1224–1242.

Gumbel, E., 1954, Statistical theory of extreme values and some practical applications: National Bureau of Standards, Applied Mathematics Series, v. 33, 40 p.

Halchuk, S., and Adams, J., 1999, Crossing the border: Assessing the differences between new Canadian and American seismic hazard maps, *in* Proceedings of the 8th Canadian Conference on Earthquake Engineering: Vancouver, 13–16 June 1999, p. 77–82.

Halchuk, S., and Adams, J., 2004, Disaggregation of seismic hazard for selected Canadian cities, *in* Proceedings of the 13th World Conference on Earthquake Engineering: Vancouver, Canada, paper 2470 on CD-ROM.

Hartzell, S., Harmsen, S., Frankel, A., and Larsen, S., 1999, Calculation of broadband time histories of ground motion: Comparison of methods and validation using strong-ground motion from the 1994 Northridge earthquake: Bulletin of the Seismological Society of America, v. 89, p. 1484–1504.

Heidebrecht, A., 2003, Overview of NBCC 2003 seismic provisions: Canadian Journal of Civil Engineering, v. 30, p. 1–10, doi: 10.1139/l02-028.

Heidebrecht, A., Basham, P., Rainer, H., and Berry, M., 1983, Engineering applications of new probabilistic seismic ground motion maps of Canada: Canadian Journal of Civil Engineering, v. 10, p. 670–680.

Johnston, A., 1996, Seismic moment assessment of earthquakes in stable continental regions: Geophysical Journal International, v. 124, p. 381–414 (Part I); v. 125, p. 639–678 (Part II); v. 126, p. 314–344 (Part III).

Johnston, A., Coppersmith, K., Kanter, L., and Cornell, C., 1994, The Earthquakes of Stable Continental Regions: Palo Alto, California, Electric Power Research Institute Report TR-102261, volume 1, 89 p.

Joyner, W., and Boore, D., 2000, Recent developments in earthquake ground motion estimation, *in* Proceedings of the 6th International Conference on Seismic Zonation: Palm Springs, California, November 12–15, 2000.

Kaka, S., and Atkinson, G., 2004, Relationships between instrumental intensity and ground motion parameters in eastern North America: Bulletin of the Seismological Society of America, v. 94, p. 1728–1736, doi: 10.1785/012003228.

Kumarapeli, P., and Saull, V., 1966, The St. Lawrence valley system: A North American equivalent of the East African Rift Valley system: Canadian Journal of Earth Sciences, v. 3, p. 639–658.

McGuire, R., 1976, FORTRAN Computer Program for Seismic Risk Analysis: U.S. Geological Survey Open-File Report 76-67, 90 p.

McGuire, R., 1977, Seismic design spectra and mapping procedures using hazard analysis based directly on oscillator response: International Journal of

Earthquake Engineering and Structural Dynamics, v. 5, p. 211–234, doi: 10.1002/eqe.4290050302.

McGuire, R., 1995, Probabilistic seismic hazard analysis and design earthquakes: Closing the loop: Bulletin of the Seismological Society of America, v. 85, p. 1275–1284.

McGuire, R., and Toro, G., 1986, Methods of Earthquake Ground Motion Estimation for the Eastern United States: Electric Power Research Institute Report RP 2556-16 (prepared by Risk Engineering, Inc), 150 p.

McGuire, R., Silva, W., and Costantino, C., 2001, Technical Basis for Revision of Regulatory Guidance on Design Ground Motions: Hazard and Risk-Consistent Ground Motion Spectra Guidelines: U.S. Nuclear Regulatory Commission Report NUREG/CR-6728, 250 p.

McGuire, R., Silva, W., and Costantino, C., 2002, Technical Basis for Revision of Regulatory Guidance on Design Ground Motions: Development of Hazard and Risk-Consistent Seismic Spectra for Two Sites: U.S. Nuclear Regulatory Commission Report NUREG/CR-6769, 300 p.

Milne, W., and Davenport, A., 1969, Distribution of earthquake risk in Canada: Bulletin of the Seismological Society of America, v. 59, p. 729–754.

Newmark, N., and Hall, W., 1982, Earthquake Spectra and Design: EERI Monographs on Earthquake Criteria, Structural Design, and Strong Motion Records: El Cerrito, California, Earthquake Engineering Research Institute, 170 p.

Pitarka, A., Somerville, P., Fukushima, Y., Uetake, T., and Irikura, K., 2000, Simulation of near-fault strong-ground motion using hybrid Green's functions: Bulletin of the Seismological Society of America, v. 90, p. 566–586, doi: 10.1785/0119990108.

Reiter, L., 1990, Earthquake Hazard Analysis: Issues and Insights: New York, Columbia University Press, 420 p.

Richter, C., 1958, Elementary Seismology: New York, W.H. Freeman and Co, 500 p.

Saikia, C., and Somerville, P., 1997, Simulated hard-rock motions in Saint Louis, Missouri, from large New Madrid earthquakes (Mw≥6.5): Bulletin of the Seismological Society of America, v. 87, p. 123–139.

Schwartz, D., and Coppersmith, K., 1984, Fault behavior and characteristic earthquakes: Examples from the Wasatch and San Andreas fault zones: Journal of Geophysical Research, v. 89, p. 5681–5698.

Somerville, P., Smith, N., Graves, R., and Abrahamson, N., 1997, Modification of empirical strong ground motion attenuation relations to include the amplitude and duration effects of rupture directivity: Seismological Research Letters, v. 68, p. 199–222.

SSHAC, 1997, Recommendations for Probabilistic Seismic Hazard Analysis: Guidance on Uncertainty and Use of Experts: Senior Seismic Hazard Analysis Committee (SSHAC): U.S. Nuclear Regulatory Commission Report NUREG/CR-6372, 256 p.

Stein, S., Tomasello, J., and Newman, A., 2003, Should Memphis build for California's earthquakes?: Eos (Transactions, American Geophysical Union), v. 84, p. 19, 177, 184–185.

Toro, G., and McGuire, R., 1987a, Calculational procedures for seismic hazard analysis and its uncertainty in the eastern United States, *in* Proceedings of the Third International Conference on Soil Dynamics and Earthquake Engineering: Princeton, New Jersey, p. 195–206.

Toro, G., and McGuire, R., 1987b, An investigation into earthquake ground motion characteristics in eastern North America: Bulletin of the Seismological Society of America, v. 77, p. 468–489.

U.S. National Research Council Panel on Seismic Hazard Analysis, Committee on Seismology, 1988, Probabilistic Seismic Hazard Analysis: Washington, D.C., National Academy Press, 200 p.

Ward, S., 1995, Area-based testing of long-term earthquake hazard estimates: Bulletin of the Seismological Society of America, v. 85, p. 1285–1295.

Wesnousky, S., 1986, Earthquakes, Quaternary faults, and seismic hazard in California: Journal of Geophysical Research, v. 91, p. 12,587–12,631.

Whitman, R., ed., 1990, Workshop on Ground Motion Parameters for Seismic Hazard Mapping: New York, NCEER (National Center for Earthquake Engineering Research), Technical Report, 200 p.

MANUSCRIPT ACCEPTED BY THE SOCIETY 29 NOVEMBER 2006

The Geological Society of America
Special Paper 425
2007

Horizontal-to-vertical ground motion relations at short distances for four hard-rock sites in eastern Canada and implications for seismic hazard assessment

Allison L. Bent

Earthquakes Canada, Natural Resources Canada, Geological Survey of Canada, 7 Observatory Crescent, Ottawa, Ontario K1A 0Y3, Canada

Emily J. Delahaye[†]

School of Earth and Ocean Sciences, University of Victoria, P.O. Box 3055, Station CSC, Victoria, British Columbia V8W 3P6, Canada

ABSTRACT

Horizontal-to-vertical (H/V) ground-motion relations were determined for four long-running, three-component broadband seismograph stations situated on hard rock in eastern Canada. We focused our attention on earthquakes of magnitude 2.5 and greater at close range (<200 km) to each station because most earthquake damage results from nearby earthquakes. An H/V value of ~2 is the general average for eastern Canada, although there are some differences from station to station. H/V in general increases with increasing frequency. There is little or no systematic variation in H/V as a function of either distance or magnitude. The H/V ratios obtained in this study are somewhat higher than previously published values. Much of the difference appears to be due to the use of different definitions of H rather than to differences in the data sets and/or methods of data processing and analysis. These results raise questions regarding the best definition for H for use in hazard assessment.

Keywords: H/V, ground motion, eastern North America.

INTRODUCTION

Because vertical ground motions have traditionally been employed most often in routine analyses (such as epicentral determination) of earthquakes, many seismograph networks consist of seismographs that record only vertical motions. However, it is the horizontal ground motions that most often contribute to damage caused directly by shaking and are therefore of most interest for engineering and hazard purposes. Expected horizontal ground motions have often been estimated from vertical seismograms using assumptions about the horizontal-to-vertical (H/V) ground-motion relations, or they have been based on empirical results from small data sets. As more three-component stations are deployed, the data set of horizontal ground motions

[†]Present address: School of Earth Sciences, Victoria University of Wellington, P.O. Box 600, Wellington, New Zealand.

Bent, A.L., and Delahaye, E.J., 2007, Horizontal-to-vertical ground motion relations at short distances for four hard-rock sites in eastern Canada and implications for seismic hazard assessment, *in* Stein, S., and Mazzotti, S., ed., Continental Intraplate Earthquakes: Science, Hazard, and Policy Issues: Geological Society of America Special Paper 425, p. 345–352, doi: 10.1130/2007.2425(22). For permission to copy, contact editing@geosociety.org. ©2007 The Geological Society of America. All rights reserved.

is augmented, which allows reliable ratios of horizontal-to-vertical ground motions to be determined. Large earthquakes are an infrequent occurrence in eastern Canada. A better understanding of the horizontal-to-vertical ground-motion ratio will allow us to make use of existing records of vertical ground motion from past earthquakes to improve hazard estimates in the short-term until three-component data from larger earthquakes become available.

While there have been some previous efforts to determine horizontal-to-vertical ground motions for eastern Canada (e.g., Boore and Atkinson, 1987, Atkinson, 1993; Atkinson and Boore, 1997a; Siddiqqi and Atkinson, 2002), these studies primarily evaluated moderate and large earthquakes and thus required the use of data recorded at relatively large distances. Earthquake damage most commonly occurs close to the epicenter, although, obviously, the stronger the earthquake, the greater the potential damage area. Thus, we focused our efforts on analyzing data recorded close to the epicenter and analyzed small earthquakes, of which there are many. We recognize that small earthquakes are unlikely to be damaging, but establishing the relation, if any, between H/V and magnitude permits us to better use small earthquakes to predict the ground motions from larger earthquakes. To ensure a reasonably large data set, we concentrated on four of the longest running eastern Canadian three-component broadband stations, all of which are situated on hard rock.

Results from this study have implications for engineering and for seismic hazard assessments. They also raise questions about the best definition for "H," or horizontal, because many

of the differences in the results among previous ground-motion studies appear to be related to the definition of H and not to differences in the data sets or processing.

DATA SELECTION AND ANALYSIS

The stations GAC (Glen Almond, Quebec), LMN (Caledonia Mountain, New Brunswick), LMQ (La Malbaie, Quebec), and SADO (Sadowa, Ontario) were selected for analysis. All are long-running, three-component broadband stations situated on hard rock. GAC and LMQ are located within active seismic zones: the west Quebec and Charlevoix seismic zones, respectively. SADO and LMN are located within the Eastern background seismic zone, a region of low-level seismicity that encompasses much of southeastern Canada, and both are near somewhat more active seismic zones. A good summary of the seismicity and seismic zones of eastern Canada may be found in Adams and Basham (1991).

Data from all earthquakes of magnitude 2.5 or greater located within 200 km of each station were extracted and evaluated (Figs. 1–3). The time period extends from the date of station installation (1992 or 1993 depending on the station) through the end of 2002. The value of 200 km was chosen as the maximum distance because most earthquake damage from shaking results from nearby earthquakes, although we recognize that large earthquakes may cause damage at greater distances. The minimum magnitude was selected to ensure a good signal-to-noise ratio

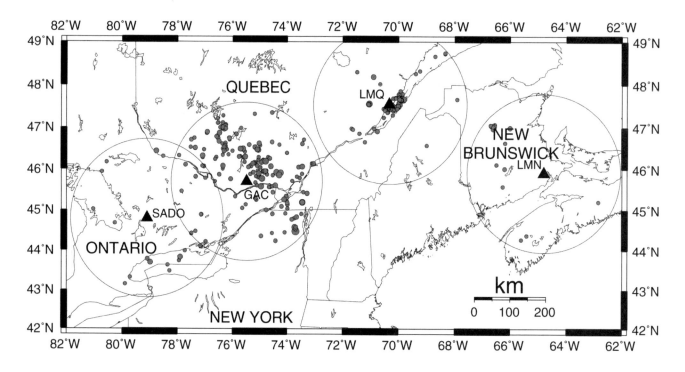

Figure 1. Map showing locations of stations evaluated in this study (black triangles) and epicenters of earthquakes analyzed (gray circles). Earthquake symbol size is scaled to magnitude. Only those earthquakes included in the study are shown. The large circles indicate the limits of the study area (200 km radius) for the station at the circle's center.

Distribution of Events by Distance

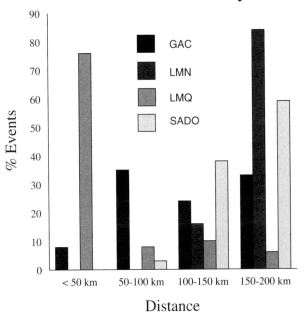

Figure 2. Data distribution by distance for the four stations studied. Note that the vertical axis represents the percentage of earthquakes in each distance range. The number of events varies from 22 for LMN to 157 for GAC.

Distribution of Events by Magnitude

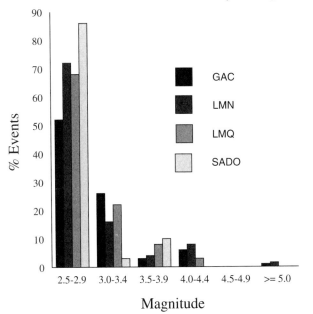

Figure 3. Data distribution by magnitude range. See comments for Figure 2.

over the distance range studied. The data set is, not surprisingly, larger for the stations LMQ and GAC than for SADO and LMN. At LMQ and GAC, the data set is also richer in terms of the percentage of earthquakes that occur at the closest distances (Fig. 2). One earthquake was added after the initial analysis had been performed. A magnitude (Ms) 5.1 earthquake occurred in the Charlevoix region on 6 March 2005, ~50 km from LMQ. Data from this earthquake were analyzed to provide data from a larger earthquake at a relatively short distance, but they did not alter the conclusions drawn from the initial analysis.

After the seismograms were extracted, they were examined visually. Any events for which there was not a clear signal were excluded from the study. Only a few events were rejected. Any DC offset and linear trend were removed from the data when applicable. Horizontal seismograms were rotated into their radial and tangential components. For the initial analysis, the peak amplitude was read from each seismogram, and the H/V ratios were computed. We define H as the vector sum of the horizontal components, as it represents the maximum horizontal ground motion.

Ratios were calculated for both velocity and acceleration at station GAC. As there was no discernible difference between the two, we subsequently worked strictly with velocity records because that is what the CNSN (Canadian National Seismograph Network) records. We also analyzed the ratio of each individual horizontal component (SH and SV) to the vertical, but focused on the combined H value since both components contribute to the ground motions. H/V for the peak amplitudes was then evaluated as a function of distance and magnitude.

To evaluate H/V as a function of frequency, we converted the data to the frequency domain using a Fourier transform of the waveform for a window beginning ~15 s before and ending ~15 s after the peak, ensuring that the same window was used for all three components. The spectra were smoothed using a root mean square (RMS) smoothing routine. Data were smoothed over a 0.06 log f window. The log f smoothing avoids some of the pitfalls of fixed window smoothing, which tends to oversmooth the long periods and/or undersmooth the high-frequency portion of the spectrum. We experimented with overlapping and nonoverlapping windows. The choice did not have a significant effect on the results, although overlapping windows generally resulted in smoother-looking spectra. The spectra shown in this paper were smoothed over a 0.06 log f window that moved in increments of 0.03 log f. That is, each frequency plotted is the center frequency (f_c) of a window with end points $f = f_c \pm 0.03 \log(f_c)$. The high-frequency end point of one window becomes the center frequency of the subsequent window. We experimented with other window widths and found that the results were not significantly affected by the choice.

We chose to use a RMS smoothing routine because it is more consistent with the laws of physics in that it does not violate the law of conservation of energy. However, log average smoothing has been used in some previous H/V studies, and there has been considerable debate about the relative merits of the two (Atkinson and Boore, 1998, 2000; Haddon, 2000). We calculated H/V for a subset of our data using log average smoothing. Although the amplitudes of the individual spectra were ~17% higher using RMS smoothing, the difference was nullified when ratios were taken. Thus, any differences between H/V calculated in this study and any previous ones employing log average smoothing do not appear to be related to the choice of spectral smoothing method.

RESULTS

Average H/V values obtained from peak amplitudes ranged from 2.0 to 2.5 (Table 1). Over the magnitude and distance ranges studied, there was no obvious correlation with H/V to either parameter (Fig. 4). Because average focal mechanisms within each of the western Quebec and Charlevoix seismic zones are similar—northwest-striking thrust for the former and northeast-striking thrust for the latter—their radiation patterns should be similar, and we also looked for systematic variations in H/V with back azimuth for the stations GAC and LMQ. No correlation was found.

At all stations, we saw a general trend of increasing H/V with increasing frequency (Fig. 5) over the range evaluated (0.5 Hz to 10 Hz). At frequencies lower than 0.5 Hz, the signal for most of the earthquakes in this data set was too small to be evaluated properly. H/V at 0.5 Hz was comparable to that at other frequencies, but the standard deviation was considerably higher, and the results could not be considered robust. Thus, the plots begin at 1 Hz. The trend of increasing H/V with frequency has been observed by other researchers, such as Siddiqqi and Atkinson (2002).

DISCUSSION

The H/V values obtained in this study are somewhat higher than those determined by previous researchers. An evaluation of the earlier studies shows that the mostly likely reason for the different results stems not from differences in the data sets used, as was suspected before this study was undertaken, or from errors in handling, but from differences in the methods for defining and calculating H or horizontal.

Horizontal ground motion is usually defined in terms of two parameters—most often either north-south and east-west or radial and transverse. When expressing horizontal motion as a single number, there does not appear to be a consistent definition in the published literature. The January/February 1997 issue of *Seismological Research Letters* was a special issue devoted to ground-motion relations. Three different definitions of H were used in the papers published therein: random horizontal component (Atkinson and Boore, 1997a, 1997b; Boore et al., 1997),

geometric mean (Abrahamson and Silva, 1997; Campbell, 1997; Sadigh et al., 1997; Somerville et al., 1997; Spudich et al., 1997; Youngs et al., 1997), and the larger of the two components (Anderson, 1997). Other methods that could conceivably be used to define H include, but are not restricted to, arithmetic mean and vector sum, the latter of which was the method used in this study. In a search of previous papers where H/V was calculated, the rationale for selecting any particular method was vague at best and most often nonexistent.

The arithmetic mean, geometric mean, and random vibration methods all use a contribution from both components but can underestimate the ground motion since the value for H will always be less than that of one of the individual components. Taking the maximum of the two components to represent the horizontal component may also underestimate the ground motion because the peak on one component does not always correspond to zero on the other. If one component is small when the other is at a maximum, the effect will be minimal. However, if the amplitude of the smaller component represents a significant proportion of the larger, the ground motion will be underestimated. The vector sum method used in this paper may sometimes overestimate the ground motion since the peak amplitudes do not always occur at the same time on the two components. An examination of the seismograms shows that at longer periods, the peaks frequently coincide, but that at higher frequencies they often do not. On the other hand, we are presenting the average ratios, so even if the average H/V overestimates the average ground motion, this value is always less than the maximum observed ratio.

If the ratios are calculated in the time domain, the results will be independent of the timing of the peaks. We attempted to do these calculations with the data but obtained suspiciously high ratios. Further examination showed that the problem was juxtaposed frequencies. That is, a high-frequency peak often occurred at the peak of a lower-frequency wave. Since many of the lower frequencies were within the range of interest, we were somewhat leery of overfiltering the data. Instead, we ran the test with synthetic seismograms. We made one seismogram and then phase-shifted it, which resulted in seismograms with peak amplitudes that were more or less the same but that did not occur at the same time. When the frequency domain analysis was repeated for the synthetic seismograms, the H/V ratio was consistently 1.4, which would be expected if all seismograms had the same peak amplitude, regardless of which phase shift was assigned to which component. In the time domain, however, the results were much more varied (Fig. 6), but the average value was ~4. Although the offset peaks means that the maximum H value will generally be smaller, it also means that the V value by which we are dividing is also smaller, and this effect appears to be dominant. These results suggest that by using the vector sum method in the frequency domain, we are not overestimating H/V, but it raises an additional question, whether H/V should be defined as H_{max}/V_{max} or $(H/V)_{max}$, similar to the (A/T) debate for magnitude. The frequency domain calculations use the former definition. As far as we have been able to ascertain, this is also true of previously published

TABLE 1. AVERAGE H/V FOR SELECTED FREQUENCIES

Station	Peak amp.	1 Hz	2 Hz	5 Hz	10 Hz
GAC	2.0	1.8	1.8	2.0	2.1
LMN	2.5	1.8	2.2	2.0	2.6
LMQ	2.3	1.9	1.9	1.8	2.9
SADO	2.0	1.4	1.7	1.9	1.9

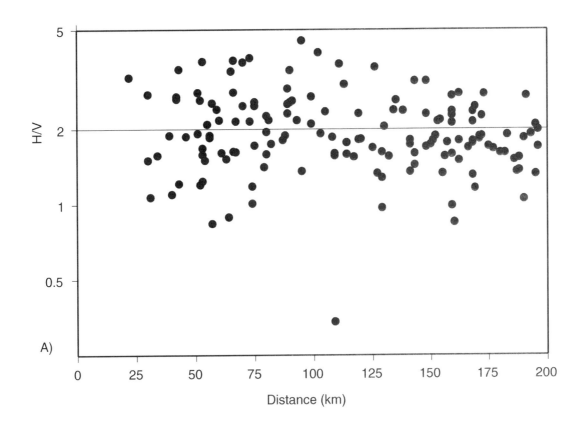

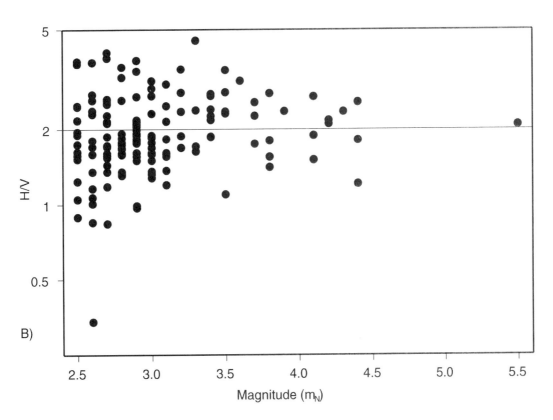

Figure 4. Horizontal-to-vertical ratio (H/V) for peak amplitude at GAC as a function of (A) distance (km) and (B) magnitude (m_N). The horizontal line shows the mean value. Note that the vertical scale is logarithmic.

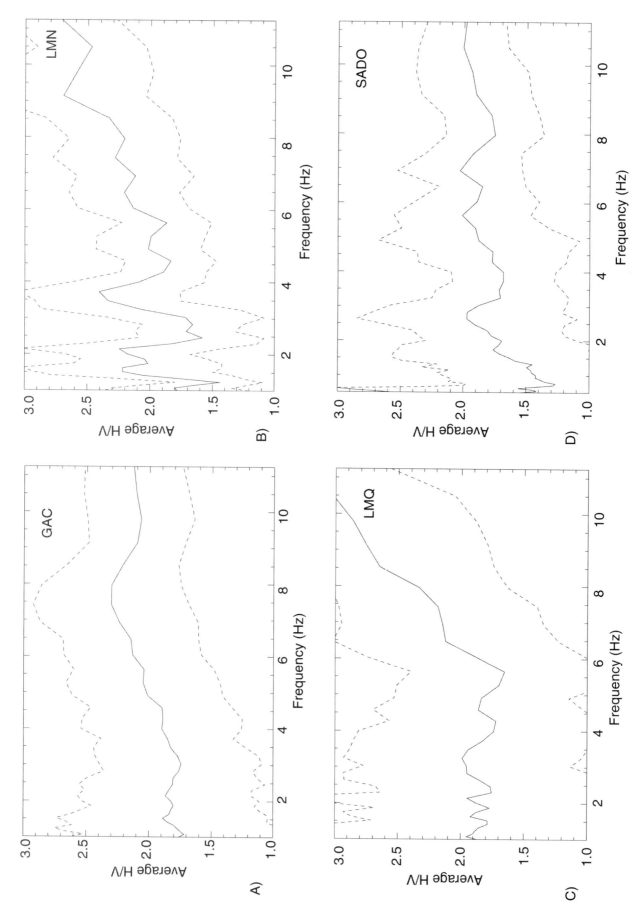

Figure 5. Horizontal-to-vertical ratio (H/V) as a function of frequency for (A) GAC, (B) LMN, (C) LMQ, and (D) SADO. Solid lines represent the mean H/V. Dashed lines show one standard deviation from the mean. Note that both axes are plotted on a linear scale.

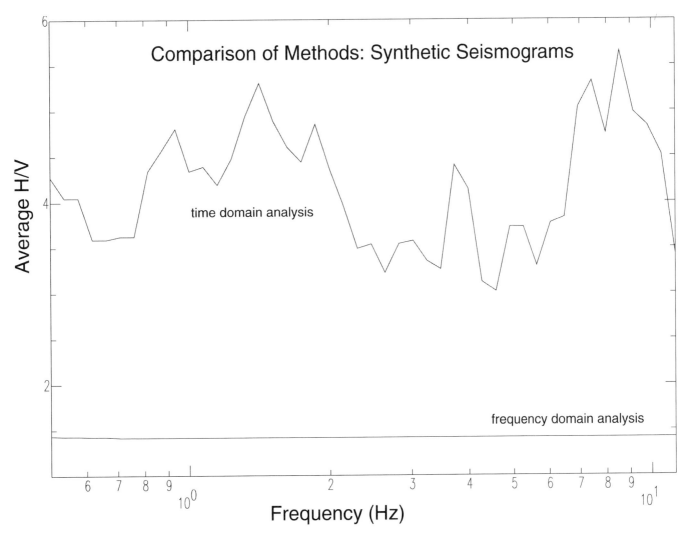

Figure 6. Horizontal-to-vertical ratio (H/V) calculated using phase-shifted synthetic seismograms and plotted as a function of frequency. The upper line shows the mean value for the time domain calculations, and the lower line shows the frequency domain. The time domain values will vary depending on which synthetic seismogram is assigned to which component, whereas the frequency domain values will be constant.

H/V studies. The time domain calculations, on the other hand, are more closely indicative of $(H/V)_{max}$.

To compare the results obtained by different methods, the value of H/V at GAC was calculated at three frequencies (1 Hz, 2 Hz, and 5 Hz) using five different methods to calculate H—arithmetic mean, geometric mean, largest of two components, random horizontal vibration, and vector sum. The results are presented in Table 2. All values have standard deviations on the order of 0.5. The arithmetic mean, geometric mean, and random vibration methods resulted in similar values of H/V at all

frequencies examined and lower values than the other methods did. As expected, the largest H/V values came from the vector sum method, and intermediate values were found using the largest horizontal component as H. The relative H/V as a function of method is consistent with what would be expected based on logic. The relative values of any two methods are consistent and independent of frequency, and therefore results from one method can be reliably converted to those that would be expected from another method if a particular application specifies how H should be defined. We note that seismic hazard assessments normally reflect ground motions only. The application of the estimated ground motions to the construction of a particular structure must take the design into consideration, which may influence the preferred definition of H.

We also note that when we selected the definition of H used by previous researchers and applied it to our data set, we obtained values for H/V similar to theirs. For example, when we applied

TABLE 2. AVERAGE H/V AT GAC FOR VARIOUS DEFINITIONS OF H

Definition of H	1 Hz	2 Hz	5 Hz
Arithmetic mean	1.2	1.3	1.4
Geometric mean	1.2	1.2	1.4
Maximum (N or E)	1.5	1.5	1.6
Random vibration	1.2	1.3	1.4
Vector sum	1.8	1.8	2.0

the random vibration method preferred by Siddiqqi and Atkinson (2002) to the GAC data, we obtained H/V values of 1.2 at 1 Hz and 1.4 at 5 Hz (Table 2), which are comparable to their values of 1.1 and 1.4.

CONCLUSIONS

We determined H/V values for four hard-rock sites in eastern Canada based on data recorded at close distances. The average value of H/V was two, although there were variations with frequency and site. The average H/V was higher than expected based on the results of previous studies. Comparisons of data sets and methods showed that the differences result primarily from variations in the definition of H when treated as a single value. These results suggest that using data recorded at several hundred kilometers will not adversely affect H/V ratios, which allows the use of larger data sets. The lack of a correlation between H/V and magnitude implies that H/V from small earthquakes can be used to predict ground motions for larger earthquakes, although we caution that there were no earthquakes of magnitude 6.0 or greater in our data set, and that for very large earthquakes, these relations may not apply due to nonlinear soil behavior. We also note that if we calculate H/V in the time domain rather than in the frequency domain, the ratio increases by about a factor of two. Using phase-shifted synthetic seismograms where each component had the same peak amplitude, we obtained an average H/V of 1.4 in the frequency domain and ~4 in the time domain. Finally, we urge that anyone using published H/V values pay close attention to the method used to calculate them and verify that the method is compatible with their application. For example, if recorded ground motions are significantly higher or lower than a design ground motion, the method of defining H may be unimportant, but if the two are close, it is important that they be calculated in the same manner in order to determine whether the shaking would exceed the design level.

ACKNOWLEDGMENTS

All maps and graphs in the paper were generated using GMT (Generic Mapping Tool) (Wessel and Smith, 1991). We thank Chris Majewski for his scripts, which facilitated data handling, and John Cassidy, Martin Chapman, and Stéphane Mazzotti for their constructive reviews.

REFERENCES CITED

Abrahamson, N.A., and Silva, W.J., 1997, Empirical response spectral attenuation relations for shallow crustal earthquakes: Seismological Research Letters, v. 68, p. 94–127.

Adams, J., and Basham, P.W., 1991, The seismicity and seismotectonics of eastern Canada, *in* Slemmons, D.B., Engdahl, E.R., Zoback, M.D., and Blackwell, D.D., eds., Neotectonics of North America: Boulder, Colorado, Geological Society of America Decade Map Volume, p. 261–276.

Anderson, J.G., 1997, Nonparametric description of peak acceleration above a subduction thrust: Seismological Research Letters, v. 68, p. 86–95.

Atkinson, G.M., 1993, Notes on ground motion parameters for eastern North America: Duration and H/V ratio: Bulletin of the Seismological Society of America, v. 83, p. 587–596.

Atkinson, G.M., and Boore, D.M., 1997a, Some comparisons between recent ground-motion relations: Seismological Research Letters, v. 68, p. 24–40.

Atkinson, G.M., and Boore, D.M., 1997b, Stochastic point-source modeling of ground motions in the Cascadia region: Seismological Research Letters, v. 68, p. 74–85.

Atkinson, G.M., and Boore, D.M., 1998, Evaluation of models for earthquake source spectra in eastern North America: Bulletin of the Seismological Society of America, v. 88, p. 917–934.

Atkinson, G.M., and Boore, D.M., 2000, Evaluation of models for earthquake source spectra in eastern North America: Reply: Bulletin of the Seismological Society of America, v. 90, p. 1339–1341, doi: 10.1785/0120000088.

Boore, D.M., and Atkinson, G.M., 1987, Stochastic prediction of ground motion and spectral response parameters at hard-rock sites in eastern North America: Bulletin of the Seismological Society of America, v. 77, p. 440–467.

Boore, D.M., Joyner, W.B., and Fumal, T.E., 1997, Equations for estimating horizontal response spectra and peak acceleration from western North American earthquakes: A summary of recent work: Seismological Research Letters, v. 68, p. 128–153.

Campbell, K.W., 1997, Empirical near-source attenuation relationships for horizontal and vertical components of peak ground acceleration, peak ground velocity, and pseudo-absolute acceleration response spectra: Seismological Research Letters, v. 68, p. 154–179.

Haddon, R.A.W., 2000, Evaluation of models for earthquake source spectra in eastern North America: Discussion: Bulletin of the Seismological Society of America, v. 90, p. 1332–1338, doi: 10.1785/0119990058.

Sadigh, K., Chang, C.-Y., Egan, J.A., Makdisi, F., and Youngs, R.R., 1997, Attenuation relationships for shallow crustal earthquakes based on California strong motion data: Seismological Research Letters, v. 68, p. 180–189.

Siddiqqi, J., and Atkinson, G.M., 2002, Ground-motion amplification at rock sites across Canada as determined from the horizontal-to-vertical component ratio: Bulletin of the Seismological Society of America, v. 92, p. 877–884, doi: 10.1785/0120010155.

Somerville, P.G., Smith, N.F., Graves, R.W., and Abrahamson, N.A., 1997, Modification of empirical strong ground motion attenuation relations to include the amplitude and duration effects of rupture directivity: Seismological Research Letters, v. 68, p. 199–222.

Spudich, P., Fletcher, J.B., Hellweg, M., Boatwright, J., Sullivan, C., Joyner, W.B., Hanks, T.C., Boore, D.M., McGarr, A., Baker, L.M., and Lindh, A.G., 1997, SEA96—A new predictive relation for earthquake ground motions in extensional tectonic regimes: Seismological Research Letters, v. 68, p. 190–198.

Wessel, P., and Smith, W.H.F., 1991, Free software helps map and display data: Eos (Transactions, American Geophysical Union), v. 72, p. 441, doi: 10.1029/90EO00319.

Youngs, R.R., Chiou, S.-J., Silva, W.J., and Humphrey, J.R., 1997, Strong ground motion attenuation relationships for subduction zone earthquakes: Seismological Research Letters, v. 68, p. 58–73.

MANUSCRIPT ACCEPTED BY THE SOCIETY 29 NOVEMBER 2006

The Geological Society of America
Special Paper 425
2007

Does it make sense from engineering and economic perspectives to design for a 2475-year earthquake?

G.R. Searer

Consultant, Wiss, Janney, Elstner Associates, Inc., 2550 North Hollywood Way, Suite 502, Burbank, California 91505, USA

S.A. Freeman

Principal, Wiss, Janney, Elstner Associates, Inc., 2200 Powell Street, Suite 925, Emeryville, California 94608, USA

T.F. Paret

Senior Consultant, Wiss, Janney, Elstner Associates, Inc., 2200 Powell Street, Suite 925, Emeryville, California 94608, USA

ABSTRACT

The code provisions for earthquake-resistant design have been substantially revised in the development of the International Building Code (IBC) and other relatively new standards. One of the most significant changes, which is likely to have significant repercussions nationwide, is the switch from the use of a 475-year mean return period earthquake (an earthquake with a 10% chance of exceedance in 50-year) to a 2475-year mean return period earthquake (an earthquake with a 2% chance of exceedance in 50-year) in design of new buildings and in the evaluation of existing buildings. According to the Federal Emergency Management Agency (FEMA), the design life span (or average life) of a normal structure is generally assumed to be ~50 years. Consequently, wind and snow loads are based on a 50-year mean return period, and flood hazard maps are generally based on 100-year and 500-year return periods. Prior to the 2000 IBC, seismic design loads generally matched the assumption of a 50-year design life and were designed with the intent to withstand a 475-year earthquake. This paper addresses the switch to a 2475-year earthquake from engineering and economic perspectives. Based on engineering and economic research, the switch does not appear to be fully justified and may have significant negative repercussions, particularly on existing building stock. Using the annualized estimated loss (AEL), the annualized estimated loss ratio (AELR), and the per capita annualized estimated loss—all based on results from the FEMA-developed *Hazards* U.S. (HAZUS) analysis program—the use of the 2475-year earthquake in the design of new structures appears to be questionable from both economic and engineering standpoints.

Keywords: structural engineering, economics, 2475-year earthquake, 475-year earthquake.

Searer, G.R., Freeman, S.A., and Paret, T.F., 2007, Does it make sense from engineering and economic perspectives to design for a 2475-year earthquake?, *in* Stein, S., and Mazzotti, S., ed., Continental Intraplate Earthquakes: Science, Hazard, and Policy Issues: Geological Society of America Special Paper 425, p. 353–361, doi: 10.1130/2007.2425(23). For permission to copy, contact editing@geosociety.org. ©2007 The Geological Society of America. All rights reserved.

INTRODUCTION

The code provisions for earthquake-resistant design have been substantially revised in the last few years. One of the most significant changes—and one likely to have significant repercussions nationwide—is the switch from the use of a 475-year mean return period earthquake (an earthquake with a 10% chance of exceedance in 50 years) to a 2475-year mean return period earthquake (an earthquake with a 2% chance of exceedance in 50 years) in design of new buildings and in the evaluation of existing buildings. Note that while an earthquake is referred to in terms of its mean return period, a return period is merely the best available scientific estimate as to the approximate recurrence interval for that size earthquake—and it is not a precise prediction of probability.

Recently developed codes and standards, such as the International Building Code (IBC), the International Existing Building Code (IEBC), American Society of Civil Engineers (ASCE) 31 *Seismic Evaluation of Existing Buildings*, and Federal Emergency Management Agency (FEMA) 356 *Prestandard and Commentary for the Seismic Rehabilitation of Buildings*, now rely on the use of the 2475-year earthquake for the design, evaluation, and upgrade of buildings. A number of recent papers have touched on this issue (Frankel, 2004; Stein, 2005). This paper addresses the switch to a 2475-year earthquake from engineering and economic perspectives.

From an engineering standpoint, use of the 2475-year earthquake represents a dramatic shift from the design philosophy used for other loads, such as wind, snow, and flood.

We argue here that use of the 2475-year earthquake in the design of new buildings appears to be questionable from an engineering perspective but may be somewhat more justified from a public policy standpoint. We further argue that use of the 2475-year earthquake in the evaluation of existing structures appears to be inappropriate based on the shortened remaining life span of these structures, and it has the potential to result in large numbers of structures improperly being deemed seismically inadequate; these structures may be upgraded or abandoned or demolished unnecessarily.

USE OF THE 2475-YEAR EARTHQUAKE

With the adoption of the IBC in at least portions of 45 states (ICC, 2005), use of an earthquake with a 2475-year return period in seismic design of new buildings is nearly ubiquitous. Furthermore, with continued pressure on earthquake engineers to use ASCE-31 *Seismic Evaluation of Existing Buildings* and FEMA 356 *Prestandard and Commentary for the Seismic Rehabilitation of Buildings*, the use of the 2475-year earthquake as the benchmark to evaluate and upgrade existing buildings is also common. For most of the United States, the use of the 2475-year earthquake generally results in substantially larger design forces than previously required or recommended by codes and standards (FEMA, 2001).

For the 1997 National Earthquake Hazards Reduction Program (NERHP), which serves as the basis for the International Building Code (IBC), the seismic design criterion was changed to two-thirds of the maximum considered earthquake (MCE) from the 475-year average return period earthquake that has been the basis for the seismic provisions of previous editions of the Uniform Building Code (UBC) and NEHRP. The MCE is defined as the lesser of the 2475-year average return period or 1.5 times the mean deterministic earthquake.

According to FEMA, the decision to use the 2475-year earthquake was made to "provide for a uniform margin against collapse" throughout the United States (FEMA, 2001, p. 45). While this approach may be a commendable goal, in reality, this decision involves what appears to be an arbitrary selection of a design return period without considering the costs and benefits of this choice. Absent such consideration, a 10,000-year earthquake or even a 100,000-year earthquake could have been selected as the design basis, just as a 100-year event could have been selected—all with the goal of "providing for a uniform margin against collapse" throughout the United States.

As mentioned already, for most of the United States, the 2475-year earthquake is significantly larger than the 475-year earthquake, and the appropriateness of designing structures for an earthquake with such a long return period is questionable, particularly when the criterion is applied to all buildings without consideration of use or expected longevity (Stein, 2005). In our opinion, the return period used for seismic design purposes is arbitrary and should be selected rationally while weighing possible risks and rewards, but there is no clear evidence that the adoption of the 2475-year earthquake in new codes involved any such rational approach.

Examples of the significant changes that resulted from switching from the 475-year average return period to the 2475-year average return period include the case of Charleston, South Carolina, where the design forces were approximately doubled, and the case of Boise, Idaho, where the design forces were reduced by ~50%. In the 1997 Uniform Building Code (as well as prior editions), the country was divided up into seismic zones that generally represented the relative magnitude of the seismic hazard (i.e., San Francisco, California, and Los Angeles, California, had an expected peak ground acceleration [PGA] of 0.4*g* to 0.6*g* and were ranked with the highest seismicity—Zone 4; Portland, Oregon, Seattle, Washington, Salt Lake City, Utah, Memphis, Tennessee, and New Madrid, Missouri, were estimated to have a somewhat lower PGA of 0.2*g* to 0.3*g* and were placed in Zone 3; and Charleston, South Carolina, and Boston, Massachusetts were estimated to have a moderate to low PGA of 0.15*g* to 0.2*g* and were grouped in seismic Zone 2A; etc.). Figure 1 is a plot that compares the new and old seismic design criteria for a typical California Zone 4 city with sites in Seattle, Washington; Salt Lake City, Utah; New Madrid, Missouri; and Charleston, South Carolina.

As stated above, in the previous editions of both the UBC and NEHRP, the design earthquake was based on the 475-year earthquake. Changing the basis from the 475-year earthquake to a 2475-year earthquake radically altered seismic design for

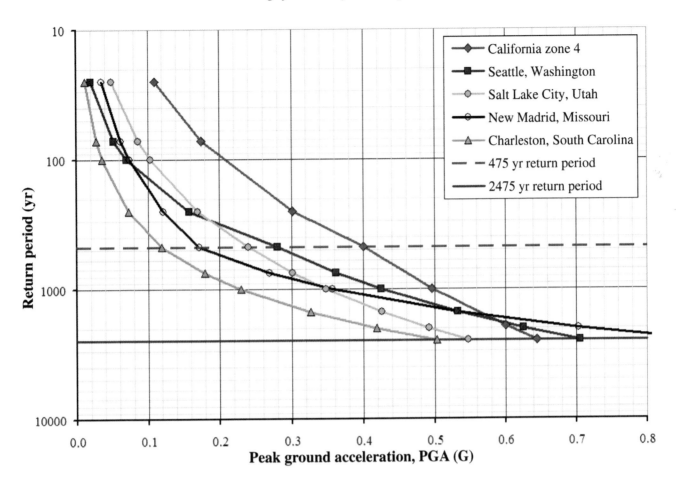

Figure 1. Peak ground acceleration (PGA) for sites in various U.S. cities (after Searer and Freeman, 2002).

numerous cities across the United States, including the above cities. As shown, use of the longer return period requires designing for much higher forces (as represented by the PGA). For example, although for the 475-year return period, the design PGA in New Madrid is substantially lower than the design PGA values for California, Seattle, or Salt Lake City, the 2475-year earthquake is substantially higher for New Madrid than the others. Furthermore, even the highest seismic areas in California are surpassed in seismicity by Seattle and New Madrid when the 2475-year earthquake is considered.

On a worldwide basis, we note that the use of a 2475-year earthquake also differs significantly from projects such as the Global Seismic Hazard Assessment Program, which bases its seismicity estimates on the size of the 475-year earthquake (GSHAP, 1999).

BRIEF INTRODUCTION TO SEISMIC DESIGN LOADS

It is important to understand that seismic design is significantly different from design for wind or snow loads. Wind and snow forces used in design are similar in magnitude to the actual forces expected from the event, so that if the design event occurs, no significant damage is generally expected. Conversely, in seismic design, the design forces are substantially *smaller* than the expected seismic forces, and if the design earthquake occurs, significant architectural and structural damage is generally expected (SEAOC, 1999). This difference in philosophy occurs because social policy makers and engineers have weighed the cost of stronger buildings against the hazard posed by infrequent earthquakes. From a design perspective, it is relatively easy to increase the design forces for earthquakes; however, the cost to actually build structures designed for much larger forces has the potential to become prohibitive. It is fairly inexpensive to increase the sizes of a few selected members (such as going from a steel column that weighs 120 pounds per foot to a column that weighs 150 pounds per foot or going from an 8-in-thick wall to a 12-in-thick wall), but a wholesale increase in design forces can have substantial effects on other portions of a structure that do not necessarily result in better performance during an earthquake (such as the foundations). Given that most structures, on the whole, have performed reasonably well during recent earthquakes in California, up until quite recently, practicing engineers have generally resisted increasing seismic design forces on a wholesale level.

CHANGE IN DESIGN PHILOSOPHY

According to FEMA, the design life span (or average life) of a normal structure is generally assumed to be approximately 50 years. FEMA apparently presumes that after 50 years, many structures will be approaching the end of their useful life, and the structures are assumed to require evaluation and possibly rehabilitation for their continued use (FEMA, 2001). We note that the concept of a 50-year lifespan for buildings is questionable, given that many buildings in the United States are much older than 50 years and that a longer design life (perhaps 100 years) would appear more realistic and defensible. Nevertheless, wind loads as well as snow loads are based on a 50-year mean return period, and flood hazard maps are generally based on 100-year and 500-year return periods. Prior to the 2000 IBC, seismic design loads generally matched the assumption of a 50-year design life and were designed with the intent to withstand a 475-year earthquake (10% chance of exceedance in 50 years). Stated this way, design of structures for a 475-year earthquake is not inconsistent with a structure's useful life expectancy of 50 years, and the acceptance of a 10% chance of exceedance does not seem to be an unreasonable risk for a normal structure. By selecting a 2475-year return period for design of typical structures (2% chance of exceedance in 50 years), the seismic codes have significantly increased the differences in design philosophy between seismic and nonseismic design criteria and have essentially reduced the risk of exceedance by 80%. We note that critical or essential facilities are often designed for higher seismic loads that, in effect, correspond to earthquakes with longer mean recurrence intervals and that have a corresponding lower annual probability of exceedance. One potential reason for designing critical or essential facilities for higher seismic loads is to reduce the potential for indirect economic and life-safety losses associated with the disruption of such facilities.

In our opinion, the selection of a 2475-year design earthquake seems extreme in light of the much shorter design life span of a typical building. As an example, consider two sites that have similar Peak Ground Acceleration (PGAs) values for the 2475-year return period: one may be subjected to large PGAs for earthquakes with a 475-year return period and the other site may be subjected to relatively minor PGAs for earthquakes with a 475-year return period. While both sites would be rated equal according to the new criteria, the latter case actually has significantly less risk of damage during a building's normal life span than the former case.

However, while the use of a 2475-year design earthquake may seem excessive on a building-by-building basis, we also note that use of a 2475-year earthquake in the design of new buildings may be reasonable when considering the risk of major community-wide disasters or when designing critical or essential structures.

ECONOMIC ASSESSMENT OF EARTHQUAKE HAZARD

With respect to expression of seismic hazard in the United States, it has been said that 39 of the 50 states "are directly vul-nerable to serious earthquakes" (EERI, 2003); this statement gives the impression that these 39 states have a relatively significant seismic risk. Another variation is that "at least 39 states are considered at risk from moderate to great earthquakes" (NEHRP, 2005). At least one basis for these statements appears to be a figure from FEMA 366, *HAZUS99 Estimated Annualized Earthquake Losses for the United States*, which, through a detailed study using the FEMA-developed Hazards U.S. (HAZUS) analysis program, estimated the losses to building stock due to all potential future ground motions along each known fault or fault system. By weighting these losses by the annual likelihood of rupture of each fault, an annualized estimated loss was produced for metropolitan areas and for each state. Figure 2 shows the annualized estimated loss from earthquakes for various regions in the United States (reproduced directly from FEMA 366).

Note that Figure 2 would seem to indicate that in absolute terms, the regional annualized seismic risk (as measured by dollars) for most of the country seems to be small, especially since 74% of the annualized earthquake loss (AEL) for the entire United States occurs in California alone and since 84% of the AEL occurs on the West Coast (California, Oregon, and Washington). In fact, Figure 2 actually gives the impression that only three states have substantially greater seismic risk as measured by AEL than the rest of the country. However, Figure 2 is not normalized by state building inventory or population and does not tell the whole story, as is discussed later.

Regarding the statement that 39 states are "directly vulnerable to serious earthquakes," the basis of this statement appears to be a figure from FEMA 366 (shown in Fig. 3 in this paper) that shows 39 states shaded with various colors. The colors represent the annualized estimated loss ratio (AELR)—the ratio of the annual estimated loss in each state divided by the replacement value of the building stock in each state. At least on the surface, this would appear to be a more meaningful measure of risk—weighting each state's estimated loss by the replacement value of the state's building stock. Indeed, we concur that this is a valuable measure; however, we note that the use of irregular ranges of data values for each color employed in the figure leads to irrational conclusions.

The key for Figure 3 shows that red-colored states have over $500 per year of seismic losses per million dollars in replacement value; yellow-colored states have AELR values of $100 to $500 per million dollars in replacement value, green-colored states have AELR values of 50–100, and blue-colored states have AELR values of 10–50. Presumably the 11 white-colored states, which have AELR values less than 10, are not "directly vulnerable to serious earthquakes." Although we cannot intuit the reasoning behind the chosen scale, it is important to note that the relative scale range for each color is not uniform. The net result of this irregular scale is that approximately one fifth of the states fall into each color category. Since four of the five groups are assigned colors other than white, it is not surprising that ~80% (39 out of 50) of the states have a color assigned to them, and thus it is not surprising that someone taking a cursory look at this

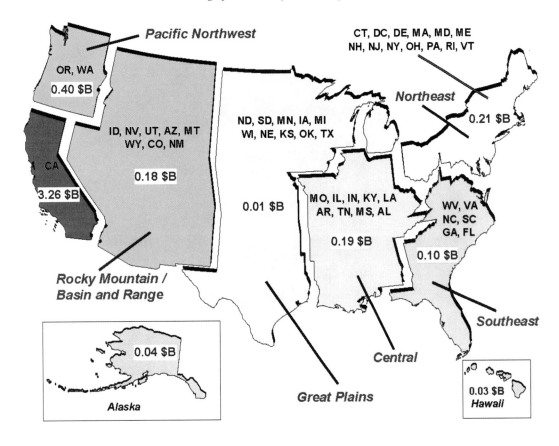

Figure 2. Figure reproduced from Federal Emergency Management Agency (FEMA) 366 showing the annualized earthquake loss for regions of the United States.

figure could conclude that 39 out of 50 states are "directly vulnerable to serious earthquakes."

We contend, however, that the same data, properly viewed, leads to an opposite conclusion—namely that only a few states have significant relative seismic risk based on the AELR (Fig. 4).

In Figure 4, we elected to use the same four colors (and white) as the figure in FEMA 366, but instead chose to assign the red color to a much higher AELR (above 2000) and to consistently reduce each subsequent color range by a multiple of two. With this AELR scale, only 10 states would qualify as "directly vulnerable to serious earthquakes," not 39 as implied by FEMA 366. Note that if we had instead used a linear scale to assign colors (such as red = AELR over 2000, yellow = 1500–2000, green = 1000–1500, blue = 500–1000, and white = less than 500), California would still be red, there would be no yellow-colored states, only Alaska and Oregon would be green-colored, and only Washington, Nevada, Utah, and Hawaii would be blue-colored, as shown in Figure 5.

In fact, California, with an AELR of 2049 is almost double the next closest state of Alaska, with an AELR of 1165. So based on either this normalized or linearized measure, California still clearly dominates the rest of the United States in terms of seismic hazard; therefore, the decision to switch to a 2475-year design

earthquake—which increases seismic forces for most areas of the country with the exception of California—appears questionable. Even when AELR values are computed on a city-by-city basis, with a few exceptions, California cities generally appear to be the most vulnerable to earthquakes, as shown in Table 1. We note that other researchers have reached similar conclusions (Stein et al., 2003).

Further evidence that the switch to the 2475-year design earthquake may be inappropriate is presented in Table 2, which again uses the annualized estimated losses due to earthquakes (FEMA, 2004) but this time normalizes the results based on population (U.S. Census Bureau, 2000). Since the number of buildings and the population in a given metropolitan area can be presumed to be correlated, we calculated an *approximate* annual per capita loss for each metropolitan area. Looked at another way, the data can be used to approximate a normalized or relative seismic risk associated with each area.

Note that this table is based on limited data and should be viewed only as an approximate measure of relative risk. However, the table shows that California cities are generally the most prone to significant damage on an annual per capita basis.

Like the AELR, using the per capita estimated annual loss results in approximately the same conclusion as using the relative seismic hazard between various metropolitan areas; i.e., the West

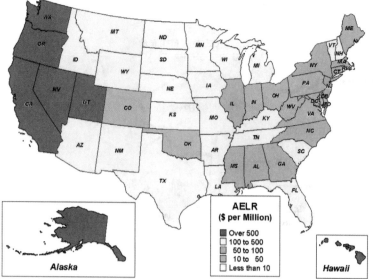

Figure 3. Figure reproduced from Federal Emergency Management Agency (FEMA) 366 showing the annualized earthquake loss ratio (AELR) for each state.

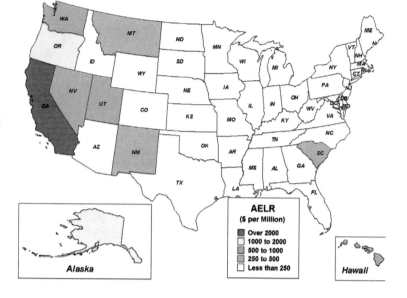

Figure 4. Federal Emergency Management Agency (FEMA) 366 figure modified to show a more uniform scale range for each color.

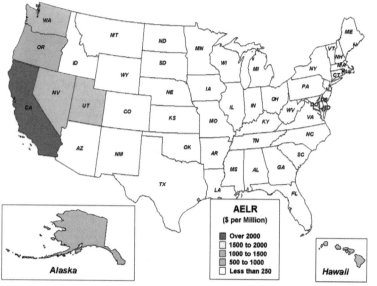

Figure 5. Federal Emergency Management Agency (FEMA) 366 figure modified to show a linear scale range for each color.

Coast of the United States—in particular California—poses the most substantial seismic risk by far, even when normalized by replacement value of the building stock or by population. Consequently, it appears that the seismic zones from the 1997 UBC may have been more appropriate than the use of a longer period earthquake.

The results of the switch to a 2475-year earthquake are clear: a number of cities will now have seismic design criteria significantly more stringent than have ever been required, and in some cases, more stringent than many areas of high seismicity in California. We believe that this jump in design criterion appears difficult to justify for all buildings given the extremely low probability of the 2475-year earthquake occurring within the lifetime of a given structure.

It is important to note that the AELR and per capita estimated annual loss are limited by the laws of probability (as well as our limited knowledge of seismicity); for example, if a city were located near a fault that ruptures once every 2475 years but that had built up stresses over several thousand years without an earthquake and was now "due" for a major earthquake, the AELR and the per capita estimated annual loss would not reflect this probability—other than by distributing the losses expected in the near term over the 2475-year return period. Of course, given the inability of seismologists to predict earthquakes even at locations where earthquakes occur at regular intervals (e.g., Parkfield, California), the reliability of any assessment that a major earthquake is "due" would appear questionable.

Also important to understand is the fact that while HAZUS can estimate losses to building stock, HAZUS cannot estimate the effects on a community's economy or the societal effects that a major disaster will have on communities in general. Thus, economic studies based on HAZUS, such as the AELR and the per capita estimated annual loss, substantially underestimate the effects of earthquakes on communities, states, and even the country.

IMPACT OF THE 2475-YEAR EARTHQUAKE ON EXISTING BUILDING STOCK

The most critical consequence of the change in seismic design criteria is that most of the existing building stock in the majority of affected cities will suddenly be perceived to be inadequate, thus creating an economic burden relating to expensive seismic upgrades that will likely provide little or no benefit during the remaining life of the structures.

For cities that have 2475-year design earthquakes that are substantially larger than the old 475-year earthquake (which, according to FEMA, is most of the United States, with California as the notable exception), the existing building stock is likely to have been designed for seismic forces substantially less than that required as a result of the 2475-year earthquake. Since national standards such as ASCE-31 and FEMA 356 use the larger forces associated with the 2475-year earthquake for the evaluation and upgrade existing buildings, the natural temptation facing owners,

building officials, and engineers is to ask that existing buildings be compared with the requirements of current code, and it is easy to see how nearly all existing buildings will suddenly be perceived to pose great risks to occupants despite the fact that the probability of a major earthquake is so low. Furthermore, the newly developed 2003 International Code for Existing Buildings (IEBC) has the potential—unlike all prior recent codes—to require full seismic upgrades for buildings as a result of only moderate damage completely unrelated to earthquakes. Consequently, existing building stock in much of the United States will be deemed to have significant deficiencies when evaluated according to ASCE-31 and to require significant upgrading according to FEMA 356 and the IEBC. Use of the 2475-year earthquake in evaluation of existing buildings has the potential to result in large numbers of structures being deemed seismically inadequate. Structures that are deemed inadequate may need to be upgraded or abandoned or demolished, and adoption of the poorly worded IEBC may make repair of damaged buildings economically infeasible.

While upgrading existing buildings to meet current seismic codes (including the 2475-year earthquake) is not necessarily a bad idea, it is not necessarily a good idea either. Governments as well as private industry have limited monetary resources, and it is simply not economically feasible to require that every existing building be upgraded to current code; owning buildings would quickly become cost-prohibitive. Funneling money into upgrading a building must naturally take away money from some other cause or objective. The only logical way to determine whether or not a building should be upgraded is to meaningfully weigh potential risks and the rewards.

As an example, consider the Veterans Administration (VA) hospital in Memphis, Tennessee. The Memphis VA hospital, which is located within the New Madrid fault zone, was the subject of a recent upgrade project. The structure, which was built in 1967, was evaluated and found to be deficient according to modern seismic standards. According to recent research, the New Madrid fault has a 7%–10% chance of experiencing a major earthquake in the next 50 years (Gomberg, 2003), so at least at first blush, upgrade of the structure seems prudent.

The resultant upgrade involved the demolition of the top 9 stories of this 14 story structure, strengthening of the bottom five stories, and construction of a new tower. In all, the upgrade cost was approximately $107 million, which was funded by the federal government (Charlier, 2003). At approximately the same time that the VA hospital in Memphis was undergoing this massive upgrade, a seismic upgrade project involving a VA hospital complex in San Francisco was cancelled due to lack of adequate funding; this particular VA hospital is located less than 6 km from the San Andreas fault. According to the U.S. Geological Survey (USGS), the San Francisco Bay area has a 62% chance of experiencing a major earthquake in the next 30 years, and the San Andreas fault alone has a 21% chance of rupturing within that time (USGS, 2002). In this light, the decision to upgrade the hospital in Memphis—and to not upgrade the at-risk San Francisco

VA facility—appears somewhat questionable given the apparently greater probability of a major earthquake affecting the VA hospital complex in San Francisco.

One might question whether the $107 million could have been spent in other ways that would also have had significant benefits, such as finding ways to make structures more fire-resistant; fires kill nearly 4000 people and injure nearly 18,000 people every year (Karter, 2005). The money could also have been spent strengthening the levees around New Orleans, possibly precluding the catastrophic flooding that occurred as a result of Hurricane Katrina. Likewise, if $107 million were "spent" as tax incentives in California to retrofit seismically weak apartment buildings with tuck-under parking—such as the Northridge Meadows apartment building that killed 16 of the approximately 60 people killed during the 1994 Northridge earthquake—at an approximate cost of $100,000 per building, over 1000 such buildings could be upgraded for the cost of this single VA hospital upgrade. Other more cost-effective ways of saving lives were proposed by Stein et al. (2003).

Similarly, the switch to the 2475-year earthquake in evaluation of existing buildings can have a significant impact on privately owned structures. With the greater forces associated with the 2475-year earthquake, greater numbers of privately owned buildings are likely to be deemed inadequate, even though they may be unlikely to experience a major earthquake within their remaining lifetime. The true cost of this change may not be realized for many years.

DISCUSSION OF ENGINEERING AND ECONOMICS

Engineering and economics have always been inextricably linked. According to *Webster's II New Collegiate Dictionary*, engineering is defined as "the application of mathematics and scientific principles to practical ends, as the design, construction, and operation of economical and efficient structures, equipment, and systems." A less formal saying—but just as apt—is that an engineer can do for one dollar what any fool can do for two.

In the case of design forces for new buildings, studies have shown that using national seismic provisions in design and construction adds between 0.5% and 3.3% to the total cost of construction in jurisdictions that did not previously have seismic provisions and adds approximately 0.9% to the cost of construction in jurisdictions that already had seismic provisions (Weber, 1985). So, assuming that implementation of the 2475-year earthquake adds approximately 1%–2% to the cost of a new building for most of the country, one might look to the AELR to determine, based on economics, which states should implement this change for new buildings. In the case of the AELR, assuming that the cost of incorporating seismic loads in the design adds 1%–2% to the cost of a new building, the "break even" AELR is approximately 1%–2% of the replacement value (normalized in millions of dollars) divided by 100 years (which might be a more realistic expected life span of a new building) to produce the cutoff AELR, which equals approximately 100–200. However, as

mentioned already, because the AELR does not capture the economic and societal effects of a major community-wide disaster, perhaps the limit should be even lower.

The question of whether requirements for existing buildings should be based on the demands from the 2475-year earthquake is far less complicated. Existing buildings, by definition, are older than new buildings and therefore generally have a shorter remaining life expectancy than new buildings. Consequently, with a shorter remaining life span, the probability that a given building will experience a given earthquake is generally substantially reduced, and the time over which the cost of the upgrade can be financed or depreciated is substantially smaller. When added to the fact that seismic upgrades can be enormously cost-prohibitive due to the constraints posed by working on an existing—and often occupied—structure, the decision to upgrade an existing building in a geographic area with infrequent earthquakes is economically very difficult to justify.

If the change to a 2475-year earthquake has any legitimacy, it is only to protect in the long term against community-wide catastrophe—not to limit risk associated with individual buildings.

CONCLUSIONS

Our primary points are as follows:

From an engineering standpoint, use of the 2475-year earthquake represents a dramatic shift from the design philosophy used for other loads, such as wind, snow, and flood.

In our opinion, the return period used for seismic design purposes is arbitrary and should be selected rationally while weighing potential risks and rewards; there is no clear evidence that the adoption of the 2475-year earthquake in new codes involved any such rational approach, and in fact, the selection of such a rare earthquake for design purposes may have been based on improper interpretation of the available data.

Use of the 2475-year earthquake in the design of new buildings appears to be questionable from an engineering perspective but may be somewhat more justified from a public policy standpoint with respect to long-term protection against community-wide catastrophe.

Use of the 2475-year earthquake in new design will generally increase design forces and detailing requirements; these increases will increase construction costs for new structures. Given that there exists a limited amount of funding for construction projects, the use of the 2475-year earthquake in design has some potential to cause increased expenditures in portions of the country that are less likely to experience a significant earthquake at the expense of other areas of the country (such as the West Coast), where structures are much more likely to experience a significant earthquake during their useful life.

Use of the 2475-year earthquake in the evaluation of existing structures appears to be inappropriate based on the shortened remaining life span of these structures.

Use of the 2475-year earthquake in the evaluation of existing buildings has the potential to result in large numbers of structures improperly being deemed seismically inadequate; these structures may be upgraded or abandoned or demolished unnecessarily.

Adoption of standards such as ASCE-31, FEMA 356, and the poorly worded IEBC—all of which use the 2475-year earthquake—may make upgrading existing structures cost prohibitive in large portions of the country and make the repair of damaged buildings economically infeasible.

FEMA recently concluded that some regions with low seismic hazard actually have "high seismic risk" and provided the examples of New York City and Boston (FEMA, 2000). Given that Boston and New York City have extremely low hazards based on both the computation of annualized estimated loss ratio and per capita annualized loss (Tables 1 and 2), we are somewhat skeptical of this claim. While it might be reasonable for new buildings in these cities to be designed for higher forces than previously required, we believe that it would be unwise to judge older buildings in these cities according to these new standards. Instead of using an extremely rare 2475-year earthquake in the evaluation of these buildings, earthquakes with much shorter return periods should be used, and the return period should be based on the expected remaining life span of these buildings. In the case of existing buildings, we believe that FEMA may be missing the larger picture—that there is a cost to these overly conservative policies, that the funding available in both the public and private arenas is limited, and that these policies must necessarily result in less funding for other projects.

Given that 74% of the annualized earthquake loss for the entire United States occurs in California and that 84% of the AEL occurs on the West Coast (California, Oregon, and Washington), it appears that reduction of seismic risk is most economically and efficiently addressed by concentrating limited available resources in these areas, particularly when it comes to upgrading existing building stock. Even when seismic risk is normalized on a per-capita basis or replacement cost basis, California far outpaces other states in terms of earthquake hazard.

We do recognize, however, as Benjamin Disraeli once said, "There are three kinds of lies: lies, damned lies, and statistics," and that the same data presented in this paper can lead to many different conclusions. Using our knowledge of engineering and economics, we have attempted to analyze the available data regarding seismic risks; this paper is presented in the hopes that a dialogue can be opened up between policy makers (such as FEMA), engineers, and seismologists.

REFERENCES CITED

Charlier, T., 2003, Quake-Proofing Costly, Difficult: New Building Code Could Shake Skyline: Memphis, Tennessee, GoMemphis.com (last accessed May 2003).

Earthquake Engineering Research Institute (EERI), 2003, Sample NEHRP Letter to Members of Congress: Oakland, California, Earthquake Engineering Research Institute, p. 1.

Federal Emergency Management Agency (FEMA), 2000, HAZUS99 Estimated Annualized Earthquake Losses for the United States (FEMA 366): Washington, D.C., FEMA, p. 1–32.

Federal Emergency Management Agency (FEMA), 2001, NEHRP Recommended Provisions for Seismic Regulations for New Buildings and Other Structures: Part 2. Commentary, 2000 Edition: Washington, D.C., FEMA, 377 p.

Federal Emergency Management Agency, (FEMA), 2004, Ranking of U.S. metropolitan areas with future estimated earthquake losses of more than $10 million per year: http://www.fema.gov/nwz01/nwz01_13a.shtm (August 2005).

Frankel, A., 2004, How can seismic hazard in the New Madrid seismic zone be similar to that in California?: Seismological Research Letters, v. 75, no. 5, p. 574–585.

Global Seismic Hazard Assessment Program (GSHAP), 1999, http://www.seismo.ethz.ch/GSHAP (January 2006).

Gomberg, J., 2003, Research dampens earthquake speculation: http://memphis.bizjournals.com/memphis/stories/2003/01/06/daily21.html (August 2005).

International Code Council (ICC), 2005, International code adoptions: http://www.iccsafe.org/government/adoption.html (August 2005).

Karter, M.J., 2005, Fire Loss in the United States during 2004: Quincy, Massachusetts, National Fire Protection Association, p. ii.

National Earthquake Hazards Reduction Program (NEHRP), 2005, About the NEHRP: http://www.fema.gov/hazards/earthquakes/nehrp/about.shtm (August 2005).

Searer, G.R., and Freeman, S.A., 2002, Unintended consequences of code modification, in Proceedings of the Seventh U.S. National Conference on Earthquake Engineering (Boston, Massachusetts): Oakland, California, Earthquake Engineering Research Institute, CD-ROM.

Stein, S., 2005, How can seismic hazard in the New Madrid seismic zone be similar to that in California?: Comment: Seismological Research Letters, v. 76, no. 3, p. 364–365.

Stein, S., Tomasello, J., and Newman, A., 2003, Should Memphis build for California's earthquakes?: Eos (Transactions, American Geophysical Union), v. 84, p. 177, 184–185.

Structural Engineers Association of California (SEAOC), 1999, Recommended Lateral Force Requirements and Commentary (7th edition): Sacramento, California, Structural Engineers Association of California, 83 p.

U.S. Census Bureau, 2000, Ranking tables for metropolitan areas: Population in 2000 and population change in 2000 (PHC-T-3): http://www.census.gov/population/cen2000/phc-t3/tab01.xls (August 2005).

U.S. Geological Survey (USGS), 2002, San Francisco Bay region earthquake probability: http://quake.wr.usgs.gov/research/seismology/wg02/images/percmap-lrg.jpg (August 2005).

Weber, S.F., 1985, Cost impact of the NEHRP recommended provisions on the design and construction of buildings, in Societal Implications: Washington, D.C., Selected Readings, Building Seismic Safety Council, p. 1-1–1-19.

Manuscript Accepted by the Society 29 November 2006

The Geological Society of America
Special Paper 425
2007

Seismic hazard and risk assessment in the intraplate environment: The New Madrid seismic zone of the central United States

Zhenming Wang[†]

Kentucky Geological Survey, 228 Mining and Mineral Resources Building, University of Kentucky, Lexington, Kentucky 40506, USA

ABSTRACT

Although the causes of large intraplate earthquakes are still not fully understood, they pose certain hazard and risk to societies. Estimating hazard and risk in these regions is difficult because of lack of earthquake records. The New Madrid seismic zone is one such region where large and rare intraplate earthquakes (M = 7.0 or greater) pose significant hazard and risk. Many different definitions of hazard and risk have been used, and the resulting estimates differ dramatically. In this paper, seismic hazard is defined as the natural phenomenon generated by earthquakes, such as ground motion, and is quantified by two parameters: a level of hazard and its occurrence frequency or mean recurrence interval; seismic risk is defined as the probability of occurrence of a specific level of seismic hazard over a certain time and is quantified by three parameters: probability, a level of hazard, and exposure time. Probabilistic seismic hazard analysis (PSHA), a commonly used method for estimating seismic hazard and risk, derives a relationship between a ground motion parameter and its return period (hazard curve). The return period is not an independent temporal parameter but a mathematical extrapolation of the recurrence interval of earthquakes and the uncertainty of ground motion. Therefore, it is difficult to understand and use PSHA. A new method is proposed and applied here for estimating seismic hazard in the New Madrid seismic zone. This method provides hazard estimates that are consistent with the state of our knowledge and can be easily applied to other intraplate regions.

Keywords: New Madrid seismic zone, seismic hazard, seismic risk, probabilistic seismic hazard analysis, seismic hazard assessment.

[†]E-mail: zmwang@uky.edu.

Wang, Z., 2007, Seismic hazard and risk assessment in the intraplate environment: The New Madrid seismic zone of the central United States, *in* Stein, S., and Mazzotti, S., ed., Continental Intraplate Earthquakes: Science, Hazard, and Policy Issues: Geological Society of America Special Paper 425, p. 363–374, doi: 10.1130/2007.2425(24). For permission to copy, contact editing@geosociety.org. ©2007 The Geological Society of America. All rights reserved.

INTRODUCTION

Although most damaging earthquakes occur along plate boundaries, such as the subduction zones around the Pacific Ocean and the San Andreas fault in California, some large earthquakes have occurred in intraplate regions. For example, the 1811–1812 New Madrid earthquakes (M 7.0–8.0) and the 1886 Charleston, South Carolina, earthquake (~M 7.3) both occurred in intraplate regions. Geologic records (paleoliquefaction data) also show that large earthquakes have occurred in other intraplate regions in eastern North America, such as the Wabash Valley (Obermeier et al., 1991; Obermeier, 1998). The causes of these large intraplate earthquakes are not well understood (Braile et al., 1986; Zoback, 1992; Newman et al., 1999; Kenner and Segall, 2000), and they pose hazards and risk because of their proximity to population centers.

The New Madrid seismic zone, located in northeastern Arkansas, western Kentucky, southeastern Missouri, and northwestern Tennessee, is a seismically active intraplate region in the central United States. It is so named because the town of New

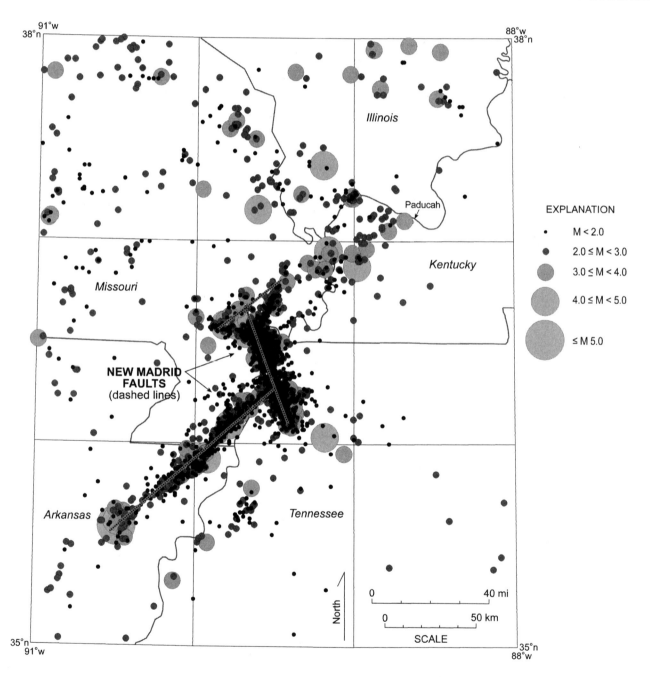

Figure 1. Seismicity in the New Madrid seismic zone of the central United States between 1974 and 2004 (CERI, 2004).

Madrid, Missouri, was the closest settlement to the epicenters of the 1811–1812 earthquakes. Between 1811 and 1812, at least three large earthquakes, with magnitudes estimated between M = 7.0 and 8.0, occurred during a 3 mo period (Nuttli, 1973). Instruments were installed in and around the seismic zone in 1974 to closely monitor seismic activity. Figure 1 shows locations of earthquakes with magnitude equal to or greater than 2.0 that occurred in the New Madrid seismic zone and the surrounding areas between 1974 and 2004 (CERI, 2004). The low seismicity and lack of strong-motion recordings from large earthquakes (M > 6.0) make estimating seismic hazard and risk difficult.

In this paper, I first review probabilistic seismic hazard analysis (PSHA), the most commonly used method for estimating seismic hazard and risk. I then develop a new method, called seismic hazard assessment (SHA), and apply it to the New Madrid seismic zone.

PROBABILISTIC SEISMIC HAZARD ANALYSIS

PSHA was originally developed by Cornell in 1968 for estimating engineering risk in comparison with the analogous flood or wind problem. A similar method was also developed by Milne and Davenport (1969) for estimating seismic risk in Canada. In 1971, Cornell extended his method to incorporate the possibility that ground motion at a site could be different (i.e., ground motion uncertainty) for different earthquakes of the same magnitude at the same distance because of differences in site conditions or source parameters. This method (Cornell, 1971) was coded into a FORTRAN algorithm by McGuire (1976) and became a standard PSHA (Frankel et al., 1996, 2002). It should be noted that there is a fundamental difference between the formulations in Cornell (1968) and those in Cornell (1971), i.e., the former does not include ground-motion uncertainty, whereas the latter does.

Following Cornell's (1971) and McGuire's (1995, 2004) formula for multiple sources, an annual probability of exceedance (γ) of a ground-motion amplitude y is

$$\gamma(y) = \sum_j v_j P_j [Y \geq y]$$
$$= \sum_j v_j \iint P_j [Y \geq y \mid m, r] f_{M,j}(m) f_{R,j}(r) dm dr, \quad (1)$$

where v_j is the activity rate for seismic source j; $f_{M,j}(m)$ and $f_{R,j}(r)$ are earthquake magnitude and source-to-site distance density functions, respectively; and $P_j(Y > y|m,r)$ is the probability ground motion Y exceeds a specific level y conditioned at a given m and r. The conditional exceedance probability $P_j(Y > y|m,r)$ is equal to the exceedance probability of the ground-motion uncertainty (a log-normal distribution) as

$$P_j [Y \geq y \mid m, r] =$$
$$1 - \int_0^y \frac{1}{\sqrt{2\pi}\sigma_{\ln,y}} \exp\left(-\frac{(\ln y - \ln y_{mr})^2}{2\sigma_{\ln,y}^2}\right) d(\ln(y)), \quad (2)$$

where y_{mr} and $\sigma_{\ln,y}$ are the median and standard deviation (log) determined by the ground-motion attenuation relationships (Campbell, 1981, 2003). Earthquakes in the intraplate regions are rare and can be described as a characteristic: the large and damaging earthquakes repeat regularly with few or no moderate and small earthquakes. For characteristic seismic sources, we have

$$\gamma(y) =$$
$$\sum_j \frac{1}{T_j} \left\{ 1 - \int_0^y \frac{1}{\sqrt{2\pi}\sigma_{\ln,y}} \exp\left(-\frac{(\ln y - \ln y_{mr})^2}{2\sigma_{\ln,y}^2}\right) d(\ln(y)) \right\}, \quad (3)$$

where T_j is the average recurrence interval of the characteristic earthquake for source j. As shown in Equations 1 and 3, PSHA generally involves many seismic sources, ground-motion attenuation relationships, recurrence intervals, and associated uncertainties. No matter how complicated the parameters are, however, the end results from PSHA are simple, total hazard curves, which give a range of annual probability of exceedance versus a range of ground-motion values (Frankel et al., 1996, 2002).

As shown in Equation 3, the annual probability of exceedance, γ, is a function of average recurrence interval of earthquake and ground-motion uncertainty. This can be illustrated through an example for a single characteristic source,

$$\gamma(y) =$$
$$\frac{1}{T} \left\{ 1 - \int_0^y \frac{1}{\sqrt{2\pi}\sigma_{\ln,y}} \exp\left(-\frac{(\ln y - \ln y_{mr})^2}{2\sigma_{\ln,y}^2}\right) d(\ln(y)) \right\}. \quad (4)$$

Figure 2 shows a peak ground acceleration (PGA) hazard curve (A) and probability density of PGA for a hypothetical characteristic earthquake of M = 7.5 with an average recurrence interval of 500 yr at a point 20 km from the epicenter. According to Equation 4, annual probability of exceedance (hazard) is the product of the annual occurrence rate, 0.002 (1/500), and the probability that PGA exceeds a given value. For example, for a PGA of 0.3g, the probability of exceedance is 0.5, which results in an annual probability of exceedance of 0.001 (0.002 × 0.5). For an annual probability of exceedance of 0.0004 (or return period of 2500 yr), a PGA of 0.5g can be obtained using the curves in Figure 2. The annual probability of exceedance of 0.0004 is equal to 0.002 (annual occurrence rate) × 0.2 (probability of PGA exceeding 0.5g). This example demonstrates the basic function of PSHA, i.e., a mathematical extrapolation from the time-domain characteristics of earthquakes and the spatial characteristics of ground motion (uncertainty).

The inverse of annual probabilities of exceedance (1/γ), called return period (T_p), is also often used (Frankel et al., 1996, 2002),

$$T_P(y) = \frac{T}{1 - \int_0^y \frac{1}{\sqrt{2\pi}\sigma_{\ln,c}} \exp\left(-\frac{(\ln y - \ln y_c)^2}{2\sigma_{\ln,c}^2}\right) d(\ln(y))}. \quad (5)$$

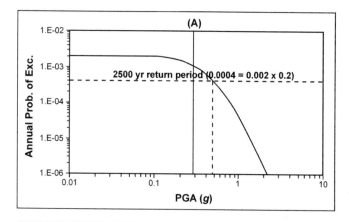

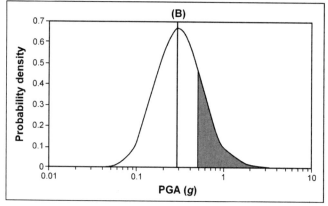

Figure 2. (A) Hazard (annual probability of exceedance) curve for a hypothetical characteristic earthquake of M = 7.5 with average recurrence interval of 500 yr at a point 20 km from the epicenter. (B) Probability density (median peak ground acceleration [PGA] of 0.3*g* and a standard deviation [log] of 0.6 are assumed).

For example, a 2500 yr return period is the inverse of annual probabilities of exceedance of 0.0004. As shown in Figure 2, return periods range between 500 and 1 million years, and they can reach infinity because there is no upper boundary on the log-normal distribution (Fig. 2B). Moreover, ground motion with a return period derived from PSHA has been communicated and used as the ground motion that will occur in that return period, for example, the ground motion with a 2500 yr return period (Frankel et al., 1996, 2002; Frankel, 2005). As shown in Figure 2, it is assumed that there is only one characteristic earthquake with an average recurrence interval of 500 yr (input). The ground motion will not occur in 2500 yr because it is a consequence of the earthquake; rather, it will have a 20% probability of being exceeded when the earthquake occurs in 500 yr. Similarly, for multiple sources, Wang and Ormsbee (2005) showed that ground motion with a particular return period does not mean that that ground motion will occur in that return period; rather, there are certain probabilities that the ground motion will be exceeded when all the considered earthquakes occur. The return period is a number extrapolated from the recurrence intervals of earthquakes and the probability of ground motions. Hence, using the return period to

communicate seismic hazard is not only inappropriate, but it also results in a fundamental change of PSHA, i.e., from a probable occurrence to a certain occurrence of a ground motion.

It is difficult to explain the physical meaning of ground motion derived from PSHA. The first thorough review of PSHA was conducted by a committee chaired by K. Aki, at the National Research Council (NRC, 1988). One of the conclusions reached by the Aki Committee was that "the aggregated results of PSHA are not always easily related to the inputs" (NRC, 1988, p. 5). In other words, "the concept of a 'design earthquake' is lost; i.e., there is no single event (specified, in simplest terms, by a magnitude and distance) that represents the earthquake threat at, for example, the 10,000-yr ground-motion level" (McGuire, 1995, p. 1275). Wang et al. (2003) and Wang and Ormsbee (2005) also demonstrated that it is difficult to explain the physical meaning of ground motion derived from PSHA for a single or three characteristic sources.

Frankel (2005, p. 474) offered a physical explanation for ground motion with a 2500 yr return period from a characteristic earthquake with a 500 yr recurrence interval. He stated "one of the five earthquakes expected to occur over the 2500 years will produce ground motions at that site greater than the 2% PE in 50 years (2500-year return period) value." This explanation contradicts the basics of PSHA, i.e., probability of ground-motion occurrence. As shown in Figure 2, the probability that PGA exceeds 0.5*g* is 0.2 if the characteristic earthquake occurs. The probability of PGA exceeding 0.5*g* after five characteristic earthquakes (in 2500 yr) is ~0.67 ($p \approx 1 - [1 - 0.2]^5$), not 1.0. This means that the PGA with a 2500 yr return period may not occur. An explanation similar to Frankel's was offered by Holzer (2005) for ground motion with a 2500 yr return period from three characteristic earthquakes. Holzer's explanation also contradicts the basics of PSHA (Wang, 2005).

As pointed out by Hanks (1997, p. 369), "PSHA is a creature of the engineering sciences, not the earth sciences, and most of its top practitioners come from engineering backgrounds." The main problem with PSHA is how it is being used in engineering risk analysis, particularly in regard to return period. Three risk levels, ground motions with 10%, 5%, and 2% probability of exceedance (PE) in 50 yr, are commonly considered in engineering design. In engineering risk analysis, a ground motion with 10%, 5%, or 2% PE in 50 yr means that a particular ground motion (an event) will occur at least once in 500, 1000, or 2500 yr (recurrence intervals) (Cornell, 1968; Milne and Davenport, 1969; Wang and Ormsbee, 2005; Wang et al., 2005). As shown by Frankel et al. (1996, 2002) and Frankel (2004), the ground motion with 2% PE in 50 yr is equivalent to the ground motion with a return period of 2500 yr (or annual probability of exceedance of 0.0004) derived from PSHA. As discussed earlier, the ground motion with a 2500 yr return period does not mean it will occur in 2500 yr; rather, it has certain probabilities of being exceeded when all the considered earthquakes occur. In other words, the return period defined in PSHA is not equivalent to the recurrence interval defined in engineering risk analysis. Hence, using PSHA for engineering risk analysis is not appropriate (Wang and Ormsbee, 2005).

SEISMIC HAZARD ANALYSIS

Seismic Risk Estimation

It is necessary to briefly review the definition of seismic risk because the purpose of seismic hazard analysis is to provide parameters for estimating risk (Cornell, 1968; Milne and Davenport, 1969). Although risk has different meanings among different professions, it can generally be quantified by three terms: probability, hazard (loss or other measurements), and time exposure. For example, in health sciences, risk may be defined as the probability of getting cancer if an average daily dose of a hazardous substance (hazard) is taken over a lifetime (70 yr on average). In the financial market, risk may be defined as the probability of losing a certain amount of money (loss) over a period of time. In seismology, risk may be defined as the probability of earthquakes with a certain magnitude or greater striking at least once in a region during a specific period of time. Therefore, a clear definition of risk is necessary in any discussion and communication of the risk.

In earthquake engineering, risk is defined as the probability that ground motion at a site of interest exceeds a specific level (hazard) at least once in a period of time (Cornell, 1968; Milne and Davenport, 1969). This definition is similar to those defined in hydraulic engineering (Gupta, 1989) and wind engineering (Sacks, 1978). In fact, seismic risk was originally defined from analogous flood and wind risks (Cornell, 1968; Milne and Davenport, 1969). Seismic risk estimation is based on a Poisson model, which assumes that earthquake occurrence is independent of time and independent of the past history of occurrences or nonoccurrences. Although the Poisson model fails to incorporate the most basic physics of the earthquake process, whereby the tectonic stress released when a fault fails must rebuild before the next earthquake can occur at that location (Stein and Wysession, 2003; Working Group on California Earthquake Probabilities, 2003), it is the standard model for seismic risk analysis, as well as for other risk analyses, such as for flood and wind. In the Poisson model (Cornell, 1968; Stein and Wysession, 2003), the probability of n earthquakes of interest in an area or along a fault occurring during an interval of t years is

$$p(n,t,\tau) = \frac{e^{-t/\tau}(t/\tau)^n}{n!},\qquad(6)$$

where τ is the average recurrence interval (or average recurrence rate, $1/\tau$) of earthquakes with magnitudes equal to or greater than a specific size. The probability that no earthquake will occur in an area or along a fault during an interval of t years is

$$p(0,t,\tau) = e^{-t/\tau}.\qquad(7)$$

The probability of one or more (at least one) earthquakes with magnitudes equal to or greater than a specific size occurring in t years is

$$p(n \geq 1,t,\tau) = 1 - p(0,t,\tau) = e^{-t/\tau} = 1 - (1 - 1/\tau)^t.\qquad(8)$$

Equation 8 can be used to calculate the risk, expressed as $x\%$ PE in Y years, for a given recurrence interval (τ) of earthquakes with a certain magnitude or greater. For example, the U.S. Geological Survey (2002) estimated a 7%–10% probability of a repeat of the 1811–1812 New Madrid earthquakes (M 7.5–8.0) in 50 yr in the New Madrid region. This estimate was determined from Equation 8 and an average recurrence interval of ~500 yr, which was inferred from interpretation of paleoliquefaction records (Tuttle et al., 2002). Equation 8 can also be used to calculate the average recurrence interval (τ) of earthquakes with a certain magnitude or greater for a given risk level. For example, 10%, 5%, and 2% PE in 50 yr are commonly used in earthquake engineering (BSSC, 1998; ICC, 2000). According to Equation 8, these risk levels are equivalent to 500, 1000, and 2500 yr recurrence intervals for earthquakes. For comparison, 1% PE in 1 yr and 2% PE in 1 yr are being considered for building designs for flood and wind, respectively (ICC, 2000). These risk levels are equivalent to 100 and 50 yr recurrence intervals for floods (100 yr flood) and wind storms, respectively.

In practice, knowledge of the consequences of earthquakes (i.e., ground motions or modified Mercalli intensity [MMI]) at a point or in a region of interest is desirable. For example, PGA and response acceleration (SA) in a given period are common measurements needed for a site. This is similar to the situation in flood and wind analyses whereby knowledge of the consequences of floods and winds, such as peak discharge and 3-s-gust wind speed, is desired for specific sites. The ground motions (consequences of earthquake) and their recurrence intervals (τ), hazard curves, are determined through seismic hazard analyses.

Seismic Hazard Assessment

The hazard curves used in seismic risk analysis describe relationships between a ground-motion parameter and its recurrence interval. As discussed earlier, the hazard curves derived from PSHA describe relationships between a ground-motion parameter and its return period, and the return period is not equal to the recurrence interval. Therefore, the hazard curves derived from PSHA are not appropriate for seismic risk analysis. A new method, seismic hazard assessment (SHA), is proposed here for developing a relationship between a ground-motion parameter and its recurrence interval (i.e., seismic hazard curve).

In seismology, the number of earthquakes that occur yearly can be represented by a magnitude-frequency relationship or Gutenberg-Richter relationship:

$$\text{Log}(N) = a - bM \text{ or } N = 10^{(a-bM)},\qquad(9)$$

where N is the cumulative number of earthquakes with magnitude equal to or greater than M occurring yearly, and a and b are constants. As discussed earlier, the average recurrence rate ($1/\tau$) of earthquakes with magnitudes equal to or greater than a specific size (M) in Equation 8 has the same meaning as N. Therefore,

$$1/\tau = N = e^{2.303a - 2.303bM} \text{ or } \tau = 1/N = e^{-2.303a + 2.303bM}. \quad (10)$$

Estimations of the expected ground motion at a site are given by assuming a ground-motion attenuation relationship, which describes a relationship between a ground-motion parameter (Y) and magnitude of an earthquake (M) and epicentral distance (R) (Campbell, 1981, 2003). Generally, the attenuation relationship follows the functional form of

$$\text{Ln } Y = a_0 + f(M, R) + \varepsilon, \quad (11)$$

where ε is uncertainty (a_0 is a constant). The uncertainty (ε) can be modeled using a log-normal distribution with a standard deviation (σ). From Equation 11, M can be expressed as a function of R, ln Y, and ε:

$$M = f(R, \ln Y, \varepsilon). \quad (12)$$

Combining Equations 10 and 12 results in:

$$1/\tau = e^{2.303a - 2.303bf(R, \ln Y, \varepsilon)} \text{ or } \tau = e^{-2.303a + 2.303bf(R, \ln Y, \varepsilon)}. \quad (13)$$

Equation 13 describes a relationship between the ground motion (ln Y) with an uncertainty (ε) and its annual recurrence rate ($1/\tau$) or recurrence interval (τ) at a distance (R), i.e., a hazard curve. Equation 13 can be used to estimate ground motion at a site or in a region.

SEISMIC HAZARD AND RISK IN THE NEW MADRID SEISMIC ZONE

Seismicity in the New Madrid seismic zone is quite low. Table 1 lists instrumental and historical earthquakes with M ≥ 4.0 known to have occurred in the New Madrid seismic zone (Bakun and Hopper, 2004). Two M 4.0 earthquakes that occurred in 2003 have also been included in Table 1. As shown in the table, there is only one event with M = 6.0 since the last 1811–1812 events, the 1843 Marked Tree, Arkansas, earthquake. This earthquake catalog is too short to be sufficient for constructing a reliable Gutenberg-Richter curve, as illustrated in Figure 3, which shows the Gutenberg-Richter curve for earthquakes with magnitudes between 4.0 and 5.0 in the New Madrid seismic zone (Stein and Newman, 2004). The a and b values are estimated to be ~3.15 and 1.0, respectively. The b value of 1.0 is consistent with that used in the national seismic hazard maps (Frankel et al., 1996, 2002). Figure 3 also shows that recurrence intervals for large earthquakes (M ≥ 6.0) would be quite long, ~700 yr for M 6.0, 7000 yr for M 7.0, and 70,000 yr for M 8.0, if these a and b values are assumed to be applicable for large earthquakes in the New Madrid seismic zone. This is not consistent with paleoseismic interpretations by Tuttle et al. (2002): an average recurrence interval of ~500 yr was inferred from the interpretation of the paleoliquefaction records for large earthquakes similar to the 1811–1812 New Madrid events. These large earthquakes were treated as characteristic events (Frankel et al., 1996, 2002), even though it is difficult to determine that they are characteristic because of the lack of data (Stein and Newman, 2004).

I assume that (1) the a and b values could be applied to earthquakes with magnitudes up to M 5.5 (Fig. 2), and (2) the large earthquake (M 7.6) is characteristic. For $a = 3.15$ and $b = 1.0$:

$$1/\tau = e^{7.254 - 2.303M} \text{ for } 4.0 \le M \le 5.5. \quad (14)$$

Equation 14 describes a hazard curve in terms of earthquake magnitude and its annual recurrence rate. For $M = 4.85$, Equation 14 results in an annual recurrence rate ($1/\tau$) of ~0.02 or a

TABLE 1. EARTHQUAKES WITH MAGNITUDE EQUAL TO OR GREATER THAN 4.0 IN THE NEW MADRID SEISMIC ZONE (FROM BAKUN AND HOPPER, 2004)

Date	Latitude (°N)	Longitude (°W)	M
16 December 1811	36.00	89.96	7.6
16 December 1811 "dawn"	36.25	89.50	7.0
23 January 1812	36.80	89.50	7.5
05 January 1843	35.90	89.90	6.2
17 February 1843	35.90	89.90	4.2
17 August 1865	35.54	90.40	4.7
19 November 1878	35.65	90.25	5.0
11 January 1883	36.80	89.50	4.2
04 November 1903	36.59	89.58	4.7
28 October 1923	35.54	90.40	4.1
07 May 1927	35.65	90.25	4.5
17 September 1938	35.55	90.37	4.4
02 February 1962	36.37	89.51	4.2
03 March 1963	36.64	90.05	4.7
17 November 1970	35.86	89.95	4.1
25a March 1976	35.59	90.48	4.6
25b March 1976	35.60	90.50	4.2
04 May 1991	36.56	89.80	4.1
30 April 2003	35.920	89.920	4.0
06 June 2003	36.87	88.98	4.0

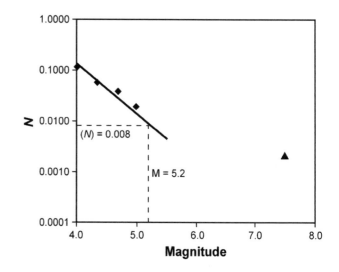

Figure 3. Magnitude-frequency (Gutenburg-Richter) curve for the New Madrid seismic zone. Diamond—historical rate, triangle—geological (paleoliquefaction) rate.

recurrence interval (τ) of 50 yr, which means that at least one earthquake with magnitude equal to or greater than 4.85 would be expected to occur in 50 yr. Similarly, Equation 14 results in an annual recurrence rate of ~0.01 or a recurrence interval of 100 yr if $M = 5.15$. Hence, according to Equation 8, we can calculate risks for the New Madrid area; i.e., there is about a 63% PE in 50 yr that the area will be hit by at least one earthquake with $M = 4.85$ or greater, and about a 39% PE in 50 yr that the area will be hit by at least one earthquake with $M = 5.15$ or greater.

The estimated risk of a large earthquake (~M 7.5) hitting the New Madrid area is ~10% PE in 50 yr (USGS, 2002). Figure 4 is the earthquake probability (risk) map for the New Madrid area generated from the U.S. Geological Survey earthquake hazard Web site (eqint.cr.usgs.gov/eq/html/eqprob.html).

Campbell (2003) found that in the central and eastern United States, ground motion on very hard rock (Vs of 2.8 km/s) follows the relationship

$$\text{Ln } Y = c_1 + f_1(M) + f_2(M, r_{\text{rup}}) + f_3(r_{\text{rup}}) + \varepsilon_a + \varepsilon_e, \quad (15)$$

where r_{rup} is the closest distance to fault rupture, ε_a is aleatory (randomness) uncertainty, and ε_e is epistemic uncertainty. For $r_{\text{rup}} \leq 70$ km, PGA of 0.2, and SA of 1.0 s:

$$\ln(\text{PGA}) = 0.0305 + 0.633M - 0.0427(8.5 - M)^2$$
$$- 1.591 \ln R + (-0.00428 + 0.000483M)r_{\text{rup}} \quad (16)$$
$$+ \varepsilon_a + \varepsilon_e,$$

$$\ln(\text{SA}_{0.2\,s}) = -0.4328 + 0.617M - 0.0586(8.5 - M)^2$$
$$- 1.320 \ln R + (-0.00460 + 0.000337M)r_{\text{rup}} \quad (17)$$
$$+ \varepsilon_a + \varepsilon_e,$$

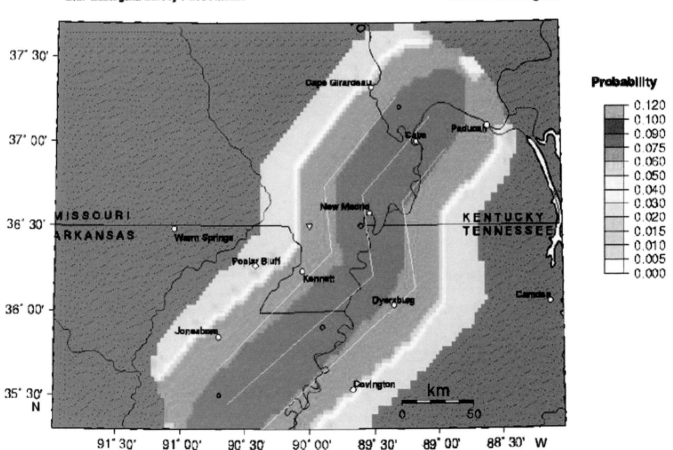

Figure 4. Earthquake probability map of the New Madrid seismic zone (USGS, 2005).

$$\ln\left(SA_{1.0\ s}\right) = -0.6104 + 0.451M - 0.2090\left(8.5 - M\right)^2$$
$$-1.158 \ln R + \left(-0.00255 + 0.000141M\right)r_{rup}$$
$$+ \varepsilon_a + \varepsilon_e, \tag{18}$$

and

$$R = \sqrt{r_{rup}^2 + \left[c_7 \exp\left(c_8 M\right)\right]^2} . \tag{19}$$

The standard deviation ($\sigma_{\ln Y}$) of ε_a is magnitude dependent and equal to

$$\sigma_{\ln Y} = \begin{cases} c_{11} + c_{12}M & \text{for } M < 7.16 \\ c_{13} & \text{for } M \geq 7.16 \end{cases}. \tag{20}$$

The coefficients c_7, c_8, c_{11}, c_{12}, and c_{13} are listed in Table 2. The standard deviation of ε_e depends on earthquake magnitude and the rupture distance as listed in Campbell (2003).

By combining the ground-motion attenuation relationships (Equations 16, 17, and 18) and the Gutenburg-Richter relationship (Equation 14), we can derive seismic hazard curves in terms of ground motions and their annual recurrence rates for a site at a certain distance from the source. Figures 5, 6, and 7 show the median ($\varepsilon = 0.0$) hazard curves for PGA, 0.2 s SA, and 1.0 s SA at a site 30 km from the source. As shown already, there is significant uncertainty ($\sigma \approx 0.66$–0.90) in the predicted ground motions, and the uncertainty depends on magnitude and distance. The uncertainty can be estimated in the hazard analysis by adding a total uncertainty ($\varepsilon \neq 0.0$) to the attenuation relationship. Also shown in Figures 5, 6, and 7 are the hazard curves with 16% and 84% confidence levels (i.e., $\pm 1\sigma$). These hazard curves (Figs. 5, 6, and 7) are similar to those derived in flood-frequency analysis (Gupta, 1989; Wang and Ormsbee, 2005) and wind-frequency analysis (Sacks, 1978). Points on the hazard curves have a similar meaning. For example, the median PGA of ~0.07g has an annual recurrence rate of 0.008, or recurrence interval of 125 yr. This PGA (0.07g) could occur at least once in a 125 yr period because it is a consequence of an earthquake with magnitude equal to 5.2 or greater (Fig. 3).

As shown in Figures 5–7, the median ground motions with the annual recurrence rate of 0.002 are significant: 0.44g PGA, 0.59g 0.2 s SA, and 0.26g 1.0 s SA, respectively. According to these results, the characteristic earthquake (M 7.0–8.0) is of safety concern in the New Madrid area. The risk posed by the characteristic earthquake is ~10% PE in 50 yr. There is no knowledge on large earthquakes or ground motions generated by the earthquakes that have recurrence intervals much longer than 500 yr in the New Madrid area. In another words, there is no information on the earthquakes or ground motions with PE much less than 10% in 50 yr, such as 2% PE or less in 50 yr, in the New Madrid area. However, PSHA has derived the ground motions with 2% or less PE in 50 yr (Frankel et al., 1996, 2002; Frankel,

TABLE 2. COEFFICIENTS c_7, c_8, c_{11}, c_{12}, AND c_{13} OF CAMPBELL'S (2003) ATTENUATION

Coefficients	PGA	0.2 s SA	1.0 s SA
c_7	0.683	0.399	0.299
c_8	0.416	0.493	0.503
c_{11}	1.030	1.077	1.110
c_{12}	−0.0860	−0.0838	−0.0793
c_{13}	0.414	0.478	0.543

Note: PGA—peak ground acceleration; SA—response acceleration.

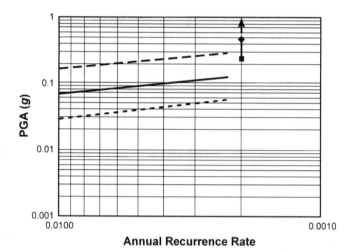

Figure 5. Peak ground acceleration (PGA) hazard curves at a site 30 km from the New Madrid faults. Diamond—median (mean) PGA, square—PGA with 16% confidence, and triangle—PGA with 84% confidence from the characteristic earthquake of M = 7.5.

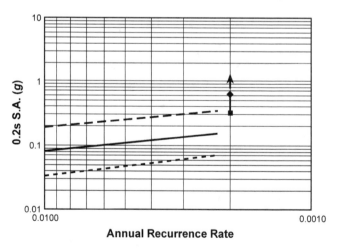

Figure 6. Hazard curves for 0.2 s response acceleration (SA) at a site 30 km from the New Madrid faults. Diamond—median (mean) 0.2 s SA, square—0.2 s SA with 16% confidence, and triangle—0.2 s SA with 84% confidence from the characteristic earthquake of M = 7.5.

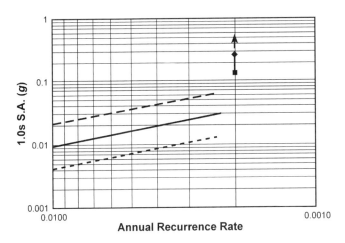

Figure 7. Hazard curves for 1.0 s response acceleration (SA) at a site 30 km from the New Madrid faults. Diamond—median (mean) 1.0 s SA, square—1.0 s SA with 16% confidence, and triangle—1.0 s SA with 84% confidence from the characteristic earthquake of M = 7.5.

2005). These ground motions are numerically created by using the ground-motion uncertainty.

The ground-motion maps corresponding to a specific annual recurrence rate or a PE in *Y* years can also be generated from the hazard curves at grid points according to Equation 13. For example, for the annual recurrence rate of 0.002 or 10% PE in 50 yr, PGA and SA can be generated according to Equation 13 using a ground-motion attenuation relationship, such as Campbell's (2003) attenuation relationship. Figure 8 shows median PGA, 0.2 s SA, and 1.0 s SA maps for the New Madrid area.

DISCUSSION

Estimations of seismic hazard and risk depend both on the definition of hazard and the definition of risk. In general terms, the hazard is the intrinsic natural occurrence of earthquakes and the resulting ground motion and other effects, whereas the risk is the danger the hazard poses to life and property. Because many different definitions of hazard and risk can be used, the resulting estimates can differ dramatically. For example, seismic risk was originally defined in terms of the probability of a given level of strong shaking occurring in a year or a time interval (Cornell, 1968; Milne and Davenport, 1969). This definition of seismic risk has become the definition of seismic hazard in PSHA (Frankel, 2004, 2005), however. Hence, a clear definition of hazard and risk is needed in any discussion of hazard and risk.

In this paper, seismic risk is defined as the probability of the occurrence of one or more (at least one) earthquakes with magnitudes equal to or greater than a specific size, or ground motion generated by the earthquakes, in a certain period of time; seismic hazard is defined as one or more (at least one) earthquakes with magnitudes equal to or greater than a specific size, or ground

motion generated by the earthquakes, recurring in a time interval. These definitions are consistent with those of Cornell (1968) and Milne and Davenport (1969). These definitions are also consistent with those defined in hydraulic engineering (Gupta, 1989) and wind engineering (Sacks, 1978). Although PSHA has been widely used in seismic hazard and risk assessments, the return period derived from PSHA is not an independent temporal parameter but a mathematical extrapolation of the recurrence interval of earthquakes and the uncertainty of ground motion. Thus, PSHA is not appropriate for use in seismic hazard and risk assessments (Wang and Ormsbee, 2005).

A new method (SHA) for estimating seismic hazards (ground motions) at a point of interest is proposed here. SHA is similar to the procedure described by Cornell (1968), but there is one important difference: Cornell (1968) treated the uncertain focal distance (distance between the focus and site) as an independent term with a probability density function and incorporated the uncertainty directly into hazard analysis, but in our procedure, this uncertainty (at least part of it) is implicitly included in the ground-motion attenuation relationships (Atkinson and Boore, 1995; Frankel et al., 1996; Toro et al., 1997; Somerville et al., 2001; Campbell, 2003). For example, the uncertainty in focal depth was treated as an aleatory uncertainty in the attenuation relationship of Toro et al. (1997). The uncertainty (epistemic uncertainty) in the attenuation relationship of Campbell (2003) depends on the rupture distance. The uncertainty of the focal distance may be counted twice in the hazard calculation if the uncertainty is explicitly included (Klügel, 2005). Therefore, it would be more appropriate to directly use the ground-motion attenuation relationship to estimate the hazards (ground motions) at a point of interest.

For the New Madrid area, there are at least 13 ground-motion attenuation relationships available (EPRI, 2003), and all of them were developed from theoretical models with or without calibration from limited ground-motion records from small earthquakes (M < 6.0). There is no unique way to use these attenuation relationships in seismic hazard analysis (SSHAC, 1997). SHA can be easily applied to any one or all of them. No matter how these ground-motion attenuation relationships are used, as either a single one or multiple ones with assigned weights (logic-tree), SHA will explicitly provide hazard estimates with associated uncertainties.

The hazard curves derived through SHA are similar to those derived through flood-frequency and wind-frequency analyses and have the same meaning. Therefore, use of SHA in risk analysis is appropriate. SHA also provides hazard (ground-motion) estimates that are consistent with the state of knowledge. The U.S. Geological Survey (2002) estimated the probability of a repeat of the 1811–1812 earthquakes with magnitude of 7.5–8.0 to be 7–10% PE in 50 yr (risk). This estimate was based on an average recurrence interval of ~500 yr, interpreted from paleoliquefaction records (Tuttle et al., 2002). The SHA method results in risk estimates (Fig. 8) that are consistent with the estimates of the U.S. Geological Survey (2002).

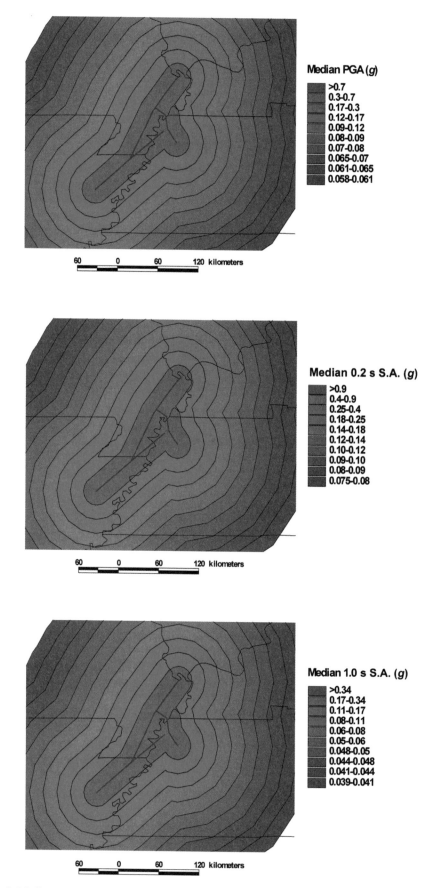

Figure 8. Median peak ground acceleration (PGA) (top), 0.2 s response acceleration (SA) (middle), and 1.0 s SA (bottom) with 10% PE in 50 yr for the New Madrid seismic zone. The New Madrid faults of Johnston and Schweig (1996) and attenuation relationship of Campbell (2003) were used.

ACKNOWLEDGMENTS

This research was in part supported by a grant from the U.S. Department of Energy, contract no. DE-FG05-03OR23032. We thank Meg Smath of the Kentucky Geological Survey for editorial help. We appreciate comments and suggestions from the editors, Seth Stein and Stéphane Mazzotti, and two anonymous reviewers, which helped to improve the manuscript greatly.

REFERENCES CITED

Atkinson, G.M., and Boore, D.M., 1995, Ground motion relations for eastern North America: Bulletin of the Seismological Society of America, v. 85, p. 17–30.

Bakun, W.H., and Hopper, M.G., 2004, Historical seismic activity in the central United States: Seismological Research Letters, v. 75, p. 564–574.

Braile, L.W., Hinze, W.J., Keller, G.R., Lidiak, E.G., and Sexton, J.L., 1986, Tectonic development of the New Madrid rift complex, Mississippi Embayment: North America: Tectonophysics, v. 131, p. 1–21, doi: 10.1016/0040-1951(86)90265-9.

Building Seismic Safety Council (BSSC), 1998, NEHRP Recommended Provisions for Seismic Regulations for New Buildings (1997 ed.): Federal Emergency Management Agency (FEMA) 302, 337 p.

Campbell, K.W., 1981, Near-source attenuation of peak horizontal acceleration: Bulletin of the Seismological Society of America, v. 71, p. 2039–2070.

Campbell, K.W., 2003, Prediction of strong ground motion using the hybrid empirical method and its use in the development of ground-motion (attenuation) relations in eastern North America: Bulletin of the Seismological Society of America, v. 93, p. 1012–1033, doi: 10.1785/0120020002.

Center for Earthquake Research and Information (CERI), 2004, New Madrid Earthquake Catalog: www.folkworm.ceri.memphis.edu/catalogs/html/cat_nm.html (last accessed 13 January 2005).

Cornell, C.A., 1968, Engineering seismic risk analysis: Bulletin of the Seismological Society of America, v. 58, p. 1583–1606.

Cornell, C.A., 1971, Probabilistic analysis of damage to structures under seismic loads, *in* Howells, D.A., Haigh, I.P., and Taylor, C., eds., Dynamic Waves in Civil Engineering: Proceedings of the Society for Earthquake and Civil Engineering Dynamics Conference: New York, John Wiley, p. 473–493.

Electric Power Research Institute (EPRI), 2003, CEUS Ground Motion Project, Model Development and Results: Electric Power Research Institute Report 1008910, 67 p.

Frankel, A., 2004, How can seismic hazard around the New Madrid seismic zone be similar to that in California?: Seismological Research Letters, v. 75, p. 575–586.

Frankel, A., 2005, Reply to "Comment on 'How can seismic hazard around the New Madrid seismic zone be similar to that in California?' by Arthur Frankel," by Zhenming Wang, Baoping Shi, and John D. Kiefer: Seismological Research Letters, v. 76, p. 472–475.

Frankel, A., Mueller, C., Barnhard, T., Perkins, D., Leyendecker, E., Dickman, N., Hanson, S., and Hopper, M., 1996, National seismic Hazard Maps: Documentation June 1996: U.S. Geological Survey Open-File Report 96-532, 110 p.

Frankel, A.D., Petersen, M.D., Mueller, C.S., Haller, K.M., Wheeler, R.L., Leyendecker, E.V., Wesson, R.L., Harmsen, S.C., Cramer, C.H., Perkins, D.M., and Rukstales, K.S., 2002, Documentation for the 2002 Update of the National Seismic Hazard Maps: U.S. Geological Survey Open-File Report 02-420, 33 p.

Gupta, R.S., 1989, Hydrology and Hydraulic Systems: Englewood Cliffs, New Jersey, Prentice-Hall, 739 p.

Hanks, T.C., 1997, Imperfect science: Uncertainty, diversity, and experts: Eos (Transactions, American Geophysical Union), v. 78, p. 369, 373, 377, doi: 10.1029/97EO00236.

Holzer, T.L., 2005, Comment on "Comparison between probabilistic seismic hazard analysis and flood frequency analysis" by Zhenming Wang and Lindell Ormsbee: Eos (Transactions, American Geophysical Union), v. 86, p. 303, doi: 10.1029/2005EO330004.

International Code Council (ICC), 2000, International Building Code: Falls Church, Virginia, International Code Council, 678 p.

Johnston, A.C., and Schweig, E.S., 1996, The enigma of the New Madrid earthquakes of 1811–1812: Annual Review of Earth and Planetary Sciences, v. 24, p. 339–384, doi: 10.1146/annurev.earth.24.1.339.

Kenner, S.J., and Segall, P., 2000, A mechanical model for intraplate earthquakes: Application to the New Madrid seismic zone: Science, v. 289, p. 2329–2332, doi: 10.1126/science.289.5488.2329.

Klügel, J.-U., 2005, Reply to the comment on "Problems in the application of the SSHAC probability method for assessing earthquake hazards at Swiss nuclear power plants," by R.J. Budnitz, C.A. Cornell, and P.A. Morris: Engineering Geology, v. 82, p. 79–85, doi: 10.1016/j.enggeo.2005.09.010.

McGuire, R.K., 1976, FORTRAN Computer Program for Seismic Risk Analysis: U.S. Geological Survey Open-File Report 76-67, 68 p.

McGuire, R.K., 1995, Probabilistic seismic hazard analysis and design earthquakes: Closing the loop: Bulletin of the Seismological Society of America, v. 85, p. 1275–1284.

McGuire, R.K., 2004, Seismic Hazard and Risk Analysis: Earthquake Engineering Research Institute Report MNO-10, 240 p.

Milne, W.G., and Davenport, A.G., 1969, Distribution of earthquake risk in Canada: Bulletin of the Seismological Society of America, v. 59, p. 729–754.

National Research Council (NRC), 1988, Probabilistic seismic hazard analysis: Report of the Panel on Seismic Hazard Analysis: Washington, D.C., National Academy Press, 97 p.

Newman, A., Stein, S., Weber, J., Engeln, J., Mao, A., and Dixon, T., 1999, Slow deformation and low seismic hazard at the New Madrid seismic zone: Science, v. 284, p. 619–621, doi: 10.1126/science.284.5414.619.

Nuttli, O.W., 1973, The Mississippi Valley earthquakes of 1811 and 1812: Intensities, ground motion and magnitudes: Bulletin of the Seismological Society of America, v. 63, p. 227–248.

Obermeier, S.F., 1998, Liquefaction evidence for strong earthquakes of Holocene and latest Pleistocene ages in the states of Indiana and Illinois, USA: Engineering Geology, v. 50, p. 227–254, doi: 10.1016/S0013-7952(98)00032-5.

Obermeier, S.F., Bleuer, N.K., Munson, C.A., Munson, P.J., Martin, W.S., McWilliams, K.M., Tabaczynski, D.A., Odum, J.K., Rubin, M., and Eggert, D.L., 1991, Evidence of strong earthquake shaking in the lower Wabash Valley from prehistoric liquefaction features: Science, v. 251, p. 1061–1063, doi: 10.1126/science.251.4997.1061.

Sacks, P., 1978, Wind Forces in Engineering (2nd ed.): Elmsford, New York, Pergamon Press, 400 p.

Senior Seismic Hazard Analysis Committee (SSHAC), 1997, Recommendations for Probabilistic Seismic Hazard Analysis: Guidance on Uncertainty and Use of Experts: Lawrence Livermore National Laboratory Report NUREG/CR-6372, 81 p.

Somerville, P., Collins, N., Abrahamson, N., Graves, R., and Saikia, C., 2001, Ground motion attenuation relations for the central and eastern United States: Final report to U.S. Geological Survey, 16 p.

Stein, S., and Newman, A., 2004, Characteristic and uncharacteristic earthquakes as possible artifacts: Applications to the New Madrid and Wabash seismic zones: Seismological Research Letters, v. 75, p. 173–187.

Stein, S., and Wysession, M., 2003, An Introduction to Seismology, Earthquakes, and Earth Structure: Maldem, Massachusetts, Blackwell Publishing, 498 p.

Toro, G.R., Abrahamson, N.A., and Schneider, J.F., 1997, Model of strong ground motions from earthquakes in central and eastern North America: Best estimates and uncertainties: Seismological Research Letters, v. 68, p. 41–57.

Tuttle, M.P., Schweig, E.S., Sims, J.D., Lafferty, R.H., Wolf, L.W., and Haynes, M.L., 2002, The earthquake potential of the New Madrid seismic zone: Bulletin of the Seismological Society of America, v. 92, p. 2080–2089, doi: 10.1785/0120010227.

U.S. Geological Survey (USGS), 2002, Earthquake hazard in the heart of the homeland: Fact Sheet FS-131-02, 4 p.

U.S. Geological Survey, 2005, The Earthquake Probability Mapping (EPM) page: www.eqint.cr.usgs.gov/eq-men/html/eqprob.html (last accessed 15 February 2005).

Wang, Z., 2005, Reply to "Comment on 'Comparison between probabilistic seismic hazard analysis and flood frequency analysis' by Zhenming Wang and Lindell Ormsbee" by Thomas L. Holzer: Eos (Transactions, American Geophysical Union), v. 86, p. 303.

Wang, Z., and Ormsbee, L., 2005, Comparison between probabilistic seismic hazard analysis and flood frequency analysis: Eos (Transactions, American Geophysical Union), v. 86, p. 45, 51–52.

Wang, Z., Woolery, E.W., Shi, B., and Kiefer, J.D., 2003, Communicating with uncertainty: A critical issue with probabilistic seismic hazard analysis: Eos (Transactions, American Geophysical Union), v. 84, p. 501, 506, 508.

Wang, Z., Woolery, E.W., Shi, B., and Kiefer, J.D., 2005, Comment on "How can seismic hazard around the New Madrid seismic zone be similar to that in California?" by Arthur Frankel: Seismological Research Letters, v. 76, p. 466–471.

Working Group on California Earthquake Probabilities, 2003, Earthquake probabilities in the San Francisco Bay region: 2002–2031: U.S. Geological Survey Open-File Report 03-214, 235 p.

Zoback, M.L., 1992, Stress field constraints on intraplate seismicity in eastern North America: Journal of Geophysical Research, v. 97, p. 11,761–11,782.

MANUSCRIPT ACCEPTED BY THE SOCIETY 29 NOVEMBER 2006

The Geological Society of America
Special Paper 425
2007

Policy development and uncertainty in earthquake risk in the New Madrid seismic zone

Jay H. Crandell, P.E.[†]

5095 Sudley Rd., West River, Maryland 20778, USA

ABSTRACT

This paper addresses policy issues related to uncertainty in earthquake risk in the New Madrid seismic zone and similar environments. I evaluate various sources of data and policy implications related to earthquake hazard, vulnerability, and risk (economic and life-safety consequences). Findings support my belief that, despite the significant uncertainties, opportunities exist to address earthquake risk through properly focused policy. These include: (1) assigning high priority to disaster-response emergency services that also address more routine life-safety risks; (2) focusing life-safety mitigation efforts on the most threatening conditions, such as unreinforced masonry construction; (3) ensuring that mitigation of potential economic loss is strengthened by policy that encourages and does not interfere with a competitive and affordable insurance market based on risk-consistent rates; and (4) requiring that public awareness efforts include realistic information on what is known and not known about the hazard to promote a rational basis for individual decision-making and public policy debate. I also show how proper qualification of "lessons learned" from past earthquakes can improve understanding of policy needs and potential impacts.

Keywords: earthquake, risk, policy, regulation, New Madrid, disaster.

INTRODUCTION

Four features influence policy issues regarding seismic hazard and risk in the New Madrid seismic zone or similar areas:

1. Very long time between the "big ones": Large (M > 7) earthquakes are infrequent on human time scales and occur only once in many generations, or about every 500 yr, according to a recent generalization (Atkinson et al., 2000).

2. Subdued seismic activity: Moderate (M > 5.5) earthquakes, which serve to maintain public awareness and expose vulnerabilities, are uncommon: the last occurred in 1968. Further-more, the last major (M > 6) earthquake was in 1895, more than 110 yr ago.

3. Low risk perception: Accordingly, the population generally perceives earthquake hazard to be low and may not appreciate that earthquake-risk exposure varies substantially within the New Madrid seismic zone.[1]

4. Accumulation of vulnerability: These factors have contributed to vulnerable conditions, notably the use of unreinforced masonry construction for more than 150 yr after the 1811–1812

[†]E-mail: jcrandell@aresconsulting.biz.

[1]Risk (i.e., probability of damage to structures or people) depends both on the magnitude of the hazard (potential for a damaging level of ground motion) as well as vulnerabilities (potential for a particular building type to collapse or soil to subside).

Crandell, J.H., 2007, Policy development and uncertainty in earthquake risk in the New Madrid seismic zone, *in* Stein, S., and Mazzotti, S., ed., Continental Intraplate Earthquakes: Science, Hazard, and Policy Issues: Geological Society of America Special Paper 425, p. 375–386, doi: 10.1130/2007.2425(25). For permission to copy, contact editing@geosociety.org. ©2007 The Geological Society of America. All rights reserved.

New Madrid earthquake series (Ginsburg and Stark, 2002; American Re, 2003).

This situation poses a serious challenge for development of policies to mitigate earthquake risk. Uncertainty in knowing the earthquake hazard causes added difficulty in considering and implementing appropriate policies. Policymakers must somehow formulate understandable and actionable policies that achieve desired effects in politically acceptable ways. Effective policy also must strive to maximize benefits to society by addressing failures in the free market (or existing regulation) without creating adverse or unintended impacts that erode the net value of the policy. Furthermore, policy development occurs in a political atmosphere influenced by divergent expert opinions, special interests, and varied public perspectives. This challenge would be daunting even in an ideal condition where earthquake hazard in the New Madrid seismic zone were understood with a high level of confidence.

Hence, this paper begins by reviewing objectives of earthquake-risk policy and then specific federal and state policies. Next, I explore policy issues dealing with hazard, vulnerability, and risk management related to economic and life-safety consequences. Finally, I suggest specific policies to better help society understand and cope with the uncertain but potentially significant seismic hazard in the New Madrid seismic zone.

POLICY OBJECTIVES

In general, public policy related to the risk of earthquakes or other natural or manmade hazards addresses one or more of the following objectives:
- awareness,
- preparedness,
- response,
- recovery,
- insurance,
- mitigation, and/or
- research.

Awareness, which can be defined as public education about specified risks such as earthquakes, is one of the first considerations in policy development. Awareness efforts include programs for school children, seminars for professionals and government officials, and general dissemination of information. They may be aimed at specific concerns, such as alerting the public to especially vulnerable forms of construction like unreinforced masonry buildings. For example, the state of California requires such buildings to display warning placards. Another example is the state of Arkansas' requirement that insurers notify property owners of the availability of earthquake-hazard insurance.

Awareness efforts face difficulties if they are unreinforced by the routine occurrence of moderately damaging events. Because this is the case in the New Madrid seismic zone, I recommend focusing awareness efforts toward conditions that represent the greatest relative risk. For example, earthquake-risk exposure varies significantly in the New Madrid seismic zone, in a large part

due to different vulnerabilities (e.g., living in a multistory unreinforced masonry building versus a wood-frame home). I have not found data regarding effectiveness of awareness efforts in the New Madrid seismic zone. Hence, I recommend periodic surveys of public awareness level and perception of seismic hazard. For example, the Missouri Department of Health's consumer risk survey could be modified to collect data on perception of seismic risk relative to variation in risk exposure. Such data may help direct earthquake awareness activities to groups with highest relative risk exposure.

The content of awareness information is also important. It should explain what is known and not known about the hazard, identify expected vulnerabilities, and suggest reasonable actions to address them with available resources. This consideration is particularly relevant to communicating uncertainties in seismic hazard in the New Madrid seismic zone (Newman et al., 2001). Realistic awareness efforts can help to educate the public and promote a rational debate on policy considerations. However, even with the best awareness strategies, creating a knowledgeable electorate is difficult (Somin, 2004).

Preparedness is closely tied with awareness because it involves informing the public of appropriate measures to prepare for an impending event. An adequately prepared public would reduce demands on emergency services after a major earthquake. Typical actions include storing provisions and securing building contents to prevent injury or collateral damages (e.g., broken gas lines resulting in fire). Preparedness information can be readily found on the Web (e.g., www.fema.gov, www.cusec.org). However, I found no information with which to assess the public's preparedness in the New Madrid seismic zone.

Response and recovery, which deal predominantly with the aftermath of a disaster, are generally considered to be the most important aspect of public policy related to natural hazards. Even with exceptional success, other policy measures cannot remove natural hazards or completely mitigate vulnerabilities.

Response and recovery involve executing coordinated plans immediately following a natural disaster. Coordination of response and recovery is especially important in the New Madrid seismic zone because relatively low attenuation of ground motion will cause a larger area to be affected than in California for earthquakes of comparable magnitude (Atkinson et al., 2000). Disaster response and recovery plans provide for adequate emergency services (medical, police, fire/rescue, national guard, etc.); shelter, food, and water for displaced individuals; financial assistance; damage assessment of buildings and infrastructure; clean up and debris removal; and other forms of relief and recovery assistance. These activities generally involve states working with federal agencies, such as the Federal Emergency Management Agency (FEMA), as well as regional entities, such as the Central United States Earthquake Consortium (CUSEC). In addition, private-sector groups partner or contract with the public sector to provide emergency services in the event of a disaster.

Insurance is generally considered to be the ideal instrument to finance recovery from disasters. It also can provide incentives to mitigate economic loss if risk-consistent rates correctly differentiate classes of vulnerability. However, for insurance to maximize its potential role, several conditions must exist. First, the risk must be considered insurable, and the market demand must be such that insurance is widely used to reduce uninsured economic loss. Furthermore, the supply of earthquake insurance to meet that demand must be covered by risk-consistent premiums and sufficient financial reserves to fulfill all policy obligations. Ideally, when most economic losses are insured by a viable hazard insurance market, the economic impact of an event is essentially transformed into a necessary and beneficial form of commerce. Automobile insurance is an example, although it has different policy issues, and it addresses a much higher risk, which is considered to be more insurable, and comprises 40% of the $300 billion in property-liability premiums collected annually (Cummins, 2002). However, such ideal conditions for natural-disaster insurance appear difficult to attain due primarily to the nature of low-probability, high-consequence events with highly correlated losses (i.e., many claims for each rare event).

State governments regulate the insurance industry with varying approaches. For example, about half of the states require insurance rates (premiums) to be approved, whereas others allow rates to be set by the competitive market (Kunreuther, 1998). Interestingly, this variation occurs despite studies showing no significant public benefit from insurance rate regulation (Klein, 1998). Unfortunately, regulatory troubles are common in establishing risk-consistent hazard insurance rates (Stelzer, 1999) and can also result in legal disputes (DeLollis, 2000). These troubles are sometimes entangled with pressure on politically appointed state insurance commissioners to protect various short-term special interests (e.g., constituents in areas where rate increases may occur). As a result, insurance regulation in some states may actually contribute to natural-disaster insurance problems. For example, insurance regulations can prevent the use of insurance rates that are consistent with variation in actual risk. Such controls result in cross-subsidy of risk whereby those in low-risk categories are not rewarded and those in higher-risk categories are. This creates a disincentive for appropriate risk-management decisions.

Regulatory policy failures can eventually lead to difficulties such as the insurance crises (and significant losses to the insurance industry) following Hurricane Andrew in Florida and the Northridge earthquake in California (WSSPC, 1998). Such crises have prompted state government interventions like the creation of government-administered hazard-insurance programs.

Several approaches might reduce the problems. In states where natural hazard insurance rates are regulated, the establishment of risk-consistent hazard insurance rates should be isolated from political interference. Where rate adjustments are required, phasing in rate changes may improve the public's willingness to accept them. This concept is particularly relevant in the New Madrid seismic zone because the long time between major earthquakes would allow such transitions to occur over a reasonable length of time. Other policy actions include using tax relief to reduce the cost of holding huge financial reserves required to cover highly correlated earthquake losses (CATO Institute, 2004). Although the reduction in tax revenue raises other concerns, such a policy could promote more affordable and widespread use of natural disaster insurance to reduce uninsured disaster losses, which would otherwise burden society through taxpayer funding of disaster relief. The state of California has granted this type of tax exemption for its government-operated California Earthquake Authority, justifying similar action for private-sector insurers. Many other insurance policy ideas can be found in the literature (e.g., Cummins, 2002; WSSPC, 1998).

Research for this paper did not identify data on the proportion of total exposed property value in the New Madrid seismic zone not covered by earthquake insurance. However, one source indicated that 42% of homeowners in Missouri, and 59% of those in the most seismic areas, purchased earthquake coverage in 2002, primarily due to low rates (e.g., 36¢ to 76¢ per $1000 of coverage) (III, 2005). In contrast, available data suggest that the majority (as much as 75%–85%) of residential properties in California are uninsured against earthquake losses as a result of recent rate increases and changing market risk perception with time from the last major event (Klein, 1998; PIFC, 2001; III, 2005).

About $12.5 billion of the $20–$30 billion in total losses caused by the 1994 Northridge earthquake was covered by insurance (Dunmoyer, 1998). This resulted in significant losses to the insurance industry because the claims dramatically exceeded the roughly $4 billion in hazard insurance premiums collected over the prior 26 yr considered by Dunmoyer (1998). This imbalance was in part related to economically unsound insurance underwriting and rate-setting regulations combined with less controllable factors, such as a dramatic increase in purchases of earthquake-hazard insurance prompted by the 1989 Loma Prieta earthquake. Thus, for many claims, premiums had only been collected for a few years. This situation may be similar in the New Madrid seismic zone, particularly if rates are below actuarial levels. Thus, development of insurance policy should be a high priority for states affected by the New Madrid seismic zone.

Mitigation involves reducing economic and life-safety impacts of natural disasters through structural and nonstructural practices. Structural practices include adoption of building codes, public infrastructure regulations, retrofitting of vulnerable existing structures, and other actions. Nonstructural practices include land-use regulations and buy-out strategies. Land-use regulations may include identification of areas subject to liquefaction and limiting land use to low-consequence activities or improvements. Buy-out strategies, used most commonly to address repetitive flood losses, involve government purchase of an at-risk or damaged property with the consent of the property owner. Subsequently, restrictions are placed on the property to limit future use and improvement and avoid repeated losses.

Mitigation can also involve incentives for safer construction than the minimum required by building codes. A key consideration includes insurance rate credits, such as rate categories based on variation in vulnerability as offered by the Florida Windstorm

Underwriting Association. In addition, low-interest loans, grants, or property-tax incentives may be used to encourage mitigation, particularly for projects with major economic or life-safety implications (e.g., major highway bridges, hospitals, etc.). In the National Flood Insurance Program, access to federal disaster relief requires local adoption of various mitigation requirements for new and existing construction in mapped flood-hazard zones. Thus, federal flood insurance (the only option for flood insurance) is offered as an incentive to adopt specific mitigation actions. However, in the National Earthquake Hazard Reduction Program (NEHRP), similar mandates are not present, and different strategies are used to promote mitigation. Innovative ways to promote mitigation should also be considered. For example, tort liability associated with unmitigated, vulnerable buildings may motivate some property owners to mitigate risk (Ginsburg and Stark, 2002). This motivation could be strengthened by legislation to provide liability protections to property owners who implement a reasonable level of mitigation to vulnerable structures.

The major downside of mitigation is its up-front cost. Thus, a central question arises: what is an acceptable level of mitigation in view of variable and uncertain risk-reduction benefits? I explore this question in greater detail later in this paper (see Policy Issues).

Research and research-related policy seeks to learn from past disasters. Research strives to improve understanding of hazards and consequences (risk) and expand knowledge of cost-effective mitigation methods. At the federal level, the Stafford Act and the NEHRP program fund research and development of new or improved mitigation technologies. Several states also support research, usually by designating a state university as a center for earthquake research and monitoring. I address appropriate uses of research to identify and apply "lessons learned" from past disasters later under Policy Issues.

POLICY REVIEW

Various earthquake-risk policies of federal and state governments address several or all of the policy objectives discussed in the previous section.

The Stafford Act provides a national program, administered through FEMA, for assistance and relief to states after a disaster that invokes a presidential declaration. On average, ~30 disasters per year have been so recognized (Sylves, 2004). However, the number of declared disasters has increased with each administration and now exceeds more than 100 per year as a reaction to political pressure and growing expectations for federal aid (Olasky, 2005). FEMA obligations under the Stafford Act range from a normal low of about $500 million in quiet years to a high of $12 billion in 2001 as a result of terrorist attacks (Sylves, 2004). By one measure, federal taxpayers provide more than $7 billion per year in disaster assistance (CATO Institute, 2001).

Public support for disaster relief is in part motivated by compassion for those who suffer significant loss or injury, which may partly explain the increasing trend in disaster declarations (no administration wants to be viewed as being less compassionate than their predecessor). The Stafford Act also minimizes the disruption of commerce due to disasters. The difficult question is how to address the need without the policy becoming an entitlement that "contributes to disaster losses by reducing the incentives for hazard mitigation and preparedness" (Olasky, 2005, p. 19). Critical policy analysts have described the Stafford Act as a financial disaster created as an act of Congress by intervening into and crowding out market-based alternatives (CATO Institute, 2001). Such reviews call for improved control in the declaration of a disaster and distribution of public funds.

The National Earthquake Hazards Reduction Program (NEHRP) is the federal government's primary program to reduce the risk to life and property from future earthquakes (Ginsburg and Stark, 2002; FEMA 383, 2003). The program involves multiple government agencies with FEMA in the lead, a role recently transferred to National Institute of Standards and Technology, Department of Commerce. The program includes earthquake hazard analysis, risk modeling, monitoring of earthquakes, and research to better understand earthquake risk and means to reduce it. The program also sponsors regional groups, such as the Central United States Earthquake Consortium. One of its primary outputs has been NEHRP "Recommended Provisions for Seismic Regulations for New Buildings and Other Structures" (FEMA 368, 2001). This document, which incorporates technical outputs of NEHRP as well as private-sector research, has a significant influence on national model building codes and design standards (ICC, 2003; ASCE, 2002) adopted by state and local governments.

State policies in the New Madrid seismic zone in regard to earthquake hazard are quite varied (Ginsburg and Stark, 2002). They are influenced by federal policy in only a limited fashion, such as coordination of disaster preparedness plans with FEMA (e.g., MSSC, 1997). States regulate insurance and also regulate construction through adoption or creation of building codes. Within a state, the institution and administration of building codes vary from complete control by the state to complete control by local jurisdictions.

Missouri has enacted five key pieces of earthquake legislation since 1990. They include the following provisions:

• They require local jurisdictions to adopt earthquake-resistant requirements for new construction based on a specified national model building code where predicted ground shaking is equivalent to a modified Mercalli intensity of VII or above due to a magnitude 7.6 earthquake on the "New Madrid fault." The requirement does not apply to existing buildings or to private structures with a total area of 10,000 ft^2 or less. Political jurisdictions that do not comply are not eligible to receive state aid, assistance, grants, etc.

• They establish earthquake emergency procedures in schools expected to experience ground motion of a modified Mercalli intensity of VII or above due to a magnitude 7.6 earthquake along the "New Madrid fault."

• They promote volunteer emergency services without incurring liability except for willful misconduct or gross negligence

• They establish the Missouri Seismic Safety Commission to prepare for a major earthquake, set goals and priorities for mitigation, monitor earthquake activity, and assist in promoting earthquake and disaster safety.

• They require insurers to notify property insurance applicants of the availability of earthquake insurance

Arkansas is considered to have the most comprehensive earthquake legislation among states in the New Madrid seismic zone. Relevant statutes are summarized as follows:

• The Earthquake Preparedness Act requires the Department of Emergency Management to educate government officials, coordinate government officials in response and recovery, and disseminate public information for awareness and preparedness. It also amends the Interstate Civil Defense and Disaster Compact to be consistent with the Central United States Earthquake Consortium (CUSEC).

• The Earthquake Resistant Design for Public Structures requires that all public structures be designed and constructed to resist earthquake forces in three seismic damage zones based on expected ground accelerations.

• The Arkansas Earthquake Authority (AEA) Act makes earthquake insurance available to homeowners through the Market Assistance Program (MAP). If no approved insurer is available in the MAP, or if rates exceed those which could be offered by the AEA, then consumers can purchase coverage up to $100,000 through the AEA. Insurers are also required to notify existing and new customers of their eligibility for residential earthquake insurance through the MAP or AEA. Those who elect not to use it must reject such coverage in writing.

• The Earthquake Activity Act established a center for Earthquake Education and Technology Transfer at the University of Arkansas to operate an observatory for monitoring earthquake activity and collaborating with seismic monitoring programs at St. Louis University and the University of Memphis.

Policy in Tennessee covers issues similar to that in Arkansas. Illinois has the least amount of earthquake-related policy of the states considered. Illinois statutes focus on awareness, preparedness, and insurance. In contrast, California has enacted 16 different policies since 1933 that address various seismic hazard concerns, including awareness, preparedness, response, recovery, mitigation (new and existing buildings), insurance, and research.

POLICY ISSUES

Earthquake-risk policy addresses four interrelated topics: hazard, vulnerability, economic consequences, and life-safety consequences. The combined effect of hazard and vulnerability defines the risk in terms of potential economic and life-safety consequences (i.e., risk = f[hazard, vulnerability]). Risk management seeks to define acceptable levels of risk, identify unacceptable risks, and efficiently allocate available resources accordingly.

Earthquake Hazard

Characterization of seismic hazard is the first step in establishing a rational basis for earthquake-risk policy. Unfortunately, seismic hazard in the New Madrid seismic zone is not easily characterized because the underlying cause of the earthquakes is unknown. Moreover, important hazard parameters are poorly known, including the magnitude of the largest earthquakes, the attenuation of ground motions, site amplification, location of faults on which major earthquakes may originate, tectonic mechanisms responsible for current and past seismicity, and future activity of these mechanisms (Newman et al., 2001; Stein et al., 2003). Due to the limited earthquake record, key parameters such as the recurrence interval and magnitude of major earthquakes are inferred through indirect methods with significant uncertainty. For example, maximum earthquake magnitude estimates vary within the range of M 7.4 to M 8.1 (HUD, 2003), with recent studies converging on the lower half of this range (Newman et al., 1999; Hough et al., 2000; Newman et al., 2001; HUD, 2003; Kochkin and Crandell, 2004; Bakun and Hopper, 2004). Uncertainty in the largest earthquake magnitude and other parameters translates directly into significant uncertainty or potential bias in predicting ground motion associated with a given probability of occurrence. Hence, the choice of an appropriate level and cost of mitigation is similarly uncertain and subject to bias.

Because of the significant uncertainty, the magnitude and extent of mapped probabilistic ground motions (hazard) in the New Madrid seismic zone primarily represent the judgment of those participating in NEHRP. These judgments result in seismic hazard maps found in NEHRP's "Recommended Provisions for Seismic Regulations for New Buildings and Other Structures" (FEMA 368, 2001). The maps have generated some controversy and worthy criticisms. For example, plausible treatments of the key uncertain parameters in hazard modeling can result in ground motions that, depending on location within the New Madrid seismic zone, vary by factors of 2–13 at the probability levels chosen for earthquake-risk mitigation in recent model building codes (Newman et al., 2001). These uncertainties reflect differences in the assumed ground-motion attenuation functions and estimates of magnitude and recurrence interval for "characteristic" New Madrid earthquakes. As such, the NEHRP seismic hazard maps (as also represented in national model building codes) represent only one of a wide range of plausible representations of earthquake hazard for the New Madrid seismic zone. The implications of this uncertainty are staggering for policy decision-making and the expected value or impact of any particular policy decision. In considering practical limits, however, a minimum representation of hazard should at least result in adequate mitigation of the nature and extent of the major known vulnerabilities, such as damage to unreinforced masonry construction, revealed in historic evidence from the 1811–1812 New Madrid earthquakes (HUD, 2003; Kochkin and Crandell, 2004). As one example, such practical considerations have influenced modifications of the NEHRP seismic hazard maps and related mitigation provisions in the state of Indiana's residential building code (http://www.in.gov/dhs/osbc/techserv/res_code.pdf).

This discussion has addressed only the uncertainty in prediction of ground motion or hazard at a prescribed probability level (i.e., 2500 yr recurrence interval in the NEHRP maps for the New Madrid seismic zone). However, there are broader impacts related to the selection of a 2500 yr recurrence interval for hazard mapping. When newer probabilistic seismic hazard maps became available from the U.S. Geologic Survey (USGS) in 1996, NEHRP convened a small group of experts to determine how to use them (FEMA 369, 2001). Those experts initially judged that a 500 yr return period hazard level generally represented past successful design practice in California. However, they also felt that design ground motions at this level of probability were too low for the rest of the country. Conversely, they also felt that the 500 yr probabilistic ground motions were too high for the most hazardous parts of California. Others felt that a 500 yr basis for mapped hazard resulted in some areas of California having design ground motions that were too low (e.g., much lower than prior maps). To accommodate these divergent "expert feelings," some circuitous judgments were imposed on the original probabilistic seismic hazard maps that involved shifting to a nonuniform probabilistic basis for NEHRP seismic hazard maps. As a final result, the NEHRP map represents a 2500 yr recurrence interval in the eastern United States, including the New Madrid seismic zone, but a lower hazard probability level in the most-hazardous parts of California. A more rational and equitable approach would have been to calibrate the probabilistic hazard levels for the entire map to be consistent with the lower probability (e.g., 500 yr) used in the most hazardous areas of California, where experience may be deemed suitable for establishing a baseline for acceptable minimum practice and, thus, acceptable maximum risk. Such an approach would tend to address one of the major intuitive problems in the NEHRP seismic hazard maps, namely, that the design ground motions in the New Madrid seismic zone are greater than those of California.

Vulnerability

In addition to uncertainty in characterizing seismic hazard, vulnerability is another major source of uncertainty. For example, uncertainty in seismic design theory's ability to predict actual building performance in an earthquake can be quite substantial (Crandell and Kochkin, 2003). Key design factors related to seismic response and building performance are still largely established by expert opinion and judgment, not scientific methods. Furthermore, the use of safety or reliability concepts to establish appropriate strength levels for design are calibrated in a limited fashion to past successful construction or design practices (Galambos et al., 1982). This process of defining acceptable safety introduces potential biases and uncertainties, particularly in regard to common forms of construction that were not included in the safety calibration exercises, such as conventional wood frame construction (HUD, 2000, 2001). As with earthquake

hazard in the New Madrid seismic zone, safety concepts used in modern codification of structural design rules are not rigidly or consistently constrained by science or other means.

Despite the uncertainties in predicting vulnerabilities or building performance, it seems that at least one notable vulnerability has accumulated in the New Madrid seismic zone since the early 1800s. For example, the 1811–1812 New Madrid earthquakes "threw down a few chimneys" and caused "a few stone houses to split" in St. Louis, which may then have had a greater proportion of buildings on softer, more vulnerable soils (HUD, 2003). Thus, historic damage reports, which represent the most direct indication of vulnerabilities in the New Madrid seismic zone, tend to show that most forms of damage were related to unreinforced masonry construction at distances of more than 50 miles from the epicenter (Kochkin and Crandell, 2004). Because such construction continued up to the 1980s (Ginsburg and Stark, 2002), ~30% or more of the buildings in the New Madrid seismic zone today are made of unreinforced masonry construction (American Re, 2003). In contrast, because the vulnerability of unreinforced masonry construction has been repeatedly revealed in California, where moderately damaging events are considerably more frequent, less than 3% of buildings in Los Angeles are unretrofitted unreinforced masonry buildings (American Re, 2003). Thus, mitigation efforts in the New Madrid seismic zone should focus on such vulnerable construction and other conditions of similar or greater consequence. This point is further reinforced in a later discussion on life-safety consequences.

Understanding of the vulnerability of structures has been a primary focus in studies of past earthquakes. But, many mitigation policy decisions are based on expert opinion from anecdotal assessments of building performance immediately following a disaster or damaging earthquake. Such anecdotal damage assessments tend to accurately support only the most obvious findings and misrepresent "interesting" but infrequent forms of damage. Fortunately, scientific principles can be applied to postdisaster damage assessments in much the same way that scientific experiments are conducted to limit human bias and test hypotheses (Crandell and Kochkin, 2005). I have conducted several studies using this method to assess residential construction performance following recent natural disasters (HUD, 1993, 1994; Crandell, 1996, 2002). Such studies provide data for statistical analysis of building performance and cause-effect relationships that help to guide important policy decisions and advance engineering science. For example, studies of conventional single-family homes in the near-field of the 1994 Northridge earthquake (HUD, 1993; McKee and Crandell, 1999; Crandell and Kochkin, 2003; Crandell, 2004) found that:

• Primary damage was related to cracking of exterior (stucco) and interior (gypsum wall board or plaster) finishes. Cracking of finishes occurred in ~50% of the homes and was mostly limited to small or hairline cracking at points of high stress concentration.

• Significant structural damage to walls and foundations was limited to ~2% of the housing stock, generally ones with extenuating circumstances (e.g., building grossly out of compliance with code, owner modifications, poor or steep-sloped site conditions, localized site effects such as liquefaction and fissuring, etc.).

• Variation in building damage was not correlated with short-period ground motion, the primary hazard parameter used for seismic design of light-frame buildings. Instead, variation in damage correlated with long-period ground motion.

• Increasing amounts of wall bracing provided diminishing returns for reducing damage probability. For example, as wall bracing increased from 20% of a wall length to 40%, the rate of cracking in stucco wall finishes decreased from ~70% to 55% (i.e., doubling of bracing amount reduced damage probability by ~20%). Further increases in bracing from 40% to 80% (theoretically again doubling strength) reduced stucco cracking probability only from 55% to 50% (i.e., a 10% reduction in damage probability). For such buildings, modern building codes require a minimum 40% braced wall length, so further increases appear to yield significantly diminished benefits in exchange for the increased cost and design impacts. Unfortunately, codified seismic design theory suggests that the benefits accrue linearly with increased strength of bracing because of a failure to account for an offsetting strength-stiffness relationship on seismic load magnitude

These "lessons learned" have a number of implications for policy. First, the amount of bracing required for conventional residential construction in modern building codes seems to be adequate, and further increases yield diminishing returns. Thus, reductions in economic loss might be better sought via cost-effective wall finish materials that are less likely to exhibit cracking. There is also a need for research to define a design method and ground-motion representation that correlates more closely with actual residential building performance. Finally, most of the worst damage experienced by less than 2% of the affected housing stock was associated with "extenuating circumstances." We should thus improve the ability to identify and mitigate these "extenuating circumstances" while limiting the impact of such policy to affect only those circumstances that lead to a higher-than-normal risk. Unfortunately, anecdotal damage studies and data are frequently used in reactive policy development after a major disaster. While not all policy created this way is faulty, it runs that risk. Therefore, local and state policy in regard to how lessons are to be learned and applied following natural disasters should be more carefully considered. Findings from scientifically robust damage assessment methods should be favored over anecdotal methods that rely heavily on expert opinion and interpretation.

Economic Consequences (Risk)

Economic consequences are a complex and significant concern that require careful consideration in regard to mitigation and insurance policy issues. For a repeat of the 1811–1812 New Madrid earthquakes, damage loss estimates range from $60 to $120 billion depending on the risk model and assumptions used (Klein, 1998; Ginsburg and Stark, 2002). Even with the progress made in California, large vulnerabilities for economic loss continue. The 1994 Northridge and 1989 Loma Prieta earthquakes caused an estimated $20–$30 billion and $7 billion in total losses, respectively.

For the New Madrid seismic zone, the most significant economic risk is concentrated in the St. Louis and Memphis metropolitan areas (FEMA 366, 2001). The estimated annualized earthquake loss ratios in California are generally an order of magnitude greater than those in the New Madrid seismic zone, despite the greater vulnerability of the building stock in the New Madrid seismic zone (FEMA 366, 2001). This difference in economic risk is due to the lower frequency of moderately damaging earthquake events in the New Madrid seismic zone. Hence, policy decisions and priorities for the New Madrid seismic zone should differ somewhat from those in California.

This point is illustrated in one preliminary estimate of the cost impacts of adopting a newer building code, such as the International Building Code (IBC), in Memphis, Tennessee. Stein et al. (2003) estimated that the new code could cost an additional $200 million per year in terms of added construction and design cost, which is more than 10 times FEMA's estimate of $17 million in annualized earthquake losses (AEL). The negative return on investment is even worse because differences in risk-mitigation levels between building-code choices may only affect a modest change in the actual AEL.

A similar cost-benefit analysis can be made for the national-scale housing market. For example, the cost impact of increased seismic building-code requirements could approach or exceed $1 billion per year for new homes in higher-seismic-risk regions of the United States.[2] Yet, with a very liberal estimate of the benefits in the new code, the earthquake loss reductions would be modest, at about $20 million per year, and increase over a period of ~100 yr to $600 million per year as the housing stock is slowly replaced.[3] Thus, over the first 100 yr after implementation, $100 billion dollars would be spent by consumers to save an uncertain $31 billion in losses. Consideration of housing in the only New Madrid seismic zone would produce a similar result due to the lower AEL in the New Madrid seismic zone relative to more active seismic zones of the United States. This lack of favorable cost-benefits for additional mitigation expenditures indicates the importance of concentrating policy on efforts to address expected losses by a competitive and widely used supply of hazard insurance. It also reinforces the need to identify and concentrate on the most critical and least costly mitigation technologies or improvements.

This discussion raises an important question: what is a loss? As discussed previously, losses that are insured by a sustainable insurance market are actually a form of necessary and beneficial

---

[2]The $1 billion cost impact was roughly estimated by assuming 20% of 1,500,000 annual housing starts in moderate- to high seismic-hazard areas with a median sale price of $200,000, of which 50% is associated with the building cost and 3% of the building cost is attributed to seismic design cost increase. Thus, $0.20 \times 1{,}500{,}000 \times \$200{,}000 \times 0.50 \times 0.03 = \$1$ billion.

[3]Loss reduction benefits for the $1 billion dollar per year cost impact are roughly estimated by assuming that 75% of the $4 billion/yr of total earthquake losses is associated with housing and that increased seismic design requirements will result in about a 20% reduction in losses for new one- and two-family homes added to the housing stock each year (e.g., 1.5 million new, 50 million existing). Thus, for the first year, the loss reduction benefit was estimated as follows: $0.75 \times \$4$ billion $\times 0.20 \times 1.5/50 = \$20$ million.

commerce. Losses that are not insured, and thus require disaster relief via financial assistance, are a loss to the taxpayers who provide the relief. Losses that are not insured may be a consequence of a property owner's decision to retain risk or the result of economic disadvantage (e.g., no choice but to retain risk). Losses also include business disruption (which also may be insured or uninsured), loss of income, and liability. In addition, other losses may be intangible (e.g., loss of irreplaceable keepsakes, etc.). These loss conditions have very different implications with respect to policy development. Yet, natural-disaster loss data seem to be rarely focused on what the losses really mean in relation to appropriate policy considerations. Fortunately, some efforts are beginning to address this problem.

To gain public support of various policies aimed at reducing the economic impact of natural disasters, it is commonly argued that natural-disaster losses are increasing. However, when historic economic loss data are normalized for population growth and corrected for inflation, a different perspective unfolds. Thus, gross disaster loss values should be viewed with caution as a basis for policy. One recent study using appropriately normalized data shows that hurricane damage losses have been relatively constant over the past century, even considering the $20 billion losses from Hurricane Andrew (Pielke and Landsea, 1998). Thus, the commonly noted exponential growth in gross damage losses results not so much from potential long-term change in hazard or frequency of events, but rather from population and property value growth in hazardous areas. However, the losses on per unit value exposed basis have been more or less stable. Because insurance premiums are determined on a per unit value basis, a properly functioning insurance market should be capable of addressing the economic risk, even in the absence of additional mitigation (see earlier discussion on *Insurance* under Policy Objectives). Moreover, risk-consistent insurance rates would place pressures in the market to properly temper growth in hazardous areas. A similar study normalizing historic earthquake losses was not found in preparing this paper.

The two basic policy questions involved in this discussion on economic consequences are (Ament, 1998):

• Who pays?

• What is the best vehicle to provide for this payment?

Ideally, those who accept a particular risk should pay. However, this approach, taken to its extreme, leaves little room for compassion, particularly for those least capable of coping with economic risk. Thus, policy ideas on these questions vary widely and are often mutually exclusive. This situation also holds true for policies related to maximizing the utility of insurance as a means to pay for losses. Ultimately, a policy that addressed economic consequences must balance the following options (Hamburger, 1998):

• retain the risk (i.e., accept responsibility for the risk); or

• transfer some or all of the risk (e.g., purchase insurance); and/or

• mitigate the risk (e.g., take measures to reduce loss or vulnerability).

Invariably, some risk may be retained, some may be insured, and some may be mitigated. The amount of risk to place in each category requires significant policy consideration. For example, the amount of mitigation (e.g., stringency of building codes) should be coordinated with the amount of risk transferred via insurance and the rates paid for that transfer. An optimum balance may require coordination of various policies and departments of government. Unfortunately, mitigation policy (building codes) and insurance policy are rarely coordinated, and related policy decisions are often made in isolation.

Life-Safety Consequences (Risk)

Life-safety consequences are generally viewed as the most critical public policy concern. Defining an acceptable level of risk on a cost-benefit basis is controversial because it requires assigning value to life. Yet, any policy involving life-safety risks inherently involves valuing life. Therefore, objective approaches to this issue have been contemplated. For example, one approach assesses the value for life through actual market response to various life-safety risks (Wilson, 2000). However, other means are generally preferred, because they avoid directly valuing lives.

Modern structural-safety design provisions have been established via a calibration to historically accepted norms for risk as determined from a relative comparison of failure probability represented by successful past practice (Galambos et al., 1982). This approach avoids directly valuing life but implicitly does so. In this process, judgments are made as to what constitutes a "norm" of past practice, and safety levels based on these historically accepted norms are frequently altered (usually increased) due to changing perceptions of risk. Furthermore, such changes are often influenced by special interest. For example, reducing life-safety risk may be used to justify an increase in mitigation aimed at reducing economic loss (or future insurance obligations), although the cost-benefits of increased mitigation may be unfavorable. Manufacturers who sell products associated with a higher level of mitigation and higher cost often support or encourage such policy changes. Even the engineering community has a conflict of interest because such policy tends to increase the market for and cost of engineering services. As with mitigation of economic loss, mitigation of life-safety risk also faces diminishing returns due in part to countervailing life-safety risk effects.[4] However, such effects are difficult to quantify and are subject to significant uncertainty. Therefore, arguments for relative increases in safety often appear compelling, and arguments to maintain or reduce current safety levels are often perceived as imprudent or politically ugly. To the degree that uncertainty

---

[4]Countervailing life-safety risks are those risks that are affected as a result of addressing one particular risk of a similar or different nature. For example, increasing the cost of new construction to mitigate (reduce) future earthquake risk may cause some increase in risk exposure associated with the increased or extended use of less costly but more vulnerable existing construction as a normal market response.

also exists and is not fully understood, the ensuing confusion also favors the faulty logic that "you can never be too safe."

People are faced with a multitude of "commonplace" life-safety risks that vary in probability and perceived importance. It is important to place any specific risk in this context, as illustrated by Table 1.

In general, a probability threshold exists where risk of death is viewed as an "act of God" (Otway et al., 1970; Melchers, 1999). This threshold has been estimated at ~10^{-6} annual probability of death. According to Table 1, lightening, flood, tornado, hurricane, earthquakes, and mud slides all fall below this threshold. Above an annual probability of 10^{-5}, people are inclined to at least acknowledge the risk. At higher risk probabilities, people are inclined to spend money to reduce the risk (e.g., homicide, auto accidents, cancer, etc.).

In society, significant deviations from a rational response to risk occur due to the personal nature of risk perception in addition to variation in actual risk exposure. These deviations can be explained by facets of human behavior such as "dread," personal experience, or reaction to the occurrence of a recent major event (May et al., 1999; Ginsburg and Stark, 2002). For this reason, major policy actions in regard to low-probability and high-consequence risks usually occur after a major event when risk perception may be momentarily elevated and perhaps "over appreciated." Whether any perception of a particular risk at a point in time is designated as an under- or over-appreciation of the risk is difficult to judge. For example, one person may view earthquake risk as greater than automobile risk, regardless of the actual risks. On the other hand, if people are inclined to spend money to reduce risk of death in an automobile accident rather than by earthquake, their behavior is reasonably consistent with the actual risk. However, risks vary with specific circumstances: people living in a multistory unreinforced masonry building in Memphis or St. Louis are at greater earthquake risk than ones in safer construction. Thus, statements in regard to a particular risk being under- or over-appreciated need to be carefully qualified. The challenge is that the public's level of risk perception provides a rational basis for policy decisions only to the extent that it represents the actual risk, knowledge of uncertainties in actual risk, and an understanding of policy implications (see earlier discussion on the policy objective of *Awareness*).

It is important to note that not all of the risks in Table 1 can be addressed by specific mitigation policies. For example, of the 25 deaths per year related to earthquakes, many are associated with indirect causes (e.g., heart attack, etc.). Of the estimated 40 deaths per year related to hurricanes, 75% are related to flooding and indirect causes. Therefore, increasing wind damage mitigation may only affect a portion of 25% of the problem (HUD, 2000). In comparison, more than 60,000 deaths per year are related to automobile accidents. While any loss of life is tragic, such factors should be considered in policy decisions given society's limited resources to address risks. For example, the costs to society of a particular level of mitigation should be weighed against other opportunities, such as improving emergency services to assist in disaster response. These same services carry a significant benefit in reducing other commonplace and greater life-safety risks (Table 1) in the decades or centuries between major earthquakes.

The correct balance of such policy considerations involves the specific circumstances. This point can be illustrated by considering at-risk populations in the New Madrid seismic zone, such as in Memphis, where ~1.2 million people live. Within the bounds of uncertainty, the ground motions in Memphis from a repeat of the 1811–1812 New Madrid earthquakes may be comparable to those experienced in Charleston during the 1886 earthquake (HUD, 2003; Kochkin and Mays, 2005). In Charleston at that

TABLE 1. COMMONPLACE RISKS (UNITED STATES)

Type of risk	Estimated annual risk (mortality)	Estimated annual average mortality	Estimated annual injuries
Smoking	3.6×10^{-3}	1,000,000	?
Cancer	2.8×10^{-3}	800,000	?
Auto accidents	2.4×10^{-4}	66,000	?
Medical errors	1.6×10^{-4}	44,000	?
Homicide	1.0×10^{-4}	27,400	?
Fires	1.4×10^{-5}	3800	?
Building collapse	1.0×10^{-6}	Probability applies only to damage of structure	
Lightning	5.0×10^{-7}	136	800
Floods (nonhurricane)	3.9×10^{-7}	106	1700
Tornadoes	2.9×10^{-7}	80	1500
Hurricanes and tropical storms (wind and flooding)	1.5×10^{-7}	40 (10 wind; 30 flooding)	60
Thunderstorm	1.3×10^{-7}	35	500
Other high winds	1.1×10^{-7}	30	120
Earthquakes	3.3×10^{-8}	25[†]	?
Hail or mud slide	1.8×10^{-9}	0.5	70

Note: Data are from various sources listed in HUD (2000), and mortality rates are based on 1999 U.S. population estimate of 273,800,000.

[†]For earthquake, others estimate a mortality rate of 9 lives per year (Stein and Wysession, 2003).

time, a similar proportion of unreinforced masonry construction existed as is found today in the New Madrid seismic zone, and ~100 people of a 50,000 population were killed, the vast majority by falling bricks (Dutton, 1890; Harlan and Lindbergh, 1988). If the analogy holds, then the expected death toll in Memphis (at recent population levels) may approach 2400 in a repeat of the 1811–1812 New Madrid earthquakes. This number corresponds to roughly 5 deaths per year assuming a 500 yr recurrence interval for 1811–1812–style New Madrid seismic zone earthquakes. Focusing resources on mitigation of particularly vulnerable existing building construction may substantially reduce this estimated death toll. In fact, existing unreinforced masonry buildings are still a life-safety concern in California based on outcome of the 2003 San Simeon earthquake (CSSC, 2004). Despite the low death toll of two people, both were killed by falling brick while attempting to evacuate an unretrofitted unreinforced masonry building.

CONCLUSIONS AND RECOMMENDATIONS

Despite the uncertainties of earthquake hazard in the New Madrid seismic zone, I believe that opportunities exist to effectively address earthquake risk. While some of these opportunities are similar to those in more seismically active parts of the United States, important differences in the emphasis of policy should be considered for the New Madrid seismic zone. I thus offer the following conclusions and recommendations:

Life-Safety Consequences (Risk)

1. *Mitigation* efforts should focus on identifying and addressing the greatest vulnerabilities to earthquakes (e.g., retrofitting of unreinforced masonry buildings and chimneys, critical facilities, identification and mitigation of landslide hazards, etc.). Incentives to promote retrofitting should be encouraged, such as low-interest loans, property tax breaks, or liability protections for owners who retrofit vulnerable buildings using prequalified measures.

2. *Disaster response and recovery* efforts should continue to be given a high priority. In the many years between major earthquakes, the required resources (e.g., emergency response personnel, facilities, and equipment) address more commonplace and greater life-safety risks. Thus, preparing to respond to a major earthquake will save many more lives by routine use of those resources.

3. Efforts to instill public *awareness* of earthquake risk should focus on specific conditions where the risk is highest (e.g., unreinforced masonry buildings, buildings subject to landslide hazards, buildings with gas services, etc.). General awareness and preparedness efforts should continue and also include educational content that adequately discloses the implications of uncertainty in hazard predictions. Surveys of earthquake-risk perception should be used to monitor progress and identify where awareness efforts are needed most.

Economic Consequences (Risk)

1. *Insurance* should be the primary tool to address the economic consequences of natural disasters. Regulations that prevent or deter competitive and risk-consistent hazard insurance rate setting and categorization by vulnerability should be identified and corrected. In addition, earthquake hazard insurance may be made more affordable and available via tax relief for the large financial reserves required to cover obligations in the event of a major earthquake.

2. *Mitigation* of economic consequences from earthquake risk should focus on the most significant existing vulnerabilities and on practices that are the most cost-effective to implement. Wide-sweeping policy changes that increase the level and cost of mitigation to reduce uncertain economic loss should be carefully scrutinized and also coordinated with insurance policy considerations.

General

1. Uncertainties in seismic hazard and vulnerabilities suggest a wide range of plausible or appropriate hazard levels for the purpose of establishing earthquake-risk mitigation policy. Local and state governments in the New Madrid seismic zone should carefully consider available resources, such as national model building codes and NEHRP seismic design provisions, and use or modify them with a reasonable and careful understanding of the implications as described in this study.

2. NEHRP seismic hazard maps as used in national model building codes should be reevaluated to establish a uniform probabilistic basis for ground-motion hazard and risk throughout the United States, including the New Madrid seismic zone and California. This approach would resolve some of the controversy surrounding the creation and use of newer earthquake hazard maps in the New Madrid seismic zone.

3. Given the many uncertainties and judgments involved in the current codification of seismic design "rules," engineers and building designers should be given reasonable latitude in making "best value" judgments for new building design and seismic retrofit applications. Simplified and effective seismic design guidelines also should be developed because the level of uncertainty in seismic hazard and building response does not appear to justify the level of detail and complexity found in modern building codes for many routine construction applications. Such actions will encourage mitigation by making it more practical and cost-effective to implement (CSSC, 2004).

4. Research should continue to address major uncertainties in understanding earthquake hazard in the New Madrid seismic zone and building performance in earthquakes. Research and development aimed at reducing the cost of mitigation also should continue.

5. Damage assessments using scientific methods, as opposed to typical anecdotal studies, should be used more extensively and be more widely considered in policy development. Such studies

provide an objective means to verify impacts of past policy decisions and explore the need for and effectiveness of future policy considerations.

6. Policy reviews and analyses on earthquake risk in the New Madrid seismic zone appear to be limited in quantity and diversity. A greater diversity of policy analyses should be encouraged because they would lead to a broader perspective and deepened understanding of the many issues involved.

ACKNOWLEDGMENTS

I thank Seth Stein, Nicholas Pinter, Vladimir Kochkin, and Mark Nowak for their assistance, critical review, and encouragement.

REFERENCES CITED

Ament, J., 1998, Earthquake insurance: Public policy analysis: Palo Alto, California, Public Policy Perspectives from the Western United States Earthquake Insurance Summit, Western States Seismic Policy Council (WSSPC), www.wsspc.org (last accessed 12/29/2004).

American Re, 2003, Annual review of North American natural catastrophes 2003: Princeton, New Jersey, American Re-Insurance Company, 44 p.

American Society of Civil Engineers (ASCE), 2002, Minimum Design Loads for Buildings and Other Structures: ASCE Standard 7–02: Reston, Virginia, American Society of Civil Engineers, 376 p.

Atkinson, G., et al., 2000, Reassessing the New Madrid seismic zone: Eos (Transactions, American Geophysical Union), v. 81, no. 35, p. 397, 402–403.

Bakun, W.H., and Hopper, M.G., 2004, Magnitudes and Locations of the 1811–1812 New Madrid, Missouri, and the 1886 Charleston, South Carolina Earthquakes: Bulletin of the Seismological Society of America, v. 94, no. 1, p. 64–75, doi: 10.1785/0120020122.

California Seismic Safety Commission (CSSC), 2004, Findings and Recommendations from the San Simeon Earthquake of December 22, 2003: California Seismic Safety Commission report 04-02, 7 p.

CATO Institute, 2001, CATO Handbook for Congress, Policy recommendations for the 107th Congress: Washington, D.C., CATO Institute, 680 p.

CATO Institute, 2004, CATO Handbook for Congress, Policy recommendations for the 108th Congress: Washington, D.C., CATO Institute, Chapter 37, p. 385–396.

Crandell, J.H., 1996, Assessment of Damage to Homes Caused by Hurricane Opal: Upper Marlboro, Maryland, NAHB (National Association of Home Builders) Research Center, Inc, 39 p.

Crandell, J.H., 2002, Housing Performance Assessment Report: F-4 La Plata Tornado of April 28, 2002: Upper Marlboro, Maryland, NAHB (National Association of Home Builders) Research Center, Inc, 26 p.

Crandell, J.H., 2004, Using system-based design principles for affordable, durable, and disaster-resistant housing, *in* Conference on Wood-Frame Housing Durability and Disaster Issues, Las Vegas, Nevada (October 4–6, 2004): Madison, Wisconsin, Forest Products Society, 29 p.

Crandell, J.H., and Kochkin, V.G., 2003, Common engineering issues in conventional construction: Wood Design Focus, v. 13, no. 3, p. 13–23.

Crandell, J.H. and Kochkin, V.G., 2005, Scientific Damage Assessment Methodology and Practical Applications, Proceedings of the 2005 Structures Congress and the 2005 Forensic Engineering Symposium, New York, 20–24 April 2005: Reston, Virginia, American Society of Civil Engineers, Structural Engineering Institute.

Cummins, J.D., 2002, Deregulating Property-Liability Insurance: Restoring Competition and Increasing Market Efficiency: Washington, D.C., AEI-Brookings Joint Center, www.aei-brookings.org.

DeLollis, B., 2000, Court rejects challenge to storm rates: Wind insurance could skyrocket if ruling stands: The Miami Herald, 17 June 2000, www.newslibrary.com (last accessed 22 June 2000).

Dunmoyer, D., 1998, Earthquake hazards and insurance, *in* Earthquake Insurance: Public Policy Perspectives from the Western United States Earthquake Insurance Summit: Palo Alto, California, Western States Seismic Policy Council, www.wsspc.org (last accessed 29 December 2004).

Dutton, C.E., 1890 (reprinted 1979), The Charleston earthquake of August 31, 1886: Washington, D.C., U.S. Geological Survey Ninth Annual Report, 1887–88, p. 203–528.

Federal Emergency Management Agency (FEMA) 366, 2001, HazUS 99 Estimated Annualized Earthquake Losses for the United States: Washington, D.C., Federal Emergency Management Agency, February 2001, 33 p.

FEMA 368, 2001, NEHRP Recommended Provisions for Seismic Regulations for New Buildings and Other Structures (Part I—Provisions): Washington, D.C., Federal Emergency Management Agency, 374 p.

FEMA 369, 2001, NEHRP Recommended Provisions for Seismic Regulations for New Buildings and Other Structures (Part II—Commentary): Washington, D.C., Federal Emergency Management Agency, 444 p.

FEMA 383, 2003, Expanding and Using Knowledge to Reduce Earthquake Losses: The National Earthquake Hazards Reduction Program Strategic Plan 2001–2005: Washington, D.C., Federal Emergency Management Agency, 66 p.

Galambos, T.V., Ellingwood, B., MacGregor, J.G., and Cornell, C.A., 1982, Probability based load criteria: Assessment of current design practice: Journal of the Structural Division, v. 108, no. ST5, p. 959–977.

Ginsburg, T.B., and Stark, T.D., 2002, The law and policy of earthquake hazard in the central United States: Champaign-Urbana, Institute of Government and Public Affairs, University of Illinois, http://www.igpa.uiuc.edu (last accessed January 2005).

Hamburger, R.O., 1998, Earthquake insurance—Incentive or disincentive for mitigation?, *in* Earthquake Insurance: Public Policy Perspectives from the Western United States Earthquake Insurance Summit: Palo Alto, California, Western States Seismic Policy Council (WSSPC), www.wsspc.org (last accessed 12/29/2004).

Harlan, M.R., and Lindbergh, C., 1988, An earthquake vulnerability analysis of the Charleston, South Carolina, Area,: Charleston, South Carolina, The Citadel, Department of the Civil Engineering Report CE-88-1.

Hough, S.E., Armbruster, J.G., Seeber, L., and Hough, J.F., 2000, On the modified Mercalli intensities and magnitudes of the 1811–1812 New Madrid earthquakes: Journal of Geophysical Research, v. 105, no. B10, p. 23,839–23,864, doi: 10.1029/2000JB900110.

Department of Housing and Urban Development (HUD), 1993, Assessment of Damage to Single-Family Homes Caused by Hurricanes Andrew and Iniki: Washington, D.C., U.S. Department of Housing and Urban Development, 112 p.

HUD, 1994, Assessment of Damage to Residential Buildings Caused by the Northridge Earthquake: Washington, D.C., U.S. Department of Housing and Urban Development, 76 p.

HUD, 2000, Residential Structural Design Guide: 2000 Edition: Washington, D.C., U.S. Department of Housing and Urban Development, 396 p.

HUD, 2001, Studies on Probability-Based Design for Residential Construction: Washington, D.C., U.S. Department of Housing and Urban Development, 34 p.

HUD, 2003, New Madrid Seismic Zone: Overview of Earthquake Hazard and Magnitude Assessment based on Fragility of Historic Structures: Washington, D.C., U.S. Department of Housing and Urban Development, March 2003, 48 p.

International Code Council (ICC), 2003, International Building Code: Falls Church, Virginia, International Code Council, 656 p.

Insurance Information Institute (III), 2005, Earthquakes: Risk and Insurance Issues: New York, New York, Insurance Information Institute, http://iiidev.iii.org/media/hottopics/insurance/earthquake (last accessed 29 January 2005).

Klein, R.W., 1998, Regulating catastrophe insurance: Issues and options: Palo Alto, California, Earthquake Insurance: Public Policy Perspectives from the Western United States Earthquake Insurance Summit, Western States Seismic Policy Council (WSSPC), www.wsspc.org (last accessed 29 December 2004).

Kochkin, V.G., and Crandell, J.H., 2004, Survey of historic buildings predating the 1811–1812 New Madrid earthquakes and magnitude estimation based on structural fragility: Seismological Research Letters, v. 72, no. 6, p. 22–35.

Kochkin, V.G., and Mays, T., 2005, Estimation of Ground Motions during the 1886 Charleston, South Carolina, Earthquake based on the Performance of Structures: Upper Marlboro, Maryland, NAHB Research Center, Inc., www.nahbrc.org (last accessed January 2005).

Kunreuther, H., 1998, Insurance as an integrating policy tool for disaster management: The role of public-private partnerships, *in* Earthquake Insurance:

Public Policy Perspectives from the Western United States Earthquake Insurance Summit: Palo Alto, California, Western States Seismic Policy Council, www.wsspc.org (last accessed 29 December 2004).

May, P.J., Feeley, T.J., Wood, R., and Burby, R.J., 1999, Adoption and enforcement of earthquake risk-reduction measures: Berkeley, California, Pacific Earthquake Engineering Research Center, University of California, 84 p.

McKee, S.P., and Crandell, J.H., 1999, Evaluation of Housing Performance and Seismic Design Implications in the Northridge Earthquake: Washington, D.C., U.S. Department of Housing and Urban Development, 13 p.

Melchers, R.E., 1999, Structural Reliability Analysis and Prediction: New York, John Wiley & Sons, 437 p.

MSSC, 1997, A Strategic Plan for Earthquake Safety in Missouri, Missouri Seismic Safety Commission (MSSC): Jefferson City, Missouri, State Emergency Management Agency, 24 p.

Newman, A., Stein, S., Weber, J., Engeln, J., Mao, A., and Dixon, T., 1999, Slow deformation and low seismic hazard at the New Madrid seismic zone: Science, v. 284, p. 619–621, doi: 10.1126/science.284.5414.619.

Newman, A., Schneider, J., Stein, S., and Mendez, A., 2001, Uncertainties in seismic hazard maps for the New Madrid seismic zone and implications for seismic hazard communication: Seismological Research Letters, v. 72, no. 6, p. 647–663.

Olasky, M., 2005, Disasters R Us, A history of disaster relief: World Magazine, v. 20, no. 46, p 19.

Otway, H.J., Battat, M.E., Lohrding, R.K., Turner, R.D., and Cubit, R.L., 1970, A risk analysis of the Omega West reactor: Los Alamos, New Mexico, Los Alamos Scientific Laboratory Report No. LA 4449, 34 p.

Personal Insurance Federation of California (PIFC), 2001, Can California Survive the Next Major Earthquake?: Sacramento, California, Personal Insurance Federation of California, www.pifc.org (last accessed 17 May 2007).

Pielke, R.A., and Landsea, C.W., 1998, Normalized hurricane damages in the United States: 1925–1995: Weather and Forecasting, v. 13, p. 621–631, doi: 10.1175/1520-0434(1998)013<0621:NHDITU>2.0.CO;2.

Somin, I., 2004, When Ignorance Isn't Bliss: How Political Ignorance Threatens Democracy: Washington, D.C., Policy Analysis, no. 525, CATO Institute, 27 p.

Stein, S., and Wysession, M., 2003, An Introduction to Seismology, Earthquakes, and Earth Structure: Malden, Massachusetts, Blackwell Publishing, 498 p.

Stein, S., Tomasello, J., and Newman, A., 2003, Should Memphis build for California's earthquakes?: Eos (Transactions, American Geophysical Union), v. 84, no. 19, p. 13, doi: 10.1029/2003EO190002.

Stelzer, C.D., 1999, On Shaky Ground: Riverfront Times, December 15, 1999: www.riverfronttimes.com (last accessed 29 January 2005).

Sylves, R., 2004, Floods and Presidential Declarations of Disaster: Washington, D.C., Disaster Roundtable No. 10, March 2, 2004, The National Academies, http://dels.nas.edu/dr/docs/dr10/sylves.ppt (last accessed 17 May 2005).

Western States Seismic Policy Council (WSSPC), 1998, Earthquake Insurance: Public Policy Perspectives from the Western United States Earthquake Insurance Summit: Palo Alto, California, Western States Seismic Policy Council, www.wsspc.org (last accessed 29 December 2004).

Wilson, R., 2000, Regulating environmental hazards: Regulation, v. 23, no. 1, http://www.cato.org/pubs/regulation/regv23n1/reg23n1.html (last accessed January 2005).

MANUSCRIPT ACCEPTED BY THE SOCIETY 29 NOVEMBER 2006

The Geological Society of America
Special Paper 425
2007

Disasters and maximum entropy production

Cinna Lomnitz
Department of Seismology, Institute of Geophysics, Universidad Nacional Autónoma de México (UNAM), Mexico City 04510, Mexico

Heriberta Castaños
Department of Social Science, Institute of Economics Research, Universidad Nacional Autónoma de México (UNAM), Mexico City 04510, Mexico

ABSTRACT

Disasters have a structure. They self-organize and "pessimize" the production of damage. In response, society develops technology at the interface between nature and society. The cost of earthquake disasters has been rising exponentially since the 1960s, when probabilistic seismic hazard assessment began to be introduced. Why is it so difficult to learn from disasters? A rational strategy of disaster prevention must be based on better science, meaning disaster physics as well as social science. The principles involved are different. Disaster physics is based on the maximum entropy principle. Social disaster science is based on a principle of least resistance. For example, poor designs survive because it is easier to blame the operators, or the government. Plate-boundary disasters occur because the hazard is high; intraplate disasters occur because it is low. In both cases, the entropy is maximized. Examples of recent disasters are discussed.

Keywords: disasters, entropy, cascades of failures, disasters and society.

INTRODUCTION

Disasters are the result of a freakish chain of events in complex, socionatural systems. We discuss seven features of disasters that distinguish them from accidents. Disasters are

- high-energy events,
- nonscaleable,
- multicausal (Perrow chains),
- nonrepeatable, "unique,"
- nonlinear, entropy-maximizing,
- "tragic" or socially controlled, and they
- feature unexpected couplings.

The list is not exhaustive. Disaster science offers challenging and largely unexplored opportunities for cooperation among engineers, physical scientists, and social scientists. We propose a broad preliminary reflection on these seven distinguishing features of disasters. We introduce an argument of maximization of entropy production, and we discuss a tentative critique of proba-

Lomnitz, C., and Castaños, H., 2007, Disasters and maximum entropy production, *in* Stein, S., and Mazzotti, S., ed., Continental Intraplate Earthquakes: Science, Hazard, and Policy Issues: Geological Society of America Special Paper 425, p. 387–396, doi: 10.1130/2007.2425(26). For permission to copy, contact editing@geosociety.org. ©2007 The Geological Society of America. All rights reserved.

bilistic assessment of hazard and other topics of disaster prevention based on the underlying assumption that risk equals hazard multiplied by vulnerability.

MAXIMUM ENTROPY PRODUCTION

An open planetary system with many degrees of freedom may convect in a steady state far from equilibrium. A steady state is defined as a balance between work produced and heat dissipated. However, this balance can be attained along many different and equally probable paths, and there is a steady state that has a higher output of work and attenuation. This state is more probable within a given set of dynamical constraints (Lorenz, 2005; Dewar, 2005). This general result is known as the principle of maximum entropy production (MEP).

Modern disasters appear to follow the principle of maximum entropy production. This observation can be especially meaningful to the social scientist because it may provide a common basis of analysis with the physical scientist. From the point of view of social science, MEP is familiar: it has been extensively discussed by philosophers of the Enlightenment. Maximum entropy production is reminiscent of Voltaire's critique of the follies of optimism and reductionism, and his related reflections on the "indifference" of the natural world.

More recently, Dombrowsky (1973) defined disaster as a negation of progress and a slap in the face of scientific hubris. In the context of any orderly civilization, there should not be any disasters at all, but they do happen. And they are getting worse. Losses from earthquakes have not only been rising exponentially since the mid-1960s, they increasingly affect the more developed nations.

Even without considering terrorism or other manmade disasters, science stands relatively helpless in the face of disaster. The public comments made after the Indian Ocean tsunami of 26 December 2004 seem to suggest that things have changed little since the Lisbon earthquake of 1 November 1755. British Foreign Secretary Jack Straw even argued that "we are at the mercy of natural forces, no matter how much we think these days, particularly in the West, that we have greater control of our own lives through the great advances of science and technology" (Radio 4's *Today* programme).

Earthquakes can produce a peak ground acceleration of up to $2g$, or twice the mean acceleration of gravity at sea level. This is hardly a cause for alarm: a commercial automobile can take higher accelerations. When the driver suddenly applies the brakes, the car does not fall to pieces, because the vehicle has been *designed and tested* to resist at least $5g$. Is it reasonable, given the "great advances of science and technology," to feel safer inside our cars than we do inside our homes?

The principle of maximum entropy production is familiar to engineers under the name of Murphy's Law: *If anything can go wrong, it will.* It is the basis of defensive design. More recently, it has reappeared in the shape of "performance-based engineering." John A. Blume (1909–2002), an American earthquake engineer, spoke of "capacity design." The principle is the same. After the

1994 Northridge, California, and the 1995 Kobe, Japan, earthquakes—two moderate earthquakes that produced damage in excess of $20 billion each—it was realized that relative displacement, not lateral force, is the cause of most damage. "Damage is related to *material strain* (e.g., yield of reinforcement or structural steel, cracking or crushing of concrete)... [while] non-structural damage is typically related to drift. Both can be integrated to provide a direct relationship between damage and displacement" (Priestley, 2004, p. iii).

However, this is not what the building codes recommend. The concept of "equivalent lateral force" was first adopted in Japan and the United States as a rule of thumb to improve the resistance of masonry structures to earthquakes. It was later given a physical meaning, and it became the basis for elaborate building codes and regulations all over the world. In Mexico, damage to structures is still attributed to horizontal ground acceleration. A lateral force coefficient C of around 10% of the weight of the building, or $0.1g$, has been specified. Around 1940, narrow-band instruments called strong-motion accelerographs were installed in buildings in order to improve our understanding of the forces unleashed by an earthquake.

A building code is defined as a set of legal requirements intended to ensure that a building will not pose a significant threat to the life and safety of its occupants. It is not expected to eliminate property loss. The building itself is not tested; when an earthquake occurs, it is too late to introduce needed improvements. Building codes are adopted by state or local governments, but their enforcement depends on standards of legal practice. In the United States, the responsibility of engineers is limited by law. Engineering standards are legally defined as "that level or quality of service ordinarily provided by other normally competent practitioners of good standing in that field, contemporaneously providing similar services in the same locality and under the same circumstances" (*Paxton v. County of Alameda*, 119 C.A. 2d 393, 398, 259 P. 2d 934, 1953). In other words, the responsibility of an engineer is measured by the standards of the local engineering community at the time. If something goes wrong, the law makes allowances for the fact that the engineer must use judgment "gained from experience and learning" in "situations where a certain amount of unknown or uncontrollable factors are common" (*City of Mounds View v. Walijarvi*, 263 N.W. 2d 420, 424 [Minn. 1978]). Indeed, when you hire an engineer, you better remember that you "purchase service, not insurance" (*Gagne v. Bertran*, California Supreme Court, 43 C2d 481, 1954).

This is a telling comment: insurance is paid by the public. It is a hidden form of taxation levied by governments supposed to protect us from disaster. The 2005 Katrina hurricane caused the highest losses of any disaster in history. The federal agencies in charge of disaster management had relinquished their responsibility to private contractors under a system known as outsourcing. Insurance picked up the tab.

Similar trends have been noticed in other types of disasters. The economic losses from earthquakes in the United States and Japan have been rising (National Research Council, 2003). These losses now account for the major share of worldwide damage from disasters. The impact on the world economy has been severe.

DISASTERS, ACCIDENTS, AND COMPLEXITY

Is science responsible for this lag? As Petroski (2005) has pointed out, science may be the theater, but engineering is the action on the stage. Losses from disasters are climbing everywhere, and our knowledge base is not particularly solid, especially in terms of disasters and their effects on society. It is estimated that a repeat of the 1906 San Francisco earthquake would cause damage on the order to $200 billion: never mind that engineering technologies have advanced considerably over the last two decades (National Research Council, 2003). Let us see why this is happening.

The Kobe earthquake (M 7.2) occurred at 5:46 a.m. on 17 January 1995. About 92,800 buildings and houses collapsed and ~192,700 were damaged. This earthquake killed more than 5500 people, ~90% of them inside their homes. The Northridge earthquake (M 6.7) occurred at 4:31 a.m. on 17 January 1994. It killed 51 people, about the same number as in the 1989 Loma Prieta, California, earthquake, and it injured more than 9000. It surprised engineers with extensive brittle failure of welds in nominally ductile-steel moment-frame buildings (National Research Council, 2003). In plain language, the beam-to-column joints were torn apart by the earthquake because of weak welds. They had been designed in accordance with the local building code. Welds can be made stronger. They are about three times as strong in mechanical moving parts (e.g., in automobiles) as compared to welds in steel buildings.

When the engineers who designed the Torre Mayor Building in Mexico City (Fig. 1) decided to apply the principle of performance-based engineering to this important structure, they included 98 fluid-viscous dampers for earthquake protection. Laboratory testing used some leftover dampers from military production. The results were dramatic: adding 20% fluid damping to the structure increased the earthquake resistance by a factor of three. During the design procedure, floors were added one by one. As the structure grew from 40 to 55 stories, its total weight was not increased because the dampers absorbed energy, which thus produced a sleeker and tougher structure. The result was a safer and more elegant building that performed very satisfactorily in the Colima earthquake of 22 January 2003 (M 7.6).

Experience suggests that the principles of defensive design can lead to more efficient and less expensive solutions. Disaster planning cannot be ignored as part of the design procedure. Disasters are critical phenomena that unfold on specific space-time scales. Some examples are earthquakes, hurricanes, terrorist attacks, stock-market crashes, and the irreversible decline of world oil reserves. Disasters tend to strike at the interface between society and nature. This interface is the specific concern of technology. Disasters are often caused by supercomplex dynamics involving interactions among large numbers of components (Bunde et al., 2002; Castaños and Lomnitz, 2003).

Figure 1. Torre Mayor Building in Mexico City (2004): 55 floors, 738 ft high, fully damped, tallest building in Latin America.

Complexity may be measured by the ability of a system to display long-range coherence in space and time and to undergo transitions between different states. An often-cited example is the case of market fluctuations (Sornette et al., 2002), with some important reservations. Playing the futures on the stock exchange is a game—techniques borrowed from game theory may be applicable, because we are interested in maximizing our profit and minimizing our personal risk. The player knows that his win is someone else's loss: The zero-sum game applies as a model.

There are no winners in a disaster. It is not a game. On Sunday, 26 December 2004, at 00:58 UTC, a magnitude 9.3 earthquake occurred off the northwest coast of Sumatra. This earthquake generated a water wave that caused an estimated 250,000 deaths and extensive damage due to runup, landward inundation, and wave-structure interactions (Lomnitz and Nilsen-Hofseth, 2005). Three months later, on 28 March 2005, another earthquake (M 8.7) occurred near the same location, but this time, no tsunami disaster occurred. In both cases, direct earthquake damage was comparatively minor. Disaster experts tended to attribute the high casualty figures to underdevelopment, but primitive human populations survived in the epicentral area and suffered no significant losses. Casualties and damage were highest in areas of intermediate stages of development.

The International Decade for Natural Disaster Reduction 1990–1999 spent millions of dollars on learning about disasters and reducing disaster vulnerability. "Never have more nongovernmental organizations, both local and international, been working on natural disaster risk," wrote Tucker (2004, p. 696) a few weeks before the Indian Ocean tsunami. The U.S. Geological Survey (2004, Web page, U.S. Geological Survey Earthquake Hazards Program, 2005, FAQ, Magnitude 9.0 Sumatra-Andaman Earthquake) was quick to point out that "the only way to know for certain if a tsunami has been generated is to directly measure the height and propagation of the ensuing wave." The speed of propagation of a tsunami wave on the open ocean is about the same as that of a commercial airliner, and air traffic can be monitored in time to warn passengers about delays in flight arrivals and departures. Forty-three military and commercial airports were operating in the coastal area of the Indian Ocean at the time of the tsunami. A stream of weather and traffic reports was reaching these airports, yet no information on abnormal wave activity was transmitted or received.

Mexico City has been flooded in every rainy season since before the Spanish conquest in 1521. Some of the early floods were very severe, but they were not unexpected, they were recurring climatic features, not disasters. Eventually, they were controlled by a series of drainage systems.

THREE-MILE ISLAND

It has been suggested that the surprise element in disasters might be caused by a cascading sequence of coupled failures in a multi–degree-of-freedom system. "No one dreamed that when X failed, Y would also be out of order and the two failures would interact so as to both start a fire and silence the fire alarm. Furthermore, no one can figure out the interaction at the time and thus know what to do. The problem is just something that never occurred to the designers" (Perrow, 1999, p. 4). Other well-known examples of cascading failures are discussed by Moscow (1981) and Chiles (2002).

Figure 2 shows a diagram of the Three-Mile Island nuclear power plant. A minute-by-minute account of the disaster of 28 March 1978 can be found in Perrow (1999). There were several unforeseen sequential failures: (1) the condensate polisher line clogged, possibly because of some minor leak, which caused two pumps to stop operating. This failure automatically shut down the turbine, and the emergency feed-water pumps turned themselves on. (2) The valves to the steam generator had been left accidentally in the closed position two days before the disaster. They refused to admit cooling water to the core. The core was starved for water, and its temperature began to rise. The operators noticed it but did not understand the reason. (3) The steam generator quickly boiled

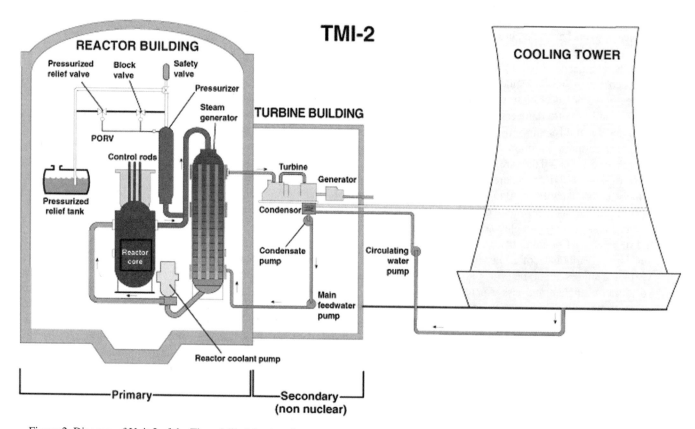

Figure 2. Diagram of Unit 2 of the Three-Mile Island nuclear power plant in Pittsburgh, Pennsylvania. PORV—pilot-operated relief valve.

dry, and the reactor scrammed automatically. This means that the graphite control rods dropped into place and reduced the flow of neutrons to let the reactor cool down, but this takes time. (4) The pilot-operated relief valve (PORV) opened automatically and began to relieve the pressure by releasing steam into the containment building. (5) The operators commanded the valve to close again, but it jammed and refused to close. This valve had an indicator circuit that was supposed to activate a warning light on the control panel, but the indicator circuit failed, and the control panel erroneously showed the valve in the closed position. (6) The time was thirteen seconds after the beginning of the emergency, and the operators thought the situation was under control. (7) Eight minutes later, there was a change of the operating shift. The new operators checked the control panel, which erroneously showed the pilot-operated relief valve in the closed position. The previous operators had reported no malfunction. The operators were baffled by the behavior of the plant and thought they must have a pipe break, but all their corrective measures worsened the problem.

The defective pilot-operated relief valve remained stuck for two hours and twenty minutes expelling radioactive steam into the containment building. This was a feared "loss-of-coolant-accident," but there was no way for the operators to understand the true cause of the problem. The disaster was not scaleable because it only takes seconds for a reactor to go into failure mode, but it may take hours or days to control it. Three different alarms were sounding as the new shift came on. The new operators decided to block the escape of steam, a decision which caused a hydrogen explosion in the reactor. By pure chance, a total meltdown did not occur, but much of the core was already gone, and a small amount of radioactive material managed to leak out of the containment building and into the environment.

The experts blamed the disaster on operator error on the grounds that no operator could recall having left the feed valves in a closed position. One of the indicator lights on the control panel had burned out and the other had a service tag hanging over it, but the feed-water situation was discovered and corrected eight minutes later. According to several experts, the initial operator error did not make much difference. Nevertheless, several operators were fired and no new nuclear power plant has been cleared for construction in the United States since that time. The cause of the disaster was not operator error, though blaming the operators was the easy way out. It was excessive complexity of design of a system, which did not permit an adequate corrective action to be taken during an emergency. The operators did follow the prescribed instructions, but they were unable to gather timely information on the state of the system. They never found out what was wrong.

The cascade of malfunctions at Three-Mile Island may seem highly improbable. The pilot-operated relief valve had been tested and was known to get stuck once in ten operations, on average. This was not regarded as serious because the valve was a part of a secondary circuit. Also, it was monitored from the control panel. Unfortunately, the control panel of the nuclear power plant contained 1600 lights and the system is so complex (Perrow, 1999).

THE 1985 MEXICO EARTHQUAKE

The Three-Mile Island disaster was typical of modern disasters in general. It was not scaleable, nonrepeatable, nonlinear, high-energy, multicausal, and socially controlled. Consider now the case of the Mexico City disaster of 19 September 1985. There is agreement about the general cause: an earthquake of magnitude 8.1 struck the west coast of Mexico at a distance of 400 km from Mexico City. However, there is still no consensus on what went wrong.

Large subduction earthquakes are known to occur at the megathrust boundary between the North America and Cocos plates. They can contain a high amount of energy—as reflected by their magnitude—yet their ground accelerations can remain quite low. Only two people died in the epicentral area—both in a highway accident caused by loss of steering—but 371 high-rise buildings collapsed and more than 10,000 people died within a small area of downtown Mexico City. A cascade of unusual and still poorly understood factors contributed toward making this distant earthquake the worst disaster in Mexican history.

Mexico City is located on a high-altitude plateau near the center of the continent, halfway between Veracruz on the Gulf of Mexico and Acapulco on the Pacific coast. This is an intraplate location. The local seismicity is low—as in the central United States. Mexico City has no record of any local earthquakes above magnitude 4, but distant subduction earthquakes were known to produce serious damage.

The city had been founded by Aztecs on a low-lying island within a large lake ringed by active volcanoes. The present urban area has extended over most of the mud flats formerly occupied by the lake and beyond. Mexico Lake mud has an extremely low shear-wave velocity of 50 m/s and a density of only 1100 kg/m^3. Structural damage and casualties are always confined to the area on soft ground. This fact was reflected in the zoning provisions of the Mexico City building code, which was modeled after California seismic regulations. In 1985, more than 16% of reinforced concrete-frame structures in the 7- to 18-story range collapsed on lake mud. Damage was highly selective: structures designed under earthquake regulations collapsed, while Colonial churches and palaces survived.

Baffled scientists have suggested various explanations. The geological situation is unusual. Mexico City mud, usually described as a sensitive clay, is actually a black organic soil or clayey muck (histosol), which developed in a high-altitude lacustrine system surrounded by active volcanoes (Cowardin et al., 1979). The mud layer is only ~30 m thick, but its high fertility gave rise to the original agricultural technology of artificially reclaimed mud flats called *chinampas*, which contributed to the rise of Mesoamerican cultures. Seismic signals from distant sources propagate through Mexico City mud as characteristic, long-duration, coherent monochromatic wave trains with a frequency of ~0.4 Hz. The damage is attributed to these wave trains.

Why was the damage so selective, and why was it limited to modern reinforced-concrete frame construction on soft mud? One possibility is that the mud layer acted as a waveguide that trapped incoming shear waves with a frequency similar to the eigenfrequency of the layer. This is possible because of the presence of the *Lg* layer, a waveguide of ~2 km thickness that conveys seismic energy in the 0.2–0.5 Hz frequency band from the Pacific coast to the interior. In large-magnitude earthquakes, the two waveguides couple at the resonant frequency of the mud layer, and the damping ratio of the soft mud may be offset by the large influx of seismic energy from a distant epicenter.

Some authors talked about a double or triple resonance, meaning that the structures collapsed when their resonant frequency peaked in the neighborhood of 0.4 Hz, while the design response spectrum in the building code was flat-topped in the range of 0.3–1.3 Hz. Engineers were incredulous about the disastrous effects of the earthquake, and the mayor insisted for three days that there was nothing to worry about: the buildings were safe against earthquakes!

The Mexico earthquake failed to confirm the received wisdom, or prejudice, that blames disaster on underdevelopment. The disaster struck one of the most highly developed areas in Latin America. Low-income housing suffered no appreciable damage, but the impact on the urban middle class was tremendous. The earthquake occurred at 07:19 a.m. local time on a working day. Several modern hospital buildings collapsed. Some local industries had started operating, especially in the garment trade, where sweatshop practices had long been tolerated. The working girls were entombed under concrete slabs in multistory buildings. Victims were hard to reach by rescue squads. A few victims were extracted alive four to seven days after the disaster.

As in Three-Mile Island, the evidence pointed to a combination of unexpected adverse factors. After the disaster, the first reaction was one of denial and disbelief. The local building code, regarded as one of the more advanced in the world, had been written by engineers schooled at American universities (Suh, 1985).

"Bad luck" is a way of blaming coincidences for a mishap, but does bad luck exist? Murphy's Law ("If anything can go wrong, it will") is beginning to be recognized as having a basis in thermodynamics as well as in fact. If an open nonequilibrium system is perturbed or displaced from its steady state, the resulting transient state will tend to evolve along a path that maximizes the rate of entropy production. The result is solidly based in statistical mechanics (Dewar, 2005). The system behaves as if its very complexity provides multiple opportunities of failure.

In Mexico City, seismic energy was trapped in a shallow waveguide. It could not escape except by causing structural damage. This explains the unusually long duration of the monochromatic wave train (up to five minutes) and the selectivity of the damage. It also suggests a possible way out. Damage was inflicted basically through resonance: therefore it could be controlled by preventing resonance. The resonant frequency is known in advance, since the thickness and elastic properties of the mud layer are known. Thus, fluid-viscous dampers can be extremely effective in stabilizing a multi–degree-of-freedom engineering structure. The Torre Mayor Building in Mexico City (Fig. 1), the first high-rise structure that incorporates 98 fluid-viscous dampers for resonance control, was inaugurated in 2004.

Several important questions about this disaster still remain unanswered. Why did disaster strike 400 km from the coast when there was no significant damage closer to the epicenter? Why did the building code prove inadequate? Why was structural collapse confined to 371 engineered structures? Again, it was easier to blame shoddy workmanship. In large disasters, some essential factors frequently may remain hidden forever.

RISK, HAZARD, AND PERROW CHAINS

A widely accepted theory of disasters, attributed to the American geographer Gilbert F. White (1974), proposes that the risk, R, or probable financial loss in a disaster, is the product of the hazard, H, times the vulnerability, V:

$$R = H \times V, \qquad (1)$$

where hazard is defined as the probability of recurrence of a disaster, and vulnerability is the unit loss likely to be inflicted on a specific physical and social environment. Most current concepts of disaster prevention, including probabilistic seismic hazard assessment (PSHA), invoke this model. However, if H and V are probabilities, Equation 1 is correct if, and only if, hazard is independent of vulnerability. Indeed, most disastrous earthquakes seem to occur in developing countries.

Thus, it seems to follow from Equation 1 that losses in disasters might be to some extent the fault of the victims. If you got hurt because you were vulnerable, vulnerability was up to you.

However, insurance companies know better. In terms of losses, the great 1985 Mexican earthquake cost the insurance industry about one-fifth of the damages paid out after a hailstorm in Munich the preceding year. Could Munich have been more vulnerable than Mexico City? Surely it was not the hazard that was higher than for the worst earthquake in the history of Mexico, with thousands of people killed in the capital. The hailstorm came and went without causing any casualties, and there was barely a ripple in the news. Germany was hardly lagging behind Mexico in terms of economic development.

White's Equation 1 must contain some logical flaw. Social science comes to the rescue: the problem may be traced to faulty logic in the definitions of hazard or vulnerability, or both. If we look closer, the vulnerability factor V refers to completely different losses in each case. Damage in Munich was mainly in cracked automobile windshields; in Mexico City, it concerned multistory reinforced concrete-frame construction. This example is not unusual or farfetched. Reinsurance companies cannot always tell, because they sell protection in bulk, as a package deal. In principle, even the local insurer cannot tell how much of his liability is in earthquake risk and how much corresponds to automobile insurance.

On the other hand, anyone can tell the difference between Munich and Mexico City. Reinsurance companies are not naïve. They rank among the oldest, wisest, and most solid international corporations. The chief geophysicist of a leading reinsurance corporation visited Mexico City five years before the earthquake. He took a leisurely walk in the downtown area and jotted down a few notes. His damage assessment for a potential earthquake proved to be correct within 5%. When the earthquake came, the insurance man was ready. He arrived in Mexico City a few days after the disaster, and he paid out his claims without any argument.

This procedure helps local insurance companies at a time when they are most grievously in need of cash—and it saves money for the reinsurance people too. Equation 1 was not used at any time in this operation, nor was any method of probabilistic hazard assessment. Kenneth Hewitt (1999) pointed out that a linear, independent definition of vulnerability tends to put the onus upon the victims. Behind the apparent neutrality of a mathematical expression, the thrifty ant was virtuously slamming the door shut in the face of the lazy improvident cricket.

For what was vulnerability if not a structural weakness neglected and left unattended? Actually, as Hewitt suggests, developing countries are not just unlucky: they endure a permanent state of disaster. Foreign assistance often adds insult to injury. Foreign aid to people in the hot tropics consists of a generous supply of blankets. Worse, much of the assistance will never reach the people for whom it was intended. It will lie for weeks under the rain on remote airfields, or it will disappear in warehouses to be resold. We are to blame for callous political and economic policies. Disasters merely bring the situation into the open.

Parameters such as "hazard" and "vulnerability" are notoriously difficult to evaluate. Disasters are nonscaleable and nonrepeatable: how could the hazard of the Indian Ocean tsunami have been estimated? There was no precedent, and there will be no repetition as social change in the Indian Ocean region is notoriously rapid. Society is a moving target. This applies also to the vexing problem of estimating the value of human life. Every life is unique, so there is no replacement value.

Humans are also uniquely vulnerable. They must somehow be made part of the risk equation if we want to know how much should be spent on telemetric buoys in the Indian Ocean. Each buoy costs $260,000 to make and install plus $200,000 a year to maintain. According to Graham and Vaupel (1981), human life should be valued at around $300,000 each, which is roughly the cost of a buoy. Should we entrust the lives of future Scandinavian tourists on Thai beaches to such a hit-or-miss device?

The Negrito tribes in the Andaman Islands executed a timely withdrawal from coastal areas. They registered no human or material losses. No one gave them credit for it: they had just followed the wildlife. Perhaps a lot of money could be saved by learning from the birds and the bees—not to mention elephants!

The proliferation of disasters is an emerging crisis of civilization. An exponential rise of losses from earthquakes or hurricanes is not sustainable. It is time to reexamine the rationality of our technology.

The assessment of hazard—that is, the probability of occurrence of a disaster—is the subject of emerging methodologies such as PSHA, which tend to be based on game theory. In game theory, we assume that the system has certain invariable features or "rules," which are known to the players. We also know in advance how much we may expect to win and how much we can afford to lose. Disasters cannot be treated in this way because both systems involved (nature and society) interact closely and change rapidly in terms of a human lifetime.

Some of the worst earthquake disasters occur in intraplate regions where the incidence of previous disasters has been lowest. In the 1976 Tangshan earthquake (M 7.8), most of the local housing lacked minimum earthquake resistance because there was no special reason for fearing such a disaster. China has the longest written earthquake history in the world, yet there was no record of any previous earthquake activity in Tangshan—possibly because Tangshan City is only a century old. Not every town in China can be traced back to the Ming Dynasty. Geophysical evidence was there all the time. Because of the underground coal-mining activity, the local geology was especially well mapped. The Tangshan fault was well known to geologists, but it had been believed to be inactive since Oligocene times. The city was built around the mining activity—most of the population worked in the mines. Housing was provided by the company. No one foresaw a need to provide increased earthquake protection at this intraplate location (Lomnitz, 1994).

Arguably, even a lesser earthquake might have caused damage. However, we still lack a reliable methodology for predicting earthquake damage at a specific location. The 1985 Mexico earthquake struck selectively at structures that had been designed against earthquakes, while historical masonry structures remained undamaged.

Perrow made the point that the causes of disasters are multiple but need to be faced singly. The 2004 Indian Ocean disaster was a result of an amazing string of coincidences. It occurred on Boxing Day (December 26), a holiday in countries of the British Commonwealth. It also fell on a Sunday, when most offices were closed. Christmas time is tourist season in the tropics, including the Indian Ocean and the Andaman coasts. At nine or ten in the morning, the weather was perfect, and the beaches were crowded. Significant tsunamis are quite rare in this part of the world. None of the coastal nations was a member of the International Tsunami Information Center, a UNESCO-sponsored club of nations bordering on the Pacific Ocean. Poorly understood waveguide phenomena may have played a significant role and may have originated unexpectedly high amplitudes along the unguarded coasts of Sri Lanka. Finally, coastal resorts have attracted many international tourists to the region, especially over the last decade. They also have attracted service personnel and vulnerable migrants housed in precarious shacks near the ocean.

Such chains of coincidences, which we call Perrow chains, are typical of disasters. Perrow (1999) attributed them to complexity combined with "tight" coupling of failure modes. They are consistent with the maximum entropy principle: "if something can go wrong, it will." This principle, however, should be understood as a statistical result: not all potential mishaps combine to cause a disaster. Earthquake disasters may be defined as large earthquakes plus Perrow chains. They are due to ground motion and also to the deficient quality of our homes and other social factors. Disasters cannot be neatly classified as either natural or social—they are both.

In the case of the Three-Mile Island disaster, if all had gone well, no questions would have been asked, and no one would have been the wiser. Such simple incidents are called *accidents*, not disasters. The same is true for natural events: most of them are foreseen and do not lead to a disaster or an investigation. St. Philastrius (ca. 384 A.D.) claimed that earthquake disasters were always caused by the wrath of God. Attributing them to natural forces, he wrote, was heresy. This is, indeed, a misreading of the way nature and societies operate.

Engineering is about ongoing process, even in the absence of existing knowledge (Petroski, 2005). If things go seriously wrong, as in the Three-Mile Island disaster, the final report of the fact-finding commission may blame the disaster on "inappropriate actions by those who were operating the plant." It is generally recognized that the design of nuclear power plants, precisely because of the risk of earthquakes, was excessively complex around that time (Fig. 1), but engineers cannot be expected to anticipate every possible malfunction of a complex system or systems.

Suppose we built every home to resist $2g$—the maximum observed lateral ground acceleration in an earthquake (Brune, 1993). Military engineers call this the "maxi-maxi approach." All military vehicles and installations can stand better than $2g$, which is not really considered such a high acceleration for an engineering structure. Commercial automobiles can take more than $2g$, and the requirement does not seem to make the price of a vehicle prohibitive. The rationale for building homes to $2g$ is actually flawless: in the face of uncertainty, this is not an excessive price to pay. It would spell the end of earthquake hazard: no need to worry about building codes. In case of an earthquake, passersby could confidently take refuge in the nearest building.

HOW MUCH WOULD IT COST TO MAKE OUR HOMES EARTHQUAKE-PROOF?

The point we are trying to make is quite simple. We face rising losses in disasters because our societies evolve rapidly and are getting so increasingly complex that we cannot foresee all factors that might conspire against our safety. It is not just the threat of terrorism. A few decades ago "earthquake country" was just California to Americans, but today we realize that intraplate earthquakes may be just as hazardous. We cannot really tell and there is argument about it, but we cannot afford to risk citizens' lives.

In conclusion, all homes and other structures used by people must be safe against earthquakes, but how far are we from safe housing? Since the 1985 earthquake the Mexico City Building Code has a flat-topped maximum at $0.4g$ in the design spectrum.

This is an improvement on the earlier version, which peaked at $0.24g$, but the recorded response spectrum of the 1985 earthquake peaked at $1g$, so we know that the current requirement is not safe enough. Rosenblueth et al. (1989) argued that the deficiency in the building code would be compensated by damping in the structures, but the damping factor in most of Mexico City construction does not exceed 5% of critical damping. In the case of a repeat of the 1985 earthquake, this amount of damping would hardly protect the population of Mexico City. Resonance remains a distinct problem and a probable cause of damage. Earthquake damage probably involves coupling of resonant modes in waveguides, but it will take years to find out, and we cannot afford to wait.

How much would it cost to build for $2g$? Nobody knows: it has never been tried. In the meantime, at least four million citizens of Mexico City can now afford cars. No one would have guessed this a few decades ago, but cars have become a part of our way of life. An acceleration of $2g$ won't release an airbag. Is it really beyond reach? If I were to tell an engineer that I wanted my home to be designed like my car, I might get a blank stare.

How does the automobile industry do it? Cars are unitized structures. They are standardized and mass-produced. Every model is test-driven over potholes. Automobile engineers worry a lot about safety. This is because consumers demand it.

Our cars are safer than our homes. That is a fact, but we do not have to live with it. We may not have to go all the way to $2g$ to be safe: the point is not to accept living in homes that are expensive and untested. An automobile without damping is not acceptable, nor is a computer without resistors—the electrical analog of dampers. In the case of the Torre Mayor Building (Fig. 1), the addition of safety in the form of 98 fluid-viscous dampers was cost-effective (Douglas Taylor, 2004, personal commun.). The engineers found that by adding dampers, a substantial saving on structural steel could be achieved, so that the height could be increased from 45 to 55 floors without additional load to the foundations. When resonance plays a major role in earthquake hazard, this is a logical evolutionary design change.

There is no particular reason why such technological advances could not be made commercially feasible in the present state of scientific and technological knowledge. As engineers and as scientists, we stand to reap significant benefits by paying more attention to things that have not yet been tried. There is a market out there for added safety. It may cost less than we think.

The Torre Mayor Building did not have to be designed to withstand horizontal ground accelerations as high as $2g$. It was not necessary, as we knew in advance what the ground motion would be. The structure was designed to withstand an earthquake of the same size and characteristics of the 1985 earthquake (M 8.1). The requirements of the Mexico City Building Code were amply met. If there had been no prior knowledge of the seismic peak ground accelerations, $2g$ might be a good first guess. Hardly any laboratory work has been done on soft ground: the wavelengths are too long to fit in a normal laboratory, and the available seismic instrumentation has not changed much since the 1930s.

THE TRAGIC CYCLE

The seismic behavior of structures on soft ground is like the behavior of terrorists: it remains poorly understood. So is the behavior of society in disasters. Earthquake-proof design ought to be available at a reasonable price, anywhere on earth.

Earthquake disasters are often called tragic. *Tragoedia*, in the original Greek, means literally a "goat-song," a ritual performed at religious ceremonies where a he-goat was sacrificed to placate the gods. Later, an actor was provided to reply to the chorus. Man was portrayed in ritual as a powerless plaything manipulated by the gods.

This is essentially the same thing the British Foreign Secretary was saying: "We are at the mercy of natural forces" (Radio 4's *Today* programme). If true, this is an indictment of science. However, scientists are not innocent bystanders. It is tragic that not enough high-quality research is being done in disasters. Knowledge is not a game of chance, or a contract to be outsourced by government agencies. It is the supreme challenge of the human mind.

In ancient Greek tragedies, the actor who questioned the chorus was called *hypocrite*, as his role was to cast doubt and suspicion on the motives of the gods. Oedipus was blinded and tormented on the stage because of some secret affairs and petty jealousies of the gods. While a chorus droned on in praise of the Olympian powers, a hypocrite prepared the way for critical science.

The roles have become reversed. The chorus of public opinion now demands better disaster physics, safer construction methods, more effective prevention, and rightly so. Science has been successful in unraveling far deeper mysteries, but scientists are lagging behind the expectations of the public.

In the 1989 Loma Prieta earthquake, a one-mile section of the Nimitz Freeway collapsed in the heart of the city of Oakland, California. Sixty people were crushed to death in their cars. The collapse did not occur along the entire length of the freeway: it affected one small section built to earthquake specifications on top of a soft layer of San Francisco Bay mud. Sections of the freeway built on hard ground did not collapse. Yet the final report of the experts was inconclusive and noncommittal about the cause of the disaster. No broadband seismic stations existed on Bay mud; more than fifteen years after the disaster there still aren't any. The basic data needed for solving the problem of earthquake hazard on soft ground are still unavailable. So how can we claim to make progress in disaster research?

The 1994 Northridge, California, earthquake was the worst disaster in history in terms of cost. Exactly one year later, the losses were topped by the Kobe, Japan, earthquake. Reinsurance companies began to envisage going out of business, yet some of the freakish features in both disasters pointed to unattended areas of vulnerability. Kobe had been razed in the air attacks of March 1945, so that no housing in Kobe was more than fifty years old, yet nearly 5000 people were killed in one- to two-story homes. In the Northridge earthquake, steel structures failed because of inadequate welding at joints. The welds conformed to building code specifications. If they had been designed as moving parts, the welds might have stood the test of the earthquake.

In the 1985 Mexico earthquake, the Perrow chain is far from clear. A large earthquake occurred at a distance of 400 km and generated seismic energy at all frequencies, but it does not follow that the relevant features of propagation were the same as for a small earthquake. A regional waveguide in the upper continental crust, which propagates energy in the 0.1–1 Hz frequency range to large distances, was largely ignored. An imbedded soft waveguide under downtown Mexico City coupled into the regional waveguide and trapped 0.4 Hz energy. In small earthquakes, nothing happens, but it does not follow that we understand the causes of earthquake disasters in Mexico City.

Gravity-coupled seismic waves have been sighted in large earthquakes. They have also been observed in jelly and in soft condensed matter. Matuzawa (1925) described "frozen" gravity waves in the 1923 Tokyo earthquake. They were also observed in the 1985 Mexico earthquake, but we have not developed instruments to observe them.

In terms of earthquake-risk prevention, we do less than we should. The 1989 collapse of the Nimitz Freeway might have been due to an unanticipated, coherent, longitudinal mode of vibration. Other multi–degree-of-freedom structures, such as the San Francisco Bay Bridge and the Hanshin Expressway in Kobe, exhibited similar failure modes. Why were they vulnerable?

Disasters have a history: That of the 1985 Mexico earthquake disaster may be traced back to 1325, when an Aztec tribal council decided their people should settle on an uninhabited flat island in the middle of Anahuac Lake. The reason was an omen in the shape of an eagle sitting on a nopal (*Opuntia ficus indica*), and devouring a rattlesnake, supposedly sent by tribal god Huitzilopochtli, or Left-Handed Hummingbird. The island was a good place for an eagle to hunt for snakes, but perhaps less so for founding a large city. Aztec homes were made of light materials. They were practically earthquake-proof, but after the fall of Tenochtitlan in 1521, the victorious Spanish began to use heavy stone-and mortar construction—as in Spain. Severe earthquake losses occurred at the rate of one or two disasters per century. Reinforced concrete-frame apartment buildings were introduced—and promptly damaged—after 1944.

Tangshan City in Hebei Province, China, was founded in the 1880s by the British to exploit the rich coal seams under the area. They controlled the mines and the city until 1954. A national program of earthquake hazard prevention was started in China after the 1966 Xingtai earthquake, yet housing conditions in Tangshan were not significantly improved, even after the British were evicted. The hazard was thought to be low. By 1976, when the earthquake struck, more than a million people—mostly coal miners and their families—lived there. The earthquake struck at night and demolished the entire city. At least 242,000 people were killed in seconds.

China has produced many brilliant thinkers over the centuries. An ancient classical sage was convinced that wars and other disasters could be understood if technologists could be persuaded to join hands with philosophers, the social scientists of the time. He wrote:

> Old-time warriors first made themselves invulnerable:
> Then they would wait for the enemy to become vulnerable.
> Invulnerability is up to you:
> The vulnerability of an enemy depends on the enemy.
> Sun Tzu, *Art of War* (1964). Chapter 4. Fifth Century B.C.

Being invulnerable to disasters is up to us.

ACKNOWLEDGMENTS

We thank Seth Stein for valuable criticism and useful suggestions. We are indebted to three anonymous reviewers for constructive comments.

REFERENCES CITED

Brune, J.N., 1993, The Seismic Hazard at Tehri Dam, Proceedings of the International Workshop: "New Horizons in Strong Motion: Seismic Studies and Engineering Practice," Santiago, June 4–7, 1991: Tectonophysics, v. 218, p. 281–286.

Bunde, A., Kropp, J., and Schellnhuber, H.J., 2002, Introductory remarks, *in* Bunde, A., Kropp, J., and Schellnhuber, H.J., eds., The Science of Disasters: Heidelberg, Springer, p. VII–X.

Castaños, H., and Lomnitz, C., 2003, Disasters at the interface of nature and society provoke thought: Eos (Transactions, American Geophysical Union), v. 84, p. 521.

Chiles, J.R., 2002, Inviting Disaster: Lessons from the Edge of Catastrophe: New York, Harper Business, 368 p.

Cowardin, L.M., Carter, V., Golet, F.C., and LaRoe, E.T., 1979, Classification of wetlands and deepwater habitats of the United States: U.S. Department of the Interior, Fish and Wildlife Service, FWS/OBS-79/31, 131 p.

Dewar, R.C., 2005, Maximum entropy production and non-equilibrium statistical mechanics, *in* Kleidon, A., and Lorenz, R.D., eds., Non-Equilibrium Thermodynamics and the Production of Entropy: Heidelberg, Springer, p. 41–55.

Dombrowsky, W.R., 1979, Katastrophenbekaempfung und Katastrophenprophylaxe: Zivilverteidigung, v. 3, p. 62–64.

Graham, J.D., and Vaupel, J.W., 1981, Value of a life, what difference does it make?: Risk Analysis, v. 1, p. 89–95.

Hewitt, K., 1999, Regions of Risk: Hazards, Vulnerability and Disasters: London, Pearson-Longman, 389 p.

Lomnitz, C., 1994, Fundamentals of Earthquake Prediction: New York, John Wiley & Sons, 326 p.

Lomnitz, C., and Nilsen-Hofseth, S., 2005, The Indian Ocean disaster: Tsunami physics and early warning dilemmas: Eos (Transactions, American Geophysical Union), v. 86, no. 65, p. 70.

Lorenz, R.D., 2005, Entropy production in the planetary context, *in* Kleidon, A., and Lorenz, R.D., eds., Non-Equilibrium Thermodynamics and the Production of Entropy: Heidelberg, Springer, p. 147–159.

Matuzawa, T., 1925, On the possibility of gravitational waves in soils and allied problems: Journal of Institute of Astronomy and Geophysics, Tokyo, v. 3, p. 161–174.

Moscow, A., 1981, Collision Course: The Classic Story of the Collision of the *Andrea Doria* and the *Stockholm*: New York, Putnam, 316 p.

National Research Council, 2003, Living on an Active Earth: Washington, D.C, National Academies Press, 418 p.

Perrow, C., 1999, Normal Accidents: Princeton, Princeton University Press, 451 p.

Petroski, H., 2005, Technology and the humanities: American Scientist, v. 93, p. 304–307.

Priestley, M.J.N., 2004, Guest editor's note: Indian Society of Earthquake Technology Journal of Earthquake Technology, v. 41, p. iii–vii.

Radio 4's *Today* programme, http://www.bbc.co.uk/radio4/today/.

Rosenblueth, E., Ordaz, M., Sánchez-Sesma, F.J., and Singh, S.K., 1989, Design spectra for Mexico's Federal District: Earthquake Spectra, v. 5, p. 258–272.

Sornette, D., Stauffer, D., and Takayasu, H., 2002, Market fluctuations II, *in* Bunde, A., Kropp, J., and Schellnhuber, H.J., eds., The Science of Disasters: Heidelberg, Springer, p. 411–435.

Suh, N.P., 1985, Statement before the Senate Subcommittee for Science, Research and Technology, Congressional Record: Washington D.C., Proceedings and Debates, 99th Congress, October 3, 1985.

Sun Tzu, 1963, The Art of War: Oxford, Oxford University Press, 197 p.

Tucker, B.E., 2004, Trends in Global Urban Earthquake Risk: A Call to the International Earth Science and Earthquake Engineering Communities: Seismological Research Letters, v. 75, p. 695–700.

U.S. Geological Survey, 2004, FAQ—Magnitude 9.0 off W coast of northern Sumatra, Earthquake Hazards Program: http://earthquake.usgs.gov/eqinthenews/2004/usslav/neic_slav_faq.html (25 December 2004).

White, G.F., 1974, Natural hazards research: Concepts, methods, and policy implications, *in* White, G.F., ed., Natural Hazards: Local, Natural and Global: New York, Oxford University, p. 3–15.

Manuscript Accepted by the Society 29 November 2006

Index